Handbook of Microplastic Pollution in the Environment

In this timely handbook, one of a series of three, leading contributors from around the world offer practical insight into the challenges and opportunities for using various technologies to tackle microplastic pollution and improve microplastic management in agricultural soil.

Through this book, readers will gain a deep understanding of microplastic pollution in soils and its effects and knowledge of strategies and technologies to identify, monitor, and combat this. Early chapters describe microplastics in the rural environment, their accumulation, and their effects on soil biological processes and biodiversity, before moving on to address identification, analysis, and monitoring. Finally, they address innovative solutions for soil remediation, such as microbial strain engineering and other techniques. An introduction is provided at the beginning of each chapter for those interested in a brief synopsis, and copious references are provided for those wishing to study each chapter topic in greater detail. This book furnishes readers with the knowledge to identify and monitor microplastic levels in soil and reduce these to mitigate their harmful effects on soil biodiversity and ecosystems in agricultural soil, improving environmental and human health and minimizing economic losses. For a wider perspective, readers are encouraged to refer to the other two titles in this series, subtitled *Microplastic Pollution in Aquatic Environments* and *Monitoring and Treatment of Microplastics and Policy Perspectives.*

In its exploration of the relationships among the characteristics of microplastics, their mobility, transport pathways, and treatment, this handbook represents a vital practical guide for academics, industry-based researchers, and policymakers worldwide to address soil biodiversity and ecological health.

Handbook of Microplastic Pollution in the Environment

Microplastic Pollution in the Soil

Edited by
Tonni Agustiono Kurniawan and
Abdelkader Anouzla

CRC Press
Taylor & Francis Group
Boca Raton London New York

CRC Press is an imprint of the
Taylor & Francis Group, an **informa** business

Designed cover image: Roman Mikhailiuk\shutterstock

First edition published 2025
by CRC Press
2385 NW Executive Center Drive, Suite 320, Boca Raton FL 33431

and by CRC Press
4 Park Square, Milton Park, Abingdon, Oxon, OX14 4RN

CRC Press is an imprint of Taylor & Francis Group, LLC

ISBN: 9781032783734 (hbk)
ISBN: 9781032783772 (pbk)
ISBN: 9781003487579 (ebk)

DOI: 10.1201/9781003487579

Typeset in Minion
by codeMantra

Contents

Editors

Tonni Agustiono Kurniawan is an Associate Professor at Xiamen University, China. He obtained his Bachelor's Degree in Chemistry from Bogor Agricultural University, Indonesia, his Master's Degree in Environmental Technology from Thammasat University, Thailand, and his Ph.D. in Applied Biology and Chemical Technology from The Hong Kong Polytechnic University, China. He has been named as one of the top 2% of highly cited scientists in the field of Environmental Sciences by Stanford University.

Abdelkader Anouzla is with the Laboratory of Process Engineering and Environment, Faculty of Science and Technology, Hassan II University, Morocco. He attained his Ph.D. in Science and Technology at Hassan II University. He has published almost 100 peer-reviewed articles and ten books.

Contributors

Areeha Afsheen
Faculty of Life Sciences, Department of Microbiology and Molecular Genetics
University of Okara
Okara, Punjab, Pakistan

Manzoor Ahmad Ahanger
Department of Civil Engineering
National Institute of Technology
Srinagar, Jammu and Kashmir, India

Rashid Ali
Faculty of Life Sciences, Departments of Zoology
Government College University Faisalabad
Faisalabad, Punjab, Pakistan

Lander de Jesus Alves
Postgraduate Program in Biology and Biotechnology of Microorganisms
State University of Santa Cruz (UESC)
Ilhéus, Bahia, Brazil

Tayba Ameen
Faculty of Life Sciences, Departments of Zoology
Government College University Faisalabad
Faisalabad, Punjab, Pakistan

Kassian T.T. Amesho
Institute of Environmental Engineering
National Sun Yat-Sen University
Kaohsiung, Taiwan
and
Center for Emerging Contaminants Research
National Sun Yat-Sen University
Kaohsiung, Taiwan
and
Faculty of Management of Sciences
Tshwane School for Business and Society, Tshwane University of Technology
Pretoria, South Africa
and

Centre for Environmental Studies, Main Campus
The International University of Management
Windhoek, Namibia
and
Research and Development Division
Destinies Biomass Energy and Farming Pty Ltd
Swakopmund, Namibia

Khadija Amjad
Atta-Ur-Rahman School of Applied Biosciences
National University of Sciences and Technology
Islamabad, Pakistan

Moruf Afolabi Asani
Department of Urban and Regional Planning
Ladoke Akintola University Technology
Ogbomoso, Nigeria

Uzair Ashraf
Faculty of Physical Sciences, Departments of Chemistry
Government College University Faisalabad
Faisalabad, Punjab, Pakistan

Muhammad Asrar
Faculty of Life Sciences, Departments of Zoology
Government College University Faisalabad
Faisalabad, Punjab, Pakistan

Baiju V
Department of Mechanical Engineering
TKM College of Engineering
Kollam, Kerala, India

Damilola Gbadebo Bobola
Department of Civil Engineering
Obafemi Awolowo University
Ile-Ife, Nigeria

Olanrewaju Rotimi Bodede
Department of Mechanical Engineering
Rufus Giwa Polytechnic
Owo, Ondo State, Nigeria

Naveen Chand
Environment and Biofuel Research Lab (EBRL), Hydro and Renewable Energy Department
Indian Institute of Technology (IIT)
Roorkee, Uttarakhand, India

Pallavi Chattopadhyay
International Rice Research Institute South Asia Regional Centre
Varanasi, Uttar Pradesh, India

Manish Chaudhary
School of Environment & Natural Resources
Doon University
Dehradun, Uttarakhand, India

Suchith Chellappan
Department of Civil Engineering
UKF College of Engineering and Technology
Kollam, India

Chingakham Chinglenthoiba
Centre for Research Impact and Outcome (CRIO)
Chitkara University
Rajpura, Punjab, India

Dele Adedapo Daramola
Dept. Works and Maintenance Services
Bamidele Olumilua University of Education, Science and Technology
Ikere Ekiti, Nigeria

Era Juliet Das
Department of Environmental Science Bangladesh
University of Professionals Mirpur
Cantonment, Dhaka, Bangladesh

Cláudia Cseko Nolasco de Carvalho
Department of Biology
State University of Alagoas
Santana do Ipanema, Alagoas, Brazil

Ankur Dwivedi
Chemical Engineering Department
Institute of Technology, Nirma University
Ahmedabad, Gujarat, India

Adeyemi Babayemi Fajobi
Civil Engineering Department
Obafemi Awolowo University
Ile-Ife, Nigeria

Asif Farooq
Department of Civil Engineering
National Institute of Technology
Srinagar, Jammu and Kashmir, India

Noor Fatima
Atta-Ur-Rahman School of Applied Biosciences
National University of Sciences and Technology
Islamabad, Pakistan

José Rodrigues de Souza Filho
Academic Department
Federal Institute Baiano (IFBAIANO)
Catu, Bahia, Brazil

Anthony Fulford
International Rice Research Institute South Asia Regional Centre
Varanasi, UP, India

Khalid Muzamil Gani
Department of Civil Engineering
National Institute of Technology
Srinagar, Jammu and Kashmir, India

Abdul Ghaffar
Entomological Research Institute
Ayub Agricultural Research Institute
Faisalabad, Punjab, Pakistan

Zakia Gueboudji
Department of Microbiology and Biochemistry
Faculty of Natural and Life Sciences
University of Batna 2 (Mostefa Ben Boulaid)
Batna, Algeria
and
Biotechnology, Water, Environment and Health Laboratory
Abbes Laghrour University of Khenchela
Khenchela, Algeria

Amal I. Hassan
Radioisotope Department
Nuclear Research Center, Egyptian Atomic Energy Authority
Cairo, Egypt

Dilbar Hussain
Entomological Research Institute
Ayub Agricultural Research Institute
Faisalabad, Punjab, Pakistan

Zeeshan Javed
Faculty of Life Sciences, Departments of Zoology
Government College University Faisalabad
Faisalabad, Punjab, Pakistan

Milind Joshipura
Chemical Engineering Department
Institute of Technology, Nirma University
Ahmedabad, Gujarat, India

Mohd Yahya Fadzli Jusoh
School of Educational Studies
Universiti Sains Malaysia
Penang, Malaysia

Kenza Kadi
Biotechnology, Water, Environment and Health Laboratory
Abbes Laghrour University of Khenchela
Khenchela, Algeria
and
Faculty of Natural and Life Sciences
Abbes Laghrour University of Khenchela
Khenchela, Algeria

Muhammad Saleem Khan
Faculty of Life Sciences, Department of Zoology
University of Okara
Okara, Punjab, Pakistan

Zeeshan Khan
Atta-Ur-Rahman School of Applied Biosciences
National University of Sciences and Technology
Islamabad, Pakistan

Naeema Khanum
Atta-Ur-Rahman School of Applied Biosciences
National University of Sciences and Technology
Islamabad, Pakistan

Sruthi S Krishna
Department of Civil Engineering
UKF College of Engineering and Technology
Kollam, Kerala, India

Mohd Nizam Lani
Faculty of Fisheries and Food Science
Universiti Malaysia Terengganu
Kuala Nerus, Terengganu, Malaysia
and
Food Security Research Cluster
Universiti Malaysia Terengganu
Kuala Nerus, Terengganu, Malaysia
and
Microplastic Research Interest Group (MRIG)
Universiti Malaysia Terengganu
Kuala Nerus, Terengganu, Malaysia

Sabrina Lekmine
Biotechnology, Water, Environment and Health Laboratory
Abbes Laghrour University of Khenchela
Khenchela, Algeria
and
Faculty of Natural and Life Sciences
Abbes Laghrour University of Khenchela
Khenchela, Algeria

Mahnoor Malik
Atta-Ur-Rahman School of Applied Biosciences
National University of Sciences and Technology
Islamabad, Pakistan

Fatima Malik
Atta-Ur-Rahman School of Applied Biosciences
National University of Sciences and Technology
Islamabad, Pakistan

Muhammad Asad Mangat
Faculty of Life Sciences, Departments of Zoology
Government College University Faisalabad
Faisalabad, Punjab, Pakistan

Sara Mechaala
Department of Microbiology and Biochemistry
Faculty of Natural and Life Sciences
University of Batna 2 Mostefa Ben Boulaid
Batna, Algeria

Ajay Kumar Mishra
International Rice Research Institute South Asia Regional Centre
Varanasi, Uttar Pradesh, India

Sajida Mustafa
Department of Zoology, Fisheries and Wildlife
University of Agriculture
Faisalabad, Punjab, Pakistan

Somrita Nag
International Rice Research Institute South Asia Regional Centre
Varanasi, UP, India

Saba Naz
Faculty of Life Sciences, Departments of Zoology
Government College University Faisalabad
Faisalabad, Punjab, Pakistan
and
Department of Zoology, Fisheries and Wildlife
University of Agriculture Faisalabad
Faisalabad, Punjab, Pakistan

Hong Nam Nguyen
University of Science and Technology of Hanoi
Vietnam Academy of Science and Technology
Cau Giay, Hanoi, Vietnam

Syeda Gul Nisa
Atta-Ur-Rahman School of Applied Biosciences
National University of Sciences and Technology
Islamabad, Pakistan

Fábio Carvalho Nunes
Academic Department
Federal Institute Baiano (IF BAIANO)
Alagoinhas, Bahia, Brazil

Isaiah Adesola Oke
Civil Engineering Department
Obafemi Awolowo University
Ile-Ife, Nigeria

Indu M Sasidharan Pillai
Department of Civil Engineering
TKM College of Engineering
Kollam, Kerala, India

Shibu Pillai
Chemical Engineering Department
Institute of Technology, Nirma University
Ahmedabad, Gujarat, India

M.N.V. Prasad
Department of Plant Sciences
School of Life Sciences, University of Hyderabad (An Institution of Eminence)
Hyderabad, Telangana, India

Priya K L
Department of Civil Engineering
TKM College of Engineering
Kollam, Kerala, India

Muhammad Sohail Qadir
Entomological Research Institute, Ayub Agricultural Research Institute
Faisalabad, Punjab, Pakistan

Fizza Qamar
Department of Zoology, Fisheries and Wildlife
University of Agriculture Faisalabad
Faisalabad, Punjab, Pakistan

Nimra Qureshi
Atta-Ur-Rahman School of Applied Biosciences
National University of Sciences and Technology
Islamabad, Pakistan

Md. Mostafizur Rahman
Department of Environmental Science Bangladesh
University of Professionals Mirpur
Cantonment, Dhaka, Bangladesh

Megha R Raj
Department of Civil Engineering
TKM College of Engineering
Kollam, Kerala, India

Bilal Rasool
Faculty of Life Sciences, Departments of Zoology
Government College University Faisalabad
Faisalabad, Punjab, Pakistan

Sabeen Sabri
Faculty of Life Sciences, Department of Microbiology and Molecular Genetics
University of Okara
Okara, Punjab, Pakistan

Mamta Saiyad
Chemical Engineering Department
Institute of Technology, Nirma University
Ahmedabad, Gujarat, India

Mubshar Saleem
Department of Plant Pathology
PMAS Arid Agriculture University
Rawalpindi, Punjab, Pakistan

Muhammad Saleem
Entomological Research Institute, Ayub Agricultural Research Institute
Faisalabad, Punjab, Pakistan

Usama Saleem
Faculty of Life Sciences, Departments of Zoology
Government College University Faisalabad
Faisalabad, Punjab, Pakistan

Hosam M. Saleh
Radioisotope Department
Nuclear Research Center, Egyptian Atomic Energy Authority
Cairo, Egypt

Fareeha Sattar
Faculty of Life Sciences, Departments of Zoology
Government College University Faisalabad
Faisalabad, Punjab, Pakistan

Selvam S
Department of Geology
V.O. Chidambaram College
Tuticorin, Tamil Nadu, India

Nimish Shah
Chemical Engineering Department
Institute of Technology, Nirma University
Ahmedabad, Gujarat, India

Tariq Shah
Faculty of Crop Production Sciences, Department of Agronomy
The University of Agriculture
Peshawar, Pakistan

Mashal Shahzadi
Faculty of Life Sciences, Departments of Zoology
Government College University Faisalabad
Faisalabad, Punjab, Pakistan

Sheetal Sharma
International Rice Research Institute South Asia Regional Centre
Varanasi, Uttar Pradesh, India

Sadia Sikder
Department of Environmental Science
Bangladesh University of Professionals
Mirpur, Dhaka, Bangladesh

Vineet Singh
Wildlife Toxicology Research Group
Wildlife Institute of India
Dehradun, Uttarakhand, India

Mohamad Nor Azzimi Sohedein
Faculty of Science, Functional Omics and Bioprocess Development Lab
Institute of Biological Sciences, Universiti Malaya
Kuala Lumpur, Malaysia
and
Faculty of Science, Biomass Energy Laboratory, Environmental Science and Management Program
Institute of Biological Sciences, Universiti Malaya
Kuala Lumpur, Malaysia

Surindra Suthar
School of Environment & Natural Resources, Doon University
Dehradun, Uttarakhand, India

Tahsin Tabassum
Department of Environmental Science
Bangladesh University of Professionals
Mirpur, Dhaka, Bangladesh

Elham Taghavi
Faculty of Fisheries and Food Science
Universiti Malaysia Terengganu
Kuala Nerus, Terengganu, Malaysia
and
Food Security Research Cluster
Universiti Malaysia Terengganu
Kuala Nerus, Terengganu, Malaysia

Mohammad Toha
Department of Environmental Science
Bangladesh University of Professionals
Mirpur, Dhaka, Bangladesh

Ali Umar
Faculty of Life Sciences, Department of Zoology
University of Okara
Okara, Punjab, Pakistan

Ahmad Waheed
Faculty of Life Sciences, Department of Zoology
University of Okara
Okara, Punjab, Pakistan

Arooj Zaib Warraich
Department of Zoology, Fisheries and Wildlife
University of Agriculture Faisalabad
Faisalabad, Punjab, Pakistan

Ufuoma Phoebe Williams Bello
Federal Ministry of Works and Housing Highway Construction & Rehabilitation Department (South-South Zone)
Mabushi, Abuja, Nigeria

Malja Imaan Yousaf
Faculty of Life Sciences, Departments of Zoology
Government College University
Faisalabad, Punjab, Pakistan

Zahwa Zaffar
Atta-Ur-Rahman School of Applied Biosciences
National University of Sciences and Technology
Islamabad, Pakistan

CHAPTER 1

Identifying Knowledge Gaps and Research Needs in Microplastic Pollution Research

Mohammad Toha and Md. Mostafizur Rahman

1.1 INTRODUCTION

Since the industrial production of plastics started in the 1950s, a massive amount of plastic has been used globally due to easy manufacturing processes, socio-economic growth, urbanization, expanding populations, and modernization, which has already become an emerging concern as they break down and are converted into microplastics (MPs) (<5 mm) in our environment (Thacharodi et al., 2024). Over the past years, plastic production has increased by 3% each year, estimated to amount to 400 million tons per year by 2025 (Akdogan & Guven, 2019). Moreover, if this plastic production rate is maintained, billions of tons of plastic will be assembled in our nature by 2050, as only 9% of these plastics are recycled, indicating an issue of potential threat and an active research hotspot (Belkhiri et al., 2022; Myszka et al., 2023). On top of that, this increasing trend of MPs has been predominantly found in environmental compartments, has potential

DOI: 10.1201/9781003487579-1

carcinogenic ability, and negatively affects aquatic and terrestrial ecosystems (Zhang et al., 2020; Huang et al., 2023). MPs can be generated from both primary and secondary sources (O'Brien et al., 2023). In primary sources, MPs enter directly into the environment as plastic beads, used as raw materials in the plastic manufacturing sector and personal care and hygiene products (Beaurepaire, 2021). On the other hand, in secondary sources, microplastics are converted into MPs in our environment due to photocatalytic breakdown, oxidation, and mechanical weathering (Wang et al., 2020; Qiu et al., 2024), for example, fibers from synthetic textiles (Shafi et al., 2024). Moreover, secondary sources of MPs are considered majorly responsible for microplastic pollution, as many macroplastic wastes enter our environment daily. However, according to a recent study, MPs can be entered into our environmental compartments from all steps of plastic production to waste management (Thacharodi et al., 2024). After discharging into the environment, these rapid amounts of MPs are distributed in our environmental compartments due to the help of various environmental factors such as wind, runoff, heat, pressure, and other chemical and biological factors (Chen et al., 2024; Kitahara and Nakata, 2020). Consequently, MPs have already been found in various media of soils, oceans, rivers, shorelines, and swamps across the continents (North America, Europe, Africa, and Asia) (Zhang et al., 2020; Chen et al., 2024). Moreover, the mean concentration of MPs was found to be $11{,}742.64 \pm 31{,}143.09$ items/m^3 in lake environments, where Michigan Lake contained the lowest levels (1.8 items/m^3) and Rawal Lake contained the highest levels (141,236 items/m^3) (Chen et al., 2024). Moreover, many MPs were found in the lakes of developing countries such as India, Bangladesh, and Pakistan (Banik et al., 2024). Globally, in sediment samples, the MPs range varied from 14.44 to 18,088.08 items/kg, with an average of 1,072.67 items/kg (Chen et al., 2024). MPs type fibers were found to be most predominant in water and sediment samples, constituting 34.02% and 44.12%, respectively (Chen et al., 2024). On the other hand, in water and sediment samples, fragments and pellets were 23.40%, 26.05%, 24.37%, and 14.23%, respectively. Green (28.14%) and blue (22.59%) were found to be the most abundant colors, while the rest of the colors were transparent, black, and red (Chen et al., 2024; Thacharodi et al., 2024). Due to their small size, these MPs are very easy for aquatic organisms to consume, which could result in chronic toxicity and death (Wang et al., 2020). The pollution hazard index (PHI) in lake water bodies was low (11%), medium (36%), high (50%), and ultra-high (4%), indicating the serious potential risk (Nakano et al., 2024;

Chen et al., 2024). Like water bodies, in sediment samples, the PHI was found to be 47%, 16%, 28%, and 9%, respectively (Chen et al., 2024). As MPs can be transferred from one trophic level to another trophic level, they can disrupt the growth and productivity of phytoplankton, alter the nutrient cycle, and bioaccumulate and biomagnify the food chain (Saikumar et al., 2024). Moreover, these MPs can transport various vectors with the help of their relatively large surface area, hydrophobic composition, and the dynamism of the aquatic body, showing the complex level of aquatic alteration (Rahman et al., 2021; Stang et al., 2022). After that, these can enter the human body by ingesting various aquatic fish and other organisms, indicating a potential threat to human health (Shao et al., 2022). On the other hand, it has been found that the occurrence of MPs in soils is 4–23 times higher than in the ocean (Yang et al., 2021; Sridharan et al., 2021). Most importantly, when rainfall occurs, these MPs in soils are transferred into groundwater (Eberhard et al., 2024; Wang et al., 2021a). Hence, according to the existing literature, although most of the MP research has been conducted focusing on the aquatic environment and soil, the issue of MPs in the atmospheric environment lags behind. Suspended particles are moved from one place to another using the atmospheric environment, an essential transportation medium (Ageel et al., 2022; Enyoh et al., 2019). Recent studies have demonstrated that atmospheric MPs can also be transported from one area to another (from ocean surface air to remote areas) through the help of wind speed, wind directions, precipitation, convection lift, and turbulence mechanisms (Zhu et al., 2024; Rao et al., 2024). Notably, atmospheric MPs contribute significantly (1.4×103 tons) seasonally in the ocean systems, indicating that atmospheric pathway systems can act as a source-sink dynamic of plastic pollution in marine and terrestrial environments (Zhao et al., 2023). Moreover, carbonyl bonds (C=O) can emerge due to MPs' long-term aging and adsorption processes in air, seawater, and freshwater with the combination of high temperatures (75°C). As this C=O bond is highly reactive, it modifies the surface texture, adsorption behavior, and internal characteristics of MPs (Thacharodi et al., 2024). Various toxic pollutants (heavy metals, polycyclic aromatic hydrocarbons, polychlorinated biphenyls, DDT, PBDEs, PFAS, and pharmaceuticals) can be carried by MPs, and humans can easily be exposed to these mixed pollutants (Can-Güven, 2021; Wright et al., 2021). It must have to be said that MPs are an emerging component of air pollution. Moreover, meteorological variables could greatly influence the distribution of MPs in the atmospheric medium. Various sampling methods, atmospheric circulation patterns, and boundary layer height might have also affected this complex correlation (Huang et al., 2023). However, the majority of published studies focus on atmospheric deposition, where a passive collection system was used to take deposited material at a particular location (Beaurepaire, 2021; Thacharodi et al., 2024; Zhao et al., 2023; Zhang et al., 2020). Moreover, MPs' morphological characteristics (size, shape, abundance, color, and components) have been identified and reported for urban, remote, and suburban areas (Chen et al., 2024; Zhang et al., 2020). The long-term distributional pattern and morphological characteristics of MPs in various sizes of atmospheric material are yet to be discovered. Moreover, the health risks of atmospheric MPs are still paid less attention. In these circumstances, it is crucial to understand the meteorological variables and their influence on the atmospheric MPs transportation system, the atmospheric MPs settlement process in the aquatic bodies, their dispersion patterns in various sizes of atmospheric materials, and the health risks of the atmospheric MPs on the human body. Prompted by this context, this chapter aims to overview the current state of microplastic pollution and collate and compare the current atmospheric MPs research findings with MPs in soil and aquatic environments. Moreover, this chapter presents the potential pathways of atmospheric MPs, their identification techniques, and their impacts, as well as identifying future research needs in this issue. Lastly, this chapter will focus on the standard management of microplastic pollution in environmental compartments.

1.2 SOURCES, FATE, AND TRANSPORT OF MICROPLASTICS (MPS) IN ENVIRONMENTAL COMPARTMENTS

Most studies on MPs' sources, fate, and transport have focused on soil and water environmental compartments. Furthermore, although few studies have considered the atmospheric environment, information on the sources, fate, and transport of MPs in atmospheric compartments is haphazardly distributed in the literature. Hence, this section presents the sources, fate, and transport of MPs in atmospheric compartments.

1.2.1 Existing Research on the Sources of Microplastics (MPs) in Air

Although assessing the sources of airborne MPs is still largely unknown, it is crucial to understand the sources and transport of atmospheric MPs for establishing legislation and guidance. Based on the sources, MPs can enter the atmospheric environment through multiple pathways. However, the synthetic fiber in textiles is considered one of the prominent sources of atmospheric MPs (O'Brien et al., 2023). Microplastic fibers have been detected in various areas of Paris, France, and China from atmospheric fallout, where 17% contained pure synthetic fibers (PET and PE) and 12% were mixtures of multiple materials (Zhao et al., 2023). Notably, atmospheric fallout in the Paris agglomeration assembled 3–10 tons of these fibers (Akdogan & Guven, 2019). It has been predicted that the atmospheric MP fiber will be generated at 0–36.64 g/km^2 within a few years (Zhao et al., 2023). Moreover, polyesters were the most abundant synthetic polymers in indoor environments (Ageel et al., 2022; Can-Güven, 2021). Fragments of fibers are generally created due to anthropogenic activities such as wearing, washing, movement, and other activities (Beaurepaire, 2021; Zhao et al., 2023). During these anthropogenic activities, the physical breakdown of fibers occurs when fragments are emitted from bed sheets, pillows, blankets, and folding (O'Brien et al., 2023). Moreover, drying these household products under the sun contributes to the fragmentation of fibers (Zhao et al., 2023). Research studies have shown that a single person can generate 1×10^9 microplastic fibers annually by wearing 1 kg of synthetic dress for 8 hours daily (O'Brien et al., 2023). Furthermore, the length and diameters of this fiber range between 15 and 19 μm (O'Brien et al., 2023). Mechanical dryers were found to emit fibers in the air with >1 fiber/m^3 when drying a blanket for 20 minutes (Brahney et al., 2020). However, fiberglass and rubber fibers are other sources of fibers in atmospheric compartments (Kannankai & Devipriya, 2024). Road paint tires and brake wear-offs are significant sources of MPs in the atmospheric environment (Shao et al., 2022). Due to the brake abrasion, fine particles of synthetic rubber core and metal additives are emitted into the air (Zhu et al., 2024). In the atmospheric environment, 5% of tire wear-off emissions are airborne, and their concentrations generally range from 2 to 5 μg/m^3 (Myszka et al., 2023; O'Brien et al., 2023). Research studies have shown that almost 146 k tons of MPs in tire wear are being emitted per year from both developed (Netherlands et al., Australia, USA, and Japan) and developing nations (India and China) (O'Brien et al., 2023). Furthermore, 2,907 k tons of MPs in tire wear are estimated to be emitted globally each year, where PM10 contains 288 k, and the rest are PM2.5 (Evangeliou et al., 2020). Lastly, the most horrible thing is that 34% of these wearing particles finally enter the ocean by deposition (Zhao et al., 2023). The sources of MPs are illustrated in Figure 1.1.

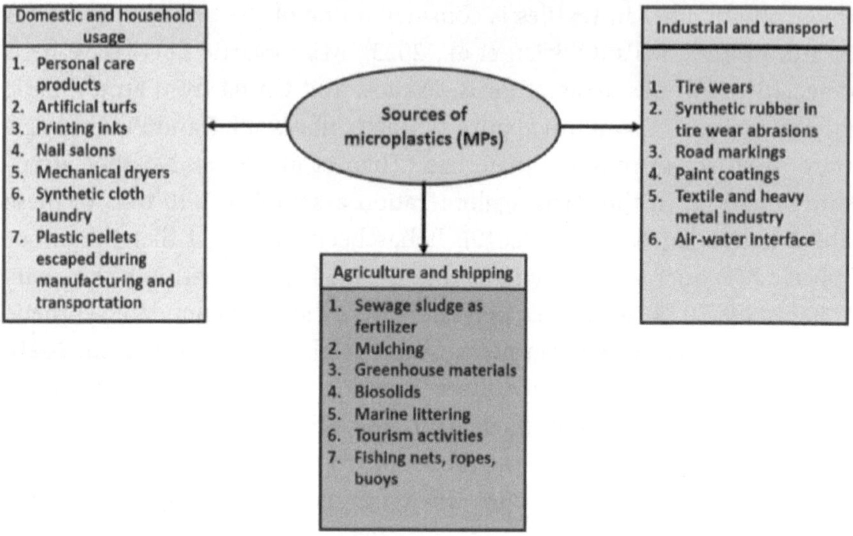

FIGURE 1.1 Various sources of microplastics (MPs) in environmental compartments.

Research studies stated that industrial activities, waste incineration, interior decoration, building materials, landfills, and unique wood and metal products are potential sources of MPs in the atmospheric environment (Can-Güven, 2021; Zhao et al., 2023). It has been noticed that industries generated various morphological characteristics of MPs in Iran, which are almost three times greater than those in other urban areas based on their abundance (Abbasi et al., 2019). According to research, 1 kg of incineration plant ash contains 171 MP particles, indicating incomplete combustion, and another burning process can convert large plastic materials into finer particles (Singh et al., 2023; Shen et al., 2022). Notably, the rate of atmospheric MPs in landfill environments is 0.16 MPs/t of waste/h (Hu et al., 2022). However, the chimneys of incineration plants can directly emit MPs into the atmospheric compartments. Various coating materials, especially alkyd resins and epoxy, are generally used in wooden or metal products. These coating materials are gradually fragmented, break down, and finally emitted into the air (O'Brien et al., 2023). Moreover, plastic bags, gloves, masks, and personal care products are gradually degraded, producing MPs in our environment (Shao et al., 2022). These MPs can be suspended into the air with the help of wind at a rate of 0.08–1.48 mg/m^2/min (O'Brien et al., 2023). Notably, approximately 2×10^{12} MPs were suspended in the air from soil media by a dust storm, where 1.5×10^{12} MPs were deposited within the area (Abbasi et al., 2022). However, nail salons can also generate MPs in indoor environments with a rate of 18 MPs/m^3) consisting of acrylic (27%), rubber (21%), and polyurethane (13%) composition (Chen et al., 2022). Scientists have discovered that air-water interfaces are potential sources of aerosolized MPs (Ferrero et al., 2022). Furthermore, more than >30% of atmospheric particles above the ocean were found to be plastic particles (Trainic et al., 2020). Notably, scientists proved in a laboratory environment that this interface transferred smaller nanoplastics with <350 nm into the surrounding atmosphere by bubble-bursting mechanisms (O'Brien et al., 2023). Hence, it can be concluded that MPs are spreading in the air from various sources.

1.2.2 Existing Research on the Fate and Transport of Microplastics (MPs) in Air

The fate and transport of airborne MPs are interlinked with soil, marine, freshwater, and terrestrial environmental compartments, as these compartments together create a diverse network of source, pathway, and sink connections (Zhu et al., 2024; Wang et al., 2020). The fate and transport of MPs are illustrated in Figure 1.2.

Currently, the atmosphere is considered to be one of the significant transportation routes of MPs that can cause MPs to accumulate in terrestrial or marine environments. The morphological characteristics (size and shape) of MPs and meteorological factors such as wind velocity, wind direction, precipitation, snow, temperature, humidity, and air pressure have a significant influence on the fate and transport of MPs in the air (Zhao et al., 2023; Shao et al., 2022). In air medium, the particle size of MPs is <0.5 mm,

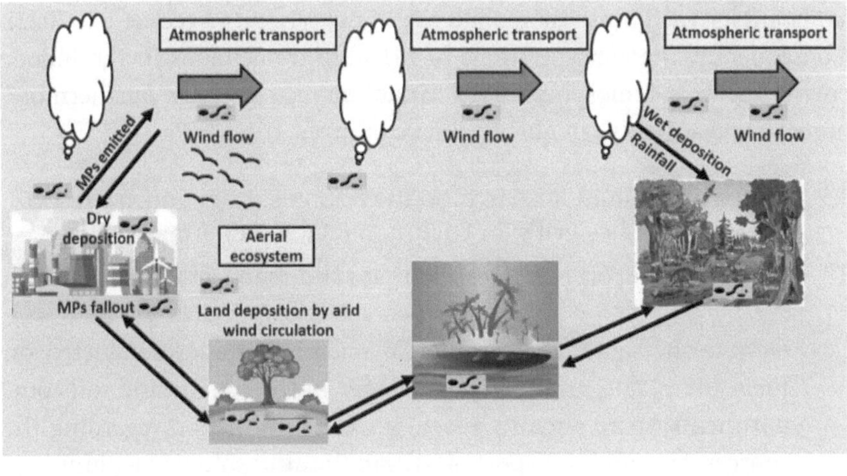

FIGURE 1.2 The fate and transport of atmospheric microplastics (MPs).

which is less than the size of water <4.975 mm, and soil <2 mm (Wright et al., 2020; Shao et al., 2022). Moreover, most airborne MPs are fibers, indicating that they are easily suspended in the air (Beaurepaire, 2021). However, higher-density MPs (PC, nylon, PVC, and PET) sink smoothly, whereas PE, PP, and PS tend to be buoyant in atmospheric medium (Sharma, 2024; Zhao et al., 2023). Generally, the intensity of wind and its direction are the main drivers of the transport of the MPs in the air. According to the research, the duration of MP particles in an air medium is 1 hour to 6.5 days, showing a favorable situation to move from one place to another through the wind (Brahney et al., 2021). Moreover, MPs have already been discovered in the Arctic, Antarctica, and Himalayan mountainous regions (Zhao et al., 2023). These states proved that MPs are transported by wind and dispersed in nature. MPs can be moved long distances in an atmospheric environment before they settle down. According to the Hybrid Single-Particle Lagrangian Integrated Trajectory (HYSPLIT), MPs can be transported in the atmospheric environment more than 95 km (Shao et al., 2022). Furthermore, Wang et al. (2020) stated that low-density MPs could be moved more than 1,000 km in the air medium. After that, these MPs are settled in various compartments of the environment, especially on roads and road edges during dry weather. These deposited MPs are moved into nearby soil and aquatic bodies through runoff when heavy rainfall occurs. Consequently, according to a recent study, every year, 140,000 tons of MPs from road traffic are deposited into the ocean via the atmosphere, producing a devastating impact on marine ecology (Zhao et al., 2023; Kannankai & Devipriya, 2024). On the other hand, the fate and transport of indoor airborne MPs are comparatively different from the outdoor atmospheric MPs, as in indoor mediums, air mobility is impoverished (Ageel et al., 2022). Consequently, the concentrations of MPs are tremendous in the indoor environment, as atmospheric MPs cannot be moved easily. Furthermore, humans and animals can quickly inhale or ingest these MPs.

1.2.3 Gaps and Future Research on the Sources, Fate, and Transport of Microplastics (MPs) in Environmental Compartments

The following is a list of critical research gaps and changes that are required:

- Despite the significant number of scientific studies conducted on the sources, fate, and transport of MPs in the aquatic and soil compartments, there remains a lack of clear information regarding the sources, fate, and transport of MPs in the atmospheric medium.

- Current MPs in air-related studies just focus on a few sources, such as synthetic textiles and road dust, while other known and unknown sources are yet to be explored. Furthermore, the contribution percentage of various sources cannot be measured, indicating that particle type, occurrence, and concentration-related data are scarce. In these circumstances, it is crucial to focus on all anthropogenic and natural sources and identify their contribution percentage.
- Future research should focus on establishing a pollution source localization model that can analyze the physical, chemical, and biological factors and explore the weathering, photo-transformation, and microbial degradation processes. Moreover, this model would make airborne MP source inventories and identify the amount of MP emission from anthropogenic sources.
- Research focusing on the environmental and anthropogenic causes of the fate and transport of MPs in the atmospheric medium is limited. The role of meteorological factors on the transport of airborne MPs in local, regional, and global areas requires more attention as they are transboundary pollutants. Investigating how meteorological factors contribute to the transport and settlement of MPs in environmental settings is essential for developing effective strategies to mitigate the environmental dissemination of airborne MPs. Furthermore, it is essential to identify and quantify atmospheric pollutants (aerosols, organic, and inorganic toxic pollutants) adsorbed onto MPs and make a quantitative relationship between MPs and these contaminants.
- Lastly, our environment follows a route of dynamism where breaking and making is a spontaneous event. Hence, the sources, fate, and transport of MPs in the atmospheric compartments should be continuously investigated to identify and monitor these pollutants' exact temporal and spatial patterns.

1.3 OCCURRENCE AND DISTRIBUTION OF MICROPLASTICS (MPS) IN ENVIRONMENTAL COMPARTMENTS

Over the past years, MPs have been detected in air, water, and soil environmental compartments. However, MPs in atmospheric environment-related studies are still limited compared to those in water and soil environments. Hence, this section presents the occurrence of MPs in the atmospheric environment and compares it with those in aquatic and soil mediums.

1.3.1 Occurrence and Distribution of Microplastics (MPs) in Atmospheric Compartments

In recent years, MPs have been observed in urban, suburban, rural, and indoor environments (Qiu et al., 2024). The concentration and distribution of MPs firmly depend on various factors such as environmental (vertical concentration gradient, rainfall, temperature, and wind flow), geographical locations, and anthropogenic factors (Huang et al., 2023). Among the European countries, 29–280 MPs/m^2/day were first detected in atmospheric fallout in Paris, France (Dris et al., 2015). Following that, various studies regarding airborne MPs study have been conducted in Europe, where a significant level of airborne MPs was detected with the concentration (MPs/m^2/day) of 275, 366, 700, (3–128), and (550–874) in Hamburg, Germany; the Pyrenees Mountains; Helsinki, Finland; Nottingham, UK; and Central London, UK, respectively (Klein & Fischer, 2019; Allen et al., 2019; Zhang et al., 2020; Stanton et al., 2019; Wright et al., 2020). On the other hand, the presence of airborne MPs has already been detected in the Asian countries of Iran, China, India, and Indonesia (Sharma, 2024). The concentrations of MPs were (88–605) items/30g, (0.3–1.10) items/m^3, (175–313) items/m^2/day, (115–602) items/m^2/day, 4.18 items/m^3, and (132.75–174.97) items/m^3 in Tehran, Iran; Asaluyeh, Iran; Dongguan, China; Yantai, China; Shanghai, China; and Surabaya, Indonesia, respectively (Dehghani et al., 2017; Abbasi et al., 2019; Cai et al., 2017; Zhou et al., 2017; Liu et al., 2019a; Zhang et al., 2020). Furthermore, the concentrations of Indian Ocean and South China Sea were found to be 0.077 MPs/m^3 and (0.01–0.06) MPs/m^3, whereas the concentrations of (0.10–1.37) MPs/m^3, (0.10–1.37) MPs/m^3, and (7.70±6.10) MPs/m^3 were observed in Pacific Ocean, North Atlantic Ocean, and French Shoreline, respectively (Wang et al., 2020; Wang et al., 2021b; Liu et al., 2019a; Yang et al., 2021). However, most of the studies (43%) measured the concentrations of MPs in the atmospheric medium were ≤2 (items/m^3) whereas, 33%, 10%, and 10% studies results showed that the concentration of MPs in the atmosphere was (5–100) (items/m^3), (100–200) (items/m^3), and >1,000 (items/m^3), respectively (Wang et al., 2020). However, the atmospheric MPs have found various shapes such as fiber, fragment, foam, and film. Fibers were found to be the most dominant shape (>60%) of MPs in the atmospheric fallout samples of China (Donggum, Shanghai, and Paris) and France (Paris) (Zhang et al., 2020). On the other hand, atmospheric samples from Hamburg, Germany, showed that 95% of their MPs were in fragmented shape. However, O'Brien et al. (2023) analyzed existing published papers and found that fibers (55%) were the most abundant shape of MPs. Notably, the maximum fiber lengths have been noticed with >1,475 μm in China, 1,650 μm in France, 1,928 μm in Portugal, 2,061 μm in the USA, and 2,181 μm in China (Zhao et al., 2023). Furthermore, the mean measurements of fiber length were found to be 330, 429, and 616 ± 537 μm in Portugal, China, and the USA, respectively (Zhao et al., 2023). The concentration of MPs in indoor environments was 6–6.5 times higher than in outdoor environments (O'Brien et al., 2023). It has been found that the concentrations of MPs in indoor environments were 1,586–11,130 particles/m^2/day in Paris, France whereas, in Shanghai, China, and Sydney, Australia, it has been detected with the concentration (particles/m^2/day) of (500–29,000) and (22–6,169), respectively (Dris et al., 2017; Zhang et al., 2022; Gaston et al., 2020). According to the various studies, fibers' maximum lengths were noticed in the range of 2,181 μm to >5,000 μm (Zhao et al., 2023). Furthermore, the average length of fibers was found to be 177, 237, 250, 330, and 383 μm, whereas the diameter was reported to be 26–30 μm wide (Zhao et al., 2023). The composition and shape of indoor airborne MPs are yet to be properly defined. Few existing studies reported that fibers were found to be the most dominant shape of MPs, such as the indoor environment of Iran containing 85% fibers, whereas China showed 80%–92% (Kashfi et al., 2022). Lastly, the concentration of airborne MPs in the indoor environment depends on the ventilation systems, which have a dilution effect, as well as the vertical height of the environment (Table 1.1).

1.3.2 Microplastics (MPs) in Aquatic and Soil Compartments and Its Comparison with the Atmospheric Environment

The distribution of MPs in soil and aquatic environments depends on the geographic location of aquatic bodies, seasonal variations, meteorological factors, sampling, and analysis methods (Huang et al., 2023). The occurrence and distribution of MPs are demonstrated in Figure 1.3. Notably, the abundance of MPs was significantly influenced by the geographic locations of aquatic bodies, indicating that MPs abundance was potentially affected by the local economic growth, anthropogenic activities, and lack of waste management practices (Chen et al., 2024). Generally, developing nations are moving towards economic development and using the environment haphazardly, due to which massive numbers of MPs are spreading in our nearby aquatic bodies and soil compartments (Zhao et al., 2023). Consequently, the concentration of MPs in aquatic and soil environments is increasing with a scale-up (Rahman et al., 2021). Recent studies have shown that the concentration of MPs in the surface water of the Buriganga River

TABLE 1.1 The Prevalence of Microplastics (MPs) in Atmospheric Environments

Study Area	Type of Samples	Abundance of Airborne MPs	Size of the MPs (μm)	References
Paris, France	Atmospheric fallout	(29–280) particles/m^2/day	(500–1,000)	Dris et al. (2015)
Paris, France	Atmospheric fallout	(2.10–355.40) particles/m^2/day	(200–1,400)	Dris et al. (2016)
Paris, France	Atmospheric fallout	(1,586–11,130) particles/m^2/day	(50–450)	Dris et al. (2017)
London, UK	Atmospheric fallout	(771 ± 167) particles/m^2/day	(75–180)	Wright et al. (2020)
Madrid, Spain	Atmospheric fallout	12/sample	—	Zhao et al. (2023)
Sakarya, Turkey	Outdoor air	(9,057–30,793) particles/L	(50–500)	Kaya et al. (2018)
Edinburgh, UK	Atmospheric fallout	(5 ± 33) particles/L	—	Catarino et al. (2018)
West Pacific Ocean	Ocean air	(0–1.37) particles/m^3	(16.14–2,086.99)	Liu et al. (2019a)
Trent Catchment, UK	Atmospheric fallout	(2.90–128.42) particles/m^2/day	—	Stanton et al. (2019)
Italian Alps	Supraglacial debris	74.40 particles/kg	—	Ambrosini et al. (2019)
Artic	Snow (wet deposition)	($0–14.40 \times 10^3$) particles/L	(11–250)	Bergmann et al. (2019)
Pyrenees Mountains, Europe	Atmospheric fallout	366 particles/m^2/day	(10–5,000)	Allen et al. (2019)
Hamburg, Germany	Atmospheric fallout	275 particles/m^2/day	<63	Klein and Fischer (2019)
Helsinki, Finland	Wet deposition	700 particles/m^2	—	Ambrosini et al. (2019)
USA	Atmospheric fallout	132 particles/m^2/day	—	Brahney et al. (2020)

(Continued)

TABLE 1.1 (*Continued*) The Prevalence of Microplastics (MPs) in Atmospheric Environments

Study Area	**Type of Samples**	**Abundance of Airborne MPs**	**Size of the MPs (μm)**	**References**
Melbourne, Australia	Dry deposition	1,130 items/kg		Monira et al. (2021)
Nagpur, India	Dry deposition	(50–120) items/day	—	Narmadha et al. (2020)
Lahore, Pakistan	Dry deposition	241.45 items/m^2	—	Aslam et al. (2022)
Dongguan, China	Atmospheric fallout	(175–313) particles/m^2/day	(200–700)	Cai et al. (2017)
Yantai, China	Atmospheric fallout	(115–602) particles/m^2/day	(100–300)	Zhou et al. (2017)
Surabaya, Indonesia	Atmospheric fallout	(132.75–174.97) items/m^3	(1,000–1,500)	Zhang et al. (2020)
Asaluyeh, Iran	Atmospheric fallout	(0.30–1.10) items/m^3	(100–1,000)	Abbasi et al. (2019)
Shanghai, China	Indoor air	500–29,000 particles/m^2/day	—	Zhang et al. (2022)
California, USA	Indoor air	(15.90±10.90) items/m^3	—	Zhang et al. (2020)
Sydney, Australia	Indoor air	(22–6,169) particles/m^2/day	—	Soltani et al. (2021)
Eastern Coast, China	Indoor air	(1,583±1,180) items/m^3	—	Liao et al. (2021)
Bushehr port, Iran	Indoor air	90.8 MPs/mg	—	Kashfi et al. (2022)
Barcelona, Spain	Indoor air	(8±1.9) items/m^3	—	Torres-Agullo et al. (2022)

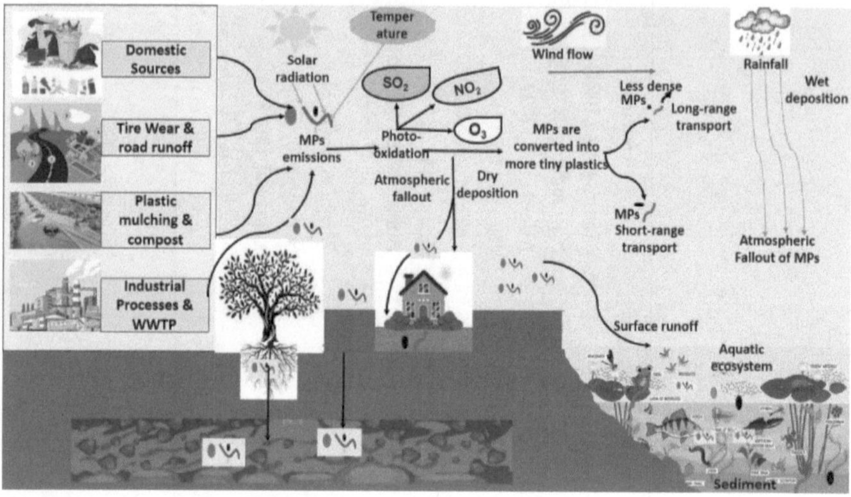

FIGURE 1.3 The occurrence and distribution of microplastics (MPs) in environmental compartments.

and Ganges River Basin, Bangladesh, was (0.25–0.12) particles/mL and (55.26±25.40) particles/L. In contrast, the concentrations were 7.89 particles/g and (3,307.22±1,732.85) particles/kg for the sediments of the Buriganga River and the Ganges River Basin (Haque et al., 2023; Alam et al., 2023). Furthermore, (0.01–95.80 particles/L), (3.67–10.70 particles/L), (17±4 items/L), and (5.85±3.28 particles/L) were observed in the surface water of the Berlin River, Wei River, Manas River, and Ciwalengke River, respectively (Schmidt et al., 2018; Ding et al., 2019; Wang et al., 2021c; Alam et al., 2019). However, a significant concentration of MPs was observed in the sediment samples of the Amazon River (417–8,178 particles/kg), Pearl River (80–9,597 items/kg), Brahmaputra River (531–3,485 particles/kg), Indus River (525–1,752 particles/kg), and Ganga River (99.27–409.86 particles/kg) (Alam et al., 2023; Gerolin et al., 2020; Lin et al., 2018). In China, 7,579 and 48 particles/kg MPs were found in the soil samples of tropical humid and temperate semi-arid regions, respectively (Haque et al., 2023). Furthermore, a significant concentration of MPs in agricultural farmland soil in Bangladesh was discovered with ($2.13 \times 10^4 \pm 0.13 \times 10^4$) particles/kg, whereas 0.34 particles/kg were discovered in Middle Franconia, Germany (Hossain et al., 2023; Piehl et al., 2018). However, massive concentrations of MPs in soil were also measured in the Swiss (593 particles/kg) and industrial areas of Australia (300–67,500 mg/kg) (Scheurer & Bigalke, 2018; Guo et al., 2020), as atmospheric compartments, fibers, and fragments were found to be the most dominant shape of MPs in the samples of freshwater, sediments, marine, and soil. PET, PP, PE, PS, and PVC polymers were found in air, water, and soil compartments (Zhang et al., 2020). As these types of polymers are generally produced through clothes and textiles, it is predicted that indoor MPs could be transported and deposited in outdoor environments (Ageel et al., 2022). After that, these MPs are distributed in all respects of environmental compartments with the help of environmental and meteorological factors. The mean size of MPs was <700 μm in aquatic environments, whereas most of the fibers were in 100–500 μm, and the fragments were <100 μm in atmospheric compartments (Zhang et al., 2020). Recent studies have also predicted that air, water, and soil compartments could be a great interconnection in terms of the characteristics of MPs and their presence (Tables 1.2 and 1.3).

1.3.3 Gaps and Future Research on the Occurrence and Distribution of Microplastics (MPs) in Environmental Compartments

The crucial gaps in research and necessary modifications are listed below:

- Despite the significant number of scientific studies conducted on the occurrence and distribution of MPs in environmental compartments, there remains a lack of clear information regarding the standardization of data, sampling techniques, and standard units. Hence, future research needs to focus on maintaining standard sampling procedures, uniform units, and efficient analyzing techniques to avoid overestimating or underestimating the abundance of MPs in environmental compartments.
- While many studies have examined the abundance and morphological characteristics of MPs in environmental compartments, there is a need for more comprehensive investigations into the spatial and temporal distribution of MPs in the various environmental compartments. Moreover, sediments act as a reservoir of MPs in aquatic bodies and conserve significant information regarding anthropogenic activities. Hence, vertical distribution of MPs-related research must be necessary to assess the MP's life cycle and historical record of MP pollution of a particular aquatic body.
- Most research focuses on European and Asian nations, indicating primary limits to MP study in other parts of the world. Therefore, more research and consistent reporting on the issue of MPs in environmental compartments is necessary for the continents of South America, Africa, and Australia.

TABLE 1.2 The Prevalence of MPs in Water Compartments

Study Area	Type of Samples	Abundance of Aquatic MPs	Size of the MPs	References
Miri River	Surface water	(10.7–14.3) particles/L	—	Liong et al. (2021)
Magdalena River	Surface water	(0.11–0.163) particles/L	—	Silva and Nanny (2020)
Dongting Lake, China	Surface water	(616.7–2316.7) particles/m^3	<5 mm	Jiang et al. (2018)
Hong Lake, China	Surface water	685.5 particles/m^3	(0.05–5) mm	Wang et al. (2021)
Vembanad Lake, India	Sediment	252.80 particles/m^2	<5 mm	Shruti et al. (2018)
River Kelvin, UK	Sediment	(161–432) particles/kg	(0.063–2.80) mm	Blair et al. (2019)
Ofanto River, Italy	Surface water	(0.90–13) particles/m^3	(0.03–5) mm	Campanale et al. (2020)
Ganges River Basin, Bangladesh	Surface water	55.26±25.40 particles/L	—	Alam et al. (2023)
Ganges River Basin, Bangladesh	Sediment	(3,307.22±1,732.85) particles/kg	—	Alam et al. (2023)
West River downstream	Surface water	(2.99–9.87) particles/L	—	Huang et al. (2021)
Amazon River	Sediment	(417–8,178) particles/kg	—	Gerolin et al. (2020)
Nakdong River	Sediment	(1,970±62 particles/kg)	—	Eo et al. (2019)
Jiaozhou Bay, China	Seawater	(20–120) items/m^3	<4 mm	Zheng et al. (2019)
Qatar, Arabian Gulf	Seawater	(0.0438–1.46) particles/m^2	<5 mm and >5mm	Abayomi et al. (2017)
Mobile Bay, USA	Sediment	(5–117) particles/m^2	<5 mm	Wessel et al. (2016)
Sea ports in Australia	Sediment	(83–350) particles/kg	(0.10–4.9) mm	Jahan et al. (2019)
Baltic Sea	Seawater	400 particles/m^3	(0.50–5) mm	Bagaev et al. (2018)

TABLE 1.3 The Prevalence of MPs in Soil Compartments

Study Area	Type of Samples	Abundance of MPs in Soil (Particles/kg)	Size of the MPs (mm)	References
Chai River valley, China	Agricultural soil	7,100–42,960	0.05–10	Zhang and Liu (2018)
Shandong province, China	Coastal soil	1.30–14,712.5	<5	Zhou et al. (2018)
Nankai University, Tianjin, China	Urban soil	75	0.10–3.2	Han et al. (2021)
Hangzhou Bay, China	Agricultural soil	0–2,760	1–3	Zhou et al. (2020)
Swiss	Floodplain agricultural soil	593	<5	Scheurer and Bigalke (2018)
Middle Franconia, Germany	Agricultural soil	0.34	1–5	Piehl et al. (2018)
Bangladesh	Agricultural soil	$(2.13\times10^4 \pm 0.13\times10^4)$	—	Hossain et al. (2023)

- Research on the correlation between marine, terrestrial, and atmospheric MPs is still limited. Moreover, the impact of natural organic matter on MPs in soil compartments is yet to be discovered. However, as natural organic matter plays a vital role in the environmental behavior of pollutants, future research should consider natural organic matter and its impact on the mobility and abundance of MPs in soil.
- Future research should be focused on the construction of a mathematical model for better understanding the occurrence and distribution of MPs in environmental compartments.

1.4 MICROPLASTICS (MPS) DETECTION AND ANALYSIS PROCEDURE

Despite the massive number of MPs already found in environmental compartments, no standard methods for detecting MPs have yet to be discovered. Hence, it is very challenging to identify and quantify the contamination level of MPs in environmental compartments. However, this section aims to present the existing MP detection and analysis procedure, identify research gaps, and recommend future research on this issue.

1.4.1 Sample Collection

Currently, the scientific communities are facing a lack of standardized sampling collection systems. However, it has been found that most of the research on atmospheric MPs focused on passive sampling collection

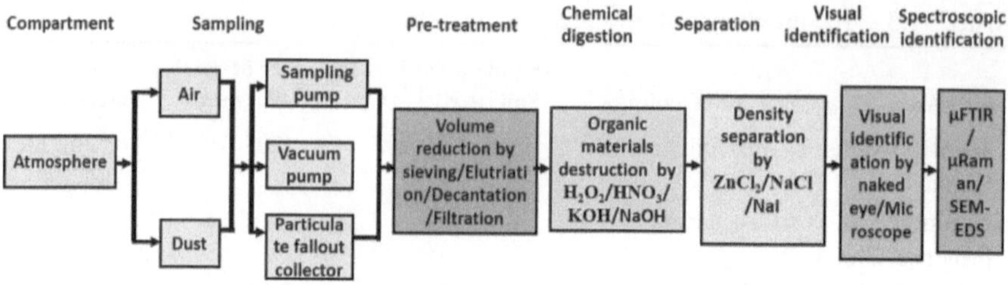

FIGURE 1.4 The detection and analysis procedure of airborne microplastics (MPs).

systems (Huang et al., 2023; Abbasi et al., 2019). The MPs detection and analysis procedures are shown in Figure 1.4. In this sampling collection system, dry or wet atmospheric particles are collected in a glass container through a funnel that is made of stainless steel or glass and has a smooth surface (Wang et al., 2022; Shao et al., 2022). The instrument must be physically covered during the dry sample collection periods to protect it from rain or snow. However, water-soluble pollution can contaminate the wet deposition collection techniques. Another passive sampling technique is collecting airborne particles in a petri dish with a glass slide (Zhang et al., 2020). In recent years, the Norwegian Institute for Air Research has designed a metallic/glass standardized collection system, considered a plastic-free standardized method (Zhang et al., 2020). However, although passive sampling can collect an extensive range of particle-size MPs, it cannot collect suspended MPs, showing significant drawbacks of this collection system. On the other hand, active sampling methods generally function based on pumping sampler systems. This technique pumps a specific amount of air within a certain time frame (Zhao et al., 2023). In this procedure, particulate matter is deposited when air passes through the sampler. Previously, this technique was used to collect 10 PM2.5 and PM1 in the air, but currently, it is also used to study MPs. However, this technique has a significant advantage in accurately collecting atmospheric MPs from outdoor or indoor environments within a certain period.

1.4.2 Sample Preparation

The particles collected by passive or active sampling collection systems contain a mixture of microplastic, natural, and anthropogenic particles. Hence, separating microplastic particles from the mixed mixture of natural and anthropogenic particles is crucial. Although there are no recognized standardized techniques for sample preparation, published previous research has suggested density separation and chemical digestion that can be used to remove non-microplastic organic matter. It is found that organic substances are generally eliminated using hydrogen peroxide (H_2O_2), hydrogen nitrate (HNO_3), potassium hydroxide (KOH), and sodium hydroxide (NaOH) (Patchaiyappan et al., 2021; Qiu et al., 2020). In this digestion system, collected samples, ultrapure water, and the solution of H_2O_2/HNO_3/KOH/NaOH are introduced to the beaker (Wang et al., 2022; Shao et al., 2022). However, most research regarding atmospheric MPs has used a 30% H_2O_2 solution. Then, this beaker is kept on a heated plate at 75°C in a fume hood to initiate digestion while a magnetic stirrer is utilized to achieve the mixture's homogenous blending. The procedure will be continued until there are no visible bubbles in the beaker. The mixture would then be covered with aluminum foil and settled for a few days. Inside the walls of the beaker, deionized water is generally sprayed to remove particulates adhering to the walls. On the other hand, various salt solutions based on density, such as $ZnCl_2$ (1.60 g/cm^3), NaCl (1.20 g/cm^3), and NaI (1.80 g/cm^3), are generally used for density separation (Abbasi et al., 2019; Myszka et al., 2023). However, $ZnCl_2$, with a density of 1.60 g/cm^3, has been rapidly used in most studies due to its low cost and environmental friendliness. In the density separation process, it is necessary to prevent the physical damage of MPs. Therefore, the solution should be kept moving constantly to avoid turbulence.

1.4.3 Instrumental Analysis of Microplastics (MPs)

After the chemical digestion and density separation process, a vacuum pump and filter paper are used for filtration (Shen et al., 2022; Myszka et al., 2023). After filtration, filter paper should be dried to identify and quantify MPs' physical and chemical characteristics. Stereomicroscopes with various models, scanning electron microscopy (SEM), and energy dispersive X-ray spectroscopy (EDS) are used for identifying the MPs' physical characteristics (color, shape, and size) and chemical composition.

1.4.4 Gaps and Future Research on the Detection and Analysis of Microplastics (MPs)

- There remains a lack of clear information regarding the standard MPs detection and analysis techniques. As a result, it will not be possible to compare the MP's results from one place to another. Hence, future research needs to focus on making a uniform standard MPs detection and analysis technique.

- Stable carbon isotope ratio mass spectrometry (IRMS), along with FTIR and SEM-EDS, should be used for the identification of sources as well as physical and chemical properties. Moreover, future research needs to focus on making technologies more efficient to analyze MPs.

1.5 EFFECTS OF MICROPLASTICS ON ECOSYSTEMS AND HUMAN HEALTH

Over the past decades, humans and aquatic and terrestrial ecosystems have been exposed to MPs, which pose a severe risk to human health and ecosystems. Moreover, research studies only focus on the direct effects of MPs on the environment and human health. Notably, after entering into the environment, MPs can undergo several degradation processes and adsorb with other toxic pollutants, indicating that MPs can also have some indirect effects. Hence, it is crucial to discuss this issue. By prompting this context, this section presents the effects of MPs on human health and ecosystems.

1.5.1 Effects of Microplastics (MPs) on Ecosystems

After releasing the environment, MPs are distributed around the environmental compartments through runoff, precipitation, wind flow, and diffusion (Huang et al., 2023). However, nature's dynamism can transport them from urban areas to remote areas (Monira et al., 2021). Once MPs enter the aquatic and soil compartments, it slowly begins to cause damage (Wang et al., 2020). Recent studies have shown that atmospheric MPs can create a new niche with airborne microbes (Song et al., 2024). After that, these microbes can enter our surface by atmospheric deposition, making the plants more vulnerable. Moreover, biofilms containing various bacterial phyla can be present in MPs (Zhang et al., 2020). However, airborne MPs could absorb and scatter radiation from the sun and the ground, which can affect the Earth's climate and the photosynthesis of plants (Zhao et al., 2023). It has been found that impaired reproduction, bioaccumulation, biomagnification, disruptions in organisms' behavior, and ecological implications occurred due to the massive presence of MPs in aquatic ecosystems (Rahman et al., 2021). Sometimes, fish, zooplankton, and invertebrates can mistakenly consume MPs, creating physical damage, blockages in the digestive system, reduced nutrient absorption, and impaired feeding capacity (Thacharodi et al., 2024). After entering the organism's body, MPs can interact with cellular components of organisms and trigger various toxicological responses (Yang et al., 2021). However, this cellular toxicity can occur through some mechanisms. First, MPs create physical interactions with cells, creating oxidative stress and inflammation (Enyoh et al., 2019). Then, chemical additives of MPs can lead to toxic effects on cells. Lastly, the large surface area of MPs can enhance the bioavailability of toxic pollutants and generate toxicity to organism cells (Zhang et al., 2020). Moreover, due to this cellular toxicity, DNA damage can occur in aquatic organisms (Wang et al., 2020). MPs can act as a barrier to critical cellular processes and oxidative stress and impair coral immune responses (Shao et al., 2022). Moreover, it can modify the coral's surface area, volume, and calcification rate, leading to tissue necrosis and impaired reproduction (Sridharan et al., 2021). As MPs have already been distributed with a massive concentration, moving towards effective reduction techniques is crucial.

1.5.2 Effects of Microplastics (MPs) on Human Health

MPs are generally exposed to the human body through inhalation, ingestion, and dermal contact (Zhao et al., 2023). The effects of MPs on human health are illustrated in Figure 1.5. Recently, MPs have been detected in various foods such as seafood, sea salt, honey, sugar, beer, drinking water, fast food, frozen food, and fish (Liu et al., 2024). Thus, people easily contaminate by taking these types of food. It has been found that 142, 113, 126, and 106 MPs/day were consumed by American male adults, female adults, male children, and female children, respectively, whereas European adults were exposed to MPs with 37–100 persons/year through the consumption of table salt (Shao et al., 2022; Ageel et al., 2022). Moreover, according to

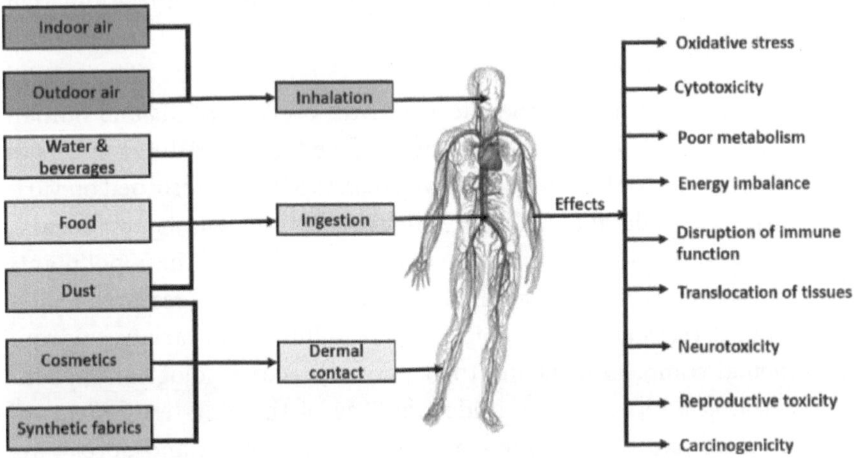

FIGURE 1.5 The effects of microplastics (MPs) on human health.

research, on average, a person ingests 4,000 MPs/year from drinking water, 11,000 MPs/year from shellfish, and (37–1,000) MPs/year from edible sea (Zhao et al., 2023). On the other hand, inhalation has recently become a significant route of human exposure to MPs. Moreover, a considerable abundance of MPs has been consistently detected in indoor and outdoor environments (Ageel et al., 2022). Research studies found that male and female MPs inhalation rates were 170/day and 132/day, respectively, while children were exposed to 110 MPs/day and 97 MPs/day MPs (Ageel et al., 2022). However, dermal contact is considered a less significant pathway of MPs exposure (Zhang et al., 2020). In this pathway, particle size requires <100 nm for direct absorption, showing that only nanoplastics enter the skin through this pathway (Zhang et al., 2020). After entering the human body, MPs rapidly spread throughout the body by circulatory systems, causing detrimental effects on the organs and whole systems of the body (Shao et al., 2022). Moreover, it can enhance the glutathione *S*-transferase activity, generate reactive oxygen species, increase mitogen-activated protein kinase signaling pathways, and reduce acetylcholinesterase inhibition and lipid digestion in human body systems (Rahman et al., 2021). As a result, the human body moves towards oxidative stress, cytotoxicity, poor metabolism, energy imbalance, disruption of immune function, translocation to tissues, neurotoxicity, reproductive toxicity, and carcinogenicity.

1.5.3 Gaps and Future Research on the Effects of Microplastics on Human Health and Ecosystems

The following is a list of important research gaps and modifications that are needed:

- Research focusing on the effects of MPs on ecosystems and human health is limited. Most of the studies focused on the direct effects of MPs on human health and ecosystems. As MPs are adsorbed on various toxic pollutants (POPs, chemical additives, and heavy metals), it is crucial to understand the combined effects of these pollutants along with MPs. Hence, there is a need for additional research to better understand the persistence of these pollutants in various environmental compartments and their potential ecotoxicological impacts. Gaining a more precise understanding of the long-term effects of MPs will aid in assessing their risks to ecosystems and developing appropriate mitigation measures.

- Despite the significant number of scientific studies conducted on the effects of MPs in environmental compartments, there remains a lack of clear information regarding the toxicity of microplastics in soil biota (earthworms, nematodes, and ciliates) and terrestrial plants (crops). Understanding the impact of MPs on soil biota and terrestrial plants is crucial for assessing their potential risks to ecosystems and human health.
- The persistence and ecotoxicological risks associated with MPs require further investigation. While existing studies have shed light on the presence and effects of MPs, there is a need for additional research to better understand the risks of various polymers in various environmental compartments and their reference value.
- The impacts of nanoplastics on human health and ecosystems are yet to be discovered. Hence, future studies need to investigate the role of nanoplastics in disrupting ecosystems and human health. Moreover, the impact of MPs and nanoplastics on human health and ecosystems should be investigated based on various places and people to assess their risks precisely in a particular area.
- In vivo, experiments and chemometric models should be conducted to assess and interpret ecological risks in a particular area.

1.6 MANAGEMENT OF MICROPLASTICS IN ENVIRONMENTAL COMPARTMENTS

The contamination of MPs significantly impacts environmental compartments as they can lead to ecotoxicological effects in our mother nature. Therefore, it is essential to adopt appropriate management strategies that can effectively reduce the entry of MPs into the environment. Thus, this section presents MPs' source reduction, mitigation, and remediation strategies in our environment.

1.6.1 Source Reduction

Source reduction is a vital approach to lessening the entry of MPs into environmental compartments (Thacharodi et al., 2024). In this strategy, regulatory measures, sustainable product design, and public awareness initiatives are taken to decrease the input of microplastics and prevent their release into environmental compartments (Kabir et al., 2023). Many countries worldwide have already taken significant steps and implemented various policies regarding the usage of plastic bags, plastic straws, and plastic packaging materials. Moreover, countries are trying to promote sustainable waste management practices, education, and awareness campaigns that can play a pivotal role in promoting behavioral changes among consumers to reduce plastic usage (Zhao et al., 2023). However, comprehensive waste management infrastructure and robust enforcement mechanisms act as irreplaceable components for reducing this emerging pollution.

1.6.2 Mitigation Approaches

Despite the high removal efficiency, conventional wastewater treatment techniques have some significant drawbacks, especially since these plants can discharge significant amounts of effluent. However, in recent years, a massive development has occurred in the field of wastewater treatment plants. Various advanced treatment processes are demonstrated in the following:

i. **Membrane Filtration Methods**

   Membrane filtration technologies such as dynamic membranes (DM), membrane bioreactors (MBRs), nanofiltration, ultrafiltration, and microfiltration have been favored to efficiently remove micro-or nanoplastics, as well as operational flexibility, low energy consumption, enhanced stability, and the potential to process large volumes of wastewater (Thacharodi et al., 2024). The performance of this technique depends on various factors, such as membrane material, pore size, MP concentration, and particle size distribution (Gandhi et al., 2022). Notably, the combination of MBRs and activated sludge processes has shown high removal efficiencies of 99.5% (Kumar et al., 2023). Various enzymes can also increase the removal of microplastics. To enhance efficiency and sustainability, membrane processes can be integrated with reverse osmosis (RO). Generally, this RO can filter out various impurities, contaminants, heavy metals, and salts (Lofty et al., 2022). According to the research, ultrafiltration (UF) can successfully remove 86% of MPs and is recognized as the most efficient and cost-effective method (Kumar et al., 2023). Furthermore, pilot-scale UF-MBR and RO-MBR hybrid treatments have also shown excellent removal efficiency at 90% and 99%, respectively (Thacharodi et al., 2024). The efficiency of MPs' mitigation approaches is shown in Table 1.4.

TABLE 1.4 Efficiency of Mitigation Approaches of MPs

Process	Size	Efficiency (%)	References
Electrocoagulation	300–355 μm	>90 (pH 3–10) and 99.24 (pH 7.5)	Dey et al. (2021); Padervand et al. (2020)
Filtration with biochar	10 μm	>95	Wang et al. (2020)
Microalgal biomass	20 μm	94.50	Sundaek et al. (2018)
Coagulation	<0.5 mm	45.34±3.93	Ma et al. (2020)
Reverse osmosis	0.1–1 μm	90	Mengesha and Sahu (2022)
Magnetic extraction	10–20 μm <1 mm	92 & 93	Dey et al. (2021)
Sol-gel method	1 mm	>95	Sturm et al. (2021)
Membrane filtration	20–300 μm	>94	Pizzichetti et al. (2021)
Metal-organic framework	—	95.5±1.2	Chen et al. (2020)

ii. **Coagulation**

Coagulation is widely recognized due to its cost-effectiveness and high efficiency (Ma et al., 2020). Iron- or aluminum-based salts are generally used to generate flocs by trapping the particles (Kumar et al., 2023). The aluminum-based method has higher removal efficiency, minor clogs, and a higher surface area, making it preferable to the Iron-based method (Hanif et al., 2021). However, as coagulants cannot be able to separate the particles less than 1 μm and produce a tremendous amount of residual sludge, the modern world has started to use bio-coagulants instead of coagulants (Govindaraj et al., 2022). Over the years, electrocoagulation has been considered an effective removal of micro and nanoplastics (Thacharodi et al., 2024). This technique generates coagulants electrically, which destabilizes pollutants and removes contaminants. However, this technique's efficiency varies between (84% and 100%) (Valentin-Reyes et al., 2022).

iii. **Magnetic Extraction**

Magnetic extraction effectively removes micro- or nanoplastics from aquatic compartments (Dey et al., 2021). This technique generally uses external magnetic fields, magnetic seeds, and acids to improve separation (Shen et al., 2020). These techniques' efficiency in removing pollutants is (up to 98.3%) (Kumar et al., 2023). Iron nanoparticles are mainly chosen for this extraction due to their low cost, large surface area, and ferromagnetic properties (Goeppert & Goldscheider, 2021).

Lastly, biochar, sol-gel, advanced oxidation process, and metal-organic framework are some of the upgrading removal techniques and mitigation approaches that could be potential strategies for removing MPs from environmental compartments.

1.6.3 Remediation Techniques

Although source reduction and mitigation techniques are critical for controlling MPs pollution, remediation techniques are essential for removing existing MPs from the environmental compartments (Shen et al., 2022). However, MPs are pretty impossible to remove from environmental compartments, but it can be possible to reduce their concentrations by taking various remediation approaches. Despite the available conventional techniques, research regarding removing MPs and nanoplastics by upgrading techniques is still limited. Thus, it is indispensable to promote and implement the upgrading removal strategies. However, over the past years, various types of bacteria and fungi have been successfully used in the MPs removal process (Kumar et al., 2023; Thacharodi et al., 2024). It has been found that certain bacteria and fungi can utilize the hydrocarbons in plastic and degrade plastic fragments (Zhai et al., 2023). Microorganisms can participate in the plastic degradation process, where they decompose and consume various plastic molecules by active enzymes. Microorganisms can use plastic products as substrates under aerobic conditions and produce carbon dioxide, water, and biomass (Nabi et al., 2023). On the other hand, microorganisms can convert the carbon chain of plastics into acetic acid. After that, methanogens can further decompose and generate methane (Shen et al., 2022). According to the research, Bacillus sp. (strain 27) can modify the surface structures of MPs (Thacharodi et al., 2024)—*Bacillus* sp. YP1 can degrade 10.7% of the plastic polymers of PE within 2 months (Yang et al., 2014). Several strains of bacteria, such as *Pseudomonas*, *Staphylococcus*, and *Streptococcus*, are capable of degrading PET and PS (Thacharodi et al., 2024). Furthermore, several strains of the fungi (*Aspergillus*, *Penicillium*, and *Trichoderma* genera) and other types of organisms (*Acremonium license*, *Fusarium redolent*, and *Verticillium lecanii*) can efficiently degrade PE and PET (Nabi et al., 2023; Xiang et al., 2023). Bacteria and fungi capable of degrading MPs are given in Table 1.5.

TABLE 1.5 The Efficiency of Various Bioremediation Techniques

Strain	Degrading Polymers	Efficiency (%)	Time Span (Days)	References
Pseudomonas lysinibacillus	HDPE	6.4	40	Zhai et al. (2023)
Rhodococcus ruber	PE	0.86	7	Zhai et al. (2023)
Paenibacillus amylolyticus (TB-13)	PLA	—	14	Nabi et al. (2023)
Pseudomonas aeruginosa	PS-PLA	10	21	Sharma and Neelam (2023)
Enterobacter asburiae (strain YT1)	PE	6.10	28	Nabi et al. (2023)
Comamonas acidovorans (TB-35)	PUR	100	8	Rajan and Vijayalakshmi (2023)
Aspergillus sp. (VRKPT1)	PE	6.02	30	Viel et al. (2023)
Gloeophyllum trabeum	PS	50	20	Nabi et al. (2023)
Pestalotiopsis microspora	PUR	—	14	Xiang et al. (2023)
Penicillium pinophilum	LDPE	0.64	930	Nabi et al. (2023)
Penicillium simplicissimum (YK)	PE	0.50	90	Nabi et al. (2023)

Land treatment technology is considered one of the most upgraded MP removal technologies (Shen et al., 2022). Combining vegetation and microorganisms removes MPs in the constructed wetland technique (land treatment technology) (Shen et al., 2022). MPs can efficiently capture and break down by this technique. In this technology, aquatic vascular plants play a vital role in stabilizing, capturing, and degrading plastics within the surrounding environment (Dovidat et al., 2020). Generally, MPs tend to float on water surface area and interact with vegetation in this technique. This approach has successfully removed diverse pollutants, including dissolved contaminants and engineered nanoparticles.

1.6.4 Gaps and Future Research on the Source Reduction, Mitigation, and Remediation Techniques of Microplastic Pollution

- Research focusing on the remediation of MPs techniques is still limited. Investigating how microorganisms contribute to removing MPs from environmental compartments is essential for developing upgrading technologies. Moreover, it is crucial to implement microplastic remediation technologies at appropriate sources to prevent contamination in different ecosystems and atmospheres effectively.

- Research should focus on designing hybrid remediation techniques that can help achieve sustainable development goals (SDGs) and more efficiency.
- Despite many countries taking various legislative approaches, standard legislative approaches regarding microplastic pollution are still limited. Hence, establishing practical policy and management tools could help reduce the impact of MPs on environmental compartments.

1.7 CONCLUSION

From terrestrial and aquatic environments to the Himalayan mountainous regions, airborne MPs have been found everywhere and seriously threaten the ecosystem and human health. However, there has been limited research specifically dedicated to studying airborne MPs compared to MPs found in marine, freshwater, and terrestrial ecosystems. Hence, this chapter aimed to identify knowledge gaps and establish the criteria for further research in the domain of microplastic pollution. According to the analysis of existing research papers, the contribution percentage of MPs from various data sources, data standardization, sampling techniques, standard uniform units, and remediation technologies still need to be improved. Moreover, as airborne MPs are transboundary pollutants, the influence of meteorological factors on their transport at the local, regional, and international levels must be addressed further. Investigating how meteorological conditions influence the transit and settlement of MPs in environmental contexts is critical for creating effective measures to reduce the ecological impact of airborne MPs. Furthermore, it is necessary to detect and quantify air pollutants (aerosols, organic, and inorganic hazardous pollutants) adsorbed on MPs and establish a quantitative link between MPs and these contaminants. Nevertheless, most of the research has concentrated on examining the direct impacts of MPs on human health and ecosystems. As MPs can adsorb harmful pollutants such as persistent organic pollutants (POPs), chemical additives, and heavy metals, it is imperative to comprehend the cumulative impacts of these pollutants in conjunction with MPs. While several studies have examined MPs' morphological traits and abundance in different environmental compartments, more research is still needed to determine the temporal and spatial distribution. Research on the vertical distribution of MPs is vital to evaluating the life cycle of MPs and the historical record of MP pollution in a particular aquatic body. However, it has been observed that bioremediation technology has many advantages, such as lower costs, better environmental compatibility, and more societal acceptance. The advancements in molecular biology and nanotechnology have facilitated the development of novel approaches for the purification of MPs. Lastly, standard rules and the prioritization of environmentally friendly technology are required to reduce the risks associated with MPs adequately.

REFERENCES

Abayomi, O. A., Range, P., Al-Ghouti, M. A., Obbard, J. P., Almeer, S. H., & Ben-Hamadou, R. (2017). Microplastics in coastal environments of the Arabian Gulf. *Mar. Pollut. Bull.* 124(1), 181–188.

Abbasi, S., Keshavarzi, B., Moore, F., Turner, A., Kelly, F. J., Dominguez, A. O., & Jaafarzadeh, N. (2019). Distribution and potential health impacts of microplastics and microrubbers in air and street dusts from Asaluyeh County, Iran. *Environ. Pollut.* 244, 153–164. https://doi.org/10.1016/j.envpol.2018.10.039

Abbasi, S., Rezaei, M., Ahmadi, F., & Turner, A. (2022). Atmospheric transport of microplastics during a dust storm. *Chemosphere* 292, 133456.

Ageel, H. K., Harrad, S., & Abdallah, M. A. E. (2022). Occurrence, human exposure, and risk of microplastics in the indoor environment. *Environ. Sci. Process. Impacts*, 24(1), 17–31.

Akdogan, Z., & Guven, B. (2019). Microplastics in the environment: A critical review of current understanding and identification of future research needs. *Environ. Pollut.* 254, 113011.

Alam, F. C., Sembiring, E., Muntalif, B. S., & Suendo, V. (2019). Microplastic distribution in surface water and sediment river around slum and industrial area (case study: Ciwalengke River, Majalaya district, Indonesia). *Chemosphere* 224, 637–645. https://doi.org/10.1016/j.chemosphere.2019.02.188

Alam, M. J., Shammi, M., & Tareq, S. M. (2023). Distribution of microplastics in shoreline water and sediment of the Ganges River Basin to Meghna Estuary in Bangladesh. *Ecotoxicol. Environ. Saf.* 266, 115537.

Allen, S., Allen, D., Phoenix, V., Roux, G. L., Jiménez, P. D., Simonneau, A., Binet, S., & Galop, D. (2019). Atmospheric transport and deposition of microplastics in a remote mountain catchment. *Nat. Geosci.* 12(5), 339–344.

Ambrosini, R., Azzoni, R. S., Pittino, F., Diolaiuti, G., Franzetti, A., & Parolini, M. (2019). First evidence of microplastic contamination in the supraglacial debris of an alpine glacier. *Environ. Pollut.* 253, 297–301. https://doi.org/10.1016/j.envpol.2019.07.005

Aslam, I., Qadir, A., & Ahmad, S. R. A. (2022). Preliminary assessment of microplastics in indoor dust of a developing country in South Asia. *Environ. Monit. Assess* 194(5), 340.

Bagaev, A., Khatmullina, L., & Chubarenko, I. (2018). Anthropogenic microlitter in the Baltic Sea water column. *Mar. Pollut. Bull.* 129(2), 918–923.

Banik, P., Anisuzzaman, M., Bhattacharjee, S., Marshall, D. J., Yu, J., Nur, A. A. U., & Hossain, M. B. (2024). Quantification, characterization and risk assessment of microplastics from five major estuaries along the northern Bay of Bengal coast. *Environ. Poll.* 342, 123036.

Beaurepaire, M. (2021). Improving the understanding of microplastic sources and fate through atmospheric deposition monitoring during lockdown. *Available at SSRN 4523820.*

Belkhiri, A. H., Carre, F., & Quiot, F. (2022). State of knowledge and future research needs on microplastics in groundwater. *J. Water Health* 20(10), 1479–1496.

Bergmann, M., Mützel, S., Primpke, S., Tekman, M. B., Trachsel, J., & Gerdts, G. (2019). White and wonderful? Microplastics prevail in snow from the Alps to the Arctic. *Science Advances* 5(8), eaax1157.

Blair, R. M., Waldron, S., Phoenix, V. R., & Gauchotte-Lindsay, C. (2019). Microscopy and elemental analysis characterisation of microplastics in sediment of a freshwater urban river in Scotland, UK. *Environ. Sci. Pollut. Res.* 26, 12491–12504.

Brahney, J., Hallerud, M., Heim, E., Hahnenberger, M., & Sukumaran, S. (2020). Plastic rain in protected areas of the United States. *Science* 368(6496), 1257.

Brahney, J., Mahowald, N., Prank, M., Cornwell, G., Klimont, Z., Matsui, H., & Prather, K. A. (2021). Constraining the atmospheric limb of the plastic cycle. *Proc. Natl. Acad. Sci.* 118.

Cai, L., Wang, J., Peng, J., Tan, Z., Zhan, Z., Tan, X., & Chen, Q. (2017). Characteristic of microplastics in the atmospheric fallout from Dongguan city, China: Preliminary research and first evidence. *Environ. Sci. Pollut. Res.* 24, 24928–24935. https://doi. org/10.1007/s11356-017-0116-x

Campanale, C., Stock, F., Massarelli, C., Kochleus, C., Bagnuolo, G., Reifferscheid, G., & Uricchio, V. F. (2020). Microplastics and their possible sources: The example of Ofanto River in southeast Italy. *Environ. Pollut.* 258, 113284.

Can-Güven, E. (2021). Microplastics as emerging atmospheric pollutants: A review and bibliometric analysis. *Air Qual. Atmos. Health* 14(2), 203–215.

Catarino, A. I., Valeria, M., William, G. S., Richard, C. T., & Theodore, B. H. (2018). Low levels of microplastics (MP) in wild mussels indicate that MP ingestion by humans is minimal compared to exposure via household fibres fall-out during a meal. *Environ. Pollut.* 237, 675–684. https://doi.org/10.1016/j.envpol.2018.02.069

Chen, G., Fu, Z., Yang, H., & Wang, J. (2020). An overview of analytical methods for detecting microplastics in the atmosphere. *TrAC Trends Anal. Chem.* 130, 115981.

Chen, L., Zhou, S., Zhang, Q., Su, B., Yin, Q., & Zou, M. (2024). Global occurrence characteristics, drivers, and environmental risk assessment of microplastics in lakes: A meta-analysis. *Environ. Pollut.* 344, 123321.

Chen, Y., Li, X., Zhang, X., Zhang, Y., Gao, W., Wang, R., & He, D. (2022). Air conditioner filters become sinks and sources of indoor microplastics fibers. *Environ. Pollut.* 292, 118465.

Dehghani, S., Moore, F., & Akhbarizadeh, R. (2017). Microplastic pollution in deposited urban dust, Tehran metropolis. *Iran. Environ. Sci. Pollut. Res.* 24, 20360–20371. https://doi.org/10.1007/s11356-017-9674-1

Dey, T. K., Uddin, M. E., & Jamal, M. (2021). Detection and removal of microplastics in wastewater: evolution and impact. *Environ. Sci. Pollut. Res.* 28(14), 16925–16947. https://doi.org/10.1007/s11356-021-12943-5

Ding, L., Mao, R. F., Guo, X., Yang, X., Zhang, Q., & Yang, C. (2019). Microplastics in surface waters and sediments of the Wei River, in the northwest of China. *Sci. Total Environ.* 667, 427–434. https://doi.org/10.1016/j.scitotenv.2019.02.332

Dovidat, L. C., Brinkmann, B. W., Vijver, M. G., & Bosker, T. (2020). Plastic particles adsorb to the roots of freshwater vascular plant Spirodela polyrhiza but do not impair growth. *Limnol. Oceanogr. Lett.* 5, 37–45.

Dris, R., Gasperi, J., Mirande, C., Mandin, C., Guerrouache, M., Langlois, V., & Tassin, B. (2017). A first overview of textile fibers, including microplastics, in indoor and outdoor environments. *Environ. Pollut.* 221, 453–458. https://doi.org/10.1016/j.envpol.2016.12.013

Dris, R., Gasperi, J., Rocher, V., Saad, M., Renault, N., & Tassin, B. (2015). Microplastic contamination in an urban area: A case study in greater Paris. *Environ. Chem.* 12, 592. https://doi.org/10.1071/en14167

Dris, R., Gasperi, J., Saad, M., Mirande, C., & Tassin, B. (2016). Synthetic fibers in atmospheric fallout: A source of microplastics in the environment? *Mar. Pollut. Bull.* 104, 290–293. https://doi.org/10.1016/j.marpolbul.2016.01.006

Eberhard, T., Casillas, G., Zarus, G. M., & Barr, D. B. (2024). Systematic review of microplastics and nanoplastics in indoor and outdoor air: Identifying a framework and data needs for quantifying human inhalation exposures. *J. Exposure Sci. Environ. Epidemiol.*, 34, 1–12.

Enyoh, C. E., Verla, A. W., Verla, E. N., Ibe, F. C., & Amaobi, C. E. (2019). Airborne microplastics: A review study on method for analysis, occurrence, movement and risks. *Environ. Monit. Assess.* 191, 1–17.

Eo, S., Hong, S. H., Song, Y. K., Han, G. M., & Shim, W. J. (2019). Spatiotemporal distribution and annual load of microplastics in the Nakdong River, South Korea. *Water Res.* 160, 228–237. https://doi.org/10.1016/j.watres.2019.05.053

Evangeliou, N., Grythe, H., Klimont, Z., Heyes, C., Eckhardt, S., Lopez-Aparicio, S., & Stohl, A. 2020. Atmospheric transport is a major pathway of microplastics to remote regions. *Nat. Commun.* 11(1), 3381.

Ferrero, L., Scibetta, L., Markuszewski, P., Mazurkiewicz, M., Drozdowska, V., Makuch, P., Jutrzenka-Trzebiatowska, P., Zaleska-Medynska, A., Andò, S., Saliu, F., Douglas Nilsson, E., & Bolzacchini, E. (2022). Airborne and marine microplastics from an oceanographic survey at the Baltic Sea: An emerging role of air-sea interaction? *Sci. Total Environ.* 824, 153709.

Gandhi, T. P., Sampath, P. V., & Maliyekkal, S. M. (2022). A critical review of uranium contamination in groundwater: Treatment and sludge disposal. *Sci. Total Environ.* 825, 153947. https://doi.org/10.1016/j.scitotenv.2022.153947

Gaston, E., Woo, M., Steele, C., Sukumaran, S., & Anderson, S. (2020). Microplastics differ between indoor and outdoor air masses: Insights from multiple microscopy methodologies. *Appl. Spectrosc.* 74, 1079–1098. https://doi.org/10.1177/0003702820920652

Gerolin, C. R., Pupim, F. N., Sawakuchi, A. O., Grohmann, C. H., Labuto, G., & Semensatto, D. (2020). Microplastics in sediments from Amazon Rivers, Brazil. *Sci. Total Environ.* 749, 141604. https://doi.org/10.1016/j.scitotenv.2020.141604

Goeppert, N., & Goldscheider, N. (2021). Experimental field evidence for transport of microplastic tracers over large distances in an alluvial aquifer. *J. Hazard. Mater.* 408, 124844. https://doi.org/10.1016/j.jhazmat.2020.124844

González-Pleiter, M., Edo, C., Aguilera, Á., Viúdez-Moreiras, D., Pulido-Reyes, G., González-Toril, E., Osuna, S., de Diego-Castilla, G., Leganés, F., Fernández-Piñas, F., & Rosal, R. (2021). Occurrence and transport of microplastics sampled within and above the planetary boundary layer. *Sci. Total. Environ.* 761, 143213.

Govindaraj, V., Manoharan, K., & Govindarajan, M. (2022). Cost-effective treatment for polluted groundwater using eco-friendly bio-coagulants 2022. https://doi.org/10.21203/rs.3.rs-1375291/v1

Guo, J. J., Huang, X. P., Xiang, L., Wang, Y. Z., Li, Y. W., Li, H., & Wong, M. H. (2020). Source, migration and toxicology of microplastics in soil. *Environ. Int.* 137, 105263.

Han, Y., Cheng, J., He, L., Zhang, M., Ren, S., Sun, J., & Tang, Z. (2021). Polybrominated diphenyl ethers in soils from Tianjin, North China: Distribution, health risk, and temporal trends. *Environ. Geochem. Health* 43, 1177–1191.

Hanif, M. A., Ibrahim, N., Dahalan, F. A., Ali, U. F. M., Hasan, M., & Jalil, A. A. (2021). Microplastics and nanoplastics: Recent literature studies and patents on their removal from aqueous environment. *Sci. Total Environ.* 810, 152115.

Haque, M. R., Ali, M. M., Ahmed, W., Siddique, M. A. B., Akbor, M. A., Islam, M. S., & Rahman, M. M. (2023). Assessment of microplastics pollution in aquatic species (fish, crab, and snail), water, and sediment from the Buriganga River, Bangladesh: An ecological risk appraisals. *Sci. Total Environ.* 857, 159344.

Hossain, M. N., Rahman, M. M., Afrin, S., Akbor, M. A., Siddique, M. A. B., & Malafaia, G. (2023). Identification and quantification of microplastics in agricultural farmland soil and textile sludge in Bangladesh. *Sci. Total Environ.* 858, 160118.

Hu, T., He, P., Yang, Z., Wang, W., Zhang, H., Shao, L., & Lü, F. (2022). Emission of airborne microplastics from municipal solid waste transfer stations in downtown. *Sci. Total Environ.* 828, 154400.

Huang, D., Li, X., Ouyang, Z., Zhao, X., Wu, R., & Zhang, C. (2021). The occurrence and abundance of microplastics in surface water and sediment of the West River downstream, in the south of China. *Sci. Total Environ.* 756, 143857. https://doi.org/10.1016/j.scitotenv.2020.143857

Huang, X., Chen, Y., Meng, Y., & Liu, G. (2023). Mitigating airborne microplastics pollution from perspectives of precipitation and underlying surface types. *Water Res.* 243, 120385.

Jahan, S., Strezov, V., Weldekidan, H., Kumar, R., Kan, T., Sarkodie, S. A., & Wilson, S. P. (2019). Interrelationship of microplastic pollution in sediments and oysters in a seaport environment of the eastern coast of Australia. *Sci. Total Environ.* 695, 133924.

Jiang, C., Yin, L., Wen, X., Du, C., Wu, L., & Long, Y. (2018). Microplastics in sediment and surface water of west Dongting Lake and south dongting lake: Abundance, source and composition. *Int. J. Environ. Res. Public Health* 15(10), 2164. https://doi.org/10.3390/ijerph15102164

Kabir, M. S., Wang, H., Luster-Teasley, S., Zhang, L., & Zhao, R. (2023). Microplastics in landfill leachate: Sources, detection, occurrence, and removal. *Environ. Sci. Ecotechnol.* 16, 100256.

Kannankai, M. P., & Devipriya, S. P. (2024). Atmospheric microplastic deposition in a coastal city of India: The influence of a landfill source on monsoon winds. *Sci. Total Environ.* 908, 168235.

Kashfi, F. S., Ramavandi, B., Arfaeinia, H., Mohammadi, A., Saeedi, R., De-la-Torre, G. E., & Dobaradaran, S. (2022). Occurrence and exposure assessment of microplastics in indoor dusts of buildings with different applications in Bushehr and Shiraz cities. *Iran. Sci. Total Environ.* 829, 154651. https://doi.org/10.1016/j.scitotenv.2022.154651

Kaya, A. T., Meral, Y., & Senem, C. B. (2018). Ubiquitous exposure to microfiber pollution in the air. *Eur. Phys. J. Plus* 133(488), 1–10.

Kitahara, K. I., & Nakata, H. (2020). Plastic additives as tracers of microplastic sources in Japanese road dusts. *Sci. Total Environ.* 736, 139694.

Klein, M., & Fischer, E. K. (2019). Microplastic abundance in atmospheric deposition within the Metropolitan area of Hamburg, Germany. *Sci. Total Environ.* 685, 96–103.

Kumar, V., Singh, E., Singh, S., Pandey, A., & Bhargava, P. C. (2023). Micro-and nano-plastics (MNPs) as emerging pollutant in ground water: Environmental impact, potential risks, limitations and way forward towards sustainable management. *Chem. Eng. J.* 459, 141568.

Liao, Z., Ji, X., Ma, Y., Lv, B., Huang, W., Zhu, X., Fang, M., Wang, Q., Wang, X., Dahlgren, R., & Shang, X. (2021). Airborne microplastics in indoor and outdoor environments of a coastal city in eastern China. *J. Hazard. Mater.* 417, 126007. https://doi.org/10.1016/j.jhazmat.2021.126007

Lin, L., Zuo, L. Z., Peng, J. P., Cai, L. Q., Fok, L., & Yan, Y. (2018). Occurrence and distribution of microplastics in an urban river: A case study in the Pearl River along Guangzhou City, China. *Sci. Total Environ.* 644, 375–381. https://doi.org/10.1016/j.scitotenv.2018.06.327

Liong, R. M. Y., Hadibarata, T., Yuniarto, A., Tang, K. H. D., & Khamidun, M. H. (2021). Microplastic occurrence in the water and sediment of Miri River Estuary, Borneo Island. *Water Air Soil Pollut.* 232(8), 1–12.

Liu, C., Li, J., Zhang, Y., Wang, L., Deng, J., Gao, Y., Yu, L., Zhang, J., & Sun, H. (2019a). Widespread distribution of PET and PC microplastics in dust in urban China and their estimated human exposure. *Environ. Int.* 128, 116–124. https://doi.org/10.1016/j.envint.2019.04.024

Liu, K., Wu, T., Wang, X., Song, Z., Zong, C., Wei, N., & Li, D. (2019b). Consistent transport of terrestrial microplastics to the ocean through atmosphere. *Environ. Sci. Technol.* 53, 10612–10619.

Liu, Q., Wu, D., Chen, Y., Chen, Z., Yuan, S., Yu, H., & Yao, W. (2024). Interaction and mechanistic studies of thiram and common microplastics in food and associated changes in hazard. *J. Hazard. Mater.* 461, 132464.

Lofty, J., Muhawenimana, V., Wilson, C. A. M. E., & Ouro, P. (2022). Microplastics removal from a primary settler tank in a wastewater treatment plant and estimations of contamination onto European agricultural land via sewage sludge recycling. *Environ. Pollut.* 304, 119198. https://doi.org/10.1016/jenvpol2022119198

Ma, H., Pu, S., Liu, S., Bai, Y., Mandal, S., & Xing, B. (2020). Microplastics in aquatic environments: Toxicity to trigger ecological consequences. *Environ. Pollut.* 261, 114089.

Mengesha, A., & Sahu, O. (2022). Sustainability of membrane separation technology on groundwater reverse osmosis process. *Cleaner Eng. Technol.* 7, 100457.

Monira, S., Bhuiyan, M. A., Haque, N., Shah, K., Roychand, R., Hai, F. I., & Pramanik, B. K. (2021). Understanding the fate and control of road dust-associated microplastics in stormwater. *Process Saf. Environ. Prot.* 152, 47–57.

Myszka, R., Enfrin, M., & Giustozzi, F. (2023). Microplastics in road dust: A practical guide for identification and characterisation. *Chemosphere* 315, 137757.

Nabi, I., Zaheer, M., Jin, W., & Yang, L. (2023). Biodegradation of macro-and micro-plastics in environment: A review on mechanism, toxicity, and future perspectives. *Sci. Total Environ.* 858, 160108.

Nakano, H., Alfonso, M. B., Jandang, S., Imai, K., & Arakawa, H. (2024). Microplastic pollution indexes in the coastal and open ocean areas around Japan. *Reg. Stud. Mar. Sci.* 69, 103287.

Narmadha, V. V., Jose, J., Patil, S., Farooqui, M. O., Srimuruganandam, B., Saravanadevi, S., & Krishnamurthi, K. (2020). Assessment of microplastics in roadside suspended dust from urban and rural environment of Nagpur, India. *Int. J. Environ. Res.* 14, 629–640.

O'Brien, S., Rauert, C., Ribeiro, F., Okoffo, E. D., Burrows, S. D., O'Brien, J. W., & Thomas, K. V. (2023). There's something in the air: A review of sources, prevalence and behaviour of microplastics in the atmosphere. *Sci. Total Environ.* 874, 162193.

Padervand, M., Lichtfouse, E., Robert, D., & Wang, C. (2020). Removal of microplastics from the environment. A review. *Environ. Chem. Lett.* 18(3), 807–828. https://doi.org/10.1007/s10311-020-00983-1

Patchaiyappan, A., Dowarah, K., Ahmed, S. Z., Prabakaran, M., Jayakumar, S., Thirunavukkarasu, C., & Devipriya, S. P. (2021). Prevalence and characteristics of microplastics present in the street dust collected from Chennai metropolitan city, India. *Chemosphere* 269, 128757.

Piehl, S., Leibner, A., Löder, M. G., Dris, R., Bogner, C., & Laforsch, C. (2018). Identification and quantification of macro-and microplastics on an agricultural farmland. *Sci. Rep.* 8(1), 17950.

Pizzichetti, A. R. P., Pablos, C., Alvarez-Fernández, C., Reynolds, K., Stanley, S., Marugán, J. (2021). Evaluation of membranes performance for microplastic removal in a simple and low-cost filtration system. *Case Studies Chem. Environ. Eng.* 3, 100075.

Qiu, R., Song, Y., Zhang, X., Xie, B., & He, D. (2020). Microplastics in urban environments: Sources, pathways, and distribution. In D. He, Y. Luo (Eds.), *Microplastics in Terrestrial Environments: Emerging Contaminants and Major Challenges*, pp. 41–61. Springer.

Qiu, X., Wang, X., Pan, J., Ding, L., Liang, X., & Guo, X. (2024). Significant contribution of different sources of particulate organic matter to the photoaging of microplastics. *Water Res.* 121173.

Rahman, A., Sarkar, A., Yadav, O. P., Achari, G., & Slobodnik, J. (2021). Potential human health risks due to environmental exposure to nano-and microplastics and knowledge gaps: A scoping review. *Sci. Total Environ.* 757, 143872.

Rajan, A., & Vijayalakshmi, S. (2023). New insights on polyurethane biodegradation. *AIP Conf. Proc.* 2690, 020060. AIP Publishing.

Rao, W., Fan, Y., Li, H., Qian, X., & Liu, T. (2024). New insights into the long-term dynamics and deposition-suspension distribution of atmospheric microplastics in an urban area. *J. Hazard. Mater.* 463, 132860.

Saikumar, S., Mani, R., Ganesan, M., Dhinakarasamy, I., Palanisami, T., & Gopal, D. (2024). Trophic transfer and their impact of microplastics on estuarine food chain model. *J. Hazard. Mater.* 464, 132927.

Scheurer, M., & Bigalke, M. (2018). Microplastics in Swiss floodplain soils. *Environ. Sci. Technol.* 52(6), 3591–3598.

Schmidt, L. K., Bochow, M., Imhof, H. K., & Oswald, S. E. (2018). Multi-temporal surveys for microplastic particles enabled by a novel and fast application of SWIR imaging spectroscopy – Study of an urban watercourse traversing the city of Berlin, Germany. *Environ. Pollut.* 239, 579–589. https://doi.org/10.1016/j.envpol.2018.03.097

Shafi, M., Lodh, A., Khajuria, M., Ranjan, V. P., Gani, K. M., Chowdhury, S., & Goel, S. (2024). Are we underestimating stormwater? Stormwater as a significant source of microplastics in surface waters. *J. Hazard. Mater.* 465, 133445.

Shao, L., Li, Y., Jones, T., Santosh, M., Liu, P., Zhang, M., Xu, L., Li, W., J., Yang, C.-X., Zhang, D., Feng, X., & BéruBé, K. (2022). Airborne microplastics: A review of current perspectives and environmental implications. *J. Clean. Prod.* 347, 131048.

Sharma, H., & Neelam, D. K. (2023). Understanding challenges associated with plastic and bacterial approach toward plastic degradation. *J. Basic Microbiol.* 63, 292–307.

Sharma, M. (2024). Microplastics in the atmosphere. In S. Mondal, P. Das, A. Mondal, S. Paul, J. K. Pandey, T. K. Das (Eds.), *Remediation of Plastic and Microplastic Waste*, pp. 60–70. CRC Press.

Shen, M., Song, B., Zhu, Y., Zeng, G., Zhang, Y., Yang, Y., Wen, X., Chen, M., & Yi, H. (2020). Removal of microplastics via drinking water treatment: Current knowledge and future directions. *Chemosphere* 251, 126612. https://doi.org/10.1016/j.chemosphere.2020.126612

Shen, M., Xiong, W., Song, B., Zhou, C., Almatrafi, E., Zeng, G., & Zhang, Y. (2022). Microplastics in landfill and leachate: Occurrence, environmental behavior and removal strategies. *Chemosphere* 305, 135325.

Shruti, V. C., Jonathan, M. P., Rodriguez-Espinosa, P. F., & Rodríguez-González, F. (2018). Microplastics in freshwater sediments of Atoyac River basin, Puebla City, Mexico. *Sci. Total Environ.* 654, 154–163. https://doi.org/10.1016/j.scitotenv.2018.11.054

Silva, P. M., & Nanny, M. A. (2020). Impact of microplastic fibers from the degradation of nonwoven synthetic textiles to the Magdalena River water column and river sediments by the City of Neiva, Huila (Colombia). *Water* 12(4), 1210. https://doi.org/10.3390/ W12041210

Singh, S., Malyan, S. K., Maithani, C., Kashyap, S., Tyagi, V. K., Singh, R., & Pandey, R. P. (2023). Microplastics in landfill leachate: Occurrence, health concerns, and removal strategies. *J. Environ. Manage.* 342, 118220.

Soltani, N. S., Taylor, M. P., & Wilson, S. P. (2021). Quantification and exposure assessment of microplastics in Australian indoor house dust. *Environmental Pollution* 283, 117064.

Song, H., Xiao, S., Zhou, X., Li, Y., Tao, M., Wu, F., & Xu, X. (2024). Temporal dynamics of bacterial colonization on five types of microplastics in a freshwater lake. *Sci. Total Environ.* 913, 169697.

Sridharan, S., Kumar, M., Singh, L., Bolan, N. S., & Saha, M. (2021). Microplastics as an emerging source of particulate air pollution: A critical review. *J. Hazard. Mater.* 418, 126245.

Stang, C., Mohamed, B. A., & Li, L. Y. (2022). Microplastic removal from urban stormwater: Current treatments and research gaps. *J. Environ. Manage.* 317, 115510.

Stanton, T., Johnson, M., Nathanail, P., MacNaughtan, W., & Gomes, R. L. (2019). Freshwater and airborne textile fibre populations are dominated by 'natural', not microplastic, fibres. *Sci. Total Environ.* 666, 377–389.

Sturm, M. T., Horn, H., & Schuhen, K. (2021). Removal of microplastics from waters through agglomeration-fixation using organosilanes—Effects of polymer types, water composition and temperature. *Water* 13(5), 675. https://doi.org/10.3390/w13050675

Sundaek, K. B., Koch, I. D. W., Villaro, C. G., Rasmussen, N. S., Holdt, S. L., & Hartmann, N. B. (2018). Sorption of fluorescent polystyrene microplastic particles to edible seaweed Fucus vesiculosus. *J. Appl. Phycol.* 30(5), 2923–2927.

Thacharodi, A., Meenatchi, R., Hassan, S., Hussain, N., Bhat, M. A., Arockiaraj, J., & Pugazhendhi, A. (2024). Microplastics in the environment: A critical overview on its fate, toxicity, implications, management, and bioremediation strategies. *J. Environ. Manage.* 349, 119433.

Torres-Agullo, A., Karanasiou, A., Moreno, T., & Lacorte, S. (2022). Airborne microplastic particle concentrations and characterization in indoor urban microenvironments. *SSRN Electron. J.* 308, 119707. https://doi.org/10.2139/ssrn.4058106

Trainic, M., Flores, J., Pinkas, I., Pedrotti, M., Lombard, F., Bourdin, G., Gorsky, G., Boss, E., Rudich, Y., Vardi, A., & Koren, I. (2020). Airborne microplastic particles detected in the remote marine atmosphere. *Commun. Earth Environ.* 1(1), 64.

Valentin-Reyes, J., Trejo, D. B., Coreno, O., & Nava, J. L. (2022). Abatement of hydrated silica, arsenic, and coexisting ions from groundwater by electrocoagulation using iron electrodes. *Chemosphere* 297, 134144. https://doi.org/10.1016/j.chemosphere.2022.134144

Viel, T., Manfra, L., Zupo, V., Libralato, G., Cocca, M., & Costantini, M. (2023). Biodegradation of plastics induced by marine organisms: Future perspectives for bioremediation approaches. *Polymers* 15, 2673.

Wang, C., Zhao, J., & Xing, B. (2021a). Environmental source, fate, and toxicity of microplastics. *J. Hazard. Mater.* 407, 124357.

Wang, G., Lu, J., Li, W., Ning, J., Zhou, L., & Tong, Y. (2021c). Seasonal variation and risk assessment of microplastics in surface water of the Manas River Basin, China. *Ecotoxicol. Environ. Saf.* 208, 111477. https://doi.org/10.1016/j.ecoenv.2020.111477

Wang, T., Niu, S., Wu, J., & Yu, J. (2022). Seasonal and daily occurrence of microplastic pollution in urban road dust. *J. Clean. Prod.* 380, 135025.

Wang, Y., Huang, J., Zhu, F., & Zhou, S. (2021b). Airborne microplastics: A review on the occurrence, migration and risks to humans. *Bull. Environ. Contamin. Toxicol.* 107(4), 657–664.

Wang, Z., Sedighi, M., & Lea-Langton, A. (2020). Filtration of microplastic spheres by biochar: removal efficiency and immobilisation mechanisms. *Water Res.* 184, 116165.

Wessel, C. C., Lockridge, G. R., Battiste, D., & Cebrian, J. (2016). Abundance and characteristics of microplastics in beach sediments: Insights into microplastic accumulation in northern Gulf of Mexico estuaries. *Mar. Pollut. Bull.* 109(1), 178–183.

Wright, S. L., Gouin, T., Koelmans, A. A., & Scheuermann, L. (2021). Development of screening criteria for microplastic particles in air and atmospheric deposition: Critical review and applicability towards assessing human exposure. *Microplastics Nanoplastics* 1(1), 1–18.

Wright, S. L., Ulke, J., Font, A., Chan, K. L. A., & Kelly, F. J. (2020). Atmospheric microplastic deposition in an urban environment and an evaluation of transport. *Environ. Int.* 136, 105411.

Xiang, P., Zhang, T., Wu, Q., & Li, Q. (2023). Systematic review of degradation processes for microplastics: Progress and prospects. *Sustainability* 15, 12698.

Yang, H., He, Y., Yan, Y., Junaid, M., & Wang, J. (2021). Characteristics, toxic effects, and analytical methods of microplastics in the atmosphere. *Nanomaterials* 11(10), 2747.

Yang, J., Yang, Y., Wu, W.-M., Zhao, J., & Jiang, L. (2014). Evidence of polyethylene biodegradation by bacterial strains from the guts of plastic-eating waxworms. *Environ. Sci. Technol.* 48, 13776–13784.

Zhai, X., Zhang, X.-H., & Yu, M. (2023). Microbial colonization and degradation of marine microplastics in the plastisphere: A review. *Front. Microbiol.* 14, 1127308.

Zhang, G. S., & Liu, Y. F. (2018). The distribution of microplastics in soil aggregate fractions in southwestern China. *Sci. Total Environ.* 642, 12–20.

Zhang, Q., Du, F., Liang, W., Chen, Q., Meng, J., & Shi, H. (2022). Microfiber fall-out during dining and potential human intake. *J. Hazard. Mater.* 430, 128477. https://doi.org/10.1016/j.jhazmat.2022.128477

Zhang, Y., Kang, S., Allen, S., Allen, D., Gao, T., & Sillanpää, M. (2020). Atmospheric microplastics: A review on current status and perspectives. *Earth Sci. Rev.* 203, 103118.

Zhao, X., Zhou, Y., Liang, C., Song, J., Yu, S., Liao, G., & Wu, C. (2023). Airborne microplastics: Occurrence, sources, fate, risks and mitigation. *Sci. Total Environ.* 858, 159943.

Zheng, Y., Li, J., Cao, W., Liu, X., Jiang, F., Ding, J., & Sun, C. (2019). Distribution characteristics of microplastics in the seawater and sediment: A case study in Jiaozhou Bay, China. *Sci. Total Environ.* 674, 27–35.

Zhou, B., Wang, J., Zhang, H., Shi, H., Fei, Y., Huang, S., & Barceló, D. (2020). Microplastics in agricultural soils on the coastal plain of Hangzhou Bay, east China: Multiple sources other than plastic mulching film. *J. Hazard. Mater.* 388, 121814.

Zhou, Q., Tian, C., & Luo, Y. (2017). Various forms and deposition fluxes of micro-plastics identified in the coastal urban atmosphere. *Chin. Sci. Bull.* 62, 3902–3909. https://doi.org/10.1360/n972017-00956

Zhou, Q., Zhang, H., Fu, C., Zhou, Y., Dai, Z., Li, Y., & Luo, Y. (2018). The distribution and morphology of microplastics in coastal soils adjacent to the Bohai Sea and the Yellow Sea. *Geoderma* 322, 201–208.

Zhu, J., Xu, A., Shi, M., Su, Y., Liu, W., Zhang, Y., & Qi, S. (2024). Atmospheric deposition is an important pathway for inputting microplastics: Insight into the spatiotemporal distribution and deposition flux in a mega city. *Environ. Pollut.* 341, 123012.

CHAPTER 2

Microplastics in the Environment

Sources and Impacts on Human Health

Usama Saleem, Mashal Shahzadi, Aqsa Nisar, Zeeshan Javed, Arooj Zaib Warraich, Fizza Qamar, Dilbar Hussain, Rashid Ali, Mubshar Saleem, Saba Naz, and Muhammad Asrar

2.1 INTRODUCTION

Due to their toxic effects on the aquatic environment and soil matrix, microplastics (MPs) have lately been recognized as an emergent class of contaminants. Microplastics are formed from primary sources (such as plastic pellets and beads in cosmetics) or secondary sources (like the deterioration of everyday plastics and washing of clothing and textiles) and are less than 5 mm in size. Due to the ecotoxicological threats they offer to soil and aquatic creatures as well as to people, MPs are a growing source of worry. The usage of MPs by various species can result in bioaccumulation and biomagnification throughout the food chain. Microplastics, which are less than 5 mm in size, are small enough for marine species to

DOI: 10.1201/9781003487579-2

easily consume them (Okeke et al., 2022). As a result, the food chain will be thrown off since the main consumers will absorb the MPs, pass them on to their secondary consumers, and eventually end up on human tables. Additionally, because to the hydrophobic properties of their surfaces. Microplastics, for example, can act as carriers for heavy metals and other pollutants such as polycyclic aromatic hydrocarbons (PAHs) and polychlorinated biphenyls (PCBs) (Yang et al., 2021a).

Depending on their weight and density, MPs can be carried through air or water currents in the environment. The three main processes that could move MPs from the pedosphere to the hydrosphere are rainfall, surface runoff, and ocean circulation. Due to ocean circulation, MPs can move not only from water to land but also from land to water (Zhang et al., 2019). Additionally, lighter and smaller MPs can be airborne MPs that travel to far-off locations like glacier zones and high mountains. Denser MPs could build up or be buried in the pedosphere (soil), Lighter microplastics, on the other hand, can be carried about the pedosphere by wind. Microplastics can be transported to surface waters (the hydrosphere) by heavy rainfall and surface runoff from urban and agricultural areas. According to studies, agricultural practices that use domestic sewage sludge as a soil amendment or plastic mulches to boost crop growth may introduce MPs to the soil. In addition, stormwater runoff transports MPs to nearby surface waters as a result of tire wear on the road (Horton et al., 2017). In addition, airborne MPs, which are composed of light fibers derived from trash incineration, landfills, and textiles, may be carried long distances to remote regions and left behind as atmospheric fallout.

2.2 MICROPLASTIC IN RUNOFF

In comparison to control and manure-amended plots, runoff from plots with land-applied biosolid had higher concentrations of MPs. The most found plastic in runoff was fibers and fragments, which decreased in surface roughness (Roychand et al., 2021). Although pathogens, heavy metals, and trace organic compounds may be present in biosolids, which have the potential to affect downstream water quality, they also contribute valuable nutrients. MPs have recently been discovered in biosolids at concentrations as high as 286 particles/g of sludge (dry weight) (Mohajerani and Karabatak, 2020).

MPs have been found in feces from both humans and animals as well as in the air, soil, and drinking water. MPs capacity to sorb pollutants like heavy metals and susceptibility to biofilm growth are both environmental and health concerns (Dong et al., 2021). For instance, MPs have the capacity to absorb nickel and Cr (VI), two of the most dangerous pollutants (Luqman et al., 2021). Additionally, MPs may be slow the growth of helpful soil microorganisms and alter the community structure and functions of terrestrial organisms like fungi (Fan et al., 2022).

MPs have been demonstrated to be transported through urban and suburban stormwater runoff (Mak et al., 2020). In retention ponds, the primary source of MP was found to be runoff (Liu et al., 2019a). MPs can enter wastewater treatment facilities (WWTP) via stormwater in locations with integrated sewage and stormwater collecting systems (Yang et al., 2021b). Over 95% of MPs removed from municipal WWTPs are finally collected as biosolids (Koutnik et al., 2021).

There aren't many studies that looked into how MPs move into agricultural soils. According to research, agricultural roots transport MPs upward, maintaining their presence in upper soil layers, and that MPs move vertically in soil with vegetation (Li et al., 2021). Smaller polymers with lower densities demonstrated greater mobility. Han et al. (2022) conducted a study on the influence of rain-induced runoff on the horizontal mobility of plastic particles. Municipal biosolids applied to land can transport MPs to surface water in a less well-known manner. Despite the fact that biosolids have the potential to be a substantial source of MP pollution, only a few studies have investigated MPs accumulation in soil from land-applied biosolids and WWTP sludges (Schell et al., 2022).

2.3 NATURAL AND ANTHROPOGENIC IMPACTS ON SOIL MICROPLASTICS

2.3.1 Impact of Biosolids on MPs

The recent discovery of microplastic presence in most ecosystems is increasing worldwide health concerns, but the destiny of MPs, particularly in the terrestrial environment, remains unknown. Microplastic particle concentrations in biosolids range from 506 to 15,385 per kg of biosolids. Microplastic numbers range from 18 to 6.9 10^5 particles/kg of soil after biosolid application on land, depending on soil makeup. Soil organic carbon content, Fe and Al oxide concentrations, and soil ionic strength all improve microplastic retention.

Crossman et al. investigated the impact of applying biosolids on the presence of MPs in soil and reported MP concentrations of 8.7 and 14 particles/g soil. They gathered soil samples both from and without a history of biosolid application, and they concluded that MP was exported to the soil at all sampling locations as a result of biosolid applications (Crossman et al., 2020). The objectives of this study were to compare biosolids and manure as two soil amendments that could be sources of MPs, to determine how applying biosolids to the land affects the concentration of MPs in agricultural runoff, to investigate preferential transport of various MP morphologies, and to estimate the potential for MP transport from corn and soybean production regions in the United States.

So far, little study has been conducted on the degradation of MPs in soil, which is a very slow process that predominantly results from biodegradation, while the impacts of soil environment, mechanical crushing, high-temperature oxidation, and UV radiation appear to have a minimal impact on degradation. Researchers discovered that plastic mulching film (PMF) PE MP fragments are resistant to soil degradation, persisting for years or even decades before finally forming small MP residues. It was discovered that after 800 days for PE and 1 year for PP, the weight loss of MPs in soil was only 0.1%–0.4%. MPs may remain in soils for up to 100 years because to low light and oxygen availability, which gradually limit photo-oxidative destruction. Furthermore, inefficient biodegradation of biodegradable polymers (oxo-biodegradable and compostable plastics used as carrier bags) is a key source of MPs in leachate generated by landfills and unregulated dumpsites (Schell et al., 2022).

2.3.2 Wastewater Discharge

Wastewater discharges and biosolids/SS, which are substantial byproducts of WWTPs, have been identified as major causes of MP contamination (Sun et al., 2019).

Particularly in soils where sludge compost has been applied, in coastal soils, in forested soils near farmlands, and in landfill refuse and leachates, MPs have been found. In addition to the previously mentioned methods, MPs also enter soil ecosystems through wastewater irrigation and agricultural mulching layers (van den Berg et al., 2020). Other sources of runoff include dumping grounds, beach trash, and agricultural, industrial, and urban runoff. Numerous studies have shown that the migration of MPs through the food chain profoundly modifies the biological community and that as MP particle sizes decrease, their accumulation and transmission in food chains rise until they finally reach people (Hurley & Nizzetto, 2018).

2.3.3 Application of Organic Fertilizers, and Irrigation

Organic fertilizers and composts derived from diverse wastes such as unsegregated municipal solid waste (MSW), SS, and animal manure are a possible source of MPs on agricultural areas because to the nature of the original feedstock for composting/digestion. It seems that the majority of studies in this area that have been published have looked into SS application to agricultural soils. According to estimates, each year between 2,800 and 19,000, 63,000–430,000, and 44,000–300,000 t of MPs are released into Australian, European, and North American agricultural fields, respectively (Ng et al., 2018). In agricultural soils that have not been amended with composts or organic fertilizers, the presence of MPs has also been noted; this finding may be related to the breakdown of PMFs. MP concentrations in 16 agricultural fields before and after SS application were compared in studies that demonstrated that SS application on land could increase MP abundance in soil and discovered that when 22–25 t/ha of SS with an average MP abundance of about 5.0×10^4 particles/kg sludge was applied, MP abundance in the top 30 cm soil layer increased by about 7.1×10^2 particles/kg soil. By taking into account MP abundances in SS and the amount of sludge applied to agricultural land, the amount of MPs transferred from SS sources to agricultural soil for the 13 countries under consideration could be estimated to range between 1.84×10^{11} and 7.76×10^{13} MP particles/annum, with the limits of this range corresponding to Ireland and Spain, respectively. Therefore, it makes sense to assume that the majority of MP particles that have accumulated in SSs eventually end up in soils. According to studies, up to 40% of MPs can be found in the effluent discharged by WWTPs, some of which enter agricultural soils through irrigation (Conley et al., 2019).

2.3.4 Mulching Activities

The widespread use of plastic as a covering material in agriculture uses up 2% of the world's annual plastic production, or 6.96 Mt. PMFs are a water-saving agricultural practice that enhances crop growth, decreases weeds and pests, maintains moisture and soil nutrients, and raises crop yields (Bläsing and Amelung, 2018). According to reports, PMFs cover more than 128,652 km^2 of agricultural land worldwide, which, as a result of its poor recovery from agricultural fields, has produced significant amounts of PMF residue. Since PE makes up the majority of PMF, soil does not readily degrade any remaining PMF (Kasirajan and Ngouajio, 2012). The plasticizing agents, like the carcinogenic and extensively researched soil contaminant phthalate, would leach from the mulch film as soon as the PE-based PMF residue (or MPs) began to degrade even slightly (Wang et al., 2018). As a result, the prevalence of MPs and their harmful equivalents may pose a serious danger to the sustainability of critical ecosystem services in agricultural settings. These pollutants with MPs have the potential to expand into surface and groundwater sources via plant absorption and runoff. For example, MPs smaller than 0.5 m accumulate in biota, such as filamentous bacteria and yeast, which do not 'ingest' their food. This might point to potential MP multiplication in the soil and food chain (Hodson et al., 2017).

2.3.5 Landfills and Landfill Leachate

Without identifying their sources, three recent investigations found MPs in landfill-mined soil-like fractions (LFMSFs) derived from landfill mining and landfill leachate (He et al., 2019). Flooding, rag-picking, surface runoff, wind-driven movement, and animal scavenging are all key environmental transport pathways for landfilled plastic debris. The leachates contained 17 different types of MPs, with concentrations ranging from 0.42 to 24.58 particles/L, including PE, PP, PVC, PS, PET, acrylonitrile butadiene styrene (ABS), and polyurethane (PUR). He et al. (2019) identified these MPs in the leachates. A slight modification to the National Oceanic and Atmospheric Administration laboratory procedures allowed for the extraction and identification of the majority (roughly 77%) of the MP particles, which were in the 0.1–1.0 mm size range (van den Berg et al., 2020). No study appears to have evaluated the depth to which MPs penetrate landfill base or cover liners or the soils below landfill sites. Therefore, from the perspective of environmental geotechnics, it is now necessary to quantify the sources of plastic in landfills, the processes of plastic fragmentation in waste, and the degree of MP migration into the wider environment surrounding landfills.

2.3.6 Common Causes of Microplastics

The geology, geomorphology, and vegetation of the coast, the amount and direction of rainfall, the prevailing wind speed and direction, tides, ocean currents, and sea temperature are just a few of the variables that influence MP deposition along coastal shorelines. Plastics may be transported to the sea or the land depending on these variables. UV radiation exposure causes plastic to fragment, which is then subject to abrasion and/or biodegradation on the seafloor (Kershaw et al., 2019).

2.3.7 Impacts of Microplastics on the Processes in Soil

MP particle characteristics (size, shape, density, and polymer type), climate (wind, rainfall, and temperature), soil physics and biochemistry (e.g., soil biota), and other environmental factors (e.g., mechanical disturbance) all influence MP migration, partitioning, and degradation in soil environments (Bläsing and Amelung, 2018). Furthermore, the weathering of friction with soil particles and the oxidation of sunlight to form an electric charge may result in positively or negatively charged MP surfaces, which will impact the MPs' characteristics and migratory processes (Mammo et al., 2020).

The process of MP migration and transformation can be significantly impacted by soil microbes. These are carried from the surface of the soil downward by soil biota, such as earthworms and local fungi, affecting the different soil horizons. Earthworms can migrate MPs from the soil's surface into its pores both horizontally and vertically; the easier the migration, the smaller the MP particle size (Zhu et al., 2018). MPs may also bind to pathogens and other microorganisms in the soil ecosystem within the soil matrix. Higher retention of MPs in deeper soil layers results from slower native microorganism decomposition/degradation of MPs due to the smaller microbial population in deeper parts of the soil profile. Native fungal mycelia may also aid in the downward migration of MPs through air-filled pores (Mammo et al., 2020).

The material cycle and energy flow of terrestrial ecosystems are significantly impacted by MPs. Since entering MPs have adsorption properties, they also take in organic pollutants. The material cycle and energy flow of terrestrial ecosystems are significantly impacted by MPs. In addition to absorbing organic pollutants due to their adsorption properties, entering MPs also act as carriers of heavy metals, increasing the latter's bioavailability and their accumulation in the food chain through animal consumption (Beckingham and Ghosh, 2017). High MP levels may change soil chemistry by interfering with the degradation of soil organic matter (SOM) (Liu et al., 2017).

2.3.8 Impact of Microplastics with Other Pollutant in Soil

Because of their tiny particle size, large specific surface area (SSA), and significant hydrophobicity, MPs can boost the mobility of other contaminants in the terrestrial subsurface by a factor of two. By absorbing and transporting them, soil properties are altered, resulting in increased release and transport. On the surfaces of MPs, heavy metals, POPs, and antibiotics have all been found. When used as a vehicle for these substances, MPs have a complex impact on the soil environment, which can contain a variety of clay minerals, metal oxides and hydroxides, humus, microorganisms, and other naturally occurring elements that can mix with pollutants. MPs can also cause changes in the soil's characteristics that influence how other soil pollutants in the geoenvironment are absorbed, changing their chemical forms and bioavailability. To produce reliable evidence regarding these interactions and their effects on the environment, more study is still needed (Hartmann et al., 2017). There are two categories of potential factors that might influence MPs' propensity to draw heavy metals:

Environmental factors (such as dissolved organic matter, ionic concentration and salinity of soil pore water, soil solids properties, contact time, pH value, temperature, outdoor weathering effects, and biofilm formation on MP surfaces) as well as those inherent to MP particles (e.g. size, surface properties, porosity, morphology, material type, manufacturing process, pigments/masterbatch used for color, and natural ageing of the MP material) (Hodson et al., 2017).

2.3.9 Impact of Microplastics on Biota

The biotic component of the terrestrial ecosystem is changed by the presence of MPs in the soil. The biogeochemical cycle and the enzymatic activity of soil microorganisms that interact with MPs change the metabolic pathways that these organisms follow. The surface charges on MPs cause them to interact electrostatically with microorganisms; for example, a positively charged MP's surface could adsorb onto a negatively charged microbe's surface, or vice versa, and then be consumed when these microbes are consumed by their predators (de Souza Machado et al., 2018). Microplastics in soils have been shown in studies to reduce a plant's ability to root, to suppress photosynthesis, and to inhibit plant growth (Huang et al., 2022).

Through a number of pathways, MPs are likely to have an impact on soil fauna. The proliferation and operation of decomposers are disrupted, growth is negatively impacted, and MPs are easily ingested by soil-dwelling organisms (Okeke et al., 2022). Second, as some microplastic types degrade in specific soil environments, they may release hazardous plastic additives (Zhang et al., 2020). Third, pathogenic bacteria, fungi, and bacteria resistant to antibiotics may spread through MPs (Martínez-Campos et al., 2022). It is likely that MPs will have indirect effects on soil animals, which are essential for providing a variety of ecological functions and services. Microplastics can reduce habitat quality, movement, and the potential prey (including plant roots, root exudates, litters, and soil microbes) of soil animals. Only a few studies have looked at the ecological risks of microplastic pollution on the diversity, community structure, and ecological functioning of soil animals, particularly those that occupy multiple trophic groups, despite the fact that the majority of studies have focused on the effects of microplastic pollution on the soil microbial community and the fitness of soil animals (Lin et al., 2020).

2.4 HOW TO REDUCE MICROPLASTICS IN SOIL

Every ecosystem on Earth is contaminated by microplastic contamination. It can be found in all oceans, the land, the water we drink, the food we consume, and our own bodies. This massive problem will not be solved fast, and the consequences for human health are not fully understood.

2.4.1 Avoid Synthetic Materials

Some typical garden items are possible microplastic contamination sources. Plastics in the form of synthetic textiles, such as horticultural fleece, might shed particles more than common products, such as plastic pots and bags. Synthetic rope or netting might degrade in the environment as well. Some garden paints can potentially release microscopic particles into the surrounding environment. In the grand scope of things, they may appear to be minor and trivial activities (and, of course, larger-scale answers are required). However, everything we can do to reduce MPs in our gardens' soil implies better soil and fewer particles coming up on our plates.

2.4.2 Avoid Commercial Compost

There are no tools or approaches for determining the amount of MPs in synthetic and commercial compost, however, MPs have been identified in municipal compost. As a result, it is preferable to manufacture your own compost. Home composts can also act as a conduit for MPs to enter the ecosystem, so be mindful of what goes into your compost to help combat this issue.

2.4.3 Filter Runoff

When car tires brush against road surfaces, they create microplastic-containing dust, which is another important cause of microplastic pollution. Taking runoff and diverting it into vegetated swales and rain garden systems can assist to minimize the distribution of those particles in a minor amount.

2.5 STATISTICAL ANALYSIS AND PATTERN RECOGNITION

The identification of MPs, ubiquitous pollutants in various environmental compartments, demands robust analytical techniques. Statistical analysis and pattern recognition play pivotal roles in processing and interpreting the vast datasets generated during microplastic studies. These methodologies enable researchers to discern patterns, classify particles, and extract valuable information from complex samples.

Descriptive statistics provide fundamental insights into the size distribution, shape diversity, and abundance of MPs within a sample (Gündoğdu et al., 2021). Measures like mean size, standard deviation, and skewness aid in summarizing key characteristics, offering a snapshot of the microplastic population. Hypothesis testing and multivariate analyses are employed to discern sources of MPs. By comparing the composition and characteristics of MPs across different environments, researchers can infer potential origins, whether from anthropogenic sources like industrial processes or natural sources like weathering of geological formations. Regression models are utilized to investigate relationships between microplastic properties and environmental variables (Kooi et al., 2017). This aids in understanding factors influencing microplastic distribution, transport, and accumulation in various ecosystems.

Machine learning algorithms, particularly convolutional neural networks (CNNs), have shown promise in automating the identification of MPs based on visual characteristics (Ng et al., 2019). This technology can enhance efficiency and accuracy in large-scale microplastic studies. Techniques like Raman and FTIR spectroscopy, coupled with chemometric analyses, facilitate the identification of MPs based on their spectral signatures (Gündoğdu et al., 2021). These methods enable discrimination between polymers and other organic or inorganic materials. Unsupervised learning approaches, including hierarchical clustering and k-means clustering, aid in categorizing MPs based on size, shape, and other morphological features (Hidalgo-Ruz et al., 2012). This assists in understanding the distribution of different microplastic types within a sample.

The synergy between statistical analysis and pattern recognition amplifies the effectiveness of microplastic identification. For instance, combining spectral data from techniques like Raman spectroscopy with machine learning algorithms allows for rapid and accurate classification of MPs (Ren et al., 2023). Furthermore, the integration of data from different analytical methods, such as microscopy and spectroscopy, provides a more comprehensive understanding of microplastic characteristics.

2.6 REMOTE SENSING AND IMAGING TECHNIQUES

Remote sensing and imaging techniques have emerged as powerful tools in the identification and monitoring of MPs in various environmental compartments. These technologies leverage advanced sensors and imaging platforms to detect and characterize microplastic particles, providing valuable insights into their distribution and abundance. Satellite-based remote sensing has revolutionized our ability to assess large-scale patterns of microplastic pollution, particularly in marine environments. Platforms like Sentinel-2, which employ multispectral imaging, enable the detection of floating macroplastic debris and large aggregations of MPs on the ocean surface (Sannigrahi et al., 2022). Additionally, synthetic aperture radar (SAR) on satellites like Sentinel-1 can penetrate the water's surface, allowing for the detection of subsurface MPs (Lavender, 2022). These technologies provide a bird's-eye view of the spatial distribution of MPs over vast areas.

Aerial photography using drones equipped with high-resolution cameras offers a versatile and cost-effective approach to monitor microplastic pollution in coastal and inland waters (Taddia et al., 2021). Drones can capture detailed images of surface water and shoreline areas, allowing for the identification of larger microplastic items and hotspots of accumulation. Furthermore, drone-based imaging can provide valuable contextual information for ground-truthing studies.

Underwater imaging technologies, such as remotely operated vehicles (ROVs) equipped with cameras, have proven invaluable for studying MPs in the water column and benthic habitats (Woodall et al., 2014). ROVs can capture high-resolution images and videos, enabling researchers to identify and quantify MPs in their natural underwater environments. This approach is particularly crucial for assessing the vertical distribution and behavior of MPs.

Hyperspectral imaging involves capturing a wide range of spectral bands, allowing for detailed characterization of materials based on their unique spectral signatures. This technique has shown promise in discriminating MPs from natural materials in complex environmental samples (Primpke et al., 2020a and b). By analyzing the reflectance patterns, hyperspectral imaging aids in the precise identification and classification of MPs.

2.7 COMPUTATIONAL MODELING AND SIMULATION

Computational models simulate the movement of MPs in natural systems, considering factors like water currents, wind patterns, and sedimentation rates. These models use principles of fluid dynamics to predict the distribution of MPs in oceans, rivers, and other aquatic environments (Bujaczek et al., 2021). By inputting parameters such as particle size, shape, and density, researchers can estimate the trajectories and potential accumulation zones of MPs. Models predict the fate of MPs over time, taking into account degradation processes, such as photodegradation, biodegradation, and fragmentation (Enders et al., 2015). These simulations consider environmental conditions, like temperature, UV exposure, and microbial activity, to estimate the persistence and transformation of MPs in different habitats.

Computational models help predict the bioaccumulation of MPs in aquatic organisms and their potential transfer along food webs. These models integrate information on the physical and chemical properties of MPs with biological parameters of organisms, allowing for assessments of potential ecological impacts (Liu et al., 2019b). They are crucial in understanding the implications of microplastic ingestion on higher trophic levels. Computational tools aid in identifying potential sources and pathways of MPs in the environment. Source apportionment models use statistical techniques to trace back the origins of MPs based on their characteristics and distribution patterns (Geyer et al., 2017). These models are instrumental in distinguishing between natural and anthropogenic sources.

Simulation models explore the interactions between MPs and environmental components, including sediments, biota, and water columns. These models help elucidate processes such as sedimentation rates, aggregation behavior, and microbial colonization on microplastic surfaces (Zhai et al., 2023). Understanding these interactions is essential for predicting the long-term fate of MPs in different ecosystems.

2.8 DATA FUSION AND MULTIVARIATE ANALYSIS

Data fusion and multivariate analysis are critical techniques in the identification of MPs, allowing researchers to integrate information from multiple sources and analyze complex datasets. By combining various types of data and employing advanced statistical methods, these approaches enhance our ability to accurately detect and characterize MPs in different environmental matrices. Combining imaging and spectroscopic data provides a comprehensive characterization of MPs. Imaging techniques offer morphological information, while spectroscopy reveals chemical composition. Multivariate analysis techniques, such as combining PCA with cluster analysis, enable researchers to correlate these data sets and gain a more nuanced understanding of microplastic properties (Fang et al., 2021).

Data fusion involves integrating information from different sensors or measurement techniques to obtain a more comprehensive understanding of microplastic characteristics. For instance, combining data from optical microscopy, spectroscopy, and elemental analysis techniques provides a holistic view of MPs, including their size, shape, composition, and surface features (Blettler et al., 2018). This synergistic approach increases the reliability and accuracy of microplastic identification. Spectroscopic techniques, such as Raman and FTIR spectroscopy, generate complex spectral data that can be challenging to interpret. Chemometric analysis applies advanced statistical methods to extract meaningful information from these spectra, allowing for the identification and classification of MPs based on their chemical composition (Song et al., 2018). Principal component analysis (PCA) and partial least squares-discriminant analysis (PLS-DA) are commonly used chemometric techniques in microplastic research.

Multivariate analysis includes unsupervised clustering methods like hierarchical clustering and k-means clustering. These techniques group MPs based on similarities in their morphological or spectral features (Hidalgo-Ruz et al., 2012). Clustering algorithms aid in identifying distinct categories or classes of MPs within a sample, providing valuable information about their sources and origins. PCA is employed to reduce the dimensionality of complex datasets while retaining essential information. In microplastic studies, PCA can be applied to streamline data from different analytical techniques, allowing for more efficient processing and interpretation. By extracting the most informative components, PCA facilitates a clearer visualization and classification of MPs.

2.9 CASE STUDIES: COMPARATIVE ANALYSIS OF SOURCE IDENTIFICATION METHODOLOGIES

One notable case study (Kappler et al., 2015) compared the efficacy of spectroscopic techniques (Raman and FTIR spectroscopy) with traditional microscopy for polymer identification in marine sediments. The study found that while microscopy allowed for detailed morphological characterization, spectroscopic techniques provided more accurate and rapid identification of polymer types, highlighting the complementary nature of these methods.

A case study (Lebreton et al., 2018) focused on mapping the global distribution of floating MPs by integrating data from remote sensing technologies (e.g., satellite imagery) with in-situ sampling. The study demonstrated the feasibility of combining remote sensing with traditional sampling methods to obtain a comprehensive understanding of microplastic abundance and distribution in marine environments. A comparative case study (Lin et al., 2022) assessed the performance of machine learning algorithms in automating microplastic identification from sediment samples. The study compared the accuracy and efficiency of machine learning models with manual identification methods. Results indicated that machine learning significantly expedited the identification process without compromising accuracy.

A case study (Huang et al., 2023) utilized multivariate analysis techniques to apportion sources of MPs in a coastal environment. By combining data from different analytical methods (e.g., FTIR, microscopy, and elemental analysis), the study successfully differentiated between natural and anthropogenic sources, providing valuable information for targeted mitigation efforts. Another case study (Felsing et al., 2018) focused on evaluating different sample preparation techniques for isolating MPs from water samples. The study compared density-based separation methods with chemical digestion approaches. Results indicated that a combination of techniques, including density separation and enzymatic digestion, yielded higher microplastic recovery rates.

2.10 CHALLENGES AND FUTURE DIRECTIONS

One of the main challenges in microplastic identification arises from the diverse range of sizes and morphologies exhibited by microplastic particles. They can be irregularly shaped fragments, fibers, spheres, or films, making them difficult to differentiate from natural particles with similar characteristics (Löder et al., 2017). Environmental matrices, such as sediments, soils, and biological tissues, often contain a complex mixture of organic and inorganic materials. Isolating and identifying MPs from these matrices requires robust sample preparation techniques and analytical methods capable of distinguishing MPs from background materials (Zhang et al., 2020).

The distribution of MPs in natural environments is highly heterogeneous, influenced by factors like water currents, wind patterns, and sedimentation rates. Conducting comprehensive and representative sampling to capture this variability is challenging, and standardization of sampling protocols remains a critical issue (Hidalgo-Ruz et al., 2012). The identification of nanoplastics (particles <1 μm) and sub-micron particles presents a significant technical challenge due to their size. Conventional analytical techniques may struggle to detect and characterize particles at this scale, necessitating the development of specialized methodologies (Cheng et al., 2021). While several analytical techniques are available for microplastic identification, there is a need for standardized protocols and methodologies. Comparative studies and inter-laboratory validation exercises are crucial for establishing best practices in microplastic analysis (Löder et al., 2017).

Continued development and refinement of analytical techniques, such as hyperspectral imaging, X-ray computed tomography, and advanced spectroscopy, hold promise for more accurate and efficient microplastic identification. These technologies offer the potential to enhance the resolution and specificity of microplastic analysis (Faltynkova et al., 2021). The application of machine learning algorithms and artificial intelligence (AI) in microplastic identification is an emerging area of research. These technologies have shown potential in automating the identification process, reducing reliance on manual sorting and analysis (Xu et al., 2022). Efforts to standardize sample collection, processing, and analysis protocols are essential for ensuring consistency and comparability across studies. Inter-laboratory studies and proficiency tests can help validate and refine analytical methods. Given the increasing recognition of nanoplastics as a potential environmental concern, future research should focus on developing specialized techniques for the identification and characterization of particles at the nanoscale. This may involve the adaptation of existing methodologies or the development of novel approaches (Cheng et al., 2021).

2.11 UNCERTAINTY AND ERROR ANALYSIS

Variability in sampling techniques, including the choice of sampling equipment, spatial coverage, and depth profiles, can introduce uncertainty in microplastic abundance estimations (Löder et al., 2017). Cross-contamination during sample handling, storage, and analysis can lead to false positives or overestimations of microplastic concentrations. It's crucial to implement rigorous protocols to minimize the risk of contamination (Bogdanowicz et al., 2021). The establishment of clear and standardized criteria for microplastic identification is essential. Ambiguities in distinguishing MPs from natural particles or other debris can introduce uncertainty in the results (Hidalgo-Ruz et al., 2012). The precision and accuracy of microplastic identification methods should be evaluated through replication and inter-laboratory studies. This helps assess the variability in results and provides a measure of measurement error. The use of reference materials, such as certified microplastic standards, allows for the validation and calibration of analytical methods. This helps establish a known baseline for error assessment (Song et al., 2021). Applying statistical methods, including confidence intervals, regression analysis, and hypothesis testing, aids in quantifying uncertainty and assessing the significance of results (Smith et al., 2018).

Continued efforts to establish standardized protocols for sample collection, processing, and analysis are essential. These protocols should include detailed guidelines for quality control and quality assurance procedures (Löder et al., 2017). Intercomparison studies involving multiple analytical techniques and laboratories can provide valuable insights into the variability and comparability of results obtained using different methods. Ongoing research into the development and certification of microplastic reference materials will enhance the accuracy and reliability of microplastic identification methods. Combining data from different analytical techniques, such as microscopy, spectroscopy, and chemical analysis, can help mitigate uncertainties by cross-validating results and providing a more comprehensive characterization of MPs (Song et al., 2021).

2.12 EMERGING TECHNOLOGIES AND APPROACHES

Advancements in technology are instrumental in enhancing the accuracy and efficiency of microplastic identification. Hyperspectral imaging technology enables the acquisition of high-resolution images with a broad range of spectral bands. This allows for the precise characterization of materials based on their unique spectral signatures. In microplastic identification, hyperspectral imaging aids in distinguishing MPs from natural particles in complex environmental samples (Primpke et al., 2020). The detailed spectral information provides a powerful tool for accurate identification.

With the increasing recognition of nanoplastics as a potential environmental concern, researchers have developed specialized techniques for the identification and characterization of particles at the nanoscale. These methods may involve advanced electron microscopy, nanoparticle tracking analysis (NTA), and other nanoscale-specific analytical tools (Cheng et al., 2021). These technologies address the challenges associated with detecting particles below the micrometer range. Machine learning algorithms and AI applications are being integrated into microplastic identification processes. These technologies have shown promise in automating the identification process, reducing reliance on manual sorting and analysis. Machine learning models can be trained to recognize microplastic features based on large datasets, significantly expediting the identification process (Lin et al., 2022).

Raman and Fourier-transform infrared (FTIR) spectroscopy continue to be essential tools for microplastic analysis. Recent advancements in these techniques include improved spectral resolution, faster data acquisition, and enhanced software capabilities for more accurate and rapid identification of polymer types (Mikulec et al., 2023). These developments have significantly enhanced the capabilities of spectroscopic methods. Enzymatic digestion methods have been developed to target MPs specifically. Enzymes can selectively break down organic materials, leaving MPs intact and facilitating their isolation from environmental samples (Mbachu et al., 2021). This approach improves the recovery rate of MPs and reduces the potential for contamination. The development of certified microplastic reference materials provides a standardized baseline for microplastic identification and quantification. These reference materials mimic real-world microplastic particles in terms of size, shape, and composition, allowing for the calibration and validation of analytical methods.

2.13 POLICY IMPLICATIONS AND DECISION-MAKING PROCESSES

The identification of MPs plays a crucial role in informing policies and decision-making processes aimed at mitigating their environmental impacts. Rigorous identification of MPs allows for comprehensive risk assessments. These assessments form the basis for regulatory frameworks aimed at controlling microplastic pollution. Scientific evidence on the sources, distribution, and potential ecological and human health risks guides the development of policies to limit microplastic emissions (Catarino et al., 2021). Accurate identification of MPs is essential for establishing baseline data on their presence in different ecosystems. This data serves as a foundation for ongoing monitoring efforts and helps track changes in microplastic abundance over time. It also enables policymakers to assess the effectiveness of mitigation measures (Belioka and Achilias, 2023).

Identification of the sources of MPs (natural vs. anthropogenic) is pivotal for implementing targeted interventions. Understanding the origins of MPs allows policymakers to focus on reducing specific sources, whether they are from industrial processes, consumer products, or natural degradation (Henderson and Green, 2020). Microplastic identification supports the transition towards a circular economy, where products are designed for reusability and recyclability. Policies based on microplastic research promote sustainable production, consumption, and waste management practices, thereby reducing the generation of plastic waste (Smith et al., 2018).

Identifying specific products that contribute significantly to microplastic pollution allows for the development of legislation to regulate or phase out their production or use. This might involve restrictions on microplastic ingredients in personal care products, packaging materials, or industrial processes (Löder et al., 2017). Clear communication of microplastic research findings to the public is essential for raising awareness and fostering behavioral change. Policy efforts to combat microplastic pollution often involve public outreach and education initiatives, which rely on accurate identification and compelling scientific evidence (Hidalgo-Ruz et al., 2012). Microplastic pollution is a global issue that requires international cooperation. Accurate identification of MPs and sharing of research findings across borders facilitate the development of multinational agreements and protocols aimed at reducing plastic waste and pollution (Lebreton et al., 2018).

2.14 CONCLUSION

The identification of MPs is fundamental for understanding their prevalence, sources, and potential impacts on the environment. Over the years, significant progress has been made in developing and refining various analytical techniques. From spectroscopic methods like Raman and FTIR spectroscopy to advanced imaging technologies and machine learning algorithms, researchers have a growing toolbox at their disposal. However, this field is not without its challenges. Variability in particle size, complexity in environmental matrices, and the emergence of nanoplastics are ongoing areas of concern. Standardization of protocols and rigorous error analysis are essential for ensuring the reliability and comparability of results.

Emerging technologies like hyperspectral imaging and nanoscale analytical methods are pushing the boundaries of what is possible in microplastic identification. These approaches offer unprecedented levels of precision and sensitivity, enabling researchers to delve deeper into the microplastic problem. Policy implications are a natural extension of accurate microplastic identification. This scientific foundation empowers policymakers to make informed decisions about regulatory measures, risk assessments, and targeted interventions. It drives the transition towards a circular economy and informs public awareness campaigns. Looking ahead, international collaboration and continued research are paramount. By working together, scientists and policymakers can develop effective strategies to mitigate the environmental impacts of MPs and move towards a more sustainable future.

2.15 SUMMARY OF KEY FINDINGS

Spectroscopic methods, such as Raman and FTIR spectroscopy, have emerged as powerful tools for identifying MPs. They offer rapid and accurate identification based on the unique spectral signatures of different polymer types (Mikulec et al., 2023). Statistical analysis and pattern recognition techniques have been employed to streamline the identification process. These approaches, combined with spectroscopic methods, improve the efficiency and reliability of microplastic identification.

Remote sensing and imaging techniques, including satellite imagery and aerial surveys, have been utilized to map the distribution of MPs in large water bodies. These technologies provide valuable insights into the spatial patterns of microplastic pollution (Lebreton et al., 2018). Computational modeling and simulation have been employed to apportion sources of MPs in complex environmental systems. These models integrate data from various sources to differentiate between natural and anthropogenic contributions.

Data fusion techniques, which combine information from multiple analytical methods, have proven effective in improving the accuracy of microplastic identification. By integrating complementary data, researchers can achieve a more comprehensive understanding of microplastic characteristics. Comparative case studies have played a crucial role in evaluating the effectiveness of different identification methodologies. These studies have highlighted the strengths and limitations of various techniques, helping to guide researchers in selecting the most appropriate approach for specific environmental contexts (Mbachu et al., 2021).

Recognizing and quantifying uncertainties associated with microplastic identification is essential for producing robust and reliable results. Error analysis techniques, including validation with reference materials and statistical approaches, contribute to the accuracy of findings (Tagg et al., 2015). Emerging technologies, such as hyperspectral imaging, nanoscale analytical methods, and machine learning, are revolutionizing microplastic identification. These cutting-edge approaches offer unprecedented levels of precision and sensitivity, enabling researchers to delve deeper into the microplastic problem (Cheng et al., 2021).

2.16 RECOMMENDATIONS FOR FUTURE RESEARCH

As the field of microplastic identification continues to evolve, there are several critical avenues for future research that can further advance our understanding of microplastic pollution and its impacts. Given the increasing recognition of nanoplastics as a potential environmental concern, there is a pressing need for the development and refinement of specialized techniques for the identification and characterization of particles at the nanoscale (Cheng et al., 2021). Future research should focus on adapting existing methodologies or devising novel approaches to address this challenge. Efforts to standardize sample collection, processing, and analysis protocols remain crucial for ensuring consistency and comparability across studies. Future research should prioritize collaborative inter-laboratory studies and proficiency tests to validate and refine analytical methods.

The application of machine learning algorithms and AI in microplastic identification is an emerging area of research with significant potential (Xu et al., 2022). Future studies should explore and refine the integration of these technologies to automate and enhance the identification process. Continued development and refinement of analytical techniques, such as hyperspectral imaging, X-ray computed tomography, and advanced spectroscopy, hold promise for more accurate and efficient microplastic identification (Primpke et al., 2020). Future research should focus on pushing the boundaries of these technologies to improve resolution and specificity. The creation and certification of microplastic reference materials are essential for establishing a standardized baseline for microplastic identification methods (Song et al., 2021). Future research should prioritize the development and validation of these reference materials to enhance the accuracy and reliability of microplastic analysis.

Combining data from multiple analytical techniques, such as microscopy, spectroscopy, and chemical analysis, can provide a more comprehensive characterization of MPs. Future studies should explore the synergistic benefits of employing multimodal approaches for improved identification accuracy. Research should focus on identifying and characterizing emerging contaminants, such as microfibers from textiles and microbeads from cosmetics, to stay ahead of potential environmental threats (Osman et al., 2023). This includes developing specialized techniques tailored to specific types of microplastic pollutants.

Conducting long-term monitoring studies to track temporal trends in microplastic abundance and distribution is crucial for understanding the dynamics of microplastic pollution over time (Phan and Luscombe, 2023). Future research should prioritize establishing consistent monitoring programs in various environments. Further research is needed to investigate the interactions between MPs and organisms, as well as the potential ecological impacts on ecosystems (Saud et al., 2023). Understanding these dynamics is critical for assessing the broader implications of microplastic pollution.

REFERENCES

Beckingham, B., & Ghosh, U. (2017). Differential bioavailability of polychlorinated biphenyls associated with environmental particles: Microplastic in comparison to wood, coal and biochar. *Environmental Pollution, 220*, 150–158.

Belioka, M. P., & Achilias, D. S. (2023). Microplastic pollution and monitoring in seawater and harbor environments: A meta-analysis and review. *Sustainability, 15*(11), 9079.

Bläsing, M., & Amelung, W. (2018). Plastics in soil: Analytical methods and possible sources. *Science of the Total Environment, 612*, 422–435.

Blettler, M. C. M., Abrial, E., Khan, F. R., Sivri, N., & Espinola, L. A. (2018). Freshwater plastic pollution: Recognizing research biases and identifying knowledge gaps. *Water Research, 143*, 416–424.

Bogdanowicz, A., Zubrowska-Sudol, M., Krasinski, A., & Sudol, M. (2021). Cross-contamination as a problem in collection and analysis of environmental samples containing microplastics—A review. *Sustainability, 13*(21), 12123.

Bujaczek, T., Kolter, S., Locky, D., & Ross, M. S. (2021). Characterization of microplastics and anthropogenic fibers in surface waters of the North Saskatchewan River, Alberta, Canada. *Facets, 6*, 26–43.

Catarino, A. I., Kramm, J., Völker, C., Henry, T. B., & Everaert, G. (2021). Risk posed by microplastics: Scientific evidence and public perception. *Current Opinion in Green and Sustainable Chemistry, 29*, 100467.

Cheng, Y., Mai, L., Lu, X., Li, Z., Guo, Y., Chen, D., & Wang, F. (2021). Occurrence and abundance of poly-and perfluoroalkyl substances (PFASs) on microplastics (MPs) in Pearl River Estuary (PRE) region: Spatial and temporal variations. *Environmental Pollution, 281*, 117025.

Conley, K., Clum, A., Deepe, J., Lane, H., & Beckingham, B. (2019). Wastewater treatment plants as a source of microplastics to an urban estuary: Removal efficiencies and loading per capita over one year. *Water Research X, 3*, 100030.

Crossman, J., Hurley, R. R., Futter, M., & Nizzetto, L. (2020). Transfer and transport of microplastics from biosolids to agricultural soils and the wider environment. *Science of the Total Environment, 724*, 138334.

de Souza Machado, A. A., Kloas, W., Zarfl, C., Hempel, S., & Rillig, M. C. (2018). Microplastics as an emerging threat to terrestrial ecosystems. *Global Change Biology, 24*(4), 1405–1416.

Dong, S., Xia, J., Sheng, L., Wang, W., Liu, H., & Gao, B. (2021). Transport characteristics of fragmental polyethylene glycol terephthalate (PET) microplastics in porous media under various chemical conditions. *Chemosphere, 276*, 130214.

Enders, K., Lenz, R., Stedmon, C. A., & Nielsen, T. G. (2015). Abundance, size and polymer composition of marine microplastics ≥10 μm in the Atlantic Ocean and their modelled vertical distribution. *Marine Pollution Bulletin, 100*(1), 70–81.

Faltynkova, A., Johnsen, G., & Wagner, M. (2021). Hyperspectral imaging as an emerging tool to analyze microplastics: A systematic review and recommendations for future development. *Microplastics & Nanoplastics, 1*, 13.

Fan, P., Tan, W., & Yu, H. (2022). Effects of different concentrations and types of microplastics on bacteria and fungi in alkaline soil. *Ecotoxicology and Environmental Safety, 229*, 113045.

Fang, C., Luo, Y., Zhang, X., Zhang, H., Nolan, A., & Naidu, R. (2021). Identification and visualisation of microplastics via PCA to decode Raman spectrum matrix towards imaging. *Chemosphere, 286*, 131736.

Felsing, S., Kochleus, C., Buchinger, S., Brennholt, N., Stock, F., & Reifferscheid, G. (2018). A new approach in separating microplastics from environmental samples based on their electrostatic behavior. *Environmental Pollution, 234*, 20–28.

Geyer, R., Jambeck, J. R., & Law, K. L. (2017). Production, use, and fate of all plastics ever made. *Science Advances, 3*(7), e1700782.

Gündoğdu, S., Eroldoğan, O. T., Evliyaoğlu, E., Turchini, G. M., & Wu, X. G. (2021). Fish out, plastic in: Global pattern of plastics in commercial fishmeal. *Aquaculture, 534*, 736316.

Han, N., Zhao, Q., Ao, H., Hu, H., & Wu, C. (2022). Horizontal transport of macro-and microplastics on soil surface by rainfall induced surface runoff as affected by vegetations. *Science of the Total Environment, 831*, 154989.

Hartmann, N. B., Rist, S., Bodin, J., Jensen, L. H., Schmidt, S. N., Mayer, P., Meibom, A., & Baun, A. (2017). Microplastics as vectors for environmental contaminants: Exploring sorption, desorption, and transfer to biota. *Integrated Environmental Assessment and Management, 13*(3), 488–493.

He, P., Chen, L., Shao, L., Zhang, H., & Lü, F. (2019). Municipal solid waste (MSW) landfill: A source of microplastics. Evidence of microplastics in landfill leachate. *Water Research, 159*, 38–45.

Henderson, L., & Green, C. (2020). Making sense of microplastics? Public understandings of plastic pollution. *Marine Pollution Bulletin, 152*, 110908.

Hidalgo-Ruz, V., Gutow, L., Thompson, R. C., & Thiel, M. (2012). Microplastics in the marine environment: A review of the methods used for identification and quantification. *Environmental Science & Technology, 46*(6), 3060–3075.

Hodson, M. E., Duffus-Hodson, C. A., Clark, A., Prendergast-Miller, M. T., & Thorpe, K. L. (2017). Plastic bag derived-microplastics as a vector for metal exposure in terrestrial invertebrates. *Environmental Science & Technology, 51*(8), 4714–4721.

Horton, A. A., Walton, A., Spurgeon, D. J., Lahive, E., & Svendsen, C. (2017). Microplastics in freshwater and terrestrial environments: Evaluating the current understanding to identify the knowledge gaps and future research priorities. *Science of the Total Environment, 586*, 127–141.

Huang, D., Wang, X., Yin, L., Chen, S., Tao, J., Zhou, W., ... & Xiao, R. (2022). Research progress of microplastics in soil-plant system: Ecological effects and potential risks. *Science of the Total Environment, 812*, 151487.

Huang, Z., Hu, B., & Wang, H. (2023). Analytical methods for microplastics in the environment: A review. *Environmental Chemistry Letter, 21*(1), 383–401.

Hurley, R. R., & Nizzetto, L. (2018). Fate and occurrence of micro (nano) plastics in soils: Knowledge gaps and possible risks. *Current Opinion in Environmental Science & Health, 1*, 6–11.

Käppler, A., Windrich, F., Löder, M., Malanin, M., Fischer, D., Labrenz, M., Eichhorn, K., & Voit, B. (2015). Identification of microplastics by FTIR and Raman microscopy: A novel silicon filter substrate opens the important spectral range below 1300 cm^{-1} for FTIR transmission measurements. *Analytical and Bioanalytical Chemistry, 407*(22), 134–145.

Kasirajan, S., & Ngouajio, M. (2012). Polyethylene and biodegradable mulches for agricultural applications: A review. *Agronomy for Sustainable Development, 32*, 501–529.

Kershaw, P. J., Turra, A., & Galgani, F. (2019). Guidelines for the monitoring and assessment of plastic litter and microplastics in the ocean. http://www.gesamp.org/publications/guidelines-for-the-monitoring-and-assessment-of-plastic-litter-in-the-ocean.

Kooi, M., van Nes, E. H., Scheffer, M., & Koelmans, A. A. (2017). Ups and downs in the ocean: Effects of biofouling on vertical transport of microplastics. *Environmental Science & Technology, 51*(14), 7963–7971.

Koutnik, V. S., Alkidim, S., Leonard, J., DePrima, F., Cao, S., Hoek, E. M., & Mohanty, S. K. (2021). Unaccounted microplastics in wastewater sludge: Where do they go? *ACS ES&T Water, 1*(5), 1086–1097.

Lavender, S. (2022). Detection of waste plastics in the environment: Application of Copernicus earth observation data. *Remote Sensing, 14*(19), 4772.

Lebreton, L., Slat, B., Ferrari, F., Sainte-Rose, B., Aitken, J., Marthouse, R., Hajbane, S., Cunsolo, S., Schwarz, A., Levivier, A., Noble, K., Debeljak, P., Maral, H., Schoeneich-Argent, R., Brambini, R., & Reisser, J. (2018). Evidence that the Great Pacific Garbage Patch is rapidly accumulating plastic. *Scientific Reports, 8*(1), 1–15.

Li, H., Lu, X., Wang, S., Zheng, B., & Xu, Y. (2021). Vertical migration of microplastics along soil profile under different crop root systems. *Environmental Pollution, 278*, 116833.

Lin, D., Yang, G., Dou, P., Qian, S., Zhao, L., Yang, Y., & Fanin, N. (2020). Microplastics negatively affect soil fauna but stimulate microbial activity: Insights from a field-based microplastic addition experiment. *Proceedings of the Royal Society B, 287*(1934), 20201268.

Lin, J., Liu, H., & Zhang, J. (2022). Recent advances in the application of machine learning methods to improve identification of the microplastics in environment. *Chemosphere, 307*(4), 136092.

Liu, F., Olesen, K. B., Borregaard, A. R., & Vollertsen, J. (2019a). Microplastics in urban and highway stormwater retention ponds. *Science of the Total Environment, 671*, 992–1000.

Liu, H., Yang, X., Liu, G., Liang, C., Xue, S., Chen, H., Ritsema, C. J., & Geissen, V. (2017). Response of soil dissolved organic matter to microplastic addition in Chinese loess soil. *Chemosphere, 185*, 907–917.

Liu, K., Wang, X., Fang, T., Xu, P., Zhu, L., & Li, D. (2019b). Source and potential risk assessment of suspended atmospheric microplastics in Shanghai. *Science of the Total Environment, 675*, 462–471.

Löder, M. G., Gerdts, G., & Moore, C. J. (2017). Methodology used for the detection and identification of microplastics—A critical appraisal. In M. G. J. Löder, G. Gerdts (Eds.), *Marine Anthropogenic Litter*, pp. 201–227. Springer International Publishing.

Luqman, A., Nugrahapraja, H., Wahyuono, R. A., Islami, I., Haekal, M. H., Fardiansyah, Y., Putri, B. Q., Amalludin, F. I., Rofiqa, E. A., Götz, F., & Wibowo, A. T. (2021). Microplastic contamination in human stools, foods, and drinking water associated with Indonesian coastal population. *Environments, 8*(12), 138.

Mak, C. W., Tsang, Y. Y., Leung, M. M. L., Fang, J. K. H., & Chan, K. M. (2020). Microplastics from effluents of sewage treatment works and stormwater discharging into the Victoria Harbor, Hong Kong. *Marine Pollution Bulletin, 157*, 111181.

Mammo, F. K., Amoah, I. D., Gani, K. M., Pillay, L., Ratha, S. K., Bux, F., & Kumari, S. (2020). Microplastics in the environment: Interactions with microbes and chemical contaminants. *Science of the Total Environment, 743*, 140518.

Martínez-Campos, S., Pissaridou, P., Drakou, K., Shammas, C., Andreou, K., González-Pleiter, M., Fernández-Piñas, F., Leganes, F., Rosal, R., Koutinas, M., Konstantinos Kapnisis, K., & Vasquez, M. I. (2022). Evolution of prokaryotic colonisation of greenhouse plastics discarded into the environment. *Ecotoxicology and Environmental Safety, 232*, 113213.

Mbachu, O., Jenkins, G., Pratt, C., & Kaparaju, P. (2021). Enzymatic purification of microplastics in soil. *MethodsX, 8*, 101254.

Mikulec, V., Adamovi, P., Cvetkovi, Ž., Iveši, M., & Gajdoš Kljusuri, J. (2023). Green techniques for detecting microplastics in marine with emphasis on FTIR and NIR spectroscopy—Short review. *Processes, 11*, 2360.

Mohajerani, A., & Karabatak, B. (2020). Microplastics and pollutants in biosolids have contaminated agricultural soils: An analytical study and a proposal to cease the use of biosolids in farmlands and utilise them in sustainable bricks. *Waste Management, 107*, 252–265.

Ng, E. L., Lwanga, E. H., Eldridge, S. M., Johnston, P., Hu, H. W., Geissen, V., & Chen, D. (2018). An overview of microplastic and nanoplastic pollution in agroecosystems. *Science of the Total Environment, 627*, 1377–1388.

Ng, W., Minasny, B., & Mcbratney, A. B. (2019). Convolutional neural network for soil microplastic contamination screening using infrared spectroscopy. *The Science of the Total Environment, 702*(6883), 134723.

Okeke, E. S., Okoye, C. O., Atakpa, E. O., Ita, R. E., Nyaruaba, R., Mgbechidinma, C. L., & Akan, O. D. (2022). Microplastics in agroecosystems-impacts on ecosystem functions and food chain. *Resources, Conservation and Recycling, 177*, 105961.

Osman, A. I., Hosny, M., Eltaweil, A. S., Omar, S., Elgarahy, A. M., Farghali, M., Yap, P. S., Wu, Y. S., Nagandran, S., Batumalaie, K., Gopinath, S. C. B., John, O. D., Sekar, M., Saikia, T., Karunanithi, P., Hatta, M. H. M., & Akinyede, K. A. (2023). Microplastic sources, formation, toxicity and remediation: A review. *Environmental Chemistry Letters, 21*, 2129–2169.

Phan, S., & Luscombe, C. K. (2023). Recent trends in marine microplastic modeling and machine learning tools: Potential for long-term microplastic monitoring. *Journal of Applied Physics, 133*(2), 020701.

en, L., Liu, S., Huang, S., Wang, Q., Lu, Y., Song, J., & Guo, J. (2023). Identification of microplastics using a convolutional neural network based on micro-Raman spectroscopy. *Talanta, 260*, 124611.

Roychand, R., Patel, S., Halder, P., Kundu, S., Hampton, J., Bergmann, D., Surapaneni, A., Shah, K., & Pramanik, B. K. (2021). Recycling biosolids as cement composites in raw, pyrolyzed and ashed forms: A waste utilisation approach to support circular economy. *Journal of Building Engineering, 38*, 102199.

Sannigrahi, S., Basu, B., Basu, A. S., & Pilla, F. (2022). Development of automated marine floating plastic detection system using Sentinel-2 imagery and machine learning models. *Marine Pollution Bulletin, 178*, 113527.

Saud, S., Yang, A., Jiang, Z., Ning, D., & Fahad, S. (2023). New insights in to the environmental behavior and ecological toxicity of microplastics. *Journal of Hazardous Materials Advances, 10*, 100298.

Schell, T., Hurley, R., Buenaventura, N. T., Mauri, P. V., Nizzetto, L., Rico, A., & Vighi, M. (2022). Fate of microplastics in agricultural soils amended with sewage sludge: Is surface water runoff a relevant environmental pathway? *Environmental Pollution, 293*, 118520.

Smith, M., Love, D. C., Rochman, C. M., & Neff, R. A. (2018). Microplastics in seafood and the implications for human health. *Current Environmental Health Reports, 5*, 375–386.

Song, Y. K., Hong, S. H., Eo, S., Jang, M., Han, G. M., Isobe, A., & Shim, W. J. (2018). Horizontal and vertical distribution of microplastics in Korean coastal waters. *Environmental Science & Technology, 52*(21), 12188–12197.

Song, Y. K., Hong, S. H., Eo, S., & Shim, W. J. (2021). A comparison of spectroscopic analysis methods for microplastics: Manual, semi-automated, and automated Fourier transform infrared and Raman techniques. *Marine Pollution Bulletin, 173*(Pt B), 113101.

Sun, J., Dai, X., Wang, Q., Van Loosdrecht, M. C., & Ni, B. J. (2019). Microplastics in wastewater treatment plants: Detection, occurrence and removal. *Water Research, 152*, 21–37.

Taddia, Y., Corbau, C., Buoninsegni, J., Simeoni, U., & Pellegrinelli, A. (2021). UAV approach for detecting plastic marine debris on the beach: A case study in the Po River Delta (Italy). *Drones, 5*(4), 140.

Tagg, A. S., Sapp, M., Harrison, J. P., & Ojeda, J. J. (2015). Identification and quantification of microplastics in wastewater using focal plane array-based reflectance micro-FT-IR imaging. *Analytical Chemistry, 87*(12), 6032–6040.

van den Berg, P., Huerta-Lwanga, E., Corradini, F., & Geissen, V. (2020). Sewage sludge application as a vehicle for microplastics in eastern Spanish agricultural soils. *Environmental Pollution, 261*, 114198.

Wang, F., Wong, C. S., Chen, D., Lu, X., Wang, F., & Zeng, E. Y. (2018). Interaction of toxic chemicals with microplastics: A critical review. *Water Research, 139*, 208–219.

Woodall, L. C., Sanchez-Vidal, A., Canals, M., Paterson, G. L., Coppock, R., Sleight, V., Calafat, A., Rogers, A. D., Narayanaswami, B. E., & Thompson, R. C. (2014). The deep sea is a major sink for microplastic debris. *Royal Society Open Science, 1*(4), 140317.

Xu, R. Z., Cao, J. S., Ye, T., Wang, S. N., Luo, J. Y., Ni, B. J., & Fang, F. (2022). Automated machine learning-based prediction of microplastics induced impacts on methane production in anaerobic digestion. *Water Research, 223*, 118975.

Yang, H., Chen, G., & Wang, J. (2021a). Microplastics in the marine environment: Sources, fates, impacts and microbial degradation. *Toxics, 9*(2), 41.

Yang, L., Zhang, Y., Kang, S., Wang, Z., & Wu, C. (2021b). Microplastics in freshwater sediment: A review on methods, occurrence, and sources. *Science of the Total Environment, 754*, 141948.

Zhai, X., Zhang, X. H., & Yu, M. (2023). Microbial colonization and degradation of marine microplastics in the plastisphere: A review. *Frontiers in Microbiology, 14*, 495.

Zhang, B., Yang, X., Chen, L., Chao, J., Teng, J., & Wang, Q. (2020). Microplastics in soils: A review of possible sources, analytical methods and ecological impacts. *Journal of Chemical Technology & Biotechnology, 95*(8), 2052–2068.

Zhang, J., Wang, L., & Kannan, K. (2020). Microplastics in house dust from 12 countries and associated human exposure. *Environment International, 134*, 105314.

Zhang, Y., Gao, T., Kang, S., & Sillanpää, M. (2019). Importance of atmospheric transport for microplastics deposited in remote areas. *Environmental Pollution, 254*, 112953.

Zhu, D., Bi, Q. F., Xiang, Q., Chen, Q. L., Christie, P., Ke, X., Wu, L.-H., & Zhu, Y. G. (2018). Trophic predator-prey relationships promote transport of microplastics compared with the single Hypoaspis aculeifer and Folsomia candida. *Environmental Pollution, 235*, 150–154.

CHAPTER 3

Source Identification of Microplastics via Natural and Anthropogenic Impacts

Era Juliet Das, Mohammad Toha, Sadia Sikder, and Md. Mostafizur Rahman

3.1 INTRODUCTION

A common way to depict human history is in terms of the materials needed to produce tools and other essentials. The Stone Age, Bronze Age, and Iron Age are the most well-known of these eras. At present, one could say that we live in the Age of Plastic (Napper & Thompson, 2020). Plastics are made up of a sequence of molecular monomers that are bound together to form macromolecules with a plethora of uses in modern civilization (Proshad et al., 2018). Since the 1950s, the manufacture of plastics has increased dramatically. Each year more than 350 million metric tons of plastic waste is generated, of which 40% comes from various forms of packaging, with consumer products and textiles each making up a further 10%–15% each. Sanitary landfills collect approximately 46% of global plastic waste, controlled incineration accounts for a further 17% and recycling collects 15%.

DOI: 10.1201/9781003487579-3

The remaining 22% of plastic waste is mismanaged – a total of almost 80 million metric tons a year (Purcell, 2023). In 2022, global plastic manufacturing reached an astounding 400.3 million metric tons, representing a rise of almost 1.6% over the prior year (Jaganmohan, 2024). According to a report by Li et al. (2023), there comes a point every year when the volume of plastic waste production exceeds the capacity of waste management systems to handle it. This is known as Plastic Overshoot Day, and on July 28, 2023, the world community passed this tipping point.

With time larger plastics brittle down through weathering, UV radiation, biodegradation, and physical-mechanical processes and create smaller particles known as microplastics (MPs) (Wicaksono, 2021; Yao et al., 2023). The word "microplastic" is described in a paper as particles generated as an outcome of the deformation of plastic materials subjected to high amounts of stress that was released by the US Air Force Materials Laboratory in 1968. In 1972, researchers worldwide first encountered micro-sized plastic fragments in the surface water of the Sargasso Sea (Carpenter & Smith, 1972). However, the proper size of these so-called MPs was still unknown in that phase. In 2004, for the first time, descriptions of MPs were given by Richard Thompson, a marine ecologist at the University of Plymouth, UK. According to his definition, plastic particles smaller than 5 mm (0.2 inches) wide are called MPs. When frequently MPs were apparent in numerous locations, the term "microplastics," was more broadly used in the scientific community (Thompson et al., 2004).

Usually, there are two ways by which plastic particles are converted into MPs. One is primary MPs, and the other is secondary MPs (Gabrielli, 2023). Primary MPs are found in a wide range of items, including toothpaste, body wash, scrubs, and cosmetics which directly enter the environment as raw materials of products (Bhatia et al., 2024; Thiele et al., 2023). Secondary MPs are produced when plastic products break down due to physical, chemical, and biological processes (Bhatia et al., 2024; Sutkar et al., 2023). For instance, fragments of large plastic bottles (Gabrielli, 2023). MPs have been found in all the spheres of Earth from aquatic to atmospheric (de Almeida et al., 2023). Various anthropogenic and natural sources such as dust from land-based sources, fertilizers, synthetic grass, highways, landfills, aviation transportation, textiles, tourism, ships, and aquaculture release MPs into the environment (Lamichhane et al., 2023). In the recent years, MPs have been reported to be in the soil (Tunali et al., 2023), sea (Kanhai et al., 2020), polar ocean (Huang et al., 2023), road dust (Yang et al., 2023), atmosphere (Sharma, 2024), indoor air (Fang et al., 2024a), lakes (Chen et al., 2024), rivers (Dris et al., 2024), rainwater (Kong et al., 2024), plants (Azeem et al., 2021), fish (Thiele et al., 2021), marine organisms (Miller et al., 2020), seafood (Al Mamun et al., 2023), etc.

MPs have been demonstrated to interact with a broad range of contaminants, including antibiotics, heavy metals, pesticides, and persistent organic pollutants (Nizzetto et al., 2016; Bradney et al., 2019; Bank & Hansson, 2019). They may also serve as vectors for the transmission of toxins in marine and terrestrial settings. MPs and chemicals are linked to having the ability to travel via food webs and accumulate at different trophic levels, which could harm the health of both people and other living things (Dissanayake et al., 2022). According to Li et al. (2018) and Prokić et al. (2019), plastics and MPs can have physical, chemical, or biological effects on the environment. Animals being physically impacted suggest both entrapment and swallowing, with entanglement occurring far more frequently. Entanglement can result in malnutrition, asphyxia, drowning, or strangling (Allsopp et al., 2006). Numerous studies have documented the physical harm inflicted upon animals by swallowed plastic pieces, which have resulted in stomach ulcers, intestinal system ruptures, and asphyxia (Cole et al., 2011; Prokić et al., 2019). Lethal entanglement and ingestion of plastic pieces by marine organisms have increased by 40% in the last 10 years, according to a Convention on Biological Diversity report (GEF, 2012) whereas data on terrestrial animals is scarcer. Many species are experiencing increasing effects over time; for instance, it is predicted that by 2050, almost all species of seabirds will be affected (Schuyler et al., 2014). Due to various chemicals being adsorbed to plastic surfaces, MPs have a chemical impact on the environment (Prokić et al., 2019). As plastics are composed of extremely hydrophobic components, chemical pollutants are found in or on the surfaces of plastics. Additionally, given MPs operate as ecological storage tanks for harmful chemicals, they can also act as a vector for hydrophobic pollutants (Bradney et al., 2019). The biological effects of MPs on organisms are uncountable. The psychological functioning of both plants and animals is biologically impacted by high MP concentrations in ecosystems (Browne et al., 2013; Oberbeckmann et al., 2015). According to studies, fish are the group of organisms that are most frequently studied in laboratory and field settings, including the impact of MPs (44%). Following the most studied worms (6%), crustaceans (21%) were studied (De Sá et al., 2018). Research on other classes of organisms was somewhat limited, especially when it came to terrestrial ones (Prokić et al., 2019). In humans, MPs can enter through the food chain and create substantial risks (Stapleton et al., 2023; Bhatia et al., 2024). Exposure to MPs negatively impacts gut health, lowers immunity, and can increase the risk of the development of diseases like Type 2 diabetes, cardiac diseases, and several inflammatory diseases (Song et al., 2016; Pattanayak, 2022). Worldwide there are many methods employed by researchers to tackle MP pollution like magnetic separation (Zhang et al., 2023), sediment dredging (Ji et al., 2021), surface skimming (Bydalek et al., 2023), biochar and activated carbon-based adsorbents (Thacharodi et al., 2024), electrocoagulation (Elkhatib et al., 2021), membrane filtration (Krishnan et al., 2023), membrane filtration with reverse osmosis (Westphalen & Abdelrasoul, 2018), ultrafiltration (Tadsuwan & Babel, 2021), microbial degradation of MPs by algae, fungi, bacteria (Thacharodi et al., 2024) etc. However, source reduction is one of the most effective methods of reducing the number of MPs in the environment (Kabir et al., 2023).

MP accumulations differ depending on the location, kind of polymer, and distance from the source. Uncertainties remain regarding the dynamics of abundance and distribution, the determinants of MP distribution, and the movement of fluxes that may have an impact on the distribution. To estimate the ecotoxicological risks and suitable management methods to reduce their influence in natural environments, it is crucial to comprehend the nature and geographic distribution dynamics of MPs in global environment pools. Therefore, this chapter aims to address the types of MPs, their probable sources, and their impact on biotic and abiotic environments. Finally, the potential methods that are used globally to reduce MPs from the environment both marine and terrestrial surroundings.

3.2 TYPES OF MICROPLASTICS

Broadly, there are two types of MPs found in the ecosystem which are primary MPs and secondary MPs. In the plastics industry, small-sized plastic pellets are utilized as feedstock for a wide range of products known as primary MPs (Almeida et al., 2024). Some instances of this include the utilization of microbeads in cosmetics and shedding MP fibers from synthetic fabrics after washing, etc. (Habib et al., 2022). Secondary MPs are formed when bigger plastics break down into smaller pieces due to a variety of factors, including wind, tillage, UV radiation, microbial activity, and chemical and mechanical disintegration (Waldman & Rillig, 2020; Guo et al., 2020; de Almeida et al., 2024). Based on the chemical composition of MPs, they are classified into six categories: polyethylene, polystyrene, polypropylene, polyurethane, polyvinyl chloride, and polyethylene terephthalate (He et al., 2022) (Figure 3.1).

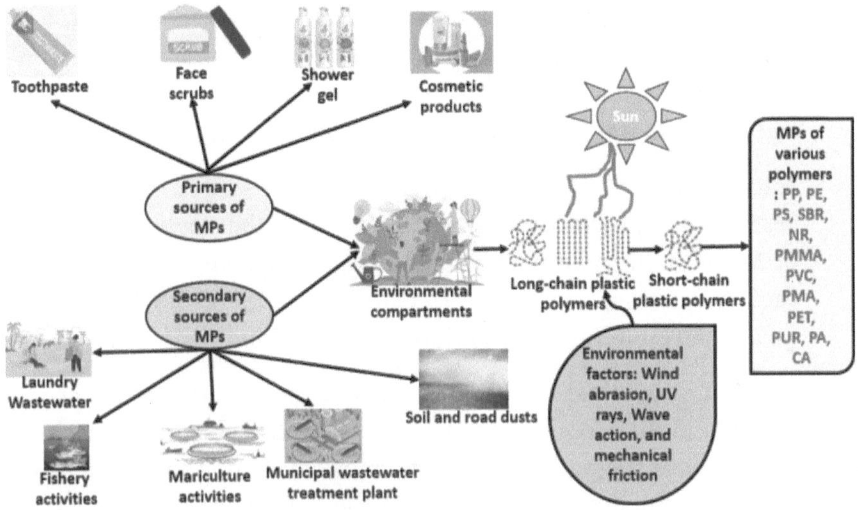

FIGURE 3.1 Source and penetration of microplastics in environmental compartments.

More specifically, MPs are of seven types and depend on the shapes, but they are categorized into five major categories according to Anderson et al. (2017), which are as follows:

1. **Microbeads:** Tiny plastic beads, usually less than 1 mm in diameter, are called microbeads. These round plastic particles are frequently found in toothpaste, cosmetics, and exfoliating scrubs, among other personal care goods. They are included in these goods to give them abrasion and texture.
2. **Microfibers:** Microfibers are very small fibers, usually less than one denier (10 µm), which is a unit of measurement for how thin a fiber is. Human hair, by contrast, has a diameter of approximately 75 µm. Microfibers are utilized extensively in numerous industries, such as the manufacturing of cleaning products, clothes, and household appliances, because of their remarkable softness and smoothness. Microfibers are widely utilized in the fashion sector to make jackets, sweaters, and other sophisticated materials.
3. **Fragmented Plastics:** Small pieces of plastic that come from the physical disintegration of bigger plastic objects, such as bottles, containers, and packaging materials, fall into this category. These little fragments can be produced in several ways, such as by weathering

naturally over time or by being exposed to harmful substances, sunshine, or wave action, among other processes. These pieces can be consumed by marine and terrestrial animals and are dispersed by the wind and water, harming ecosystems and species alike.

4. **Microplastic Pellets (Nurdles):** This plastic pellet is made by processing used plastic waste into tiny particles that have a size range of 1–5 mm. The manufacturing of new plastic items frequently uses these MPs as raw materials. They may, however, leak into the environment and result in serious pollution issues if improperly controlled.
5. **Foamed Microplastics:** Some plastics are made to be lightweight and foamy, and they are utilized in insulation and packaging. These substances have the potential to decompose over time into finer foam particles, which could then pollute the environment as secondary MPs (*All Types of Microplastics You Should Know*, n.d.).

3.2.1 Sources of Microplastics (MPs)

From point and non-point or diffuse sources of land and ocean, MPs can accumulate in the ecosystem (Osman et al., 2023; Wang et al., 2024) (Table 3.1 and Figure 3.1). Unintentional spills, integrated sewage overflows, and the release of wastewater are common indicators of point sources that are easily identifiable (Gatidou et al., 2019). Non-point sources are those origins that are hard to figure out and these include household plastic waste and emissions from agricultural plastic film, urban runoff, etc. (Zhou et al., 2023). 80%–90% of the MPs released into the aquatic environment originate from land-based sources, with the other 10%–20% from the ocean (Osman et al., 2023). Land-based sources include clothes, building materials, containers, personal care items, plastic bags, straws, food, and beverage packaging waste, etc. (Osman et al., 2023). From plastic burning facilities MPs can also be generated on land (Yang et al., 2021b). Other likely sources of MPs include sewage effluent and industrial processes, especially those that use granules and small epoxy pellets (Rolsky et al., 2020). MPs can drift at various depths in the ocean and are largely sourced from lost or discarded fishing gear, such as nylon nets and plastic monofilament lines. The issue is exacerbated by the fact that over 600,000 tons of fishing gear are discarded into the ocean annually (Osman et al., 2023). Additionally, a substantial amount of plastic trash from offshore industries, such as petrochemicals, is being dumped into marine environments (Calero et al., 2021).

TABLE 3.1 Common Sources of Microplastics to Enter the Environment

Microplastic Types	Potential Source	Location	Microplastic Shapes	Polymer Types	References
Primary microplastics	Face scrubs	Local shop, Thailand	Microbeads	Polyethylene	Shalumon and Ratanatamskul (2023)
	Toothpaste	Supermarkets, India	Fibers, fragments	Cellophane, polypropylene, polyvinyl chloride, polyamide	Madhumitha et al. (2022)
	Cosmetic products	local markets, United Arab Emirates	Irregular, granular, spherical	Polyethylene	Habib et al. (2020)
	Shower gel	Beijing, China supermarkets	Irregular	Polyethylene	Lei et al. (2017)
Secondary microplastics	Mariculture activities	Maowei Sea, China	Fragments, fiber, foam	Polyester, polypropylene, polyethylene, polyamide (nylon), polystyrene, polyoxymethylene, polyetherurethane, polybutylene terephthalate	Anderson et al. (2017)
	Laundry wastewater	Chulalongkorn University, Bangkok	Microfibers	Polyethylene	Shalumon and Ratanatamskul (2023)
	Fishery activities and human domestic sewage/ building industry	Nanxun Reef in Nansha Islands, South China Sea	Fibers, pellets, films, fragments	Polyvinylchloride, polyethylene, polyamide	Zhang et al. (2021)
	Sludge and wastewater treatment plants	The Persian Gulf	Fragment, fiber, film, granule	Polyamide (nylon), polyethylene, polypropylene	Xiang et al. (2022)
	Soil and road dust	Poland	Fragments	Rubber, polypropylene, acrylic, nylon	Worek et al. (2022)
	Anthropogenic activity	Three Gorges Reservoir, China	Fiber, Styrofoam, fragment, film, pellet	Polystyrene, polyethylene, polypropylene	Bui et al. (2020)
	Municipal wastewater treatment plant	Gyeonggi province, Korea	Fragments, fibers	Polyethylene, polypropylene, polystyrene, and polyethylene terephthalate	Lee et al. (2023)

3.3 IMPACTS OF MICROPLASTICS ON ENVIRONMENTAL COMPARTMENTS

MPs enter into the environment through a multitude of pathways and have an enormous impact on both biotic and abiotic components. Depending on their size, shape, kind of polymer, length, and varied environmental conditions, MPs have different outcomes. In this section, its impact will be discussed based on compartments.

3.3.1 Terrestrial Compartment

MPs can move vertically via the interstitial gaps in the soil to deeper layers (Li et al., 2021). There are two potential sources of MPs contaminating the terrestrial environment and they are domestic and agricultural activities. Domestic activities such as tourism result in the disposal of plastic products and single-use plastics that accumulate on land as plastic litter. As per the data, plastic litter predominantly consists of at least 90 million daily grocery bags, which end up as mismanaged plastic waste (Li & Richter, 2015). Once exposed to the environment, plastic litter undergoes deterioration and releases MPs into the soil due to UV rays, mechanical attrition, thermal decay, the action of bacteria, and contact with soil colloids (Dissanayake et al., 2022) (Figure 3.2). According to estimates, the soil contains 300 million tons of MPs. Fertilizer applications involving polymer coatings, slow-release fertilizers, composting, organic fertilizers, polyethylene-based plastic mulches, and irrigation water containing MPs (typically synthetic fiber) are some other agricultural practices that may contribute to MP contamination in the soil environment (Lebreton et al., 2017).

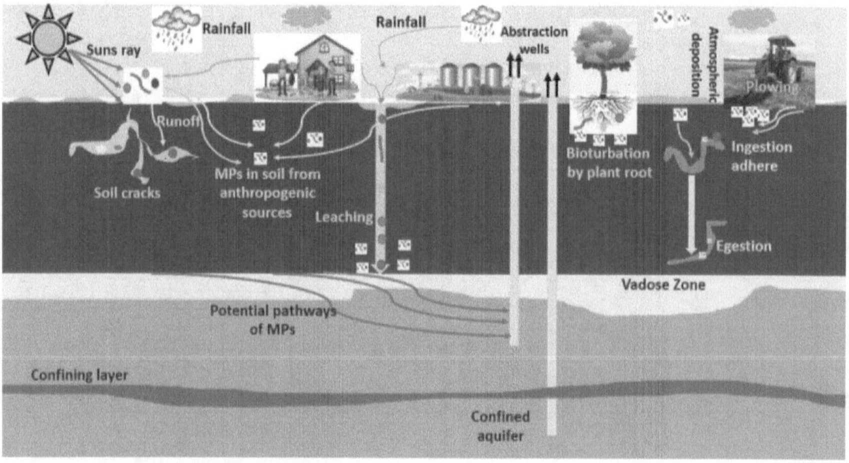

FIGURE 3.2 Fate of microplastics in terrestrial compartments.

3.3.1.1 Effects of Microplastics on Soil Physical Properties

When MPs land on soil surfaces, extrinsic factors including bioturbation processes and human activity instantly integrate them into the soil matrix. The inclusion may cause variations in the texture and structure of the soil (Dissanayake et al., 2022). According to Liu et al. (2014), large-scale MPs in the soil reduce irrigation and rainfall infiltration, which negatively affects the soil's ability to hold water and may result in oxygen deficiency. Additionally, these contaminants prevent evaporation, which may lead to an increase in soil moisture levels (Qin et al., 2015). Some writers claim that MPs decrease the water in the soil's permeability (porousness) and aeration (entry of oxygen-containing air) (Zhang et al., 2018). According to Siwek et al. (2015), MPs may eventually alter the physical characteristics and structure of soil by increasing the proportion of stable water aggregates. The negative impacts mostly depend on the shape of the MPs because soil clumping can be enhanced by linear-shaped polymers because they can blend more uniformly into the structure of the soil than non-linear plastics (Dissanayake et al., 2022). Numerous research has been conducted to date on the impact of MPs on the physical qualities of soil, with both positive as well as negative outcomes noted (Table 3.2). For example, in a 5-week garden test in loamy sand soil, de Souza Machado et al. (2018) studied four common MP types: polyacrylic fibers, polyamide beads, polyester fibers, and polyethylene fragments. They discovered that the soil contaminated with polyester (PES) fibers increased in bulk density, water-holding capacity, and water-stable aggregates. In contrast, the bulk densities of all other polluted soils decreased. Furthermore, there was a significant decrease in the creation of water-stable aggregates in soils contaminated with polystyrene (PS) and PES. However, clay soil impacted differently than sandy soil; it shrinks and cracks as evaporation occurs (Kirstein et al., 2016) and the author concluded that cracks formed due to the direct movement of water and solutes in the soil (Sajjad et al., 2022).

3.3.1.2 Effects of Microplastics on Soil Chemical Properties

While plastic tends to be resistant to breaking down in soil, MPs that have been severely degraded are labile and have strong chemical associations with soil particles. Depending on the polymer content and kind of soil, several erratic and obscure processes lead to chemical reactions after MPs break down into smaller pieces (Dissanayake et al., 2022). For instance, the ecosystem's nutrient transfer, nitrogen cycle, and soil organic carbon (SOC) are all negatively impacted by MP particles in the soil (Kumar et al., 2023).

TABLE 3.2 Effects of Microplastics on Soil Physical Properties

Polymer Type	Soil Type	Concentration	Duration	Type of Experiment	Effects	References
Polyethylene powders (<35, <125, <500 μm)	Silty clay loam and clay loam	0, 0.1, 1.0, 10 wt %	75 days	Field experiment	1. Aggregate size and MP size and concentration affected the number of MPs incorporated into the aggregates	Fang et al. (2024b)
Polyester fibers and the polypropylene granules	Silt loam	0%, 0.06%, 0.12%, 0.24%, 0.49%, 0.98% w/w for polyester fibers and 0%, 0.30%, 0.61%, 1.22%, 2.44%, 4.88% w/w for polypropylene granules	48 days	Field experiment	1. Bulk density, water holding capacity decreased 2. Contact angle and saturated hydraulic conductivity decreased 3. Aggregate stability unaffected	Yu et al. (2023)
Polyester	Vertiso, Entisol, Alfisol	0.5% w/w of the dry soil weight	6 months	Pot experiment	1. MP fibers limit the neoformation of soil aggregates 2. Polyester microplastic fibers reduced soil loss and sediment concentration 3. Affect soil bulk density, water stability unaffected	Ingraffia et al. (2022)
Polyacrylic fibers, polyethylene high-density fragments, polyamide beads, polyester fibers	Loamy sand	0.05%, 0.10%, 0.20%, and 0.40% (soil dry weight basis) for polyamide and polyethylene 0.25%, 0.50%, 1.00%, and 2.00% (soil dry weight basis) for polyester and polyacrylic	5 weeks	Pot experiment	1. The bulk density of the soil was impacted by all particles and polyester fibers reduced in a concentration-dependent manner. 2. Water holding capacity increased by water-stable aggregates	de Souza MacHado et al. (2018)

(*Continued*)

TABLE 3.2 (*Continued*) Effects of Microplastics on Soil Physical Properties

Polymer Type	Soil Type	Concentration	Duration	Type of Experiment	Effects	References
Polyester (PES) high-density polyethylene (HDPE), polypropylene (PP) polystyrene (PS), and polyethylene terephthalate (PET)	Loamy sand	PES 0.2% HDPE, PS, PP, PET 2%	2 months	Pot experiment	1. The bulk density of the soil was reduced by PS, PEHD, PES, PET, and PP 2. There were notable reductions in water-stable aggregates in soils containing PA, PES, and PS 3. Depending on the type of microplastic, all treatments changed the soil's structure, however the effects varied in severity	de Souza MacHado et al. (2019)
Synthetic fibers, high-density polyethylene (HDPE), polyacrylic (PLA)	Sandy clay loam	HDPE, PLA 0.1%, synthetic fibers 0.001%	30 days	Pot experiment	1. The mean weight diameters of the soils in the control treatments were 28%, 35%, and 24% larger than the soils in the HDPE, PLA, and fiber addition treatments, respectively 2. There were 60% and 53% more big macroaggregates (>2,000 μm) in the control soil than in the HDPE and PLA-treated soil, respectively 3. On the other hand, compared to control soil, soil exposed to all forms of microplastics had noticeably more micro-aggregates (250–63 μm)	Boots et al. (2019)
Polyester microfiber	Clay loam	0%, 0.1%, and 0.3%	75 days	Pot experiment	1. No differences in soil bulk density 2. Higher contents of large macroaggregates (>2 mm) (30.8%) were found in the PMF treatments after incubation 3. NPMF treatment significantly increased microaggregate (0.25–0.05 mm)	Zhang et al. (2019)

As reported by Palansooriya et al. (2022), the total exchangeable cations (Na^+, K^+, Mg^{2+}, and Ca^{2+}) significantly decreased at higher concentrations of low-density polyethylene (LDPE) derived MPs (3%–7%), while the electrical conductivity of soil contaminated by LDPE MPs increased when the LDPE MP concentrations were between 1% and 7%. Another study by Qi et al. (2020) showed that mulch films made of biodegradable plastic and LDPE raised the pH of the soil and decreased its electric conductivity after 2 months of incubation. Thus, the effect varies depending on the polymer type, concentration, incubation time, soil type, and other factors. Table 3.3 shows studies that are focused on the effects of MPs on the chemical properties of soil.

3.3.1.3 Effects of Microplastics on Terrestrial Plants

Based on recent research, MPs can be absorbed by plants, and it has been suggested that flexible cell wall holes may allow MP particles to enter more easily (Ullah et al., 2021). Plant root features, growth, and nutrient uptake can all be affected by changes in soil parameters caused by MPs, including moisture, density, structure, and uptake of nutritional content (de Souza Machado et al., 2018). Different plant species have different rates of uptake, transport, and accumulation of organic pollutants based on their anatomical and physiological characteristics (Ng et al., 2018). For example, de Souza Machado et al. (2018), found that MPs (PES fibers, polyamide beads, polyethylene (PE), PES terephthalate, polypropylene (PP), and PS) changed spring onions (*Allium fistulosum*) behavior in terms of their roots and leaves, total biomass, and leaf composition, which included nitrogen content and the carbon-to-nitrogen (C:N) ratio. For the consequences of MPs in terrestrial ecosystems, they put out a stochastic model in which changes in the biophysical environment of soil cause a series of cascades that impact onion development. However, a different study by Verla et al. (2020) concluded that LDPE, polypropylene, and polystyrene macro- and microplastic residues in soils had an impact on the vegetative growth of juvenile lime trees (*Citrus aurantium*) in both positive and negative ways. Juvenile lime trees' vegetative development was positively impacted by a soil mixture containing LDPE, PP, and PS, but growth was adversely affected most by LDPE. MPs have been shown to cause genotoxicity and ecotoxicity in plants, according to Jiang et al. (2019). They found that fava bean (*Vicia faba*) growth was significantly reduced when exposed to 100 mg/L of 100 mm polystyrene fluorescent microplastics (PS-MPs).

TABLE 3.3 Effects of Microplastics on Soil Physical Properties

Microplastics Type	Soil Type	Concentration	Duration	Type of Experiment	Effects	References
Low-density polyethylene and biodegradable plastic mulch films	Sandy soil	1%	4 months	Pot experiment	Increase in pH and decrease in EC	Qi et al. (2020)
Polyamide (PA), polycarbonate (PC), polyethylene (PE), polyester (PES), polyethylene terephthalate (PET), polypropylene (PP), polystyrene (PS), and polyurethane (PU)	Loamy sandy soil	0.4% w/w	31 days	Field experiment	Soil pH increased	Zhao et al. (2021)
Polyethylene (PE)	Loamy soil	Film residues (0–200 cm^2)	1–25 year	Field experiment	Soil pH increased; organic matter decreased	Hegan et al. (2015)
Polyethylene (PE)	Purple soil	Sizes (i.e., 300 and 600 μm)	4 weeks	Pot experiment	Smaller-sized PE MPs significantly reduced the content of dissolved organic compound while significantly increasing the cation exchange capacity compared with larger-sized PE MPs	Ma et al. (2023)

Likewise, several plant species have reported that MPs have an inhibiting influence on plant development and performance (Table 3.4).

3.3.2 Effects of Microplastics on Soil Organisms

Soil is a vast ecosystem for different types of organisms. Earthworms, mycorrhizae, insects, mites, and nematodes are some examples of soil creatures that are crucial to conserve and maintain a healthy ecosystem and functions of soil. Nonetheless, MPs may present a serious risk to these creatures (Qi et al., 2020). MP ecotoxicity is influenced by several variables, such as kind, size, form, color, shape, density, number, physical behavior, chemical composition, age, and biological interactions. For example, it has been found that tiny MPs are more harmful than bigger ones since the former may impede an organism's internal structure and produce abrasions. Smaller MPs can potentially breach the cell membrane and protective wall, inflicting damage to living things (Dissanayake et al., 2022). According to Zimmermann et al. (2020), the chemical makeup of plastics varies, leading to variations in MP ecotoxicity. Since weak van der Waals forces bind most of the compounds included in plastics, these additives can leak into the environment by aging, and these can harm living things. Numerous times, MPs are taken up as food by soil biota, which reduces the amount of carbon biomass ingested and can cause energy depletion, stunted development, and even death (da Costa et al., 2016). Studies showed that MPs can change the bacterial and fungal populations in soil by either actively or passively influencing the soil's microenvironmental circumstances. For instance, 1% polyethylene and polyvinyl chloride (PVC) reduced the variety and quantity of the soil bacterial community and limited the activity of specific hydrolysis enzymes in acidic soil (Fei et al., 2020). MPs that have large surface areas provide niches for soil fungi and bacteria, which subsequently changes the movement of soil microorganisms and affects the distribution and composition of soil microbial communities (Prata et al., 2020). By influencing some functional microorganisms in the soil, MPs can also change biogeochemical cycles and soil ecology functions. These include interfering with the soil's carbon and nitrogen cycles, releasing greenhouse gases, and degrading organic matter (Rong et al., 2021). In a 28-day trial, Rodriguez-Seijo et al. (2017) discovered that earthworms exposed to soil with polyethylene pellets (0–1,000 mg/kg) did not experience any significant effect on survival, growth, and reproduction. However, after exposing earthworms to a minimum of 125 mg/kg of MPs, they saw gastrointestinal harm and histopathological damage was evident at a dose as low as 62.5 mg/kg, and that the damage intensified at greater

TABLE 3.4 Effects of Microplastics on Plant Development and Performance

Plant Species	Types of MPs	MPs Amount	Incubation Time	Experiment Type	Effects	References
Cucumber (*Cucumis sativus* L.)	Polystyrene (PS)	5 and 0.1 μm	Greenhouse	60 days	Increased plant height, root length, and dry weight of the aboveground and underground parts	Liu et al. (2024)
Lettuce (*Lactuca sativa* L.)	Polystyrene, polyethylene, and polypropylene	0.01% and 0.1% w/w	Pot experiment	48 days	Aboveground length and fresh weight drastically reduced	Lian et al. (2024)
Cotton plants	Carboxyl-modified polystyrene microplastics (PS-COOH)	100, 300, and 500 mg/L	Hydroponic culture system	15 days	Dose-dependent phenotypic change occurred; oxidative stress observed	Li et al. (2024)
Rice (*Oryza sativa* L.)	PS	Low, medium, high	Suspension and greenhouse + field trials	21 days	Higher doses of PS caused ≈40% decrease in shoot biomass, compared with that of 25.9% in field trials	Wu et al. (2020)
Perennial ryegrass (*Lolium perenne* L.)	Biodegradable polylactic acid and HDPE	1 g/kg	Soil	30 days	Reduced germination of seeds when exposed to PLA. Additionally, compared to PE, the root biomass was lower with PLA. The amount of chlorophyll did not vary significantly	Boots et al. (2019)

(*Continued*)

TABLE 3.4 (*Continued*) Effects of Microplastics on Plant Development and Performance

Plant Species	Types of MPs	MPs Amount	Incubation Time	Experiment Type	Effects	References
Zea mays (L. var. Jubilee)	PE	0.0125, 100 mg/L	Hydroponic	10–15 days	The tested MP did not go to the shoots; instead, it collected on the surface of the roots. PE hampered the maize plant's ability to absorb nutrients and water	Urbina et al. (2020)
Spring onion (*Allium fistulosum*)	Polyester (PES) high-density polyethylene (HDPE), polypropylene (PP) polystyrene (PS), and polyethylene terephthalate (PET)	PES 0.2% HDPE, PS, PP, PET 2%	1.5 months	Pot experiment	1. PES and PS increased root biomass 2. PA decreased the ratio between root and leaf-dry biomass 3. All tested microplastics significantly increased total root length and decreased root average diameter	de Souza MacHado et al. (2019)
Wheat (*Triticum aestivum*)	PHBV	1%	25 days	Greenhouse	The plants died after 25 days due to the addition of PHBV. It raised the dissolved organic C content and biomass of the soil	Zhou et al. (2021)

concentrations. The research in this area is still in progress. As different species have different effects at varying concentrations. Thus, for a better comprehensive understanding of the impacts, more research in this area is required.

3.3.3 Aquatic Compartment

According to estimates, between 4.8 and 12.7 million tons of plastic were dumped into the ocean, and this number is rising daily (Gola et al., 2021). An increase in the number of human population density has been shown to positively correlate with MP abundance, which may contribute to the accumulation of plastic trash in the aquatic environment (Figure 3.3).

The distribution and accumulation of MPs can be influenced by a variety of factors, including the physicochemical properties of the particles (such as size, specific density, charge, and chemical composition), hydrodynamic factors, and environmental factors (such as temperature, wind, turbidity, velocity, water currents, and density of water mass), etc. (Guzzetti et al., 2018). Apart from the coastal region and surface water, MP has been also found in deep-sea water samples, the polar region, etc. (Gola et al., 2021). Almost 700 aquatic species worldwide, including sea turtles, penguins, and various crustaceans, were negatively impacted by the encroaching tide of MPs (Marn et al., 2020).

3.3.3.1 Effects on Aquatic Fauna

In water bodies, MP is exposed to and consumed by aquatic animals. Since there is no enzyme available to break down the synthetic polymer into

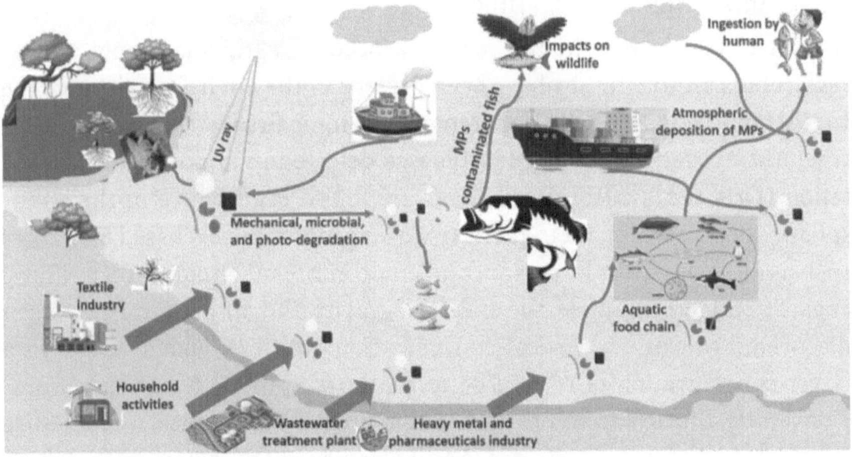

FIGURE 3.3 Microplastics in aquatic compartments.

simpler, non-toxic chemical compounds, it does not undergo any enzymatic breakdown via the metabolic process once it is consumed. In addition, the absorption of various organic and inorganic chemicals (such as heavy metals, dyes, and pesticides), hydrocarbons, and antibiotics, among others, increases the toxicity of MPs (Gola et al., 2021). Fish can absorb MPs through their gills. Some organisms may absorb MPs directly through their internal surfaces or via the epithelial tissues of their digestive systems. Once consumed or absorbed, MPs can move throughout an organism's body. Studies have demonstrated that MPs can find their way into the liver, stomach, kidneys, and even the circulatory system, among other tissues and organs, and ultimately create negative effects on the organism (Thacharodi et al., 2024; Wang et al., 2024) (Table 3.5).

3.3.3.2 Effects on Aquatic Flora

MP toxicity affects aquatic plants in two ways – direct and indirect. Direct impact includes physical entanglement or covering plant surfaces, obstructing the exchange of gases and light. Aquatic plant development and health can be further hampered by chemical additions and pollutants linked to MPs. The indirect impacts of MPs on aquatic plants include changes to the pH, dissolved oxygen content, and nutritional availability of the water. The delicate balance necessary for the best possible growth and survival of plants may be affected by these changes. According to studies, MPs can build up in seagrass beds and negatively impact the health and functionality of the plants (Thacharodi et al., 2024; Ge et al., 2021) (Table 3.6).

3.3.4 Atmospheric Compartment

In recent years, MPs in the air have become a popular topic among researchers although they have been present in the environment since the Industrial Revolution. The research on this topic first started back in 2015 when the existence of airborne MPs was detected in atmospheric precipitation (Dris et al., 2015). While MPs are found everywhere in the atmospheric compartment, their quantity varies geographically based on factors such as altitude, latitude, season, and environmental circumstances (Yang et al., 2021a). For example, the effects of gravity and air pressure have been discovered to cause a vertical concentration gradient, which results in a larger concentration of MPs close to the ground than at high elevations. Conversely, the upward motion of MPs is aided by the vertical temperature gradient since the air is less dense and warmer in the lower region, which causes it to rise. As cold air at a lower altitude is heavier and cannot rise,

TABLE 3.5 Impacts of Microplastics on Aquatic Fauna

Organisms	Types of MPs	MPs Amount/ Concentration	Effects	References
Zebrafish	High-density polyethylene (HDPE) small (14.12 μm), medium (80.32 μm), and large (120.97 μm)	20 mg/L	Large MPs affected fish's lateral line system's mechanosensory receptors and hair cells; gastrointestinal system exhibited morphological alterations because of larvae consuming MPs smaller than 100 μm; all larvae exposed to MPs displayed altered behavioral patterns correlated with variations in size	Kim et al. (2022)
Grass Carp	80 nm and 8 μm microplastics	5, 15, and 45 mg/L	Fish may re-consume excreted microplastics and store them in the mouth cavity, whereas the nasal cavity may retain 1 μm of MPs	Zhang et al. (2022)
Shrimp (*Litopenaeus vannamei*)	Polystyrene, 1.1 μm	40 and 400 μg MP/kg of ration	Damage to DNA and lipids was visible in the hepatopancreas. A histological examination showed edema in the muscle, hepatopancreas, and gut. There was granuloma formation in the gut and hepatopancreas with hemocyte infiltration	Seta et al. (2023)
Clam (*Ruditapes decussatus*)	Polyethylene (PE) (40–48 μm)	10 μg/L (low), 100 μg/L (medium), and 1,000 μg/L (high)	Increased quantity of hemocyte cells; clams' inability to filter water; clams' overall weight decline; substantial decline in hemoglobin viability	Abidli et al. (2023)
Japanese quail (*Coturnix japonica*)	Polypropylene and polyethylene	600 mg	The antioxidant enzymes glutathione-*S*-transferase, glutathione peroxidase, and catalase were all significantly stimulated by exposure to small MP. Females exposed to substantial MP showed a drop in 17β-estradiol levels and an increase in aspartate aminotransferase (a liver parameter). Ingesting MP did not directly change body mass, however, quail exposed to small MP and a combination of large and small MP grew at a different rate than quail in the control group	Monclús et al. (2022)
Coral (*Goniopora columna*)	Polyethylene (40–48 μm)		Affect the antioxidant system, impair the activity of the enzymes that detoxify G. column, raise the concentration of lipid peroxide, and harm tissue by building up on the coral mesenteric	Chen et al. (2022)

TABLE 3.6 Effects of Microplastics on Different Kinds of Aquatic Plants

Plant Species	Types of MPs	MPs Amount/ Concentration	Incubation Time	Effects	References
Spirodela polyrhiza, Salvinia natans and Phragmites australis	1-μm-sized fluorescent microplastics (FMPs)	0.1% and 0.01%	3 weeks	*S. polyrhiza* and *P. australis* showed a significant decrease in harvested biomass, confirming the phytotoxicity of FMPs; however, *S. natans* showed no variation in harvested biomass or chlorophyll contents between treatments	Song et al. (2023)
Lemna minor	Polyethylene microbeads (4–12 μm)	0, 10, 50, and 100 mg/L	7 days	The root development was notably impacted by mechanical obstruction	Kalčíková et al. (2017)
Lemna minuta	Polystyrene	50, 100 mg/L	28 days	Chlorophyll levels and photosynthetic capacity were reduced; after 28 days of exposure, growth (biomass, RGR) was most affected at the highest dosage. Adsorption of 1–5 μm MPs occurred mostly on plant surfaces in close association with the contaminated suspension	Ceschin et al. (2023)
Thalassiosira pseudonana, Skeletonema grethae, Phaeodactylum tricornutum, Dunaliella tertiolecta	0.055, 1 and 6 μm polystyrene	10^{-4} to 250 mg/L	2 days	Marine phytoplankton is more sensitive to smaller-sized plastics, it actively changes the chemical composition of EPS to withstand the strain of pollution. It was discovered that the release of EPS, which is rich in proteins, helped plastic particles form aggregates and modify their surfaces, which in turn affected the fate and colonization of the particles	Shiu et al. (2020)
Utricularia vulgaris	1, 2, and 5 μm polystyrene	15, 70. and 140 m/L	7 days	Substantial ecotoxicity and oxidative damage observed, the growth rate and functional characteristics of leaves were substantially hindered	Yu et al. (2020)
Spirodela polyrhiza	Polyvinyl chloride (PVC)	0, 10, 100, and 1,000 mg/L		Exhibited significant effects on the nitrogen fate of aquatic plant habitats, inhibited plant growth and reproduction, altered plant nutrient metabolism, and decreased the spontaneous root elongation ratio by 41.68% and the leaf multiplication ratio by 61.03%. These reductions also led to a decrease in anthocyanin and nitrogen contents to 63.45% and 84.21% of the control group, respectively	Wang et al. (2023)

it can also block MPs in the light layers of the atmosphere during temperature inversions, hindering the dispersion of atmospheric pollution and trapping it close to the surface (López et al., 2023).

MPs have been found in a variety of human organs and biological fluids, demonstrating their capacity to enter the body through a variety of entrance points (Table 3.7). Plastic particles can contaminate food for humans either directly from environmental exposure or indirectly through the movement of MPs across the ecosystem. MPs have a significant impact on the ecosystem as a whole by delivering pollutants that pose a health risk to consumers through food. MPs are easily ingested by

TABLE 3.7 Human Exposure to Microplastics: Evidence from Tissues and Fluids

Sample	Type of MPs	Polymer Type	Size	References
Human blood	---	Polyethylene terephthalate, polypropylene, polyethylene, and polymers of styrene	≥700 nm	Leslie et al. (2022)
Human placentas	Fragments	Polypropylene	5–10 μm	Ragusa et al. (2021)
Human urine	Fragments with irregular shapes	Polyethylene vinyl acetate (PVA), polyvinyl chloride (PVC), polypropylene (PP), and polyethylene (PE)	4–15 μm size	Pironti et al. (2022)
Human semen	Spheric or irregular shapes	Polypropylene (PP), polyethylene (PE), polyethylene terephthalate (PET), polystyrene (PS), polyvinylchloride (PVC), polycarbonate (PC), polyoxymethylene (POM) and acrylic	2–6 μm	Montano et al. (2023)
Human feces	---	Polypropylene and polyethylene terephthalate prominent one	50–500 μm	Schwabl et al. (2019)
Human sputum	---	Polyurethane, polyester, chlorinated polyethylene, and alkyd varnish	20–500 μm	Huang et al. (2022)
Human breastmilk	Irregular fragments and spheres	Polyethylene, polyvinyl chloride, and polypropylene,	2–12 μm	Ragusa et al. (2022)
Human lung	Fibre, fragment, and film	Polypropylene and polyethylene terephthalate fibers were the most abundant.	3 μm	Jenner et al. (2022)

--- means not given

marine and terrestrial creatures, which allow them to enter the human food chain (Al Mamun et al., 2023). Ingestion, inhalation, and dermal contact are the three primary pathways through which MPs can enter the body's interior (Wu et al., 2022). The most frequent way that MPs enter the body is by ingestion since so far, a wide variety of foods including fish, table salt, sugar, vegetables, rice, and others are detected to contain significant amounts of plastic particles (Cox et al., 2019; Pironti et al., 2021; Al Mamun et al., 2023). It is estimated that an adult may consume up to 4.1×10^4 MP particles from drinking water and 5.1×10^3 MP particles from table salts on an annual basis (Al Mamun et al., 2023). The airborne route has been found as another crucial pathway by which MPs can enter the human body. It was initially detected in the atmosphere close to Paris, France, with an average daily fallout of 118 MPs (ranging from 29 to 280 items·$m^{-2}.d^{-1}$) (Dris et al., 2015). These particles may enter the alveolar areas of the human lungs through inhalation, where they may then move into epithelial layers as a result of gas exchange between the alveoli and capillaries (Barlow et al., 2017). Another recognized method of exposure to MPs is dermal contact, however, it appears to be less common than ingesting and inhalation. Wearing clothes and cleaning the body with products like toothpaste, shower gel, and scrubs can expose MPs to the skin. The dermal barrier may be penetrated by nanoplastics (NPs) (less than 100 nm), synthetic fibers (less than 25 nm), monomers, and additives since skin pores range in size from 40 to 80 μm (Wu et al., 2022).

Once within, they may move and gather in various tissues and organs. The physical attributes of MPs (size, shape, and length), their chemical composition (kind of polymer and presence of additives), their concentration level, and the potential existence of microbial biofilms are all responsible for their hazardous effects. The components release and age over time when they are in the environment, which is why it harms health. Furthermore, MPs aging in the atmosphere due to direct photo-irradiation, causing surface changes (roughness and hydrophobicity) that worsen their cytotoxicity, adsorption of pollutants, and microbe movement. MPs-induced cytotoxicity increases oxidative stress by producing free radicals from ROS, which could interfere with cell homeostasis. When MPs encounter the interior of the body, they can cause inflammatory lesions, alter metabolism, and raise the risk of cancer, among other harmful health effects (López et al., 2023; Al Mamun et al., 2023).

FIGURE 3.4 Mitigation approaches for microplastic pollution.

3.3.5 Mitigation Approaches

The removal of existing plastics from the environment, as well as the reduction of future pollution, are the two main goals of the holistic approach to mitigate MP. These tactics cover behavioral, technical, and regulatory shifts. An outline of the principal methods for reducing MPs is provided below in Figure 3.4.

According to Jambeck et al. (2015), the absence of comprehensive waste management alone is the reason behind the yearly entry of 4.8–12.7 million tons of plastics into the oceans which over time converted to MPs. Therefore, improving waste management is one of the prime things to do. This could be accomplished by separating the waste according to its type and degree of degradation. This separation guarantees the best possible handling of various waste while also streamlining the recycling and disposal procedures. The implementation of sophisticated recycling systems, which guarantees that commodities like plastics and microfibers are properly recycled, is crucial for the effective management of plastic waste and strengthens this strategy even more (Rostami et al., 2021). The next approach is regulatory measures which include numerous laws, regulations, and voluntary commitments (Chen 2015). Implementation of taxes and levies is crucial in increasing public awareness of plastic consumption and promoting the decrease of plastic waste production. By reducing the amount of plastic that enters the environment, these coordinated efforts seek to lessen the negative effects of plastic pollution. Technological innovation helps to remove already-existing plastic particles (Thacharodi et al., 2024). Researchers throughout the world have employed a variety of techniques that have demonstrated effectiveness in eliminating MPs (Table 3.8). Advanced techniques like reverse osmosis, membrane bioreactor, biofilter, coagulation/flocculation, etc. are particularly applied in wastewater treatment systems to remove pollutants as well as MPs. Conventional methods of primary and secondary treatment could remove up to 99% of MPs. Most of the removal of MPs from wastewater occurs in the early stages of treatment. However, because of the large amounts of effluent they discharge, traditional wastewater treatment plants remain potentially a significant contributor to MPs even with this high removal efficiency (Thacharodi et al., 2024). Besides, utilizing bacteria, fungi, and algae is gaining importance day by day as they can degrade tremendous amounts of MPs. For instance, according to Das et al. (2012), several bacteria, including *Achromobacter, Bacillus*, and *Pseudomonas*, are capable of degrading PVC. In another study, it was noted that within 120 days *Pseudomonas* can degrade 20% of high-density polyethylene (HDPE) (Balasubramanian et al., 2010). Bacteria in constrained medium possess the ability to utilize polymer materials as their primary carbon source and can modify the chemical characteristics, tensile strength, average molecular weight, and structural integrity of the polymer and ultimately produce various catalytic enzymes that contribute to the adsorption, degradation, and desorption of MPs (Thacharodi et al., 2023). Further efficiency of the breakdown of MPs increases when bacteria consortia, groups of microorganisms, collaborate symbiotically (Yuan et al., 2020). Fungi are an excellent option for degrading a variety of plastic polymers due to their extensive metabolic capability, which includes the capacity to replicate extracellular multienzyme complexes. It was discovered that a number of strains belonging to the genera *Aspergillus, Penicillium*, and *Trichoderma*, such as *Aspergillus niger, A versicolor, Penicillium pinophilum, P. chrysogenum, P. frequentan* and *P. oxalicum* were degraders of PE. Other organisms that may break down polyethylene are *Verticillium lecanii, Fusarium redolens, Acremonium kiliense*, and *Gliocladium virens*

TABLE 3.8 The Advantages and Disadvantages of Several Methods for Removing Microplastics from the Environment

Method	Number of MPs Eliminated/Efficiency	Advantages	Disadvantages	References
Biofilter	79%	Large MP particles have been retained effectively by the biofilter, which was able to dramatically reduce the MP abundance in the treated effluent	Deeper layers did not guarantee the full retention of microplastics and only produced a small increase in treatment effectiveness. The final effluent from the filter did not contain any particles larger than 100 μm	Liu et al. (2020)
Coagulation/flocculation	99.4%	Can remove smaller MPs; cost-effective; high-efficiency	Chemicals are required	Rajala et al. (2020); Thacharodi et al. (2024)
Photocatalytic degradation	97% average mass loss	No chemicals required; environment-friendly	Produced a modest level of by-products (less than 10 mg/L COD)	Lee et al. (2020)
Membrane bioreactor	98.3%	Using combined advanced treatment methods with porous membranes	Shape dependency of the removal percentage, membrane fouling	Lares et al. (2018)
Constructed wetland	>89%	Operational cost affordable; green water treatment facility	Large area required; maintenance costly; risk of contamination	Zhou et al. (2022)
Reverse osmosis	86%	Most economical and efficient	Microplastics having irregular shapes can abrade surfaces and decrease filtration efficiency	Tadsuwan and Babel (2021)
Biodegradation	Depends on microbes	Versatility, low running costs, practically applicable in many contexts, ease and reliability for large-scale application	Difficult-to-control environmental factors, challenges in conducting large-scale analysis, and difficulty in identifying the right microbial community	Arossa et al. (2019); Hall et al. (2015)

(Thacharodi et al., 2024). In the fields of waste management and environmental conservation public awareness, research and development, and international cooperation play an important part. Fostering widespread environmental stewardship begins with educating the public on the significance of implementing sustainable waste management methods, such as waste reduction, recycling, and responsible disposal. To find creative solutions, it is crucial to fund research and development projects that enhance waste management technologies. To lessen environmental damage, these initiatives can involve creating innovative recycling methods and investigating biodegradable materials. Furthermore, it is tough to understate the importance of international cooperation. Nations can coordinate their efforts to address urgent environmental problems, including MP pollution, by partnering together globally. By establishing standardized approaches for evaluating and addressing environmental issues, collaborative initiatives open the door to practical solutions that cut across national boundaries. As a result, these actions will aid in building a strong framework for controlling MP contamination, encouraging more environmentally friendly behaviors, and minimizing environmental impact.

3.5 CONCLUSION

In conclusion, one of the biggest environmental problems of today is the problem caused by MP contamination, which is also an indicator of a larger problem related to the world's dependence on plastics. A concentrated and prompt response is required to lessen the effects of the extensive manufacturing and use of plastics, which have led to several environmental nuisances and serious threats to human health. The decomposition of bigger plastic objects has resulted in the introduction of MPs into environmental compartments and that affects biotic and abiotic components. Therefore, it is vital to address their existence and stop additional pollution. As a first step in addressing the problem of MP pollution, this chapter emphasizes the significance of source identification. Stakeholders can create more focused plans to mitigate the intrusion of MPS into the environment by gaining knowledge about the origins of these pollutants. A thorough framework for addressing this issue is provided for academics, policymakers, and environmental managers through the description of the several types of MPs, their characteristics, and the advanced methods for reducing existing MPs. Apart from that bioremediation by different organisms was also discussed. Moreover, this chapter emphasizes the necessity for researchers to work together to address the MP issue, acknowledging the pressing need for additional study to determine MP threshold concentrations and to implement practical regulatory measures. It is through such committed investigation and policy creation that we may hope to limit the consequences of MP contamination, saving both our environment and public health for years to come.

REFERENCES

Abidli, S., Zaidi, S., Ben Younes, R., Lahbib, Y., & Trigui El Menif, N. (2023). Impact of polyethylene microplastics on the clam Ruditapes decussatus (Mollusca: Bivalvia): Examination of filtration rate, growth, and immunomodulation. *Ecotoxicology*, *32*(6), 746–755. https://doi.org/10.1007/s10646-023-02683-2

Al Mamun, A., Prasetya, T. A. E., Dewi, I. R., & Ahmad, M. (2023). Microplastics in human food chains: Food becoming a threat to health safety. *Science of the Total Environment*, *858*, 159834. https://doi.org/10.1016/j.scitotenv.2022.159834

All types of microplastics you should know. (n.d.). EuroPlas. Retrieved February 8, 2024, from https://europlas.com.vn/en-US/blog-1/all-types-of-microplastics-you-should-know

Allsopp, M., Walters, A., Santillo, D., & Johnston, P. (2006). *Plastic Debris in the World's Oceans*. Greenpeace.

Almeida, C. M. R., Perdigão, R., Correia, B. R., Van Der Gracht, H., Dias, S., Magalhães, C., ... & Ramos, S. (2024). Potential of fishing nets for adsorption of inorganic (Cu and Pb) and organic (PAHs) pollutants. *Marine Pollution Bulletin, 209,* 117291.

Anderson, P. J., Warrack, S., Langen, V., Challis, J. K., Hanson, M. L., & Rennie, M. D. (2017). Microplastic contamination in Lake Winnipeg, Canada. *Environmental Pollution*, *225*, 223–231. https://doi.org/10.1016/j.envpol.2017.02.072

Arossa, S., Martin, C., Rossbach, S., & Duarte, C. M. (2019). Microplastic removal by red sea giant clam (*Tridacna maxima*). *Environmental Pollution*, *252*, 1257–1266. https://doi.org/10.1016/j.envpol.2019.05.149

Azeem, I., Adeel, M., Ahmad, M. A., Shakoor, N., Jiangcuo, G. D., Azeem, K., Ishfaq, M., Shakoor, A., Ayaz, M., Xu, M., & Rui, Y. (2021). Uptake and accumulation of nano/microplastics in plants: A critical review. *Nanomaterials*, *11*(11), 2935. https://doi.org/10.3390/nano11112935

Balasubramanian, V., Natarajan, K., Hemambika, B., Ramesh, N., Sumathi, C. S., Kottaimuthu, R., & Rajesh Kannan, V. (2010). High-density polyethylene (HDPE)-degrading potential bacteria from marine ecosystem of Gulf of Mannar, India. *Letters in Applied Microbiology*, *51*(2), 205–211. https://doi.org/10.1111/j.1472-765X.2010.02883.x

Bank, M. S., & Hansson, S. V. (2019). The plastic cycle: A novel and holistic paradigm for the anthropocene. *Environmental Science & Technology*, *53*, 7177–7179. https://doi.org/10.1021/acs.est.9b02942

Barlow, C. A., Grespin, M., & Best, E. A. (2017). Asbestos fiber length and its relation to disease risk. *Inhalation Toxicology, 29*(12–14), 541–554. https://doi.org/10.1080/08958378.2018.1435756

Bhatia, S. K., Kumar, G., & Yang, Y. H. (2024). Understanding microplastic pollution: Tracing the footprints and eco-friendly solutions. *Science of the Total Environment*, 169926. https://doi.org/10.1016/j.scitotenv.2024.169926

Boots, B., Russell, C. W., & Green, D. S. (2019). Effects of microplastics in soil ecosystems: Above and below ground. *Environmental Science & Technology, 53*(19), 11496–11506. https://doi.org/10.1021/acs.est.9b03304

Bradney, L., Wijesekara, H., Palansooriya, K. N., Obadamudalige, N., Bolan, N. S., Ok, Y. S., Rinklebe, J., Kim, K.-H., & Kirkham, M. B. (2019). Particulate plastics as a vector for toxic trace-element uptake by aquatic and terrestrial organisms and human health risk. *Environment International, 131*, 104937. https://doi.org/10.1016/j.envint.2019.104937

Browne, M. A., Niven, S. J., Galloway, T. S., Rowland, S. J., & Thompson, R. C. (2013). Microplastic moves pollutants and additives to worms, reducing functions linked to health and biodiversity. *Current Biology, 23*(23), 2388–2392. https://doi.org/10.1016/j.cub.2013.10.012

Bui, X. T., Nguyen, P. T., Nguyen, V. T., Dao, T. S., & Nguyen, P. D. (2020). Microplastics pollution in wastewater: Characteristics, occurrence and removal technologies. *Environmental Technology & Innovation, 19*, 101013. https://doi.org/10.1016/j.eti.2020.101013

Bydalek, F., Ifayemi, D., Reynolds, L., Barden, R., Kasprzyk-Hordern, B., & Wenk, J. (2023). Microplastic dynamics in a free water surface constructed wetland. *Science of the Total Environment, 858*, 160113. https://doi.org/10.1016/j.scitotenv.2022.160113

Calero, M., Godoy, V., Quesada, L., & Martín-Lara, M. Á. (2021). Green strategies for microplastics reduction. *Current Opinion in Green and Sustainable Chemistry, 28*, 100442. https://doi.org/10.1016/j.cogsc.2020.100442

Carpenter, E. J., & Smith, K. L. (1972). Plastics on the Sargasso Sea Surface. *Science, 175*(4027), 1240–1241. https://doi.org/10.1126/science.175.4027.1240

Ceschin, S., Mariani, F., Di Lernia, D., Venditti, I., Pelella, E., & Iannelli, M. A. (2023). Effects of Microplastic contamination on the aquatic plant Lemna minuta (least duckweed). *Plants, 12*(1), 207. https://doi.org/10.3390/plants12010207

Chen, C. L. (2015). Regulation and management of marine litter. In M. Bergmann, L. Gutow, M. Klages (Eds.), *Marine Anthropogenic Litter* (pp. 395–428). SpringerOpen.

Chen, L., Zhou, S., Zhang, Q., Su, B., Yin, Q., & Zou, M. (2024). Global occurrence characteristics, drivers, and environmental risk assessment of microplastics in lakes: A meta-analysis. *Environmental Pollution*, 123321. https://doi.org/10.1016/j.envpol.2024.123321

Chen, Y. T., Ding, D. S., Lim, Y. C., Singhania, R. R., Hsieh, S., Chen, C. W., Hsieh, S. L., & Dong, C. D. (2022). Impact of polyethylene microplastics on coral Goniopora columna causing oxidative stress and histopathology damages. *Science of the Total Environment, 828*, 154234. https://doi.org/10.1016/j.scitotenv.2022.154234

Cole, M., Lindeque, P., Halsband, C., & Galloway, T. S. (2011). Microplastics as contaminants in the marine environment: A review. *Marine Pollution Bulletin, 62*(12), 2588–2597. https://doi.org/10.1016/j.marpolbul.2011.09.025

Cox, K. D., Covernton, G. A., Davies, H. L., Dower, J. F., Juanes, F., & Dudas, S. E. (2019). Human consumption of microplastics. *Environmental Science & Technology, 53*(12), 7068–7074. https://doi.org/10.1021/acs.est.9b01517

da Costa, J. P., Santos, P. S., Duarte, A. C., & Rocha-Santos, T. (2016). (Nano) plastics in the environment–sources, fates and effects. *Science of the Total Environment, 566*, 15–26. https://doi.org/10.1016/j.scitotenv.2016.05.041

Das, G., Bordoloi, N. K., Rai, S. K., Mukherjee, A. K., & Karak, N. (2012). Biodegradable and biocompatible epoxidized vegetable oil modified thermostable poly (vinyl chloride): Thermal and performance characteristics post biodegradation with Pseudomonas aeruginosa and Achromobacter sp. *Journal of Hazardous Materials, 209*, 434–442. https://doi.org/10.1016/j.jhazmat.2012.01.043

de Almeida, M. P., Gaylarde, C. C., Baptista Neto, J. A., Delgado, J. D. F., Lima, L. D. S., Neves, C. V., Pompermayer, L. L. D. O., Vieira, K., & da Fonseca, E. M. (2023). The prevalence of microplastics on the earth and resulting increased imbalances in biogeochemical cycling. *Water Emerg. Contam. Nanoplast, 2*, 7. https://doi.org/10.20517/wecn.2022.20

De Sá, L. C., Oliveira, M., Ribeiro, F., Rocha, T. L., & Futter, M. N. (2018). Studies of the effects of microplastics on aquatic organisms: What do we know and where should we focus our efforts in the future? *Science of the Total Environment, 645*, 1029–1039. https://doi.org/10.1016/j.scitotenv.2018.07.207

de Souza Machado, A. A., Lau, C. W., Kloas, W., Bergmann, J., Bachelier, J. B., Faltin, E., ... & Rillig, M. C. (2019). Microplastics can change soil properties and affect plant performance. *Environmental Science & Technology, 53*(10), 6044–6052.

de Souza Machado, A. A., Lau, C. W., Till, J., Kloas, W., Lehmann, A., Becker, R., & Rillig, M. C. (2018). Impacts of microplastics on the soil biophysical environment. *Environmental Science & Technology, 52*(17), 9656–9665. https://doi.org/10.1021/acs.est.8b02212

Dissanayake, P. D., Kim, S., Sarkar, B., Oleszczuk, P., Sang, M. K., Haque, M. N., Ahn, J. H., Bank, M. S., & Ok, Y. S. (2022). Effects of microplastics on the terrestrial environment: A critical review. *Environmental Research, 209*, 112734. https://doi.org/10.1016/j.envres.2022.112734

Dris, R., Beaurepaire, M., Bouzid, N., Stratmann, C., Nguyen, M. T., Bordignon, F., Gasperi, J., & Tassin, B. (2024). Sampling and analyzing microplastics in rivers: What methods are being used after a decade of research? In E. Y. Zeng (Ed.), *Microplastic Contamination in Aquatic Environments* (pp. 65–91). Elsevier. https://doi.org/10.1016/B978-0-443-15332-7.00013-2

Dris, R., Gasperi, J., Rocher, V., Saad, M., Renault, N., & Tassin, B. (2015). Microplastic contamination in an urban area: A case study in Greater Paris. *Environmental Chemistry, 12*(5), 592–599. https://doi.org/10.1071/EN14167

Elkhatib, D., Oyanedel-Craver, V., & Carissimi, E. (2021). Electrocoagulation applied for the removal of microplastics from wastewater treatment facilities. *Separation and Purification Technology, 276*, 118877. https://doi.org/10.1016/j.seppur.2021.118877

Fang, C., Awoyemi, O. S., Saianand, G., Xu, L., Niu, J., & Naidu, R. (2024a). Characterising microplastics in indoor air: Insights from Raman imaging analysis of air filter samples. *Journal of Hazardous Materials, 464*, 132969. https://doi.org/10.1016/j.jhazmat.2023.132969

Fang, Z., Sallach, J. B., & Hodson, M. E. (2024b). Size-and concentration-dependent effects of microplastics on soil aggregate formation and properties. *Journal of Hazardous Materials, 465*, 133395. https://doi.org/10.1016/j.jhazmat.2023.133395

Fei, Y., Huang, S., Zhang, H., Tong, Y., Wen, D., Xia, X., Wang, H., Luo, Y., & Barceló, D. (2020). Response of soil enzyme activities and bacterial communities to the accumulation of microplastics in an acid cropped soil. *Science of the Total Environment, 707*, 135634. https://doi.org/10.1016/j.scitotenv.2019.135634

Gabrielli, A. (2023). *Microplastics*. Retrieved October 31, 2023, from https://education.nationalgeographic.org/resource/microplastics

Gatidou, G., Arvaniti, O. S., & Stasinakis, A. S. (2019). Review on the occurrence and fate of microplastics in Sewage Treatment Plants. *Journal of Hazardous Materials, 367*, 504–512. https://doi.org/10.1016/j.jhazmat.2018.12.081

Ge, J., Li, H., Liu, P., Zhang, Z., Ouyang, Z., & Guo, X. (2021). Review of the toxic effect of microplastics on terrestrial and aquatic plants. *Science of the Total Environment, 791*, 148333. https://doi.org/10.1016/j.scitotenv.2021.148333

GEF. (2012). *Impacts of marine debris on biodiversity: Current status and potential solutions*. Secretariat of the Convention on Biological Diversity. Retrieved from https://www.cbd.int/

Gola, D., Tyagi, P. K., Arya, A., Chauhan, N., Agarwal, M., Singh, S. K., & Gola, S. (2021). The impact of microplastics on marine environment: A review. *Environmental Nanotechnology, Monitoring & Management, 16*, 100552. https://doi.org/10.1016/j.enmm.2021.100552

Guo, J. J., Huang, X. P., Xiang, L., Wang, Y. Z., Li, Y. W., Li, H., Cai, Q. Y., Mo, C. H., & Wong, M. H. (2020). Source, migration and toxicology of microplastics in soil. *Environment International, 137*, 105263. https://doi.org/10.1016/j.envint.2019.105263

Guzzetti, E., Sureda, A., Tejada, S., & Faggio, C. (2018). Microplastic in marine organism: Environmental and toxicological effects. *Environmental Toxicology and Pharmacology, 64*, 164–171. https://doi.org/10.1016/j.etap.2018.10.009

Habib, R. Z., Abdoon, M. M. S., Al Meqbaali, R. M., Ghebremedhin, F., Elkashlan, M., Kittaneh, W. F., Cherupurakal, N., Mourad, A. H. I., Thiemann, T., & Al Kindi, R. (2020). Analysis of microbeads in cosmetic products in the United Arab Emirates. *Environmental Pollution, 258*, 113831. https://doi.org/10.1016/j.envpol.2019.113831

Habib, R. Z., Aldhanhani, J. A., Ali, A. H., Ghebremedhin, F., Elkashlan, M., Mesfun, M., Kittaneh, W., Al Kindi, R., & Thiemann, T. (2022). Trends of microplastic abundance in personal care products in the United Arab Emirates over the period of 3 years (2018–2020). *Environmental Science and Pollution Research, 29*(59), 89614–89624. https://doi.org/10.1007/s11356-022-21773-y

Hall, N., Berry, K., Rintoul, L., & Hoogenboom, M. (2015). Microplastic ingestion by scleractinian corals. *Marine Biology*, 162(3), 725–732. https://doi.org/10.1007/s00227-015-2619-7

He, S., Jia, M., Xiang, Y., Song, B., Xiong, W., Cao, J., Peng, H., Yang, Y., Wang, W., Yang, Z., & Zeng, G. (2022). Biofilm on microplastics in aqueous environment: Physicochemical properties and environmental implications. *Journal of Hazardous Materials, 424*, 127286. https://doi.org/10.1016/j.jhazmat.2021.127286

Hegan, D., Tong, L., Zhiquan, H., Qinming, S., & Ru, L. (2015). Determining time limits of continuous film mulching and examining residual effects on cotton yield and soil properties. *Journal of Environmental Biology*, *36*(3), 677.

Huang, L., Zhang, S., Li, L., Zhang, S., Wang, J., Liu, X., & Zhang, W. (2023). Research progress on microplastics pollution in polar oceans. *Polar Science*, *251*, 100946. https://doi.org/10.1016/j.polar.2023.100946

Huang, S., Huang, X., Bi, R., Guo, Q., Yu, X., Zeng, Q., Huang, Z., Liu, T., Wu, H., Chen, Y., Xu, J., Wu, Y., & Guo, P. (2022). Detection and analysis of microplastics in human sputum. *Environmental Science & Technology*, *56*(4), 2476–2486. https://doi.org/10.1021/acs.est.1c03859

Ingraffia, R., Amato, G., Bagarello, V., Carollo, F. G., Giambalvo, D., Iovino, M., Lehmann, A., Rillig, M. C., & Frenda, A. S. (2022). Polyester microplastic fibers affect soil physical properties and erosion as a function of soil type. *Soil*, *8*(1), 421–435. https://doi.org/10.5194/soil-8-421-2022

Jaganmohan, M. (2024). *Global plastic production*. Statista. Retrieved January 30, 2024, from https://www.statista.com/statistics/282732/global-production-of-plastics-since-1950/

Jambeck, J. R., Geyer, R., Wilcox, C., Siegler, T. R., Perryman, M., Andrady, A., Narayan, R., & Law, K. L. (2015). Plastic waste inputs from land into the ocean. *Science*, *347*(6223), 768–771. https://doi.org/10.1126/science.1260352

Jenner, L. C., Rotchell, J. M., Bennett, R. T., Cowen, M., Tentzeris, V., & Sadofsky, L. R. (2022). Detection of microplastics in human lung tissue using μFTIR spectroscopy. *Science of the Total Environment*, *831*, 154907. https://doi.org/10.1016/j.scitotenv.2022.154907

Ji, X., Ma, Y., Zeng, G., Xu, X., Mei, K., Wang, Z., Chen, Z., Dahlgren, R., Zhang, M., & Shang, X. (2021). Transport and fate of microplastics from riverine sediment dredge piles: Implications for disposal. *Journal of Hazardous Materials*, *404*, 124132. https://doi.org/10.1016/j.jhazmat.2020.124132

Jiang, X., Chen, H., Liao, Y., Ye, Z., Li, M., & Klobučar, G. (2019). Ecotoxicity and genotoxicity of polystyrene microplastics on higher plant Vicia faba. *Environmental Pollution*, *250*, 831–838. https://doi.org/10.1016/j.envpol.2019.04.055

Kabir, M. S., Wang, H., Luster-Teasley, S., Zhang, L., & Zhao, R. (2023). Microplastics in landfill leachate: Sources, detection, occurrence, and removal. *Environmental Science and Ecotechnology*, *16*, 100256. https://doi.org/10.1016/j.ese.2023.100256

Kalčíková, G., Gotvajn, A. Ž., Kladnik, A., & Jemec, A. (2017). Impact of polyethylene microbeads on the floating freshwater plant duckweed Lemna minor. *Environmental Pollution, 230*, 1108–1115. https://doi.org/10.1016/j.envpol.2017.07.050

Kanhai, L. D. K., Gardfeldt, K., Krumpen, T., Thompson, R. C., & O'Connor, I. (2020). Microplastics in sea ice and seawater beneath ice floes from the Arctic Ocean. *Scientific Reports, 10*(1), 5004. https://doi.org/10.1038/s41598-020-61948-6

Kim, S. A., Kim, L., Kim, T. H., & An, Y. J. (2022). Assessing the size-dependent effects of microplastics on zebrafish larvae through fish lateral line system and gut damage. *Marine Pollution Bulletin, 185*, 114279. https://doi.org/10.1016/j.marpolbul.2022.114279

Kirstein, I. V., Kirmizi, S., Wichels, A., Garin-Fernandez, A., Erler, R., Löder, M., & Gerdts, G. (2016). Dangerous hitchhikers? Evidence for potentially pathogenic Vibrio spp. on microplastic particles. *Marine Environmental Research, 120*, 1–8. https://doi.org/10.1016/j.marenvres.2016.07.004

Kong, J., Lee, J., & Jeong, S. (2024). Distribution of microplastics in rainfall and their control by a permeable pavement in low-impact development facility. *Journal of Environmental Management, 351*, 119710.

Kumar, A., Yu, Z. G., & Thakur, T. K. (2023). Microplastic pollutants in terrestrial and aquatic environment. *Environmental Science and Pollution Research, 30*(49), 107296–107299.

Krishnan, R. Y., Manikandan, S., Subbaiya, R., Karmegam, N., Kim, W., & Govarthanan, M. (2023). Recent approaches and advanced wastewater treatment technologies for mitigating emerging microplastics contamination–A critical review. *Science of the Total Environment, 858*, 159681. https://doi.org/10.1016/j.scitotenv.2022.159681

Lamichhane, G., Acharya, A., Marahatha, R., Modi, B., Paudel, R., Adhikari, A., Raut, B. K., Aryal, S., & Parajuli, N. (2023). Microplastics in environment: Global concern, challenges, and controlling measures. *International Journal of Environmental Science and Technology, 20*(4), 4673–4694. https://doi.org/10.1007/s13762-022-04261-1

Lares, M., Ncibi, M. C., Sillanpää, M., & Sillanpää, M. (2018). Occurrence, identification and removal of microplastic particles and fibers in conventional activated sludge process and advanced MBR technology. *Water Research, 133*, 236–246. https://doi.org/10.1016/j.watres.2018.01.049

Lebreton, L. C., Van Der Zwet, J., Damsteeg, J. W., Slat, B., Andrady, A., & Reisser, J. (2017). River plastic emissions to the world's oceans. *Nature Communications, 8*(1), 15611. https://doi.org/10.1038/ncomms15611

Lee, J. H., Kim, M. J., Kim, C. S., Cheon, S. J., Choi, K. I., Kim, J., Jung, J., Yoon, J. K., Lee, S. H., & Jeong, D. H. (2023). Detection of microplastic traces in four different types of municipal wastewater treatment plants through FT-IR and TED-GC-MS. *Environmental Pollution, 333*, 122017. https://doi.org/10.1016/j.envpol.2023.122017

Lee, J. M., Busquets, R., Choi, I. C., Lee, S. H., Kim, J. K., & Campos, L. C. (2020). Photocatalytic degradation of polyamide 66; evaluating the feasibility of photocatalysis as a microfibre-targeting technology. *Water, 12*(12), 3551. https://doi.org/10.3390/w12123551

Lei, K., Qiao, F., Liu, Q., Wei, Z., Qi, H., Cui, S., Yue, X., Deng, Y., & An, L. (2017). Microplastics releasing from personal care and cosmetic products in China. *Marine Pollution Bulletin*, *123*(1–2), 122–126. https://doi.org/10.1016/j.marpolbul.2017.09.016

Leslie, H. A., Van Velzen, M. J., Brandsma, S. H., Vethaak, A. D., Garcia-Vallejo, J. J., & Lamoree, M. H. (2022). Discovery and quantification of plastic particle pollution in human blood. *Environment International*, *163*, 107199. https://doi.org/10.1016/j.envint.2022.107199

Li, F., Boucher, J., Gaboury, A., Voirin, N., Gallato, M., & Puppi, R. (2023). *Plastic Overshoot Day July 28, 2023*. EA – Environmental Action, Lausanne, Switzerland. Retrieved from www.plasticovershoot.earth

Li, J., Liu, H., & Chen, J. P. (2018). Microplastics in freshwater systems: A review on occurrence, environmental effects, and methods for microplastics detection. *Water Research*, *137*, 362–374. https://doi.org/10.1016/j.watres.2017.12.056

Li, S., Wang, T., Guo, J., Dong, Y., Wang, Z., Gong, L., & Li, X. (2021). Polystyrene microplastics disturb the redox homeostasis, carbohydrate metabolism and phytohormone regulatory network in barley. *Journal of Hazardous Materials*, *415*, 125614. https://doi.org/10.1016/j.jhazmat.2021.125614

Li, W., Zhao, J., Zhang, Z., Ren, Z., Li, X., Zhang, R., & Ma, X. (2024). Uptake and effect of carboxyl-modified polystyrene microplastics on cotton plants. *Journal of Hazardous Materials*, *466*, 133581. https://doi.org/10.1016/j.jhazmat.2024.133581

Li, Z. G., & Richter, J. S. (2015). Problem and countermeasure on promoting the plastic bag ban of USA. *Applied Mechanics and Materials*, *768*, 787–796. https://doi.org/10.4028/www.scientific.net/AMM.768.787

Lian, Y., Shi, R., Liu, J., Zeb, A., Wang, Q., Wang, J., Yu, M., Li, J., Zheng, Z., Ali, N., & Liu, W. (2024). Effects of polystyrene, polyethylene, and polypropylene microplastics on the soil-rhizosphere-plant system: Phytotoxicity, enzyme activity, and microbial community. *Journal of Hazardous Materials*, *465*, 133417. https://doi.org/10.1016/j.jhazmat.2023.133417

Liu, B., Li, R., Zhuang, H., Lin, Z., & Li, Z. (2024). Effects of polystyrene microplastics on the phenylpropane metabolic pathway in cucumber plants. *Environmental and Experimental Botany*, *220*, 105671. https://doi.org/10.1016/j.envexpbot.2024.105671

Liu, F., Nord, N. B., Bester, K., & Vollertsen, J. (2020). Microplastics removal from treated wastewater by a biofilter. *Water*, *12*(4), 1085. https://doi.org/10.3390/w12041085

Liu, J., Bu, L., Zhu, L., Luo, S., Chen, X., & Li, S. (2014). Optimizing plant density and plastic film mulch to increase maize productivity and water-use efficiency in semiarid areas. *Agronomy Journal*, *106*(4), 1138–1146. https://doi.org/10.2134/agronj13.0582

López, A. P. A., Trilleras, J., Arana, V. A., Garcia-Alzate, L. S., & Grande-Tovar, C. D. (2023). Atmospheric microplastics: Exposure, toxicity, and detrimental health effects. *RSC Advances*, *13*(11), 7468–7489. https://doi.org/10.1039%2Fd2ra07098g

Ma, J., Xu, M., Wu, J., Yang, G., Zhang, X., Song, C., Long, L., Chen, C., Xu, C., & Wang, Y. (2023). Effects of variable-sized polyethylene microplastics on soil chemical properties and functions and microbial communities in purple soil. *Science of the Total Environment, 868*, 161642. https://doi.org/10.1016/j.scitotenv.2023.161642

Madhumitha, C. T., Karmegam, N., Biruntha, M., Arun, A., Al Kheraif, A. A., Kim, W., & Kumar, P. (2022). Extraction, identification, and environmental risk assessment of microplastics in commercial toothpaste. *Chemosphere, 296*, 133976. https://doi.org/10.1016/j.chemosphere.2022.133976

Marn, N., Jusup, M., Kooijman, S. A., & Klanjscek, T. (2020). Quantifying impacts of plastic debris on marine wildlife identifies ecological breakpoints. *Ecology Letters, 23*(10), 1479–1487. https://doi.org/10.1111/ele.13574

Miller, M. E., Hamann, M., & Kroon, F. J. (2020). Bioaccumulation and biomagnification of microplastics in marine organisms: A review and meta-analysis of current data. *PLoS One, 15*(10), e0240792. https://doi.org/10.1371/journal.pone.0240792

Monclús, L., McCann Smith, E., Ciesielski, T. M., Wagner, M., & Jaspers, V. L. (2022). Microplastic Ingestion Induces Size-Specific Effects in Japanese Quail. *Environmental Science & Technology, 56*(22), 15902–15911. https://doi.org/10.1021/acs.est.2c03878

Montano, L., Giorgini, E., Notarstefano, V., Notari, T., Ricciardi, M., Piscopo, M., & Motta, O. (2023). Raman Microspectroscopy evidence of microplastics in human semen. *Science of the Total Environment, 901*, 165922. https://doi.org/10.1016/j.scitotenv.2023.165922

Napper, I. E., & Thompson, R. C. (2020). Plastic debris in the marine environment: History and future challenges. *Global Challenges, 4*(6), 1900081. https://doi.org/10.1002/gch2.201900081

Ng, E. L., Lwanga, E. H., Eldridge, S. M., Johnston, P., Hu, H. W., Geissen, V., & Chen, D. (2018). An overview of microplastic and nano plastic pollution in agroecosystems. *Science of the Total Environment, 627*, 1377–1388. https://doi.org/10.1016/j.scitotenv.2018.01.341

Nizzetto, L., Futter, M., & Langaas, S. (2016). Are agricultural soils dumps for microplastics of urban origin? *Environmental Science & Technology, 50*, 10777. https://doi.org/10.1021/acs.est.6b04140

Oberbeckmann, S., Löder, M. G., & Labrenz, M. (2015). Marine microplastic-associated biofilms–A review. *Environmental Chemistry, 12*(5), 551–562. https://doi.org/10.1071/EN15069

Osman, A. I., Hosny, M., Eltaweil, A. S., Omar, S., Elgarahy, A. M., Farghali, M., Yap, P. S., Wu, Y. S., Nagandran, S., Batumalaie, K., Gopinath, S. C., & Akinyede, K. A. (2023). Microplastic sources, formation, toxicity and remediation: A review. *Environmental Chemistry Letters, 21*(4), 2129–2169. https://doi.org/10.1007/s10311-023-01593-3

Palansooriya, K. N., Sang, M. K., El-Naggar, A., Shi, L., Chang, S. X., Sung, J., Zhang, W., & Ok, Y. S. (2022). Low-density polyethylene microplastics alter chemical properties and microbial communities in agricultural soil. *Scientific Reports, 13*(1), 16276. https://doi.org/10.1038/s41598-023-42285-w

Pattanayak, S. (2022). Plastics and their additives reached the blood and tissue spaces: What are the possible consequences? *Exploratory Animal and Medical Research, 12*, 128. https://doi.org/10.52635/eamr/12.2.128-133

Pironti, C., Notarstefano, V., Ricciardi, M., Motta, O., Giorgini, E., & Montano, L. (2022). First evidence of microplastics in human urine, a preliminary study of intake in the human body. *Toxics, 11*(1), 40. https://doi.org/10.3390/toxics11010040

Pironti, C., Ricciardi, M., Motta, O., Miele, Y., Proto, A., & Montano, L. (2021). Microplastics in the environment: Intake through the food web, human exposure and toxicological effects. *Toxics, 9*(9), 224. https://doi.org/10.3390/toxics9090224

Prata, J. C., da Costa, J. P., Lopes, I., Duarte, A. C., & Rocha-Santos, T. (2020). Environmental exposure to microplastics: An overview on possible human health effects. *Science of the Total Environment, 702*, 134455. https://doi.org/10.1016/j.scitotenv.2019.134455

Prokić, M. D., Radovanović, T. B., Gavrić, J. P., & Faggio, C. (2019). Ecotoxicological effects of microplastics: Examination of biomarkers, current state and future perspectives. *TrAC Trends in Analytical Chemistry, 111*, 37–46. https://doi.org/10.1016/j.trac.2018.12.001

Proshad, R., Kormoker, T., Islam, M. S., Haque, M. A., Rahman, M. M., & Mithu, M. M. R. (2018). Toxic effects of plastic on human health and environment: A consequence of health risk assessment in Bangladesh. *International Journal of Health, 6*(1), 1–5. https://www.sciencepubco.com/index.php/IJH

Purcell, J. (2023). *Plastic pollution: Pathways to net zero*. Credit Suisse Center for Sustainability (CfS). Retrieved from https://www.credit-suisse.com/media/assets/corporate/docs/news-and-expertise/articles/2023/06/plastic-pollution-pathways-to-net-zero-report-en.pdf

Qi, Y., Ossowicki, A., Yang, X., Lwanga, E. H., Dini-Andreote, F., Geissen, V., & Garbeva, P. (2020). Effects of plastic mulch film residues on wheat rhizosphere and soil properties. *Journal of Hazardous Materials, 387*, 121711. https://doi.org/10.1016/j.jhazmat.2019.121711

Qin, W., Hu, C., & Oenema, O. (2015). Soil mulching significantly enhances yields and water and nitrogen use efficiencies of maize and wheat: A meta-analysis. *Scientific Reports, 5*(1), 16210. https://doi.org/10.1038/srep16210

Ragusa, A., Notarstefano, V., Svelato, A., Belloni, A., Gioacchini, G., Blondeel, C., Zucchelli, E., De Luca, C., D'Avino, S., Gulotta, A., Carnevali, O., & Giorgini, E. (2022). Raman microspectroscopy detection and characterisation of microplastics in human breastmilk. *Polymers, 14*(13), 2700. https://doi.org/10.3390/polym14132700

Ragusa, A., Svelato, A., Santacroce, C., Catalano, P., Notarstefano, V., Carnevali, O., Papa, F., Rongioletti, M. C. A., Baiocco, F., Draghi, S., D'Amore, E., Rinaldo, D., Matta, M., Giorgini, E., & Giorgini, E. (2021). Plasticenta: First evidence of microplastics in human placenta. *Environment International, 146*, 106274. https://doi.org/10.1016/j.envint.2020.106274

Rajala, K., Grönfors, O., Hesampour, M., & Mikola, A. (2020). Removal of microplastics from secondary wastewater treatment plant effluent by coagulation/flocculation with iron, aluminum and polyamine-based chemicals. *Water Research, 183*, 116045. https://doi.org/10.1016/j.watres.2020.116045

Rodriguez-Seijo, A., Lourenço, J., Rocha-Santos, T. A. P., Da Costa, J., Duarte, A. C., Vala, H., & Pereira, R. (2017). Histopathological and molecular effects of microplastics in Eisenia andrei Bouché. *Environmental Pollution*, *220*, 495–503. https://doi.org/10.1016/j.envpol.2016.09.092

Rolsky, C., Kelkar, V., Driver, E., & Halden, R. U. (2020). Municipal sewage sludge as a source of microplastics in the environment. *Current Opinion in Environmental Science & Health*, *14*, 16–22. https://doi.org/10.1016/j.coesh.2019.12.001

Rong, L., Zhao, L., Zhao, L., Cheng, Z., Yao, Y., Yuan, C., Wang, L., & Sun, H. (2021). LDPE microplastics affect soil microbial communities and nitrogen cycling. *Science of the Total Environment*, *773*, 145640. https://doi.org/10.1016/j.scitotenv.2021.145640

Rostami, S., Talaie, M. R., Talaiekhozani, A., & Sillanpää, M. (2021). Evaluation of the available strategies to control the emission of microplastics into the aquatic environment. *Environmental Science and Pollution Research*, *28*, 18908–18917. https://doi.org/10.1007/s11356-021-12888-9

Sajjad, M., Huang, Q., Khan, S., Khan, M. A., Liu, Y., Wang, J., Lian, F., Wang, Q., & Guo, G. (2022). Microplastics in the soil environment: A critical review. *Environmental Technology & Innovation*, *27*, 102408. https://doi.org/10.1016/j.eti.2022.102408

Schuyler, Q., Hardesty, B. D., Wilcox, C., & Townsend, K. (2014). Global analysis of anthropogenic debris ingestion by sea turtles. *Conservation Biology*, *28*(1), 129–139. https://doi.org/10.1111/cobi.12126

Schwabl, P., Köppel, S., Königshofer, P., Bucsics, T., Trauner, M., Reiberger, T., & Liebmann, B. (2019). Detection of various microplastics in human stool: A prospective case series. *Annals of Internal Medicine*, *171*(7), 453–457. https://doi.org/10.7326/M19-0618

Seta, A. S., Müller, L., Tavella, R., da Silva Júnior, F. M. R., Pedrosa, V., Romano, L. A., Wasielesky, W., Josende, M. E., & Ventura-Lima, J. (2023). Oxidative effects of consuming microplastics in different tissues of white shrimp Litopenaeus vannamei. *Marine Pollution Bulletin*, *193*, 115137. https://doi.org/10.1016/j.marpolbul.2023.115137

Shalumon, C. S., & Ratanatamskul, C. (2023). A novel simplified method for extraction of microplastic particles from face scrub and laundry wastewater. *Scientific Reports*, *13*(1), 14168. https://doi.org/10.1038/s41598-023-41457-y

Sharma, M. (2024). Microplastics in the atmosphere. In S. Mondal, P. Das, A. Mondal, S. Paul, J. K. Pandey, T. K. Das (Eds.), *Remediation of Plastic and Microplastic Waste* (pp. 60–70). CRC Press.

Shiu, R. F., Vazquez, C. I., Chiang, C. Y., Chiu, M. H., Chen, C. S., Ni, C. W., Gong, G. C., Quigg, A., Santschi, P. H., & Chin, W. C. (2020). Nano-and microplastics trigger secretion of protein-rich extracellular polymeric substances from phytoplankton. *Science of the Total Environment*, *748*, 141469. https://doi.org/10.1016/j.scitotenv.2020.141469

Siwek, P., Kalisz, A., & Domagala-Swiatkiewicz, I. (2015). The influence of degradable polymer mulches on soil properties and cucumber yield. *Agrochimica*, *59*, 108–123.

Song, U., Kim, J., & Rim, H. (2023). Assessing phytotoxicity of microplastics on aquatic plants using fluorescent microplastics. *Environmental Science and Pollution Research*, *30*(29), 74186–74195. https://doi.org/10.1007/s11356-023-27621-x

Song, Y., Chou, E. L., Baecker, A., You, N. C., Song, Y., & Sun, Q. (2016). Endocrine disrupting chemicals, risk of type 2 diabetes, and diabetes-related metabolic traits: A systematic review and meta-analysis. *Journal of Diabetes*, *8*(4), 516–532. https://doi.org/10.1111/1753-0407.12325

Stapleton, M. J., Ansari, A. J., & Hai, F. I. (2023). Antibiotic sorption onto microplastics in water: A critical review of the factors, mechanisms and implications. *Water Research*, *233*, 119790. https://doi.org/10.1016/j.watres.2023.119790

Sutkar, P. R., Gadewar, R. D., & Dhulap, V. P. (2023). Recent trends in degradation of microplastics in the environment: A state-of-the-art review. *Journal of Hazardous Materials Advances*, *11*, 100343. https://doi.org/10.1016/j.hazadv.2023.100343

Tadsuwan, K., & Babel, S. (2021). Microplastic contamination in a conventional wastewater treatment plant in Thailand. *Waste Management & Research*, *39*(5), 754–761. https://doi.org/10.1177/0734242X20982055

Thacharodi, A., Hassan, S., Singh, T., Mandal, R., Khan, H. A., Hussain, M. A., & Pugazhendhi, A. (2023). Bioremediation of polycyclic aromatic hydrocarbons: An updated microbiological review. *Chemosphere*, *328*, 138498. https://doi.org/10.1016/j.chemosphere.2023.138498

Thacharodi, A., Meenatchi, R., Hassan, S., Hussain, N., Bhat, M. A., Arockiaraj, J., Ngo, H. H., Le, Q. H., & Pugazhendhi, A. (2024). Microplastics in the environment: A critical overview on its fate, toxicity, implications, management, and bioremediation strategies. *Journal of Environmental Management*, *349*, 119433. https://doi.org/10.1016/j.jenvman.2023.119433

Thiele, C. J., Grange, L. J., Haggett, E., Hudson, M. D., Hudson, P., Russell, A. E., & Zapata-Restrepo, L. M. (2023). Microplastics in European sea salts–An example of exposure through consumer choice and of interstudy methodological discrepancies. *Ecotoxicology and Environmental Safety*, *255*, 114782. https://doi.org/10.1016/j.ecoenv.2023.114782

Thiele, C. J., Hudson, M. D., Russell, A. E., Saluveer, M., & Sidaoui-Haddad, G. (2021). Microplastics in fish and fishmeal: An emerging environmental challenge? *Scientific Reports*, *11*(1), 2045. https://doi.org/10.1038/s41598-021-81499-8

Thompson, R. C., Olsen, Y., Mitchell, R. P., Davis, A., Rowland, S. J., John, A. W., McGonigle, D., & Russell, A. E. (2004). Lost at sea: Where is all the plastic? *Science*, *304*(5672), 838–838. https://doi.org/10.1126/science.1094559

Tunali, M., Adam, V., & Nowack, B. (2023). Probabilistic environmental risk assessment of microplastics in soils. *Geoderma*, *430*, 116315. https://doi.org/10.1016/j.geoderma.2022.116315

Ullah, R., Tsui, M. T. K., Chen, H., Chow, A., Williams, C., & Ligaba-Osena, A. (2021). Microplastics interaction with terrestrial plants and their impacts on agriculture. *Journal of Environmental Quality*, *50*(5), 1024–1041. https://doi.org/10.1002/jeq2.20264

Urbina, M. A., Correa, F., Aburto, F., & Ferrio, J. P. (2020). Adsorption of polyethylene microbeads and physiological effects on hydroponic maize. *Science of the Total Environment, 741*, 140216. https://doi.org/10.1016/j.scitotenv.2020.140216

Verla, A. W., Enyoh, C. E., Obinna, I. B., Verla, E. N., Qingyue, W., Chowdhury, M. A. H., Enyoh, E. C., & Chowdhury, T. (2020). Effect of macro-and micro-plastics in soil on growth of Juvenile Lime Tree (Citrus aurantium). *AIMS Environmental Science, 7*(6), 526–541. https://doi.org/10.3934/environsci.2020033

Waldman, W. R., & Rillig, M. C. (2020). Microplastic research should embrace the complexity of secondary particles. *Environmental Science & Technology, 54*(13), 7751–7753. https://doi.org/10.1021/acs.est.0c02194

Wang, T., Li, B., Shi, H., Ding, Y., Chen, H., Yuan, F., Liu, R., & Zou, X. (2024). The processes and transport fluxes of land-based macroplastics and microplastics entering the ocean via rivers. *Journal of Hazardous Materials, 466*, 133623. https://doi.org/10.1016/j.jhazmat.2024.133623

Wang, Y., Bai, J., Wen, L., Wang, W., Zhang, L., Liu, Z., & Liu, H. (2023). Phytotoxicity of microplastics to the floating plant Spirodela polyrhiza (L.): Plant functional traits and metabolomics. *Environmental Pollution, 322*, 121199. https://doi.org/10.1016/j.envpol.2023.121199

Westphalen, H., & Abdelrasoul, A. (2018). Challenges and treatment of microplastics in water. *Water Challenges of an Urbanizing World, 5*, 71–82.

Wicaksono, K. B., Patria, M. P., & Suryanda, A. (2021, March). Microplastic ingestion in the black sea cucumber *Holothuria leucospilota* (Brandt, 1835) was collected from Rambut Island, Seribu Islands, Jakarta, Indonesia. In *IOP Conference Series: Materials Science and Engineering* (Vol. 1098, No. 5, p. 052049). IOP Publishing.

Worek, J., Badura, X., Białas, A., Chwiej, J., Kawoń, K., & Styszko, K. (2022). Pollution from transport: Detection of tyre particles in environmental samples. *Energies, 15*(8), 2816. https://doi.org/10.3390/en15082816

Wu, P., Lin, S., Cao, G., Wu, J., Jin, H., Wang, C., Wong, M. H., Yang, Z., & Cai, Z. (2022). Absorption, distribution, metabolism, excretion and toxicity of microplastics in the human body and health implications. *Journal of Hazardous Materials, 437*, 129361. https://doi.org/10.1016/j.jhazmat.2022.129361

Wu, X., Liu, Y., Yin, S., Xiao, K., Xiong, Q., Bian, S., S., Liang, S., Hou, H., Hu, J., & Yang, J. (2020). Metabolomics revealing the response of rice (*Oryza sativa* L.) exposed to polystyrene microplastics. *Environmental Pollution, 266*, 115159. https://doi.org/10.1016/j.envpol.2020.115159

Xiang, Y., Jiang, L., Zhou, Y., Luo, Z., Zhi, D., Yang, J., & Lam, S. S. (2022). Microplastics and environmental pollutants: Key interaction and toxicology in aquatic and soil environments. *Journal of Hazardous Materials, 422*, 126843. https://doi.org/10.1016/j.jhazmat.2021.126843

Yang, C., Niu, S., Xia, Y., & Wu, J. (2023). Microplastics in urban road dust: Sampling, analysis, characterization, pollution level, and influencing factors. *TrAC Trends in Analytical Chemistry, 168*, 117348. https://doi.org/10.1016/j.trac.2023.117348

Yang, H., He, Y., Yan, Y., Junaid, M., & Wang, J. (2021a). Characteristics, toxic effects, and analytical methods of microplastics in the atmosphere. *Nanomaterials, 11*, 2747.

Yang, Z., Lü, F., Zhang, H., Wang, W., Shao, L., Ye, J., & He, P. (2021b). Is incineration the terminator of plastics and microplastics? *Journal of Hazardous Materials, 401*, 123429. https://doi.org/10.1016/j.jhazmat.2020.123429

Yao, S., Ni, N., Li, X., Wang, N., Bian, Y., Jiang, X., Song, Y., Bolan, N. S., Zhang, Q., & Tsang, D. C. (2023). Interactions between white and black carbon in water: A case study of concurrent aging of microplastics and biochar. *Water Research, 238*, 120006. https://doi.org/10.1016/j.watres.2023.120006

Yu, H., Zhang, X., Hu, J., Peng, J., & Qu, J. (2020). Ecotoxicity of polystyrene microplastics to submerged carnivorous Utricularia vulgaris plants in freshwater ecosystems. *Environmental Pollution, 265*, 114830. https://doi.org/10.1016/j.envpol.2020.114830

Yu, Y., Battu, A. K., Varga, T., Denny, A. C., Zahid, T. M., Chowdhury, I., & Flury, M. (2023). Minimal impacts of microplastics on soil physical properties under environmentally relevant concentrations. *Environmental Science & Technology, 57*(13), 5296–5304. https://doi.org/10.1021/acs.est.2c09822

Yuan, J., Ma, J., Sun, Y., Zhou, T., Zhao, Y., & Yu, F. (2020). Microbial degradation and other environmental aspects of microplastics/plastics. *Science of the Total Environment, 715*, 136968. https://doi.org/10.1016/j.scitotenv.2020.136968

Zhang, C., Zuo, Z., Wang, Q., Wang, S., Lv, L., & Zou, J. (2022). Size effects of microplastics on embryos and observation of toxicity kinetics in Larvae of Grass Carp (Ctenopharyngodon idella). *Toxics, 10*(2), 76. https://doi.org/10.3390/toxics10020076

Zhang, G. S., Zhang, F. X., & Li, X. T. (2019). Effects of polyester microfibers on soil physical properties: Perception from a field and a pot experiment. *Science of the Total Environment, 670*, 1–7. https://doi.org/10.1016/j.scitotenv.2019.03.149

Zhang, K., Hamidian, A. H., Tubić, A., Zhang, Y., Fang, J. K., Wu, C., & Lam, P. K. (2021). Understanding plastic degradation and microplastic formation in the environment: A review. *Environmental Pollution, 274*, 116554. https://doi.org/10.1016/j.envpol.2021.116554

Zhang, S., Shi, C., Nie, Y., Xing, B., Wen, X., & Cheng, S. (2023). Separation experiment and mechanism study on PVC microplastics removal from aqueous solutions using high-gradient magnetic filter. *Journal of Water Process Engineering, 51*, 103495. https://doi.org/10.1016/j.jwpe.2023.103495

Zhang, S., Yang, X., Gertsen, H., Peters, P., Salánki, T., & Geissen, V. (2018). A simple method for the extraction and identification of light density microplastics from soil. *Science of the Total Environment, 616*, 1056–1065. https://doi.org/10.1016/j.scitotenv.2017.10.213

Zhao, T., Lozano, Y. M., & Rillig, M. C. (2021). Microplastics increase soil pH and decrease microbial activities as a function of microplastic shape, polymer type, and exposure time. *Frontiers in Environmental Science, 9*, 675803. https://doi.org/10.3389/fenvs.2021.675803

Zhou, J., Gui, H., Banfield, C. C., Wen, Y., Zang, H., Dippold, M. A., Charlton, A., & Jones, D. L. (2021). The microplastisphere: Biodegradable microplastics addition alters soil microbial community structure and function. *Soil Biology and Biochemistry, 156*, 108211. https://doi.org/10.1016/j.soilbio.2021.108211

Zhou, X., Luo, Z., Wang, H., Luo, Y., Yu, R., Zhou, S., Wang, Z., Hu, G., & Xing, B. (2023). Machine learning application in forecasting tire wear particles emission in China under different potential socioeconomic and climate scenarios with tire microplastics context. *Journal of Hazardous Materials, 441*, 129878. https://doi.org/10.1016/j.jhazmat.2022.129878

Zhou, X., Zhao, Y., Pang, G., Jia, X., Song, Y., Guo, A., Wang, A., Zhang, S., & Ji, M. (2022). Microplastic abundance, characteristics and removal in large-scale multi-stage constructed wetlands for effluent polishing in northern China. *Chemical Engineering Journal, 430*, 132752. https://doi.org/10.1016/j.cej.2021.132752

Zimmermann, L., Göttlich, S., Oehlmann, J., Wagner, M., & Völker, C. (2020). What are the drivers of microplastic toxicity? Comparing the toxicity of plastic chemicals and particles to Daphnia magna. *Environmental Pollution, 267*, 115392. https://doi.org/10.1016/j.envpol.2020.115392

CHAPTER 4

Micro/Nanoplastics in Agricultural Soils and Associated Hazard

Md. Mostafizur Rahman, Tahsin Tabassum, Mohammad Toha, and Sadia Sikder

4.1 INTRODUCTION

Over the past decades, plastics have been extensively utilized in various applications due to their affordability, resilience, and high plasticity (Hossain et al., 2023). Following their production and utilization, most of these plastics are discharged into environmental compartments, where they are subject to the influence of diverse environmental factors and undergo accelerated degradation (Tang, 2023; Iqbal et al., 2023). Due to this expeditious degradation process, macroplastics are transforming microplastics (MPs) and nanoplastics (NPs). NPs are particles with a typical size below 100 nm (Shafea et al., 2023), whereas MPs are particles smaller than 5 mm in at least one dimension. The modern global community has witnessed an abundance of MP varieties, leading to the MP pollution problem (Bethanis & Golia, 2024). The ubiquity of different types of persistent pollutants in environmental compartments and their potential threats and ecotoxicological hazards have made this emerging pollutant a trendy topic of discussion worldwide (Chukwuemeka et al., 2024). Like other environmental compartments, MPs have been identified in

DOI: 10.1201/9781003487579-4

agricultural soils at impermissible limits (Ma et al., 2024). MPs in agricultural soils are typically derived from primary and secondary sources (Hossain et al., 2023). Some of the primary contributors to plastic pollution in agricultural soil are plastic mulching films, irrigation pipes and equipment, greenhouses and tunnels, fertilizers, the textile industry, and domestic waste disposal (Bethanis & Golia, 2024). Notably, North American and European wastewater treatment facilities annually release an estimated 44,000–300,000 tons and 63,000–430,000 tons of MPs into the soil, respectively (Alimi et al., 2018). Furthermore, studies conducted in Bangladesh in 2023 identified 2.92×10^4 items/kg of MPs in textile effluent samples (Hossain et al., 2023). On the contrary, previous research has noted that the mean concentration of MPs in Chinese agricultural soil was 1,076 items/kg in 2020, compared to a range of 308 ± 138.10 MPs/kg in 2009 (Büks & Kaupenjohann, 2020; Arthur et al., 2009). Recently, it has been determined that the concentration of MPs in China's farmland varied between 1,430 and 3,410 items/kg. This state unequivocally demonstrates the upward trajectory of MP concentration in China. Hossain et al. (2023) identified nine distinct varieties of plastic polymers in agricultural soil close to an industrial zone in Bangladesh, with a total detection limit of 2.13×10^4 items/kg. Moreover, MP concentrations in the agricultural soils of Japan and Korea were 664 items/kg and 144 mg/kg, respectively (Katsumi et al., 2020; Sa'adu & Farsang, 2023). On the other hand, the concentrations of MPs in the European nations of Germany, Greece, Spain, and the Netherlands were recorded as follows: 14.60 items/g, 301 ± 140 items/kg, $2,242 \pm 984$ items/kg, and 888 ± 500 items/kg (Sa'adu & Farsang, 2023). Significantly, loamy soils exhibited greater MPs than clay soils across European nations (Büks & Kaupenjohann, 2020). Nevertheless, the number of investigations conducted in Africa is quite limited. In agricultural soils, Tunisia detected MPs ranging from 0.21 to 1.50 items/g, with most plastic polymers identified as polyethylene terephthalate (PET), LPE, high-density polyethylene (HDPE), and polystyrene (PS) (Boughattas et al., 2021). A considerable quantity of MPs was also detected in the farmlands of Mexico and Chile over time (Boughattas et al., 2021). The quantities measured were 306 ± 360 and 400–2,000 items/kg, respectively. The majority of plastic polymers found in agricultural soils are PET, polypropylene (PP), polyvinyl chloride (PVC), PS, low-density polyethylene (LDPE), and HDPE, according to recent research (Hossain et al., 2023; Sa'adu & Farsang, 2023). Furthermore, it is worth noting that the distribution of MP concentrations in agricultural land is highly variable and is substantially influenced by atmospheric conditions and geographic locations (Ma et al., 2024; Bethanis & Golia, 2024). Hydrophobic pollutants, heavy metals (HMs), and complex contaminants can be effectively absorbed by MPs owing to their substantial surface area, volume ratio, and hydrophobic characteristics. The ingestion of MPs by soil organisms may present a developmental and survival hazard (Tang, 2023). Moreover, MPs may enter the human body via the food chain, presenting possible health hazards. MP exposure has been associated with adverse effects on gastrointestinal health, compromised immune function, and an increased susceptibility to various inflammatory diseases, Type 2 diabetes, and cardiovascular disorders (Pattanayak, 2022). Research findings suggest that microorganisms residing in the rhizosphere of plants may benefit from hydroxy-valerate copolymer (PHBV), a biodegradable MP, as a carbon source, notwithstanding the hazards associated with agricultural utilization (Wang et al., 2021a). Additionally, the structure of bacterial colonies is modified at various classification levels by PHBV (Zhou et al., 2021). As a result, by stimulating functional microbial enzymes in crop rhizospheres, biodegradable plastics may influence the biogeochemical cycle of carbon and other nutrients and improve soil ecological function (Sana et al., 2020). As a result, while the influence of MPs on agroecosystems is not wholly detrimental, further extensive investigation is required to evaluate their intricate consequences in conjunction with soil microorganisms and other pollutants. The current worldwide goal for managing plastic waste is to achieve a 35% recycling rate by the year 2050 (ESCAP, 2019). The "Plastics Strategy" is an initiative the European Union implemented to establish a circular economy and reduce the use of single-use plastic bags (Patrício Silva et al., 2021). Furthermore, the "Biotechnology Innovation in Environmental Restoration" initiative is developing biotechnologies capable of decomposing persistent and biodegradable plastics. Implementing China's Soil Pollution Prevention and Control Law in 2018 prioritizes the improvement of agricultural film waste recycling (Union, 2018). Political measures are also supported to reduce plastic waste on farmland progressively, as are initiatives to address residual mulching film and plastic packaging for pesticides and fertilizers. Furthermore, the scientific community has already developed numerous treatment systems for mitigating MP contamination. Developing countries have yet to adopt comprehensive strategies to address the issue of MP contamination in agricultural soil, despite the implementation of diverse approaches in countries such as China, Europe, Canada, the United States, and Japan. Therefore, developing nations are increasingly confronted with this issue of pollution, which is exacerbated by its persistent characteristics, carcinogenic attributes, toxicity, and bioaccumulation, in addition to the absence of adequate treatment systems and strategies. Furthermore, numerous studies have identified MPs in the food webs of developing countries, demonstrating the severe contamination of food chain systems. Under these conditions, knowledge of MPs' origins, destiny, conveyance, dispersion, and impacts in agricultural soils is critical. Therefore, the objective of this chapter is to furnish essential details about the origins, destiny, and conveyance of MPs within agricultural soils. Furthermore, this chapter will elaborate on the influencing factors, impacts on ecosystems and human health, and worldwide dispersion of MPs in agricultural soils.

4.2 SOURCES AND TYPES OF MICRO/NANOPLASTICS (MNPS) IN AGRICULTURAL ENVIRONMENTS

MPs are increasingly being found in agricultural environments, raising concerns about their potential impacts on soil health, crop yields, and food safety. As the presence of MPs in agricultural environments is a growing concern, more research is needed to understand their source.

4.2.1 Sources of Micro/Nanoplastics (MNPs) in Agricultural Soils

4.2.1.1 Agricultural Plastics

Plastic Mulching Films: These are widely used to suppress weeds, retain moisture, and regulate soil temperature. However, they can fragment over time due to weathering and agricultural practices, releasing MPs into the soil (He et al., 2021).

Irrigation Pipes and Equipment: Due to wear and tear, MPs can leach from irrigation pipes and equipment (Kokalj et al., 2024).

Greenhouses and Tunnels: Plastic sheeting used in greenhouses and tunnels can also degrade into MPs (Ma et al., 2024).

Agrotextiles: These are fabrics made from synthetic fibers used for various agricultural purposes, such as protecting crops from frost and insects. They can also shed MP fibers (He et al., 2021).

4.2.1.2 Sludge Recycle

Wastewater treatment can concentrate MPs in sludge, with estimates ranging from 4,000 to 15,000 items/kg (Willén et al., 2017). Applying this sludge as fertilizer can transfer these MPs to the soil. Consequently, it has been found that millions of tons of MPs are entering European and North American soils annually (Büks & Kaupenjohann, 2020).

However, although incineration can reduce them, it has some drawbacks. In this situation, hydrothermal reaction at 200°C–300°C can be used as an alternative way, but managing the produced gas is crucial to avoid further pollution (Sharma & Sharma, 2024). This indicates the need for practical solutions to manage MPs throughout the wastewater treatment process and prevent their spread into the environment.

4.2.1.3 Surface Runoff and Sewage Irrigation

Rivers and rainwater runoff contain alarming levels of MPs, ranging from 1.6 to 12 items/L in the Yangtze River and up to 300 million/m^2 in Tijuana (de Jesus Piñon-Colin et al., 2020). Despite wastewater treatment, MPs persist in effluents at (0.40–1) items/L, potentially entering the soil through increased use of untreated or partially treated wastewater for irrigation (de Jesus Piñon-Colin et al., 2020). While research explores the distribution and dangers of MPs, their removal remains a challenge. Improved coagulation technology using magnesium hydroxide shows promise in reducing water contamination and, consequently, soil MP pollution.

4.2.1.4 Application of Organic Fertilizer

While organic fertilizer boasts eco-friendly benefits, it surprisingly contributes to soil MP pollution. MPs from animal feed packaging, processing tools, and manure end up in organic fertilizers, potentially reaching plants and animals. Studies confirm the presence of plastic fragments in different sizes within various organic fertilizers globally (Zhao et al., 2023). Notably, the main consumer of organic fertilizer is China, and the application of organic fertilizer MPs in soil can reach 52.4 t/year (Zhang et al., 2022).

4.2.1.5 Atmospheric Deposition

MP polymers have been found both in indoor medium (1–60) particles/m^3 and outside environments (0.300–1.50) particles/m^3, but it is more concentrated in indoor environments (Ageel et al., 2022; Ambrosini et al., 2019). It travels long distances through the air, even reaching remote areas like the Arctic. According to the research, MPs enter the soil from atmospheric compartments through rainfall, snow, and stormwater runoff. However, more research is needed to identify the other unknown sources.

4.2.1.6 Landfill Plastic Waste Seepage

Untreated landfill leachate is a possible source of MPs in soils. As plastic production and use are being increased, MPs would be released into the environment. However, it has been found that MPs in landfill leachate can range from 0.420 to 24.60 items/L (Rahman et al., 2023). As conventional treatment methods cannot effectively remove MPs from leachate, managing landfills should be the first and foremost choice for developing nations.

4.2.1.7 Others

Various sources contribute to the presence of MPs in soil. Improper disposal, illegal waste dumping, and ocean tides contribute to the accumulation of plastic products in soils (Rahman et al., 2023). The wear and tear of plastic items and vehicle tires is a significant source of MPs which is exacerbated by illegal garbage dumping in the environment. Moreover, tidal activity disperses MPs widely in the ocean, leading to their deposition in coastal soils. In some regions, substantial amounts of tire debris are emitted annually with 0.810 kg/capita (Jia, 2022). Additionally, in Europe, snow containing over 2,000 MPs/L has been reported (Ambrosini et al., 2019). Thus, these diverse sources contribute to a substantial accumulation of MPs in soils, raising concerns about the environmental impact and potential hazards associated with their presence (Junhao, 2021).

4.2.2 Types and Characteristics of Micro/Nanoplastics (MNPs)

Micro/nanoplastics (MNPs) in agricultural soil are an emerging environmental concern due to their potential negative impacts on soil health, crop yields, and food safety. These tiny plastic fragments, ranging in size from 1 mm to 100 nm, can enter the soil through various sources like agricultural plastics, sewage sludge, and atmospheric deposition (Tian, 2022). Here is a closer look at the different types of MNPs found in agricultural soil:

- **Polyethylene (PE):** PE is the most common type used in mulching films, irrigation pipes, and greenhouses (Hossain et al., 2023). It is highly resistant to degradation, leading to long-term persistence in soil. It can absorb organic pollutants like pesticides and herbicides, potentially affecting soil fertility and plant health.
- **Polypropylene (PP):** PP is a common type of polymers used in agro-textiles and fertilizer bags. It has similar properties to PE in terms of persistence and potential pollutant absorption (Gaylor, 2013; Fuller and Gautam, 2016).

- **Polyethylene Terephthalate (PET):** PET is found in beverage bottles and can enter the soil through littering or sewage sludge application (Tang, 2023). It has less persistence than PE and PP but can still fragment into smaller MPs over time.
- **Polyvinyl Chloride (PVC):** PVC is used in irrigation pipes, plumbing and construction materials, liners, cosmetic containers, commercial cling wraps, siding, and greenhouse covers (Iqbal et al., 2023). It can leach harmful additives like phthalates into the soil, posing potential risks to soil organisms and plant health.
- **Nylon:** Nylon is used in fishing nets and ropes and can fragment into MPs through weathering and agricultural practices (Liu et al., 2024). However, it can adsorb pollutants and potentially harm soil biota.
- **Bioplastics:** Bioplastics have been promoted as biodegradable alternatives, but their degradation rate in soil can be slow and dependent on specific conditions (Ma et al., 2024). Some bioplastics might contain MP additives, negating their environmental benefits.

TABLE 4.1 Type and Characteristics of Microplastic Polymers in Agricultural Soil

Polymer Types	Usage
High-density polyethylene (HDPE)	Shampoo and detergent bottles, juice cans, freezer bags and trash bags, furniture, squeeze bottles, standard cling wrap, and shopping bags
Low-density polyethylene (LDPE)	Video cases, disposable cups, egg cartons, cutlery foam polystyrene, and hot beverage cups
Polystyrene (PS) & expanded polystyrene (EPS)	Food takeaway trays and pellets of protective packaging
Polyoxymethylene (POM)	Gears, fan wheels, and screws
Acrylonitrile butadiene styrene (ABS)	Wheel covers, Legos, and auto parts
Polyamide (PA)	Thread, clothing, and airbags
Polyester (P)	Debris, rope
Polyetheretherketone (PEEK)	Pumps, bearings, and pistons
Polyethylene (PE)	Toys, housewares, and mulch
Bioplastic (PLA)	Syringes, packaging, and textiles
Polysulfone (PSU)	Parts for appliances, filters
Polytetrafluoroethylene (PTFE)	Teflon foams, adhesives, and coatings
Polyurethane (PU)	Teflon foams, adhesives, and coatings
Styrene acrylonitrile (SAN)	Housewares and consumer goods

References: (Hüffer & Hofmann, 2021; Hossain et al., 2023; Chukwuemeka et al., 2024; Ma et al., 2024).

4.3 SOIL INTERACTION MECHANISMS AND THE FATE AND TRANSPORT OF MICROPLASTICS (MPS) IN AGRICULTURAL SOILS

Several factors contribute to the movement of MPs within soil profiles. Attributes of MPs such as size, density, and shape, alongside external climatic conditions like wind and rain, soil organisms such as earthworms and *Orchesella cincta*, and other external forces, collectively influence their migration (Ma et al., 2024). According to O'Connor et al. (2019), the depth of MPs' migration increased significantly as the number of dry-wet cycles increased. However, individual precipitation events and surface MP concentrations had little effect on migration depth (de Jesus Piñon-Colin et al., 2020). Additionally, earthworm activity has been observed to facilitate the movement of MPs through interactions with various particles and pollutants present in agricultural soil, which are the primary determinants of MPs' transportation routes. MPs inevitably engage in interactions with other pollutants when they come into contact with the soil environment, owing to their minute particle dimensions, substantial specific surface area, and high hydrophobicity (Boughattas et al., 2021). MPs are significant contributors to transforming and transporting pollutants within the soil ecosystem. Antibiotics, persistent organic pollutants (POP), and HMs have been detected on the surfaces of MPs. The combined impact of MPs, which serve as effective carriers for these substances, on the soil environment and organisms has generated academic interest (Liu et al., 2024). Researchers in the field have investigated the synergistic effects of MPs and additional pollutants (Guo et al., 2020). Moreover, microorganisms, clay minerals, metal oxides, hydroxides, and humus can all react with environmental pollutants. Researchers anticipated that the limited presence of MPs in the soil environment may diminish their significance as carriers of pollutants (Guo et al., 2022). They can move from the soil surface (Figure 4.1) into their burrows and deeper into the soil. Their study indicated the presence of PE particles in the middle and lower soil layers after 21 days, suggesting earthworms' role in promoting MP migration through the soil profile, with smaller particle sizes exhibiting greater ease of migration. The relationship between particle size and LDPE migration shows that smaller particles cause more biological problems. Biodegradation, mechanical abrasion, high-temperature oxidation, and exposure to ultraviolet (UV) radiation are some of the processes by which MPs found in soil undergo modifications in their chemical composition, including the disproportionation

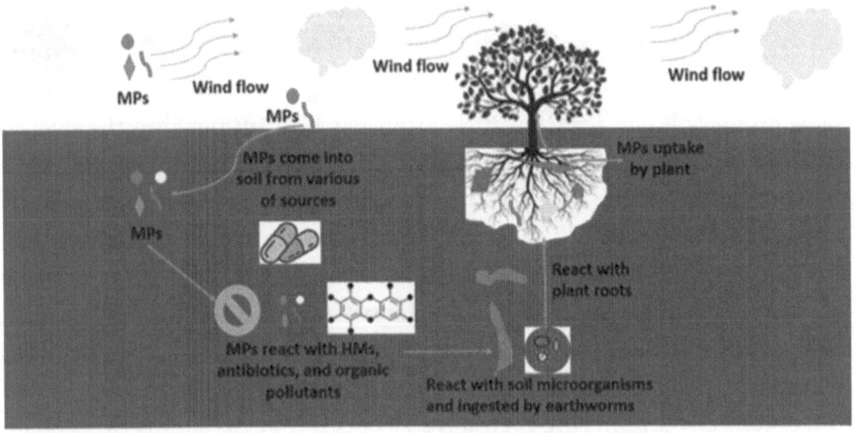

FIGURE 4.1 The distribution of microplastics (MPs) in soil compartments

and breakage of molecular bonds (Feng et al., 2022). These alterations ultimately result in the development of MNPs characterized by diminished particle diameters. Although biodegradation is the predominant mechanism by which MPs are degraded in soil, its impact is hindered by various factors such as soil conditions, mechanical forces, high temperatures, and UV radiation. MPs display an exceptionally sluggish rate of degradation in soil environments (Hartmann et al., 2017). For example, agricultural plastic film fragments, specifically PE, exhibit substantial resistance to degradation in farmland soil. These small fragments can stay in the ground for years or even decades before breaking down into smaller pieces of MP that stay in a relatively stable soil matrix (Gao et al., 2019).

4.3.1 Interaction between Microplastics (MPs) and Persistent Organic Pollutants

MPs, like silent ships in the soil, act as potent carriers for harmful pollutants. They readily adsorb POPs like PAHs, PCBs, pesticides, and herbicides, influencing their distribution and posing a synergistic threat to ecosystems (Boughattas et al., 2021). However, this adsorption is not a simple hitchhiking operation rather it is a complex dynamics where the type of MP matters. Research studies showed a strong correlation between MP hydrophobicity and its adsorption capacity for hydrophobic pollutants (Akutsu et al., 2022). In other words, more hydrophobic plastics attract more hydrophobic organic pollutants. Additionally, it depends on the organic pollutants' polarity. After that, they tend to bind more strongly and create toxic environments (Hämer et al., 2014). However, the adsorption process is not always harmonious. Different pollutants competing for space on an MP can lead to "adsorption battles," influencing the overall distribution of POPs in the soil (Guo et al., 2020). Research studies have shown that this competitive adsorption favors hydrophobic compounds over natural substances like caffeine or phenanthrene (Hartmann et al., 2017). Moreover, adsorbance of MPs greatly depends on the factors like pH, temperature, moisture, and water flow (Feng et al., 2022). For example, lower pH environments favor stronger adsorption. Scientists assumed that MPs act as Trojan horses, transporting and concentrating POPs in the soil, potentially amplifying their adverse effects on ecosystems (Kim et al., 2021). Lastly, it is clear that knowledge regarding the interaction between MPs and POPs is still limited. Hence, understanding the factors governing this adsorption process is crucial to assessing the true risks posed by MP pollution and developing effective mitigation strategies. Moreover, continuous research is needed to navigate the complex world of MPs and their impact on our soil health.

4.3.2 Interaction between Microplastics (MPs) and Heavy Metals (HMs)

In recent years, research studies have examined the impacts of MPs and HMs, revealing that their release into soil environments triggers geochemical processes that can lead to soil degradation. Researchers have explored that the surface of MPs becomes charged in soil environments through abrasion, enabling the adsorption of metal cations (Pang et al., 2023). The rate of adsorption follows nonlinear equations, and MPs make it easier for HMs to stick to surfaces on land. Various studies have already been assessed the adsorption rates of HMs such as Cd and Pb on MP stems. Consequently, adsorption rates for Cd and Pb onto MPs are reported at 6.9% and 7.5%, respectively (Lambert et al., 2017; Cai et al., 2023; Guo et al., 2024; Chen et al., 2023). Additionally, researchers noticed that sunlight oxidation and weathering alter the surface structure of MPs, facilitating the acquisition of electrical charges for metal ion adsorption to achieve charge balance (Chen et al., 2023). Factors such as soil pH and the residence time of MPs significantly influence their adsorption capacity for metal ions. Extended UV exposure increases copper and zinc adsorption capacities after 2,000 hours (Lei et al., 2018).

Moreover, metal cations form complexes with organic compounds by binding with polar regions or oxygen anions on the plastic surface (Bethanis & Golia, 2024). Despite existing studies on MPs' interaction with HMs, further investigation is needed to fully understand the mechanisms underlying MPs' adsorption of HMs.

4.3.3 Interaction between Microplastics (MPs) and Antibiotics

Antibiotics are extensively utilized in veterinary practice, and their coexistence with MPs in the natural environment has garnered attention. The interaction between MPs and antibiotics has been examined through research studies investigating the adsorption of five distinct antibiotics onto five different types of MPs (polyethylene, polystyrene, polypropylene, polyamide, and polyvinyl chloride) (Wang et al., 2021b). Findings revealed polyamide (PA) to exhibit the highest antibiotic adsorption capacity, attributed primarily to pore structure formation and hydrogen bond formation. Factors such as MP polarity, rubbery state, and crystallinity were noted to influence antibiotic adsorption, while environmental parameters like pH, ionic strength, and temperature showed negligible effects on tetracycline adsorption (de Souza Machado et al., 2018). The combined interaction of MPs and antibiotics may lead to heightened detrimental impacts on soil ecosystems, as this interaction can lead to help the emergence of antibiotic resistance genes (ARGs) (Xu et al., 2024). Notably, it has been observed that the joint effect of MPs and antibiotics on ARG abundance in *E. crypticus* surpassed that of individual MP treatments alone (Alimi et al., 2018). Additionally, antibiotics in soil can modify the degradation dynamics of MPs. The combined effect of MPs and other pollutants as well as influencing factors are shown in Table 4.2.

However, the combined effects of MPs and antibiotics need to be further investigated for better understanding the ecological risks they may pose on environmental compartments.

4.4 DETECTION AND ANALYSIS OF MICRO/NANOPLASTICS (MNPS) IN AGRICULTURAL SOILS

4.4.1 Sampling Techniques

MNPs pose a significant threat to agricultural soils, demanding accurate and representative sampling techniques to understand their distribution and impact (Jia, 2022). Here's a deeper dive into the key methods:

TABLE 4.2 The Combined Effect of Micro Plastics and Other Pollutants

Contaminant Type	Classification	Microplastic Type	Particle Size	Influences
Phthalic acid esters	POP	PVC, PE, PS	<75 µm	Adsorption is highly linear
Polybrominated biphenyls	POP	PET	<75 µm	Alter soil physicochemical properties
Chloroacetate	POP	LDPE	25 µm	Absorb more pesticides alter soil water content
4 (2,4-dichlorophenoxy) butyrate (atrazine)	POP	PE	250 µm	Reduce soil adsorption capacity
Polycyclic aromatichydro carbons, polychlorinated bi phenyls	POP	PE, PS	250 µm, 300 µm	When microplastics are added, the tissue concentration of PAHs and PCBs decreases
Chlorpyrifos	POP	LDPE	5 mm	MPs can transfer more chlorpyrifos into the soil matrix
As	Heavy metals	PVC	0.25–1 mm	Low toxicity to earthworms
Cu	Heavy metals	PA, PE, PS, PET, PVC, PMMA	70–350 µm	The adsorption of Cu^{2+} is affected by the type of MPs
Zn	Heavy metals	HDPE	<5 mm	Bioavailability of zinc increase
Sulfadiazine, amoxicillin, tetracycline, ciprofloxacin, Star, Methoxy benzidine	Antibiotics	PE, PS, PP, PA, PVC	75–180 µm	Adsorption capacity depends on the types of antibiotic, MPs, and environmental conditions
Tetracycline	Antibiotics	PE	<1 mm	Interfere the degradation process and spread of tetracycline

References: (Brodhagen et al., 2019; Espí et al., 2006; Feng et al., 2022).

Note: POP, PVC, PE, PS, PET, LDPE, PA, PMMA, and HDPE denote persistent organic pollutants, polyvinyl chloride, polyethylene, polystyrene, polyethylene terephthalate, low-density polyethylene, polyamide, polymethyl methacrylate, and high-density polyethylene, respectively.

4.4.1.1 Grab Sampling

Pros: Simple, rapid, cost-effective.

Cons: Highly susceptible to spatial variability, potentially missing hotspots or underestimating contamination. Suitable only for preliminary assessments or specific targeted areas (Kumar et al., 2020).

4.4.1.2 Composite Sampling

Pros: Provides a more representative picture of the field by combining multiple sub-samples from different locations.

Cons: Cannot capture detailed spatial variations within the field. Requires careful planning to ensure representative sub-sample selection.

4.4.1.3 Stratified Sampling

Pros: Offers better spatial resolution by dividing the field into smaller zones based on factors like soil type, topography, or past agricultural practices and taking samples from each zone.

Cons: More complex and time-consuming than grab or composite sampling. Requires prior knowledge of potential spatial variations.

4.4.1.4 Core Sampling

Pros: Crucial for understanding the depth distribution of MNPs, allowing assessment of migration patterns and potential contamination of deeper soil layers.

Cons: Mostly labor-intensive and expensive method. Requires specialized equipment and expertise.

4.4.1.5 Additional Considerations

- **Sampling Depth:** Depends on research objectives and suspected MNPs distribution. Surface samples may not be enough if studying deeper migration.
- **Number of Samples:** Determined by field size, heterogeneity, and desired level of accuracy. More samples provide more detailed information but increase cost and analysis time.
- **Replicates:** Taking multiple samples from each location or zone increases confidence in the results and allows for statistical analysis.

- **Equipment:** Sterile tools and containers are crucial to avoid contamination. Sampling equipment should be chosen based on soil type and sampling depth.
- **Documentation:** Recording detailed information about sampling locations, depth, date, and any relevant field observations is essential for data interpretation.

4.4.1.6 Emerging Techniques

Automated Sampling: Robots or drones can collect samples across large areas, improving efficiency and spatial coverage (Junhao, 2021).

Proximal Sensing: Techniques like hyperspectral imaging can identify areas with potential MNP contamination, guiding targeted sampling.

4.4.2 Analytical Methods

4.4.2.1 Unveiling the Plastic Invaders

- MNPs pose a significant threat to agricultural soil, and understanding their presence is crucial for mitigating their impact (Hossain et al., 2023). This requires efficient analytical methods to identify and characterize these elusive invaders. Various analytical techniques are discussed below:

4.4.2.2 Microscopy

- **Optical Microscopy:** Offers a low-cost and readily available option for visualizing larger MNPs (>10 µm). Techniques like stereomicroscopy and brightfield/darkfield microscopy can provide information on size, shape, and color (Koyuncuoğlu & Erden, 2021).
- **Scanning Electron Microscopy (SEM):** Provides high-resolution images of MNPs down to 1 µm, allowing for detailed morphological analysis and elemental composition through Energy-Dispersive X-ray spectroscopy (EDS) (Hossain et al., 2023).
- **Transmission Electron Microscopy (TEM):** Enables visualization of even smaller MNPs (down to 10 nm) with superior resolution, but requires specialized expertise and instrumentation.

4.4.2.3 Spectroscopy

- **Fourier Transform Infrared Spectroscopy (FTIR):** Identifies MNPs based on their unique chemical "fingerprint" created by molecular vibrations. Offers rapid analysis and can differentiate between different plastic types.
- **Raman Spectroscopy:** Similar to FTIR, analyzes molecular vibrations but uses laser light, offering higher spatial resolution and potential for in-situ analysis.
- **Near-Infrared Spectroscopy (NIR):** Provides rapid screening and differentiation between organic and inorganic materials, but may not be sensitive enough for specific MNP identification.

4.4.2.4 Separation Techniques

- **Density Separation:** Utilizes differences in density between MNPs and soil components. Techniques like floatation with salt solutions or centrifugation effectively separate MNPs for further analysis (Hossain et al., 2023).
- **Floatation:** Organic matter and lighter soil particles float in a solution like sodium chloride or sucrose, while denser MNPs sink for easier collection.
- **Centrifugation:** Spinning the soil suspension forces MNPs to settle at the bottom based on their density, allowing for isolation.
- **Filtration:** It can be used to separate MNPs based on size, but pore size selection is crucial to avoid losing smaller nanoplastics.

4.4.2.5 Other Techniques

- **Pyrolysis-Gas Chromatography/Mass Spectrometry (Py-GC/MS):** Breaks down MNPs into characteristic fragments and identifies them based on their mass spectra. Offers detailed information on polymer type and potential additives.
- **Thermal Analysis:** Techniques like Differential Scanning Calorimetry (DSC) and Thermogravimetric Analysis (TGA) can help identify specific plastic types based on their thermal properties.

The best analytical method depends on several factors, including:

- **Available Resources:** Consider cost, expertise, and instrumentation availability.
- **Sample Characteristics:** Soil type, organic matter content, and expected MNP size distribution can influence method selection.

4.4.2.6 Combining Techniques

A combination of various methods is necessary for comprehensive analysis. For example, microscopy can provide initial identification, followed by FTIR or Raman for confirmation and type differentiation. Pyrolysis-GC/MS might be used for detailed polymer analysis on isolated MNPs.

4.4.2.7 Emerging Innovations

- **Flow Cytometry:** Offers rapid analysis of individual MNPs based on their size and fluorescence properties.
- **Imaging Techniques:** Combining microscopy with Raman or fluorescence microscopy allows for simultaneous visualization and identification of individual MNPs.

4.4.3 Challenges and Limitations

While uncovering the secrets of MNPs in agricultural soil is crucial, the journey is riddled with obstacles (Junhao, 2021). Navigating these challenges and limitations is essential to developing effective detection and analysis strategies.

Sample Representativeness: Isolating these tiny invaders from vast soil samples is like finding a needle in a haystack. Traditional sampling methods often miss hidden pockets or underestimate MNP presence.

Extraction Efficiency: Separating MNPs from intricate soil matrices is tricky. Existing techniques might leave some behind, leading to incomplete pictures of the contamination landscape.

Size Differentiation: Distinguishing between MPs (>1 μm) and NPs (<1 μm) requires highly specialized equipment and expertise, limiting accessibility and potentially underestimating the presence of nanoparticles.

False Positives: Soil components like organic matter, minerals, and even microorganisms can resemble MNPs under certain analytical techniques, leading to misleading results.

Specificity vs. Sensitivity: Striking the right balance between detecting a wide range of MNPs and avoiding false positives can be challenging. Techniques optimized for sensitivity might catch natural interferences, while overly specific methods might miss certain plastic types.

Method Inconsistency: The lack of standardized protocols for sampling, extraction, and analysis creates discrepancies between studies, hindering data comparison and interpretation.

Data Comparability: Different laboratories might use varied techniques and equipment, making it difficult to combine findings and assess the full picture of MNP contamination across geographic regions or agricultural practices.

Evolving Landscape: As new types of plastics and additives emerge, existing detection methods might not be able to identify them, requiring continuous adaptation and validation of protocols.

Financial Burden: Advanced analytical techniques like pyrolysis-GC/MS or sophisticated microscopy require expensive equipment and skilled personnel, limiting their accessibility for smaller research groups or resource-constrained regions.

Time Investment: Meticulous analysis procedures can be time-consuming, hindering rapid screening and potentially delaying informed decision-making on mitigation strategies.

Balancing Resources: Allocating resources between developing new detection methods, optimizing existing ones, and implementing large-scale monitoring programs requires careful consideration and collaboration among stakeholders.

4.5 DISTRIBUTION AND ACCUMULATION PATTERNS

As MPs are infiltrating agricultural soil, posing a growing threat to food safety, ecosystem health, and soil sustainability, understanding their distribution and accumulation patterns is critical for combating this pervasive issue.

4.5.1 Uneven Spread

Researchers have observed the uneven spread of MPs in agricultural fields. Their distribution is a patchwork quilt influenced by various factors:

- **Agricultural Practices:** Tillage, irrigation, and plastic mulching practices can unwittingly fragment and disperse MPs throughout the soil profile, turning the plow into a weapon of plastic proliferation (Chukwuemeka et al., 2024).
- **Soil Composition:** Sandy soils, with their looser structure, become easy targets for MP migration compared to their clay counterparts. The finer particles in clay soils tend to hold onto MPs more tightly (He et al., 2021).
- **Field Topography:** Sloping landscapes add another layer of complexity. Runoff and erosion act as unwitting transporters, carrying MPs downhill, and creating pockets of higher concentration in low-lying areas (Bethanis & Golia, 2024).
- **Proximity to Plastic Sources:** Fields bordering bustling urban areas, industrial zones, or wastewater treatment plants become unwilling neighbors, more susceptible to plastic contamination due to aerial deposition, wastewater irrigation, or biosolids application (Fuller and Gautam, 2016).

4.5.2 Beyond the Surface

MPs aren't content with just littering the surface. They embark on a clandestine journey deeper into the soil (Zhang et al., 2022).

- **Leaching:** Rainwater and irrigation water act as Trojan horses, carrying MPs down through the soil layers, potentially contaminating groundwater and affecting deeper root systems.
- **Bioturbation:** Earthworms and other soil dwellers, the tireless ecosystem engineers, inadvertently become plastic transporters. As they burrow and mix the soil, they can unknowingly transport MPs to deeper depths.
- **Root Growth:** Plants themselves can unwittingly become accomplices. Plant roots can adhere to MPs, even absorbing them in some cases, creating a pathway for the plastic to enter the plant and potentially accumulate in edible portions.

4.5.3 Hotspots of Contamination

Within the field itself, certain areas become hotspots for MP accumulation (Zhao et al., 2023).

- **Mulch Remnants:** Improperly disposed-of mulching film, a common agricultural practice, fragments into MPs that concentrate in the surrounding soil, creating localized zones of contamination.
- **Low-Lying Areas:** As water flows downhill, it carries MPs with it, depositing them in depressions and valleys, transforming these areas into plastic sinks.
- **Intense Agricultural Activity:** Fields subjected to frequent tillage and irrigation experience increased MP fragmentation and dispersion, leading to higher concentrations compared to less intensively managed areas.

However, the global distribution of MPs is given (Table 4.3) below:

TABLE 4.3 Global Status of Microplastics (MPs)

City	Country	Concentration (Items/kg)	Composition	References
Yunnan	China	9,780.60	PP, PE, PS, PET	Huang et al. (2021)
Xishuangbanna	China	6,043.80	PES, PP, PE	Xu et al. (2022)
Pingquan	China	2,261.60	PET, PES	Zhang et al. (2021)
Hebei	China	4,143.90	PVC, PS	Wang et al. (2021c)
Shandong	China	4,292.4	PS, PES, PVC	Wang et al. (2021c)
Dhaka	Bangladesh	2.13×10^4	EVA, PVC, LDPE, HDPE	Hossain et al. (2023)
Narayanganj	Bangladesh	151	PA, PE, PU	Tajwar et al. (2023)
Sagian	Pakistan	876±194	PET, PP, PE	Luqman et al. (2023)
Tamil Nadu	India	1,500	PP, PET, PE	Karthika et al. (2024)
Yeoju	Korea	664	PE, PP	Choi et al. (2020)
Arusha	Tanzania	Up to 1,500	PET, PS, HDPE	Kundu et al. (2021)
—	Mauritius	231	PP, PA	Ragoobur et al. (2021)
—	Tunisia	852.24±124.2	PEVA, PP	Boughattas et al. (2021)
Llia, Western Greece	Greece	301±140	PE	Isari et al. (2021)
Murcia region	Spain & Netherland	2,242±984	—	Schothorst et al. (2021)
Mellipilla	Chile	360	PP, PE	Corradini et al. (2019)
Ontario	Canada	4,144	PP, PE, PS	Crossman et al. (2020)

4.6 EFFECTS ON SOIL HEALTH AND MICROBIAL COMMUNITIES

4.6.1 Effects on Soil Physical Properties

MPs have a significant impact on both the physical/chemical and biological properties of soil (Saeedi, 2024). This includes alterations in water retention capacity, pore characteristics, hydraulic properties, soil conductivity, and the soil microbial community. In a loamy sand soil, polyester, polyacrylic, and polyethylene MPs, even at environmentally relevant concentrations (2%), led to changes in soil bulk density and structural and functional modifications within a short span of 5 weeks (de Souza Machado et al., 2018). The variations in bulk density primarily result from the lower density of MPs compared to many natural soil minerals. MPs based on PS also affected the chemical properties and functional groups in aromatic rings of dissolved organic matter. Specifically, PP MPs, at concentrations of up to 6% and varying sizes (20, 200, and 500 μm), substantially reduced the saturated hydraulic conductivity (K_s) values in loam, clay, and sandy soils, with the most significant impact observed in clay soils (Ren et al., 2021). Moreover, the presence of MPs caused a more pronounced decrease in water retention capacity in clay soils compared to loamy and sandy soils (Pang et al., 2023). Additionally, clay soils experienced a notable increase in the slope of the water characteristic curve (SWCC) and a reduction in saturated water content (θ_s) and residual water content (θ_r) due to the influence of MPs (Guo et al., 2022). It has been noteworthy that compared to larger MPs, smaller MPs cause a notable decrease in soil porosity and aeration (Akutsu et al., 2022). The quantity of MPs present in soil exhibits an inverse relationship with the number of micropores. It has been observed that mixing MPs with soil led to a reduction in pore-size distribution, resulting in decreased hydraulic conductivity of saturated soils. MPs can alter the distribution of water-stable soil aggregates, hindering water infiltration and reducing soil stability (Liang et al., 2019). This impairment in soil permeability and stability can negatively affect the vertical growth of plant roots and consequently impact plant yield. Recent investigations have highlighted the negative impacts of MPs on various soil physical and chemical properties (Huang et al., 2024). These effects encompass changes in soil structure, porosity, bulk density, water content, pH, organic matter, and nutrients. Different types of MPs, such as polyacrylic acid, polyamide, polyethersulfone, and various soil types, have been studied under different experimental conditions (Liang et al., 2019). The data indicate that MPs play a significant role in altering soil fertility by influencing factors like water stability, bulk density, soil moisture, water-holding capacity, vertical water flow, pH, nutrients, dissolved organic carbon (DOC), and water-soluble organic carbon (WSOC). However, conflicting reports exist, with some investigations suggesting no significant impact of MPs on soil properties (Lozano et al., 2021). These varied outcomes are attributed to factors such as the type, dose, size, and shape of MPs, as well as the properties of the tested soil. Upon entering the soil matrix, MPs undergo changes in their properties. Hetero-aggregation occurs between MPs and soil inorganic elements like Fe, Mg, Si, and Al, enhancing the density and zeta potential of MPs (Wan et al., 2019). The positive charge of Fe oxides facilitates electrostatic attraction with negatively charged MPs. The interaction between MPs and soil minerals and organic matter (OM) can affect the density and zeta potential of MPs. The size, shape, exposure time, polymer type, and the "rhizosphere effect" due to plant presence contribute to the varying effects of MPs on soil pH and other properties (Kim et al., 2021). The abundance and exposure time of MPs in agricultural ecosystems also play a crucial role, with the impact changing from non-significant to significant levels over time (Feng et al., 2022).

4.6.2 Effects on Nutrient Cycling

Nutrient cycling keeps essential elements like carbon, nitrogen, and phosphorus moving and changing forms within an ecosystem. In healthy ecosystems, these amounts stay balanced. However, human activities like MP pollution can disrupt this balance by affecting how nutrients move and are available to living things. This can have serious consequences for the health of natural systems (Fuller and Gautam, 2016).

4.6.2.1 Soil Nitrogen Cycling

The nitrogen cycle is a crucial process for life on Earth, involving the conversion of atmospheric nitrogen (N_2) into usable forms and its movement through various ecosystems (Hopkins et al., 2017). The key stages within the soil are included below:

- **Nitrogen Fixation:** Transforming N_2 into reactive forms usable by plants and microbes.
- **Mineralization:** Releasing organic nitrogen back into the soil as mineral nitrogen.
- **Immobilization:** Conversion of mineral nitrogen into organic forms by plants and microbes.

- **Nitrification:** Oxidizing ammonium (NH^{4+}) to nitrate (NO^{3-}) for plant uptake.
- **Denitrification:** Converting nitrates back to atmospheric N_2 gas by bacteria.

4.6.2.2 Soil Carbon Cycle

MPs in soil act as significant reservoirs for non-plant-derived carbon, affecting the soil carbon cycle in various ways (Gaylor, 2013).

Become Immobilized: Most retained MPs bind with minerals and organic matter, forming aggregates that alter the natural carbon cycle.

Influence Microbial Activity: MP-derived carbon sources influence microbial growth and functions, affecting decomposition, plant growth, and carbon allocation from plants (Zhou et al., 2021).

Impact Native Soil Organic Matter (SOM): MPs can inhibit native SOM storage (Chen et al., 2023) while promoting MP-derived DOC and SOM. However, the effect on native SOM decomposition varies depending on the MP type:

- Biodegradable polymers accelerate native SOM decomposition by providing microbial fuel (Zhou et al., 2021).
- Other MPs can suppress SOM decomposition through a "negative priming" effect due to increased DOC (Rillig et al., 2021).

Influence Carbon Flow: MPs affect above and below-ground carbon flow by impacting CO_2 assimilation and emission. Dose-dependent effects are observed:

- Polythene and PVC increase root carbon allocation while decreasing soil allocation.
- Low-dose PVC powder enhances rhizodeposition and root growth, leading to increased soil carbon (Zang et al., 2020).
- MP presence can influence colonization of Arbuscular Mycorrhizal Fungi (AMF), influencing major carbon fluxes like rhizodeposition, photosynthesis, and organic carbon sequestration (Zhou et al., 2021).

Limited Impact on Specific Processes: Studies suggest minimal effects of PVC and PE MPs on low-molecular-weight carbon mineralization (Zang et al., 2020; Yu et al., 2022).

Dose-Dependent Effects on SOM Decomposition: Low MP doses can stimulate SOM decomposition, while high doses can suppress it.

Negligible Effect on Exogenous Carbon Mineralization: Addition of MPs seems to have minimal impact on the breakdown of external carbon sources like glucose and rice straw (Xiao et al., 2021).

4.6.2.3 Phosphorus Cycle

MPs can have diverse effects on the total phosphorus content in soil, depending on their type and composition. Studies show:

Reduced Total Phosphorus: MPs derived from PVC and PS can significantly decrease the overall phosphorus content in soil (Yu et al., 2021; Gao et al., 2019). This suggests they might bind or immobilize phosphorus, making it less available for plants and soil organisms.

Increased Available Phosphorus: MP polymer PVC can increase the availability of phosphorus in acidic red soils. This might be due to their interaction with soil properties or the release of plasticizer additives, potentially affecting phosphorus cycling in these specific soil types.

4.6.3 Influence on Microbial Diversity

The complex interaction between microorganisms and MPs in soil, paves the way for potential bioremediation strategies. Microbial responses to MPs vary depending on surface properties and concentration. Studies show LDPE MPs (76 mg/kg) altered bacterial community composition in Cinnamon soil after 90 days, while LDPE and PVC (1%–5%) reduced bacterial richness and diversity after 50 days, affecting specific bacterial families (increased Burkholderiaceae, decreased Sphingomonadaceae and Xanthobacteraceae) (Boughattas et al., 2021). Enrichment of Acidobacteria and Bacteroidetes was observed in soils dosed with 2% PE and PP for 29 days, accompanied by depletion of Deinococcus-Thermus and Chloroflexi populations. In an acidic cropped soil, MP presence significantly impacted bacterial richness and diversity. Co-amendment with PE (1%) and ciprofloxacin in Ustic Cambosol reduced bacterial diversity within 35 days but enriched Serratia and Achromobacter populations. Ammonia-oxidizing bacteria and nitrite reductase gene copies were reduced in clay loam soil treated with varying LDPE concentrations (0.10%–18%) for 30 days. Despite these negative impacts, positive effects on soil microbial communities have also been reported, highlighting the dynamic and intricate nature of these interactions (ESCAP, 2019). Studies investigating the impact of MNPs on soil microorganisms have yielded mixed results (Wang et al., 2022; Yu et al., 2022). At low concentrations (0.01%–1%), polyurethane (PU) did not significantly affect bacterial diversity, but increased specific bacterial groups like Firmicutes, Bacteroidetes, and Verrucomicrobia (Zhang et al., 2024). Similarly, polyethersulfone (PES) at 0.40% increased arbuscular mycorrhizal colonization, while PS nanoparticles mitigated the negative effects of sulfamethazine on bacterial communities (Gao et al., 2019). However, soil microbial communities drive diverse biofilm-mediated processes like penetration, leaching, and enzymatic action to transform and degrade microplastics and nanoplastics (MPs/NPs). This process is influenced by soil properties (organic matter, minerals, pH) and environmental factors (temperature, rainfall, sunlight) (Huang et al., 2024). These combined biological and abiotic processes dictate the MPs' fate, including dispersion, accumulation, and lifespan within agricultural soil. Scientists have identified around 16,170 plastic-degrading traits in different bacterial and fungal species by analyzing available microbial genomes (Han et al., 2024). Studies have shown promising results, with weight loss of metal-blended PP reaching 79% after a year of fungal treatment (Katsumi et al., 2020). Even some species can utilize specific polyesters as their sole carbon source. Additionally, polyester microfibers have been observed to increase aboveground biomass and even promote the colonization of beneficial arbuscular mycorrhizal fungi. Beyond individual microbe analysis, researchers have constructed models utilizing experimentally verified enzymes and metagenomic data to identify over 30,000 non-redundant enzyme homologs linked to plastic degradation (Zhang et al., 2022). Notably, they observed an enrichment of these enzymes within specific bacterial groups and even a potential correlation between a country's plastic pollution and its microbiome's degradation potential, suggesting possible adaptation to global plastic pressures. Understanding these novel enzymes and the diverse plastic-degrading microbial communities is crucial for developing innovative approaches to plastic waste management. This ongoing research holds promise for mitigating the growing threat of MP pollution in our agricultural soils and beyond (Brodhagen et al., 2019; Fuller and Gautam, 2016; Liang et al., 2019). Moreover, these findings highlight the potential for both positive and negative impacts of MPs on soil microbiota. However, current research limitations include:

- **Laboratory-Based Studies:** Most investigations occur in controlled laboratory settings, using pure MNPs in specific soil types for short-term periods (weeks to a year).
- **Limited Environmental Realism:** Real-world scenarios involve complex MNP mixtures associated with agricultural inputs, requiring investigation of their combined effects on both biological and non-biological processes.

Therefore, extending research to field-scale, long-term experiments using MPs-laden agricultural inputs is crucial to gain a more comprehensive understanding of their true impact on soil microbial activity and ecosystem health. This knowledge is essential for developing informed strategies to mitigate potential negative consequences and harness any potential benefits of MPs in agricultural systems.

4.6.4 Effects on Invertebrates

Earthworms, crucial players in soil ecosystems, are at the forefront of research examining the impact of MPs on soil animals (Zhou et al., 2021). Studies on species like *Lumbricus terrestris, Eisenia fetida*, and *Eisenia Andrei* show they can ingest and digest MPs, leading to varied effects on growth, reproduction, and mortality (Boughattas et al., 2021). These effects are dose-dependent, with even 0.10% (w/w) MPs concentration impacting growth and survival. Research on Lumbricus terrestris exposed to varying polyethylene MP concentrations in litter (7%–60%) confirms this, with higher concentrations leading to increased mortality and decreased growth rates (Gaylor, 2013). The impact of MPs extends beyond individual organisms, influencing various soil animal functions. Interestingly, studies show that the environment itself can influence toxicity. Kim et al. (2021) observed that polystyrene nanoparticles (PS NPs) were more harmful to invertebrates in soil than in liquid media at the same concentration (10 mg/kg or 10 mg/L). Furthermore, *Caenorhabditis elegans* exhibited greater sensitivity to larger PS MPs (530 nm) compared to smaller ones

(42 nm) (Huang et al., 2024). This size-dependent trend is also seen in other organisms. Smaller PS particles (50 nm) were more toxic to rotifer fecundity and lifespan than larger ones (0.50–6 µm), while smaller PE particles (10–27 µm) were more detrimental to *Chironomus tepperi* and *C. elegans* compared to larger counterparts (Zhu et al., 2018). This size-based difference likely arises from easier accumulation, ingestion, and digestion of smaller particles by organisms (Han et al., 2024). Understanding these impacts on earthworms and other soil animals is crucial for assessing the broader consequences of MP pollution on soil ecosystems (Palansooriya et al., 2022). The dose-dependent nature and size-dependent variations highlight the complexity of this issue, emphasizing the need for further research and mitigation strategies to protect these vital components of our soil health. Invertebrates like mealworms and wax moth larvae can harbor gut microbes capable of degrading plastics like PS and PE (Zhang et al., 2024). These microbes, including *Bacillus sp.* YP1 and members of *Citrobacter* and *Kosakonia*, likely aid in the grinding and breakdown of plastic pieces by the larvae (Zhou et al., 2018). However, MPs also negatively impact soil invertebrates, affecting their reproduction, feeding behavior, stress response, movement, and gene expression. Studies show varying ecotoxicology effects of MPs due to differences in both the MPs (size, shape, polymer) and the invertebrates (species, exposure dose, duration) (Zhao et al., 2023). For example, particle-to-mouth size ratio, specific to each species, influences their ability to ingest MPs, leading to diverse responses across different soil invertebrates. Recent research by Yang et al. (2023) highlights the combined threat of MPs and other contaminants. Their study found that co-exposure to PS nanoparticles (100 nm, 1,000 mg/kg soil) and tetracycline (20 mg/kg soil) in *Enchytraeus crypticus* enhanced the antibiotic's toxicity and enriched antibiotic resistance genes (Zhang et al., 2022). This suggests that the combined ecotoxicity of multiple contaminants can pose a significant threat to soil microbiomes, invertebrates, and ultimately, the health of agroecosystems.

4.6.5 Effects on Plants

Plant organisms have been observed to absorb materials or particles ranging from 3–4 to up to 40–50 nm in size. Research indicates that tobacco BY-2 cells were reported to uptake PS nanoparticles sized between 20 and 40 nm via endocytosis (Lei et al., 2018). Magnetic nanoparticles (MPs) present in terrestrial ecosystems have been found to be accessible to plants, leading to phytotoxic effects upon root absorption and subsequent translocation (Huang et al., 2024). This translocation poses concerns for food safety as MPs can enter the food chain. Recent studies have confirmed the uptake and translocation of MPs into edible tissues of various food crop species including wheat, carrot, cucumber, rice, maize, and lettuce (Saeedi, 2024). Investigations utilizing PS MPs observed their accumulation in the root cap cells of Arabidopsis and wheat, though not their uptake into roots directly (de Souza Machado et al., 2018). The ability of plants to adsorb or internalize MPs depends on factors such as MP size, charge, and the nature of plant tissues. A quantitative analysis method involving pyrolysis gas chromatography-mass spectrometry (Py-GC/MS) has been developed to detect MPs, with detection levels ranging from 2.31 to 4.15 μg/g for PS and 3.87 to 8.20 μg/g for polymethyl methacrylate (PMMA) (Ahmed et al., 2023). Cucumber plants grown hydroponically and exposed to 100 nm PS MNPs for 7 and 14 days showed MNP accumulation ranging from 0 to 6,893 μg PS/g. Both Py-GC/MS and scanning electron microscopy confirmed the presence of PS in dried cucumber, indicating uptake, translocation, and accumulation of MPs (Saeedi, 2024). However, challenges remain in quantifying diverse plastic particles such as PE and PVC in field-grown plants, particularly non-spherical particles. The impact of MPs on crop plants poses significant concerns for food security, affecting photosynthesis, antioxidant systems, and morphology in plants like maize, lettuce, wheat, and cress. Confocal microscopic observations further confirmed PS accumulation on root surfaces and entry into various tissues, particularly the xylem of rice seedling roots. MNPs can deform cell walls, creating larger pores facilitating their penetration (Wang et al., 2022). A higher transpiration rate enhances root uptake of MPs, while apoplastic and symplastic transports play crucial roles in translocating MPs to aerial tissues. Despite these findings, research on the genetic level to identify the molecular mechanisms of MP transport in plants remains scarce. Addressing these gaps in understanding is crucial for comprehensively assessing the impact of MPs on plant health and food security. Numerous detrimental impacts of magnetic nanoparticles (MPs) on plant functions have been documented (Anik et al., 2021; Huang et al., 2024). These include decreased root length, plant fresh weight, and chlorophyll content, as well as diminished shoot/root ratio, genetic alterations, reduced seed setting, and altered metabolic pathways. MPs also impede seed germination rate, dry biomass, and plant height, while interfering with photosynthetic metabolism and mineral nutrition in roots, stems, and leaves. Additionally, MPs can indirectly affect plant growth by altering soil physical/chemical properties, soil microbiome, and invertebrates. The charge of MNPs is a crucial factor influencing phytotoxicity. Positively charged PS NPs (PS–NH_2) exhibit lower accumulation in Arabidopsis thaliana root tips but cause greater accumulation of reactive oxygen species (ROS), resulting in more severe effects on plant growth and seed development compared to negatively charged PS NPs (sulfonic-modified; PS-SO_3H) (Kumari et al., 2022). Conversely, negatively charged PS NPs show higher accumulation in the apoplast and xylem. Research findings indicate that PS NPs significantly alter gene expression patterns in Triticum aestivum L. in a tissue-specific manner. In hydroponic conditions, PS NPs (100 nm) at concentrations of 0.01–10 mg/L disrupt various plant functions including carbon metabolism, amino acid biosynthesis, mitogen-activated protein kinase (MAPK) signaling, plant hormone signal transduction, and plant–pathogen interaction pathways (Boughattas et al., 2021). Cress seed germination decreases in a dose-dependent manner with MNPs at concentrations ranging from 10^{-3} to 10^{-7} particles/mL. Phytotoxicity worsens when MNPs are combined with co-pollutants (He et al., 2021). For instance, the combination of PS NPs (0.20 g/L) and As(III) (4.0 mg/L) more adversely affects photosynthesis and antioxidant activities in rice than the As(III) treatment alone. Despite these negative effects, there are also positive and non-significant impacts of MPs on plant functionalities, complicating conclusions regarding their overall impact. Magnetic nanoparticles exhibit varying impacts on biometric, biochemical, and physiological characteristics, encompassing cytotoxic and genotoxic effects, contingent upon particle properties, plant types, and duration of exposure (Hämer et al., 2014). The conflicting findings regarding MP effects on plants necessitate forthcoming investigations aimed at elucidating mechanisms governing bioaccumulation, phytotoxicity, and trophic transfer. Although demonstrating trophic transfers of MPs from plants to humans might pose challenges, humans, being significant recipients of MPs, are gradually being recognized as such, evidenced by reports indicating the presence of phthalates in pregnant women's urine samples, MPs in the feces of healthy human volunteers and infants, and polyethylene microparticles in the blood samples of healthy adults (Yu et al., 2022). These insights strongly suggest that MPs in agricultural soils are susceptible to root uptake and subsequent translocation into food crops, posing a substantial threat to the terrestrial food chain within farmland ecosystems. However, technical obstacles in quantifying MPs impede efforts to mitigate their ingress into farmland soils and subsequent dispersion within agroecosystems.

4.7 RISK ASSESSMENT AND HAZARD EVALUATION

4.7.1 Impacts on the Soil Properties and Characteristics

Soil properties, including structure, porosity, pH, microbial community, bulk density, and enzymatic activities, are significantly influenced by pollutants such as MPs (Wang et al., 2022). These plastic particles, predominantly composed of carbon, adversely impact soil functions. Previous studies indicate that MPs can detrimentally affect soil structure, organic matter, nutrient availability, and microbial activities (Kokalj et al., 2024). They adhere to root surfaces, penetrate root structures, move through plant xylem, and interfere with plant growth, causing stunted roots, altered nutrient uptake, and changes in plant physiology. MPs also impact soil porosity, influencing water and nutrient retention, and can create soil cracks that increase porosity (Maddela, 2023). While soil bulk density is a crucial indicator of soil health, studies suggest that, although soil contamination with polyester MPs may not significantly affect bulk density, it can impact the soil biogeochemical cycle and nutrient availability to crops (Han et al., 2024). The concentration of MPs smaller than 100 μm is influenced by soil composition (sand, silt, and clay), with higher silt and clay content hindering migration into deeper soil layers (Bethanis & Golia, 2024). Conversely, higher sand content and greater porosity enable vertical migration but reduce water and fertilizer retention. Overall, MP contamination adversely affects soil physical, chemical, and biological properties, consequently impacting plant growth and productivity (Anik et al., 2021). This includes negative effects on antioxidant enzyme activity, reactive oxygen species, and proteomics profiling, influencing crop yields. Thus, MP contamination poses multifaceted challenges to soil health and agricultural ecosystems.

4.7.2 Potential Risks to Plant Health

Just like humans, plants experience environmental stress in different ways and adopt unique strategies to cope. This also applies to their susceptibility to MPs. Studies show lettuce and corn are particularly vulnerable to root development disruption by polystyrene MPs, while radish and wheat seem more resilient (Saeedi, 2024; Boughattas et al., 2021). Even if MPs can't penetrate roots, they can still wreak havoc by clinging to the surface, clogging pores, or damaging cell walls. This can lead to oxidative stress, nutrient and water uptake blockage, and even detached root cells, as observed in wheat and maize exposed to MNPs. The damage does not stop MPs travel from roots to leaves, potentially impacting photosynthesis and overall plant growth (Huang et al., 2024). While roots seem to bear the brunt of the physical damage, leaves experience reduced oxidative stress due to ROS. Studies have documented various negative effects on crops like tomatoes and rice, including reduced photosynthesis, chlorophyll content, biomass, and growth. Additionally, MPs can induce oxidative stress and DNA damage in leaves, increasing ROS levels and lipid peroxidation in maize and Vigna radiata plants (Anik et al., 2021). Overall, the impact of MPs on plants varies depending on the species and the specific stress caused, highlighting the complexity of this issue in the plant kingdom.

- **Stunt Plant Growth:** Tomato plants exposed to MPs exhibited reduced growth, root length, height, and weight. Similar effects were observed in cereal crops like wheat and rice, with lower germination rates, stunted growth, and decreased yields (Anik et al., 2021).
- **Contaminate Edible Parts:** Vegetables can take up MPs, raising concerns about potential health risks for consumers. This can also affect the nutritional content, flavor, and market value of these vegetables (Chukwuemeka et al., 2024).
- **Harm Fruit Crops:** Studies on strawberries and tomatoes suggest MPs might negatively impact fruit size, nutrient composition, and ripening. Additionally, the accumulation of MPs on fruit surfaces poses a food safety threat (Alimi et al., 2018).
- **Threaten Native Plants:** While research on crops is limited, early findings suggest MPs may similarly affect native plants, potentially impacting ecosystem function and biodiversity (Han et al., 2024).

4.7.3 Implications for Human Health

MPs are a growing concern due to their potential harm to both wildlife and humans (Anik et al., 2021; Ma et al., 2024). They can enter the food chain through various sources, including:

- Breaking down of larger plastic pieces in the environment.
- Direct release from products like synthetic textiles and rubber tires.
- Leaching from food packaging and other plastic items.

Once ingested by organisms, MPs can biomagnify, meaning they become more concentrated as they move up the food chain. This can lead to negative effects on both the animals themselves and humans who consume them. The main routes of MP exposure for humans are inhalation, ingestion, and skin contact (Tang, 2023). MPs sneak into our bodies through the air we breathe, making inhalation the second most common exposure route (Wright and Kelly, 2018). These tiny invaders, especially from indoor materials and outdoor sources like ocean spray, easily penetrate our lungs' thin barrier and vast surface area, potentially spreading throughout the body (Ahmed et al., 2023). Inhaling them can trigger various health problems, from particle and chemical toxicity to disease spread and even cancer. Studies on animals suggest higher occupational exposure increases the risk of lung inflammation (Guo et al., 2020). On the other hand, body and face scrubs used in the beauty industry are a potential source of NPs entering the body. While research hasn't conclusively proven the impact, smaller particles and compromised skin are thought to increase penetration. However, no direct studies specifically address how well NPs pass through the outer skin layer (stratum corneum). While this outer layer acts as a protective barrier, contact with contaminated water or beauty products could still leave plastic particles on the skin. Additionally, even if direct absorption through the stratum corneum is unlikely, NPs might enter the body through sweat ducts, wounds, or hair follicles due to their hydrophobic nature. This highlights the need for further research to better understand the potential risks of NPs in skincare products and their interaction with the human body.

4.8 MITIGATION AND REMEDIATION STRATEGIES

4.8.1 Prevention at the Source

A comprehensive strategy is required to mitigate and remediate MP contamination in agricultural soil, with prevention at the source being a critical component.

4.8.1.1 Manufacturing/Packaging

Primary and secondary MP pollution is becoming a significant and expanding issue globally. Reducing the use of plastic packaging while increasing the recycling and reuse of some plastics can help with this global problem (Brodhagen et al., 2019). While boosting the proportion of recycled materials, facilities should be urged to reduce the amount of plastic packaging they produce.

4.8.1.2 Improving Production Efficiency

Plastic use can be reduced by (1) using alternative (for example, glass), recycled, or nonplastic biodegradable materials; (2) improving product design to reduce plastic use, extend product life, allow repair and reuse, and improve recyclability by limiting the number of polymers, additives, and mixtures; and (3) prohibiting certain types of single-use plastics (Hodson et al., 2017).

4.8.1.3 Reducing the Consumption of Plastics

The global MP contamination issue can only be addressed by incorporating responsible and sustainable practices into the production, use, and disposal of various plastic materials (Hopkins et al., 2017). To address plastic pollution, governments, corporations, and consumers must work together to implement solutions such as educating consumers about the impact of MPs and their source materials.

4.8.2 Soil Amendments and Remediation Techniques

To address the issue of MP contamination in agricultural soil, various soil amendments and remediation techniques can be employed (Anik et al., 2021). These strategies aim to mitigate the presence of MPs and improve soil health. Here are some key approaches:

i. **Biochar Application:**
Utilizing biochar, a carbon-rich material produced from organic matter through pyrolysis, as a soil amendment. Biochar has been shown to enhance soil structure, reduce leaching of contaminants, and potentially adsorb MPs (Elbasiouny & Elbehiry, 2023).

ii. **Compost Addition:**
Incorporating compost into the soil to improve organic matter content and enhance microbial activity. Microbial degradation facilitated by compost may contribute to the breakdown of certain types of MPs (Elbasiouny & Elbehiry, 2023).

iii. **Soil Microbial Communities Enhancement:**
Promoting the growth of beneficial soil microorganisms that have the potential to degrade or assimilate MPs. Strategies may include

the application of microbial inoculants or fostering conditions that support natural microbial communities (Kumari et al., 2022).

iv. **Phytoremediation:**
Using plants with the ability to accumulate and tolerate certain contaminants, including MPs. Plants with hyperaccumulation capabilities may uptake MPs from the soil, offering a potential avenue for remediation (Elbasiouny & Elbehiry, 2023).

v. **Agrochemical Amendments:**
Exploring the use of certain agrochemicals or soil conditioners that may facilitate the breakdown or sequestration of MPs. Research is ongoing to identify suitable chemicals without causing adverse effects on soil and crop health (Ahmed et al., 2023).

vi. **Soil Cover and Mulching:**
Implementing soil cover and mulching practices to reduce direct exposure of soil to environmental sources of MPs. Mulching can also contribute to moisture retention and overall soil health (Kumari et al., 2022).

vii. **Advanced Filtration and Filtration Systems:**
Developing and implementing advanced filtration systems in stormwater drains and agricultural runoff to capture MPs before they reach the soil. Installing barriers and filters in water channels to trap MPs (Ahmed et al., 2023).

viii. **Soil Monitoring and Assessment:**
Establishing routine monitoring programs to assess the concentration and distribution of MPs in agricultural soils. Regular soil testing helps track the effectiveness of remediation efforts and informs further actions (Anik et al., 2021).

4.9 CONCLUSION

The presence of MPs in the soils of agricultural land presents a substantial hazard, as it causes imbalances in soil composition, obstructs the absorption of nutrients and water, and may even impair the development of plants. In addition, they disrupt microbial communities and discharge hazardous chemicals, which may adversely affect human health and soil quality. Furthermore, developing countries are experiencing severe consequences due to inadequate treatment infrastructure and a lack of knowledge regarding MPs' origin, destiny, and transportation. In agricultural fields, the literature indicates that fertilizers are applied haphazardly by farmers in developing nations, which poses a potential risk of MP pollution. Developing countries must continuously investigate the pathways and interconnections of MPs in agricultural fields to address this issue. Additionally, they must increase awareness and implement strategies such as improved farm management, decreased plastic use, alternative materials, and enhanced wastewater treatment. However, it has been observed that there are no particular laws, policies, or regulatory frameworks regarding MPs in developing nations. Therefore, standard rules and regulations regarding this matter must be established in developing countries. Considering the effectiveness and cost-effectiveness of biological treatment systems in removing MPs, these techniques should be investigated and implemented in developing nations. Overall, mitigating MP pollution requires a multi-pronged approach to ensure sustainable agricultural development.

REFERENCES

Ageel, H. K., Harrad, S., & Abdallah, M. A. E. (2022). Occurrence, human exposure, and risk of microplastics in the indoor environment. *Environmental Science: Processes & Impacts*, 24(1), 17–31.

Ahmed, A. S. S., Billah, M. M., Ali, M. M., Bhuiyan, M. K. A., Guo, L., Mohinuzzaman, M., Belal Hossain, M., Safiur Rahman, M., Islam, M. S., Yan, M., & Cai, W. (2023). Microplastics in aquatic environments: A comprehensive review of toxicity, removal, and remediation strategies. *Science of the Total Environment*, 876, 162414.

Akutsu, Y., Nakajima-Kambe, T., Nomura, N., & Nakahara, T. (2022). Purification and properties of a polyester polyurethane-degrading enzyme from comamonas acidovorans. *Applied and Environmental Microbiology*, 64, 62–67.

Alimi, O., Farner Budarz, J., Hernandez, L., & Tufenkji, N. (2018). Microplastics and nanoplastics in aquatic environments: Aggregation, deposition, and enhanced contaminant transport. *Environmental Science & Technology*, 52, 1704–1724.

Ambrosini, R., Azzoni, R. S., Pittino, F., Diolaiuti, G., Franzetti, A., & Parolini, M. (2019). First evidence of microplastic contamination in the supraglacial debris of an alpine glacier. *Environmental Pollution*, 253, 297–301. https://doi.org/10.1016/j.envpol.2019.07.005

Anik, A. H., Hossain, S., Alam, M., Sultan, M. B., Hasnine, M. T., & Rahman, M. M. (2021). Microplastics pollution: A comprehensive review on the sources, fates, effects, and potential remediation. *Environmental Nanotechnology, Monitoring & Management*, 16, 100530.

Arthur, C., Baker, J., & Bamford, H. (2009). *Proceedings of the International Research Workshop on the Occurrence, Effects and Fate of Microplastic Marine Debris*, September 9–11, 2008, National Oceanic and Atmosphere Administration.

Bethanis, J., & Golia, E. E. (2024). Micro-and nano-plastics in agricultural soils: A critical meta-analysis of their impact on plant growth, nutrition, metal accumulation in plant tissues and crop yield. *Applied Soil Ecology*, 194, 105202.

Boughattas, I., Hattab, S., Zitouni, N., Mkhinini, M., Missawi, O., Bousserrhine, N., & Banni, M. (2021). Assessing the presence of microplastic particles in Tunisian agriculture soils and their potential toxicity efects using eisenia andrei as bioindicator. *Science of the Total Environment*, 796, 148959.

Brodhagen, M., Goldberger, J. R., Hayes, D. G., Inglis, D. A., Marsh, T. L., & Miles, C. (2019). Policy considerations for limiting unintended residual plastic in agricultural soils. *Environmental Science and Policy*, 69, 81–84.

Büks, F., & Kaupenjohann, M. (2020). Global concentrations of microplastics in soils–A review. *Soil*, 6(2), 649–662.

Cai, L., Zhao, X., Liu, Z., & Han, J. (2023). The abundance, characteristics and distribution of microplastics (MPs) in farmland soil—Based on research in China., 876, 162782.

Chen, Q., Zhao, H., Liu, Y., Jin, L., & Peng, R. (2023). Factors affecting the adsorption of heavy metals by microplastics and their toxic effects on fish. *Toxics*, 11(6), 490.

Choi, Y. R., Kim, Y., Yoon, J., Dickson, N., & Kim, K. (2020). Plastic contamination of forest, urban, and agricultural soils: A case study of Yeoju in the republic of Korea. *Journal of Soils and Sediments*, 21, 1962–1973.

Chukwuemeka, I. S., Li, G., Mo, Y., & Jacques, K. J. (2024). Impacts of microplastics and urbanization on soil health: An urgent concern for sustainable development. *Green Analytical Chemistry*, 8, 100095.

Corradini, F., Meza, P., Eguiluz, R., Casado, F., Huerta-Lwanga, E., & Geissen, V. (2019). Evidence of microplastics accumulation in agricultural soils from sewage sludge disposal. *Science of the Total Environment*, 671, 411–420.

Crossman, J., Hurley, R. R., Futter, M., & Nizzetto, L. (2020). Transfer and transport of microplastics from biosolids to agricultural soils and the wider environment. *Science of the Total Environment*, 724, 138334. https://doi.org/10.1016/j.scitotenv.2020.138334. PMID: 32408466.

de Jesus Piñon-Colin, T., Rodriguez-Jimenez, R., Rogel-Hernandez, E., Alvarez-Andrade, A., & Wakida, F. T. (2020). Microplastics in stormwater runoff in a semiarid region, Tijuana, Mexico. *Science of the Total Environment*, 704, 135411.

de Souza Machado, A., Lau, C., Till, J., Kloas, W., Lehmann, A., Becker, R., & Rillig, M. (2018). Impacts of microplastics on the soil biophysical environment. *Environmental Science & Technology*, 52, 9656–9665.

Elbasiouny, H., & Elbehiry, F. (2023). Addressing the microplastic dilemma in soil and sediment with focus on biochar-based remediation techniques. *Soil Systems*, 7(4), 110.

ESCAP. (2019). A Call for Healthy "Blue Oceans" in Asia and the Pacific. Bangkok.

Espí, E., Salmerón, A., Fontecha, A., García, Y., & Real, A. I. (2006). Plastic films for agricultural applications. *Journal of Plastic Film and Sheeting*, 22, 85–102.

Feng, X., Wang, Q., Sun, Y., Zhang, S., & Wang, F. (2022). Microplastics change soil properties, heavy metal availability and bacterial community in a Pb-Zn-contaminated soil. *Journal of Hazardous Materials*, 424, 127364.

Fuller, S. G., & Gautam, A. (2016). A procedure for measuring microplastics using pressurized fluid extraction. *Environmental Science and Technology*, 50, 5774–5780.

Gao, H., Yan, C., Liu, Q., Ding, W., Chen, B., & Li, Z. (2019). Effects of plastic mulching and plastic residue on agricultural production: A meta-analysis. *Science of the Total Environment*, 651, 484–492.

Gaylor, M. O., Harvey, E., & Hale, R. C. (2013). Polybrominated diphenyl ether (PBDE) accumulation by earthworms (Eisenia fetida) exposed to biosolids-, polyurethane foam microparticle-, and penta-BDE-amended soils. *Environmental Science and Technology*, 47, 13831–13839.

Guo, J., Du, Y., Yang, L., Luo, Y., Zhong, G., Zhao, H. M., & Liu, J. (2024). Effects of microplastics on the environmental behaviors of the herbicide atrazine in soil: Dissipation, adsorption, and bioconcentration. *Journal of Hazardous Materials*, 465, 133085.

Guo, J. J., Huang, X.-P., Xiang, L., Wang, Y.-Z., Li, Y.-W., Li, H., Cai, Q.-Y., Mo, C.-H., & Wong, M.-H. (2020). Source, migration and toxicology of microplastics in soil. *Environment International*, 137, 105263.

Guo, Z., Li, P., Yang, X., Wang, Z., Lu, B., Chen, W., … & Xue, S. (2022). Soil texture is an important factor determining how microplastics affect soil hydraulic characteristics. *Environment International*, 165, 107293.

Hämer, J., Gutow, L., Köhler, A., & Saborowski, R. (2014). Fate of microplastics in the marine isopod Idotea emarginata. *Environmental Science and Technology*, 48, 13451–13458.

Han, Y., Teng, Y., Wang, X., Wen, D., Gao, P., Yan, D., & Yang, N. (2024). Biodegradable PBAT microplastics adversely affect pakchoi (Brassica chinensis L.) growth and the rhizosphere ecology: Focusing on rhizosphere microbial community composition, element metabolic potential, and root exudates. *Science of the Total Environment*, 912, 169048.

Hartmann, N. B., Rist, S., Bodin, J., Jensen, L. H., Schmidt, S. N., Mayer, P., Meibom, A., & Anders Baun, A. (2017). Microplastics as vectors for environmental contaminants: Exploring sorption, desorption, and transfer to biota. *Integrated Environmental Assessment and Management*, 13, 488–493.

He, D., Zhang, Y., & Gao, W. (2021). Micro (nano) plastic contaminations from soils to plants: Human food risks. *Current Opinion in Food Science*, 41, 116–121.

Hodson, M. E., Duffus-Hodson, C. A., Clark, A., Prendergast-Miller, M. T., & Thorpe, K. L. (2017). Plastic bag derived-microplastics as a vector for metal exposure in terrestrial invertebrates. *Environmental Science and Technology*, 51, 4714–4721.

Hopkins, D. W., Wheatley, R. E., Coakley, C. M., Daniell, T. J., Mitchell, S. M., Newton, A. C., & Neilson, R. (2017). Soil carbon and nitrogen and barley yield responses to repeated additions of compost and slurry. *Journal of Agricultural Science*, 155, 141–155.

Hossain, M. N., Rahman, M. M., Afrin, S., Akbor, M. A., Siddique, M. A. B., & Malafaia, G. (2023). Identification and quantification of microplastics in agricultural farmland soil and textile sludge in Bangladesh. *Science of the Total Environment*, 858, 160118.

Huang, J., Chen, H., Zheng, Y., Yang, Y., Zhang, Y., & Gao, B. (2021). Microplastic pollution in soils and groundwater: Characteristics, analytical methods and impacts. *Chemical Engineering Journal*, 425, 131870.

Huang, W., Zhang, J., Zhang, Z., Gao, H., Xu, W., & Xia, X. (2024). Insights into adsorption behavior and mechanism of Cu (II) onto biodegradable and conventional microplastics: Effect of aging process and environmental factors. *Environmental Pollution*, 342, 123061.

Hüffer, T., & Hofmann, T. (2021). Sorption of non-polar organic compounds by micro-sized plastic particles in aqueous solution. *Environmental Pollution*, 214, 194–201.

Isari, E. A., Papaioannou, D., Ioannis, K., Kalavrouziotis, I. K., & Karapanagioti, H. K. (2021). Microplastics in agricultural soils: A case study in cultivation of watermelons and canning tomatoes. *Water*, 13(16), 2168. https://doi.org/10.3390/w13162168

Iqbal, B., Zhao, T., Yin, W., Zhao, X., Xie, Q., Khan, K. Y., & Du, D. (2023). Impacts of soil microplastics on crops: A review. *Applied Soil Ecology*, 181, 104680.

Jia, W. (2022). Automated identification and quantification of invisible microplastics in agricultural soils. *Science of the Total Environment*, 844, 156853.

Junhao, C. X. (2021). Extraction and identification methods of microplastics and nanoplastics in agricultural soil: A review. *Journal of Environmental Management*, 294, 112997.

Karthika, S., Sivasubramanian, K., Dhevagi, P., Thavamani, P., Rajkishore, S. K., Amrutha, V. M., & Lakshmanan, A. (2024). Distribution pattern and risk assessment of microplastics contamination in different agricultural systems. *International Journal of Environmental Analytical Chemistry*, 36, 1–20.

Katsumi, N., Kusube, T., Nagao, S., & Okochi, H. (2020). Accumulation of microcapsules derived from coated fertilizer in paddy felds. *Chemosphere*, 267, 129185.

Kim, S., Liang, Y., Zhao, T., & Rillig, M. (2021). Indirect effects of microplastic-contaminated soils on adjacent soil layers: Vertical changes in soil physical structure and water flow. *Frontiers in Environmental Science*, 9, 681934.

Kokalj, A. J., Kalčíková, G., Selonen, S., Bosker, T., Drobne, D., Dvorakova, D., & van Gestel, C. A. (2024). Strategy towards producing relevant and reliable data for the hazard assessment of micro-and nanoplastics in agricultural soils. *TrAC Trends in Analytical Chemistry*, 172, 117567.

Koyuncuoğlu, P., & Erden, G. (2021). Sampling, pre-treatment, and identification methods of microplastics in sewage sludge and their effects in agricultural soils: A review. *Environmental Monitoring and Assessment*, 193, 1–28.

Kumar, M., Xiong, X., He, M., Tsang, D. C., Gupta, J., Khan, E., & Bolan, N. S. (2020). Microplastics as pollutants in agricultural soils. *Environmental Pollution*, 265, 114980.

Kumari, A., Rajput, V. D., Mandzhieva, S. S., Rajput, S., Minkina, T., Kaur, R., & Glinushkin, A. P. (2022). Microplastic pollution: An emerging threat to terrestrial plants and insights into its remediation strategies. *Plants*, 11(3), 340.

Kundu, M. N., Hans, C., & Lugomela, K. G. (2021). Analysis of macro- and microplastics in riverine, riverbanks, and irrigated farms in Arusha, Tanzania. *Archives of Environmental Contamination and Toxicology*, 82(1), 142–157. https://doi.org/10.1007/s00244-021-00897-1. PMID: 34741639.

Lambert, S., Scherer, C., & Martin Wagner, M. (2017). Ecotoxicity testing of microplastics: Considering the heterogeneity of physicochemical properties. *Integrated Environmental Assessment and Management*, 13, 470–475.

Lei, K., Qiao, F., Liu, Q., Wei, Z., Qi, H., Cui, S., Yue, X., Deng, Y., & An, L. (2018). Microplastics releasing from personal care and cosmetic products in China. *Marine Pollution Bulletin*, 123, 122–126.

Liang, Y., Lehmann, A., Ballhausen, M.-B., Muller, L., & Rillig, M. (2019). Increasing temperature and microplastic fibers jointly influence soil aggregation by saprobic fungi. *Frontiers in Microbiology*, 10, 2018.

Liu, Z., Wen, J., Liu, Z., Wei, H., & Zhang, J. (2024). Polyethylene microplastics alter soil microbial community assembly and ecosystem multifunctionality. *Environment International*, 183, 108360.

Lozano, Y. M., Aguilar-Trigueros, C. A., Onandia, G., Maaß, S., Zhao, T., & Rillig, M. C. (2021). Effects of microplastics and drought on soil ecosystem functions and multifunctionality. *Journal of Applied Ecology*, 58(5), 988–996.

Luqman, M., Shahid, T., Awan, M. U. F., Kashif, S. U. R., Arooj, F., & Awan, A. R. (2023). Quantification and characterization of microplastics (MPs) pollution in peri-uburban agricultural lands of Lahore, Pakistan. *PLoS One*, 18(10), e0291760.

Ma, X., Shan, J., Chai, Y., Wei, Z., Li, C., Jin, K., & Ji, R. (2024). Microplastics enhance nitrogen loss from a black paddy soil by shifting nitrate reduction from DNRA to denitrification and Anammox. *Science of the Total Environment*, 906, 167869.

Maddela, N. R. (2023). Do microplastics and nanoplastics pose risks to biota in agricultural ecosystems? *Soil Systems*, 7, 19.

Palansooriya, K. N., Sang, M. K., El-Naggar, A., Shi, L., Chang, S. X., Sung, J., Zhang, W., & Ok, Y. S. (2022). Low-density polyethylene microplastics alter chemical properties and microbial communities in agricultural soil. *Scientific Reports*, 13(1), 16276. https://doi.org/10.1038/s41598-023-42285-w

Pang, X., Chen, C., Sun, J., Zhan, H., Xiao, Y., Cai, J., & Yang, G. (2023). Effects of complex pollution by microplastics and heavy metals on soil physicochemical properties and microbial communities under alternate wetting and drying conditions. *Journal of Hazardous Materials*, 458, 131989.

Patrício Silva, A., Prata, J., Mouneyrac, C., Barcelò, D., Duarte, A., & Rocha-Santos, T. (2021). Risks of COVID-19 face masks to wildlife: Present and future research needs. *Science of the Total Environment*, 792, 148505.

Pattanayak, S. (2022). Plastics and their additives reached the blood and tissue spaces: What are the possible consequences? *Exploratory Animal and Medical Research*, 12(2), 128–133. https://doi.org/10.52635/eamr/12.2.128-133

Ragoobur, D., Huerta-Lwanga, E., & Somaroo, G. D. (2021). Microplastics in agricultural soils, wastewater effluents and sewage sludge in Mauritius. *Science of the Total Environment*, 798, 149326. https://doi.org/10.1016/j.scitotenv.2021.149326. PMID: 34340075.

Rahman, M. M., Sultan, M. B., & Alam, M. (2023). Microplastics and adsorbed micropollutants as emerging contaminants in landfill: A mini review. *Current Opinion in Environmental Science & Health*, 31, 100420.

Ren, X., Tang, J., Wang, L., & Liu, Q. (2021). Microplastics in soil-plant system: Effects of nano/microplastics on plant photosynthesis, rhizosphere microbes and soil properties in soil with different residues. *Plant Soil*, 462, 561–576.

O'Connor, D., Pan, S., Shen, Z., Song, Y., Jin, Y., Wu, W. M., & Hou, D. (2019). Microplastics undergo accelerated vertical migration in sand soil due to small size and wet-dry cycles. *Environmental Pollution*, 249, 527–534.

Sa'adu, I., & Farsang, A. (2023). Plastic contamination in agricultural soils: A review. *Environmental Sciences Europe*, 35(1), 13.

Saeedi, M. (2024). How microplastics interact with food chain: A short overview of fate and impacts. *Journal of Food Science and Technology*, 61(3), 403–413.

Sana, S., Dogiparthi, L., Gangadhar, L., Chakravorty, A., & Abhishek, N. (2020). Effects of microplastics and nanoplastics on marine environment and human health. *Environmental Sciences and Pollution Research International*, 27, 44743–44756.

Schothorst, B., Beriot, N., Lwanga, E. H., & Geissen, V. (2021). Sources of light density microplastic related to two agricultural practices: The use of compost and plastic mulch. *Environments*, 8(4), 3.

Shafea, L., Yap, J., Beriot, N., Felde, V. J., Okoffo, E. D., Enyoh, C. E., & Peth, S. (2023). Microplastics in agroecosystems: A review of effects on soil biota and key soil functions. *Journal of Plant Nutrition and Soil Science*, 186(1), 5–22.

Sharma, P., & Sharma, P. (2024). Micro (nano) plastics: Invisible compounds with a visible impact. *F1000Research*, 13, 69.

Tajwar, M., Hasan, M., Shreya, S. S., Rahman, M., Sakib, N., & Gazi, M. Y. (2023). Risk assessment of microplastic pollution in an industrial region of Bangladesh. *Heliyon*, 9(7), e17949.

Tang, K. H. D. (2023). Microplastics in agricultural soils in China: Sources, impacts and solutions. *Environmental Pollution*, 322, 121235.

Tian, L. C. (2022). Microplastics in agricultural soils: Sources, effects, and their fate. *Current Opinion in Environmental Science & Health*, 25, 100311.

Union, E. (2018). *Plastics Strategy*; European Union: Brussels.

Wan, Y., Wu, C., Xue, Q., & Hui, X. (2019). Effects of plastic contamination on water evaporation and desiccation cracking in soil. *Science of the Total Environment*, 654, 576–582.

Wang, X., Song, N., Xue, Y., Wang, J., Liang, L., Li, M., Liu, X., & Liu, D. (2021c). Occurrence characteristics of residual film in cultivated soil of peanut planting area in Shandong Province, China. *Journal of Agro-Environment Science*, 40, 1729–1737.

Wang, Y., Yang, Y., Liu, X., Zhao, J., Liu, R., & Xing, B. (2021b). Interaction of microplastics with antibiotics in aquatic environment: Distribution, adsorption, and toxicity. *Environmental Science & Technology*, 55(23), 15579–15595.

Wang, Z., Fu, D., Gao, L., Qi, H., Su, Y., & Peng, L. (2021a). Aged microplastics decrease the bioavailability of coexisting heavy metals to microalga Chlorella vulgaris. *Ecotoxicology and Environmental Safety*, 217, 112–199.

Willén, A., Junestedt, C., Rodhe, L., Pell, M., & Jönsson, H. (2017). Sewage sludge as fertiliser – Environmental assessment of storage and land application options. *Water Science and Technology*, 75, 1034–1050.

Wright, S. L., & Kelly, F. J. (2018). Plastic and human health: A micro issue? *Environmental Science and Technology*, 51, 6634–6647.

Xu, G., Yang, L., Xu, L., & Yang, J. (2022). Soil microplastic pollution under different land uses in tropics, southwestern China. *Chemosphere*, 289, 133176.

Xu, M., Chen, M., Pan, C., Xu, R. Z., Gao, P., Chen, H. Q., & Shen, X. X. (2024). Microplastics shape microbial interactions and affect the dissemination of antibiotic resistance genes in different full-scale wastewater treatment plants. *Science of the Total Environment*, 912, 168313.

Yang, J., Monnot, M., Sun, Y., Asia, L., Wong-Wah-Chung, P., Doumenq, P., & Moulin, P. (2023). Microplastics in different water samples (seawater, freshwater, and wastewater): Removal efficiency of membrane treatment processes. *Water Research*, 232, 119673.

Yu, J., Adingo, A., Liu, X., Li, X., Sun, J., & Zhang, X. (2022). Micro plastics in soil ecosystem – A review of sources, fate, and ecological impact. *Plant Soil and Environment*, 68, 1–17.

Zhang, J., Zou, G., Wang, X., Ding, W., Xu, L., Liu, B., & Chen, Y. (2021). Exploring the occurrence characteristics of microplastics in typical maize farmland soils with long-term plastic film mulching in northern China. *Frontiers in Marine Science*, 8, 1–13.

Zhang, J., Wang, X., Xue, W., Xu, L., Ding, W., Zhao, M., & Chen, Y. (2022). Microplastics pollution in soil increases dramatically with long-term application of organic composts in a wheat–maize rotation. *Journal of Cleaner Production*, 356, 131889.

Zhang, Y., Song, M., Zhu, Y., Li, H., Zhang, Y., Wang, G., & Yang, Q. (2024). Impact of microplastic particle size on physiological and biochemical properties and rhizosphere metabolism of Zea mays L.: Comparison in different soil types. *Science of the Total Environment*, 908, 168219.

Zhao, H., Zhou, Y., Lu, Z., Ren, X., Barcelo, D., Zhang, Z., & Wang, Q. (2023). Microplastic pollution in organic farming development cannot be ignored in China: Perspective of commercial organic fertilizer. *Journal of Hazardous Materials*, 460, 132478.

Zhou, J., Gui, H., Banfield, C., Wen, Y., Zang, H., & Dippold, M. (2021). The microplastisphere: Biodegradable microplastics addition alters soil microbial community structure and function. *Soil Biology and Biochemistry*, 156, 108211.

Zhou, Q., Zhang, H., Fu, C., Zhou, Y., Dai, Z., Li, Y., Tu, C., & Luo, Y. (2018). The distribution and morphology of microplastics in coastal soils adjacent to the Bohai Sea and the Yellow Sea. *Geoderma*, 322, 201–208.

Zhu, B. K., Fang, Y.-M., Zhu, D., Christie, P., Ke, X., & Zhu, Y.-G. (2018). Exposure to nanoplastics disturbs the gut microbiome in the soil oligochaete Enchytraeus crypticus. *Environmental Pollution*, 239, 408–415.

CHAPTER 5

Quantification and Ecotoxicological Evolution of Microplastics in Soil Ecosystem

Muhammad Asrar, Tayba Ameen, Zeeshan Javed, Usama Saleem, Malja Imaan Yousaf, Muhammad Saleem, Abdul Ghaffar, Muhammmad Usman, Muhammad Uzair Ashraf, Muhammad Sohail Qadir, and Mashal Shahzadi

5.1 PRETREATMENT AND EXTRACTION METHODS

5.1.1 Dissolution and Extraction Techniques of Soil Micro-Plastics

The identification of soil micro-plastics (MPs) needs a precise treatment of the sample. Exhibiting a concise pretreatment method for the analysis of MPs is very crucial as it influences the results of samples.

DOI: 10.1201/9781003487579-5

According to reviewed researches, the methods of pretreatment procedures varied (Lee et al., 2023). The initial step for extraction of MPs is screening (Yang et al., 2021). The procedure for screening is mainly used for particles with a size less than 5.0 mm (Perez-Reveron et al., 2022). The screening is done to untangle the investigation process. This is followed by the second procedure known as density separation (DS), with the requirement of samples to be digested chemically prior to DS (Zhang et al., 2021; Fan et al., 2023). Organic matter digestion and DS are procedures used as pretreatment protocols for freshwater soil and sediments but few researches used only DS and not organic matter digestion (Lee et al., 2023) (Figure 5.1).

In distinct studies, the pretreatment procedures performed for freshwater soil and sediments show similarities. The most performed protocols for both samples were DS and afterward organic removal. The order of reviewed researches is as follows: 55% of researches (Density separation → organic removal); 25% of researches (Density separation only); 10% of researches (organic matter removal → density separation); 5% of researches (organic removal → density separation → organic matter removal); and 5% of researches used both (density separation and organic removal)

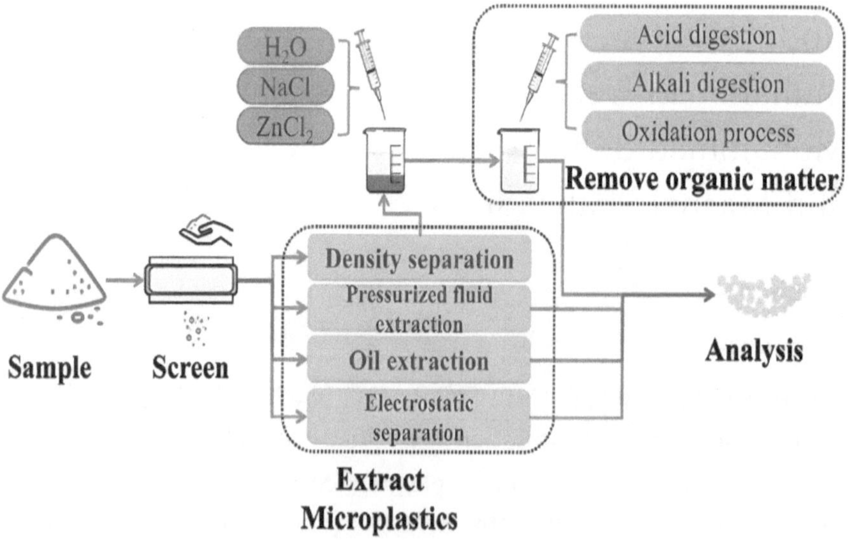

FIGURE 5.1 Micro-plastics extraction (Fan et al., 2023).

simultaneously for sediment samples of freshwater. For soil samples 50% of researches used (density separation → organic matter removal); 20% of researches used (density separation only); 20% of researches used (organic matter removal → density separation); and 10% of researches used (organic matter removal → density separation → organic matter removal). Sediment matrix samples can also be processed by using both DS and organic oxidation simultaneously. DS procedures can be repeated several times if required to enhance the efficacy of MP isolation and extraction from specimens (Simon-Sanchez et al., 2019; Lee et al., 2023). The sequences of pretreatment methods can notably change the procedures used in every step. The specimens from sediment and soil contain a large quantity of organic material due to which if organic digestion of matter is catalyzed initially then DS will require a lot of reagent for removal. For instance, H_2O_2 oxidation of organic matter will form a notable quantity of foam which may result in the loss of sample. On the contrary, if DS was catalyzed initially numerous contaminants can enter in the supernatant, it can clog the filter paper due to the presence of organic material while collecting MPs. Therefore, it is very important to construct the sequence of pretreatment methods while considering these pros and cons (Lee et al., 2023) (Figure 5.2).

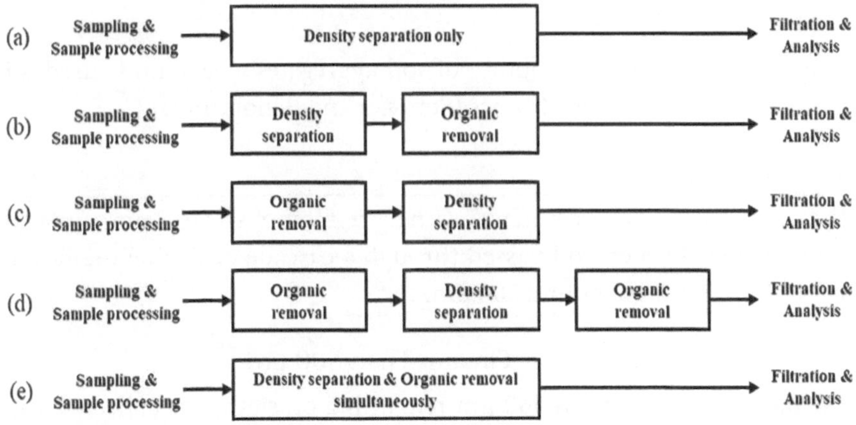

FIGURE 5.2 Flow chart representation of pretreatment protocols: (a) DS only, (b) DS comply with OR, (c) OR comply with DS, (d) OR prior and after DS, (e) DS and OR catalyzed simultaneously (Lee et al., 2023).

5.1.2 Filtration

The pretreatment process will be finalized by the filtration step. It uses numerous kinds of micro-filters, ranging in size from 0.2 to 20 μm, though some studies have also used a micro-filter with a pore size of 50 μm and larger. The selection of filters depends on the type and particle size of the MPs to be studied. Furthermore, the material of the filter paper used included nylon, cellulose nitrate, glass fiber, aluminum oxide, etc. Generally, the types of filter paper have no relationship with analytical techniques (Raman, microscopic examination, and FTIR). This is due to the fact that MPs show different patterns of adsorption on different filter materials. However, in a research conducted by Mani et al., various types of filters were used depending upon the analytical protocol. The anodisc of 0.2 μm was utilized for FTIR examination (Mani et al., 2019; Lee et al., 2023).

5.1.3 Sample and Subsample Collection

There are various sampling procedures that can be used for sampling depending on the research in question (Moller et al., 2022). The soil samples from the fields that are larger than 250 g can be homogenized (mixed entirely but not ground); soil specimens larger than 2 kg are split into subsamples of 250 g by the protocol known as cone and quarter given by Schumacher et al. (1990).

5.1.4 Dried by Freezing

In order to decrease the stability of soil aggregates a vacuum is used for subsamples for a time of 24 hours (Staricka and Benoit, 1995).

5.2 SIEVING

The subsamples can be sieved wet with a pre-filter of size (0.2 μm), washed with deionized water, and passed through a cascade of sieving meshes of 5 mm (five), 1 mm (one), and 500 μm.

5.2.1 Protocols for Particles Greater Than 500 μm

The particles larger than 500 μm remaining on sieves are collected and sorted for stereomicroscopic examination, photographed, and are reserved for FTIR analysis for MPs (Moller et al., 2022) (Figure 5.3).

5.3 EXTRACTION OF MICRO-PLASTICS

5.3.1 Density Separation

MP density ranges between 0.800 and 1.40 g/cm^3, while the density of soil levels is between 2.60 and 2.70 g/cm^3; thus MPs from soil can be extracted

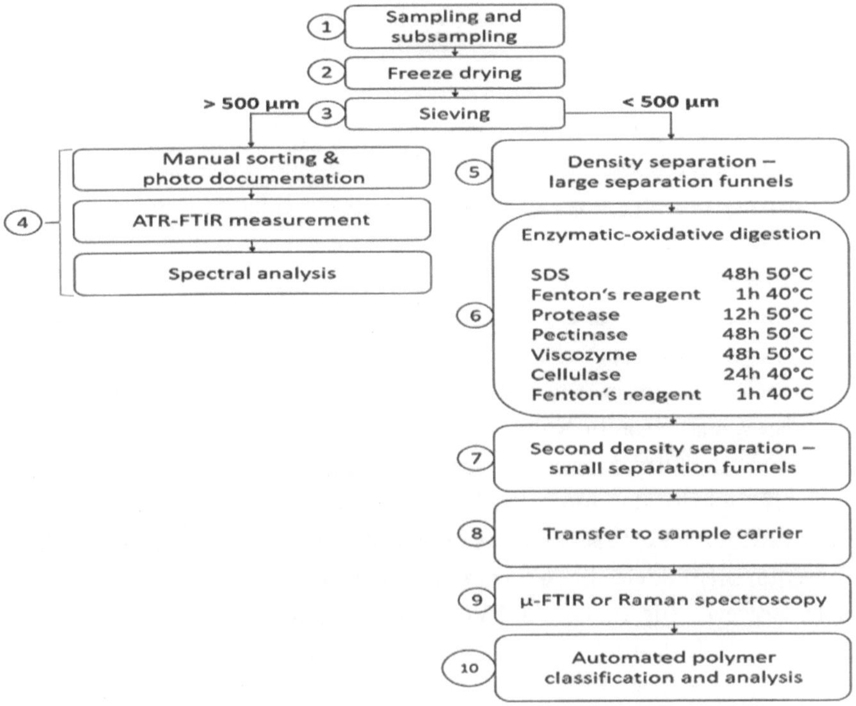

FIGURE 5.3 Diagrammatic representation for methods used in micro-plastic extraction from soil (Moller et al., 2022).

on the basis of their dissimilar densities. Most films used in agriculture are composed of polypropylene or low-density polyethylene, both having a density of <1.00 g/cm^3. Therefore, MPs can be separated in deionized water and thus become cost-effective. A high-level density solution of salt is used for floatation and extraction of those MPs present with undetermined densities (Van Cauwenberghe et al., 2015). Only saturated solutions of NaCl are used for extraction of MPs with low densities (PP, PS, and PE), the recovery time is 30 minutes (30 min/batch). However, the increase in frequency of floatation, extraction, and time of sonographic actions can be efficacious for extraction of acrylonitrile butadiene styrene (ABS), PE, polymethyl methacrylate (PMMA), PP, polycarbonate (PC), and PS). According to Plastics–The Facts, 2020(b) polyvinyl chloride (PVC) and polyethylene terephthalate (PET) encompasses 17.9% of the total world's plastic burden. Therefore, the extraction and separation of higher-density MP polymers (PVC, PMMA, and PET) remain challenging. The use of high-action solutions of salts like $ZnBr_2$ and $CaCl_2$ is assumed to increase the efficacy of the extraction process of MPs. The extraction efficacy of $CaCl_2$ is remarkably higher than that of

NaCl extraction. However, the positive charge of Ca^{+2} can form a bridge with the organic molecules comprising of a negative charge thereby, accumulating organic matter in the soil and intervening with the identification of MPs (Scheurer and Bigalke, 2018). The densities of MPs that can be extracted efficiently range between 1.60 and 1.80 g/cm^3. Sodium iodide (NaI) solution density is 1.08 g/cm^3, so it can be used for efficient removal of MPs. However, the high cost, the oxidation state for 7 days, and the use of H_2O_2 (35%) make it a less common method for MP extraction (Nuelle et al., 2014). To minimize the cost of extraction, a mixture in the ratio of 1:1 of NaI and NaCl was proposed with efficacy rates of 68%–99% for many forms of MPs). Considering the embryo-toxicity and ecological hazards of NaI for aquatic organisms, a solution of NaBr is proposed for MPs extraction in soil samples (Liu et al., 2019). Massively-parallel-signature-sequencing (MPSS) is the basic principle for designing a sediment MP unit comprising a ball valve of PVC, $ZnCl_2$, a PVC tube, and a magnetic stirrer bar (Coppock et al., 2017). This equipment device has a low price, a high rate of extraction, and is portable. However, the analysis of PVC is excluded due to the corrosion of the PVC tube. There are different density solutions used such as $ZnBr_2$, Na_2WO_4, and NaCl, and 50% of the studies used KI. Over 90% of MPs (>50.0 μm) recovery is achieved by DS. However, this method is not very suitable for extraction of smaller particles comprising 35%–90% of the entire plastic load. Additionally, MPs with related densities are difficult to isolate like PET (1.35–1.40 g cm^3) and PVC (1.37–1.42 g cm^3). For such samples, DS will be integrated with other methods, like electrostatic separation (Fan et al., 2023) (Table 5.1).

TABLE 5.1 The Extracting Agents for DS (Density Separation Method) of Micro-Plastics (Fan et al., 2023)

Extracting Agents	Relevant MPs	Retrieval Rate	References
H_2O	PE & PP	40%–90%	Quinn et al. (2017)
NaI	PVC	98%–100%	Van Cauwenberghe et al. (2013)
NaCl	PP, PC, PS, PE, ABS, PA and PMMA	69%–98%	Liu et al. (2018)
$CaCl_2$	MPs	93%–98%	Scheurer and Bigalke (2018)
NaBr	MPs	85%–100%	Liu et al. (2019)
$ZnCl_2$	MPs	85%–98%	Maes et al. (2017)
HCO_2K	PET, PE, PS and PE	No retrieval rate	Quinn et al. (2017)
$3Na_2WO_4.9WO_5.H_2O$	MPs	No retrieval rate	
NaCl + NaI	MPs	68%–99%	
$ZnBr_2$	MPs	95%–100%	

5.4 ELECTROSTATIC SEPARATION

Electrostatic separation is the segregation of MPs due to variations in their conductivity. The minerals in the soil show a greater conductivity rate than MPs. The samples from soil are often dried and frozen to remove the moisture effect on the electrostatic conductivity of matter minerals and MPs from soil. The electrostatic equipment separator is known to efficiently extract MPs from soil (Felsing et al., 2018; Enders et al., 2020). This separator method is also used to isolate non-conductive matter (like plastics) from metallic matter with a size of less than 8.00 mm. The recovery rate of MPs extracted through electrostatic separation is 90% without being affected by form, aging, density, and deposit of MPs ranging in size from 63 µm to 5.0 mm. To achieve a high rate of recovery the samples from soil should be dried thrice, each session being 3–4 h/150 g of sample specimen (Felsing et al., 2018). This method is suitable for MPs with a size higher than 60 µm and can be utilized to isolate MPs with greater size prior to DS (Enders et al., 2020).

5.5 OIL STRIPPING

The MPs that are trapped in the layer of oil can be separated through transfer-filtration. The oil on the surface can be eliminated by using ethanol. Oils with different polarities and densities (e.g., mineral oil, olive oil, rapeseed oil, and synthetic oil) can be picked to be extracted. The levels of recovery range from 90% to 97% (Scopetani et al., 2020). This method of extraction is safe, effective, simple, and inexpensive. However, residues of fat cannot be present in the sample when analyzed with spectral identification. Thus, the sample is washed with hexane to remove oil. Additionally, this process cannot eliminate intervention of organic matter present in soil.

5.6 FLUID EXTRACTION UNDER PRESSURE

Pressurized-fluid-extraction (PFE) is utilized to retrieve volatile and semi-volatile organic material from solid matter at sub-critical pressure and temperature. Pre-extraction of MPs is carried out by using methanol at 100°C to eliminate intervention from semi-volatile organic matter like grease, combined with PFE with dichloromethane at 180°C. The outcome of this method gives a retrieval rate of 84.5%–94.0% for MPs like PVC, PP, and PE present in municipal solid wastes from and soil, and MPs having a size less than 30.0 µm can be retrieved effectively (Fuller and Gautam, 2016). Nevertheless, some plastic particles can be eliminated in the pre-extraction stage causing inaccurate qualitative outcomes. To remove this inaccuracy, the pre-extraction step can be excluded and the remaining one-stage extraction of PVC, PE, PC, PMMA, PET, PS, and PP can be directly implemented using dichloromethane at 180°C. The average retrieval rate of extraction is greater than 80% (Okoffo et al., 2020). Regardless, the PFE method causes changes in the morphological aspects of MP particles, affecting physical characterization. PFE can extract MPs within 10 minutes. Therefore, it is an efficient technique to extract MPs.

5.7 ELIMINATION OF SOM (SOIL-ORGANIC-MATTER)

Organic fibers, MPs, and soil-organic-matter (SOM) show similar densities, which cause obstructions with the spectral and microscopic visual analysis of MPs (Blasing and Amelung, 2018). They can be isolated using density isolation (Scheurer and Bigalke, 2018). Thus, it is hard to isolate MPs from the rich soil of SOM. Organic fibers and SOM can be removed by digesting solutions containing H_2O_2 and strong bases and acids (Zhang et al., 2021). Therefore, SOM must be eliminated prior to and after separation by density to increase the retrieval accuracy.

5.7.1 Acid Digestion

It utilizes the action of powerful acids either in open or confined systems, for the purpose of oxidizing the compounds usually integrated with high pressure and temperature, and comparatively is a severe method of organic removal; resulting in cleavages at the molecular level causing different degrees of deterioration to MPs. At present, the powerful acids for the purpose of digestion and organic matter removal integrated into MPs are HCL and HNO_3. Of the two acids, HNO_3 is usually used and has the capacity to remove organic matter of high size in a short time with efficient digestion; however, many types of MP polymers like PA, PET, ABS, and PS are negatively affected by the use of HNO_3 (Prata et al., 2019). HNO_3 shows a retrieval rate of MPs ranging from 93% to 98% when sampled from soil (Scheurer and Bigalke, 2018). The efficacy of HCL as a digester is usually lower. High temperature through heating can result in the reduction of digestion duration. However, a powerful acid at a temperature of 60°C can cause the destruction of MPs analysis (Prata et al., 2019). Therefore, the retrieval rates and efficacy of powerful acids can be defined by using varying degrees of temperatures between 50°C and 60°C with varying concentrations of acids (Fan et al., 2023).

5.7.2 Alkaline Digestion

Contrary to digestion by acids, digestion by alkali is mild. It is best suited for the biological specimen's digestion, leading to the denaturation of protein molecules and hydrolysis of various compounds with less or no impact on the analysis of MPs (Karami et al., 2017). However, with the exception of a mild nature, it requires a high duration for digestion of usually 2–3 weeks for maximum organic matter removal (Foekema et al., 2013). Its efficacy can be increased by using high concentrations of alkali and temperature. Similarly, hot alkali solutions integrated with sodium hydroxide (NaOH) can cause deterioration of MP polymers like PC, PE, and PET similar to acid digestion (Foekema et al., 2013; Karami et al., 2017). Moreover, alkali at high temperatures combined with a KOH solution shows no degradation effect on MPs with the exception of cellulose acetate fiber (Dehaut et al., 2016). It can be used as a supplemental method with others as it is efficient for alkali-insoluble material (Fan et al., 2023).

5.7.3 Fenton's Reagent and H_2O_2 (Hydrogen-Peroxide)

H_2O_2 is a universally used oxidant that efficiently removes organic matter present in soil samples (Liu et al., 2018). Hydrogen peroxide digestion is a more strongly used solution for removing organic matter than acid and bases from plants and algae. Organic matter can be removed by using a 30% solution of H_2O_2; however, this solution can form dense foam in samples containing rich organic material. This issue can be resolved using smaller amounts of H_2O_2 integrated with high temperatures. Fenton's reagent can be used as an efficient alternative to H_2O_2 as it can reduce both temperature and time of reaction with an optimum range of PH between 3 and 5. Organic matter and inorganic compounds can be removed effectively using Fenton's reagents which are hard to remove by H_2O_2. Moreover, the MP polymers like PP, PVC, and PE are not affected in structure by this Fenton's reagent (Hurley et al., 2018). However, Fenton's reagent effects on the weathered MPs require more investigations. Further, Fenton's reagent solution is unable to remove organic matter completely so it requires some additional processes, like the addition of alkali KOH. Samples rich in organic material cannot be digested above 60°C with Fenton's reagent, because it causes thermal degradation as a result of exothermic reactions (Fan et al., 2023).

5.7.4 Enzymatic Digestion

This method is used to eliminate organic material from fluent samples. For soil samples, this method is dubious (Loder et al., 2017). Organic material from the soil samples can be removed by utilizing the effect of enzymes present for three natural constituents like biomass containing lingocellulosic (lignin, cellulose, and hemicellulose) but time and cost are two limitations. It can become faster when integrated with oxidative digestion (Fan et al., 2023).

5.8 ISOLATION, IDENTIFICATION, AND EXTRACTION

5.8.1 Visual Classification and Identification

MPs with larger sizes from the crude soil specimens are passed through a sieve of 2 mm mesh. The remaining MPs on the 0.6 mm mesh are then collected and counted visually. ATR- FTIR can be used for identification of the isolated MPs (Shim et al., 2016; He et al., 2019). MPs can also be identified by a dissecting microscope (stereo-zoom-microscope) at a magnification of 300X. The retained MPs on the filter can be morphologically observed under SEM. EDS can be used to highlight the elemental components of MPs observed in SEM (Isari et al., 2021).

5.8.2 Microscopic Characterization

5.8.2.1 Optical Microscopy

Optical microscopy (OM) is very efficient as it includes easy operation, detection of 500 µm, and low cost. $ZnCl_2$, H_2O, and NaCl solutions are used in organic matter digestion and extraction of MPs, the extracted MPs are observed under OM resulting in the successful detection of PVC, LDPE, and PES (Corradini et al., 2019). However, when the size of grain particles decreases causing increase in transparency, interference from substances will result in increased faulty positive presence of MPs. It can be utilized to view MPs with greater sizes of particles (>500 µm) (Dekiff et al., 2014; Fan et al., 2023).

5.8.2.2 Electron Microscopy

Beams of electrons are used in electron microscopy to image and expand samples. Such research uses scanning electron microscopy (SEM). SEM uses high resolution, magnification, secondary electron beam, and background scattering light to form images. MPs can be singled out from organic matter particles by utilizing high-definition particle surface images (Gong and Xie, 2020). SEM is used to configure the presence of MPs such as PET, HDPE, PS, PP, and PE in agricultural soils treated with DS separation ($CaCl_2$, NaCl) in China. Flatness and conductivity of the sample are the two limitations of SEM applications (Ding et al., 2020b). Although SEM is non-corrosive, the chemical and physical properties of MPs are retained by integrating SEM along with other methods (energy spectrum, spectroscopy, and mass spectroscopy). Images can be formed on three dimensions to view targeted areas better.

5.8.2.3 *Spectral Investigation*

The structural information, number of MPs, and chemical composition can be estimated by comparing the MP spectra with the spectra present in the library. Integrating microscopy, chemo-metric methods, and imaging techniques enables a rapid, quantitative, and qualitative analysis of MP specimens. Spectral analysis can be done by various techniques.

5.8.2.4 *Infrared Spectroscopy*

Absorption spectrum principle is used in the infrared spectrum, and data collected from electromagnetic wave absorption by molecules. The most prevalent techniques include near-infrared spectroscopy (NIR) and Fourier-transformed infrared spectroscopy (FTIR). The high powerful spectral qualities and simple operation are the advantages of FTIR which covers for quantification and in-situ detection of MPs in environmental specimens. The identified peaks (e.g., C=O, C=C, and O–H) figures between 700 and 3,700 cm^{-1} were also utilized for MPs analysis (Li et al., 2020; Ding et al., 2020a). To meet the requirements of MPs analysis FTIR can change freely modes between transmission, ATR, and reflection. MPs greater than 200 µm can be tested directly and rapidly using ATR-FTIR (Fuller and Gautam, 2016). However, for a morphological analysis of MPs FTIR must be integrated with SEM. The extraction of MPs (like PS, PP, and PE) from soil specimens in U.S wetlands digested with 0.500 M Fe(II) and (30%) H_2O_2 can be analyzed by an integrated system of SEM and ATR-FTIR (Helcoski et al., 2020).

5.8.2.5 *µ-FTIR (Fourier Transformed Micro-Infrared-Spectroscopy)*

Micro-FTIR is an apparatus that is an integrated structure of a microscope and FTIR. It is efficient for detection of MPs greater than 10 µm. However, it is very laborious because only a single polymer of plastic can be analyzed (Wang et al., 2020). FTIR integrated with a focal plane array can naturally quantify the MPs present on the filter membrane. It is the most pervasively used process for visualization, quantification, time-saving, and identification of MPs (Shim et al., 2016). The limitations of this procedure are to eliminate the abundance of non-target components. For this purpose, pretreatment protocols are very essential (Sarker et al., 2020). Additionally, FTIR must be integrated with other methods like SEM for visual and morphological characterization. Soil specimens pretreated by oxidative digestion (30% hydrogen peroxide, 50°C-48 hours) analyzed in μ-FTIR results in extraction of PA, PS, PP, PES, and PE (Wang et al., 2021). Micro-plastics and nano-plastics can be characterized by AFM-IR (Shim et al., 2016).

For the evaluation of quantitative and qualitative measurements, NIR employs infrared radiation absorption (at 700–2,500 nm) by numerous bonding (C–C, C–H, C–N, and O–H) (Galvez-Sola et al., 2010). The three-dimensional structure of plastic polymers can be analyzed using NIR-HIS (Near-Infrared Hyperspectral Imaging) (Piarulli et al., 2020). Visual-near-infrared-spectroscopy (VIS-NIR) can directly compute the number of MPs in soil specimens with the exception of the pretreatment process (Corradini et al., 2019). An algorithm known as MEDA was developed for evaluation of MPs burden in various regions along with the integration of transfer learning and NIR for quick quantification of MPs (Feng et al., 2020). Nevertheless, absence of a proper database can restrict this algorithm, so it is necessary to create a plastic database library. LDIR (Laser-Driven Ion Radiation) uses trending quantum laser innovation integrated with rapid scanning optics is the latest innovation for detection of MPs and gives clear, quick, and high-resolution images (Tian et al., 2022). In contrast to IR processes (FTIR-micro-FTIR), LDIR has the capacity to scan the specimen prior to imaging and can target the MPs resulting in improvement in accuracy of detection of various polymers thereby, minimizing duration of analysis (Dong et al., 2022). LDIR can detect MPs in soil specimens pretreated with oxidative removal of organic matter (30% hydrogen peroxide, 70°C-48 hours). LDIR integrated with density isolation is also used to calculate the concentration of MPs present in farm-land loam (Jia et al., 2022). It can also carry out automated scanning of MPS no. in the samples area and afterward gives classification. However, LDIR is unable to discover the thickness of MPs resulting in inaccurate detection of aged MP particles (Primpke et al., 2020).

5.8.2.6 Raman-Spectroscopy

The Raman spectral study uses the principle of scattering scale spectrum and utilizes the information obtained from the difference in frequency of scattering and incident light. By using spectral data, the rotational and vibrational data of particles is retrieved, and the structural chemical makeup of the focused particles is presented (Moller et al., 2022). It can detect and identify the quality peaks (like; C=O, CH_2, and C--O) ranging between 800 and 2,900 cm^{-1} of MPs like PA, PE, and PS from soil samples (Feng et al., 2020). Surface enhanced Raman spectroscopy (SERS) has the capability to detect a single MP particle because of fast characterization and maximum sensitivity toward targeted molecules (Xie et al., 2022). Pretreatment of the sample solution is necessary as trace elements can interfere with the detection by Raman spectra. It is suitable for samples containing a non-polar pattern of bonding vibration of a single atom; the signal from Raman is stronger based on the powerful molecular symmetry. For the detection of polar bonds infrared vibrations are more suited due to vibration of many different atoms and the uncertain dipole movement of molecules (Phan et al., 2022). Both IR and Raman spectroscopy can be integrated together for better precision of MP identification and quantification.

5.8.2.7 Energy Dispersion Spectrometry/Spectroscopy (EDS)

EDS is based on the principle of using various X-ray quantum (photon) energies of different elements for the conduction of the component study. In complement to micro-FTIR, EDS can also be utilized for the characterization of MPs in the soil samples for more accurate chemical data collection (Yin et al., 2022). The detection of MPs from the paddy soils, with subsequent addition of plastic debris, can be conducted using EDS integrated with (atomic force microscope) AFM (Yang et al., 2022). The combination of SEM with EDS shows the effects of polypropylene in cadmium-rich soil and production of MPs in coastal sediments from East Sea China. The quality peaks of EDS are more prominent than FTIR. It can be used as a semi-quantitative method of analysis (Cao et al., 2023; Lin et al., 2021).

5.9 MASS-SPECTROMETRY

5.9.1 TOF-SIMS or (Time-of-Flight Secondary Ion Mass Spectrum)

TOF-SIMS is a delicate and responsive technology for surface analysis. It is established on the principle of excitation of secondary ions by primary ions, thereby detecting various times of excitation of secondary ions on the surface of the sample by a detector (Sodhi, 2004). It has very low detection power and, on the contrary, high-resolution power. It was utilized for the detection of 12 samples of soil from China and identified the presence of PVC, PA, PP, and PET (Du et al., 2020a). TOF-SIMS is used in the characterization of chemical components of MPs with the help of a beam of high energy that bangs on the MPs, and the particles vanish in the pattern of fragments; it acts as an essential property of MPs specified imaging. TOF-SIMS is also used for the characterization of MPs in farmland samples, detection of particle sizes, their chemical composition, and differences in densities produced by decomposition. The MPs identified by TOF-SIMS are PP, PET, PA6, and PET from soil samples (Du et al., 2020b).

5.9.2 MALDI-TOF-MS or Matrix-Assisted-Laser-Desorption-Ionization Mass Spectrometry

This is an effective method of identifying and ionizing compounds with high molecular size (Badia et al., 2011). It is used to evaluate the oligomers present in the MPs and can arbitrarily degrade the long-chain polymers for MS analysis (Luo et al., 2022a). It is divided into oligomers which are tiny monomers, the spectra form patterns based on the difference in m/z in relation to monomer size (Luo et al., 2022b). This spectrogram library includes peaks of various polymers. MALDI-TOF-MS is very sensitive, easily operated in quantifying and identifying new arriving pollutants in the ecosystem (Badia et al., 2011). It is also utilized to identify MPs in sediments from rivers. A curve showing the quantitative correlation formed among the normal signal wave and the wave signals from the low and high molecular size plastic polymers, using coefficients for correlation higher and beyond 0.960 and 0.980; MPs (like PS and PET) can be identified on the basis of the Quantitative Correlation Curve (Wu et al., 2020). Nevertheless, the identification efficacy of this method for mixed specimens is low, leading to absence of quantitative analysis. For this reason, pretreatment of samples is necessary to avoid interference of various elements in the sample. To improve the efficacy of detection the form and concentration of the matrix should be optimized.

5.9.3 Inductively-Coupled-Plasma-Mass-Spectrometry (ICP-MS)

This is a productive method of detecting trace minerals (Wu et al., 2020). Single-particle ICP-MS has the efficacy of rapid detection of single particles by utilizing high-speed modes of detection combined with time-resolved detection (Chouchene et al., 2022). It is an upgraded procedure for directly characterizing individual particles like MPs. Luo et al. (2022a) established a procedure for MPs labeled with the help of lanthanides. If these elements are utilized for labeling MPs, the vulnerability of ICP-MS for metal substances detection makes it an accurate method for quantitative MP analysis. These labeled MPs are effective for studying the effect of MPs from plant material and soil in the laboratory with the exception of analyzing MPs in the natural environment. Additionally, ICP-MS is unable to furnish data about the morphological and chemical nature of MPs, for this reason, SEM is integrated with ICP-MS to clarify the chemical and morphological characteristics of MPs.

5.10 CONCLUSIONS

The contamination caused by MPs in the soil is becoming worse. As a result, it's critical to develop and enhance techniques for separating and analyzing soil MPs, assess their release into the environment, reduce the hazards to the environment, and methodically comprehend the negative effects of MPs on the organism's health and soil, establish environmental remediation methods. Soil MPs may be effectively separated by using DS. Removal of organic material at a density comparable to MPs is difficult because of the high quantity of SOM and its distinct binding with MPs. As a result, soil samples are usually broken down. MPs might be undervalued because of the variable degree of damage caused by different digestive techniques. Consequently, it is imperative to look at less-cost, less-time separation, and extraction techniques that maintain the properties of MPs. To prevent artificial error operation, the level of automation methods for microscopic evaluation should be increased. Pretreatment procedures need to be reduced for spectrum studies by integrating digestion protocols and efficacy optimization with numerous technologies. The development of multispectral imaging protocols in conjunction with autonomous data processing can reduce the amount of energy and time required for the evaluation process. Strengthening the establishment of pertinent databases is necessary for mass spectrometry procedure. To offer more precise data for thermal evaluation, appropriate pretreatment techniques for various plastic kinds must be established in conjunction with many detection procedures. Standard techniques for separating, extracting, identifying, and measuring MPs in soils need to be developed and reinforced going forward. A single technical specification needs to be created by comparing different quantitative methodologies. To increase the accuracy, consistency, and viability of inspecting data among areas, the technological and methodological requirements of soil MP examination should be established and refined based on comparative tests of various laboratory facilities.

REFERENCES

Badia, J. D., Stromberg, E., Ribes-Greus, A., & Karlsson, S. (2011). A statistical design of experiments for optimizing the MALDI-TOF-MS sample preparation of polymers. An application in the assessment of the thermo-mechanical degradation mechanisms of poly (ethylene terephthalate). *Analytica Chimica Acta*, *692*(1–2), 85–95.

Blasing, M., & Amelung, W. (2018). Plastics in soil: Analytical methods and possible sources. *Science of the Total Environment*, *612*, 422–435.

Cao, Y., Ma, X., Chen, N., Chen, T., Zhao, M., Li, H., & Yang, J. (2023). Polypropylene microplastics affect the distribution and bioavailability of cadmium by changing soil components during soil aging. *Journal of Hazardous Materials, 443*, 130079.

Chouchene, K., Nacci, T., Modugno, F., Castelvetro, V., & Ksibi, M. (2022). Soil contamination by microplastics in relation to local agricultural development as revealed by FTIR, ICP-MS and pyrolysis-GC/MS. *Environmental Pollution, 303*, 119016.

Coppock, R. L., Cole, M., Lindeque, P. K., Queiros, A. M., & Galloway, T. S. (2017). A small-scale, portable method for extracting microplastics from marine sediments. *Environmental Pollution, 230*, 829–837.

Corradini, F., Meza, P., Eguiluz, R., Casado, F., Huerta-Lwanga, E., & Geissen, V. (2019). Evidence of microplastic accumulation in agricultural soils from sewage sludge disposal. *Science of the Total Environment, 671*, 411–420.

Dehaut, A., Cassone, A. L., Frere, L., Hermabessiere, L., Himber, C., Rinnert, E., & Paul-Pont, I. (2016). Microplastics in seafood: Benchmark protocol for their extraction and characterization. *Environmental Pollution, 215*, 223–233.

Dekiff, J. H., Remy, D., Klasmeier, J., & Fries, E. (2014). Occurrence and spatial distribution of microplastics in sediments from Norderney. *Environmental Pollution, 186*, 248–256.

Ding, L., Mao, R., Ma, S., Guo, X., & Zhu, L. (2020a). High temperature depended on the ageing mechanism of microplastics under different environmental conditions and its effect on the distribution of organic pollutants. *Water Research, 174*, 115634.

Ding, L., Zhang, S., Wang, X., Yang, X., Zhang, C., Qi, Y., & Guo, X. (2020b). The occurrence and distribution characteristics of microplastics in the agricultural soils of Shaanxi Province, in north-western China. *Science of the Total Environment, 720*, 137525.

Dong, M., She, Z., Xiong, X., Ouyang, G., & Luo, Z. (2022). Automated analysis of microplastics based on vibrational spectroscopy: Are we measuring the same metrics? *Analytical and Bioanalytical Chemistry, 414*(11), 3359–3372.

Du, C., Liang, H., Li, Z., & Gong, J. (2020a). Pollution characteristics of microplastics in soils in southeastern suburbs of Baoding City, China. *International Journal of Environmental Research and Public Health, 17*(3), 845.

Du, C., Wu, J., Gong, J., Liang, H., & Li, Z. (2020b). ToF-SIMS characterization of microplastics in soils. *Surface and Interface Analysis, 52*(5), 293–300.

Enders, K., Tagg, A. S., & Labrenz, M. (2020). Evaluation of electrostatic separation of microplastics from mineral-rich environmental samples. *Frontiers in Environmental Science, 112.*

Fan, W., Qiu, C., Qu, Q., Hu, X., Mu, L., Gao, Z., & Tang, X. (2023). Sources and identification technology of microplastics in soil. *Soil & Environmental Health, 1*, 100019.

Felsing, S., Kochleus, C., Buchinger, S., Brennholt, N., Stock, F., & Reifferscheid, G. (2018). A new approach in separating microplastics from environmental samples based on their electrostatic behavior. *Environmental Pollution, 234*, 20–28.

Feng, S., Lu, H., Tian, P., Xue, Y., Lu, J., Tang, M., & Feng, W. (2020). Analysis of microplastics in a remote region of the Tibetan Plateau: Implications for natural environmental response to human activities. *Science of the Total Environment, 739*, 140087.

Foekema, E. M., De Gruijter, C., Mergia, M. T., van Franeker, J. A., Murk, A. J., & Koelmans, A. A. (2013). Plastic in North Sea fish. *Environmental Science & Technology, 47*(15), 8818–8824.

Fuller, S., & Gautam, A. (2016). A procedure for measuring microplastics using pressurized fluid extraction. *Environmental Science & Technology, 50*(11), 5774–5780.

Galvez-Sola, L., Moral, R., Perez-Murcia, M. D., Perez-Espinosa, A., Bustamante, M. A., Martinez-Sabater, E., & Paredes, C. (2010). The potential of near infrared reflectance spectroscopy (NIRS) for the estimation of agroindustrial compost quality. *Science of the Total Environment, 408*(6), 1414–1421.

Gong, J., & Xie, P. (2020). Research progress in sources, analytical methods, eco-environmental effects, and control measures of microplastics. *Chemosphere, 254*, 126790.

He, P., Chen, L., Shao, L., Zhang, H., & Lü, F. (2019). Municipal solid waste (MSW) landfill: A source of microplastics?-Evidence of microplastics in landfill leachate. *Water Research, 159*, 38–45.

Helcoski, R., Yonkos, L. T., Sanchez, A., & Baldwin, A. H. (2020). Wetland soil microplastics are negatively related to vegetation cover and stem density. *Environmental Pollution, 256*, 113391.

Isari, E. A., Papaioannou, D., Kalavrouziotis, I. K., & Karapanagioti, H. K. (2021). Microplastics in agricultural soils: A case study in cultivation of watermelons and canning tomatoes. *Water, 13*(16), 2168.

Jia, W., Karapetrova, A., Zhang, M., Xu, L., Li, K., Huang, M., & Huang, Y. (2022). Automated identification and quantification of invisible microplastics in agricultural soils. *Science of the Total Environment, 844*, 156853.

Lee, H., Kim, S., Sin, A., Kim, G., Khan, S., Nadagouda, M. N., & Han, C. (2023). Pretreatment methods for monitoring microplastics in soil and freshwater sediment samples: A comprehensive review. *Science of the Total Environment, 871*, 161718.

Li, J., Song, Y., & Cai, Y. (2020). Focus topics on microplastics in soil: Analytical methods, occurrence, transport, and ecological risks. *Environmental Pollution, 257*, 113570.

Lin, J., Xu, X. M., Yue, B. Y., Xu, X. P., Liu, J. Z., Zhu, Q., & Wang, J. H. (2021). Multidecadal records of microplastic accumulation in the coastal sediments of the East China Sea. *Chemosphere, 270*, 128658.

Luo, Y., Li, L., Feng, Y., Li, R., Yang, J., Peijnenburg, W. J., & Tu, C. (2022a). Quantitative tracing of uptake and transport of submicrometre plastics in crop plants using lanthanide chelates as a dual-functional tracer. *Nature Nanotechnology, 17*(4), 424–431.

Luo, Y., Sobhani, Z., Zhang, Z., Zhang, X., Gibson, C. T., Naidu, R., & Fang, C. (2022b). Raman imaging and MALDI-MS towards identification of microplastics generated when using stationery markers. *Journal of Hazardous Materials, 424*, 127478.

Mani, T., Primpke, S., Lorenz, C., Gerdts, G., & Burkhardt-Holm, P. (2019). Microplastic pollution in benthic midstream sediments of the Rhine River. *Environmental Science & Technology, 53*(10), 6053–6062.

Moller, J. N., Heisel, I., Satzger, A., Vizsolyi, E. C., Oster, S. J., Agarwal, S., & Loder, M. G. (2022). Tackling the challenge of extracting microplastics from soils: A protocol to purify soil samples for spectroscopic analysis. *Environmental Toxicology and Chemistry, 41*(4), 844–857.

Nuelle, M. T., Dekiff, J. H., Remy, D., & Fries, E. (2014). A new analytical approach for monitoring microplastics in marine sediments. *Environmental Pollution, 184*, 161–169.

Okoffo, E. D., Ribeiro, F., O'Brien, J. W., O'Brien, S., Tscharke, B. J., Gallen, M., & Thomas, K. V. (2020). Identification and quantification of selected plastics in biosolids by pressurized liquid extraction combined with double-shot pyrolysis gas chromatography–mass spectrometry. *Science of the Total Environment, 715*, 136924.

Perez-Reveron, R., Gonzalez-Salamo, J., Hernandez-Sanchez, C., Gonzalez-Pleiter, M., Hernandez-Borges, J., & Diaz-Pena, F. J. (2022). Recycled wastewater as a potential source of microplastics in irrigated soils from an arid-insular territory (Fuerteventura, Spain). *Science of the Total Environment, 817*, 152830.

Phan, S., Padilla-Gamiño, J. L., & Luscombe, C. K. (2022). The effect of weathering environments on microplastic chemical identification with Raman and IR spectroscopy: Part I. Polyethylene and polypropylene. *Polymer Testing, 116*, 107752.

Piarulli, S., Sciutto, G., Oliveri, P., Malegori, C., Prati, S., Mazzeo, R., & Airoldi, L. (2020). Rapid and direct detection of small microplastics in aquatic samples by a new near infrared hyperspectral imaging (NIR-HSI) method. *Chemosphere, 260*, 127655.

Prata, J. C., da Costa, J. P., Duarte, A. C., & Rocha-Santos, T. (2019). Methods for sampling and detection of microplastics in water and sediment: A critical review. *TrAC Trends in Analytical Chemistry, 110*, 150–159.

Primpke, S., Godejohann, M., & Gerdts, G. (2020). Rapid identification and quantification of microplastics in the environment by quantum cascade laser-based hyperspectral infrared chemical imaging. *Environmental Science & Technology, 54*(24), 15893–15903.

Sarker, A., Deepo, D. M., Nandi, R., Rana, J., Islam, S., Rahman, S., & Kim, J. E. (2020). A review of microplastics pollution in the soil and terrestrial ecosystems: A global and Bangladesh perspective. *Science of the Total Environment, 733*, 139296.

Scheurer, M., & Bigalke, M. (2018). Microplastics in Swiss floodplain soils. *Environmental Science & Technology, 9*(52), 3591–3598. https://doi.org/10.1021/acs.est.7b06003

Schumacher, B. A., Shines, K. C., Burton, J. V., & Papp, M. L. (1990). A comparison of soil homogenization techniques. In M. S. Simmons (Ed.), *Hazardous Waste Measurements*. Lewis, Chelsea, MI, pp. 1–40.

Scopetani, C., Chelazzi, D., Mikola, J., Leiniö, V., Heikkinen, R., Cincinelli, A., & Pellinen, J. (2020). Olive oil-based method for the extraction, quantification and identification of microplastics in soil and compost samples. *Science of the Total Environment, 733*, 139338.

Shim, W. J., Song, Y. K., Hong, S. H., & Jang, M. (2016). Identification and quantification of microplastics using Nile Red staining. *Marine Pollution Bulletin, 113*(1–2), 469–476.

Simon-Sanchez, L., Grelaud, M., Garcia-Orellana, J., & Ziveri, P. (2019). River Deltas as hotspots of microplastic accumulation: The case study of the Ebro River (NW Mediterranean). *Science of the Total Environment, 687*, 1186–1196.

Sodhi, R. N. (2004). Time-of-flight secondary ion mass spectrometry (TOF-SIMS):—Versatility in chemical and imaging surface analysis. *Analyst, 129*(6), 483–487.

Staricka, J. A., & Benoit, G. R. (1995). Freeze-drying effects on wet and dry soil aggregate stability. *Soil Science Society of America Journal, 59*(1), 218–223.

Tian, X., Been, F., & Bauerlein, P. S. (2022). Quantum cascade laser imaging (LDIR) and machine learning for the identification of environmentally exposed microplastics and polymers. *Environmental Research, 212*, 113569.

Van Cauwenberghe, L., Devriese, L., Galgani, F., Robbens, J., & Janssen, C. R. (2015). Microplastics in sediments: A review of techniques, occurrence and effects. *Marine Environmental Research, 111*, 5–17.

Van Cauwenberghe, L., Vanreusel, A., Mees, J., & Janssen, C. R. (2013). Microplastic pollution in deep-sea sediments. *Environmental Pollution, 182*, 495–499.

Wang, J., Li, J., Liu, S., Li, H., Chen, X., Peng, C., & Liu, X. (2021). Distinct microplastic distributions in soils of different land-use types: A case study of Chinese farmlands. *Environmental Pollution, 269*, 116199.

Wang, W., Ge, J., Yu, X., & Li, H. (2020). Environmental fate and impacts of microplastics in soil ecosystems: Progress and perspective. *Science of the Total Environment, 708*, 134841.

Wu, P., Tang, Y., Cao, G., Li, J., Wang, S., Chang, X., & Cai, Z. (2020). Determination of environmental micro (nano) plastics by matrix-assisted laser desorption/ionization–time-of-flight mass spectrometry. *Analytical Chemistry, 92*(21), 14346–14356.

Xie, L., Gong, K., Liu, Y., & Zhang, L. (2022). Strategies and challenges of identifying nanoplastics in environment by surface-enhanced Raman spectroscopy. *Environmental Science & Technology, 57*(1), 25–43.

Yang, H., Yumeng, Y., Yu, Y., Yinglin, H., Fu, B., & Wang, J. (2022). Distribution, sources, migration, influence and analytical methods of microplastics in soil ecosystems. *Ecotoxicology and Environmental Safety, 243*, 114009.

Yang, L., Zhang, Y., Kang, S., Wang, Z., & Wu, C. (2021). Microplastics in soil: A review on methods, occurrence, sources, and potential risk. *Science of the Total Environment, 780*, 146546.

Yin, M., Cao, H., Zhao, W., Wang, T., Huang, W., & Cai, M. (2022). Tide-driven microplastics transport in an elongated semi-closed bay: A case study in Xiangshan Bay, China. *Science of the Total Environment*, *846*, 157374.

Zhang, D., Fraser, M. A., Huang, W., Ge, C., Wang, Y., Zhang, C., & Guo, P. (2021). Microplastic pollution in water, sediment, and specific tissues of crayfish (Procambarus clarkii) within two different breeding modes in Jianli, Hubei province, China. *Environmental Pollution*, *272*, 115939.

CHAPTER 6

Fate, Transport Pathways, and Accumulation of Microplastics Agricultural Soil and Their Interaction with Agrochemicals

Zeeshan Javed, Mashal Shahzadi, Fareeha Sattar, Bilal Rasool, Usama Saleem, Muhammad Asad Mangat, Rashid Ali, Saba Naz, Sajida Mustafa, and Muhammad Asrar

6.1 INTRODUCTION

As plastic trash has become a greater environmental issue, attention to global plastic pollution has increased. Due to their unknown, potentially harmful effects on the environment, microplastics (MPs), which are defined

DOI: 10.1201/9781003487579-6

as particles with a size of less than 5 mm, have come to the attention of the world's governments. Because of their tiny particle size and extremely slow rate of biodegradation, they can be readily absorbed by organisms and then moved through food webs. Globally, annual plastic output already exceeds 450 million metric tons (MMT), and if no mitigating measures are adopted, it is expected to rise even more by a factor of 2–3 by 2060 (OECD, 2022). Growing production has led to an increase in plastic trash, which is expected to reach 353 MMT in 2019 (OECD, 2022), with roughly 22 MMT of that amount coming from poorly managed plastic garbage that seeped into the environment. Due to the physical and chemical diversity of MPs, effects on crops and soil ecosystems can be unpredictable. Thus, reducing this waste will be a significant issue in the coming years (Wang et al., 2022a).

Nano-plastics (NPs) are plastic particles smaller than 1 mm, and MPs are plastics smaller than 5 mm. In agricultural soils, microplastic abundances were substantially associated with soil characteristics (Piehl et al., 2018; Chen et al., 2013).

MPs were commonly detected in a variety of habitats, including freshwater, sediments, marine and soil (Cao et al., 2021).

A soil ecosystem's key functions and services may be significantly impacted if soil MNPs exceed a specific threshold (Zhou et al., 2020).

When crops are exposed to MP residues, the health and functioning of the soil ecosystems might be impacted, which poses hazards to the safety of the food chain. According to studies, MPs have the potential to have negative effects on aquatic creatures, ecosystems, and human health (Mateos-Cárdenas et al., 2021).

According to particle size, plastic waste is commonly divided into four subgroups: macro-plastics (>25 mm), meso-plastics (5–25 mm), micro-plastics (1–5 mm), and NPs (1 mm). Microplastics and macroplastics are new pollutants in agricultural environments (Reay et al., 2023).

However, these particular size ranges are not universally accepted and frequently serve as merely indicators and a starting point for future discussion (Hartmann et al., 2019).

The classification of micro-plastic and nano-plastic particles (MNPs) as primary or secondary particles is common. Primary MPs are purposefully produced in their size range, for example, for use in industrial processing, textile applications, or the fabrication of synthetic clothing (Browne et al., 2011; Lim, 2021).

In order to determine the average mass concentration of these MPs and NPs as well as the average removal efficiency of WWTPs, six polymer types that are commonly found in WWTPs were chosen. These polymer types include polymethyl methacrylate (PMMA), polypropylene (PP), polystyrene (PS), polyethylene (PE), polyethylene terephthalate (PET), and polyamide (PA) (Chang et al., 2023).

Municipal wastewater treatment plants (WWTPs) were recognized as significant urban sources releasing MPs despite the fact that there were numerous anthropogenic sources of MPs in the environment that were recorded. (Lusher, 2015; Guo et al., 2020).

Due to their small size, MPs like microbeads found in toothpaste, face scrubs, and other personal care items may pass wastewater treatment processes at WWTPs after traveling via the sewage system (Rillig, Ziersch & Hempel, 2017; Corradini et al., 2019).

Despite the fact that particulate matter treatment in WWTPs was reported to efficiently remove MPs, the rapid flux of wastewater means that a significant amount of MPs can still enter the aquatic environment.

Concerns have been expressed over agricultural practices that could pollute agricultural soils with MP (Zhang et al., 2020). Common agricultural practices, including the use of plastic water pipes for irrigation, plastic greenhouse covers, and plastic films for insulation and mulching, as well as the use of sewage sludge as fertilizer, are some potential sources of MPs contamination in agricultural areas (Steinmetz et al., 2016).

Recent investigations have shown conflicting effects on crop yield, soil health, and nutrient cycling, although degradable "bio-plastics" have been proposed as an alternative to plastic mulching (Xu et al., 2008; Browne et al., 2011).

Plasticizers and dispersants are substances added to plastic to increase its quality, change its plasticity, and reduce its viscosity (Rochman et al., 2014; de Souza Machado et al., 2018).

Plasticizers are substances that are chemically added to plastic in liquid or solid form in order to modify its physical properties. A broad set of phthalic acid (PA) esters, including butyl benzyl phthalate (BBP), dibutyl phthalate (DBP), and diethyl hexyl phthalate (DEHP), are together known as phthalate esters (PAEs).

It has been discovered that the chemical additions used to make MPs fatal to soil organisms (Prata et al., 2020).Since PAEs have a high hydrophobic nature and are easily adsorbable by SOM, their residues may be permanently left in the soil environment (Liu et al., 2014).

To prevent infections in animals, antibiotics are widely used in animal husbandry. However, after using antibiotics excessively, these substances end up in the soil through animal manure excretion. Previous studies have noted that the addition of MPs affects both the adsorption and breakdown of antibiotics. In order to increase plant growth and production, organic fertilizers are put to the soil of greenhouses, however, this practice pollutes the environment with a mix of MPs and antibiotics (Rocha-Santos & Duarte, 2015; Sun et al., 2018).

MPs are well known for acting as vectors in the soil environment, moving both organic and inorganic contaminants from one location to another. Microplastics have the potential to act as a significant vector of pathogens, especially if they have come into contact with a source of fecal contamination. However, the plastisphere can also contain bacterial and viral human pathogens capable of maintaining their virulence and infectivity.

The deadly toxins present in the environment are known as potentially toxic elements (PTEs). MPs have persistent and bio-accumulative qualities that make them susceptible to absorbing these environmental pollutants. Due to their enormous surface area and high hydrophobicity, MPs have the ability to adsorb PTEs on their outer surfaces, according to research (Peña, Rodríguez-Liébana, & Delgado-Moreno, 2023).

In this chapter we elaborate current understanding of MPs in agricultural soil and discuss the main types of MPs and the fate and transport of MPs in agriculture soil and further elaborate the interaction of MPs with agrochemicals.

6.2 BACKGROUND AND SIGNIFICANCE

Zhou et al. (2020) Bio-MPs degrade more quickly than conventional MPs, which causes bigger changes in microbial diversity and community structure. This could change methane emissions even further, with unforeseeable effects on the rate of global warming as a result. However, neither laboratory nor field investigations have yet to demonstrate the extent to which this might take place. Additionally, compared to ordinary MPs, bio-MPs have a high likelihood of converting into NPs, which could have a more harmful effect on plants.

According to Quilliam et al. (2023), plastics and MPs are now widely present in agricultural soils as a result of irrigation with sewage or contaminated surface water, plastic sheeting mulching, and the use of plastic carriers in seed coatings. Once exposed to the environment, plastic surfaces are quickly colonized by a diverse microbial community's microbial biofilm. This "plastisphere" community may also contain human diseases, especially if the plastic has been exposed to fecal contamination (such as from wastewater, organic manures, and cattle feces).

Yang et al. (2021) research has revealed MPs interaction with contaminants, effect soil function and health, and even the food chain. Microplastic contamination and its effects on the soil are influenced by its form, chemical composition, and environmental elements. We draw the conclusion that there are significant study gaps in the characterization and estimation of regional emissions of MPs in soil, variables influencing the concentration of MPs, and MPs masquerading as soil carbon storage, all of which require additional work.

According to Wang et al. (2020), the extensive and intensive use of plastics and the products developed from them has led to the pollution of the environment with plastic trash. MPs (5 mm) are fragments of large-scale plastic waste that have drawn increasing attention from the general public and scholarly groups around the world. Until recently, the majority of MPs research that was published in the literature concentrated on aquatic settings, particularly the marine environment, and there is very little knowledge on MPs contamination in terrestrial soil systems. Ecotoxicological impacts of soil microplastic contamination on terrestrial plants, soil flora, and soil microorganisms, as well as effects on soil physiochemical characteristics. Finally, we proposed a number of options for future study on MPs in soil ecosystems, based on the most recent advancements reported in this review.

Contrary to freshwater and marine habitats, agricultural soil contamination has only lately been the focus of research while accounting for a sizable portion (14%) of the world's plastic pollution. The vertical migration of plastic particles in agricultural soils and the enhanced pesticide transfer caused by plastic are both causes for worry (Horton et al., 2017; Tian et al., 2022).

Due to their pervasive contamination, MPs have become a major environmental concern since they could endanger human health and food security if they get into agricultural soils (Wanner, 2021).

Qi et al. (2020) state that it is still unclear how plastic affects terrestrial ecosystems. MPs (particles smaller than 5 mm) can enter agroecosystems either directly (through the application of biosolids, irrigation water, or atmospheric deposition) or indirectly (from the in situ decomposition of bigger plastic pieces, such as from plastic mulch films). Despite the fact that we have supported the use of plastics in agriculture for the past 50 years in an effort to increase resource efficiency and food security, the result has been that many soils are now heavily contaminated with plastic residue.

Because of the pervasive environmental pollution with MPs, both people and other biota are exposed. While the majority of studies exaggerate the direct toxicity of MPs, it is important to take into account the particle quantities, properties, and exposure settings employed in these assays. For instance, the practical significance of toxicity tests that employ doses over 100,000 times higher than those anticipated in the environment is minimal. As a result, the detrimental impacts of existing ambient concentrations on animal and human health are recognized as a knowledge gap (Prata et al., 2021)

Rillig et al. (2017) analyzed three soil layers that were each 3.5 cm deep and found that the presence of earthworms significantly increased the amount of microplastic particles at depth, with smaller PE microbeads having been transported downward to a larger extent. Earthworms can play a substantial role in the movement of MPs in soils, likely assimilating this material through casts, burrows (which might alter the hydraulics of the soil), egestion, and adhesion to the earthworm's exterior.

Ritter et al. (2022) reported that in 2015, 4.9 billion metric tons of plastic were produced, or 60% of all plastic ever made. Due to COVID-19, this growing tendency was reversed in 2020, when production fell to about 367 MMT, a 0.3% decline. However, due to the overproduction of facial masks, this rising trend has skyrocketed since the COVID. Because "single-use" packaging is more effective and chemically stable, consumers' preferences for it have accelerated and contaminated many ecosystems, damaging the environment and posing health hazards.

It has been shown through a number of studies that MPs significantly harm both public and environmental health. Due to the large number of MPs, it has been an ongoing problem that needs to be resolved as soon as possible. We anticipate that this evaluation will stimulate creative thought among scientists studying for MPs management.

6.3 MICROPLASTIC IN THE ENVIRONMENT

The primary focus of research was on the presence and potential consequences of MPs in aquatic ecosystems. However, the soil also serves as a sink for MPs, and current research has examined how MPs affect the ecosystem of the soil (Yang et al., 2019).

For better understanding of characteristics of MPs, it is necessary to have knowledge about MPs effects on the soil environment. Mulching processes leave behind plastic waste, which over time and through environmental factors are transformed into MPs. The generated MPs spread throughout the soil, where they interact with other pollutants, such as heavy metals, insecticides, and persistent organic pollutants (POPs), having cumulative harmful effects on soil flora and animals. Agricultural water runoff may eventually transport these MPs to rivers, oceans, and other water bodies, resulting in lake water contamination. The remaining plastic pieces from mulching, biosolids (sewage sludge), atmospheric deposition, municipal trash (compost), and plastic-coated fertilizers are the main sources of MPs in agricultural soil (Alimi, Farner, & Tufenkji, 2021).

The mode and amount of plastic dispersion into various environmental components is a critical issue requiring further research. Furthermore, the potential effects of degraded small chunks on the human body and the environment are of global concern (Allan et al., 2022).

Even nano-plastics derived from a single microplastic particle are expected to be a more complex issue around the world due to their small size, which allows them to pass through biological membranes easily (Yee et al., 2021).

Researchers must then create effective techniques for detecting these particles in nanoseconds to microseconds. There are several methods for detecting MPs, and some of these are visual techniques, spectroscopic techniques, and chromatographic techniques (Wagner et al., 2019).

In recent years, some research has emphasized the efficiency of microbes that break down plastic. *Ideonella sakaiensis*, on the other hand, is a bacterium that can degrade and consume plastic PET as a sole source of carbon and energy. It belongs to the genus *Ideonella* and the family *Comamonadaceae*. Similar to this, Gao and Sun isolated three different bacterial genera called *Exiguobacterium* sp., *Halomonas* sp., and *Ochrobactrum* sp. However, significantly, SEM observations on their research revealed the combination of *Exiguobacterium* sp., *Halomonas* sp., and *Ochrobactrum* sp. had a greater degradation efficiency on both PET and PE films as compared to single isolates (Gao & Sun 2021).

By educating people about the long-term effects of MPs pollution through programs and campaigns and encouraging them to reject, reduce, reuse, and recycle plastics, various governmental and non-governmental organizations and networks can lessen MPs pollution (Figure 6.1).

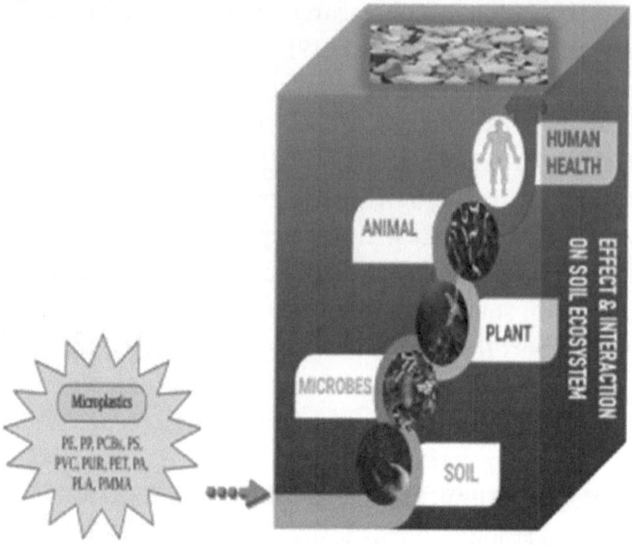

FIGURE 6.1 MPs effect & interaction on soil ecosystem.

6.4 DEFINITION AND TYPES OF MICROPLASTICS

Microplastics, as the name shows, are small pieces of plastics in lengths of less than 0.2 inches or 5 mm, smaller than the typical pearl used in jewelry, in terms of diameter. This definition frequently includes nanoscale (1 m) plastic particles. Carbon and hydrogen atoms are joined in polymer chains to form MPs. Microplastics frequently contain additional chemical additives, such as phthalates, polybrominated diphenyl ethers (PBDEs), and tetrabromobisphenol A (TBBPA), and many of these substances leak out of the plastics after they have been released into the environment (Kristanti et al., 2023).

Fragmented plastics have been accepted and given names based on their size, point of origin, and fragmentation process as a result of the extensive development of plastic materials. Studies have established their top size limitations of 1,000 or 100 nm for sub-scale plastics fragmentation, also referred to as "nano-plastics," by a number of researchers. Additionally, MPs are often known as tiny pieces of disintegrating plastic that are between 1 and 5,000 m in length (Anik et al., 2021).

Microplastics can be divided into two groups: primary and secondary. Potential sources of primary MPs include plastic pellets, scrubbers, commercial cleaning abrasives, plastic resin flakes, plastic powder, or fluff used to make plastic goods, as well as volatile particulate contaminants like micro-polyester, nano Fe_3O_4, and SiO_2 from printing toners (Sharma et al., 2021).

Similar to primary MPs, secondary MPs result from the subsequent breakdown of bigger polymers into nano, micro, and macro-sizes. The possibility for secondary MPs to be produced before being released into the environment owing to weathering includes exposure to wind, wave action, photo degradation, biodegradation, hydrolysis, and ultraviolet radiation from sunshine (Waraich & Srivastava, 2022).

In addition, tiny plastic fragments known as microbeads (size 10–500 mm) are patented for use in toothpaste and other personal care productsto exfoliate skin and remove dead skin cells (Anagnosti et al., 2021).

In order to create plastics with particular qualities (such as flexibility, roughness, resistance, and durability), multiple polymer types may be used to represent each microplastic shape. These polymer forms, however, are made up of several monomers, which can might be environmentally detrimental (Kole et al., 2017).

Plastics come in a variety of forms, and there are several types of polymers that are utilized to make plastics. However, six plastic categories dominate, including polyethylene (PE) (high and low density), polypropylene (PP), polyvinyl chloride (PVC), polystyrene (PS), polyurethane (PUR), and polyethylene terephthalate (PET) (Esterhuizen & Kim, 2022).

6.5 SOURCES OF MICROPLASTIC POLLUTION

The FAO (2021) projected that at least 12.5 MMT of plastics are used globally, with 10 MMT of this ascribed to agricultural and livestock production. However, the global use of plastic in agriculture is not thoroughly documented. A large portion of the plastic used in agriculture is single-use, which poses the risk of environmental persistence and slow degradation if improperly managed. Due to the paucity of analytical technologies that can quickly and affordably identify and quantify MPs in soils, especially smaller fractions below 5–10 mm, estimates of MnPs in soils are limited, with only a very small number of studies providing particle numbers or plastic mass (Caputo et al., 2021).

Thorough meta-study comparing global horticulture and agricultural sites on MPs in various soils. Based on 118 samples, they discovered that the amounts of microplastic in agricultural soils ranged from 1 to 530,000 particles/kg dry weight, with an average concentration of roughly 1,200 particles/kg (Büks & Kaupenjohann, 2020).

The methods employed for sampling, polymer extraction, and identification seem to have a significant impact on concentrations, which are also highly varied. With different paths depending on agricultural practices, urban influence, and hydrometeorological causes, plastic permeates agricultural soils in all size ranges. Microplastics can come from a variety of sources, including air deposition, tire wear from agricultural machines and cars, flooding incidents, and regulated aquifer recharge. Using sewage sludge or other compost as fertilizer, irrigating with wastewater, applying agrochemicals, and planting seeds that are frequently coated with polymers are other ways that MPs are directly administered to agricultural land (Yang et al., 2021).

MP emissions to soils are significantly increased by current farming methods. About 75% of all plastic used in agriculture is used as plastic mulch films, which are heavily used to increase agricultural yields and water efficiency (Qian et al., 2018).

A source of MPs in agricultural soils can also be thought of as machine wear and tear. This covers tire wear abrasion as well as abrasion of other plastic machine parts that occur while the machine is in use. Additionally, because tire wear products frequently consist of tiny particles in the low micrometer to nanometer range, road runoff from surrounding roads and air deposition have the potential to collect in agricultural areas. For instance, calculated that 67% of tire particles discharged into the environment in the Netherlands end up in the soil, making this a potentially significant source of soil MPs (Kole et al., 2017).

Plastic films with PAEs or other chemical additions that have been demonstrated to negatively impact the composition of microbial communities and soil enzymatic activity present additional issues (Liu et al., 2014).

MPs can also be purposely added to agricultural soil, for example, by using commercially produced chemical fertilizers and insecticides, which are frequently encased in a polymer shell (Weithmann et al., 2018).

Irrigation with both freshwater and wastewater, depending on the irrigation and drainage network design, is another source of MP for agricultural soil (Yang et al., 2021).

MnPs that build up in agricultural soil have the potential to be transported deeper into the groundwater over time and may consequently be present as a developing pollutant, especially in open and shallow aquifers. Direct MP intrusion into aquifers beneath agricultural areas has also lately come to light.

6.6 ENVIRONMENTAL IMPACTS OF MICROPLASTICS

Plastic particles smaller than 5 mm in size, known as "microplastics," have become a serious problem in the natural world. They may be released directly from sources like cosmetics and textiles, or they might originate in the breakdown of bigger plastic products (Yurtsever, 2019).

Microplastics have far-reaching ecological implications because they permeate both terrestrial and aquatic ecosystems once released into the environment (Huang et al., 2022). They are taken in by marine life, move up the food chain, and end up in the bellies of people who eat seafood (Mishra, Charan Rath, & Das, 2019).

This poses a risk to marine life, and there are also worries about the possible spread of harmful compounds linked to these particles (Campanale, Massarelli, Savino, Locaputo, & Uricchio, 2020).

Microplastics may be harmful to soil health and plant development on land, which might reduce agricultural output (Wang et al., 2019; Yang et al., 2020). Even worse, they may function as vectors for dangerous chemicals, transporting anything from pesticides to heavy metals (Briffa, Sinagra, & Blundell, 2020).

Our society's reliance on plastic has far-reaching and sometimes unnoticed consequences, and the spread of MPs in the environment serves as a sobering reminder of this (Franklin, Gavins, & Mehl, 2022).

6.7 FATE AND TRANSPORT OF MICROPLASTICS IN SOIL

Physical, chemical, and biological processes all have a role in the long-term destiny and movement of MPs in soil (Horton & Dixon, 2018).

These little bits of plastic, once added to the soil, may travel through the soil profiles via a number of different means. Microplastics may be spread far and wide by physical factors such as rainfall runoff and wind erosion (Rezaei, Riksen, Sirjani, Sameni, & Geissen, 2019).

Microplastics in the soil matrix may be mixed and redistributed by bioturbation by soil-dwelling creatures like earthworms and insects (Astner et al., 2023).

Microplastics' stability and mobility are affected by the chemical interactions they have with soil components such as organic matter and minerals. Some MPs may remain on the soil surface while others get trapped in soil aggregates and become immobile (Hale, Seeley, La Guardia, Mai, & Zeng, 2020).

May be degraded by microorganisms over time, although slowly. Soil type, climate, and land use practices are just a few of the variables that might affect how MPs behave in soil, highlighting the need of conducting in-depth studies to learn more about the environmental effects of these contaminants (Yuan et al., 2020; Zhai, Zhang, & Yu, 2023).

6.8 DEPOSITION AND SOURCES OF MICROPLASTICS IN AGRICULTURAL SOIL

Many different routes of deposition, from both direct and indirect sources, allow MPs to enter agricultural soil (Büks & Kaupenjohann, 2020).

Organic additives such as composts and manures may include microplastic pollutants from animal feed or plastic-based agricultural commodities (Prata et al., 2019).

Microplastics that have been carried by the wind from maritime ecosystems end up on land after being blown there from urban and industrial locations. The widespread use of plastic mulches in agriculture for weed control and water conservation may potentially contribute to microplastic pollution (Shah & Wu, 2020).

Microplastics may also enter the soil ecosystem via the decomposition of bigger plastic detritus including, films, irrigation tubing, and packing materials (Rose et al., 2023).

Microplastic pollution in agricultural soil has several causes, all of which must be addressed for the sake of protecting soil health and agricultural production (Stubenrauch & Ekardt, 2020).

6.9 TRANSPORT PATHWAYS OF MICROPLASTICS IN SOIL

The various transport routes used by MPs through soil are affected by different physical and biological factors. Microplastics, especially those with a coarse texture and high permeability, tend to flow downhill through the soil profile due to gravity settling (Boanini, 2021; Alimi et al., 2021).

Vertical advection describes the movement of MPs across an environment. Second, water movement within the soil matrix may force MPs laterally (Waldschläger et al., 2022).

Soil-dwelling creatures like earthworms and insects may also play a role in the redistribution of MPs, moving them either horizontally or vertically within the soil via a process called bioturbation (Xu et al., 2020).

Additionally, aggregates may entrap MPs, preventing them from moving (Pérez-Reverón et al., 2023).

6.10 ACCUMULATION OF MICROPLASTICS IN AGRICULTURAL SOIL

The rising problem of MPs in farmland soil has serious consequences. Tiny pieces of plastic may make their way into the soil and build up over time, usually as a result of a combination of agricultural practices and environmental factors (Itelima, Bang, Onyimba, Sila, & Egbere, 2018).

Because of their little size, they are capable of penetrating soil pores and ultimately being rooted in the ground (Huang et al., 2022).

Furthermore, organic supplements such as composts or manures are often employed in agriculture and might contribute to the accumulation of MPs since they may include microplastic contamination (Urra, Alkorta, & Garbisu, 2019; Zhao et al., 2023).

Plastic mulches, often used for weed control and water conservation, may also accumulate MPs (Steinmetz et al., 2016).

6.11 FACTORS INFLUENCING MICROPLASTIC ACCUMULATION IN SOIL

Many complex elements affect how much MPs accumulate in soil. Sandier soils, which have more permeability, may allow MPs to penetrate deeper than finer, clayey soils (Martins & Marto, 2023).

Microplastics may be introduced into the soil by land use practices, such as the use of organic amendments like composts or manures, and accumulate over time (Corradini, Casado, Leiva, Huerta-Lwanga, & Geissen, 2021).

Plastic mulches and films used in farming are another potential source of microplastic contamination (Huang, Liu, Jia, Yan, & Wang, 2020).

Extreme weather events, such as torrential downpours, may cause runoff, which can transport MPs to unaffected regions (Wakkaf et al., 2022).

Soil-dwelling creatures' digging may recaster MPs across the soil profile, among other biological activities. Soil's interaction and retention of MPs are also affected by the presence of organic matter and minerals (Ivanic et al., 2023).

It is imperative that we take a systems-level approach to learning about and fixing this environmental problem, since the patterns of microplastic buildup in soil are shaped by a complex web of interactions between natural and anthropogenic elements (Dechesne et al., 2018) (Figure 6.2).

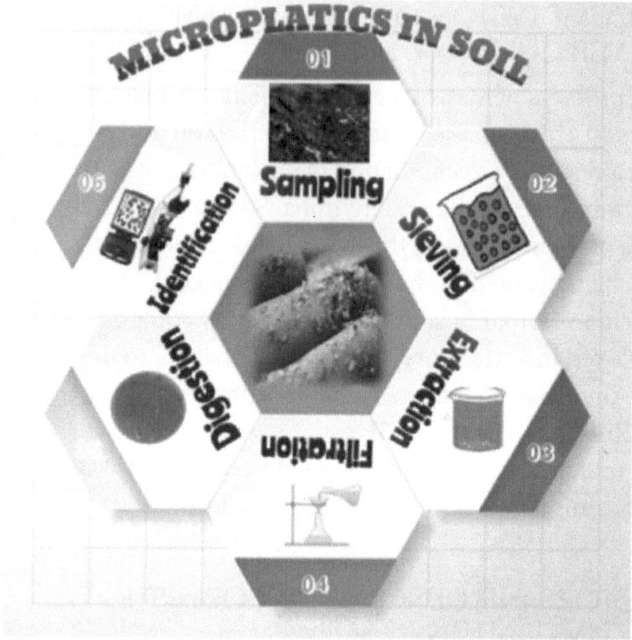

FIGURE 6.2 Microplastic buildup in soil.

6.12 MEASUREMENT TECHNIQUES FOR MICROPLASTIC CONTENT IN SOIL

Microplastic content in soil is measured using a variety of methods, each of which is optimized for detecting and quantifying these very small particles. Microplastics may be visually identified by optical microscopy due to their varied sizes, colors, and refractive indices (Li et al., 2023).

However, the size of particles that may be detected with this method may be too tiny for certain applications (Bläsing & Amelung, 2018; Tarafdar, Sharma, & Raliya, 2013).

Using the distinct absorption spectra of plastics, Fourier-transform infrared spectroscopy (FTIR) provides a chemical identification method (Comnea-Stancu, Wieland, Ramer, Schwaighofer, & Lendl, 2017).

Raman spectroscopy is a comparable technique that uses the scattering of laser light to identify the particle's composition (Qin, Chao, & Kim, 2010).

The chemical make-up of MPs may also be investigated by Pyrolysis Gas Chromatography Mass Spectrometry (Py-GCMS) (El Hayany, Rumpel, Hafidi, & El Fels, 2022).

High-resolution imaging and elemental analysis are available with more modern techniques like scanning electron microscopy (SEM) combined with energy-dispersive X-ray spectroscopy (EDS) (Hodoroaba, 2020).

6.13 SPATIAL DISTRIBUTION OF MICROPLASTICS IN AGRICULTURAL SOIL

There is a complicated web of circumstances that determines where MPs end up in farmland. Research has demonstrated that their distribution within a field may be very variable, which is consistent with the heterogeneous character of soil and the many potential contamination pathways (Binley et al., 2015; Li et al., 2020).

High microplastic concentrations have been found in places where organic amendments like composts and manures have been used, suggesting that this is a localized source of pollution (Zhang et al., 2022).

In addition, the use of plastic-based products in farming (such as mulches and films) might result in localized accumulations of MPs close to the soil's surface. Bioturbators, such as earthworms and insects, might further affect microplastic dispersal via their burrowing activities (Huerta Lwanga et al., 2016; Milosevic et al., 2024). In order to establish effective management techniques and evaluate the possible dangers posed to soil health and ecosystems, it is essential to have a firm grasp on the geographical distribution of MPs in agricultural soil (Zhou et al., 2020).

6.14 INTERACTION BETWEEN MICROPLASTICS AND AGROCHEMICALS

There are several ways in which MPs and agrochemicals might harm the ecosystem. Agrochemicals like pesticides and herbicides may be transported and stored in MPs because of their large surface area and chemical make-up (Klátyik et al., 2023).

Microplastics might pick up these substances from the air or have them enclosed in their structure (Verla, Enyoh, Verla, & Nwarnorh, 2019).

This interaction has the potential to have unexpected implications for environmental and agricultural systems by exposing soil-dwelling organisms for longer than necessary (Baho, Bundschuh, & Futter, 2021).

For instance, agrochemicals' effectiveness may be diminished and perhaps have unintended consequences if MPs were present in soil. Soil organisms may experience additive or counteracting impacts from the presence of MPs and agrochemicals, which in turn affects the organisms' behavior, reproduction, and general health (de Souza Machado et al., 2019; Oliveira et al., 2023; Subaramaniyam et al., 2023).

6.15 FATE AND BEHAVIOR OF AGROCHEMICALS IN SOIL

Agrochemical fate and behavior in soil are dynamic processes influenced by many variables. Agrochemicals, once applied, can adsorb onto soil particles and then desorb back into the soil solution, among other possible reactions (Yavari et al., 2020; Shove, 2022).

Furthermore, agrochemicals can be degraded or transformed into metabolites via chemical reactions and microbial activity, with potentially varying effects on the environment (Waldschläger et al., 2020; Ahmad et al., 2021).

Soil properties such as texture, organic matter content, and moisture levels all play a role in the distribution of agrochemicals (Wang, Cheng, Naidu, & Bowman, 2021).

Agrochemicals may leach more easily from coarse soils and enter the groundwater table, while finer soils may retain the chemicals and cause them to accumulate close to the surface (Artiola, Walworth, Musil, & Crimmins, 2019).

The presence of living organisms, such as soil microbes and plant roots, can also play a role in how agrochemicals are broken down in the environment (X. Wang et al., 2022).

6.16 ADSORPTION AND DESORPTION OF AGROCHEMICALS ON MICROPLASTICS

The interaction between agrochemicals and MPs in soil is heavily influenced by adsorption and desorption processes (Zhang et al., 2021).

Due to their high surface area and potential for chemical attraction, agrochemicals can cling to microplastic particles when they are introduced into the soil (Mubeen et al., 2023).

Effective immobilization of agrochemicals on the microplastic substrate is achieved through an adsorption phenomenon (Ricardo et al., 2021).

The rate at which ions are desorbed depends on a number of variables, including pH, temperature, and the presence of competing ions (Kończak & Huber, 2022).

Natural organic substances may compete with agrochemicals for binding sites on MPs, so the organic matter content of soil can also affect these processes (Pignatello & Nason, 2020).

6.16.1 Impacts of Microplastics on Agrochemical Availability & Mobility

Soil agrochemical mobility and availability are significantly impacted by MPs. The interaction between agrochemicals and MPs is complicated because the chemicals can adsorb onto the plastic surfaces, changing the MPs' properties (Moeck, Davies, Krause, & Schneidewind, 2023).

Agrochemicals' ability to move around in the soil and leach into groundwater or be transported off-site can be diminished by adsorption (Khalid et al., 2020; Li, Song & Cai, (2020).

On the other side, it can increase the persistence of agrochemicals in soil, which could increase their usefulness for a longer period of time (Mandal et al., 2020).

More than that, MPs can act as carriers, getting agrochemicals where they need to go in the soil (Šunta, Prosenc, Trebše, Bulc, & Kralj, 2020).

6.16.2 Effects of Microplastics on Soil Health and Crop Growth

The effects of MPs on soil quality and crop yields are complex. Their presence can change the soil's physical properties, decreasing its fertility and possibly preventing water retention and aeration (Brodhagen et al., 2017; de Jesus Duarte, Glaser, & Pellegrino Cerri, 2019).

Soil microbiome disruption by MPs has the potential to endanger beneficial soil organisms and interfere with vital nutrient cycling processes (Q. Wang, Adams, Wang, Sun, & Zhang, 2022).

6.17 EFFECTS OF MICROPLASTICS ON PLANT GROWTH AND NUTRIENT UPTAKE MITIGATION STRATEGIES AND FUTURE PERSPECTIVES

Microplastics, tiny plastic particles less than 5 mm in size, can have adverse effects on plant growth and nutrient uptake. They obstruct root development, alter nutrient balance, and can release toxic substances, hampering plant health. To mitigate these effects, reducing microplastic pollution and improving soil management are crucial. Bioremediation and the use of biodegradable alternatives can also help. Future perspectives include continued research, regulatory measures, public awareness, innovative solutions, and integrated approaches to agriculture, with the aim of minimizing the impact of MPs on plant ecosystems and ensuring long-term sustainability (Figure 6.3).

6.18 CONCLUSION

To address these challenges and uncertainties, future research is essential. A better understanding of the fate, transport, and behavior of MPs in agricultural soil, as well as their interactions with agrochemicals, will enable the development of effective mitigation strategies and regulatory measures. Furthermore, promoting sustainable agricultural practices that

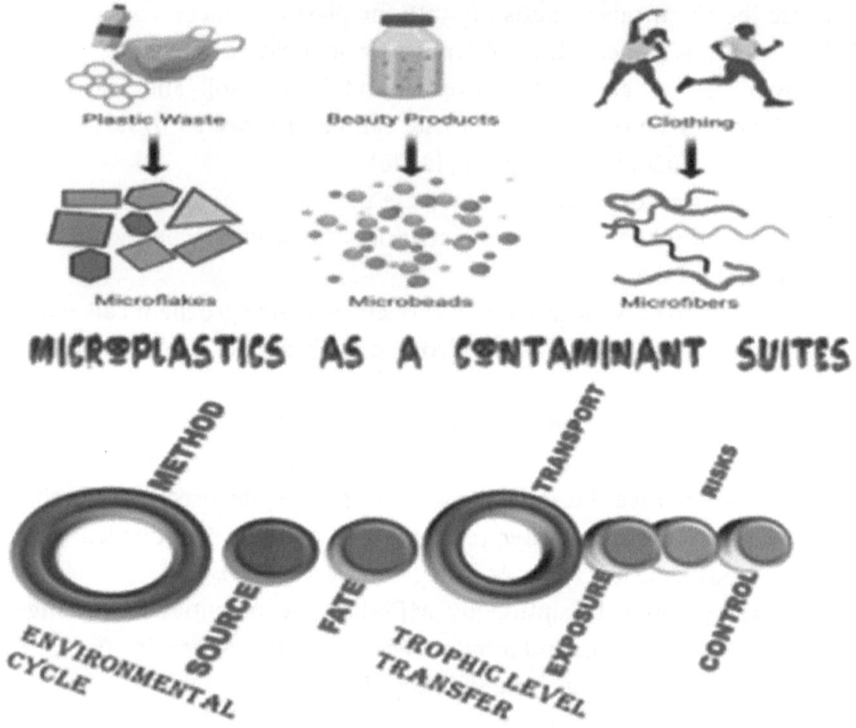

FIGURE 6.3 Microplastics in soil.

minimize the use of plastics and enhance soil health is critical to reducing the long-term impact of MPs in agricultural ecosystems. By addressing these issues, we can work toward preserving the health and productivity of agricultural soil while minimizing the environmental consequences of microplastic contamination.

REFERENCES

Ahmad, U. M., Ji, N., Li, H., Wu, Q., Song, C., Liu, Q., Ma, D., & Lu, X. (2021). Can lignin be transformed into agrochemicals? Recent advances in the agricultural applications of lignin. *Industrial Crops and Products*, 170: 113646.

Alimi, O. S., Farner, J. M., & Tufenkji, N. (2021). Exposure of nanoplastics to freeze-thaw leads to aggregation and reduced transport in model groundwater environments. *Water Research*, 189: 116533.

Allan, I. J., Samanipour, S., Manoli, K., Gigault, J., & Fatta-Kassinos, D. (2022). Examining the relevance of the microplastic-associated additive fraction in environmental compartments. *ACS ES&T Water*, 2(3): 405–413.

Anagnosti, L., Varvaresou, A., Pavlou, P., Protopapa, E., & Carayanni, V. (2021). Worldwide actions against plastic pollution from microbeads and microplastics in cosmetics focusing on European policies. Has the issue been handled effectively? *Marine Pollution Bulletin*, 162: 111883.

Anik, A. H., Hossain, S., Alam, M., Sultan, M. B., Hasnine, M. T., & Rahman, M. M. (2021). Microplastics pollution: A comprehensive review on the sources, fates, effects, and potential remediation. *Environmental Nanotechnology, Monitoring & Management*, 16:100530.

Artiola, J., Walworth, J., Musil, S., & Crimmins, M. (2019). Soil and land pollution. In M. L. Brusseau, I. L. Pepper, C. P. Gerba (Eds.), *Environmental and Pollution Science* (pp. 219–235). Elsevier.

Astner, A. F., Gillmore, A. B., Yu, Y., Flury, M., DeBruyn, J. M., Schaeffer, S. M., & Hayes, D. G. (2023). Formation, behavior, properties and impact of micro-and nanoplastics on agricultural soil ecosystems (A Review). *NanoImpact*, 31: 100474.

Baho, D. L., Bundschuh, M., & Futter, M. N. (2021). Microplastics in terrestrial ecosystems: Moving beyond the state of the art to minimize the risk of ecological surprise. *Global Change Biology*, 27(17): 3969–3986.

Binley, A., Hubbard, S. S., Huisman, J. A., Revil, A., Robinson, D. A., Singha, K., & Slater, L. D. (2015). The emergence of hydrogeophysics for improved understanding of subsurface processes over multiple scales. *Water Resources Research*, 51(6): 3837–3866.

Bläsing, M., & Amelung, W. (2018). Plastics in soil: Analytical methods and possible sources. *Science of the Total Environment*, 612: 422–435.

Boanini, C. (2021). GEAM–Geoingegneria Ambientale e Mineraria Rivista dell'Associazione Georisorse e Ambiente Anno LVIII, n. II/III, dicembre 2021 (163/164).

Briffa, J., Sinagra, E., & Blundell, R. (2020). Heavy metal pollution in the environment and their toxicological effects on humans. *Heliyon*, 6(9): e04691.

Brodhagen, M., Goldberger, J. R., Hayes, D. G., Inglis, D. A., Marsh, T. L., & Miles, C. (2017). Policy considerations for limiting unintended residual plastic in agricultural soils. *Environmental Science & Policy*, 69: 81–84.

Browne, M. A., Crump, P., Niven, S. J., Teuten, E., Tonkin, A., Galloway, T., & Thompson, R. (2011). Accumulation of microplastic on shorelines worldwide: Sources and sinks. *Environmental Science & Technology*, 45(21): 9175–9179.

Büks, F., & Kaupenjohann, M. (2020). Global concentrations of microplastic in soils, a review. *Soil Discussions*, 2020: 1–26.

Campanale, C., Massarelli, C., Savino, I., Locaputo, V., & Uricchio, V. F. (2020). A detailed review study on potential effects of microplastics and additives of concern on human health. *International Journal of Environmental Research and Public Health*, 17(4): 1212.

Cao, Y., Zhao, M., Ma, X., Song, Y., Zuo, S., Li, H., & Deng, W. (2021). A critical review on the interactions of microplastics with heavy metals: Mechanism and their combined effect on organisms and humans. *Science of the Total Environment*, 788: 147620.

Caputo, F., Vogel, R., Savage, J., Vella, G., Law, A., Della Camera, G., Hannon, G., Peacock, B., Mehn, D., & Ponti, J. (2021). Measuring particle size distribution and mass concentration of nanoplastics and microplastics: Addressing some analytical challenges in the sub-micron size range. *Journal of Colloid and Interface Science*, 588: 401–417.

Chang, J., Liang, J., Fang, W., Zhang, H., Zhang, Y., Zhao, H. & Zhang, G. (2023). Adsorption behaviors and bioavailability of tetrabromobisphenol A in the presence of polystyrene microplastic in soil: Effect of microplastics aging. *Environmental Pollution*, 334, 122156.

Chen, Y., Wu, C., Zhang, H., Lin, Q., Hong, Y., & Luo, Y. (2013). Empirical estimation of pollution load and contamination levels of phthalate esters in agricultural soils from plastic film mulching in China. *Environmental Earth Sciences*, 70: 239–247.

Comnea-Stancu, I. R., Wieland, K., Ramer, G., Schwaighofer, A., & Lendl, B. (2017). On the identification of rayon/viscose as a major fraction of microplastics in the marine environment: discrimination between natural and manmade cellulosic fibers using Fourier transform infrared spectroscopy. *Applied Spectroscopy*, 71(5): 939–950.

Corradini, F., Casado, F., Leiva, V., Huerta-Lwanga, E., & Geissen, V. (2021). Microplastics occurrence and frequency in soils under different land uses on a regional scale. *Science of the Total Environment*, 752: 141917.

Corradini, F., Meza, P., Eguiluz, R., Casado, F., Huerta-Lwanga, E., & Geissen, V. (2019). Evidence of microplastic accumulation in agricultural soils from sewage sludge disposal. *Science of the Total Environment*, 671: 411–420.

de Jesus Duarte, S., Glaser, B., & Pellegrino Cerri, C. E. (2019). Effect of biochar particle size on physical, hydrological and chemical properties of loamy and sandy tropical soils. *Agronomy*, 9(4): 165.

de Souza Machado, A. A., Kloas, W., Zarfl, C., Hempel, S., & Rillig, M. C. (2018). Microplastics as an emerging threat to terrestrial ecosystems. *Global Change Biology*, 24(4): 1405–1416.

de Souza Machado, A. A., Lau, C. W., Kloas, W., Bergmann, J., Bachelier, J. B., Faltin, E., Becker, R., Görlich, A. S., & Rillig, M. C. (2019). Microplastics can change soil properties and affect plant performance. *Environmental Science & Technology*, 53(10): 6044–6052.

Dechesne, A., Li, L., He, Z., Nesme, J., Quintela-Baluja, M., Balboa, S., Romalde, J. L., Graham, D., Sørensen, S. J., & Smets, B. F. (2018). Patterns of permissiveness towards broad host range plasmids in microbial communities across the urban water cycle in Europe. *17th International Symposium on Microbial Ecology*, Leipzig, Germany.

Esterhuizen, M., & Kim, Y. J. (2022). Effects of polypropylene, polyvinyl chloride, polyethylene terephthalate, polyurethane, high-density polyethylene, and polystyrene microplastic on Nelumbo nucifera (Lotus) in water and sediment. *Environmental Science and Pollution Research*, *29*(12): 17580–17590.

El Hayany, B., Rumpel, C., Hafidi, M., & El Fels, L. (2022). Occurrence, analysis of microplastics in sewage sludge and their fate during composting: A literature review. *Journal of Environmental Management*, 317: 115364.

FAO. 2021. Sustainable use and management of plastics in agriculture. https://www.fao.org/climate-change/areas-of-work/plastics-in-agriculture/en

Franklin, E., Gavins, J., & Mehl, S. (2022). "I don't think education is the answer": A corpus-assisted ecolinguistic analysis of plastics discourses in the UK. *Journal of World Languages*, 8(2): 284–322.

Gao, R., & Sun, C. (2021). A marine bacterial community capable of degrading poly (ethylene terephthalate) and polyethylene. *Journal of Hazardous Materials*, 416: 125928.

Guo, J.-J., Huang, X.-P., Xiang, L., Wang, Y.-Z., Li, Y.-W., Li, H., Cai, Q.-Y., Mo, C.-H., & Wong, M.-H. (2020). Source, migration and toxicology of microplastics in soil. *Environment International*, 137: 105263.

Hale, R. C., Seeley, M. E., La Guardia, M. J., Mai, L., & Zeng, E. Y. (2020). A global perspective on microplastics. *Journal of Geophysical Research: Oceans*, 125(1): e2018JC014719.

Hartmann, S., Louch, R., Zeumer, R., Steinhoff, B., Mozhayeva, D., Engelhard, C. & Witte, K. (2019). Comparative multi-generation study on long-term effects of pristine and wastewater-borne silver and titanium dioxide nanoparticles on key lifecycle parameters in Daphnia magna. *NanoImpact*, 14: 100163.

Hodoroaba, V.-D. (2020). Energy-dispersive X-ray spectroscopy (EDS). In V.-D. Hodoroaba, W. E. S. Unger, A. G. Shard (Eds.), *Characterization of Nanoparticles* (pp. 397–417). Elsevier.

Horton, A. A., & Dixon, S. J. (2018). Microplastics: An introduction to environmental transport processes. *Wiley Interdisciplinary Reviews: Water*, 5(2): e1268.

Horton, A. A., Walton, A., Spurgeon, D. J., Lahive, E., & Svendsen, C. (2017). Microplastics in freshwater and terrestrial environments: Evaluating the current understanding to identify the knowledge gaps and future research priorities. *Science of the Total Environment*, 586: 127–141.

Huang, D., Chen, H., Shen, M., Tao, J., Chen, S., Yin, L., Zhou, W., Wang, X., Xiao, R., & Li, R. (2022). Recent advances on the transport of microplastics/nanoplastics in abiotic and biotic compartments. *Journal of Hazardous Materials*, 438: 129515.

Huang, Y., Liu, Q., Jia, W., Yan, C., & Wang, J. (2020). Agricultural plastic mulching as a source of microplastics in the terrestrial environment. *Environmental Pollution*, 260: 114096.

Huerta Lwanga, E., Gertsen, H., Gooren, H., Peters, P., Salánki, T., Van Der Ploeg, M., Besseling, E., Koelmans, A. A., & Geissen, V. (2016). Microplastics in the terrestrial ecosystem: Implications for Lumbricus terrestris (Oligochaeta, Lumbricidae). *Environmental Science & Technology*, 50(5): 2685–2691.

Itelima, J., Bang, W., Onyimba, I., Sila, M., & Egbere, O. (2018). Bio-fertilizers as key player in enhancing soil fertility and crop productivity: A review. *Direct Research Journal of Agriculture and Food Science*, 6: 73–83.

Ivanic, F. M., Guggenberger, G., Woche, S. K., Bachmann, J., Hoppe, M., & Carstens, J. F. (2023). Soil organic matter facilitates the transport of microplastic by reducing surface hydrophobicity. *Colloids and Surfaces A: Physicochemical and Engineering Aspects*, 676: 132255.

Khalid, S., Shahid, M., Murtaza, B., Bibi, I., Naeem, M. A., & Niazi, N. K. (2020). A critical review of different factors governing the fate of pesticides in soil under biochar application. *Science of the Total Environment*, 711: 134645.

Klátyik, S., Simon, G., Oláh, M., Mesnage, R., Antoniou, M. N., Zaller, J. G., & Székács, A. (2023). Terrestrial ecotoxicity of glyphosate, its formulations, and co-formulants: Evidence from 2010–2023. *Environmental Sciences Europe*, 35(1): 1–29.

Kole, P. J., Löhr, A. J., Van Belleghem, F. G., & Ragas, A. M. (2017). Wear and tear of tyres: A stealthy source of microplastics in the environment. *International Journal of Environmental Research and Public Health*, 14(10): 1265.

Kończak, M., & Huber, M. (2022). Application of the engineered sewage sludge-derived biochar to minimize water eutrophication by removal of ammonium and phosphate ions from water. *Journal of Cleaner Production*, 331: 129994.

Kristanti, R. A., Hadibarata, T., Wulandari, N. F., Sibero, M. T., Darmayati, Y., & Hatmanti, A. (2023). Overview of microplastics in the environment: type, source, potential effects and removal strategies. *Bioprocess and Biosystems Engineering*, 46(3): 429–441.

Li, J., Liu, H., Liao, R., Wang, H., Chen, Y., Xiang, J., Xu, X., & Ma, H. (2023). Recognition of microplastics suspended in seawater via refractive index by Mueller matrix polarimetry. *Marine Pollution Bulletin*, 188: 114706.

Li, J., Song, Y., & Cai, Y. (2020). Focus topics on microplastics in soil: Analytical methods, occurrence, transport, and ecological risks. *Environmental Pollution*, 257: 113570.

Li, L., Luo, Y., Li, R., Zhou, Q., Peijnenburg, W. J., Yin, N., Yang, J., Tu, C., & Zhang, Y. (2020). Effective uptake of submicrometre plastics by crop plants via a crack-entry mode. *Nature Sustainability*, 3(11): 929–937.

Li, X., Chen, L., Mei, Q., Dong, B., Dai, X., Ding, G., & Zeng, E. Y. (2018). Microplastics in sewage sludge from the wastewater treatment plants in China. *Water Research*, 142: 75–85.

Lim, X. (2021). Microplastics are everywhere—But are they harmful. *Nature*, 593(7857): 22–25.

Liu, E., He, W., & Yan, C. (2014). 'White revolution' to 'white pollution'—Agricultural plastic film mulch in China. *Environmental Research Letters*, 9(9): 091001.

Lusher, A. (2015). Microplastics in the marine environment: Distribution, interactions and effects. In M. Bergmann, L. Gutow, M. Klages (Eds.), *Marine Anthropogenic Litter* (pp. 245–307). Springer.

Mandal, A., Sarkar, B., Mandal, S., Vithanage, M., Patra, A. K., & Manna, M. C. (2020). Impact of agrochemicals on soil health. In M. N. V. Prasad (Ed.), *Agrochemicals Detection, Treatment and Remediation* (pp. 161–187). Elsevier.

Martins, A. M., & Marto, J. M. (2023). A sustainable life cycle for cosmetics: From design and development to post-use phase. *Sustainable Chemistry and Pharmacy*, 35: 101178.

Mateos-Cárdenas, A., van Pelt, F. N., O'Halloran, J., & Jansen, M. A. (2021). Adsorption, uptake and toxicity of micro-and nanoplastics: Effects on terrestrial plants and aquatic macrophytes. *Environmental Pollution*, 284: 117183.

Milosevic, D., Stanković, J., Baranov, V., Jovanović, B., Krause, S., & Csabai, Z. (2024). Beneath the surface: Decoding the impact of Chironomus riparius bioturbation on microplastic dispersion in sedimentary matrix. *Science of the Total Environment*, 919: 170844.

Mishra, S., Charan Rath, C., & Das, A. P. (2019). Marine microfiber pollution: A review on present status and future challenges. *Marine Pollution Bulletin*, 140: 188–197.

Moeck, C., Davies, G., Krause, S., & Schneidewind, U. (2023). Microplastics and nanoplastics in agriculture—A potential source of soil and groundwater contamination? *Grundwasser*, 28(1): 23–35.

Mubeen, B., Hasnain, A., Wang, J., Zheng, H., Naqvi, S. A. H., Prasad, R., Rehman, A. U., Sohail, M. A., Hassan, M. Z., & Farhan, M. (2023). Current progress and open challenges for combined toxic effects of manufactured nano-sized objects (MNO's) on soil biota and microbial community. *Coatings*, 13(1): 212.

OECD. (2022). Economic survey: Guidelines for multinational enterprises and tax administrations. https://www.oecd.org/content/dam/oecd/en/publications/reports/2022/01/oecd-transfer-pricing-guidelines-for-multinational-enterprises-and-tax-administrations-2022_57104b3a/0e655865-en.pdf

Oliveira, A. M., Silva, A. L. P., Soares, A. M., Barceló, D., Duarte, A. C., & Rocha-Santos, T. (2023). Current knowledge on the presence, biodegradation, and toxicity of discarded face masks in the environment. *Journal of Environmental Chemical Engineering*, 11(2): 109308.

Peña, A., Rodríguez-Liébana, J. A., & Delgado-Moreno, L. (2023). Interactions of microplastics with pesticides in soils and their ecotoxicological implications. *Agronomy*, 13(3): 701.

Pérez-Reverón, R., Álvarez-Méndez, S. J., González-Sálamo, J., Socas-Hernández, C., Díaz-Peña, F. J., Hernández-Sánchez, C., & Hernández-Borges, J. (2023). Nanoplastics in the soil environment: Analytical methods, occurrence, fate and ecological implications. *Environmental Pollution*, 317: 120788.

Piehl, S., Leibner, A., Löder, M. G., Dris, R., Bogner, C., & Laforsch, C. (2018). Identification and quantification of macro-and microplastics on an agricultural farmland. *Scientific Reports*, 8(1): 17950.

Pignatello, J. J., & Nason, S. L. (2020). Importance of soil properties and processes on bioavailability of organic compounds. In J. J. Ortega-Calvo, J. R. Parsons (Eds.), *Bioavailability of Organic Chemicals in Soil and Sediment* (pp. 7–41). Springer.

Prata, J. C., da Costa, J. P., Lopes, I., Andrady, A. L., Duarte, A. C., & Rocha-Santos, T. (2021). A One Health perspective of the impacts of microplastics on animal, human and environmental health. *Science of the Total Environment*, 777: 146094.

Prata, J. C., da Costa, J. P., Lopes, I., Duarte, A. C., & Rocha-Santos, T. (2020). Environmental exposure to microplastics: An overview on possible human health effects. *Science of the Total Environment*, 702: 134455.

Prata, J. C., Silva, A. L. P., Da Costa, J. P., Mouneyrac, C., Walker, T. R., Duarte, A. C., & Rocha-Santos, T. (2019). Solutions and integrated strategies for the control and mitigation of plastic and microplastic pollution. *International Journal of Environmental Research and Public Health*, 16(13): 2411.

Qi, R., Jones, D. L., Li, Z., Liu, Q., & Yan, C. (2020). Behavior of microplastics and plastic film residues in the soil environment: A critical review. *Science of the Total Environment*, 703: 134722.

Qian, H., Zhang, M., Liu, G., Lu, T., Qu, Q., Du, B., & Pan, X. (2018). Effects of soil residual plastic film on soil microbial community structure and fertility. *Water, Air, & Soil Pollution*, 229: 1–11.

Qin, J., Chao, K., & Kim, M. S. (2010). Raman chemical imaging system for food safety and quality inspection. *Transactions of the ASABE*, 53(6): 1873–1882.

Quilliam, R. S., Pow, C. J., Shilla, D. J., Mwesiga, J. J., Shilla, D. A., & Woodford, L. (2023). Microplastics in agriculture–a potential novel mechanism for the delivery of human pathogens onto crops. *Frontiers in Plant Science*, 14: 1152419.

Rezaei, M., Riksen, M. J., Sirjani, E., Sameni, A., & Geissen, V. (2019). Wind erosion as a driver for transport of light density microplastics. *Science of the Total Environment*, 669: 273–281.

Ricardo, I. A., Alberto, E. A., Júnior, A. H. S., Macuvele, D. L. P., Padoin, N., Soares, C., Riella, H. G., Starling, M. C. V., & Trovo, A. G. (2021). A critical review on microplastics, interaction with organic and inorganic pollutants, impacts and effectiveness of advanced oxidation processes applied for their removal from aqueous matrices. *Chemical Engineering Journal*, 424: 130282.

Reay, M. K., Greenfield, L. M., Graf, M., Lloyd, C. E., Evershed, R. P., Chadwick, D. R., & Jones, D. L. (2023). LDPE and biodegradable PLA-PBAT plastics differentially affect plant-soil nitrogen partitioning and dynamics in a Hordeum vulgare mesocosm. *Journal of Hazardous Materials*, 447: 130825.

Rillig, M. C., Ziersch, L., & Hempel, S. (2017). Microplastic transport in soil by earthworms. *Scientific Reports*, 7(1): 1362.

Ritter, S., Ånes, E. W., Hovland, K., Stridal, A. C., Hansen, H., Henriksen, T. J., & Okkenhaug, G. (2023). Sustainable impermeable landfill barriers: The potential of using geological waste and surplus masses. *9th International Conference on Environmental Geotechnics*, Chania, Greece.

Rocha-Santos, T., & Duarte, A. C. (2015). A critical overview of the analytical approaches to the occurrence, the fate and the behavior of microplastics in the environment. *TrAC Trends in Analytical Chemistry*, 65: 47–53.

Rochman, C. M., Hentschel, B. T., & Teh, S. J. (2014). Long-term sorption of metals is similar among plastic types: Implications for plastic debris in aquatic environments. *PLoS One*, 9(1): e85433.

Rose, P. K., Jain, M., Kataria, N., Sahoo, P. K., Garg, V. K., & Yadav, A. (2023). Microplastics in multimedia environment: A systematic review on its fate, transport, quantification, health risk, and remedial measures. *Groundwater for Sustainable Development*, 20: 100889.

Shah, F., & Wu, W. (2020). Use of plastic mulch in agriculture and strategies to mitigate the associated environmental concerns. *Advances in Agronomy*, 164: 231–287.

Sharma, S., Basu, S., Shetti, N. P., Nadagouda, M. N., & Aminabhavi, T. M. (2021). Microplastics in the environment: occurrence, perils, and eradication. *Chemical Engineering Journal*, 408: 127317.

Shove, E. (2022). *Connecting Practices: Large Topics in Society and Social Theory*. Routledge.

Steinmetz, Z., Wollmann, C., Schaefer, M., Buchmann, C., David, J., Tröger, J., Muñoz, K., Frör, O., & Schaumann, G. E. (2016). Plastic mulching in agriculture. Trading short-term agronomic benefits for long-term soil degradation? *Science of the Total Environment*, 550: 690–705.

Stubenrauch, J., & Ekardt, F. (2020). Plastic pollution in soils: Governance approaches to foster soil health and closed nutrient cycles. *Environments*, 7(5): 38.

Subaramaniyam, U., Allimuthu, R. S., Vappu, S., Ramalingam, D., Balan, R., Paital, B., Panda, N., Rath, P. K., Ramalingam, N., & Sahoo, D. K. (2023). Effects of microplastics, pesticides and nano-materials on fish health, oxidative stress and antioxidant defense mechanism. *Frontiers in Physiology*, 14: 1217666.

Sun, M., Ye, M., Jiao, W., Feng, Y., Yu, P., Liu, M., Jiao, J., He, X., Liu, K., & Zhao, Y. (2018). Changes in tetracycline partitioning and bacteria/phage-comediated ARGs in microplastic-contaminated greenhouse soil facilitated by sophorolipid. *Journal of Hazardous Materials*, 345: 131–139.

Šunta, U., Prosenc, F., Trebše, P., Bulc, T. G., & Kralj, M. B. (2020). Adsorption of acetamiprid, chlorantraniliprole and flubendiamide on different type of microplastics present in alluvial soil. *Chemosphere*, 261: 127762.

Tarafdar, J., Sharma, S., & Raliya, R. (2013). Nanotechnology: Interdisciplinary science of applications. *African Journal of Biotechnology*, 12(3): 219–226.

Tian, L., Jinjin, C., Ji, R., Ma, Y., & Yu, X. (2022). Microplastics in agricultural soils: Sources, effects, and their fate. *Current Opinion in Environmental Science & Health*, 25: 100311.

Urra, J., Alkorta, I., & Garbisu, C. (2019). Potential benefits and risks for soil health derived from the use of organic amendments in agriculture. *Agronomy*, 9(9): 542.

Verla, A. W., Enyoh, C. E., Verla, E. N., & Nwarnorh, K. O. (2019). Microplastic–toxic chemical interaction: A review study on quantified levels, mechanism and implication. *SN Applied Sciences*, 1(11): 1–30.

Wakkaf, T., El Zrelli, R., Yacoubi, L., Kedzierski, M., Lin, Y.-J., Mansour, L., Bruzaud, S., & Rabaoui, L. (2022). Seasonal patterns of microplastics in surface sediments of a Mediterranean lagoon heavily impacted by human activities (Bizerte lagoon, Northern Tunisia). *Environmental Science and Pollution Research*, 29(51): 76919–76936.

Waldschläger, K., Brückner, M. Z., Almroth, B. C., Hackney, C. R., Adyel, T. M., Alimi, O. S., Belontz, S. L., Cowger, W., Doyle, D., & Gray, A. (2022). Learning from natural sediments to tackle microplastics challenges: A multidisciplinary perspective. *Earth-Science Reviews*, 228: 104021.

Waldschläger, K., Lechthaler, S., Stauch, G., & Schüttrumpf, H. (2020). The way of microplastic through the environment–Application of the source-pathway-receptor model. *Science of the Total Environment*, 713: 136584.

Wagner, J., Wang, Z. M., Ghosal, S., Murphy, M., Wall, S., Cook, A. M. & Allen, H. (2019). Nondestructive extraction and identification of microplastics from freshwater sport fish stomachs. *Environmental Science & Technology*, 53(24): 14496–14506.

Wang, F., Yang, W., Cheng, P., Zhang, S., Zhang, S., Jiao, W., & Sun, Y. (2019). Adsorption characteristics of cadmium onto microplastics from aqueous solutions. *Chemosphere*, 235: 1073–1080.

Wang, L., Cheng, Y., Naidu, R., & Bowman, M. (2021). The key factors for the fate and transport of petroleum hydrocarbons in soil with related in/ex situ measurement methods: An overview. *Frontiers in Environmental Science*, 9: 756404.

Wang, Q., Adams, C. A., Wang, F., Sun, Y., & Zhang, S. (2022). Interactions between microplastics and soil fauna: A critical review. *Critical Reviews in Environmental Science and Technology*, 52(18): 3211–3243.

Wang, W., Ge, J., Yu, X., & Li, H. (2020). Environmental fate and impacts of microplastics in soil ecosystems: Progress and perspective. *Science of the Total Environment*, 708: 134841.

Wang, X., Sial, M. U., Bashir, M. A., Bilal, M., Raza, Q.-U.-A., Ali Raza, H. M., Rehim, A., & Geng, Y. (2022). Pesticides xenobiotics in soil ecosystem and their remediation approaches. *Sustainability*, 14(6): 3353.

Wanner, P. (2021). Plastic in agricultural soils–A global risk for groundwater systems and drinking water supplies?–A review. *Chemosphere*, 264: 128453.

Waraich, A., & Srivastava, N. (2022) Microplastics: Potential risk to wildlife health. In *Xenobiotics Interaction: Consequences and Remediation* (pp. 32–52). Indu Book Services, New Delhi, India.

Weithmann, N., Möller, J. N., Löder, M. G., Piehl, S., Laforsch, C., & Freitag, R. (2018). Organic fertilizer as a vehicle for the entry of microplastic into the environment. *Science Advances*, 4(4): eaap8060.

Xu, B., Liu, F., Cryder, Z., Huang, D., Lu, Z., He, Y., Wang, H., Lu, Z., Brookes, P. C., & Tang, C. (2020). Microplastics in the soil environment: Occurrence, risks, interactions and fate–A review. *Critical Reviews in Environmental Science and Technology*, 50(21): 2175–2222.

Xu, Y., Park, J., Sorensen, V. K., Clausen, P., & Little, J. (2008). Characterizing emissions of phthalate plasticizer from vinyl flooring in a specially-designed chamber. *Epidemiology*, 19(6): S294–S295.

Yang, L., Zhang, Y., Kang, S., Wang, Z., & Wu, C. (2021). Microplastics in soil: A review on methods, occurrence, sources, and potential risk. *Science of the Total Environment*, 780: 146546.

Yang, X., Guo, X., Huang, S., Xue, S., Meng, F., Qi, Y., Cheng, W., Fan, T., Lwanga, E. H., & Geissen, V. (2020). Microplastics in soil ecosystem: Insight on its fate and impacts on soil quality. In D. He, Y. Luo (Eds.), *Microplastics in Terrestrial Environments: Emerging Contaminants and Major Challenges* (pp. 245–258). Springer.

Yang, X., Lwanga, E. H., Bemani, A., Gertsen, H., Salanki, T., Guo, X., Fu, H., Xue, S., Ritsema, C., & Geissen, V. (2019). Biogenic transport of glyphosate in the presence of LDPE microplastics: A mesocosm experiment. *Environmental Pollution*, 245: 829–835.

Yavari, S., Sapari, N. B., Malakahmad, A., Razali, M. A. B., Gervais, T. S., & Yavari, S. (2020). Adsorption–desorption behavior of polar imidazolinone herbicides in tropical paddy fields soils. *Bulletin of Environmental Contamination and Toxicology*, 104: 121–127.

Yee, M. S. L., Hii, L. W., Looi, C. K., Lim, W. M., Wong, S. F., Kok, Y. Y., & Leong, C. O. (2021). Impact of microplastics and nanoplastics on human health. *Nanomaterials*, 11(2): 496.

Yuan, J., Ma, J., Sun, Y., Zhou, T., Zhao, Y., & Yu, F. (2020). Microbial degradation and other environmental aspects of microplastics/plastics. *Science of the Total Environment*, 715: 136968.

Yurtsever, M. (2019). Glitters as a source of primary microplastics: An approach to environmental responsibility and ethics. *Journal of Agricultural and Environmental Ethics*, 32(3): 459–478.

Zhai, X., Zhang, X.-H., & Yu, M. (2023). Microbial colonization and degradation of marine microplastics in the plastisphere: A review. *Frontiers in Microbiology*, 14: 1127308.

Zhang, B., Yang, X., Chen, L., Chao, J., Teng, J., & Wang, Q. (2020). Microplastics in soils: A review of possible sources, analytical methods and ecological impacts. *Journal of Chemical Technology & Biotechnology*, 95(8): 2052–2068.

Zhang, C., Lei, Y., Qian, J., Qiao, Y., Liu, J., Li, S., Dai, L., Sun, K., Guo, H., & Sui, G. (2021). Sorption of organochlorine pesticides on polyethylene microplastics in soil suspension. *Ecotoxicology and Environmental Safety*, 223: 112591.

Zhang, S., Li, Y., Chen, X., Jiang, X., Li, J., Yang, L., Yin, X., & Zhang, X. (2022). Occurrence and distribution of microplastics in organic fertilizers in China. *Science of the Total Environment*, 844: 157061.

Zhao, H., Zhou, Y., Lu, Z., Ren, X., Barcelo, D., Zhang, Z., & Wang, Q. (2023). Microplastic pollution in organic farming development cannot be ignored in China: Perspective of commercial organic fertilizer. *Journal of Hazardous Materials*, 460: 132478.

Zhou, Y., Wang, J., Zou, M., Jia, Z., Zhou, S., & Li, Y. (2020). Microplastics in soils: A review of methods, occurrence, fate, transport, ecological and environmental risks. *Science of the Total Environment*, 748: 141368.

CHAPTER 7

Impact of Microplastics on Soil's Biodiversity and Public Health

Zakia Gueboudji, Sabrina Lekmine, Sara Mechaala, and Kenza Kadi

7.1 INTRODUCTION

Over the past seven decades, we have witnessed a substantial surge in global plastic production, leading to the pervasive presence of plastic in our environment. This proliferation has reached a level where it is apt to describe our current state as inhabiting a world dominated by plastic (Campanale et al., 2020). In our contemporary world, the issue of plastic pollution has emerged as a pressing global concern, captivating the attention of scientists, policymakers, and the general public alike. While much attention has been rightly directed towards the consequences of plastic waste in marine environments, a less conspicuous yet equally significant facet of this crisis has been gradually unfolding beneath our feet: the infiltration of microplastics into soil ecosystems (Kibria et al., 2023; Sadia et al., 2023). The proliferation of microplastics, defined as plastic particles smaller than 5 mm in size, presents an intricate challenge with far-reaching implications for soil biodiversity and public health (Fang et al., 2024; Moto et al., 2023). Over the past 10 years, microplastics have emerged as a significant global

DOI: 10.1201/9781003487579-7

environmental and health issue. These minute plastic particles originate from diverse sources, including discarded plastics, textiles, and personal care products like facial cleansers and soaps. Their diminutive size enables them to bypass water filtration methods, leading to their widespread presence in the environment (Ghosh et al., 2023; Akdogan and Guven, 2019). Microplastics are currently detected in diverse environmental mediums such as soil, water, and air. They have permeated the food chain, eventually integrating into the human diet (Emenike et al., 2023).

The ubiquity of plastics in modern society is undeniable. From the packaging of consumer goods to the fibers in our clothing, plastics have become an indispensable part of our daily lives (Gueboudji, 2024b). However, the durability that makes plastics so desirable for various applications also renders them resistant to degradation. As a result, plastic waste accumulates in terrestrial environments, where it undergoes fragmentation into increasingly smaller particles over time (Andrady, 2017). These microplastics, often imperceptible to the naked eye, are dispersed throughout the soil matrix via diverse pathways, including agricultural practices, atmospheric deposition, and the breakdown of plastic mulches (Yang et al., 2021).

The ramifications of microplastic contamination on soil biodiversity are multifaceted and complex. Soil, often regarded as the cornerstone of terrestrial ecosystems, harbors a staggering diversity of organisms, ranging from microscopic bacteria to earthworms and larger fauna (Wang et al., 2020). These organisms play pivotal roles in nutrient cycling, soil structure formation, and plant health. However, the intrusion of microplastics into soil environments can disrupt these delicate ecological balances (Li et al., 2023a). One of the primary concerns regarding the impact of microplastics on soil biodiversity is their potential to alter soil physicochemical properties (Ya et al., 2021). Studies have shown that microplastics can modify soil structure, moisture retention, and nutrient availability, thereby influencing the abundance and diversity of soil-dwelling organisms (Xu et al., 2020; Shafea et al., 2023; Feng et al., 2023). Moreover, microplastics can serve as substrates for microbial colonization, altering microbial community compositions and metabolic processes. These disruptions to soil microbial communities can reverberate throughout the broader soil food web, affecting the abundance and activity of higher trophic levels (Thacharodi et al., 2024; Zhu et al., 2023). Beyond direct ecological effects, microplastics in soil ecosystems pose a considerable risk to public health. The dynamic interactions between soil, water, and vegetation provide pathways for the transfer of microplastics into the food chain, ultimately culminating in human exposure (Vairaperumal et al., 2024; Ya et al., 2021). Recent research has detected microplastics in various food crops, including fruits, vegetables, and grains, raising concerns about potential health hazards associated with their consumption (Liang et al., 2023). Furthermore, microplastics can adsorb and transport persistent organic pollutants (POPs) and heavy metals, acting as vectors for the bioaccumulation of toxic compounds in organisms along the food chain (Okoye et al., 2022).

The objective of this research is to examine the impact of microplastics on soil biodiversity and its potential implications for public health. Through reviewing current research, the study aims to understand how microplastics affect soil-dwelling organisms and alter soil physicochemical properties, potentially disrupting nutrient cycling and overall terrestrial biodiversity. Special attention is given to exploring pathways of microplastic transfer within the soil food web and their accumulation in edible crops, highlighting potential exposure routes for humans. Addressing challenges in assessing and quantifying microplastics in soil, the research emphasizes the urgent need for standardized methodologies and comprehensive risk assessments to inform evidence-based policies for mitigating microplastic pollution and safeguarding soil ecosystems and human health.

7.2 SOURCES AND DISTRIBUTION OF MICROPLASTICS

The production of plastics has been on the rise over the years, yet their recycling rate remains low, leading to the accumulation of substantial quantities of microplastics in the environment (Zhang et al., 2023). Microplastics can originate from various anthropogenic activities, with primary sources including the degradation of larger plastic items, disintegration of synthetic textiles, and the use of personal care and cosmetic products containing microplastic particles. These sources collectively contribute to the dispersal of microplastics across terrestrial and aquatic environments (Mai et al., 2024; Zhang et al., 2023).

In soil, microplastics exhibit diverse distribution patterns influenced by factors such as land use, proximity to urban centers, and waste management practices (Rafique et al., 2020). Agricultural soils may be particularly vulnerable due to the application of plastic mulches, the breakdown of plastic films, and the use of sewage sludge as fertilizer (Gueboudji, 2024a; Kumar et al., 2020). Urban soils, on the other hand, may accrue microplastics from stormwater runoff and the accumulation of plastic waste (Qiu et al., 2020). Consequently, the intricate interplay between these sources and the myriad factors shaping microplastic distribution underscores the need for a nuanced examination of their presence in soil ecosystems (Yu et al., 2024).

7.3 IMPACTS ON SOIL BIODIVERSITY

The soil biome represents a complex network of organisms ranging from microscopic bacteria and fungi to larger invertebrates and plant roots. This intricate web of life is essential for nutrient cycling, soil structure maintenance, and overall ecosystem functioning (Voroney et al., 2024). The infiltration of microplastics into soil ecosystems introduces an additional layer of complexity, with potential repercussions for soil biodiversity (Wang et al., 2022).

Microplastics, once introduced into the soil, can interact with various components of the soil biome. For instance, soil-dwelling organisms such as earthworms, nematodes, and microorganisms may encounter microplastics in their habitats. The potential effects on these organisms, whether through physical interactions, ingestion, or alterations in soil chemistry, have yet to be comprehensively understood (Liu et al., 2023; Feng et al., 2023).

The impacts on soil microbial communities, critical drivers of nutrient cycling and ecosystem stability, merit particular attention. Microplastics may influence microbial composition, metabolic activity, and functional diversity, with potential cascading effects on overall soil health (Bouaicha et al., 2022). Moreover, the potential for microplastics to serve as vectors for contaminants and disrupt microbial processes adds a layer of complexity to the already intricate dynamics of soil ecosystems (Campanale et al., 2020).

7.4 MICROPLASTICS IN THE SOIL-WATER-PLANT NEXUS

The interconnectedness of terrestrial and aquatic ecosystems is epitomized by the soil-water-plant nexus. Microplastics, having infiltrated the soil, can traverse this nexus, potentially influencing water quality, plant health, and ultimately, human exposure. Understanding the pathways and mechanisms through which microplastics move within this nexus is imperative for comprehending their broader environmental implications (Huang et al., 2022).

In the soil-water interface, microplastics may undergo various processes such as sorption, leaching, and transport. These processes can be influenced by soil properties, land use practices, and climatic conditions. Once in the soil solution, microplastics may interact with plant roots, with the potential for uptake and translocation within the plant. The fate of microplastics within the plant, their impacts on plant physiology, and the subsequent implications for human exposure through the consumption of contaminated crops warrant meticulous investigation (Huang et al., 2022; Zhang et al., 2022).

7.5 RISKS TO PUBLIC HEALTH

Beyond the ecological realm, the potential risks to public health arising from microplastics in soil and the food chain raise critical concerns. As microplastics enter the soil-water-plant-human pathway, they may contribute to human exposure through the consumption of contaminated food and water. The health implications of this exposure, however, remain an evolving area of research.

Microplastics have been detected in various food items, including seafood, fruits, and vegetables. The potential for these particles to carry contaminants and additives, as well as their ability to accumulate in edible tissues, prompts a reevaluation of food safety standards. Furthermore, the impact of microplastic ingestion on human health, including potential effects on the gastrointestinal tract and the absorption of associated chemicals, requires thorough investigation (Sridhar et al., 2022; Campanale et al., 2020).

The presence of microplastics has been associated with potential impacts on human health according to various studies (Yee et al., 2021; Ebrahimi et al., 2022; Boskovic et al., 2023). Investigations reveal that these tiny plastic particles can infiltrate our bodies through consumption of food, water, and even inhalation of air (Table 7.1). Microplastics have been detected in tap water, bottled water, seafood, and the air we breathe (Yee et al., 2021). Upon entering the human body, they can induce inflammation, leading to a variety of health issues. Additionally, microplastics can release harmful chemicals, contributing to health problems such as cancer (Kumar et al., 2022), reproductive issues, and developmental disorders (Ghosh et al., 2023). Furthermore, microplastics have been identified as playing a role in the development of antibiotic resistance (Shi et al., 2021). This is due to their potential to serve as breeding grounds for bacteria, ultimately fostering antibiotic resistance. This poses a significant threat to global health, given the already pressing issue of antibiotic resistance (Ghosh et al., 2023).

7.6 MICROPLASTICS IN SOIL

As they infiltrate diverse environmental matrices, microplastics have been discovered not only in marine and freshwater ecosystems but also in soils, the very foundation of terrestrial life (Lwanga et al., 2023; Sajjad

TABLE 7.1 The Various Types of Plastics, Their Size Ranges, and Potential Health Effects (Ghosh et al., 2023)

Plastic Type	Size Range	Health Impacts
Polyethylene	0.06–11.06 µm	Potential ingestion through contaminated seafood, respiratory issues
Terephthalate	Less than 4 µm to greater	Disruption of hormone regulation, respiratory problems
Polystyrene	33–190 µm	Respiratory issues, potential hormone disruption, adverse neurological effects
Polypropylene	120–220 µm	Gastrointestinal problems, respiratory issues, potential immune system effects
Polyvinyl chloride (PVC)	Less than 10 nm and as much as 20 µm	Respiratory problems, potential reproductive and developmental issues
Polycarbonate	5–200 nm	Hormone disruption, potential adverse effects on neurological development
Polyethylene terephthalate (PET)	Microfiber dimensions: 12–18 µm in thickness Length: shorter than 1 mm	Ingestion through contaminated seafood, potential respiratory issues
Polyurethane	Less than 5 mm	Respiratory problems, potential skin irritation, potential effects on immune and nervous systems
Polyvinylidene fluoride (PVDF)	Less than 5 mm	Prolonged inhalation or skin contact with PVDF particles can cause irritation or sensitization in some individuals
Acrylic copolymer	80–110 nm	Develop allergic reactions or sensitization to acrylic, leading to dermatitis or respiratory issues upon exposure

et al., 2022). Soil, a dynamic and intricate ecosystem, serves as the cradle for diverse organisms and plays a pivotal role in supporting global biodiversity (Guerra et al., 2020). The introduction of microplastics into this environment raises intricate questions about their potential impacts on soil-dwelling organisms and the overall biodiversity of terrestrial ecosystems. Additionally, concerns extend beyond the confines of soil ecosystems, as microplastics have the propensity to enter the food chain, posing potential risks to human health (Yang et al., 2021).

7.6.1 Definition and Types of Microplastics

Microplastics are tiny plastic particles with a size range from 1 µm to 5 mm, making them invisible to the naked eye (Méndez Rodríguez et al., 2023). These particles can originate from various sources and come in different forms, including fragments, fibers, and microbeads. Fragmented microplastics result from the breakdown of larger plastic items, such as bottles and bags, while microbeads are intentionally manufactured for use in personal care products like exfoliating scrubs. The diverse nature of microplastics poses challenges in understanding their environmental impact and devising effective mitigation strategies (Vairaperumal et al., 2024).

7.6.2 Sources of Microplastics in Soil

Microplastics enter soil environments through several pathways, creating a complex web of contamination sources. Primary sources include the breakdown of larger plastic debris, either through mechanical processes like weathering or degradation by ultraviolet radiation. Secondary sources involve the release of microplastics from various products, such as synthetic textiles, rubber tires, and agricultural films. Urban runoff, atmospheric deposition, and improper waste disposal contribute to the dispersion of microplastics in soil, creating a widespread and pervasive issue (Gueboudji et al., 2024; Surendran et al., 2023).

7.6.3 Distribution and Persistence in Soil

Once introduced into soil ecosystems, microplastics exhibit diverse distribution patterns and persistence. They may accumulate near the surface or penetrate deeper layers, depending on factors like soil structure, particle size, and organic matter content (Xu et al., 2020). The persistence of microplastics in soil is influenced by their chemical composition, as well as environmental conditions. Some microplastics may bind to soil particles, reducing their mobility, while others may leach into water bodies, posing additional risks to aquatic ecosystems (Yu et al., 2021; Surendran et al., 2023).

Understanding the distribution and persistence of microplastics in soil is crucial for assessing their potential impact on soil health, plant growth, and overall ecosystem function (Huang et al., 2022). Research efforts are ongoing to unravel the intricate interactions between microplastics and soil organisms as well as to develop effective remediation strategies for mitigating the environmental consequences of microplastic contamination.

As the global awareness of microplastic pollution continues to grow, addressing this issue requires interdisciplinary collaboration, innovative technologies, and sustainable practices to minimize the release and accumulation of microplastics in soil ecosystems.

7.7 BIODIVERSITY IMPACT OF MICROPLASTICS

7.7.1 Effects on Soil Organisms

Microplastics can have direct and indirect effects on soil organisms (Wan et al., 2023). The alteration of soil physicochemical properties due to microplastic presence may impact microbial communities. Changes in soil moisture, aeration, and temperature resulting from microplastic accumulation can influence the abundance and diversity of bacteria, fungi, and archaea. Moreover, some soil microorganisms may exhibit the ability to break down or interact with certain types of microplastics, influencing microbial activity and community dynamics (Liu et al., 2023; Shafea et al., 2023).

The ingestion of microplastics by soil organisms can lead to adverse effects on their physiology and behavior. The potential for bioaccumulation in soil organisms raises concerns about the transfer of microplastics through different trophic levels, with implications for the broader soil food web.

7.7.2 Microbial Communities

Microplastics can act as substrates for microbial colonization, forming biofilms on their surfaces. This altered microhabitat may impact the structure and function of microbial communities (Baptista Neto et al., 2022). Changes in microbial diversity and activity can affect essential soil processes, including organic matter decomposition, nutrient cycling, and the availability of key elements for plant growth (Zifcakova, 2020). The intricate interplay between microplastics and microbial communities highlights the need for in-depth studies to elucidate the specific mechanisms driving these interactions.

7.7.3 Invertebrates

Soil invertebrates, such as earthworms and arthropods, are vital components of soil ecosystems. Microplastics can influence these organisms through various pathways (Baptista Neto et al., 2022). Ingestion of microplastics by earthworms, for example, may lead to altered feeding behaviors, reproductive effects, and changes in the overall fitness of these organisms

(Cui et al., 2022). The potential for microplastic-induced stress in soil invertebrates has implications for ecosystem functioning, as these organisms play key roles in nutrient cycling, soil aeration, and organic matter decomposition (Li et al., 2023b; Baptista Neto et al., 2022).

The effects of microplastics on soil invertebrates extend to other taxa within the soil food web, impacting predator-prey interactions and community dynamics. Understanding the nuances of these interactions is crucial for predicting the long-term consequences of microplastic contamination on soil biodiversity.

7.7.4 Plants

Microplastics can interact with plants at different stages of their life cycle, influencing germination, root development, and overall growth. The presence of microplastics in soil may affect the physical and chemical properties of the rhizosphere, potentially altering nutrient availability and microbial interactions. As plants absorb water containing microplastics, there is concern about the translocation of these particles within plant tissues, with potential implications for plant health and the quality of agricultural produce (Iqbal et al., 2023; Zhang et al., 2022).

The interaction between microplastics and plants is a complex area of study, encompassing physiological responses, nutrient dynamics, and potential impacts on ecosystem services provided by plants. Research in this field aims to unravel the specific mechanisms driving microplastic-plant interactions and their consequences for terrestrial ecosystems.

7.7.5 Ecological Consequences

The cumulative effects of microplastics on soil organisms, microbial communities, invertebrates, and plants can result in broader ecological consequences. Changes in biodiversity, nutrient cycling, and ecosystem functioning may have cascading effects on the entire ecosystem. Microplastics can alter the balance of predator-prey interactions, disrupt energy flows, and influence the resilience of ecosystems to environmental stressors (Liu et al., 2023; Ya et al., 2021).

The potential for microplastics to move through trophic levels raises concerns about the impact on higher organisms, including birds and mammals. Moreover, the persistence of microplastics in soil ecosystems underscores the need for long-term monitoring and management strategies to mitigate their ecological consequences.

7.8 PUBLIC HEALTH IMPLICATIONS OF MICROPLASTICS IN SOIL

7.8.1 Soil-to-Food Chain Transfer

The transfer of microplastics from soil to the food chain is a critical aspect of understanding the public health implications of microplastic contamination. Research indicates that microplastics can be taken up by plants through their root systems. The factors influencing this transfer include the type and size of microplastics, soil characteristics, and plant species. Some studies have shown that certain crops, such as leafy vegetables, may accumulate higher levels of microplastics than others. This soil-to-plant transfer has direct implications for human exposure to microplastics through the consumption of contaminated crops, and it underscores the need for investigations into the types of crops most susceptible to this transfer (Prata and Dias-Pereira, 2023).

7.8.2 Human Exposure Pathways

Human exposure to microplastics from soil occurs through multiple pathways. Inhalation is a potential route, especially in areas where soil disturbance is common, such as construction sites or agricultural fields. Airborne microplastics can settle on surfaces and be inhaled, potentially entering the respiratory system. Additionally, exposure can occur through the consumption of contaminated food and water. Dust from microplastic-contaminated soil can settle on crops during growth, and irrigation water may introduce microplastics into the agricultural system. Understanding these exposure pathways is crucial for assessing the overall risk to human health (Chellasamy et al., 2023).

7.8.3 Health Risks Associated with Microplastics

The potential health risks associated with microplastics are complex and multifaceted. Microplastics can carry a range of additives and contaminants, including heavy metals and persistent organic pollutants. When ingested, these substances may be released in the digestive tract, leading to concerns about the bioavailability and accumulation of potentially harmful compounds in the human body (Okoye et al., 2022). Table 7.2 delineates the heavy metals correlated with additives found in plastics, elucidating their deleterious effects on human health.

Studies have suggested that microplastics may cause physical damage to tissues and organs. The small size of microplastics allows them to traverse biological barriers, potentially reaching different organs and tissues.

TABLE 7.2 Heavy Metals Connected with Additives in Plastics and Their Detrimental Impact on Human Health (Okoye et al., 2022)

Heavy Metals	Additives	Types of Polymers	Harmful Effects on Human Health
Aluminum (Al)	Stabilizers, inorganic pigments, and flame retardants	PBT, PET, PE, PVC	Metal–estrogen; breast cancer
Antimony (Sb)	Flame retardants and biocides	Various plastics	Metal–estrogen; breast cancer
Arsenic (As)	Biocides	PVC, LDPE, and polyesters	Congenital disabilities; carcinogen: lung, skin, liver, bladder, kidneys; gastrointestinal damage; death
Barium (Ba)	Inorganic pigments and UV stabilizers	PVC	Metal–estrogen, breast cancer; cardiovascular and kidney diseases; metabolic, neurological, and mental disorders
Cadmium (Cd)	Heat stabilizers, UV stabilizers, and inorganic pigments	PVC	Changes in the metabolism of calcium, phosphorus, and bone; osteomalacia and bone fractures in postmenopausal women; lipid peroxidation and in the promotion of carcinogenesis; cellular apoptosis; DNA methylation
Mercury (Hg)	Biocides	PU	Mutagen or carcinogen; induction of the disruption of DNA molecular structure and brain damage
Mercury (Hg)	Biocides	PU	Mutagen or carcinogen; induction of the disruption of DNA molecular structure and brain damage
Cobalt (Co)	Inorganic pigments	PET bottles	Formation of reactive oxygen species (ROS); neurological (e.g., hearing and visual impairment); cardiovascular and endocrine deficits

(*Continued*)

TABLE 7.2 (*Continued*) Heavy Metals Connected with Additives in Plastics and Their Detrimental Impact on Human Health (Okoye et al., 2022)

Heavy Metals	Additives	Types of Polymers	Harmful Effects on Human Health
Chrome (Cr)	Inorganic pigments	PVC, PE, PP	Allergic reactions to the body; nasal septum ulcer; severe cardiovascular, respiratory, hematological, gastrointestinal, renal, hepatic, and neurological effects and possibly death
Lead (Pb)	Heat stabilizers, UV stabilizers, and inorganic pigments	PVC and all types of plastics, where red pigments are used	Anemia (less Hb); hypertension; miscarriages; disruption of nervous Systems; brain damage; infertility; oxidative stress and cell damage
Tin (Sn)	UV stabilizers and biocides	PU foam and PVC	Metal–estrogen; breast cancer; skin rashes; stomach complaints; nausea; vomiting, diarrhea; abdominal pain; headache and palpitations; potential clastogen
Titanium (Ti)	UV stabilizers and inorganic pigments	PVC	Cytotoxicity on human epithelial lung and colon cells

PBT, polybutylene terephthalate; PET, polyethylene-terephthalate; PE, polyethylene; PVC, polyvinylchloride; LDPE, low-density polythene; PP, polypropylene; PU, polyurethane.

The inflammatory responses triggered by microplastics and their potential hormonal effects add another layer of complexity to the assessment of health risks. Researchers are actively investigating the long-term effects of chronic exposure to microplastics, aiming to elucidate the mechanisms of toxicity and assess the potential for cumulative harm (Zhu et al., 2023; Yee et al., 2021; Campanale et al., 2020).

The identification of specific health risks associated with different types of microplastics, their additives, and the contaminants they carry is an ongoing area of research. As our understanding of the potential health impacts evolves, it is crucial to integrate this knowledge into risk assessment frameworks and public health guidelines.

In part of a study conducted by Zhu et al. (2023) about the interaction of microplastics with biological barriers and translocation into multiple organs, they find the following. Humans are subjected to substantial microplastics exposure, resulting in a significant burden. The estimated daily intake of microplastics by humans, acquired through various pathways, is approximately ($122.25 \pm 177.38 \sim 202.80 \pm 294.25$) mg per person, as reported by Yadav et al. (2022). The widespread presence of microplastics in consumer goods and diverse environments contributes to their omnipresence within human organs, as highlighted by Huang et al. (2021). Due to their diminutive size, microplastic particles can easily ingress the digestive tract and accumulate, subsequently traversing biological barriers to disseminate throughout the body, including the bloodstream, liver, kidneys, and brain, as documented by Zarus et al. (2021). Given the lifelong exposure and accumulation of microplastics in tissues and organs, there is a profound concern regarding their potential adverse health effects, as illustrated in Figure 7.1.

Concerning the pervasive impact of microplastics on various biological barriers within the human body. In brief, microplastics detected in the digestive systems of organisms, traverse the gut barrier, interacting with cells and potentially disrupting the intestinal barrier function. Larger microplastics are often expelled with stool, but chronic exposure may negatively affect the gut microbiota, leading to dysbiosis and adverse health

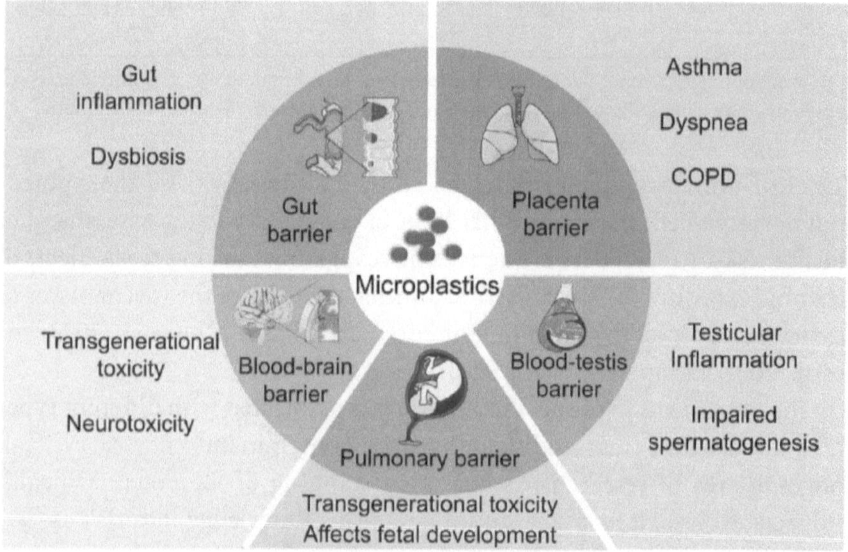

FIGURE 7.1 Interaction of microplastics with biological barriers and their subsequent translocation into various organs (Zhu et al., 2023).

effects (Enyoh et al., 2023). In the respiratory system, inhaled microplastics can trigger inflammatory responses, causing respiratory diseases, particularly chronic obstructive pulmonary disease (COPD) (Rahman et al., 2021). The blood-testis barrier is not immune, as microplastics can cross it, inducing testicular inflammation, disrupting spermatogenesis, and posing risks to male reproductive health (Rose et al., 2023). The placental barrier, vital for embryonic development, faces challenges from microplastics that can deposit in placental tissues, potentially affecting fetal health. Furthermore, there are concerns about the transgenerational transport of microplastics from mother to offspring, raising questions about its impact on fetal development (Medley et al., 2023). The blood-brain barrier is not spared, as microplastics from the food chain can cross it, inhibiting neurodevelopment-related genes and leading to neurotoxic effects, potentially causing neurological development disorders and cognitive deficits. Throughout these biological barriers, the text emphasizes the urgent need for comprehensive research to elucidate the intricate mechanisms underlying the interactions between microplastics and these barriers and to comprehend the potential long-term repercussions on human health (Zhu et al., 2023).

The public health implications of microplastics in soil involve a chain of events, from soil contamination to plant uptake and human exposure. Addressing these implications requires a multidisciplinary approach that combines environmental science, toxicology, and public health research. Developing strategies to minimize microplastic contamination in soil, monitoring human exposure pathways, and understanding the potential health risks associated with microplastics are essential steps toward safeguarding public health in the face of this emerging environmental challenge.

7.9 ANALYTICAL TECHNIQUES FOR MICROPLASTIC DETECTION IN SOIL

Detecting microplastics in soil poses unique challenges due to the complex nature of soil matrices and the small size of microplastic particles. Several analytical techniques have been developed to identify and quantify microplastics in soil samples.

7.9.1 Visual Inspection

Microplastics in soil can be initially identified through visual inspection under a microscope. However, this method is limited by the small size of microplastics and the potential for misidentification of other particles (Perez et al., 2022).

7.9.2 Density Separation

Density separation techniques involve the use of solutions with known densities to separate microplastics from soil particles. Commonly used solutions include salt solutions and heavy liquids. This method allows for the isolation of microplastics based on their buoyancy, but it may not be suitable for all soil types (Nabi et al., 2022).

7.9.3 Spectroscopy Techniques

Fourier-transform infrared (FTIR) and Raman spectroscopy are powerful techniques for identifying the chemical composition of microplastics. These methods can help distinguish between different types of plastics and identify additives associated with microplastics (Perez et al., 2022).

7.9.4 Microscopy Techniques

Scanning electron microscopy (SEM) and transmission electron microscopy (TEM) provide high-resolution imaging of microplastics. Electron microscopy allows for detailed morphological characterization, aiding in the identification of microplastics and differentiation from natural particles (Azeem et al., 2023).

7.9.5 Chemical Digestion

Chemical digestion involves treating soil samples with strong acids or enzymes to dissolve organic matter, leaving behind microplastics. This method is effective in isolating microplastics but requires careful optimization to avoid sample contamination (Liu et al., 2022).

7.9.6 Fluorescence Techniques

Some microplastics exhibit natural fluorescence, making fluorescence microscopy and spectroscopy effective tools for their detection. Fluorescent dyes can also be used to enhance the visibility of microplastics under a microscope (Liu et al., 2022).

7.10 CHALLENGES AND ADVANCES IN ANALYTICAL TECHNIQUES

According to Dong et al. (2023), Perez et al. (2022), and He et al. (2018), analyzing microplastics in soil presents several challenges, including the small size of microplastics, potential interference from soil components, and the need for standardized methods. Advances in analytical techniques aim to address these challenges and improve the accuracy and reliability of microplastic detection.

1. **Size Range:** Many conventional analytical techniques struggle to detect the smallest microplastics (<1 μm). Advanced techniques, such as nanoparticle tracking analysis (NTA) and field-flow fractionation (FFF), offer improved sensitivity for detecting smaller particles.
2. **Quantification:** Accurately quantifying microplastics in soil is challenging. Advances in image analysis software and machine learning algorithms contribute to more precise quantification by automating the identification and counting of microplastics in large datasets.
3. **Standardization:** The lack of standardized protocols for microplastic analysis in soil hinders comparability across studies. Initiatives to develop standardized methods, such as those proposed by international organizations, aim to improve the reliability and reproducibility of results.
4. **Multimethod Approaches:** Combining multiple analytical techniques, such as spectroscopy, microscopy, and chemical digestion, can provide a more comprehensive understanding of microplastic characteristics in soil. Multimethod approaches help overcome the limitations of individual techniques and enhance the reliability of results.
5. **Environmental Context:** Considering the environmental context is crucial for accurate microplastic analysis. Advances in isotopic and elemental analysis techniques allow researchers to trace the origin of microplastics and understand their fate and transport in soil environments.

The detection and analysis of microplastics in soil require a combination of established and emerging analytical techniques. Ongoing advancements aim to overcome the challenges associated with microplastic analysis, providing researchers with more robust tools to investigate the presence, distribution, and impact of microplastics in soil ecosystems.

7.11 RESEARCH GAPS AND FUTURE DIRECTIONS

While the scientific community has made significant strides in understanding the presence and potential impacts of microplastics in soil ecosystems, numerous research gaps persist. The complexity of soil biodiversity dynamics in response to microplastic exposure, the mechanisms governing the movement of microplastics within the soil-water-plant nexus, and the potential health risks associated with human exposure necessitate further exploration.

Future research efforts should prioritize the development of standardized methodologies for microplastic detection and quantification in soils. Additionally, there is a need for long-term studies that assess the chronic effects of microplastic exposure on soil organisms and ecosystems. Interdisciplinary collaborations between ecologists, microbiologists, soil scientists, and public health experts are essential to unravel the multifaceted dimensions of the microplastic conundrum (Yang et al., 2021; Akdogan and Guven, 2019).

7.12 CONCLUSION

The infiltration of microplastics into soil ecosystems represents a multifaceted environmental challenge with far-reaching implications for biodiversity and public health. The complex interplay between microplastics and soil-dwelling organisms, the intricate pathways through the soil-water-plant nexus, and the potential risks to human health underscore the urgency of continued research in this field. As we navigate the intricate web of microplastic contamination in soils, a comprehensive understanding of its ecological and health implications is paramount for developing effective mitigation strategies and informing policy decisions.

REFERENCES

Akdogan, Z., & Guven, B. (2019). Microplastics in the environment: A critical review of current understanding and identification of future research needs. *Environmental Pollution*, 254 Pt A, 113011. https://doi.org/10.1016/j.envpol.2019.113011

Andrady, A. L. (2017). The plastic in microplastics: A review. *Marine Pollution Bulletin*, 119(1), 12–22. https://doi.org/10.1016/j.marpolbul.2017.01.082

Azeem, I., Shakoor, N., Chaudhary, S., Adeel, M., Zain, M., Ahmad, M. A., … & Rui, Y. (2023). Analytical challenges in detecting microplastics and nanoplastics in soil-plant systems. *Plant Physiology and Biochemistry*, 204, 108132. https://doi.org/10.1016/j.plaphy.2023.108132

Baptista Neto, J. A., Gaylarde, C., & da Fonseca, E. M. (2022). Microplastics: A pelagic habitat for microorganisms and invertebrates. In T. Rocha-Santos, M. F. Costa, C. Mouneyrac (Eds.), *Handbook of Microplastics in the Environment* (pp. 641–665). Cham: Springer International Publishing. https://doi.org/10.1007/978-3-030-39041-9_17

Boskovic, N., Joksimovic, D., & Bajt, O. (2023). Microplastics in mussels from the Boka Kotorska Bay (Adriatic Sea) and impact on human health. *Food and Chemical Toxicology*, 173, 113641. https://doi.org/10.1016/j.fct.2023.113641

Bouaicha, O., Mimmo, T., Tiziani, R., Praeg, N., Polidori, C., Lucini, L., … & Borruso, L. (2022). Microplastics make their way into the soil and rhizosphere: A review of the ecological consequences. *Rhizosphere*, 22, 100542. https://doi.org/10.1016/j.rhisph.2022.100542

Campanale, C., Massarelli, C., Savino, I., Locaputo, V., & Uricchio, V. F. (2020). A detailed review study on potential effects of microplastics and additives of concern on human health. *International Journal of Environmental Research and Public Health*, 17(4), 1212. https://doi.org/10.3390/ijerph17041212

Chellasamy, G., Ramasundaram, S., Veerapandian, M., Chandran, M., Dhanasekaran, B., Oh, T. H., … & Yun, K. (2023). Systematic review on fate and behavior of microplastics towards the environment. *TrAC Trends in Analytical Chemistry*, 169, 117390. https://doi.org/10.1016/j.trac.2023.117390

Cui, W., Gao, P., Zhang, M., Wang, L., Sun, H., & Liu, C. (2022). Adverse effects of microplastics on earthworms: A critical review. *Science of the Total Environment*, 850, 158041. https://doi.org/10.1016/j.scitotenv.2022.158041

Dong, H., Wang, X., Niu, X., Zeng, J., Zhou, Y., Suona, Z., … & Chen, X. (2023). Overview of analytical methods for the determination of microplastics: Current status and trends. *TrAC Trends in Analytical Chemistry*, 167, 117261. https://doi.org/10.1016/j.trac.2023.117261

Ebrahimi, P., Abbasi, S., Pashaei, R., Bogusz, A., & Oleszczuk, P. (2022). Investigating impact of physicochemical properties of microplastics on human health: A short bibliometric analysis and review. *Chemosphere*, 289, 133146. https://doi.org/10.1016/j.chemosphere.2021.133146

Emenike, E. C., Okorie, C. J., Ojeyemi, T., Egbemhenghe, A., Iwuozor, K. O., Saliu, O. D., … & Adeniyi, A. G. (2023). From oceans to dinner plates: The impact of microplastics on human health. *Heliyon*, 9(10), e20440. https://doi.org/10.1016/j.heliyon.2023.e20440

Enyoh, C. E., Devi, A., Kadono, H., Wang, Q., & Rabin, M. H. (2023). The plastic within: Microplastics invading human organs and bodily fluids systems. *Environments*, 10(11), 194. https://doi.org/10.3390/environments10110194

Fang, Z., Sallach, J. B., & Hodson, M. E. (2024). Size-and concentration-dependent effects of microplastics on soil aggregate formation and properties. *Journal of Hazardous Materials*, 465, 133395. https://doi.org/10.1016/j.jhazmat.2023.133395

Feng, S., Wang, H., Wang, Y., & Cheng, Q. (2023). A review of the occurrence and degradation of biodegradable microplastics in soil environments. *Science of the Total Environment*, 904, 166855. https://doi.org/10.1016/j.scitotenv.2023.166855

Ghosh, S., Sinha, J. K., Ghosh, S., Vashisth, K., Han, S., & Bhaskar, R. (2023). Microplastics as an emerging threat to the global environment and human health. *Sustainability*, 15, 10821. https://doi.org/10.3390/su151410821

Gueboudji, Z. (2024a). Waste valorization techniques. In A. Anouzla, S. Souabi (Eds.), *Generation of Energy from Municipal Solid Waste*. Springer, Cham. https://doi.org/10.1007/978-3-031-74334-4_2

Gueboudji, Z. (2024b). From trash to treasure: Transforming landfill tyres, plastics, and glass into sustainable construction resources. In S. Souabi, A. Anouzla (Eds.), *Technical Landfills and Waste Management*. Springer Water. Springer, Cham. https://doi.org/10.1007/978-3-031-55665-4_12

Gueboudji, Z., Mahmoudi, M., Kadi, K., & Nagaz, K. (2024). Characteristics and Impacts of Municipal Solid Waste (MSW): A Review. In A. Anouzla,

S. Souabi, S. (Eds.), *Technical Landfills and Waste Management*. Springer Water. Springer, Cham. https://doi.org/10.1007/978-3-031-52633-6_4

Guerra, C. A., Heintz-Buschart, A., Sikorski, J., Chatzinotas, A., Guerrero-Ramírez, N., Cesarz, S., … & Eisenhauer, N. (2020). Blind spots in global soil biodiversity and ecosystem function research. *Nature Communications*, 11(1), 3870. https://doi.org/10.1038/s41467-020-17688-2

He, D., Luo, Y., Lu, S., Liu, M., Song, Y., & Lei, L. (2018). Microplastics in soils: Analytical methods, pollution characteristics and ecological risks. *TrAC Trends in Analytical Chemistry*, 109, 163–172. https://doi.org/10.1016/j.trac.2018.10.006

Huang, D., Tao, J., Cheng, M., Deng, R., Chen, S., Yin, L., & Li, R. (2021). Microplastics and nanoplastics in the environment: Macroscopic transport and effects on creatures. *Journal of Hazardous Materials*, 407, 124399. https://doi.org/10.1016/j.jhazmat.2020.124399

Huang, D., Wang, X., Yin, L., Chen, S., Tao, J., Zhou, W., … & Xiao, R. (2022). Research progress of microplastics in soil-plant system: Ecological effects and potential risks. *Science of the Total Environment*, 812, 151487. https://doi.org/10.1016/j.scitotenv.2021.151487

Iqbal, B., Zhao, T., Yin, W., Zhao, X., Xie, Q., Khan, K. Y., … & Du, D. (2023). Impacts of soil microplastics on crops: A review. *Applied Soil Ecology*, 181, 104680. https://doi.org/10.1016/j.apsoil.2022.104680

Kibria, M. G., Masuk, N. I., Safayet, R., Nguyen, H. Q., & Mourshed, M. (2023). Plastic waste: Challenges and opportunities to mitigate pollution and effective management. *International Journal of Environmental Research*, 17(1), 20. https://doi.org/10.1007/s41742-023-00507-z

Kumar, M., Xiong, X., He, M., Tsang, D. C., Gupta, J., Khan, E., … & Bolan, N. S. (2020). Microplastics as pollutants in agricultural soils. *Environmental Pollution*, 265, 114980. https://doi.org/10.1016/j.envpol.2020.114980

Kumar, R., Manna, C., Padha, S., Verma, A., Sharma, P., Dhar, A., ... & Bhattacharya, P. (2022). Micro(nano)plastics pollution and human health: How plastics can induce carcinogenesis to humans?. *Chemosphere*, 298, 134267. https://doi.org/10.1016/j.chemosphere.2022.134267

Li, K., Xiu, X., & Hao, W. (2023a). Microplastics in soils: Production, behavior process, impact on soil organisms, and related toxicity mechanisms. *Chemosphere*, 141060. https://doi.org/10.1016/j.chemosphere.2023.141060

Li, X., Wang, X., Ren, C., Palansooriya, K. N., Wang, Z., & Chang, S. X. (2023b). Microplastic pollution: Phytotoxicity, environmental risks, and phytoremediation strategies. *Critical Reviews in Environmental Science and Technology*, 54, 1–22. https://doi.org/10.1080/10643389.2023.2252310

Liang, Y., Cao, X., Mo, A., Jiang, J., Zhang, Y., Gao, W., & He, D. (2023). Micro (nano) plastics in plant-derived food: Source, contamination pathways and human exposure risks. *TrAC Trends in Analytical Chemistry*, 165, 117138. https://doi.org/10.1016/j.trac.2023.117138

Liu, M., Feng, J., Shen, Y., & Zhu, B. (2023). Microplastics effects on soil biota are dependent on their properties: A meta-analysis. *Soil Biology and Biochemistry*, 178, 108940. https://doi.org/10.1016/j.soilbio.2023.108940

Liu, S., Shang, E., Liu, J., Wang, Y., Bolan, N., Kirkham, M. B., & Li, Y. (2022). What have we known so far for fluorescence staining and quantification of microplastics: A tutorial review. *Frontiers of Environmental Science & Engineering*, 16, 1–14. https://doi.org/10.1007/s11783-021-1442-2

Lwanga, E. H., Van Roshum, I., Munhoz, D. R., Meng, K., Rezaei, M., Goossens, D., … & Ritsema, C. (2023). Microplastic appraisal of soil, water, ditch sediment and airborne dust: The case of agricultural systems. *Environmental Pollution*, 316, 120513. https://doi.org/10.1016/j.envpol.2022.120513

Mai, L., Bao, L. J., Wong, C. S., & Zeng, E. Y. (2024). Microplastics in the terrestrial environment. In E. Y. Zeng (Ed.), *Microplastic Contamination in Aquatic Environments* (pp. 229–247). Elsevier. https://doi.org/10.1016/B978-0-443-15332-7.00012-0

Medley, E. A., Spratlen, M. J., Yan, B., Herbstman, J. B., & Deyssenroth, M. A. (2023). A systematic review of the placental translocation of micro-and nanoplastics. *Current Environmental Health Reports*, 10, 1–13. https://doi.org/10.1007/s40572-023-00391-x

Méndez Rodríguez, K. B., Jiménez Avalos, J. A., Fernández Macias, J. C., & González Palomo, A. K. (2023). Microplastics: Challenges of assessment in biological samples and their implication for in vitro and in vivo effects. *Environmental Science and Pollution Research*, 30, 1–17. https://doi.org/10.1007/s11356-023-30853-6

Moto, E., Hossein, M., Bakari, R., Mateso, A. S., Selemani, J. R., Nkrumah, S., … & Machunda, R. L. (2023). Ecological consequences of microplastic pollution in sub-Saharan Africa aquatic ecosystems: An implication to environmental health. *HydroResearch*, 7, 39–54. https://doi.org/10.1016/j.hydres.2023.11.003

Nabi, I., Bacha, A. R., & Zhang, L. (2022). A review on microplastics separation techniques from environmental media. *Journal of Cleaner Production*, 337, 130458. https://doi.org/10.1016/j.jclepro.2022.130458

Okoye, C. O., Addey, C. I., Oderinde, O., Okoro, J. O., Uwamungu, J. Y., Ikechukwu, C. K., … & Odii, E. C. (2022). Toxic chemicals and persistent organic pollutants associated with micro-and nanoplastics pollution. *Chemical Engineering Journal Advances*, 11, 100310. https://doi.org/10.1016/j.ceja.2022.100310

Perez, C. N., Carré, F., Hoarau-Belkhiri, A., Joris, A., Leonards, P. E., & Lamoree, M. H. (2022). Innovations in analytical methods to assess the occurrence of microplastics in soil. *Journal of Environmental Chemical Engineering*, 10(3), 107421. https://doi.org/10.1016/j.jece.2022.107421

Prata, J. C., & Dias-Pereira, P. (2023). Microplastics in terrestrial domestic animals and human health: Implications for food security and food safety and their role as sentinels. *Animals*, 13(4), 661. https://doi.org/10.3390/ani13040661

Qiu, R., Song, Y., Zhang, X., Xie, B., & He, D. (2020). Microplastics in urban environments: Sources, pathways, and distribution. In D. He, Y. Luo (Eds.), *Microplastics in Terrestrial Environments: Emerging Contaminants and Major Challenges* (pp. 41–61). Springer. https://doi.org/10.1007/698_2020_447

Rafique, A., Irfan, M., Mumtaz, M., & Qadir, A. (2020). Spatial distribution of microplastics in soil with context to human activities: A case study from the urban center. *Environmental Monitoring and Assessment*, 192, 1–13. https://doi.org/10.1007/s10661-020-08641-3

Rahman, A., Sarkar, A., Yadav, O. P., Achari, G., & Slobodnik, J. (2021). Potential human health risks due to environmental exposure to nano-and microplastics and knowledge gaps: A scoping review. *Science of the Total Environment*, 757, 143872. https://doi.org/10.1016/j.scitotenv.2020.143872

Rose, P. K., Yadav, S., Kataria, N., & Khoo, K. S. (2023). Microplastics and nanoplastics in the terrestrial food chain: Uptake, translocation, trophic transfer, ecotoxicology, and human health risk. *TrAC Trends in Analytical Chemistry*, 167, 117249. https://doi.org/10.1016/j.trac.2023.117249

Sadia, M. R., Hasan, M., Islam, A. R. M. T., Jion, M. M. M. F., Al Masud, M. A., Rahman, M. N., ... & Malafaia, G. (2023). A review of microplastic threat mitigation in Asian lentic environments. *Journal of Contaminant Hydrology*, 260, 104284. https://doi.org/10.1016/j.jconhyd.2023.104284

Sajjad, M., Huang, Q., Khan, S., Khan, M. A., Liu, Y., Wang, J., ... & Guo, G. (2022). Microplastics in the soil environment: A critical review. *Environmental Technology & Innovation*, 27, 102408. https://doi.org/10.1016/j.eti.2022.102408

Shafea, L., Yap, J., Beriot, N., Felde, V. J., Okoffo, E. D., Enyoh, C. E., & Peth, S. (2023). Microplastics in agroecosystems: A review of effects on soil biota and key soil functions. *Journal of Plant Nutrition and Soil Science*, 186(1), 5–22. https://doi.org/10.1002/jpln.202200136

Shi, J., Wu, D., Su, Y., & Xie, B. (2021). Selective enrichment of antibiotic resistance genes and pathogens on polystyrene microplastics in landfill leachate. *Science of the Total Environment*, 765, 142775. https://doi.org/10.1016/j.scitotenv.2020.142775

Sridhar, A., Kannan, D., Kapoor, A., & Prabhakar, S. (2022). Extraction and detection methods of microplastics in food and marine systems: A critical review. *Chemosphere*, 286, 131653. https://doi.org/10.1016/j.chemosphere.2021.131653

Surendran, U., Jayakumar, M., Raja, P., Gopinath, G., & Chellam, P. V. (2023). Microplastics in terrestrial ecosystem: Sources and migration in soil environment. *Chemosphere*, 318, 137946. https://doi.org/10.1016/j.chemosphere.2023.137946

Thacharodi, A., Meenatchi, R., Hassan, S., Hussain, N., Bhat, M. A., Arockiaraj, J., ... & Pugazhendhi, A. (2024). Microplastics in the environment: A critical overview on its fate, toxicity, implications, management, and bioremediation strategies. *Journal of Environmental Management*, 349, 119433. https://doi.org/10.1016/j.jenvman.2023.119433

Vairaperumal, T., Selvaraj, D., & Chakraborty, P. (2024). Microplastic in the environment: Sources, workflow, identification techniques, and impacts on human health. In A. L. Srivastav, A. S. Grewal, M. Tiwari, T. D. Pham (Eds.), *Role of Green Chemistry in Ecosystem Restoration to Achieve Environmental Sustainability* (pp. 91–103). Elsevier. https://doi.org/10.1016/B978-0-443-15291-7.00002-X

Voroney, R. P., Heck, R. J., & Kuzyakov, Y. (2024). The habitat of the soil biota. In E. A. Paul, S. D. Frey (Eds.), *Soil Microbiology, Ecology and Biochemistry* (pp. 13–40). Elsevier. https://doi.org/10.1016/B978-0-12-822941-5.00002-8

Wan, L., Cheng, H., Liu, Y., Shen, Y., Liu, G., & Su, X. (2023). Global meta-analysis reveals differential effects of microplastics on soil ecosystem. *Science of the Total Environment*, 867, 161403. https://doi.org/10.1016/j.scitotenv.2023.161403

Wang, F., Wang, Q., Adams, C. A., Sun, Y., & Zhang, S. (2022). Effects of microplastics on soil properties: Current knowledge and future perspectives. *Journal of Hazardous Materials*, 424, 127531. https://doi.org/10.1016/j.jhazmat.2021.127531

Wang, W., Ge, J., Yu, X., & Li, H. (2020). Environmental fate and impacts of microplastics in soil ecosystems: Progress and perspective. *Science of the Total Environment*, 708, 134841. https://doi.org/10.1016/j.scitotenv.2019.134841

Xu, B., Liu, F., Cryder, Z., Huang, D., Lu, Z., He, Y., … & Xu, J. (2020). Microplastics in the soil environment: Occurrence, risks, interactions and fate–A review. *Critical Reviews in Environmental Science and Technology*, 50(21), 2175–2222. https://doi.org/10.1080/10643389.2019.1694822

Ya, H., Jiang, B., Xing, Y., Zhang, T., Lv, M., & Wang, X. (2021). Recent advances on ecological effects of microplastics on soil environment. *Science of the Total Environment*, 798, 149338. https://doi.org/10.1016/j.scitotenv.2021.149338

Yadav, H., Sethulekshmi, S., & Shriwastav, A. (2022). Estimation of microplastic exposure via the composite sampling of drinking water, respirable air, and cooked food from Mumbai, India. *Environmental Research*, 214, 113735. https://doi.org/10.1016/j.envres.2022.113735

Yang, L., Zhang, Y., Kang, S., Wang, Z., & Wu, C. (2021). Microplastics in soil: A review on methods, occurrence, sources, and potential risk. *Science of the Total Environment*, 780, 146546. https://doi.org/10.1016/j.scitotenv.2021.146546

Yee, M. S., Hii, L. W., Looi, C. K., Lim, W. M., Wong, S. F., Kok, Y. Y., Tan, B. K., Wong, C. Y., & Leong, C. O. (2021). Impact of microplastics and nanoplastics on human health. *Nanomaterials*, 11, 496. https://doi.org/10.3390/nano11020496

Yu, L., Zhang, J., Liu, Y., Chen, L., Tao, S., & Liu, W. (2021). Distribution characteristics of microplastics in agricultural soils from the largest vegetable production base in China. *Science of the Total Environment*, 756, 143860. https://doi.org/10.1016/j.scitotenv.2020.143860

Yu, Y., Ding, Y., Miao, C., Yang, X., & Ge, S. (2024). Exploring the influence of sludge dewatering agents on Microplastic aging under hydrothermal treatment: Insights from Polylactic Acid microplastics. *Environmental Research*, 240, 117471. https://doi.org/10.1016/j.envres.2023.117471

Zarus, G. M., Muianga, C., Hunter, C. M., & Pappas, R. S. (2021). A review of data for quantifying human exposures to micro and nanoplastics and potential health risks. *Science of the Total Environment*, 756, 144010. https://doi.org/10.1016/j.scitotenv.2020.144010

Zhang, S., Wang, W., Yan, P., Wang, J., Yan, S., Liu, X., & Aurangzeib, M. (2023). Microplastic migration and distribution in the terrestrial and aquatic environments: A threat to biotic safety. *Journal of Environmental Management*, 333, 117412. https://doi.org/10.1016/j.jenvman.2023.117412

Zhang, Z., Cui, Q., Chen, L., Zhu, X., Zhao, S., Duan, C., … & Fang, L. (2022). A critical review of microplastics in the soil-plant system: Distribution, uptake, phytotoxicity and prevention. *Journal of Hazardous Materials*, 424, 127750. https://doi.org/10.1016/j.jhazmat.2021.127750

Zhu, L., Xie, C., Chen, L., Dai, X., Zhou, Y., Pan, H., & Tian, K. (2023). Transport of microplastics in the body and interaction with biological barriers, and controlling of microplastics pollution. *Ecotoxicology and Environmental Safety*, 255, 114818. https://doi.org/10.1016/j.ecoenv.2023.114818

Zifcakova, L. (2020). Factors affecting soil microbial processes. In R. Datta, R. S. Meena, S. I. Pathan, M. T. Ceccherini (Eds.), *Carbon and Nitrogen Cycling in Soil* (pp. 439–461). Springer. https://doi.org/10.1007/978-981-13-7264-3_13

CHAPTER 8

Innovative Solutions for Soil Remediation from Microplastics Pollution

Hong Nam Nguyen

8.1 INTRODUCTION

Microplastic pollution has emerged as a pervasive and pressing environmental issue in recent years, garnering significant attention from scientists, policymakers, and the general public alike. The widespread distribution of microplastics (MPs) in various environmental compartments, including oceans, freshwater bodies, and even soils, has raised concerns about its potential impacts on ecosystem health and human well-being (Li et al., 2019; Z. Wu et al., 2024). Among these environmental compartments, soils represent a crucial component of terrestrial ecosystems, serving as the foundation for agriculture, biodiversity, and numerous ecosystem services (K. Y. Khan et al., 2024; F. Zhang et al., 2024). Many surveys around the world, especially in Asia, have highlighted the alarming situation of plastic pollution (Table 8.1).

Microplastics, minute plastic particles less than 5 mm in size, have become a ubiquitous presence in soil ecosystems worldwide. This is a

DOI: 10.1201/9781003487579-8

TABLE 8.1 Plastics Contamination in Some Asian Countries

Location	Plastic Source	Soil Type	Estimated Quantity	Composition
China (Li et al., 2022; Yang et al., 2021; Zhang et al., 2021)	Mulch	Arable land	8,885 and 2,899 pieces/kg	PE, PP, and PET
	Sludge	Upland land	149.2±52.5, 68.6±21.5, and 73.1±15.4 particles/kg	PES, PP, and PS-AC
	Plastic gauze	Paddy land	1,629.68 tons/year	PES, PP, and PS-AC
Thailand	Various	Various	12–117 items/m^2	PE, LDPE, PP, and PS
Japan	Coated fertilizer	Paddy land	144 mg/kg	PE
Korea	Various	Various	664 pieces/kg	PE and PP
Vietnam (Doan et al., 2023)	Various	Agricultural soils	1,700–38,800 items/kg	Various
Cambodia (Tun et al., 2022)	Various	Dumping site	218,182 pieces/kg	Various
Philippines (Tun et al., 2022)	Various	Dumping site	24,000 pieces/kg	Various

consequence of various anthropogenic activities, including the degradation of larger plastic items, the application of plastic mulches in agriculture, and the dispersal of microplastic-containing products (Figure 8.1).

Once introduced into soils, MPs can persist for extended periods, undergo physical and chemical transformations, and interact with soil biota and organic matter.

The pervasive presence of MPs in soil ecosystems poses significant threats to soil health, plant growth, and overall ecosystem vitality (Yang et al., 2024; S. Zhang et al., 2024). Firstly, MPs can alter soil physical properties, such as porosity and water retention capacity, thereby affecting soil structure and moisture availability for plant growth. Additionally, their presence can impede soil microbial activity and disrupt soil nutrient cycling processes, leading to diminished soil fertility and productivity. Furthermore, MPs can directly impact plant health by interfering with root development, nutrient uptake, and physiological functions. For example, MPs may physically obstruct root growth, restrict nutrient absorption, or induce stress responses in plants, ultimately compromising their growth and reproductive success (Chu et al., 2023; Yang et al., 2024). Moreover, the accumulation of MPs in plant tissues may pose risks to food

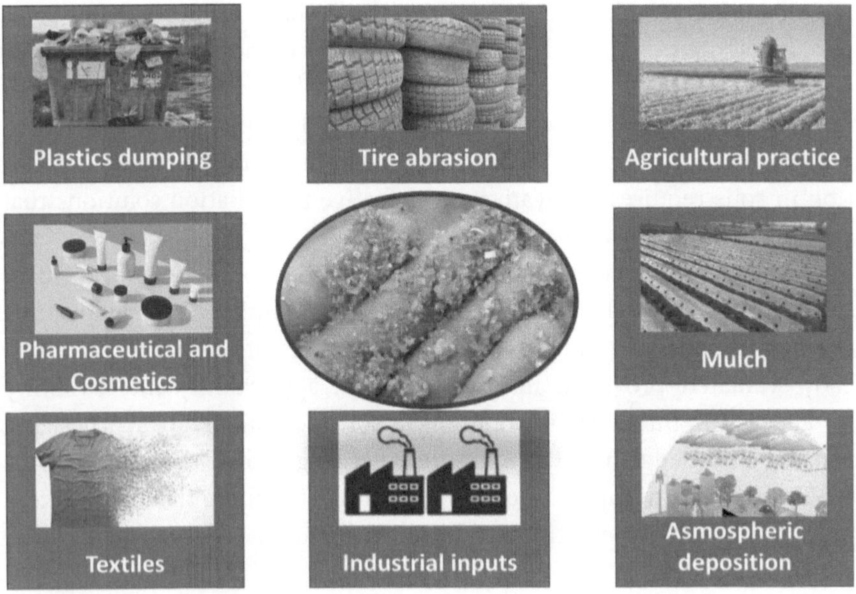

FIGURE 8.1 Different sources of microplastics pollution in soil.

safety and human health, as these particles can translocate from roots to shoots and potentially enter the food chain. Beyond their effects on soil and plant health, MPs can exert broader ecological consequences by disrupting ecosystem dynamics and functions. For instance, MPs may serve as vectors for the transport of harmful pollutants, such as pesticides and heavy metals, throughout the soil food web, leading to bioaccumulation and biomagnification in higher trophic levels. Additionally, MPs can alter soil biodiversity and community composition by influencing the abundance and activity of soil organisms, including microorganisms, invertebrates, and soil-dwelling fauna (Han et al., 2024).

The removal of plastics from soil presents greater challenges and necessitates more intricate methods compared to aquatic environments (Dong et al., 2024; I. Khan et al., 2024). Soil is a complex, heterogeneous medium with varying textures, organic matter content, and moisture levels, which can influence the distribution and mobility of plastics within it. Microplastics in soil tend to have limited mobility compared to water, as they are often trapped within pore spaces or bound to soil particles, making their removal more challenging. Microplastics in soil may interact with soil particles and organic matter through physical adsorption, making them difficult to extract using conventional methods. Moreover, MPs in soil can persist for long periods due to slower degradation rates compared to aquatic environments, posing risks to soil health and ecosystems (Adhikari et al., 2024). As a consequence, not all of MPs removal methods developed for aqueous environments are directly applicable or effective in soil remediation contexts. Addressing the challenge of microplastic pollution in soils requires innovative and effective remediation solutions that can mitigate its environmental impacts.

This chapter provides an in-depth exploration of a range of innovative solutions for soil remediation from MPs pollution. Throughout this chapter, the efficacy and limitations of each remediation solution are critically evaluated, providing insights into their applicability and feasibility for addressing microplastic pollution in soil ecosystems. By synthesizing existing knowledge and highlighting emerging technologies, this chapter aims to inform future research directions and guide the development of effective soil remediation strategies tailored to combat microplastic contamination.

8.2 PHYSICO-CHEMICAL METHODS

8.2.1 Advanced Soil Washing

Soil washing is a method of physically removing MPs from soil by capitalizing on distinctions in their density, size, or hydrophobic properties (Li et al., 2019). Usually, it utilizes water-based washing procedures, wherein soil samples are mixed with water and agitated or stirred to displace MPs (Gu et al., 2022). Subsequently, the suspension is permitted to settle, and the supernatant, which contains the separated MPs, is collected.

Various advanced soil washing technologies have been developed to achieve high removal efficiencies of MPs from soil matrices. The use of novel solvents, surfactants, or chelating agents that enhance the solubilization and mobilization of MPs has been discussed in previous studies (Liu et al., 2022; Mushtak et al., 2024). Utilization of ultrasonic or sonication techniques to enhance the disaggregation and dispersion of MPs from soil aggregates, followed by separation and concentration using physical methods such as sedimentation or centrifugation (Goli and Singh, 2023). Ultrasonic energy can disrupt intermolecular forces and weak bonds holding MPs within soil matrices, facilitating their release and subsequent removal with reduced energy and time requirements. Development of microfluidic devices or microscale reactors for soil washing applications, enabling precise control over flow conditions, mixing dynamics, and chemical interactions during the soil washing process (AlMashrea et al.,

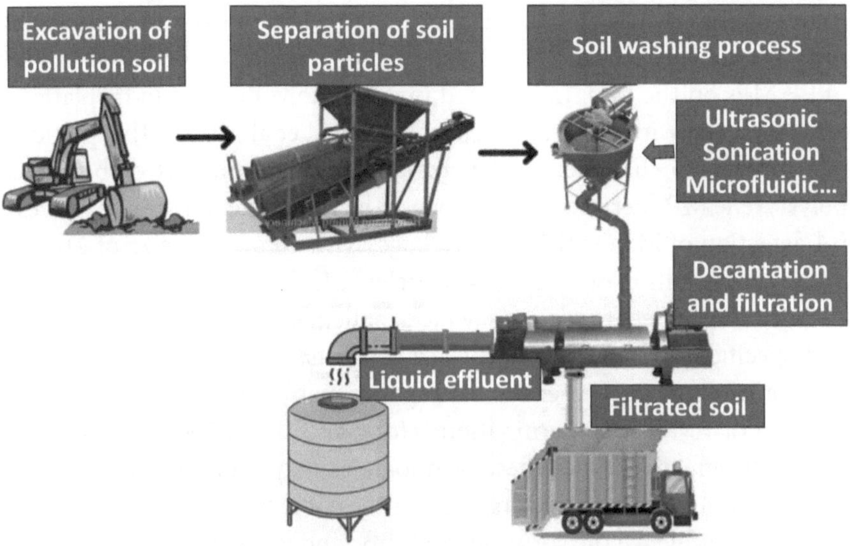

FIGURE 8.2 Main steps of an industrial soil washing process for removal of microplastics.

2024; Ece et al., 2023; Mesquita et al., 2022). Figure 8.2 summarizes the main steps of an industrial soil washing process for the removal of MPs.

For instance, the ultrasonic extraction of the common plastic additive di(2-ethylhexyl) phthalate (DEHP) from aquatic sediment proved most effective through a modified Fenton process enhanced by ultrasonic energy. This method yielded the highest average degradation efficiency of 70.74% under conditions of $C_0=10$ mg/kg dry weight and $t=60$ minutes (Kida et al., 2021). Microfluidic systems are also part of advanced soil washing technologies, which offer advantages such as reduced reagent consumption, enhanced mass transfer rates, and scalability, making them promising tools for on-site or in-situ soil remediation targeting MPs. By optimizing process parameters such as solvent concentration, temperature, and contact time, advanced soil washing technologies can effectively target specific types and sizes of MPs, leading to improved remediation outcomes. A PDMS-based microfluidic system was developed, featuring a 400 μm channel width for the detection of polyethylene microplastics (PE MPs) in soil (Wu et al., 2021). The system utilized a film whose thickness varied based on the number of capillaries, effectively capturing these impurities. The effectiveness of this detection method increased with greater film thickness. The study also explored the behavior of MPs at different flow velocities (ranging from 42 to 2,096 μm/s) and particle volume fractions. Higher flow rates and thicker films led to increased capture of PE MPs on the film. In a related investigation, a microfluidic platform was employed with integrated sensors (Pollard et al., 2020). This platform was utilized to characterize the size and concentration of carboxylated polystyrene MPs derived from tea bags. The results demonstrated successful detection of MPs within the tea bags, with an average size of 21.9 μm and a concentration of 6.52 × 104 particles/mL.

Soil washing technologies may face challenges related to the selectivity and specificity of solvents or chelating agents, which may interact with soil constituents and co-extracted organic matter, leading to incomplete removal or residual contamination (Hong and Jiang, 2005; Zhang et al., 2010). The effectiveness of these technologies may vary depending on soil properties, such as texture, organic matter content, and clay mineralogy, which can influence the sorption, mobility, and availability of MPs in soil matrices (Zheng et al., 2022). Scale-up and implementation of advanced soil washing technologies at field sites may be hindered by logistical constraints, including equipment costs, energy requirements, and the availability of suitable reagents, limiting their practical applicability and cost-effectiveness (Sutkar et al., 2023). Soil washing technologies may also generate wastewater or wash effluents containing concentrated MPs, solvents, or chemical additives, posing risks of environmental contamination and pollution if not properly managed and treated.

8.2.2 Electrokinetic Remediation

Electrokinetic remediation is reported as a novel electrochemical method utilized to extract MPs from various media, including soil (Hildebrandt et al., 2021; Thompson and Crooks, 2022). The principle of electrokinetic remediation for microplastic remediation in soil revolves around the application of an electric field to mobilize charged MPs within the soil matrix. When an electric field is applied across the soil, it induces electroosmotic flow and electrophoresis (Akarsu et al., 2021; Vocciante et al., 2021). Electroosmotic flow occurs when the electric field causes water and ions to move within the soil pores, dragging along charged particles such as MPs. Electrophoresis involves the movement of charged particles, including MPs, in response to the electric field. As a result of these processes, MPs are mobilized and transported towards the electrodes, where they can be collected and removed from the soil. This method allows for the targeted extraction of MPs from soil without the need for extensive excavation or disturbance.

Through the application of suitable electric potentials and regulation of factors like soil conductivity, pH, and electrolyte concentration, electrokinetic remediation can successfully mobilize and extract MPs from soil samples (Acar et al., 1995). As an example, a previous study reported the successful design of an ion concentration polarization-based continuous separation device using electrical repulsion in the depletion region that could facilitate the continuous separation of MPsmeasuring 4.8 μm and 9.9 μm. The 9.9 μm MPs, possessing greater mobility, are deflected farther by the ion depletion zone and electric field gradient compared to the 4.8 μm MPs. As a result, microplastic populations can be collected in distinct outlets. In a similar experiment, MPs measuring 500 and 100 nm could be separated. This technique offers advantages such as uniform treatment of soil volumes, minimal soil disturbance, and potential applicability to a wide range of soil types and contaminant distributions, enhancing its efficacy for microplastic remediation (Mustapha et al., 2024).

Electrokinetic remediation may be limited by factors such as soil heterogeneity, moisture content, and electrode placement, which can affect the distribution and mobility of MPs within soil profiles and complicate remediation efforts (Chaukura et al., 2022). The effectiveness of electrokinetic remediation may vary depending on the physicochemical properties of MPs, including their surface charge, size, and aggregation state, which can influence their response to electric fields and electroosmotic flow (Ihenetu et al., 2024). Scale-up and implementation of electrokinetic remediation at field sites may be also challenging. As the size of the remediation site increases, so does the energy required to maintain the electric field and sustain electrokinetic processes. Designing suitable electrode configurations and placement strategies to accommodate larger soil volumes while maintaining optimal electric field distribution can pose technical challenges (Kim et al., 2021). Ensuring sufficient electrode contact with the soil and minimizing electrode corrosion or degradation over time are also important considerations. The use of electric currents in soil remediation processes may also induce changes in soil pH, redox potential, and microbial activity, potentially affecting soil fertility, nutrient cycling, and ecosystem functioning, with potential long-term environmental impacts.

8.2.3 Magnetic Separation

Magnetic separation is a physical removal technique used to separate MPs from complex soil matrices by exploiting their magnetic properties. Microplastics can become magnetized through various processes, such as exposure to magnetic fields or the incorporation of magnetic nanoparticles during their production or surface modification (Grbic et al., 2019; Ramage et al., 2022). The process typically involves the application of magnetic fields or the use of magnetic nanoparticles dispersed in the soil slurry (Ramage et al., 2022). When a magnetic field is applied, magnetized MPs are attracted to the source of the magnetic field, allowing them to be captured and separated from the soil matrix. In some cases, magnetic nanoparticles specifically designed to interact with MPs are added to the soil slurry. These nanoparticles bind to the MPs, forming magnetic complexes that can be easily manipulated using magnetic fields (Rhein et al., 2022). By applying a magnetic field gradient, the magnetic complexes containing MPs are directed towards a collection point, where they can be removed from the soil slurry. The principle of magnetic separation for MPs in soil is presented in Figure 8.3.

Magnetic separation can selectively target MPs based on their magnetic properties, allowing for efficient removal while leaving other soil components unaffected. When properly implemented, magnetic separation methods can achieve high recovery rates of MPs from soil, contributing to effective pollution mitigation efforts. In a recent study, a novel extraction technique comprising three stages was devised (Ramage et al., 2022). The process begins with high-gradient magnetic separation to eliminate magnetic soil (Stage 1). Subsequently, surface-modified iron nanoparticles are employed to magnetically tag MPs (MPs) (Stage 2), followed by the high-gradient magnetic recovery of surface-modified MPs (Stage 3). The method was tested across four distinct soil types—loam, high-carbon

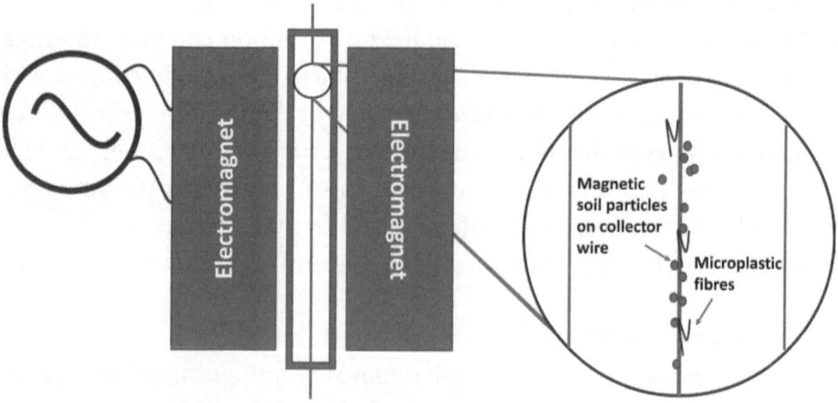

FIGURE 8.3 Principle of magnetic separation for microplastics in soil.

loamy sand, sandy loam, and high-clay sandy loam—each spiked with varying types of MPs (polyethylene, polyethylene terephthalate, and polytetrafluoroethylene) spanning a range of particle sizes from 63 to 2 mm, along with polyethylene fibers measuring 2–4 mm. The method yielded average recoveries of 96% for fibers and 92% for particles in loam, 91% for fibers and 87% for particles in high-carbon loamy sand, 96% for fibers and 89% for particles in sandy loam, and 97% for fibers and 94% for particles in high-clay sandy loam. In another study, hydrophobic Fe nanoparticles were designed to adhere to plastic, enabling magnetic retrieval (Grbic et al., 2019). The results achieved a retrieval rate of 92% for 10–20 μm polyethylene and polystyrene beads, along with 97% retrieval for MPs (>1 mm) including polyethylene, polyethylene terephthalate, polystyrene, polyurethane, polyvinyl chloride, and polypropylene, from seawater. About 84% and 78% of MPs spanning from 200 μm to 1 mm were also recovered, encompassing polyethylene, polystyrene, polyurethane, polyvinyl chloride, and polypropylene, from freshwater and sediment samples, respectively.

Nevertheless, magnetic separation technologies also face different challenges. This technique may not be effective for removing very small MPs or those with weak magnetic properties, limiting its applicability to certain types or sizes of MPs (Bakhteeva et al., 2023; Grbic et al., 2019). Soil composition and moisture content can also interfere with the magnetic separation process, potentially reducing its efficiency and effectiveness in removing MPs. The physical process of magnetic separation may disturb soil structure and biota, potentially leading to unintended ecological impacts (Mullins, 1977). The magnetic separation process may generate hazardous byproducts or secondary pollutants, particularly if the soil contains contaminants or reactive components. For example, the use of magnetic nanoparticles or additives in the separation process could result in the release of toxic substances or the formation of undesirable compounds, posing environmental and health risks if not properly managed.

8.2.4 Ultraviolet Radiation

The principle of using ultraviolet (UV) radiation for MPs in soil involves the application of specific wavelengths of UV light to induce degradation or alteration in the chemical structure of the MPs. This process is primarily based on photooxidation, where UV radiation interacts with the molecular bonds of the plastic polymers, leading to the formation of reactive oxygen species such as hydroxyl radicals (•OH), singlet oxygen ($^1O_2$), and superoxide radicals ($\bullet O_2^-$). These reactive oxygen species attack the polymer chains of MPs, breaking them into smaller fragments through processes like chain scission, cross-linking, and oxidation. As a result, the MPs undergo physical and chemical changes, ultimately leading to their degradation into smaller particles or even complete mineralization into non-toxic compounds such as carbon dioxide, water, and inorganic ions. UV radiation treatment processes can be relatively fast and efficient, allowing for the rapid removal of MPs from soil. This can be particularly beneficial for remediation projects requiring timely intervention to mitigate environmental impacts or prevent further contamination spread. This technique can also be effective for the removal of a wide range of contaminants, including various types of MPs present in soil. This versatility allows for the treatment of different types of MPs, regardless of their composition, size, or shape. The impact of UV-A and UV-C radiation, administered at 70°C, on three prevalent types of microplastics (MPs)—polypropylene (PP), polyethylene (PE), and polyethylene terephthalate (PET)—commonly encountered in sewage sludge, was investigated across three different shapes (fibers, lines, and granules) (Alavian Petroody et al., 2023). The MPs were subjected to UV radiation within biosolid suspensions. The results highlight that the combination of ultraviolet radiation and elevated temperature alters the polymer structure within wastewater biosolids, thereby expediting their degradation process.

A recent study analyzed and compared the surface degradation process and microplastic release from two types of plastic films—polyethylene (PE) and thermoplastic polyurethane (TPU)—under the influence of ultraviolet radiation (UVR), mechanical abrasion (MA), and combined UVR+MA treatments (Sun et al., 2022). It was observed that the combined UVR+MA treatment induced more pronounced surface changes in both PE and TPU films compared to individual UVR or MA treatments. This combined treatment resulted in accelerated surface degradation compared to UVR or MA alone, leading to a significant release of MPs from both PE and TPU films. Notably, the UVR+MA treatment generated the highest quantity of MPs with the smallest particle size, followed by MA and UVR treatments. UV radiation-based treatment methods can also be scaled up to treat large areas of contaminated soil. This scalability makes UV radiation a viable option for remediation projects requiring treatment of extensive soil areas affected by microplastic pollution. Another study subjected LDPE films to pretreatment with a 15-W UV lamp for 100 hours, and then placed them in a soil environment to monitor their degradation

(Tribedi and Dey, 2017). After 28 days of exposure, it was noted that the pretreated films experienced more decomposition compared to those that had not undergone pretreatment.

Nevertheless, UV radiation may have limited penetration depth into the soil matrix, especially in densely packed or clay-rich soils. This can result in uneven treatment and incomplete removal of MPs from deeper soil layers (Yang et al., 2019). Oil particles and organic matter can absorb or scatter UV radiation, reducing its effectiveness in reaching and degrading MPs embedded within the soil matrix. Additionally, soil moisture content and pH levels can affect UV radiation absorption and dispersion, further limiting its efficacy. UV radiation treatment methods may also be subject to regulatory requirements and guidelines governing the use of UV radiation for environmental remediation (Shafea et al., 2023). Compliance with safety standards and regulations, including limits on UV exposure levels and protection measures for workers and nearby communities, must be ensured to minimize potential health and safety risks.

8.2.5 Centrifugation

A centrifuge-based method has the capability to effectively eliminate a significant portion of the soil matrix while exhibiting high selectivity for the polymer (Lusher et al., 2020; Pfohl et al., 2024; Zhang et al., 2023). The principle of the centrifugation method for the removal of MPs in soil is based on the physical separation of particles of different sizes and densities through centrifugal force (Zhang et al., 2023). In this method, soil samples are mixed with a liquid solution, often containing a surfactant or density gradient medium, to create a suspension. The suspension is then subjected to high-speed rotation in a centrifuge, causing the denser soil particles and MPs to sediment at the bottom of the tube or container, while lighter particles remain suspended in the liquid. By adjusting the centrifugal force and duration of centrifugation, MPs can be effectively separated from soil particles based on their size and density. The separated MPs can then be collected from the sediment for further analysis or disposal, facilitating their removal from the soil matrix. In a recent study, a centrifugal separation process for removal of MPs in soil was developed (Grause et al., 2022). Using a $CaCl_2$ solution with a density of 1.4 g/mL, all major polymers—including polyethylene, polypropylene, polystyrene, poly (vinyl chloride), and poly (ethylene terephthalate)—were effectively recovered from thermally altered soil, with a recovery rate exceeding 94 wt%. Subsequently, this protocol was applied to agricultural soil samples collected from a site with known contamination. The highest levels of contamination were observed in close proximity to the hotspot, with 75×10^3 particles/kg, corresponding to a weight range of 20–6 mg/kg at a distance of 1 m. The particle count decreased logarithmically to 30×10^3 particles/kg, or 5–2 mg/kg, with increasing distance from the hotspot. Centrifugation and centrifugation combined with oxidation have been employed to determine the recovery rates of various types of MPs, such as expanded polystyrene (EPS), polyvinyl chloride (PVC), polyurethane (PU), polytetrafluorethylene (PTFE), polypropylene (PP), and nylon (PA), across three soil types with varying ratios of sand, silt, and clay (Perez et al., 2021). The average recovery rates for centrifugation and centrifugation with oxidation were found to be 94.4%±4.52 and 95.4%±4.85, respectively. The method can achieve high removal efficiencies, particularly for MPs with densities different from that of soil particles (Pfohl et al., 2024). It is also a relatively fast process, allowing for the processing of multiple samples in a short period and can be adapted for use with various types of soil samples and MPs.

However, centrifugation also poses various challenges in the context of microplastic removal from soil. Centrifugation necessitates specialized equipment, often unavailable in field settings, limiting its application primarily to laboratory-scale studies (Grause et al., 2022). Centrifugation may also cause disturbance to soil samples, potentially altering their properties and affecting subsequent analyses. The effectiveness of centrifugation for microplastic removal may be limited to certain size ranges, with smaller or larger particles requiring alternative methods (Zhang et al., 2023). Centrifugation also consumes a big amount of energy, particularly in high-speed or long-duration processes, contributing to operational costs and environmental impact (Perez et al., 2021).

8.3 BIO-CHEMICAL METHODS

8.3.1 Use of Surfactants

Surfactants are amphiphilic molecules that contain both hydrophilic (water-attracting) and hydrophobic (water-repelling) regions. When applied to contaminated soil, surfactants reduce the surface tension of water and increase its wetting ability (Liu et al., 2021; Paria, 2008). Common surfactants used for enhanced soil remediation include nonionic surfactants, anionic surfactants, cationic surfactants, amphoteric surfactants, and biosurfactants (Renzi et al., 2019). The types of surfactants and their related effects are presented in Table 8.2.

TABLE 8.2 Type of Surfactants and Its Related Effect

No	Type	Example	Effect
1	Nonionic surfactants	Triton X series, alkylphenol ethoxylates, and alkyl polyglucosides	Reduce surface tension and enhance solubilization of hydrophobic contaminants
2	Anionic surfactants	Linear alkylbenzene sulfonates (LAS), sodium dodecyl sulfate (SDS), and alkyl sulfates	Form stable micelles and increase the solubility of contaminants, making them suitable for soil washing
3	Cationic surfactants	Alkyltrimethylammonium salts and benzalkonium chlorides	Promote soil aggregation and reduce soil-water surface tension, enhancing the removal of organic contaminants
4	Amphoteric surfactants	Betaines and amphoterics	Effective in various soil types and pH conditions, aiding in the removal of both hydrophobic and hydrophilic contaminants
5	Biosurfactants	Rhamnolipids, sophorolipids, and lipopeptides	Enhance microorganism activities, hence accelerate microplastics degradation

Surfactant-enhanced soil remediation technologies have traditionally been utilized for addressing organic compounds and heavy metals, but they can also be adapted to tackle emerging contaminants like MPs. The unique properties of surfactants offer distinct advantages in treating these new types of contaminants (Tiwari and Tripathy, 2023). By reducing the adhesion of MPs to soil particles, surfactants facilitate their detachment and increase their mobility. This enables easier extraction or separation of MPs from the soil matrix, thereby enhancing the efficiency of the remediation process. Additionally, surfactants can play a crucial role in the transport and distribution of treatment agents for MPs in soil. Acting as carriers for other remedial agents such as sorbents, enzymes, or chemicals, surfactants assist in their delivery to the target contaminants. This not only improves the efficiency of the treatment but also enhances the contact between the treatment agents and the contaminants, facilitating their interaction and removal. Overall, surfactant-enhanced soil remediation technologies hold promise in addressing the challenges posed by microplastic contamination in soil environments (Mohanty et al., 2013). A recent study showed that the inclusion of 1,500 mg/L of surface-active compounds led to the increased degradation of PAHs observed in both the biostimulation-assisted biodegradation with the aeration system (from 61% to 100%) and the biostimulation-assisted biodegradation with the aeration and filtration system (from 54% to 86%) (Rathankumar et al., 2022). Notably, the surface-active compounds-supplemented biostimulation-assisted biodegradation with the aeration system achieved complete degradation (100%) of naphthalene and acenaphthene within 17 days. Additionally, ecotoxicity analysis indicated that the LC50 of sediment soil decreased by an average of 71% after soil remediation, demonstrating a significant reduction in toxicity. The effectiveness of surfactant-enhanced soil washing combined with adsorption for removing PAHs from real contaminated soil, as well as the feasibility of surfactant recycling, was investigated (Mirzaee and Sartaj, 2022). A synthesized magnetic granular activated carbon composite served as the adsorbent, with Tween 80 as the surfactant. It was found that only 5%–10% of Tween 80 would be adsorbed onto the soil particles, indicating a high recovery of the surfactant solution from the soil. Moreover, increasing the Tween 80 dose from 1% to 5% (v/v) resulted in a rise in the total solubility of PAHs by approximately 40%. On average, the percent removal of PAHs from the contaminated soil was 67.6%.

Nevertheless, some limitations of using surfactants for the removal of MPs in soil can be discussed. Surfactants may not be equally effective for all types of MPs or soil matrices (Jiang et al., 2021). Factors such as the size, shape, and surface properties of MPs, as well as soil texture and organic matter content, can influence the effectiveness of surfactant-based removal methods. Surfactants themselves can become contaminants if not properly managed, potentially leading to soil and groundwater pollution. Additionally, some surfactants may persist in the environment, posing long-term risks to ecosystems. The use of surfactants may have adverse effects on soil microbial communities, plant growth, and other soil organisms (Xia et al., 2020). Certain surfactants may be toxic to non-target organisms or disrupt soil ecosystems, leading to unintended ecological consequences. Moreover, surfactants may interact with soil components and other contaminants present in the soil, leading to unintended consequences such as the mobilization of other pollutants or changes in soil properties. Understanding these interactions is crucial for predicting the efficacy and potential risks of surfactant-based soil remediation methods.

8.3.2 Bioaugmentation

Bioaugmentation involves introducing specialized microbial consortia or genetically engineered microorganisms into soil to enhance the biodegradation of MPs (Auta et al., 2022; Salinas et al., 2023; Zhang and Zhang, 2022). These microorganisms may possess specific enzymes or metabolic pathways optimized for breaking down plastic polymers, thereby accelerating biodegradation rates. The principle of microbial bioaugmentation for MPs in soil revolves around the ability of microorganisms to enzymatically break down the polymer chains of MPs, leading to their degradation into smaller fragments (Barone et al., 2024). Microorganisms in soil colonize the surface of MPs through adhesion and biofilm formation. Then, microorganisms produce extracellular enzymes, such as lipases, proteases, and cellulases, which catalyze the hydrolysis of chemical bonds in the polymer chains of MPs. These enzymes break down ester, peptide, and glycosidic bonds present in MPs, leading to the cleavage of polymer chains into smaller oligomers and monomers. Microorganisms metabolize the breakdown products of MPs as carbon and energy sources. Monomers and oligomers released from the degradation of MPs can be assimilated by microorganisms and utilized in cellular metabolism, supporting microbial growth and activity. Microorganisms assimilate the degradation products of MPs into their biomass, incorporating carbon and other nutrients derived from MPs into cellular components. Some microorganisms possess metabolic pathways capable of mineralizing MPs completely, converting them into simpler, non-toxic compounds such as carbon dioxide, water, and inorganic ions (Anand et al., 2023). In a previous study, three marine bacteria from the Arabian Sea in India, *Kocuria palustris M16*, *Bacillus pumilus M27*, and *Bacillus subtilis H1584*, have been tested for MPs degradation (Raghul et al., 2014). Following 30 days of cultivation in a medium with polyethylene (PE) as the sole carbon source, Fourier Transform Infrared Spectroscopy (FTIR) analysis revealed an increase in the ketone carbonyl bond index, ester carbonyl bond index, and vinyl bond index on the surface of PE. This increase indicated the degradation of PE, resulting in weight losses of 1%, 1.5%, and 1.75%, respectively. The isolated *Vibrio parahaemolyticus* (BTTV4 and BTTN18) and *Vibrio alginolyticus* (BTTC10 and BTTC27) from marine sediments and evaluated their degradation efficiency of a polyvinyl alcohol-low linear density polyethylene (PVA-LLDPE) blend plastic film (Harshvardhan and Jha, 2013). Following 15 weeks of incubation in shake flasks at 120 rpm and 37°C, scanning electron microscopy (SEM) revealed visible cracks and grooves on the surface of the PVA-LLDPE blend film, indicating degradation. The study also noted a greater percentage of tensile strength loss in PVA-LLDPE plastic films containing 25% and 30% PVA, with tensile strength decreasing as PVA content increased. This observation suggests that the PVA component can serve as a carbon source to enhance bacterial degradation of PVA-LLDPE. Cooperative microbial consortia can enhance the efficiency and specificity of microplastic degradation by collectively producing a diverse array of enzymes targeting various chemical bonds in MPs (Cao et al., 2022).

Nevertheless, challenges for bioaugmentation are also various. This technique relies on introducing specific microbial strains capable of degrading MPs. However, identifying and obtaining suitable strains that can efficiently degrade MPs in soil may be challenging (Cai et al., 2023). Introduced microbial strains may struggle to survive and persist in the soil environment, especially if conditions such as pH, temperature, moisture, and nutrient availability are not optimal for their growth and activity. Introduced microbial strains may also face competition with native microorganisms present in the soil, which can limit their effectiveness in degrading MPs. Moreover, degradation efficiency of bioaugmentation may vary depending on factors such as the type and concentration of MPs, soil composition, and environmental conditions. In some cases, bioaugmentation may not achieve complete degradation of MPs (Tyagi et al., 2011). Introducing non-native microbial strains into the soil ecosystem for bioaugmentation purposes may pose environmental risks, such as unintended ecological disruptions or the spread of genetically modified organisms (Nuzzo et al., 2020). The long-term effects of bioaugmentation on soil health, microbial diversity, and ecosystem functioning are not fully understood. There is a need for further research to assess the potential impacts of bioaugmentation on soil ecosystems over extended periods.

8.3.3 Phytodegradation

The principle of phytodegradation for MPs in soil relies on the activities of plants and their associated rhizospheric microorganisms to degrade MPs (Sutkar et al., 2023; Figure 8.4). Plant roots exude organic compounds into the soil, creating a nutrient-rich microenvironment known as the rhizosphere. This zone harbors a diverse community of microorganisms that interact with plant roots and organic matter. Microorganisms in the rhizosphere, including bacteria and fungi, possess enzymes capable of breaking down the polymer chains of MPs (Lozano et al., 2024).

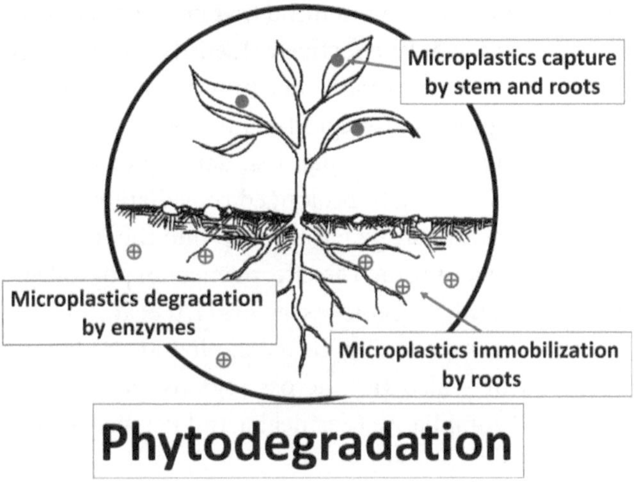

FIGURE 8.4 Principle of phytodegradation for microplastics in soil.

These enzymes catalyze the hydrolysis of chemical bonds in MPs, leading to their fragmentation into smaller oligomers and monomers. Plants can stimulate microbial activity in the rhizosphere through the release of root exudates, which serve as carbon and energy sources for soil microorganisms. In turn, microorganisms contribute to the degradation of MPs by producing extracellular enzymes and metabolizing microplastic-derived carbon compounds. The presence of plants and their associated microorganisms enhances the degradation of MPs in soil by promoting enzymatic hydrolysis and microbial metabolism (Zhong et al., 2024). Plant roots may also physically disrupt MPs in the soil matrix, facilitating their degradation and incorporation into soil organic matter. Phytodegradation of MPs contributes to the cycling of carbon and other nutrients in terrestrial ecosystems. Microbial degradation products released during phytodegradation can be assimilated by plants and soil microorganisms, supporting plant growth and ecosystem productivity. In a recent study, it has been proved that mangrove vegetation has the capacity to collect MPs in their roots, stems, and foliage by means of absorption, adsorption, and entrapment, thereby aiding in the reduction of microplastic levels in the adjacent environment (Talukdar et al., 2023). Various bacterial and fungal species have been identified in mangrove ecosystems, demonstrating the ability to break down MPs. Leveraging the bioremediation capabilities of mangrove plants presents a novel and environmentally sustainable strategy for addressing microplastic contamination.

Nevertheless, it should be kept in mind that not all plant species possess the capability to degrade MPs effectively. Identifying and utilizing plant species with appropriate enzymatic machinery for microplastic breakdown can be challenging. Phytodegradation of MPs may occur at a slow rate, especially in soils with low microbial activity or unfavorable environmental conditions, leading to prolonged remediation timelines (Arpia et al., 2021). Microplastics may penetrate deep into the soil profile, beyond the reach of plant roots, limiting the effectiveness of phytodegradation in targeting these buried microplastic reservoirs (Du et al., 2021). Regulatory frameworks and policies governing the use of phytodegradation for microplastic remediation may also vary across regions, posing challenges for implementation and compliance (Fermeglia and Perišić, 2023).

8.3.4 Enzymatic Treatment

The principle of enzyme treatment for MPs in soil revolves around utilizing specific enzymes capable of breaking down the chemical bonds present in plastic polymers (Mbachu et al., 2021). Enzymes with activity against the types of polymers present in MPs are selected. Examples include lipases, proteases, cellulases, and esterases, which target various chemical bonds found in plastics, such as ester, peptide, and glycosidic bonds (Chow et al., 2022). Enzymes are applied to soil contaminated with MPs, either directly or through enzymatic solutions or formulations. The enzymes may be added to soil as powders, liquids, or encapsulated forms, depending on the specific application method. The enzymes come into contact with microplastic particles present in soil (Ben Hmad and Gargouri, 2024). Enzymes can adhere to the surface of MPs or penetrate into the polymer matrix, depending on factors such as enzyme size, concentration, and affinity for the plastic substrate. Enzymes catalyze hydrolysis reactions, breaking down the polymer chains of MPs into smaller fragments or monomers (Jain et al., 2023). This process involves the cleavage of chemical bonds within the polymer structure, leading to the fragmentation of MPs into oligomers, dimers, and ultimately, monomers. The fragmented microplastic particles are more susceptible to physical and biological degradation processes in soil. Soil microorganisms, such as bacteria and fungi, may further metabolize the smaller plastic fragments, contributing to their eventual degradation and mineralization into harmless compounds. An innovative enzymatic purification method for MPs in soil had been studied (Mbachu et al., 2021). This method showcased remarkable efficacy by reducing organic material content by over 90% in soil samples.

Following its successful digestion, the method was applied to spiked microplastic soil samples, resulting in high recovery rates and the preservation of plastic characteristics. Hydrolase enzymes, specialized in depolymerizing polyester-based plastics, also offer a bioremediation solution applicable to agricultural soils (in situ), biofertilizers, and irrigation water (ex situ) (Palacios-Mateo et al., 2023).

However, like bioaugmentation and phytodegradation methods, enzymes used for microplastic degradation may have specificity towards certain types or compositions of MPs (Palacios-Mateo et al., 2023). This can limit the effectiveness of enzyme treatment for a wide range of microplastic contaminants. Enzyme-mediated degradation of MPs may occur at a relatively slow rate, particularly in soils with low enzyme activity or unfavorable environmental conditions. Enzymes used for microplastic degradation may degrade or become inactive over time in soil environments due to factors such as microbial degradation, chemical degradation, or denaturation (Guo et al., 2023). This can limit the long-term effectiveness of enzyme treatment. The use of enzymes for microplastic degradation may also have unintended environmental consequences, such as the release of by-products or the alteration of soil microbial communities (Palacios-Mateo et al., 2023). It is important to assess the potential environmental impact of enzyme treatment and implement appropriate mitigation measures.

8.3.5 Addition of High Sorption (Nano)Materials into Soil

The principle of adding high sorption (nano)materials into soil for MPs degradation involves incorporating materials with high sorption capacity, such as nanoparticles or other sorbents, into the soil environment (Q. Wu et al., 2024). Unlike in aquatic environments where high sorption capacities, such as activated carbon or biochar, directly adsorb MPs into their porous structures, soil amendments with high sorption capacities can adsorb MPs onto their surfaces (Siipola et al., 2022). This facilitates MPs degradation by creating a concentrated microenvironment for microbial activity and enzymatic degradation. The soil porosity and aeration will also increase, facilitating the diffusion of oxygen and nutrients to microorganisms involved in microplastic degradation. Improved soil structure also enhances microbial access to MPs and promotes the penetration of enzymes into the soil matrix. In a recent study, the combination of MPs and biochar was proved to have the potential to mitigate the adverse impacts of microplastic accumulation on sugarcane growth and soil quality (Q. Wu et al., 2024). Treatments involving the combination of MPs and biochar notably enhanced the prevalence of *Subgroup_10* in the *16S rRNA* gene, whereas treatments with microplastics alone substantially elevated the relative abundance of *Streptomyces* in the *PhoD* gene.

One limitation of high-sorption (nano)materials in soil remediation from MPs is their potential to interfere with soil nutrient availability and microbial activity. Both activated carbon and biochar possess high sorption capacities, which can lead to the adsorption of not only MPs but also essential nutrients and organic matter in the soil (Mirzaee and Sartaj, 2022). This can result in nutrient depletion and altered soil microbial communities, impacting soil fertility and overall ecosystem health. Additionally, the presence of organic matter or other soil constituents may reduce the sorption capacity of these materials, limiting their efficiency in removing MPs from the soil (Q. Wu et al., 2024). Furthermore, there is a risk that these materials may degrade over time, releasing adsorbed MPs back into the soil.

8.4 CONCLUSION

The chapter provided comprehensive insights into innovative solutions for soil remediation from MPs pollution. Physico-chemical remediation methods aim to either physically remove MPs or modify their physicochemical properties to enhance degradation or immobilization in soils. Innovative remediation methods include advanced soil washing, electrokinetic remediation, centrifugation, and ultraviolet radiation. These methods offer potential for mass removal of MPs from soil, with rapid and effective processing times. Moreover, various types of MPs can be simultaneously addressed. However, common challenges of these methods include high energy consumption and high investment costs, thus hindering their widespread application.

Bio-chemical remediation strategies harness the metabolic activities of microorganisms, plants, or other living organisms to degrade, transform, or sequester MPs in soils. Innovative remediation methods include the use of surfactants, enzymatic treatment, phytoremediation, bioaugmentation, or the addition of high sorption (nano)materials into soil. These biochemical approaches offer several advantages, such as targeted treatment of specific MPs, generally low investment costs, and applicability on a wide scale. Additionally, these solutions are relatively environmentally friendly. However, these methods often only address surface-level MPs due to limitations in penetration depth. Long-term impacts of these approaches may occur due to changes in the soil environment and its microorganisms, although further research is needed to fully understand these effects. Furthermore, regulatory barriers and policies may pose obstacles due to concerns about irreversible changes caused by these potentially impactful methods.

REFERENCES

Acar, Y.B., Gale, R.J., Alshawabkeh, A.N., Marks, R.E., Puppala, S., Bricka, M., Parker, R., 1995. Electrokinetic remediation: Basics and technology status. *Journal of Hazardous Materials, Soil Remediation: Application of Innovative and Standard Technologies* 40, 117–137. https://doi.org/10.1016/0304-3894(94)00066-P

Adhikari, K., Pearce, C.I., Sanguinet, K.A., Bary, A.I., Chowdhury, I., Eggleston, I., Xing, B., Flury, M., 2024. Accumulation of microplastics in soil after long-term application of biosolids and atmospheric deposition. *Science of the Total Environment* 912, 168883. https://doi.org/10.1016/j.scitotenv.2023.168883

Akarsu, C., Kumbur, H., Kideys, A.E., 2021. Removal of microplastics from wastewater through electrocoagulation-electroflotation and membrane filtration processes. *Water Science and Technology* 84, 1648–1662. https://doi.org/10.2166/wst.2021.356

Alavian Petroody, S.S., Hashemi, S.H., Škrlep, L., Mušič, B., van Gestel, C.A.M., Sever Škapin, A., 2023. UV light causes structural changes in microplastics exposed in bio-solids. *Polymers* 15, 4322. https://doi.org/10.3390/polym15214322

AlMashrea, B.A., Almehdi, A.M., Damiati, S., 2024. Simple microfluidic devices for in situ detection of water contamination: A state-of-art review. *Frontiers in Bioengineering and Biotechnology* 12, 1355768. https://doi.org/10.3389/fbioe.2024.1355768

Anand, U., Dey, S., Bontempi, E., Ducoli, S., Vethaak, A.D., Dey, A., Federici, S., 2023. Biotechnological methods to remove microplastics: A review. *Environmental Chemistry Letters* 21, 1787–1810. https://doi.org/10.1007/s10311-022-01552-4

Arpia, A.A., Chen, W.-H., Ubando, A.T., Naqvi, S.R., Culaba, A.B., 2021. Microplastic degradation as a sustainable concurrent approach for producing biofuel and obliterating hazardous environmental effects: A state-of-the-art review. *Journal of Hazardous Materials* 418, 126381. https://doi.org/10.1016/j.jhazmat.2021.126381

Auta, H.S., Abioye, O.P., Aransiola, S.A., Bala, J.D., Chukwuemeka, V.I., Hassan, A., Aziz, A., Fauziah, S.H., 2022. Enhanced microbial degradation of PET and PS microplastics under natural conditions in mangrove environment. *Journal of Environmental Management* 304, 114273. https://doi.org/10.1016/j.jenvman.2021.114273

Bakhteeva, I.A., Medvedeva, I.V., Filinkova, M.S., Byzov, I.V., Minin, A.S., Zhakov, S.V., Uimin, M.A., Patrakov, E.I., Novikov, S.I., Suntsov, A.Y., Demin, A.M.,

2023. Removal of microplastics from water by using magnetic sedimentation. *International Journal of Environmental Science and Technology* 20, 11837–11850. https://doi.org/10.1007/s13762-023-04776-1

Barone, G.D., Rodríguez-Seijo, A., Parati, M., Johnston, B., Erdem, E., Cernava, T., Zhu, Z., Liu, X., Axmann, I.M., Lindblad, P., Radecka, I., 2024. Harnessing photosynthetic microorganisms for enhanced bioremediation of microplastics: A comprehensive review. *Environmental Science and Ecotechnology* 20, 100407. https://doi.org/10.1016/j.ese.2024.100407

Ben Hmad, I., Gargouri, A., 2024. Stable and effective eco-enzyme cocktails in powder and liquid form of *Stachybotrys microspora* used as detergent additives. *Heliyon* 10, e25610. https://doi.org/10.1016/j.heliyon.2024.e25610

Cai, Z., Li, M., Zhu, Z., Wang, X., Huang, Y., Li, T., Gong, H., Yan, M., 2023. Biological degradation of plastics and microplastics: A recent perspective on associated mechanisms and influencing factors. *Microorganisms* 11, 1661. https://doi.org/10.3390/microorganisms11071661

Cao, Z., Yan, W., Ding, M., Yuan, Y., 2022. Construction of microbial consortia for microbial degradation of complex compounds. *Frontiers in Bioengineering and Biotechnology* 10, 1051233. https://doi.org/10.3389/fbioe.2022.1051233

Chaukura, N., Muzawazi, E.S., Katengeza, G., Mahmoud, A.E.D., 2022. Remediation technologies for contaminated soil systems, in: Gwenzi, W. (Ed.), *Emerging Contaminants in the Terrestrial-Aquatic-Atmosphere Continuum*: Elsevier, pp. 353–365. https://doi.org/10.1016/B978-0-323-90051-5.00019-5

Chow, J., Perez-Garcia, P., Dierkes, R., Streit, W.R., 2022. Microbial enzymes will offer limited solutions to the global plastic pollution crisis. *Microbial Biotechnology* 16, 195–217. https://doi.org/10.1111/1751-7915.14135

Chu, J., Zhou, J., Wang, Y., Jones, D.L., Ge, J., Yang, Y., Brown, R.W., Zang, H., Zeng, Z., 2023. Field application of biodegradable microplastics has no significant effect on plant and soil health in the short term. *Environmental Pollution* 316, 120556. https://doi.org/10.1016/j.envpol.2022.120556

Doan, T.O., Duong, T.T., Nguyen, T.M., Hoang, T.Q., Luong, T.T., Pham, P.T., Cao, T.T.N., Le, P.T., Phung, H.P.H., Le, T.P.Q., Dang, T.M.A., Bui, P.T., Duong, T.N., Bui, V.C., 2023. Preliminary results on microplastic pollution from agricultural soil in Vietnam: Distribution, characterization, and ecological risk assessment. *Vietnam Journal of Earth Sciences* 45, 405–418. https://doi.org/10.15625/2615-9783/18616

Dong, Y., Gao, M., Qiu, W., Xiao, L., Cheng, Z., Peng, H., Song, Z., 2024. Investigating the impact of microplastics on sulfur mineralization in different soil types: A mechanism study. *Journal of Hazardous Materials* 464, 132942. https://doi.org/10.1016/j.jhazmat.2023.132942

Du, H., Xie, Y., Wang, J., 2021. Microplastic degradation methods and corresponding degradation mechanism: Research status and future perspectives. *Journal of Hazardous Materials* 418, 126377. https://doi.org/10.1016/j.jhazmat.2021.126377

Ece, E., Hacıosmanoğlu, N., Inci, F., 2023. Microfluidics as a ray of hope for microplastic pollution. *Biosensors* 13, 332. https://doi.org/10.3390/bios13030332

Fermeglia, M., Perišić, M., 2023. Nature-based solution to man-made problems: Fostering the uptake of phytoremediation and low-iluc biofuels in the EU. *Journal for European Environmental & Planning Law* 20, 145–167. https://doi.org/10.1163/18760104-20020007

Goli, V.S.N.S., Singh, D.N., 2023. Effect of ultrasonication conditions on polyethylene microplastics sourced from landfills: A precursor study to establish guidelines for their extraction from environmental matrices. *Journal of Hazardous Materials* 459, 132230. https://doi.org/10.1016/j.jhazmat.2023.132230

Grause, G., Kuniyasu, Y., Chien, M.-F., Inoue, C., 2022. Separation of microplastic from soil by centrifugation and its application to agricultural soil. *Chemosphere* 288, 132654. https://doi.org/10.1016/j.chemosphere.2021.132654

Grbic, J., Nguyen, B., Guo, E., You, J.B., Sinton, D., Rochman, C.M., 2019. Magnetic extraction of microplastics from environmental samples. *Environmental Science & Technology Letters* 6, 68–72. https://doi.org/10.1021/acs.estlett.8b00671

Gu, F., Zhang, J., Shen, Z., Li, Y., Ji, R., Li, W., Zhang, L., Han, J., Xue, J., Cheng, H., 2022. A review for recent advances on soil washing remediation technologies. *Bulletin of Environmental Contamination and Toxicology* 109, 651–658. https://doi.org/10.1007/s00128-022-03584-6

Guo, C., Zhang, L.-Q., Jiang, W., 2023. Biodegrading plastics with a synthetic non-biodegradable enzyme. *Chem* 9, 363–376. https://doi.org/10.1016/j.chempr.2022.09.008

Han, L., Chen, L., Feng, Y., Kuzyakov, Y., Chen, Q., Zhang, S., Chao, L., Cai, Y., Ma, C., Sun, K., Rillig, M.C., 2024. Microplastics alter soil structure and microbial community composition. *Environment International* 185, 108508. https://doi.org/10.1016/j.envint.2024.108508

Harshvardhan, K., Jha, B., 2013. Biodegradation of low-density polyethylene by marine bacteria from pelagic waters, Arabian Sea, India. *Marine Pollution Bulletin* 77, 100–106. https://doi.org/10.1016/j.marpolbul.2013.10.025

Hildebrandt, L., Zimmermann, T., Primpke, S., Fischer, D., Gerdts, G., Pröfrock, D., 2021. Comparison and uncertainty evaluation of two centrifugal separators for microplastic sampling. *Journal of Hazardous Materials* 414, 125482. https://doi.org/10.1016/j.jhazmat.2021.125482

Hong, P.K.A., Jiang, W., 2005. Factors in the selection of chelating agents for extraction of lead from contaminated soil: Effectiveness, selectivity, and recoverability, in: Nowack, B., VanBriesen, J.M. (Eds.), *Biogeochemistry of Chelating Agents, ACS Symposium Series*: American Chemical Society, pp. 421–432. https://doi.org/10.1021/bk-2005-0910.ch025

Ihenetu, S.C., Li, G., Mo, Y., Jacques, K.J., 2024. Impacts of microplastics and urbanization on soil health: An urgent concern for sustainable development. *Green Analytical Chemistry* 8, 100095. https://doi.org/10.1016/j.greeac.2024.100095

Jain, R., Gaur, A., Suravajhala, R., Chauhan, U., Pant, M., Tripathi, V., Pant, G., 2023. Microplastic pollution: Understanding microbial degradation and strategies for pollutant reduction. *Science of the Total Environment* 905, 167098. https://doi.org/10.1016/j.scitotenv.2023.167098

Jiang, Y., Yin, X., Xi, X., Guan, D., Sun, H., Wang, N., 2021. Effect of surfactants on the transport of polyethylene and polypropylene microplastics in porous media. *Water Research* 196, 117016. https://doi.org/10.1016/j.watres.2021.117016

Khan, I., Tariq, M., Alabbosh, K.F., Rehman, A., Jalal, A., Khan, A.A., Farooq, M., Li, G., Iqbal, B., Ahmad, N., Khan, K.A., Du, D., 2024. Soil microplastics: Impacts on greenhouse gasses emissions, carbon cycling, microbial diversity, and soil characteristics. *Applied Soil Ecology* 197, 105343. https://doi.org/10.1016/j.apsoil.2024.105343

Khan, K.Y., Tang, Y., Cheng, P., Song, Y., Li, X., Lou, J., Iqbal, B., Zhao, X., Hameed, R., Li, G., Du, D., 2024. Effects of degradable and non-degradable microplastics and oxytetracycline co-exposure on soil N_2O and CO_2 emissions. *Applied Soil Ecology* 197, 105331. https://doi.org/10.1016/j.apsoil.2024.105331

Kida, M., Ziembowicz, S., Koszelnik, P., 2021. The use of an ultrasonic field in support of classical methods of oxidising component leached from microplastics in bottom sediments. *Materials* 14, 3029. https://doi.org/10.3390/ma14113029

Kim, S.-O., Jeong, J.Y., Lee, W.-C., Yun, S.-T., Jo, H.Y., 2021. Electrokinetic remediation of heavy metal-contaminated soils: Performance comparison between one- and two-dimensional electrode configurations. *Journal of Soils and Sediments* 21, 2755–2769. https://doi.org/10.1007/s11368-020-02803-z

Li, Q., Wu, J., Zhao, X., Gu, X., Ji, R., 2019. Separation and identification of microplastics from soil and sewage sludge. *Environmental Pollution* 254, 113076. https://doi.org/10.1016/j.envpol.2019.113076

Li, S., Ding, F., Flury, M., Wang, Z., Xu, L., Li, S., Jones, D.L., Wang, J., 2022. Macro- and microplastic accumulation in soil after 32 years of plastic film mulching. *Environmental Pollution* 300, 118945. https://doi.org/10.1016/j.envpol.2022.118945

Liu, J., Zhao, L., Liu, Q., Li, J., Qiao, Z., Sun, P., Yang, Y., 2022. A critical review on soil washing during soil remediation for heavy metals and organic pollutants. *International Journal of Environmental Science and Technology* 19, 601–624. https://doi.org/10.1007/s13762-021-03144-1

Liu, J.-W., Wei, K.-H., Xu, S.-W., Cui, J., Ma, J., Xiao, X.-L., Xi, B.-D., He, X.-S., 2021. Surfactant-enhanced remediation of oil-contaminated soil and groundwater: A review. *Science of the Total Environment* 756, 144142. https://doi.org/10.1016/j.scitotenv.2020.144142

Lozano, Y.M., Gordillo-Rocha, H., Waldman, W.R., Rillig, M.C., 2024. Photodegradation modifies microplastic effects on soil properties and plant performance. *Journal of Applied Ecology* 61, 13–24. https://doi.org/10.1111/1365-2664.14514

Lusher, A.L., Munno, K., Hermabessiere, L., Carr, S., 2020. Isolation and extraction of microplastics from environmental samples: An evaluation of practical approaches and recommendations for further harmonization. *Applied Spectroscopy* 74, 1049–1065. https://doi.org/10.1177/0003702820938993

Mbachu, O., Jenkins, G., Pratt, C., Kaparaju, P., 2021. Enzymatic purification of microplastics in soil. *MethodsX* 8, 101254. https://doi.org/10.1016/j.mex.2021.101254

Mesquita, P., Gong, L., Lin, Y., 2022. A low-cost microfluidic method for microplastics identification: Towards continuous recognition. *Micromachines* 13, 499. https://doi.org/10.3390/mi13040499

Mirzaee, E., Sartaj, M., 2022. The application of surfactant-enhanced soil washing process combined with adsorption using a recoverable magnetic granular activated carbon for remediation of PAH-contaminated soil. *Environmental Advances* 9, 100274. https://doi.org/10.1016/j.envadv.2022.100274

Mohanty, S., Jasmine, J., Mukherji, S., 2013. Practical considerations and challenges involved in surfactant enhanced bioremediation of oil. *BioMed Research International* 2013, e328608. https://doi.org/10.1155/2013/328608

Mullins, C.E., 1977. Magnetic susceptibility of the soil and its significance in soil science – A review. *Journal of Soil Science* 28, 223–246. https://doi.org/10.1111/j.1365-2389.1977.tb02232.x

Mushtak, F., Prakash, J., Katoch, S.S., 2024. Microplastics in complex soil matrix: Recovery, identification and removal using micro nano techniques. *Micro and Nano Engineering* 22, 100237. https://doi.org/10.1016/j.mne.2023.100237

Mustapha, S., Tijani, J.O., Elabor, R., Salau, R.B., Egbosiuba, T.C., Amigun, A.T., Shuaib, D.T., Sumaila, A., Fiola, T., Abubakar, Y.K., Abubakar, H.L., Ossamulu, I.F., Abdulkareem, A.S., Ndamitso, M.M., Sagadevan, S., Mohammed, A.K., 2024. Technological approaches for removal of microplastics and nanoplastics in the environment. *Journal of Environmental Chemical Engineering* 12, 112084. https://doi.org/10.1016/j.jece.2024.112084

Nuzzo, A., Puccio, S., Martina, C., Pietrangeli, B., Martinez, G.A., Bertin, L., Mancini, M., Fava, F., Zanaroli, G., 2020. Containment of a genetically modified microorganism by an activated sludge system. *New Biotechnology* 55, 58–64. https://doi.org/10.1016/j.nbt.2019.10.001

Palacios-Mateo, C., Meng, K., Legaz-Pol, L., Steen Redeker, E., Huerta-Lwanga, E., Blank, L.M., 2023. Enzymes for microplastic-free agricultural soils. *Ecotoxicology and Environmental Safety* 258, 114982. https://doi.org/10.1016/j.ecoenv.2023.114982

Paria, S., 2008. Surfactant-enhanced remediation of organic contaminated soil and water. *Advances in Colloid and Interface Science* 138, 24–58. https://doi.org/10.1016/j.cis.2007.11.001

Perez, S., Naderi Beni, N., Caines, S., Bartelt-Hunt, S., 2021. Methods evaluation for microplastic extraction from soil. *AGU Fall Meeting Abstracts* 2021, A45N–2037.

Pfohl, P., Roth, C., Wohlleben, W., 2024. The power of centrifugation: How to extract microplastics from soil with high recovery and matrix removal efficiency. *MethodsX* 12, 102598. https://doi.org/10.1016/j.mex.2024.102598

Pollard, M., Hunsicker, E., Platt, M., 2020. A tunable three-dimensional printed microfluidic resistive pulse sensor for the characterization of algae and microplastics. *ACS Sensors* 5, 2578–2586. https://doi.org/10.1021/acssensors.0c00987

Raghul, S.S., Bhat, S.G., Chandrasekaran, M., Francis, V., Thachil, E.T., 2014. Biodegradation of polyvinyl alcohol-low linear density polyethylene-blended plastic film by consortium of marine benthic vibrios. *International Journal of Environmental Science and Technology* 11, 1827–1834. https://doi.org/10.1007/s13762-013-0335-8

Ramage, S.J.F.F., Pagaling, E., Haghi, R.K., Dawson, L.A., Yates, K., Prabhu, R., Hillier, S., Devalla, S., 2022. Rapid extraction of high- and low-density microplastics from soil using high-gradient magnetic separation. *Science of the Total Environment* 831, 154912. https://doi.org/10.1016/j.scitotenv.2022.154912

Rathankumar, A.K., Saikia, K., Cabana, H., Kumar, V.V., 2022. Surfactant-aided mycoremediation of soil contaminated with polycyclic aromatic hydrocarbons. *Environmental Research* 209, 112926. https://doi.org/10.1016/j.envres.2022.112926

Renzi, M., Grazioli, E., Blašković, A., 2019. Effects of different microplastic types and surfactant-microplastic mixtures under fasting and feeding conditions: A case study on Daphnia magna. *Bulletin of Environmental Contamination and Toxicology* 103, 367–373. https://doi.org/10.1007/s00128-019-02678-y

Rhein, F., Nirschl, H., Kaegi, R., 2022. Separation of microplastic particles from sewage sludge extracts using magnetic seeded filtration. *Water Research* X 17, 100155. https://doi.org/10.1016/j.wroa.2022.100155

Salinas, J., Carpena, V., Martínez-Gallardo, M.R., Estrella-González, M.J., Toribio, A.J., Jurado, M.M., López-González, J.A., Suárez-Estrella, F., López, M.J., 2023. Development of plastic-degrading microbial consortia by induced selection in microcosms. *Frontiers in Microbiology* 14. https://doi.org/10.3389/fmicb.2023.1143769

Shafea, L., Yap, J., Beriot, N., Felde, V.J.M.N.L., Okoffo, E.D., Enyoh, C.E., Peth, S., 2023. Microplastics in agroecosystems: A review of effects on soil biota and key soil functions. *Journal of Plant Nutrition and Soil Science* 186, 5–22. https://doi.org/10.1002/jpln.202200136

Siipola, V., Romar, H., Lassi, U., 2022. Chapter 10: Microplastic removal from water and wastewater by carbon-supported materials, in: Mohan, D., Pittman, C.U., Mlsna, T.E. (Eds.), *Sustainable Biochar for Water and Wastewater Treatment*: Elsevier, pp. 361–393. https://doi.org/10.1016/B978-0-12-822225-6.00007-5

Sun, J., Zheng, H., Xiang, H., Fan, J., Jiang, H., 2022. The surface degradation and release of microplastics from plastic films studied by UV radiation and mechanical abrasion. *Science of the Total Environment* 838, 156369. https://doi.org/10.1016/j.scitotenv.2022.156369

Sutkar, P.R., Gadewar, R.D., Dhulap, V.P., 2023. Recent trends in degradation of microplastics in the environment: A state-of-the-art review. *Journal of Hazardous Materials Advances* 11, 100343. https://doi.org/10.1016/j.hazadv.2023.100343

Talukdar, A., Kundu, P., Bhattacharjee, S., Dey, S., Dey, A., Biswas, J.K., Chaudhuri, P., Bhattacharya, S., 2023. Microplastics in mangroves with special reference to Asia: Occurrence, distribution, bioaccumulation and remediation options. *Science of the Total Environment* 904, 166165. https://doi.org/10.1016/j.scitotenv.2023.166165

Thompson, J.R., Crooks, R.M., 2022. Electrokinetic separation techniques for studying nano- and microplastics. *Chemical Science* 13, 12616–12624. https://doi.org/10.1039/D2SC04019K

Tiwari, M., Tripathy, D.B., 2023. Soil contaminants and their removal through surfactant-enhanced soil remediation: A comprehensive review. *Sustainability* 15, 13161. https://doi.org/10.3390/su151713161

Tribedi, P., Dey, S., 2017. Pre-oxidation of low-density polyethylene (LDPE) by ultraviolet light (UV) promotes enhanced degradation of LDPE in soil. *Environmental Monitoring and Assessment* 189, 624. https://doi.org/10.1007/s10661-017-6351-2

Tun, T.Z., Kunisue, T., Tanabe, S., Prudente, M., Subramanian, A., Sudaryanto, A., Viet, P.H., Nakata, H., 2022. Microplastics in dumping site soils from six Asian countries as a source of plastic additives. *Science of the Total Environment* 806, 150912. https://doi.org/10.1016/j.scitotenv.2021.150912

Tyagi, M., da Fonseca, M.M.R., de Carvalho, C.C.C.R., 2011. Bioaugmentation and biostimulation strategies to improve the effectiveness of bioremediation processes. *Biodegradation* 22, 231–241. https://doi.org/10.1007/s10532-010-9394-4

Vocciante, M., Dovì, V.G., Ferro, S., 2021. Sustainability in electrokinetic remediation processes: A critical analysis. *Sustainability* 13, 770. https://doi.org/10.3390/su13020770

Wu, Q., Zhou, W., Chen, D., Tian, J., Ao, J., 2024. Biochar mitigates the negative effects of microplastics on sugarcane growth by altering soil nutrients and microbial community structure and function. *Plants* 13, 83. https://doi.org/10.3390/plants13010083

Wu, T., Yang, Z., Hu, R., Chen, Y., Zhong, H., Yang, L., Jin, W., 2021. Film entrainment and microplastic particles retention during gas invasion in suspension-filled microchannels. *Water Research* 194, 116919.

Wu, Z., Dong, J., Wu, Y., Zhao, Y., Wang, H., Zhao, X., Zhang, B., zhang, Y., An, L., 2024. Mass-based fates of microplastics throughout wastewater treatment processes. *Chemical Engineering Journal* 487, 150497. https://doi.org/10.1016/j.cej.2024.150497

Xia, Y., Zhou, J.-J., Gong, Y.-Y., Li, Z.-J., Zeng, E.Y., 2020. Strong influence of surfactants on virgin hydrophobic microplastics adsorbing ionic organic pollutants. *Environmental Pollution* 265, 115061. https://doi.org/10.1016/j.envpol.2020.115061

Yang, J., Cang, L., Sun, Q., Dong, G., Ata-Ul-Karim, S.T., Zhou, D., 2019. Effects of soil environmental factors and UV aging on Cu^{2+} adsorption on microplastics. *Environmental Science and Pollution Research* 26, 23027–23036. https://doi.org/10.1007/s11356-019-05643-8

Yang, M., Huang, D.-Y., Tian, Y.-B., Zhu, Q.-H., Zhang, Q., Zhu, H.-H., Xu, C., 2021. Influences of different source microplastics with different particle sizes and application rates on soil properties and growth of Chinese cabbage (*Brassica chinensis* L.). *Ecotoxicology and Environmental Safety* 222, 112480. https://doi.org/10.1016/j.ecoenv.2021.112480

Yang, S., Zhang, Y., Chen, Y., Zeng, Y., Yan, X., Tang, X., Pu, S., 2024. Studies on the transfer effect of aged polyethylene microplastics in soil-plant system. *Chemosphere* 349, 141001. https://doi.org/10.1016/j.chemosphere.2023.141001

Zhang, F., Yang, X., Zhang, Z., 2024. Effects of soil properties and land use patterns on the distribution of microplastics: A case study in southwest China. *Journal of Environmental Management* 356, 120598. https://doi.org/10.1016/j.jenvman.2024.120598

Zhang, Q., Zhao, M., Meng, F., Xiao, Y., Dai, W., Luan, Y., 2021. Effect of polystyrene microplastics on rice seed germination and antioxidant enzyme activity. *Toxics* 9, 179. https://doi.org/10.3390/toxics9080179

Zhang, S., Li, W., Bao, A., Jia, G., Lin, X., Zhang, Q., 2023. A low-cost approach for batch separation, identification and quantification of microplastics in agriculture soil. *Toxics* 11, 461. https://doi.org/10.3390/toxics11050461

Zhang, S., Li, Y., Jiang, L., Chen, X., Zhao, Y., Shi, W., Xing, Z., 2024. From organic fertilizer to the soils: What happens to the microplastics? A critical review. *Science of the Total Environment* 919, 170217. https://doi.org/10.1016/j.scitotenv.2024.170217

Zhang, T., Zhang, H., 2022. Microbial consortia are needed to degrade soil pollutants. *Microorganisms* 10, 261. https://doi.org/10.3390/microorganisms10020261

Zhang, W., Tong, L., Yuan, Y., Liu, Z., Huang, H., Tan, F., Qiu, R., 2010. Influence of soil washing with a chelator on subsequent chemical immobilization of heavy metals in a contaminated soil. *Journal of Hazardous Materials* 178, 578–587. https://doi.org/10.1016/j.jhazmat.2010.01.124

Zheng, X.-J., Li, Q., Peng, H., Zhang, J.-X., Chen, W.-J., Zhou, B.-C., Chen, M., 2022. Remediation of heavy metal-contaminated soils with soil washing: A review. *Sustainability* 14, 13058. https://doi.org/10.3390/su142013058

Zhong, Y., Zhuo, L., Lu, W., 2024. Analysis of photocatalytic degradation of polyamide microplastics in metal salt solution by high resolution mass spectrometry. *Journal of Environmental Sciences* 139, 473–482. https://doi.org/10.1016/j.jes.2023.06.018

CHAPTER 9

Mitigation of Microplastics

Impacts for Soil Remediation

Chingakham Chinglenthoiba,
Kassian T.T. Amesho,
Mohd Yahya Fadzli Jusoh,
Mohamad Nor Azzimi Sohedein,
Elham Taghavi, and Mohd Nizam Lani

9.1 INTRODUCTION: OVERVIEW OF MICROPLASTICS POLLUTION IN SOIL

Microplastic pollution has emerged as a pressing environmental issue, with widespread implications for ecosystems, wildlife, and human health (Amesho et al., 2023; Chinglenthoiba et al., 2023). Microplastics, defined as plastic particles smaller than 5 mm in diameter, originate from a variety of sources, including the breakdown of larger plastic waste, industrial processes, and household products such as cosmetics and synthetic textiles (Frias & Nash, 2019). These tiny particles have pervaded the world's oceans, rivers, and now, increasingly, our soil environments.

DOI: 10.1201/9781003487579-9

The prevalence of microplastics in soil is a relatively recent concern, attracting attention due to the critical role soil plays in global ecosystems, agriculture, and carbon cycling (Verla et al., 2020). Several pathways contribute to the accumulation of microplastics in soil. Agricultural practices, such as the use of sewage sludge as fertilizer, directly introduce microplastics into farmland (Bonyadinejad et al., 2022). Additionally, the breakdown of plastic mulch and the deposition of plastic debris from nearby urban and industrial sources contribute to soil contamination (Surendran et al., 2023). Atmospheric deposition also plays a role, with microplastics being transported by wind and rain into soil systems far removed from their original sources (Schell et al., 2022).

Once in the soil, microplastics pose significant threats to soil health and biodiversity (Kour et al., 2023). They can alter soil structure and porosity, affect water retention and nutrient cycles, and potentially introduce toxic substances into the soil ecosystem (Smedes et al., 2017). Microplastics also impact soil organisms, from microorganisms to larger soil fauna, affecting their health, behavior, and distribution, and thereby disrupting the ecological balance of soil ecosystems as shown in Figure 9.1 (Yu et al., 2022)

The ubiquity and persistence of microplastics in soil environments highlight the need for urgent action to address this growing environmental challenge (Chen et al., 2023). Understanding the sources, pathways, and impacts of microplastics on soil ecosystems is crucial for developing effective strategies to mitigate their presence and protect soil health. Soil remediation is a critical component of environmental health and sustainability, with far-reaching implications for agricultural productivity, ecosystem services, and human well-being (Bonyadinejad et al., 2022). The process involves the removal of pollutants or contaminants from soil to restore its health and function, which is essential for a variety of reasons.

Soil health is foundational to ecosystem stability and biodiversity (Saravanakumar et al., 2022). Soils serve as a habitat for a multitude of organisms, from microorganisms to larger fauna, which play vital roles in nutrient cycling, decomposition, and the regulation of greenhouse gases (Crossman et al., 2020; Wang et al., 2024). Contaminated soils can disrupt these natural processes, leading to diminished ecosystem resilience and loss of biodiversity (Song et al., 2023). Remediation helps to restore the balance, ensuring that soils can continue to support diverse life forms (Tang et al., 2024).

Agricultural productivity is heavily dependent on the quality of soil (Chen et al., 2023; Wu et al., 2021). Contaminants in the soil can inhibit plant growth, reduce crop yields, and compromise the nutritional

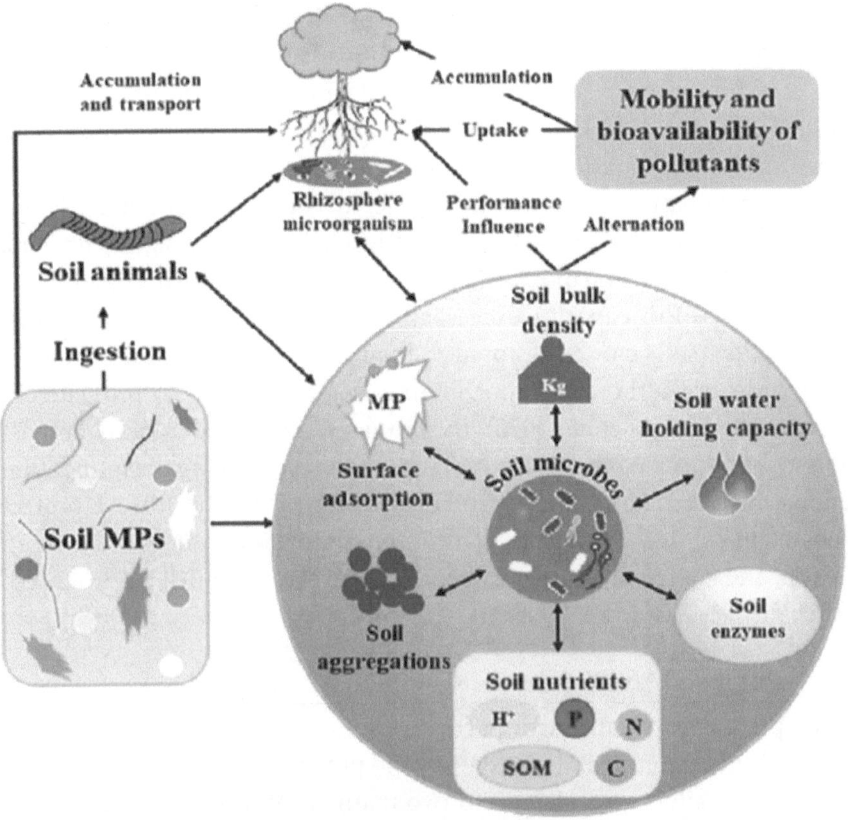

FIGURE 9.1 Interactions between microplastics and soil organisms: a conceptual overview. (With permission from Yu et al., 2022.)

value of food. This not only affects food security but also farmers' livelihoods. By removing harmful substances, soil remediation improves soil fertility, enhancing crop performance and ensuring that agricultural lands remain productive for future generations (Sima et al., 2023).

Furthermore, the health of human populations is intricately linked to soil quality. Soils contaminated with heavy metals, pesticides, or other pollutants can lead to the accumulation of toxins in the food chain, posing serious health risks to humans (Dong et al., 2020). These risks include chronic diseases, developmental disorders, and acute poisoning. Soil remediation reduces these health hazards by eliminating or neutralizing contaminants before they can enter the food chain (Guo et al., 2021).

In addition to direct health impacts, contaminated soils can affect human well-being in other ways. For example, soil pollution can degrade the aesthetic and recreational value of natural landscapes, impacting mental health and community well-being (Brožová et al., 2023). Moreover, the economic costs associated with soil degradation—through lost agricultural productivity, health care costs, and the need for environmental clean-up—can be substantial.

Soil remediation plays a crucial role in combating climate change. Healthy soils are vital for carbon sequestration, acting as a significant sink for atmospheric carbon dioxide (Xu et al., 2022). Contaminated soils, however, may lose this capacity, exacerbating the greenhouse effect. Through remediation, soils can be restored to their natural state, enhancing their ability to store carbon and contributing to global efforts to mitigate climate change (Verla et al., 2020). In summary, soil remediation is not just about cleaning up contaminated sites; it is about preserving the foundation upon which terrestrial life depends. By restoring soil health, we protect ecosystem services, ensure agricultural sustainability, safeguard human health, and contribute to the resilience of our planet against environmental challenges.

9.2 UNDERSTANDING MICROPLASTICS IN SOILS

Microplastics are small plastic particles less than 5 mm in diameter, posing significant environmental, ecological, and potentially human health risks. They are broadly categorized into two main types: primary and secondary microplastics. Primary microplastics are manufactured at a microscopic size for specific uses. These include microbeads found in personal care products like exfoliants and toothpaste, as well as pellets and powders used in industrial processes and synthetic fibers for textiles (Surendran et al., 2023). Primary microplastics are designed for particular applications, including as abrasives in cleaning products or as raw materials in the manufacturing of plastic products (Ahmed et al., 2024).

Secondary microplastics, on the other hand, result from the breakdown of larger plastic items over time due to environmental factors such as sunlight, physical abrasion, and biological degradation. Common sources of secondary microplastics include plastic bags, bottles, fishing nets, and other larger plastic debris that fragment into smaller pieces (Guerranti et al., 2019). This breakdown can occur in various environments, including marine and terrestrial ecosystems, leading to widespread distribution of microplastics (Y. Zhang et al., 2020).

Within these categories, microplastics can be further classified by their physical characteristics, such as fibers, fragments, films, and beads. Fibers, often

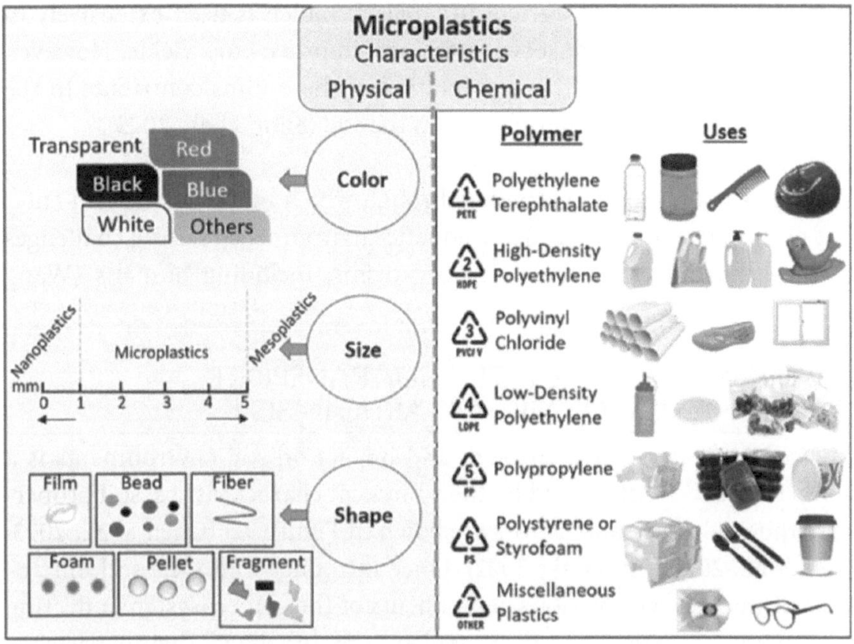

FIGURE 9.2 An illustration of the classification of microplastics based on physical characteristics. (With permission from Raza et al., 2022.)

originating from synthetic textiles like polyester and nylon, are a common type of microplastic found in the environment, often detected in marine water samples and organisms (Bhatt et al., 2021). Fragments are irregularly shaped pieces that come from the breakdown of larger plastic items. Films, typically thin and flexible, originate from plastic bags and wrap, while beads are small, manufactured spherical particles found in personal care products and industrial applications as shown in Figure 9.2 (Raza et al., 2022).

The main sources of microplastics in the environment include:

Agricultural Runoff: Microplastics enter soil and water systems through agricultural runoff, which may contain plastic fragments from mulch film, synthetic fibers from ropes and nets, and granules from fertilizers and soil conditioners (Schell et al., 2022).

Wastewater Irrigation: Wastewater treatment plants are significant sources of microplastics, which are not completely removed during treatment processes. When this treated water is used for irrigation, microplastics are introduced into agricultural soils (Cattle et al., 2020).

Plastic Mulching: In agriculture, plastic mulch is used extensively to suppress weeds, conserve water, and improve crop yields. However, the degradation and fragmentation of these films contribute to the accumulation of microplastics in the soil (Silori et al., 2023).

These sources represent pathways through which microplastics can enter and accumulate in terrestrial and aquatic environments, posing challenges to the health of ecosystems and organisms, including humans (Wang et al., 2024).

9.3 ENVIRONMENTAL FATE AND TRANSPORT MECHANISM OF MICROPLASTIC IN SOIL

Microplastic's movement through and impact on soil environments is a complex process, influenced by their physical characteristics, soil properties, and environmental factors as shown in Figure 9.3 (Chen et al., 2023; Guo et al., 2020; Lin et al., 2022). Once introduced into the soil, microplastics interact with various components of the soil ecosystem, affecting its structure, function, and the organisms that inhabit it. Their persistence

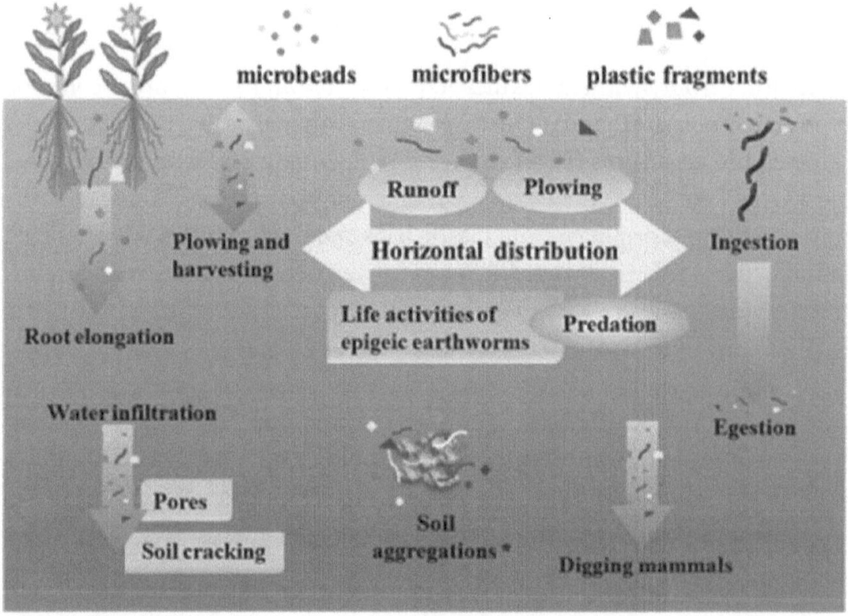

FIGURE 9.3 Environmental fate and transport mechanism of microplastic in soil. (With permission from Guo et al., 2020.)

and potential for bioaccumulation raise significant concerns regarding soil health and the broader ecological implications (Gani et al., 2024).

9.3.1 Movement Pathways of Microplastics in Soil Environments

Microplastics move through soil environments via several pathways, each influenced by different factors. Water infiltration, for instance, is a significant mechanism, where microplastics can be carried along with water as it percolates through the soil profile (Brožová et al., 2023; Schell et al., 2022). Soil tillage, including plowing and cultivation, also plays a role in redistributing microplastics, especially those located near the surface (Lin et al., 2022). Bioturbation, the disturbance of soil by organisms like earthworms and insects, further aids in the movement of microplastics, as these organisms burrow through the soil and transport particles along with them. Additionally, the structure of soil aggregates can influence the retention and transport of microplastics, with aggregated soils typically exhibiting slower movement rates compared to unaggregated soils (Wang et al., 2018).

9.3.2 Influence of Physical Characteristics on Microplastic Transport

The physical characteristics of microplastics significantly impact their transport within soil environments. Particle size is a crucial factor, with smaller microplastics, such as microfibers, having higher mobility due to their ability to infiltrate soil pores more easily (Kim et al., 2023). Conversely, larger or denser particles may remain closer to the soil surface, especially in coarse-textured soils with limited pore space. Shape also plays a role, with irregularly shaped microplastics exhibiting different transport behaviors compared to spherical particles (Wirnkor et al., 2019). Density affects the settling velocity of microplastics, influencing their vertical distribution within soil profiles. Understanding how these physical characteristics interact with soil properties helps predict the fate and transport of microplastics in different environmental settings (Surendran et al., 2023).

9.3.3 Interactions between Microplastics and Soil Components

Microplastics interact with various components of the soil ecosystem, altering soil properties and affecting soil organisms and chemical processes (Cardoso et al., 2023). Microplastics can physically alter soil structure by filling pore spaces, which can impact water infiltration and soil aeration (Arrázola-Vásquez et al., 2022). They can also serve as substrates for microbial colonization, influencing microbial community composition and activity. Chemical interactions occur as microplastics adsorb pollutants from the soil environment, potentially leading to the accumulation of toxic substances (Silori et al., 2023). These interactions have cascading effects on soil health and ecosystem functioning, highlighting the need for comprehensive assessments of microplastic impacts in soil environments (Ding et al., 2023).

9.3.4 Factors Affecting the Persistence of Microplastics in Soil

The persistence of microplastics in soil is influenced by a multitude of factors. Polymer composition is a key determinant, with some polymers being more resistant to degradation than others (Meng et al., 2022). Environmental conditions such as temperature, moisture, and pH can also affect the rate of degradation, with warmer and wetter environments typically promoting faster degradation rates (Tian et al., 2024). Microbial activity plays a significant role, as microbial communities can enzymatically degrade certain types of plastics under favorable conditions. Soil texture and organic matter content can influence the accessibility of microplastics to microbial degradation, with organic-rich soils generally exhibiting higher degradation rates. Additionally, the presence of additives and coatings on microplastics can affect their susceptibility to degradation (Tian et al., 2024). Understanding these factors is crucial for predicting the long-term fate of microplastics in soil environments and developing effective mitigation strategies.

9.3.5 Bioaccumulation Potential of Microplastics in Soil Ecosystems

Microplastics have the potential to bioaccumulate within soil ecosystems, posing risks to soil organisms and higher trophic levels (Herrera et al., 2022). Soil organisms such as earthworms, springtails, and microarthropods may ingest microplastics while feeding on organic matter or soil particles. These ingested microplastics can accumulate in the gut tissues of soil organisms over time, potentially leading to adverse effects such as physical blockages, nutrient imbalances, and compromised health (ElSafty, 2014). Furthermore, microplastics can act as vectors for transporting adsorbed pollutants into the food web, as they may be consumed by predators at higher trophic levels (Dissanayake et al., 2022). This trophic transfer of microplastics and associated contaminants can pose risks to wildlife and humans who consume contaminated food items (Herrera et al., 2023). Understanding the bioaccumulation potential of microplastics is critical for assessing their ecological impact and implementing strategies to minimize their presence in soil ecosystems (Koongolla et al., 2022).

9.3.5.1 Movement through Soil

Microplastics can be transported through soil profiles by various mechanisms, including water infiltration, soil tillage, and bioturbation by soil organisms (Dhevagi et al., 2022). The movement is influenced by the size, shape, and density of microplastics, as well as soil texture, porosity, and moisture content. Fine and lightweight microplastics, such as microfibers, can be easily transported deeper into the soil, while larger or denser particles may remain closer to the surface. This vertical distribution affects the extent to which microplastics interact with soil organisms and processes at different depths.

9.3.5.2 Effects on Soil Environments

Microplastics affect soil environments in several ways:

i. **Physical Soil Properties:** Microplastics can alter the physical structure of the soil, affecting porosity, bulk density, and water retention capacity (Aransiola et al., 2023; X. Ma et al., 2023). These changes can influence root growth, water infiltration, and soil aeration, potentially impacting plant health and agricultural productivity.

ii. **Soil Organisms:** Microplastics can harm soil biota, including microorganisms, earthworms, and other invertebrates. They may ingest microplastics, leading to physical blockages, reduced feeding activity, and energy depletion. For microorganisms, microplastics can alter community composition and function, affecting nutrient cycling and soil fertility.

iii. **Chemical Interactions:** Microplastics can adsorb and concentrate pollutants, including heavy metals and organic contaminants, from the soil environment (Tingting & Jiawei, 2022). These chemicals may desorb in the gut of soil organisms or affect the bioavailability of nutrients and toxins in the soil (Velicogna et al., 2021).

9.3.5.3 Persistence and Bioaccumulation

Microplastics are highly resistant to degradation, with some types persisting in the environment for hundreds to thousands of years (Sharma & Chatterjee, 2017). This persistence allows for the continual accumulation of microplastics in soil environments, increasing the likelihood of interaction with and ingestion by soil organisms (Mokhtarzadeh et al., 2022). The potential for bioaccumulation, particularly for smaller particles that can be readily ingested, poses a risk for trophic transfer, where microplastics and associated contaminants may move up the food chain, ultimately reaching higher organisms, including humans (Stabnikova et al., 2022). The bioaccumulation of microplastics and their effects on soil organisms and ecosystem services highlight the need for strategies to mitigate their release into the environment and to remediate contaminated soils (Jia et al., 2022; Yu et al., 2021). Understanding the dynamics of microplastics in soil environments is crucial for assessing their full ecological impact and for developing effective management and remediation approaches (C. Wang et al., 2022).

9.4 IMPACTS OF MICROPLASTICS ON SOIL HEALTH

Microplastics' presence in soil can significantly alter its physical properties, with implications for water retention, soil structure, and plant root growth (Yadav et al., 2022). These changes are primarily due to the way microplastics interact with soil particles and organic matter, affecting the soil's ability to function and support plant life (Schjønning et al., 2017). Microplastics can modify the physical structure of the soil in several ways:

i. **Soil Porosity and Bulk Density:** The incorporation of microplastics into soil can change its porosity and bulk density. Microplastics, being less dense than most soil particles, can increase soil porosity and decrease bulk density (Sadiq et al., 2021). This alteration in soil structure can affect the movement of air and water through the soil profile.

ii. **Aggregate Stability:** Soil aggregates are clumps of soil particles that are bound together, and their stability is crucial for preventing erosion and promoting water infiltration. Microplastics can disrupt the formation of these aggregates or alter their stability by physically interfering with particle-to-particle interactions and by affecting the activity of soil organisms that contribute to aggregate formation (Aksakal et al., 2020; Rieke et al., 2022).

9.4.1 Impact on Water Retention

The presence of microplastics can influence the soil's water retention capacity in complex ways. On one hand, increased porosity might improve the soil's ability to hold water, potentially benefiting plant growth under dry conditions (Wang et al., 2015). On the other hand, the altered structure could lead to changes in the distribution and availability of water to plants, with some studies suggesting that microplastics might reduce the water film thickness around soil particles, making water less available to plant roots. Moreover, microplastics may affect the hydraulic conductivity of the soil, influencing how quickly water moves through the soil profile (Hao et al., 2023). This can impact the soil's capacity to store water and supply it to plants during periods of drought (Vance & Milroy, 2022).

9.4.2 Effects on Root Growth

The physical presence of microplastics in soil can directly impact plant root growth and development (Meng et al., 2021). The alteration in soil structure and water dynamics may create a less favorable environment for root penetration and expansion (Huang et al., 2022). For instance, microplastics might physically obstruct root pathways, leading to reduced root length and branching (Ceneviva & Tsuge, 2022). This can limit the plant's ability to access water and nutrients, potentially stunting growth and reducing crop yields. Furthermore, the altered soil environment may induce stress responses in plants, affecting their overall health and productivity (Ning et al., 2023). The interaction between microplastics and soil organisms that facilitate nutrient cycling and support plant growth could also be compromised, further impacting plant development.

9.5 CHEMICAL INTERACTIONS AND CONTAMINANT TRANSPORT

Microplastics in soil environments play a significant role in the adsorption and transportation of pollutants, acting as carriers for a variety of hazardous substances (Enfrin et al., 2021). This capacity for pollutant interaction is attributed to the large surface area of microplastics relative to their size, as well as their chemical composition and the presence of functional groups on their surfaces that can bind with contaminants (Chae & An, 2020). The dynamics of this process have important implications for soil and environmental health, as well as for the organisms that inhabit these ecosystems (Hooge et al., 2023).

9.5.1 Adsorption of Pollutants

Microplastics can adsorb a wide range of soil pollutants, including heavy metals, organic pollutants (such as pesticides, polycyclic aromatic hydrocarbons (PAHs), and pharmaceuticals), and pathogens (Smedes et al., 2017). The adsorption process is influenced by several factors, including the type of microplastics (e.g., polyethylene, polypropylene, polystyrene), the chemical nature of the pollutants, and environmental conditions such as pH and temperature. The hydrophobic surfaces of many microplastics make them particularly effective at adsorbing organic pollutants, which are also hydrophobic (Cheng et al., 2022). This affinity allows microplastics to concentrate pollutants from the surrounding soil matrix, potentially leading to elevated levels of toxicity in areas with high microplastic concentrations (Cingolani et al., 2022).

9.5.2 Transportation of Pollutants

Once pollutants are adsorbed onto microplastics, they can be transported through the soil environment (Elizalde-Velázquez & Gómez-Oliván, 2021). This movement is facilitated by the same processes that transport microplastics, including water flow, soil tillage, and bioturbation. As microplastics move, they can redistribute pollutants, introducing them to new areas and possibly increasing the exposure risk for soil organisms (Chang et al., 2022). The transportation of adsorbed pollutants by microplastics can lead to several ecological consequences. For instance, when microplastics carry pollutants into deeper soil layers, they may affect groundwater quality (Z. Zhang et al., 2023). Similarly, if microplastics are ingested by soil organisms, the adsorbed pollutants can be bioaccumulated or biomagnified within the food chain, posing risks to higher organisms, including humans (Deng et al., 2023).

9.6 IMPLICATIONS FOR SOIL AND ENVIRONMENTAL HEALTH

The role of microplastics in adsorbing and transporting soil pollutants has significant implications for soil and environmental health (Acarer, 2023). The concentration and redistribution of pollutants by microplastics can exacerbate the toxicity of contaminated sites, affecting soil biota and plant health. Furthermore, the potential for microplastics to transport pollutants across different environmental compartments (e.g., from soil to water bodies) highlights the need for comprehensive management strategies to address microplastic pollution (Calabrò & Grosso, 2018). Understanding the interactions between microplastics and soil pollutants is critical for assessing the risks and developing effective remediation strategies. Research in this area can inform policies and practices aimed at reducing the release of microplastics into the environment and mitigating their impact on soil ecosystems and beyond (Tedeschi et al., 2023).

The impact of microplastics on soil biodiversity, microbial communities, and plant health is a growing concern among environmental scientists

(Ivy et al., 2023). Microplastics can have multifaceted effects on soil ecosystems, influencing the abundance, diversity, and function of soil organisms, as well as the health and growth of plants (Taleb et al., 2022). These impacts can alter ecosystem processes, nutrient cycling, and soil fertility, with potential long-term consequences for ecosystem services and agricultural productivity.

9.6.1 Impact on Soil Biodiversity

Soil biodiversity encompasses a wide range of organisms, including bacteria, fungi, protozoa, invertebrates, and plants. Microplastics can affect these organisms in various ways, potentially leading to changes in soil biodiversity. For example, microplastics can physically obstruct soil pores, altering the habitat available for microorganisms and small invertebrates. This physical disruption can lead to changes in community composition and a reduction in biodiversity, particularly for species sensitive to changes in soil structure and aeration (Gartzia-Bengoetxea et al., 2009; Kladivko, 2001).

9.6.2 Impact on Microbial Communities

Microbial communities play crucial roles in soil ecosystems, driving processes such as decomposition, nutrient cycling, and organic matter transformation (Kour et al., 2023). The presence of microplastics can influence the structure and function of these communities. Studies have shown that microplastics can affect the abundance and diversity of soil microbes, sometimes promoting the growth of specific bacterial or fungal taxa while inhibiting others (Kour et al., 2023). This can lead to shifts in microbial community composition and potentially disrupt the ecological balance of soil ecosystems. The interaction between microplastics and microbes also extends to the adsorption of pollutants, as previously discussed. Microplastics can concentrate toxic substances, which may affect microbial health and activity. Furthermore, microplastics can serve as a substrate for biofilm formation, altering microbial interactions and nutrient availability (Ma et al., 2023).

9.6.3 Impact on Plant Health

The presence of microplastics in soil can directly and indirectly affect plant health (Xueying et al., 2021). Direct effects include physical impacts, such as the obstruction of root growth and reduced seed germination rates, potentially due to the physical barrier created by microplastics or their influence on soil structure and water availability. Indirect effects involve changes in soil microbial communities that are essential for plant growth

(Bardgett & Caruso, 2020). For instance, microplastics may alter the abundance and diversity of mycorrhizal fungi, which are critical for plant nutrient uptake, water acquisition, and resistance to pathogens (Lyu et al., 2022).

Moreover, microplastics can affect the availability of nutrients in the soil, either by adsorbing nutrients onto their surfaces or by influencing the microbial processes that release or fix nutrients. This can lead to nutrient imbalances, affecting plant growth and health as shown in Figure 9.4 (Athulya et al., 2024).

9.6.4 Broader Ecological Implications

The impact of microplastics on soil biodiversity, microbial communities, and plant health has broader ecological implications. Changes in soil biodiversity and ecosystem processes can affect the stability and resilience of ecosystems, potentially leading to reduce ecosystem services such as carbon sequestration, water purification, and crop production (García et al., 2018). The disruption of plant-soil-microbe interactions can also have cascading effects on food webs and nutrient cycling, impacting wildlife and human populations that depend on these ecosystems for food, water, and other resources (Das et al., 2022). Given the pervasive nature of microplastics and their potential to affect soil ecosystems profoundly, it is crucial to further investigate their impacts and develop strategies to mitigate their presence in the environment. Understanding the interactions between microplastics, soil organisms, and plants is essential for maintaining soil health, biodiversity, and the sustainability of agricultural systems (Y. Li et al., 2023).

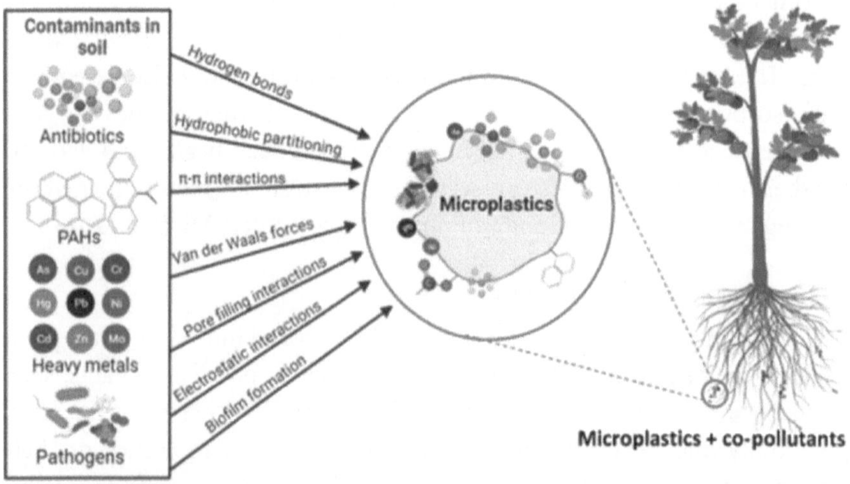

FIGURE 9.4 Alteration of soil physical properties due to microplastic presence: illustrated diagram. (With permission from Athulya et al., 2024).

9.7 CURRENT STRATEGIES FOR MICROPLASTICS REMEDIATION IN SOIL

9.7.1 Physical Removal Techniques

Sieving and flotation are two common methods used for the separation and analysis of microplastics from environmental samples, including soil and sediment (Saboor et al., 2022). These methods are part of a broader set of techniques aimed at identifying, quantifying, and characterizing microplastics in various matrices. Each method has its strengths and limitations, affecting its effectiveness in different contexts.

9.7.1.1 Sieving

Methodology: Sieving involves passing a soil or sediment sample through one or more mesh screens to separate particles based on size (do Carmo Precci Lopes et al., 2021). The choice of mesh size(s) is crucial and depends on the size range of microplastics of interest. Sieving can be performed wet or dry, although wet sieving is often preferred for soil samples to reduce dust and improve the separation of microplastics from soil particles.

Effectiveness: Sieving is particularly effective for separating larger microplastics (>1 mm) and can be a useful preliminary step in a multi-stage microplastic extraction protocol. It allows for the efficient removal of larger particles and the concentration of microplastics for further analysis.

Limitations: The main limitation of sieving is its inability to capture smaller microplastics, especially those smaller than the mesh size. This method can also be time-consuming and labor-intensive, particularly for samples with high organic matter content that may require additional steps to break down the sample matrix. Additionally, sieving does not differentiate between microplastics and other similarly sized particles, necessitating further analysis to confirm the presence of microplastics (Rajot et al., 2003; Sholokhova et al., 2021).

9.7.1.2 Flotation

Methodology: Flotation exploits the difference in density between microplastics and the surrounding matrix (soil, sediment, or water). Samples are mixed with a solution of higher density (e.g., saturated salt solutions like sodium chloride or zinc chloride) and agitated.

Microplastics, being less dense, float to the surface and can be collected for analysis (Jiang et al., 2023).

Effectiveness: Flotation is effective for isolating microplastics from denser soil particles and organic matter (Barciela-Rial et al., 2022). It is particularly useful for extracting small microplastics that pass through sieves (Ruggero et al., 2020). The method can be adapted to process large volumes of samples, making it suitable for studies requiring comprehensive microplastic inventories.

Limitations: The effectiveness of flotation can be influenced by the physical properties of the microplastics (e.g., density, shape) and the choice of flotation solution (Pooja et al., 2022). Some microplastics with densities close to or higher than the flotation solution may not separate effectively. The method may also require the use of toxic chemicals (e.g., zinc chloride), raising safety and environmental disposal concerns (Li et al., 2024). Furthermore, the process can be affected by the wetting properties of microplastics; hydrophobic particles may adhere to soil or organic matter, reducing recovery rates (Tankersley et al., 2017).

Both sieving and flotation are valuable tools in the study of environmental microplastics, offering complementary approaches to sample preparation and analysis (Chiou et al., 2013; Mäkinen et al., 2015). However, their effectiveness can be influenced by the sample type, microplastic characteristics, and the specific objectives of the study. Often, these methods are used in conjunction with other techniques (e.g., visual identification, spectroscopic analysis) to accurately quantify and characterize microplastics in environmental samples. The choice of method(s) requires careful consideration of the strengths and limitations of each approach, as well as the specific requirements of the research question at hand (Hubert et al., 2009).

9.7.2 Bioremediation Approaches

The use of microorganisms and plants to degrade or accumulate microplastics is a promising area of research, offering potential sustainable solutions to mitigate microplastic pollution (Viel et al., 2023). These biological approaches exploit the natural capabilities of certain organisms to interact with, break down, or uptake microplastics from the environment (Hemalatha et al., 2022). While still largely in the experimental or early implementation stages, these methods show significant promise for contributing to the remediation of microplastic-contaminated environments (Devi & Aravindhan, 2023).

9.7.2.1 Microorganism-Mediated Degradation

Microorganisms such as bacteria and fungi possess mechanisms that can degrade plastics into simpler, less harmful substances (G. Zhang et al., 2023). These organisms may secrete enzymes that break down the polymer chains of plastics, including microplastics. Specific examples include:

i. *Ideonella sakaiensis*: A bacterium discovered in Japan that can break down polyethylene terephthalate (PET), a common type of plastic used in bottles and packaging. It secretes enzymes known as PETases, which hydrolyze PET into terephthalic acid and ethylene glycol, substances that the bacteria can then use as a carbon and energy source (Juliana et al., 2022; Walter et al., 2022).

ii. *Aspergillus tubingensis*: A fungus found to grow on the surface of plastics, secreting enzymes that break down polyurethane, a widely used durable plastic. The fungal hyphae can penetrate the plastic, accelerating its degradation (Huang et al., 2019).

These microorganisms offer a biological approach to degrade plastics, including microplastics, potentially reducing their abundance in the environment. However, challenges remain in optimizing these processes for environmental remediation, including the efficiency of degradation in different conditions and the fate of degradation products.

9.7.2.2 Plant-Based Accumulation

Plants, through phytoremediation, can also play a role in the accumulation or stabilization of microplastics in contaminated soils (Kominko et al., 2022). While plants do not degrade microplastics in the same way microorganisms can, they can help reduce the mobility of microplastics and prevent their further dispersal into the environment. Specific mechanisms include:

i. **Root Entanglement:** Plant roots can physically trap microplastics, preventing them from being transported through soil or water runoff. This process can help localize microplastics, making them easier to manage or remove (Nakano et al., 2022).

ii. **Rhizodegradation:** The root zone (rhizosphere) is a hotbed of microbial activity. Plants can enhance the degradation of microplastics indirectly by supporting diverse microbial communities. The exudates from roots can stimulate microbial growth, including those capable of degrading plastics (de-Bashan et al., 2012).

An example of plant-based potential in interacting with microplastics is the use of fast-growing, high-biomass plants like willows and poplars, which have extensive root systems that can stabilize soil and potentially entrap microplastics.

9.7.2.3 Challenges and Considerations

While the use of microorganisms and plants offers an innovative approach to addressing microplastic pollution, there are several challenges and considerations (Crystal Thew et al., 2023). These include the scalability of such methods, the time required for significant degradation or accumulation, the environmental conditions necessary for optimal organism performance, and the potential ecological impacts of introducing non-native species for remediation purposes (Qiu et al., 2022). Despite these challenges, research into the biological remediation of microplastics is a rapidly evolving field. Further studies are needed to understand the mechanisms involved, optimize conditions for degradation or accumulation, and assess the feasibility and safety of these methods for large-scale environmental applications (Schwarz et al., 2023).

9.7.3 Advanced Techniques for Microplastic Degradation

Advanced techniques for breaking down microplastics are critical in addressing the pervasive issue of microplastic pollution (Saboor et al., 2022; Tian et al., 2023). These techniques vary in approach, from chemical to physical and biological methods, each with its potential risks and benefits (Du et al., 2021; Sadia et al., 2022). Innovations in this field offer hope for more effective management and mitigation of microplastics in the environment, but they also necessitate careful consideration of their ecological and health implications. The recent advanced techniques are described as follows:

9.7.3.1 Photocatalytic Degradation

This method involves using light-sensitive catalysts to degrade microplastics under sunlight or artificial light. Titanium dioxide (TiO_2) is a commonly used photocatalyst that, when activated by UV light, generates reactive oxygen species capable of breaking down plastic polymers. Benefits: The use of sunlight makes this method energy-efficient and potentially scalable for outdoor applications. It can degrade a variety of plastic types and reduce microplastics to harmless by-products. Risks: The release of nanoparticulate catalysts into the environment could pose ecological and health risks if not adequately contained or recovered (Naghdi et al., 2023).

9.7.3.2 Enzymatic Degradation

Leveraging specific enzymes capable of breaking down polymer chains in plastics, this method targets the chemical bonds in plastics (Kaushal et al., 2021). Engineered or naturally occurring enzymes, such as PETase for PET plastics, have shown promise in degrading microplastics. Benefits: Enzymatic degradation can be highly specific to certain types of plastics, reducing the risk of unintended environmental impacts. It operates under mild conditions, which is energy-efficient and less likely to produce harmful by-products. Risks: Scaling up enzymatic treatments to environmental levels is challenging. The production and stability of enzymes in diverse environmental conditions also pose limitations (Arnold & Gramza-Michałowska, 2022).

9.7.3.3 Microwave Pyrolysis

This technique uses microwave radiation to heat microplastics in the absence of oxygen, breaking them down into simpler chemicals, such as gases and oils. Benefits: Microwave pyrolysis can process mixed types of plastics without the need for sorting, with the potential to convert waste microplastics into useful products or energy sources (Ethaib et al., 2020). Risks: The process requires significant energy input and must be carefully controlled to avoid the formation of toxic by-products.

9.7.3.4 Potential Risks and Benefits of Advanced Microplastic Degradation Techniques

Risks: The primary risks associated with advanced microplastic degradation techniques relate to the potential release of toxic by-products, the energy requirements of certain methods, and the ecological implications of deploying these technologies in natural environments. Careful assessment and management of these risks are essential to ensure that the solutions do not introduce new environmental problems (Dey et al., 2023). Benefits: The benefits of these advanced techniques are significant. They offer the potential to reduce the environmental burden of microplastics, converting them into less harmful substances or even useful materials. By addressing microplastic pollution, these methods can contribute to the restoration of ecosystems, protect biodiversity, and reduce health risks associated with microplastic exposure (Dierkes et al., 2022).

9.8 CONCLUSION

Advanced techniques for breaking down microplastics present promising solutions to a global environmental challenge. However, the development and implementation of these methods must be accompanied by comprehensive risk assessments and strategies to mitigate potential negative impacts. The balance of risks and benefits will be crucial in determining the feasibility and sustainability of these technologies for environmental remediation. Further research is needed to deepen our understanding of microplastic's impacts on soil ecosystems and to refine remediation technologies suitable for microplastic mitigation. The global collaboration among scientists, policymakers, and industry stakeholders are important to develop effective strategies for mitigating microplastic pollution and safeguarding soil health for future generations.

REFERENCES

Acarer, S. (2023). Microplastics in wastewater treatment plants: Sources, properties, removal efficiency, removal mechanisms, and interactions with pollutants. *Water Science and Technology, 87*, 685–710. https://doi.org/10.2166/wst.2023.022

Ahmed, S. F., Islam, N., Tasannum, N., Mehjabin, A., Momtahin, A., Chowdhury, A. A., Almomani, F., & Mofijur, M. (2024). Microplastic removal and management strategies for wastewater treatment plants. *Chemosphere, 347*, 140648. https://doi.org/10.1016/j.chemosphere.2023.140648

Aksakal, E. L., Angin, I., & Sari, S. (2020). A new approach for calculating aggregate stability: Mean weight aggregate stability (MWAS). *Catena, 194*, 104708. https://doi.org/10.1016/j.catena.2020.104708

Amesho, K. T. T., Chinglenthoiba, C., Samsudin, M. S. A. B., Lani, M. N., Pandey, A., Desa, M. N. M., & Suresh, V. (2023). Microplastics in the environment: An urgent need for coordinated waste management policies and strategies. *Journal of Environmental Management, 344*, 118713. https://doi.org/10.1016/j.jenvman.2023.118713

Aransiola, S. A., Victor-Ekwebelem, M. O., Ajiboye, A. E., Zobeashia, S. S. L. T., Ijah, U. J. J., & Oyedele, O. J. (2023). Ecological impacts and toxicity of micro- and nanoplastics in agroecosystem. In N. R. Maddela, K. V. Reddy, P. Ranjit (Eds.), *Micro and Nanoplastics in Soil: Threats to Plant-Based Food*. Springer. https://doi.org/10.1007/978-3-031-21195-9_10

Arnold, M., & Gramza-Michałowska, A. (2022). Enzymatic browning in apple products and its inhibition treatments: A comprehensive review. *Comprehensive Reviews in Food Science and Food Safety 21*, 13059. https://doi.org/10.1111/1541-4337.13059

Arrázola-Vásquez, E., Larsbo, M., Capowiez, Y., Taylor, A., Sandin, M., Iseskog, D., & Keller, T. (2022). Earthworm burrowing modes and rates depend on earthworm species and soil mechanical resistance. *Applied Soil Ecology, 178*, 104568. https://doi.org/10.1016/j.apsoil.2022.104568

Athulya, P. A., Waychal, Y., Rodriguez-Seijo, A., Devalla, S., Doss, C. G. P., & Chandrasekaran, N. (2024). Microplastic interactions in the agroecosystems: Methodological advances and limitations in quantifying microplastics from agricultural soil. *Environmental Geochemistry and Health*, *46*(3), 85. https://doi.org/10.1007/s10653-023-01800-8

Barciela-Rial, M., Vardon, P. J., Van Kessel, T., Griffioen, J., & Winterwerp, J. C. (2022). Effect of composition on the compressibility and shear strength of dredged cohesive sediment. *Frontiers in Earth Science*, *10*. https://doi.org/10.3389/feart.2022.786108

Bardgett, R. D., & Caruso, T. (2020). Soil microbial community responses to climate extremes: Resistance, resilience and transitions to alternative states. *Philosophical Transactions of the Royal Society B: Biological Sciences*, *375*. https://doi.org/10.1098/rstb.2019.0112

Bhatt, P., Pathak, V. M., Bagheri, A. R., & Bilal, M. (2021). Microplastic contaminants in the aqueous environment, fate, toxicity consequences, and remediation strategies. *Environmental Research*, *200*, 111762. https://doi.org/10.1016/j.envres.2021.111762

Bonyadinejad, G., Salehi, M., & Herath, A. (2022). Investigating the sustainability of agricultural plastic products, combined influence of polymer characteristics and environmental conditions on microplastics aging. *Science of the Total Environment*, *839*, 156385. https://doi.org/10.1016/j.scitotenv.2022.156385

Brožová, K., Halfar, J., Čabanová, K., Motyka, O., Drabinová, S., Hanus, P., & Heviánková, S. (2023). The first evidence of microplastic occurrence in mine water: The largest black coal mining area in the Czech Republic. *Water Research*, *244*, 120538. https://doi.org/10.1016/j.watres.2023.120538

Calabrò, P. S., & Grosso, M. (2018). Bioplastics and waste management. *Waste Management*, 78, 800–801. https://doi.org/10.1016/j.wasman.2018.06.054

Cardoso, D. N., Gonçalves, S. F., Silva, A. R. R., Soares, A. M. V. M., Wrona, F. J., & Loureiro, S. (2023). Ecotoxicological effects of fluvial eroded bitumen sediments from the Alberta oil sands to model aquatic species. *Science of the Total Environment*, *862*, 160592. https://doi.org/10.1016/j.scitotenv.2022.160592

Cattle, S. R., Robinson, C., & Whatmuff, M. (2020). The character and distribution of physical contaminants found in soil previously treated with mixed waste organic outputs and garden waste compost. *Waste Management*, *101*, 94–105. https://doi.org/10.1016/j.wasman.2019.09.043

Ceneviva, L. V. S., & Tsuge, T. (Eds.). (2022). Environmental sustainability with polyhydroxyalkanoates (PHA) as plastic alternatives. In *Sustainable Management of Environmental Contaminants, Eco-friendly Remediation Approaches* (pp.17–49). Springer. https://doi.org/10.1007/978-3-031-08446-1_2

Chae, Y., & An, Y. J. (2020). Effects of food presence on microplastic ingestion and egestion in Mytilus galloprovincialis. *Chemosphere*, *240*, 124855. https://doi.org/10.1016/j.chemosphere.2019.124855

Chang, J., Fang, W., Liang, J., Zhang, P., Zhang, G., Zhang, H., Zhang, Y., & Wang, Q. (2022). A critical review on interaction of microplastics with organic contaminants in soil and their ecological risks on soil organisms. *Chemosphere*, *306*, 135573. https://doi.org/10.1016/j.chemosphere.2022.135573

Chen, X., Zhu, Y., Chen, F., Li, Z., Zhang, X., Wang, G., Ji, J., & Guan, C. (2023). The role of microplastics in the process of laccase-assisted phytoremediation of phenanthrene-contaminated soil. *Science of the Total Environment, 905*, 167305. https://doi.org/10.1016/j.scitotenv.2023.167305

Cheng, Y., Chen, J., Bao, M., Zhao, L., & Li, Y. (2022). The proliferation and colonization of functional bacteria on amorphous polyethylene terephthalate: Key role of ultraviolet irradiation and nonionic surfactant polysorbate 80 addition. *Chemosphere, 291*, 132940. https://doi.org/10.1016/j.chemosphere.2021.132940

Chinglenthoiba, C., Amesho, K. T. T., Reddy, D. G. C. V., Chellappan, S., & Lani, M. N. (2023). Microplastic as an emerging environmental threat: A critical review on sampling and identification techniques focusing on aquactic ecoystem. *Journal of Polymers and the Environment, 31*(5), 1725–1747. https://doi.org/10.1007/s10924-022-02716-7

Chiou, K. L., Cook, A. G., & Hastorf, C. A. (2013). Flotation versus dry sieving archaeobotanical remains: A case history from the Middle Horizon southern coast of Peru. *Journal of Field Archaeology, 38*, 38–53. https://doi.org/10.1179/0093469012Z.00000000035

Cingolani, M., Rampazzo, E., Zaccheroni, N., Genovese, D., & Prodi, L. (2022). Fluorogenic hyaluronan nanogels for detection of micro- and nanoplastics in water. *Environmental Science: Nano, 9*, 582–588. https://doi.org/10.1039/d1en00684c

Crossman, J., Hurley, R. R., Futter, M., & Nizzetto, L. (2020). Transfer and transport of microplastics from biosolids to agricultural soils and the wider environment. *Science of the Total Environment, 724*, 138334. https://doi.org/10.1016/j.scitotenv.2020.138334

Crystal Thew, X. E., Lo, S. C., Ramanan, R. N., Tey, B. T., Huy, N. D., & Chien Wei, O. (2023). Enhancing plastic biodegradation process: Strategies and opportunities. *Critical Reviews in Biotechnology, 44*, 477–494. https://doi.org/10.1080/07388551.2023.2170861

Das, P. P., Singh, K. R., Nagpure, G., Mansoori, A., Singh, R. P., Ghazi, I. A., Kumar, A., & Singh, J. (2022). Plant-soil-microbes: A tripartite interaction for nutrient acquisition and better plant growth for sustainable agricultural practices. *Environmental Research, 214*, 113821. https://doi.org/10.1016/j.envres.2022.113821

de-Bashan, L. E., Hernandez, J. P., & Bashan, Y. (2012). The potential contribution of plant growth-promoting bacteria to reduce environmental degradation - A comprehensive evaluation. *Applied Soil Ecology, 61*, 171–189. https://doi.org/10.1016/j.apsoil.2011.09.003

Deng, Y., Wu, J., Chen, J., & Kang, K. (2023). Overview of microplastic pollution and its influence on the health of organisms. *Journal of Environmental Science and Health - Part A Toxic/Hazardous Substances and Environmental Engineering, 58*, 412–422. https://doi.org/10.1080/10934529.2023.2190715

Devi, A. S., & Aravindhan, M. S. (2023). A comprehensive study of recent bioremediation approaches and techniques for a green and sustainable environment. *Environmental Quality Management, 33*, 441–449. https://doi.org/10.1002/tqem.22037

Dey, S., Anand, U., Kumar, V., Kumar, S., Ghorai, M., Ghosh, A., Kant, N., Suresh, S., Bhattacharya, S., Bontempi, E., Bhat, S. A., & Dey, A. (2023). Microbial strategies for degradation of microplastics generated from COVID-19 healthcare waste. *Environmental Research, 216*, 114438. https://doi.org/10.1016/j.envres.2022.114438

Dhevagi, P., Poornima, R., Keerthi Sahasa, R. G., Ramya, A., Karthika, S., & Sivasubramanian, K. (2022). The crux of microplastics in soil - A review. *International Journal of Environmental Analytical Chemistry, 104*, 6546–6578. https://doi.org/10.1080/03067319.2022.2148528

Dierkes, G., Lauschke, T., & Földi, C. (2022). Analytical methods for plastic (microplastic) determination in environmental samples. In *Handbook of Environmental Chemistry*. https://doi.org/10.1007/698_2021_744

Ding, J., Lv, M., Zhu, D., Leifheit, E. F., Chen, Q. L., Wang, Y. Q., Chen, L. X., Rillig, M. C., & Zhu, Y. G. (2023). Tire wear particles: An emerging threat to soil health. *Critical Reviews in Environmental Science and Technology, 53*, 239–27. https://doi.org/10.1080/10643389.2022.2047581

Dissanayake, P. D., Kim, S., Sarkar, B., Oleszczuk, P., Sang, M. K., Haque, M. N., Ahn, J. H., Bank, M. S., & Ok, Y. S. (2022). Effects of microplastics on the terrestrial environment: A critical review. *Environmental Research*, 209, 112734. https://doi.org/10.1016/j.envres.2022.112734

do Carmo Precci Lopes, A., Senfter, T., Ebner, C., Senn, M., Pillei, M., Kraxner, M., Robra, S., & Bockreis, A. (2021). Separation of biodegradable material from the low calorific fraction of municipal solid waste. *Journal of Cleaner Production, 280*, 124681. https://doi.org/10.1016/j.jclepro.2020.124681

Dong, Y., Gao, M., Liu, X., Qiu, W., & Song, Z. (2020). The mechanism of polystyrene microplastics to affect arsenic volatilization in arsenic-contaminated paddy soils. *Journal of Hazardous Materials, 398*. https://doi.org/10.1016/j.jhazmat.2020.122896

Du, H., Xie, Y., & Wang, J. (2021). Microplastic degradation methods and corresponding degradation mechanism: Research status and future perspectives. *Journal of Hazardous Materials, 418*, 126377. https://doi.org/10.1016/j.jhazmat.2021.126377

Elizalde-Velázquez, G. A., & Gómez-Oliván, L. M. (2021). Microplastics in aquatic environments: A review on occurrence, distribution, toxic effects, and implications for human health. *Science of the Total Environment, 780*, 146551. https://doi.org/10.1016/j.scitotenv.2021.146551

ElSafty, A. (2014). Health implications of heavy metal overload. *Occupational Medicine & Health Affairs, 2*, 1000145. https://doi.org/10.4172/2329-6879.1000145

Enfrin, M., Hachemi, C., Hodgson, P. D., Jegatheesan, V., Vrouwenvelder, J., Callahan, D. L., Lee, J., & Dumée, L. F. (2021). Nano/micro plastics – Challenges on quantification and remediation: A review. *Journal of Water Process Engineering, 42*, 102128. https://doi.org/10.1016/j.jwpe.2021.102128

Ethaib, S., Omar, R., Kamal, S. M. M., Biak, D. R. A., & Zubaidi, S. L. (2020). Microwave-assisted pyrolysis of biomass waste: A mini review. *Processes, 8*, 1190. https://doi.org/10.3390/PR8091190

Frias, J. P. G. L., & Nash, R. (2019). Microplastics: Finding a consensus on the definition. *Marine Pollution Bulletin, 138*, 145–147. https://doi.org/10.1016/j.marpolbul.2018.11.022

Gani, A., Hussain, A., Pathak, S., & Banerjee, A. (2024). An empirical investigation on the elimination of heavy metals using bioremediation method for selected plant species. *Physics and Chemistry of the Earth, 134*, 103568. https://doi.org/10.1016/j.pce.2024.103568

García, R. A., Franzese, J., Policelli, N., Sasal, Y., Zenni, R. D., Nuñez, M. A., Taylor, K., & Pauchard, A. (2018). Non-native pines are homogenizing the ecosystems of South America. In R. Rozzi, R. H. May Jr, F. Stuart Chapin III, F. Massardo, M. C. Gavin, I. J. Klaver, A. Pauchard, M. A. Nunez, D. Simberloff (Eds.), *From Biocultural Homogenization to Biocultural Convservation* (pp. 245–263). Springer International. https://doi.org/10.1007/978-3-319-99513-7_15

Gartzia-Bengoetxea, N., González-Arias, A., Merino, A., & Martínez de Arano, I. (2009). Soil organic matter in soil physical fractions in adjacent semi-natural and cultivated stands in temperate Atlantic forests. *Soil Biology and Biochemistry, 41*, 1674–1683. https://doi.org/10.1016/j.soilbio.2009.05.010

Guerranti, C., Martellini, T., Perra, G., Scopetani, C., & Cincinelli, A. (2019). Microplastics in cosmetics: Environmental issues and needs for global bans. *Environmental Toxicology and Pharmacology, 68*, 75–79. https://doi.org/10.1016/j.etap.2019.03.007

Guo, J.-J., Huang, X.-P., Xiang, L., Wang, Y.-Z., Li, Y.-W., Li, H., Cai, Q.-Y., Mo, C.-H., & Wong, M.-H. (2020). Source, migration and toxicology of microplastics in soil. *Environment International, 137*, 105263. https://doi.org/10.1016/j.envint.2019.105263

Guo, Q. Q., Xiao, M. R., & Zhang, G. S. (2021). The persistent impacts of polyester microfibers on soil bio-physical properties following thermal treatment. *Journal of Hazardous Materials, 420*. https://doi.org/10.1016/j.jhazmat.2021.126671

Hao, X., Sun, H., Zhang, Y., Li, S., & Yu, Z. (2023). Co-transport of arsenic and micro/nano-plastics in saturated soil. *Environmental Research, 228*. https://doi.org/10.1016/j.envres.2023.115871

Hemalatha, I., Harika, D., & Karnena, M. K. (2022). Sustainable nano-bioremediation approaches for the treatment of polluted soils. *Nature Environment and Pollution Technology, 21*. https://doi.org/10.46488/NEPT.2022.v21i04.036

Herrera, A., Acosta-Dacal, A., Pérez Luzardo, O., Martínez, I., Rapp, J., Reinold, S., Montesdeoca-Esponda, S., Montero, D., & Gómez, M. (2022). Bioaccumulation of additives and chemical contaminants from environmental microplastics in European seabass (Dicentrarchus labrax). *Science of the Total Environment, 822*, 153396. https://doi.org/10.1016/j.scitotenv.2022.153396

Herrera, A., Acosta-Dacal, A., Pérez-Luzardo, O., Martínez, I., Rapp, J., Reinold, S., Montesdeoca-Esponda, S., Montero, D., & Gómez, M. (2023). Trophic transfer of DDE, BP-3 and chlorpyrifos from microplastics to tissues in Dicentrarchus labrax. *Science of the Total Environment, 882*, 163295. https://doi.org/10.1016/j.scitotenv.2023.163295

Hooge, A., Hauggaard-Nielsen, H., Heinze, W. M., Lyngsie, G., Ramos, T. M., Sandgaard, M. H., Vollertsen, J., & Syberg, K. (2023). Fate of microplastics in sewage sludge and in agricultural soils. *TrAC - Trends in Analytical Chemistry, 166*, 117184. https://doi.org/10.1016/j.trac.2023.117184

Huang, D., Song, Y., Liu, Y., & Qin, Y. (2019). A new strain of Aspergillus tubingensis for high-activity pectinase production. *Brazilian Journal of Microbiology, 50*, 53–65. https://doi.org/10.1007/s42770-018-0032-3

Huang, G., Kilic, A., Karady, M., Zhang, J., Mehra, P., Song, X., Sturrock, C. J., Zhu, W., Qin, H., Hartman, S., Schneider, H. M., Bhosale, R., Dodd, I. C., Sharp, R. E., Huang, R., Mooney, S. J., Liang, W., Bennett, M. J., Zhang, D., & Pandey, B. K. (2022). Ethylene inhibits rice root elongation in compacted soil via ABA- and auxin-mediated mechanisms. *Proceedings of the National Academy of Sciences of the United States of America*, 119, e2201072119. https://doi.org/10.1073/pnas.2201072119

Hubert, J., Nesvorna, M., & Stejskal, V. (2009). The efficacy of sieving, filth flotation and Tullgren heat extraction for detecting various developmental stages of Tribolium castaneum and Ephestia kuehniella in samples of wheat grain, flour and semolina. *Journal of Stored Products Research, 45*, 279–288. https://doi.org/10.1016/j.jspr.2009.05.003

Ivy, N., Bhattacharya, S., Dey, S., Gupta, K., Dey, A., & Sharma, P. (2023). Effects of microplastics and arsenic on plants: Interactions, toxicity and environmental implications. *Chemosphere, 338*, 139542. https://doi.org/10.1016/j.chemosphere.2023.139542

Jia, H., Wu, D., Yu, Y., Han, S., Sun, L., & Li, M. (2022). Impact of microplastics on bioaccumulation of heavy metals in rape (Brassica napus L.). *Chemosphere, 288*, 132576. https://doi.org/10.1016/j.chemosphere.2021.132576

Jiang, H., Bu, J., Bian, K., Su, J., Wang, Z., Sun, H., Wang, H., Zhang, Y., & Wang, C. (2023). Surface change of microplastics in aquatic environment and the removal by froth flotation assisted with cationic and anionic surfactants. *Water Research, 233*, 119794. https://doi.org/10.1016/j.watres.2023.119794

Juliana, S., Parhusip, M., Simanullang, A., Tita, E., & Irawati, W. (2022). Potential of Ideonella sakaiensis bacteria in degrading plastic waste type polyethylene terephthalate. *Jurnal Biologi Tropis.*

Kaushal, J., Khatri, M., & Arya, S. K. (2021). Recent insight into enzymatic degradation of plastics prevalent in the environment: A mini-review. *Cleaner Engineering and Technology, 2*, 100083. https://doi.org/10.1016/j.clet.2021.100083

Kim, W.-K., Park, H., Ishii, K., & Ham, G.-Y. (2023). Investigation on microplastics in soil near landfills in the Republic of Korea. *Sustainability (Switzerland), 15*(15). https://doi.org/10.3390/su151512057

Kladivko, E. J. (2001). Tillage systems and soil ecology. *Soil and Tillage Research, 61*, 61–76. https://doi.org/10.1016/S0167-1987(01)00179-9

Kominko, H., Gorazda, K., & Wzorek, Z. (2022). Effect of sewage sludge-based fertilizers on biomass growth and heavy metal accumulation in plants. *Journal of Environmental Management, 305*, 114417. https://doi.org/10.1016/j.jenvman.2021.114417

Koongolla, J. B., Lin, L., Yang, C. P., Pan, Y. F., Li, H. X., Liu, S., & Xu, X. R. (2022). Microplastic prevalence in marine fish from onshore Beibu Gulf, South China Sea. *Frontiers in Marine Science, 9*. https://doi.org/10.3389/fmars.2022.964461

Kour, H., Khan, S. S., Kour, D., Rasool, S., Sharma, Y. P., Rai, P. K., Singh, S., Chaubey, K. K., Rai, A. K., & Yadav, A. N. (2023). Microbes mediated plastic degradation: A sustainable approach for environmental sustainability. *Journal of Applied Biology and Biotechnology, 11*(3), 9–19. https://doi.org/10.7324/JABB.2023.110515

Li, H., Min, F., Liu, C., Cheng, Y., & Chen, J. (2024). Experimental study on treatment of high ash coal slime by three-products flotation. *International Journal of Coal Preparation and Utilization, 44*, 67–81. https://doi.org/10.1080/19392699.2023.2170366

i, Y., Liu, Q., Junaid, M., Chen, G., & Wang, J. (2023). Distribution, sources, transportation and biodegradation of microplastics in the soil environment. *TrAC - Trends in Analytical Chemistry, 164*, 117106. https://doi.org/10.1016/j.trac.2023.117106

Lin, Y., Xie, J., Xiang, Q., Liu, Y., Wang, P., Wu, Y., & Zhou, Y. (2022). Effect of propiconazole on plastic film microplastic degradation: Focusing on the change in microplastic morphology and heavy metal distribution. *Science of the Total Environment, 822*, 153609. https://doi.org/10.1016/j.scitotenv.2022.153609

Lyu, Y., Zhou, J., Yang, Y., Zang, H., Hu, Y., & Zeng, Z. (2022). Microplastics in agroecosystem: Research status and future challenges. *Chinese Journal of Eco-Agriculture, 30*, 1–14. https://doi.org/10.12357/cjea.20210442

Ma, J.-W., Wu, Y.-Q., Xu, C.-L., Luo, Z.-X., Yu, R.-L., Hu, G.-R., & Yan, Y. (2023). Inhibitory effect of polyethylene microplastics on roxarsone degradation in soils. *Journal of Hazardous Materials, 454*, 131483. https://doi.org/10.1016/j.jhazmat.2023.131483

Ma, X., Han, F., Bai, Y., Wu, S., Wu, J., & Ma, Y. (2023). Application and environmental effects of organic materials in orchard ecosystem: Research status and prospects. *Chinese Journal of Eco-Agriculture, 31*, 1240–1255. https://doi.org/10.12357/cjea.20230207

Mäkinen, J., Bachér, J., Kaartinen, T., Wahlström, M., & Salminen, J. (2015). The effect of flotation and parameters for bioleaching of printed circuit boards. *Minerals Engineering, 75*, 26–31. https://doi.org/10.1016/j.mineng.2015.01.009

Meng, F., Yang, X., Riksen, M., & Geissen, V. (2022). Effect of different polymers of microplastics on soil organic carbon and nitrogen – A mesocosm experiment. *Environmental Research, 204*, 111938. https://doi.org/10.1016/j.envres.2021.111938

Meng, F., Yang, X., Riksen, M., Xu, M., & Geissen, V. (2021). Response of common bean (Phaseolus vulgaris L.) growth to soil contaminated with microplastics. *Science of the Total Environment, 755*, 142516. https://doi.org/10.1016/j.scitotenv.2020.142516

Mokhtarzadeh, Z., Keshavarzi, B., Moore, F., Busquets, R., Rezaei, M., Padoan, E., & Ajmone-Marsan, F. (2022). Microplastics in industrial and urban areas in South-West Iran. *International Journal of Environmental Science and Technology, 19*, 10199–10210. https://doi.org/10.1007/s13762-022-04223-7

Naghdi, S., Shahrestani, M. M., Zendehbad, M., Djahaniani, H., Kazemian, H., & Eder, D. (2023). Recent advances in application of metal-organic frameworks (MOFs) as adsorbent and catalyst in removal of persistent organic pollutants (POPs). *Journal of Hazardous Materials, 442*, 130127. https://doi.org/10.1016/j.jhazmat.2022.130127

Nakano, M., Furuichi, T., Sokabe, M., Iida, H., Yano, S., & Tatsumi, H. (2022). Entanglement of Arabidopsis seedlings to a mesh substrate under microgravity conditions in KIBO on the ISS. *Plants, 11*, 956. https://doi.org/10.3390/plants11070956

Ning, Y., Wang, X., Lu, J., Li, Y., Yang, Y., Zou, D., & Zhou, D. (2023). Study on the life maintenance mechanism of Eisenia fetida under low-density polyethylene stress: Based on path analysis and canonical correlation analysis. *Ecotoxicology and Environmental Safety, 250*, 114500. https://doi.org/10.1016/j.ecoenv.2023.114500

Pooja, G., Kumar, P. S., & Indraganti, S. (2022). Recent advancements in the removal/recovery of toxic metals from aquatic system using flotation techniques. *Chemosphere, 287*, 132231. https://doi.org/10.1016/j.chemosphere.2021.132231

Qiu, J., Zhou, Q., Ye, W., Chen, Q., & Bao, Y. J. (2022). SweepCluster: A SNP clustering tool for detecting gene-specific sweeps in prokaryotes. *BMC Bioinformatics, 23*, 19. https://doi.org/10.1186/s12859-021-04533-6

Rajot, J. L., Alfaro, S. C., Gomes, L., & Gaudichet, A. (2003). Soil crusting on sandy soils and its influence on wind erosion. *Catena, 53*, 1–16. https://doi.org/10.1016/S0341-8162(02)00201-1

Raza, M., Lee, J.-Y., & Cha, J. (2022). Microplastics in soil and freshwater: Understanding sources, distribution, potential impacts, and regulations for management. *Science Progress, 105*(3), 003685042211266. https://doi.org/10.1177/00368504221126676

Rieke, E. L., Bagnall, D. K., Morgan, C. L. S., Flynn, K. D., Howe, J. A., Greub, K. L. H., Mac Bean, G., Cappellazzi, S. B., Cope, M., Liptzin, D., Norris, C. E., Tracy, P. W., Aberle, E., Ashworth, A., Bañuelos Tavarez, O., Bary, A. I., Baumhardt, R. L., Borbón Gracia, A., Brainard, D. C., Brennan, J. R., Honeycutt, C. W. (2022). Evaluation of aggregate stability methods for soil health. *Geoderma, 428*, 116156. https://doi.org/10.1016/j.geoderma.2022.116156

Ruggero, F., Gori, R., & Lubello, C. (2020). Methodologies for microplastics recovery and identification in heterogeneous solid matrices: A review. *Journal of Polymers and the Environment, 28*, 739. https://doi.org/10.1007/s10924-019-01644-3

Saboor, F. H., Hadian-Ghazvini, S., & Torkashvand, M. (2022). Microplastics in aquatic environments: Recent advances in separation techniques. *Periodica Polytechnica Chemical Engineering, 66*, 167–181. https://doi.org/10.3311/PPch.18930

Sadia, M., Mahmood, A., Ibrahim, M., Irshad, M. K., Quddusi, A. H. A., Bokhari, A., Mubashir, M., Chuah, L. F., & Show, P. L. (2022). Microplastics pollution from wastewater treatment plants: A critical review on challenges, detection, sustainable removal techniques and circular economy. *Environmental Technology and Innovation, 28*, 102946. https://doi.org/10.1016/j.eti.2022.102946

Sadiq, M., Li, G., Rahim, N., & Tahir, M. M. (2021). Sustainable conservation tillage technique for improving soil health by enhancing soil physicochemical quality indicators under wheat mono-cropping system conditions. *Sustainability (Switzerland), 13*, 8177. https://doi.org/10.3390/su13158177

Saravanakumar, K., SivaSantosh, S., Sathiyaseelan, A., Naveen, K. V, AfaanAhamed, M. A., Zhang, X., Priya, V. V, MubarakAli, D., & Wang, M.-H. (2022). Unraveling the hazardous impact of diverse contaminants in the marine environment: Detection and remedial approach through nanomaterials and nano-biosensors. *Journal of Hazardous Materials, 433*, 128720. https://doi.org/10.1016/j.jhazmat.2022.128720

Schell, T., Hurley, R., Buenaventura, N. T., Mauri, P. V, Nizzetto, L., Rico, A., & Vighi, M. (2022). Fate of microplastics in agricultural soils amended with sewage sludge: Is surface water runoff a relevant environmental pathway? *Environmental Pollution, 293*, 118520. https://doi.org/10.1016/j.envpol.2021.118520

Schjønning, P., McBride, R. A., Keller, T., & Obour, P. B. (2017). Predicting soil particle density from clay and soil organic matter contents. *Geoderma, 286*, 83–87. https://doi.org/10.1016/j.geoderma.2016.10.020

Schwarz, A. E., Lensen, S. M. C., Langeveld, E., Parker, L. A., & Urbanus, J. H. (2023). Plastics in the global environment assessed through material flow analysis, degradation and environmental transportation. *Science of the Total Environment, 875*, 162644. https://doi.org/10.1016/j.scitotenv.2023.162644

Sharma, S., & Chatterjee, S. (2017). Microplastic pollution, a threat to marine ecosystem and human health: A short review. *Environmental Science and Pollution Research, 24*, 21530–21547. https://doi.org/10.1007/s11356-017-9910-8

Sholokhova, A., Denafas, G., & Mykhaylenko, V. (2021). The organic output from mechanical–biological treatment plants as a source of microplastics: Mini-review on current knowledge, research methodology and future study perspectives. *Waste Management and Research, 39*, 652–663. https://doi.org/10.1177/0734242X20983914

Silori, R., Shrivastava, V., Mazumder, P., Mootapally, C., Pandey, A., & Kumar, M. (2023). Understanding the underestimated: Occurrence, distribution, and interactions of microplastics in the sediment and soil of China, India, and Japan. *Environmental Pollution, 320*, 120978. https://doi.org/10.1016/j.envpol.2022.120978

Sima, J., Wang, J., Song, J., Du, X., Lou, F., Pan, Y., Huang, Q., Lin, C., Wang, Q., & Zhao, G. (2023). Dielectric barrier discharge plasma for the remediation of microplastic-contaminated soil from landfill. *Chemosphere, 317*, 137815. https://doi.org/10.1016/j.chemosphere.2023.137815

Smedes, F., Rusina, T. P., Beeltje, H., & Mayer, P. (2017). Partitioning of hydrophobic organic contaminants between polymer and lipids for two silicones and low density polyethylene. *Chemosphere, 186*, 948–957. https://doi.org/10.1016/j.chemosphere.2017.08.044

Song, X., Jin, J., Li, H., Wang, F., Liu, J., Wang, X., Huang, X., Chai, C., Song, N., & Zong, H. (2023). Kaolinite reduced Cd accumulation in peanut and remediate soil contaminated with both microplastics and cadmium. *Ecotoxicology and Environmental Safety, 266*, 115580. https://doi.org/10.1016/j.ecoenv.2023.115580

Stabnikova, O., Stabnikov, V., Marinin, A., Klavins, M., & Vaseashta, A. (2022). The role of microplastics biofilm in accumulation of trace metals in aquatic environments. *World Journal of Microbiology and Biotechnology*, *38*, 117. https://doi.org/10.1007/s11274-022-03293-6

Surendran, U., Jayakumar, M., Raja, P., Gopinath, G., & Chellam, P. V. (2023). Microplastics in terrestrial ecosystem: Sources and migration in soil environment. *Chemosphere*, *318*, 137946. https://doi.org/10.1016/j.chemosphere.2023.137946

Taleb, M., Gani, A., Saiful, J., Irham, M., Windari, W., & Erdiwansyah. (2022). An overview of the potential risks, sources, and analytical methods for microplastics in soil. *AIMS Environmental Science*, *9*, 169–200. https://doi.org/10.3934/environsci.2022013

Tang, S., Gong, J., Song, B., Cao, W., & Li, J. (2024). Remediation of biochar-supported effective microorganisms and microplastics on multiple forms of heavy metals in eutrophic lake. *Journal of Hazardous Materials*, *465*, 133098. https://doi.org/10.1016/j.jhazmat.2023.133098

Tankersley, K. B., Owen, L. A., Dunning, N. P., Fladd, S. G., Bishop, K. J., Lentz, D. L., & Slotten, V. (2017). Micro-flotation removal of coal contaminants from archaeological radiocarbon samples from Chaco Canyon, New Mexico, USA. *Journal of Archaeological Science: Reports*, *12*, 66–73. https://doi.org/10.1016/j.jasrep.2017.01.029

Tedeschi, A., Schillaci, M., & Balestrini, R. (2023). Mitigating the impact of soil salinity: Recent developments and future strategies. *Italian Journal of Agronomy*, *18*. https://doi.org/10.4081/ija.2023.2173

Tian, H., Wang, L., Zhu, X., Zhang, M., Li, L., Liu, Z., & Abolfathi, S. (2024). Biodegradation of microplastics derived from controlled release fertilizer coating: Selective microbial colonization and metabolism in plastisphere. *Science of the Total Environment*, *920*, 170978. https://doi.org/10.1016/j.scitotenv.2024.170978

Tian, W., Song, P., Zhang, H., Duan, X., Wei, Y., Wang, H., & Wang, S. (2023). Microplastic materials in the environment: Problem and strategical solutions. *Progress in Materials Science*, 132, 101035. https://doi.org/10.1016/j.pmatsci.2022.101035

Tingting, H., & Jiawei, C. (2022). A review on adsorption and transport of microplastics in soil and the effect of ageing on environmental behavior of pollutants. *Rock and Mineral Analysis*, *41*(3), 353–363. https://doi.org/10.15898/j.cnki.11-2131/td.202202180024

Vance, W. H., & Milroy, S. P. (2022). Water acquisition by roots from the subsoil: Impact of physical constraints on the dynamics of water capture. In T. S. de Oliveira, R. W. Bell (Eds.), *Subsoil Constraints for Crop Production* (pp. 323–345). Springer. https://doi.org/10.1007/978-3-031-00317-2_13

Velicogna, J. R., Schwertfeger, D., Jesmer, A., Beer, C., Kuo, J., DeRosa, M. C., Scroggins, R., Smith, M., & Princz, J. (2021). Soil invertebrate toxicity and bioaccumulation of nano copper oxide and copper sulphate in soils, with and without biosolids amendment. *Ecotoxicology and Environmental Safety*, *217*, 112222. https://doi.org/10.1016/j.ecoenv.2021.112222

Verla, A W, Enyoh, C. E., Obinna, I. B., Verla, E. N., Qingyue, W., Chowdhury, M. A. H., Enyoh, E. C., & Chowdhury, T. (2020). Effect of macro-and micro-plastics in soil on growth of juvenile lime tree (Citrus aurantium). *AIMS Environmental Science*, *7*(6), 526–541. https://doi.org/10.3934/environsci.2020033

Viel, T., Manfra, L., Zupo, V., Libralato, G., Cocca, M., & Costantini, M. (2023). Biodegradation of plastics induced by marine organisms: Future perspectives for bioremediation approaches. *Polymers*, *15*, 2673. https://doi.org/10.3390/polym15122673

Walter, A., Sopracolle, L., Mutschlechner, M., Spruck, M., & Griesbeck, C. (2022). Biodegradation of different PET variants from food containers by Ideonella sakaiensis. *Archives of Microbiology*, *204*, 711. https://doi.org/10.1007/s00203-022-03306-w

Wang, C., Tang, J., Yu, H., Wang, Y., Li, H., Xu, S., Li, G., & Zhou, Q. (2022). Microplastic pollution in the soil environment: Characteristics, influencing factors, and risks. *Sustainability (Switzerland)*, *14*, 13405. https://doi.org/10.3390/su142013405

Wang, H., Tang, X., Zhang, W., Liu, C., Guan, Z., & Xiao, L. (2015). Effects of biochar application on tilth soil hydraulic properties of slope cropland of purple soil. *Nongye Gongcheng Xuebao/Transactions of the Chinese Society of Agricultural Engineering*, *31*. https://doi.org/10.3969/j.issn.1002-6819.2015.04.016

Wang, Y., Hou, P., Liu, K., Hayat, K., & Liu, W. (2024). Depth distribution of nano- and microplastics and their contribution to carbon storage in Chinese agricultural soils. *Science of the Total Environment*, *913*, 169709. https://doi.org/10.1016/j.scitotenv.2023.169709

Wang, Z., Taylor, S. E., Sharma, P., & Flury, M. (2018). Poor extraction efficiencies of polystyrene nano- and microplastics from biosolids and soil. *PLoS One*, *13*(11), e0208009. https://doi.org/10.1371/journal.pone.0208009

Wirnkor, V. A., Ebere, E. C., & Ngozi, V. E. (2019). The importance of microplastics pollution studies in water and soil of Nigeria ecosystems. *Analytical Methods in Environmental Chemistry Journal*, *2*(3), 89–96. https://doi.org/10.24200/amecj.v2.i03.69

Wu, R.-T., Cai, Y.-F., Chen, Y.-X., Yang, Y.-W., Xing, S.-C., & Liao, X.-D. (2021). Occurrence of microplastic in livestock and poultry manure in South China. *Environmental Pollution*, *277*, 116790. https://doi.org/10.1016/j.envpol.2021.116790

Xu, B., Liu, F., Cryder, Z., Huang, D., Lu, Z., He, Y., Wang, H., Lu, Z., Brookes, P. C., Tang, C., Gan, J., & Xu, J. (2020). Microplastics in the soil environment: Occurrence, risks, interactions and fate–A review. *Critical Reviews in Environmental Science and Technology*, *50*, 2175–2222. https://doi.org/10.1080/10643389.2019.1694822

Xu, P., Zhu, X., Tian, H., Zhao, G., Chi, Y., Jia, B., & Zhang, J. (2022). The broad application and mechanism of humic acids for treating environmental pollutants: Insights from bibliometric analysis. *Journal of Cleaner Production*, *337*, 130510. https://doi.org/10.1016/j.jclepro.2022.130510

Xueying, F., Yuhuan, S., Shuwu, Z., & Fayuan, W. (2021). Ecological effects of microplastics on soil-plant systems. *Acta Pedologica Sinica, 58*, 299–313. https://doi.org/10.11766/trxb202007190402

Yadav, V., Dhanger, S., & Sharma, J. (2022). Microplastics accumulation in agricultural soil: Evidence for the presence, potential effects, extraction, and current bioremediation approaches. *Journal of Applied Biology and Biotechnology, 10*, 38–47. https://doi.org/10.7324/JABB.2022.10s204

Yu, H., Chen, Q., & Shi, H. (2021). The bioaccumulation effects of microplastics and associated organic pollutants in the aquatic environment. *Kexue Tongbao/Chinese Science Bulletin, 66*. https://doi.org/10.1360/TB-2020-1426

Yu, J., Adingo, S., Liu, X., Li, X., Sun, J., & Zhang, X. (2022). Micro plastics in soil ecosystem - A review of sources, fate, and ecological impact. *Plant, Soil and Environment, 68*(1), 1–17. https://doi.org/10.17221/242/2021-PSE

Zhang, B., Yang, X., Chen, L., Chao, J., Teng, J., & Wang, Q. (2020). Microplastics in soils: A review of possible sources, analytical methods and ecological impacts. *Journal of Chemical Technology and Biotechnology, 95*. https://doi.org/10.1002/jctb.6334

Zhang, G., Bai, J., Zhai, Y., Jia, J., Zhao, Q., Wang, W., & Hu, X. (2023). Microbial diversity and functions in saline soils: A review from a biogeochemical perspective. *Journal of Advanced Research, 59*, 129–140. https://doi.org/10.1016/j.jare.2023.06.015

Zhang, Y., Pu, S., Lv, X., Gao, Y., & Ge, L. (2020). Global trends and prospects in microplastics research: A bibliometric analysis. *Journal of Hazardous Materials, 400*, 123110. https://doi.org/10.1016/j.jhazmat.2020.123110

CHAPTER 10

Microbial Strain Engineering for Microplastics Removal in Soils

Sabeen Sabri, Areeha Afsheen, Ali Umar, Ahmad Waheed, and Muhammad Saleem Khan

10.1 INTRODUCTION TO MICROPLASTICS

Microplastics are tiny particles of plastic, approximately less than 5 mm in size. According to the literature, plastic usage started in 1600 BC. Today, plastic has become an important part of our lives and a major cause of environmental pollution (Chang et al., 2022). In 1839, after the invention of polystyrene and rubber, people started misuse or overuse of plastic by utilizing it in industries such as the food industry, agricultural industry, and packaging, product, and utensil industries (Tudor, Mocuta, Teodorescu, & Smedescu, 2019). These plastics were discarded in soil, environments, fields, etc., after their usage, which causes major environmental problems (Chia, Lee, Kim, & Jang, 2021). According to a survey, nearly 140 million tons of synthetic plastic polymers are manufactured worldwide. Plastic manufacturing has increased by 20 in the previous 50 years, leading to an expected 9,200 million metric tons (Mt) of plastic manufactured worldwide, while

DOI: 10.1201/9781003487579-10

6,900 Mt is dumped or contributes to ecological contamination (Zhang, Tan, Ji, Ma, & Li, 2022). Over the last 50 years, plastic production has increased by approximately 392 million metric tons.

Global plastic output reached 368 million metric tons (Mt) in 2019 but is expected to quadruple within the next 20 years. It is reported that 1,923 Mt of plastic garbage produced internationally in 2016 reached waterways, but this figure is expected to rise to 53 Mt/year by the year 2030. The usage of plastic products is considered a danger to human health because some plastics are poisonous in nature. According to the International Agency for Research on Cancer, some plastics, such as vinyl chloride, are carcinogenic to human health (Yongming et al., 2018).

As a result, there is an urgent need to develop polymers that are biodegradable and break down readily by microbes. Plastics are carbon-capacity resources that strengthen microbes with heterotrophic metabolism, such as fungal and bacterial species. A variety of bacteria have recently been revealed as capable of producing degradation enzymes that produce non-toxic metabolites along with dwelling organisms after the degradation of plastic (Crawford & Quinn, 2016). It is thought that this is the most secure way of removing plastic from the soil with considerably fewer hazardous byproducts and the potential for biological cycling of the substrates (Tudor et al., 2019). As a result, the present article sheds light on the role of microorganisms in plastic biodegradation and demonstrates the impact on the natural environment, particularly aquatic and soil environments (Shim, Hong, & Eo, 2017). The distribution of microplastics in soil is given below.

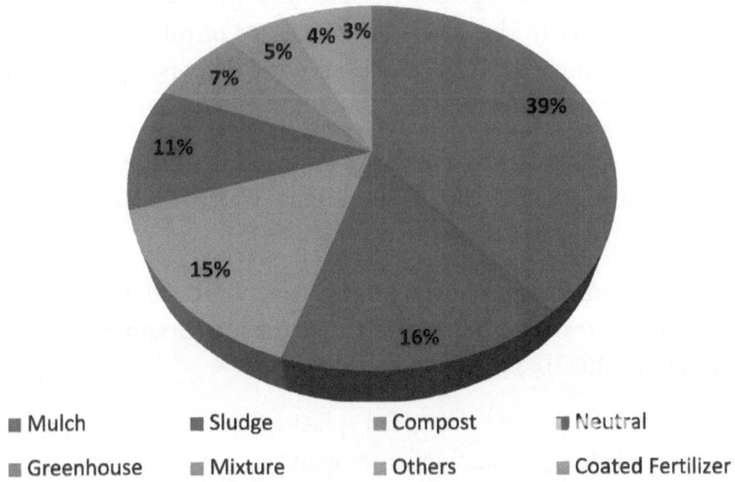

FIGURE 10.1 Distribution of microplastics in soil (Sa'adu et al., 2023).

10.2 SOURCE OF MICROPLASTICS IN SOIL

Microplastic pollution in soil comes from many different kinds of reasons, including both direct and indirect pollution caused by human actions. These factors help to introduce and accumulate microplastics in soil, resulting in extensive contamination (Huang et al., 2021). The following are some of the key sources of microplastic pollution in soil.

- Plastic compost, polyethylene films, and thermoplastic-based agricultural receptacles used in farming processes may gradually disintegrate plastic into small pieces. Small plastic fragments can reach the soil by rainfall, winds, and water drainage (He et al., 2018).
- Microplastic pollution in soil is significantly increased in urban areas. Stormwater runoff can convey microplastics from littered plastic items, manufactured textiles, and plastic debris from different places to the soil (Dioses-Salinas, Pizarro-Ortega, & De-la-Torre, 2020).
- Microplastics may end up in wastewater sludge from residential and industrial effluent. Microplastics are introduced into the soil when sewage sludge is put into crops as a fertilizer (Crawford & Quinn, 2016).
- Through exposure to the atmosphere, mechanical stress, and UV rays, macroscopic polymer products such as plastic bags, bottles, and containers can gradually break down into tiny pieces of microplastic (Chia et al., 2021).
- Improper disposal of plastic trash in garbage microplastics and landfills may result in the accidental discharge of microplastics into the soil around it via drainage as well as wind dispersal (Chia, Lee, Jang, Kim, & Kwon, 2022).
- Vehicle tires and road markings deteriorate and discharge microplastic particles into the surroundings, which can ultimately settle adjacent to soil surfaces (Chang et al., 2022).
- Garbage and organic matter coming from microplastic-contaminated sources can transfer such particles into the soil (Braun, Mail, Krupp, & Amelung, 2023).
- Some fertilizers may contain plastics used as filler materials, inadvertently introducing microplastics into the soil (Bharath, Muthulakshmi, & Usha, 2022).

- Plastic trash, including wrapping paper and disposable plastics, can enter natural habitats and gradually degrade into microplastics (Monkul & Özhan, 2021).
- The drying process can cause microplastics in the air to accumulate on the soil surface.

Each of these factors leads to the widespread occurrence of microplastics in soils around the world, impacting a variety of ecosystems, such as fields, cities, and ecological systems. Knowing these causes is critical for designing effective microplastic pollution mitigation techniques that safeguard soil wellness and ecosystem stability (Huang, Liu, Jia, Yan, & Wang, 2020).

10.3 MICROPLASTIC INTERACTION WITH SOIL MICROBES

The interaction of microplastics with soil microbiota is a complicated and diverse process involving a variety of psychological, chemical, and physiological variables. The soil microbiota, which consists of bacteria, fungi, archaea, and other kinds of microorganisms, is critical in the breakdown, transportation, and destiny of microplastics in soil (de Souza Machado et al., 2019). The following are some of the most important elements of microplastic association with soil microbiota:

1. **Adhesion and Surface Colonization:** Microplastics in soil have a high surface area-to-volume ratio, allowing microorganisms to cling to them. Microorganisms such as fungi and bacteria are able to cling to the exterior surface of microplastics, resulting in biofilm formation. The physicochemical characteristics of both microplastics and bacterial cells influence this adhesion (de Souza Machado et al., 2018).
2. **Biofilm Formation:** Biofilms are frequently formed on the surfaces of microplastics by soil microorganisms. By creating an organized and protected environment for microbial activity, these biofilms can aid in the degradation of microplastics (Helmberger, Tiemann, & Grieshop, 2020).
3. **Enzymatic Degradation:** Enzymes outside of the cell capable of breaking down polymers, including those contained in microplastics, are secreted by soil microorganisms. These enzymes have the ability

to break the chemical bonds in polymeric polymers, thereby commencing the degradation process (Rillig, Leifheit, & Lehmann, 2021).

4. **Ingestion by Microorganisms:** Microplastic particles can be consumed by some microorganisms, including particular bacterial species and protozoa. This intake can occur as a result of nutritional probing or as a consequence of immediate contact with microplastics (J. Shi, Sun, Wang, & Wang, 2022).

5. **Impact on Microbial Community Structure:** Microplastics in soil can alter both the composition and variety of soil microbial species. Several microbes can flourish in an environment full of microplastics, although others might be harmed (Sun et al., 2022).

6. **Microbial Degradation Efficiency:** Microorganisms' ability to destroy microplastics fluctuates based on factors such as the type of plastic, its dimensions, and its surroundings. Certain microplastics are more resistant to breakdown than others, while some may be more vulnerable to assault by bacteria (You, Wang, Li, Du, & Dong, 2022).

7. **Plastic-Mediated Transport of Microorganisms:** Microplastics can act as microbial cell transporters, allowing them to travel far across and within the ecosystems of soil (Zhang et al., 2021).

8. **Release of Chemical Additives:** Some microplastics carry compounds, such as polymers and flammable substances, which can leach into the soil as they degrade. These compounds may be harmful to soil microbes (Rillig, Leifheit, & Lehmann, 2021).

9. **Microplastic-Associated Pollution:** Additional contaminants from the environment, including toxic metals and organic toxins, can be absorbed, and accumulated by microplastics. When soil microbes consume microplastics, these contaminants may enter the microbe food chain (Zhou et al., 2021).

To understand the impact and fate of microplastics in soil, it is important to comprehend the relationships between soil microbes and microplastics. Additional studies need to be conducted to understand the particular biological processes that are involved in microplastic-microorganism relationships, as well as the implications for soil wellness and the conservation of the environment (Helmberger, Tiemann, & Grieshop, 2020).

10.4 EFFECTS OF MICROPLASTICS ON SOIL HEALTH AND ECOSYSTEM SERVICES

Microplastics present in soil deteriorate the quality of soil and pose a danger to soil fertility and biological processes occurring in soil. The impact of microplastics on the ecosystem is given below.

1. **Impact on Quality of Soil:** Microplastics can alter the quality of soil, such as density, water retention, and permeability. The existence of these elements may diminish soil drainage and enhance compression, hindering root development and nutrient intake by plants (J. Shi et al., 2022).
2. **Nutrient Cycling:** Modifications in microbial communities and the activity of enzymes might affect the method of organic matter breakdown, leading to variations in the availability of necessary nutrients for plants as well as soil-based organisms (You et al., 2022).
3. **Plant Growth and Health:** Microplastics in soil can either positively or negatively damage plant growth and health. Microplastic physical interactions with roots can obstruct both water and nutrient intake. Furthermore, harmful compounds released by degraded microplastics can have a deleterious impact on the development of plants (Zhou et al., 2021).
4. **Soil Biota:** Microplastics can be consumed by soil organisms such as earthworms, bacteria, and earth-dwelling insects whenever hunting for food. This can cause harm to the body, decrease nutritional efficacy, and change the behavior of soil microbes (X. Zhang et al., 2021).
5. **Biodiversity:** Variations in soil microbes and fauna caused by microplastics can cause shifts in soil biodiversity. Biodiversity changes can have an adverse impact on the functioning of ecosystems and sustainability (Zhou et al., 2021).
6. **Carbon Sequestration:** Soil microplastics can disrupt the retention of carbon systems. Lower activity of microbes and increased organic matter degradation in the soil may impede carbon storage over time (Rillig et al., 2021).
7. **Water Quality:** Microplastics that are underlying soil can enter lakes and rivers via runoff from the surface or seeping. Microplastics, after entering the water, can worsen water quality, affecting aquatic habitats and living things (de Souza Machado et al., 2019).

8. **Erosion and Sediment Transport:** Microplastics can attach to soil particles and disrupt the processes of erosion and sediment movement. This can alter sedimentary movements in waterways (de Souza Machado et al., 2018).

9. **Natural Pest Control:** Predatory insects and nematodes, for example, play an important role in naturally controlling insects. Microplastics can reduce the efficacy of alternative pest control methods by altering the biodiversity of soils.

10. **Human Health Concerns:** The presence of microplastics in soil prompts concerns about prospective exposure to humans via ingestion of crops for human consumption cultivated in toxic soil or by contact during harvesting (Rillig et al., 2021).

The total impact of microplastics on soil health and ecosystem services is still being studied, and the magnitude of these effects may vary depending on microplastic qualities, soil characteristics, and the surrounding environment. Tackling the problem of microplastic contamination in soil necessitates effective remediation techniques such as adequate disposal of waste, restriction of plastic consumption, and studies on microplastic-contaminated soil restoration (Ya et al., 2021) (Figure 10.2).

FIGURE 10.2 Sources and impact of microplastics on soil (Okoffo et al., 2021).

10.5 IMPACT AND CHALLENGES OF MICROPLASTIC ON SOIL

Impacts	Challenges
Microplastics may influence the structure of soil by diminishing pore dimensions and influencing the aeration of the soil, which can limit root growth and plant absorption of nutrients (Xu et al., 2020)	Due to their widespread occurrence in soil from agriculture field to the natural environment microplastics become challenging (Wang, Ge, Yu, & Li, 2020)
Microplastics have the potential to disturb soil microbes and their metabolic processes, hence influencing the cycling of nutrients and biological material degradation (Ya et al., 2021)	Due to their small size less than 5 mm, they become very challenging for us (Möller et al., 2022)
The presence of microplastics in soil can interfere with ecosystem functions including retention of carbon, fertility of the soil, and biological control of insects (Qi, Jones, Li, Liu, & Yan, 2020)	Due to increase in their abundance in soil microplastics become challenging (J. Shi et al., 2022)
Microplastics in soil are frequently carried to organisms that live in water by drainage, leading to water contamination and endangering aquatic life (Sun et al., 2022)	Due to their persistence in soil for a long time microplastics are challenging (Sun et al., 2022)
Microplastics can be consumed by soil organisms like earthworms and bacteria, which could lead to injuries and plastic bioaccumulation in the food chain (Xu et al., 2020)	Conventional remediation approaches are frequently inefficient at eliminating microplastics from the soil matrix (He et al., 2018)
By affecting soil physical and chemical qualities and accessibility to nutrients, microplastics can have a deleterious impact on plant growth, development, and general health (Wang et al., 2019)	Microplastics may serve as carriers for a variety of toxic contaminants, such as insecticides or toxic metals, potentially causing biological accumulation in soil-borne organisms (Ya et al., 2021)
Microplastics have the potential to disturb soil microbes and their metabolic processes, hence influencing the cycling of nutrients and biological material degradation (de Souza Machado et al., 2018)	Microplastics exist in a variety of forms, dimensions, and plastic compositions, making description and removing complicated (Möller et al., 2022)

Ultimately, the issues raised by microplastics in soil, as well as their effects on soil welfare and ecological systems, demand immediate attention and appropriate mitigation solutions. Microbial strain engineering is an attractive strategy for improving soil health and functionality by increasing the breakdown of microplastics. Researchers can design specialized microbial strains capable of efficiently eliminating microplastics from the soil, minimizing their negative environmental consequences, by understanding the genetic basis of microplastic degradation and utilizing biological engineering methods (Wang et al., 2020).

TABLE 10.1 List of Natural and Artificial Plastics and Their Degradation Mechanisms (Thakur, Singh, Singh, Angmo, & Vig, 2023)

Natural Plastic	Degradation Mechanism	Artificial Plastic	Degradation Mechanism
Polyhydroxyalkanoates	UV light	Polyethylene PE	Hydrobiodegradation, oxobiodegradation
Polypropylene PP	UV light exposure	Polyester	Hydrolytically labile ester bond
Polyvinyl chloride PVC	Phodegradation or white rot fungi	Teflon	Thermal degradation
Polystyrene PS	Thermal or chemical method	Nylon	Photolysis 3
PHB, PHBV	UV light	Epoxy	Thermal oxidative degradation

10.6 TYPES OF MICROPLASTICS

The types of microplastics depend on the primary and secondary sources. Primary sources of microplastics are fiber fabric, cosmetic products, etc. Secondary sources are particles of plastic bottles, large containers, etc. Plastic degradation depends on several factors, such as UV rays (Setälä, Norkko, & Lehtiniemi, 2016). Microplastics are further divided into two types on the basis of degradability and nondegradability. Degradable plastics are those that can be easily degraded by microorganisms such as bacteria and fungi. Nondegradable plastics are those that cannot be easily degraded by microorganisms (J. Shi et al., 2022). There are two major types of plastic: (1) natural plastic and (2) artificial plastic (Setälä et al., 2016) (Table 10.1).

10.7 MICROORGANISMS INVOLVED IN MICROPLASTIC DEGRADATION

Microorganisms play an important role in the breakdown of microplastics in a broad range of environments, particularly soil. A broad array of microorganisms with varying metabolic capacities performs the process of microplastic breakdown (Yuan et al., 2020). The following are some of the major microorganisms that contribute to microplastic degradation (Figure 10.3):

10.7.1 Bacteria

Numerous strains of bacteria have been found to be effective microplastic degraders. Bacteria have extracellular enzymes that can break down the polymer bonds of plastics, such as lipases and esterases. Microplastic degradation bacteria include representatives of the *genera Pseudomonas, Bacillus, Acinetobacter*, and *Ideonella* (Gaur et al., 2022) (Table 10.2).

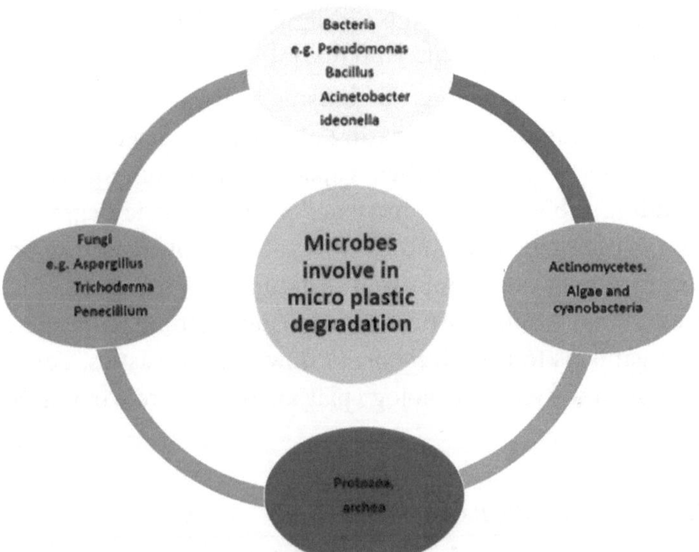

FIGURE 10.3 Microplastic degradation.

TABLE 10.2 List of Bacteria, Their Plastic-Degrading Types and Enzymes

Bacteria	Degrade Plastic Type	Plastic Degrader Enzyme
Pseudomonas	Polyethylene, polyethylene terephthalate polypropylene	PETase, cutinases, alkane hydroxylase
Bacillus	Polystyrene, polyethylene	
Ideonella	Polyethylene terephthalate	
Acinetobacter	Polyethylene, polypropylene	

TABLE 10.3 List of Fungi, Their Plastic-Degrading Types and Enzymes

Fungi	Degrade Plastic Type	Plastic Degrader Enzyme
Aspergillus	Polyurethane, polyethylene	Depolymerases, Laccases
Trichoderma	Polyethylene, polystyrene	Cutinases, Peroxidases
Penicillium	Synthetic polymer degraders	

10.7.2 Fungi

Fungi are also key contributors to the breakdown of microplastics. They secrete enzymes that can break down the complicated polymer structures of plastics, including lignin-degrading enzymes. Fungal genera reported to break down microplastics include *Aspergillus, Penicillium,* and *Trichoderma* (Yuan et al., 2020) (Table 10.3).

10.8 MICROBIAL STRAIN ENGINEERING STRATEGIES FOR MICROPLASTIC REMOVAL IN SOIL

To improve the ability of microorganisms to break down microplastics, genetic engineering offers fascinating alternatives. Scientists can change microorganisms to enhance their capacity for decomposing plastic by understanding the genetic components of microplastic degradation and the action of essential genes and enzymes. The following are some genetic engineering strategies for improved degradation (Gaur et al., 2022). The goal of microbial strain engineering strategies is to increase the capacity of microorganisms to effectively break down microplastics. For example, CRISPR-Cas9 and synthetic biology play an essential role in manipulating microbial strains.

10.9 SYNTHETIC BIOLOGY

In the science of "synthetic biology," organisms are modified to serve new functions by being given the capacity to produce new biological structures, gadgets, and components. Synthetic biology enables scientists to develop microbes with improved plastic-degrading abilities, which is relevant to the decomposition of microplastics. The following are the essential steps in synthetic biology for microbial strain engineering (Nguyen, Huang, Collins, Collins, & Lu, 2023).

10.9.1 Identification of Genes

Experts have found important genes in microbes that possess the ability to break down plastics. The genes involved in the production of enzymes PETase and cutinases are important in the breakdown of plastic.

10.9.2 Synthesis of Genes

After the synthesis of the identified gene, the sequences of the genes are capable of being adjusted for enhanced expression and productivity in the target microorganism (Nguyen et al., 2023).

10.9.3 Promoter Modification

After the synthesis of the desired gene, scientists are able to regulate the expression of genes that break down plastic through modifications to promoters, which boost enzyme production (Gaur et al., 2022).

10.9.4 Optimization and Synthesis of Genes

To stimulate the synthesis of plastic-degrading genes in a selected microbe, the genes and their optimized promoters have been inserted into expression vectors, which are DNA molecules (Thakur, Mathur, Patel, & Paital, 2022).

10.9.5 Validation and Screening

The modified genes are incorporated into the target microorganism's genome once the expression vectors have been incorporated into it. The improved capacity for breaking down plastic is tested in the altered microorganisms. Improved degrading microbes are further characterized to understand their activity (Nguyen et al., 2023).

10.10 APPLICATION OF CRISPR–CAS9 FOR TARGETED GENETIC MODIFICATIONS IN MICROPLASTIC DEGRADERS

The topic of microbial strain engineering for microplastic breakdown may be extensively upgraded by the cutting-edge gene-editing technique CRISPR-Cas9. Investigators may boost the ability of microbes to degrade plastic in a more regulated and effective way by employing CRISPR-Cas9, which enables accurate and selective alterations of microbial genomes. The implication method of CRISPR-Cas9 is given below (Prasad & Elchuri, 2022).

10.10.1 Target Gene Identification

Before using CRISPR-Cas9, investigators have to determine particular genes in the selected microorganism responsible for microplastic breakdown. PETases, cutinases, and other plastic-degrading enzymes might have been synthesized by these genes.

10.10.2 Guide RNA (gRNA)

To selectively target the identified gene(s) in the genome of microbes, a guide RNA (gRNA) is synthesized. By serving as a molecular "address," the gRNA leads the enzyme known as Cas9 to the exact region in the desired protein.

Gene editing by using CRISPR-Cas9 plasmids or other delivery systems is commonly used to deliver the system known as CRISPR-Cas9 into microbes. When the enzyme Cas9 enters the cell, it utilizes the gRNA to find the gene of interest and makes a split in both strands.

10.10.3 Repair Mechanism

Following the double-strand break, the cell's inherent repair systems try to repair the DNA damage. These repair processes are capable of being guided by experts to generate particular modifications to genes. In this case, they could alter the function of the target gene by deleting or introducing specific regions of DNA.

10.10.4 Validation and Selection

Once the successful modification of genes has been verified, the modified microbes are tested and validated. The selection of microorganisms for further testing and applications is based on desirable characteristics, such as improved plastic-degrading capacities (Prasad & Elchuri, 2022).

Powerful tools are available for designing microbes with improved plastic-degrading capabilities, including synthetic biology and CRISPR-Cas9. The development of effective and long-lasting methods for reducing microplastic contamination in various ecosystems could be considerably aided by these efforts. To ensure the safe and legal implementation of these approaches, it is essential to carry out comprehensive risk evaluations and follow appropriate research and application practices (Thakur et al., 2022).

10.11 MICROBIAL STRAIN ENGINEERING PATHWAYS OF MICROPLASTIC DEGRADATION

For increased degradation and removal of microplastics from soils, some potential genetic engineering techniques are as follows.

10.11.1 Plastic Degradation Mechanism

The degradation of plastic is a method in which plastic polymers are broken down into monomers by changing their chemical and physical properties through microorganisms. This process of degradation is mediated by enzymatic and nonenzymatic hydrolysis of fungi and bacteria. Plastic degradability depends on physical, chemical, and environmental factors. Plastic degradation occurs under aerobic and anaerobic conditions (Urso & Pumera, 2022).

10.11.1.1 Degradation under Aerobic Conditions

Aerobic degradation is a process in which aerobic microorganisms utilize oxygen as an electron carrier and utilize CO_2 and water metabolites to break down organic molecules into smaller organic compounds. Aerobic biodegradation, often known as aerobic respiration, is a key aspect of natural pollutant elimination in many toxic waste sites (Bano et al., 2017).

10.11.1.2 Degradation under Anaerobic Conditions

Anaerobic biodegradation is the degradation of organic pollutants utilizing bacteria in the absence of oxygen, and it is also an important component of natural pollution removal at toxic waste sites (Kaushal, Khatri, & Arya, 2021). As electron acceptors, some anaerobic bacteria employ iron, nitrate, manganese, sulfate, and CO_2 to break down large biological molecules into smaller organic compounds, such as CH_4, CO_2, H_2O, C residual, biomass, and C plastic.

The biodegradability of a polymer depends upon its chemical and physical properties, which are given below (De Gisi et al., 2022) (Figure 10.4).

1. The density and molecular weight of the polymer
2. Presence of functional groups of polymers
3. Morphology of polymer
4. Molecular composition of polymer
5. The structure of the polymer
6. Hardness of polymer
7. Structural complexity of polymers
8. Presence of chemical bonds in the polymer

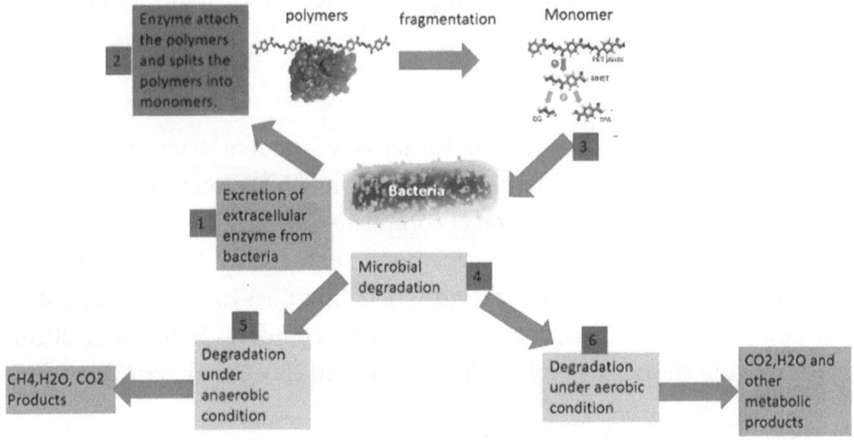

FIGURE 10.4 Schematic representation of microplastic degradation by microbes under aerobic conditions.

10.11.1.3 Bioaugmentation

Bioaugmentation is a modern technique for microplastic removal in soil. The process of introducing living organisms into the soil to break down particular groundwater as well as soil pollutants is known as bioaugmentation (Sharma & Jain, 2020). This phenomenon is used to accelerate the enzymatic activity of microbes to degrade microplastics from soil. The following steps are involved in this process (Baskaran & Sathiavelu, 2022).

1. Identification and isolation of plastic-degrading enzyme-producing microbes.
2. Isolated microorganisms were cultured in the laboratory by providing suitable nutrients and environmental conditions.
3. Cultured microbes are inserted into contaminated soil directly or indirectly.
4. The ability to survive and the stability of the inserted microbes in the soil are tracked over time.
5. After persistence and survival, the efficacy of plastic-degrading microbes is evaluated.

When plastic-degrading enzymes are absent in contaminated soil, this method is used to insert plastic-degrading microbes into contaminated soil. The effectiveness of this process relies on the ability of inserted microbes to interact with existing microbes in the soil (Sharma & Jain, 2020).

10.11.1.4 Biostimulation

The change in the environment to activate resident bacteria that are able to carry out bioremediation is known as biostimulation (Baskaran & Sathiavelu, 2022). This can be accomplished by adding various speed-limiting elements and electron acceptors, such as C, O, N, and P. This method is also used to enhance the growth of microbes that produce plastic-degrading enzymes (Cosgrove, McGeechan, Handley, & Robson, 2010). The following steps are involved in this method (Jang et al., 2023).

1. The organic nutrients necessary for the survival of microbes are inserted into the soil directly or indirectly in the form of supplements, such as compost.

2. The quantity of oxygen in the soil is vital for the survival of microbes in the soil.
3. To encourage the degrading activity of microorganisms as well as water retention, water content is retained in the substrate.
4. Ensuring the optimal temperature can boost the degrading activity of microorganisms, particularly in adverse circumstances in ecosystems.
5. After persistence and survival, the efficacy of plastic-degrading microbes is evaluated.

The effectiveness of this process relies on the existence of microbes that are able to produce plastic-degrading enzymes. This method is more efficient and rapid than the bioaugmentation method. To mitigate the challenges and improve the viability of synthetic microbes with existing soil microbes, extensive study and experimental research are needed prior to use in the field (Jang et al., 2023). Furthermore, taking into account the specialists in environmental microbiology, microbial biodiversity, and risk evaluation in an interactive way could assure the ethical and effective application of engineered microbes for the removal of microplastics from soil (Baskaran & Sathiavelu, 2022) (Figure 10.5).

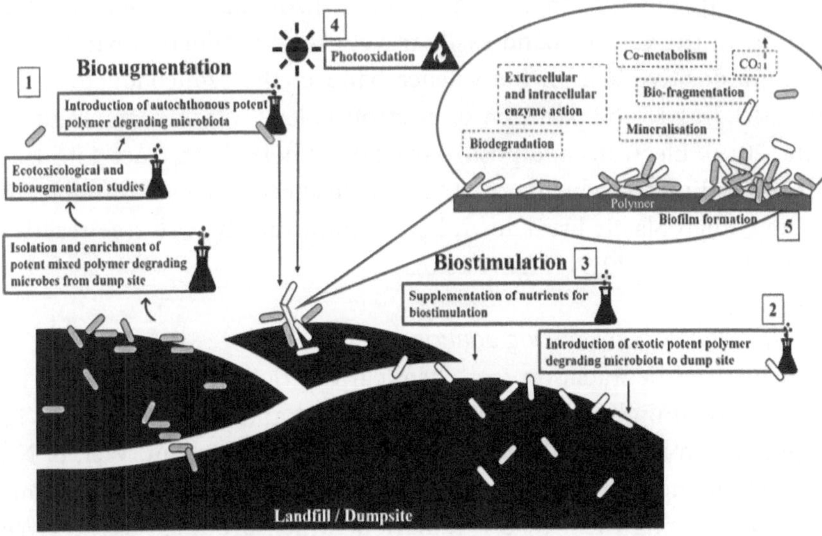

FIGURE 10.5 Plastic degradation by the bioaugmentation and biostimulation process (Baskaran & Sathiavelu, 2022).

10.12 BROAD CATEGORY OF MICROBIAL STRAIN ENGINEERING MECHANISMS

The degradation of microplastics is broadly categorized into two methods: (1) the oxidative degradation of microplastics and (2) the biological degradation of microplastics (Thakur et al., 2023).

10.12.1 Oxidative Degradation

Oxidative degradation consists of different techniques, including

- Photocatalytic degradation
- Photochemical degradation
- Electrochemical degradation

10.12.1.1 Photocatalytic Degradation

Photocatalytic degradation is a sophisticated oxidation process in which light energy is used to drive pollutant degradation in photocatalytic degradation. This process can be utilized to destroy contaminants that have a high concentration, complexity, and limited biodegradability. Light energy is used to drive pollutant degradation in photocatalytic degradation (Uheida, Mejía, Abdel-Rehim, Hamd, & Dutta, 2021). Decomposition begins in this process when the photon energy of semiconductor components exceeds the difference in energy of semiconductor substances. The electrons in the valence band easily travel to the conduction band, thus causing a positive hole in the valence band of the semiconductor substance, leading to the partition of electron holes (Tofa, Kunjali, Paul, & Dutta, 2019). Electrons and positively charged holes both react with free hydroxyl radicals, producing reactive oxygen species that are unstable. In this way, microplastic breakdown begins immediately with free reactive oxygen (Bratovcic, 2019) (Figure 10.6).

10.12.1.2 Photochemical Degradation

Photochemical degradation is a chemical reaction that is triggered by exposure to sunlight, such as ultraviolet rays, and is the key factor in the photochemical degradation process (X. Shi, Chen, Liu, Wei, & Ni, 2022). Photochemical processes are used to degrade organic contaminants by exposing them to UV radiation, which produces oxygen-free radicals and bonds between the long chain of polymers (Song et al., 2017) (Figure 10.7).

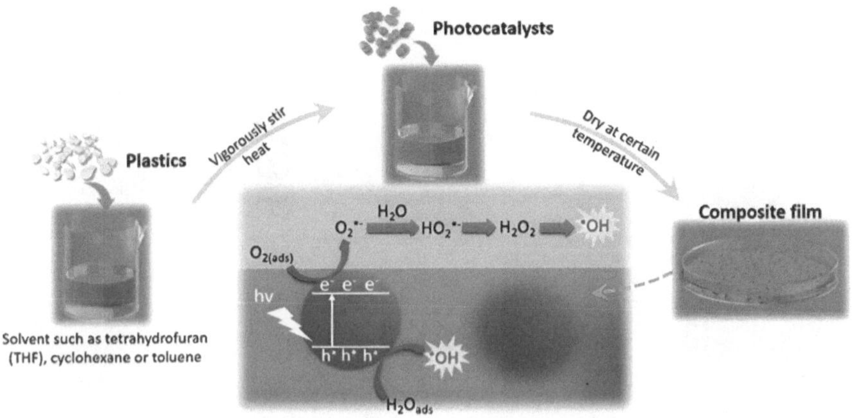

FIGURE 10.6 shows the mechanism of photocatalytic degradation of microplastics (Chu et al., 2022).

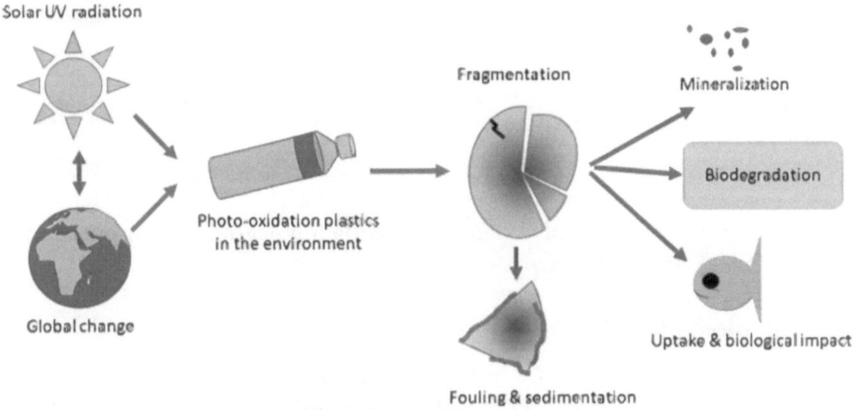

FIGURE 10.7 shows the mechanism of photochemical degradation of microplastics (Andrady et al., 2022).

10.12.1.3 Electrochemical Degradation

Electrochemical degradation (ECD) is a method that uses oxidative or reductive reactions to detoxify organic and inorganic pollutants (Martic, Tabobondung, Gao, & Lewis, 2022).

Pollutants are oxidized in a cell with electrodes via direct transport of electrons toward the anode or indirect oxidation via diverse radicals produced at the anode. The electron-Fenton method and oxygen free radicals accomplish cathodic degradation. It is produced by Fe^{+} and serves as a catalyst for the degradation of microplastics (Lu, Hou, Wang, Zhou, & Yuan, 2022) (Figure 10.8).

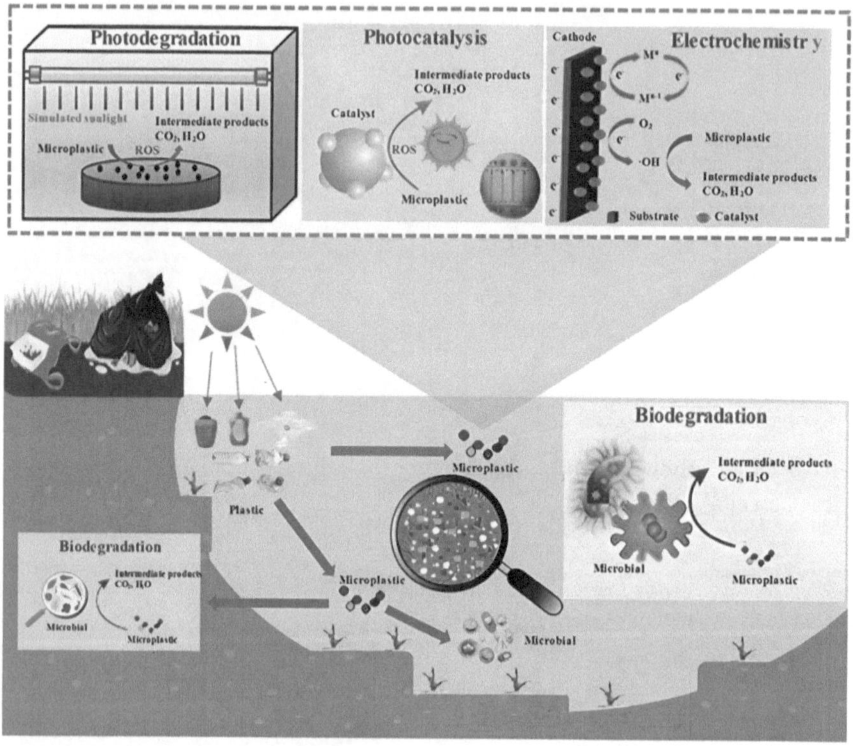

FIGURE 10.8 Mechanism of action of the oxidation degradation of microplastics (Du, Xie, & Wang, 2022).

10.12.2 Biological Degradation

Biological degradation is the breakdown of the polymer by microorganisms during the repository. Although biopolymers tend to be more vulnerable to biological attack, the degradation of synthetic polymers is mediated by biological degradation (Haben Fesseha, 2019).

Biological degradation is categorized into three main types

- Biodeterioration
- Biofragmentation
- Mineralization

10.12.2.1 Biodeterioration

Biodeterioration is a process in which plastic polymers break down into monomers by the action of enzymes produced by microbes. Enzymes act on the microplastic during this phase of degradation, breaking down the microplastic polymer into monomers (Tareen, Saeed, Iqbal, Batool, &

Jamil, 2022). Microplastic polymer disruption occurs both within and outside of the plastic. The enzymatic breakdown of microplastics relies on a variety of enzymes produced by microbes, which include the urease enzyme, cutinase enzyme, protease enzyme, and lipase enzyme (Cappitelli, Principi, & Sorlini, 2006).

10.12.2.2 Biofragmentation

Biofragmentation refers to the physiological splitting or disintegration of a microplastic material by enzymes produced by microorganisms. Enzymes such as hydrolase and oxidoreductase undergo hydrolysis in the polymers of plastics at this step (El Menofy & Khattab, 2023). Additional enzymes mediate the various oxidation reactions leading to the formation of free radicals. The polar nature of the polymer obtained by oxidation occurs in free radicals and in the formation of carbonyl and hydroxyl functional groups (Dussud & Ghiglione, 2014).

10.12.2.3 Assimilation

During this phase of degradation, microbes absorb monomers to form biomass from bacteria, CO_2, and CH_4. The aforementioned monomers are employed as a source of carbon energy and are used for microbial cell development. Some plastic monomers are difficult to ingest due to the semipermeability of cell membranes (Vaksmaa et al., 2023). Through the biotransformation process, microbial cells use nonassimilated monomers. Nonetheless, nonabsorbed monomers of polymers are employed by microbes in a process known as transformation, in which secreted enzymes of microorganisms generate the transformation of chemical compounds into an end product that can be absorbed by comparable microbes (Amobonye, Bhagwat, Singh, & Pillai, 2021) (Figure 10.9).

10.12.3 Extracellular Enzymatic Degradation

Microplastic microbial degradation pathways involve numerous mechanisms by which microorganisms break down plastic polymers into tiny pieces or monomers. Extracellular enzymatic degradation is a fundamental pathway in which microbes produce particular enzymes beyond their cells to catalyze the breakdown of polymeric materials. The following steps are involved in this process (Bano et al., 2017).

Attachment and Adhesion: In this step, some microbes, such as fungi and bacteria, stick on the outside of the microplastic after attachment, and these microbes start producing extracellular polymeric

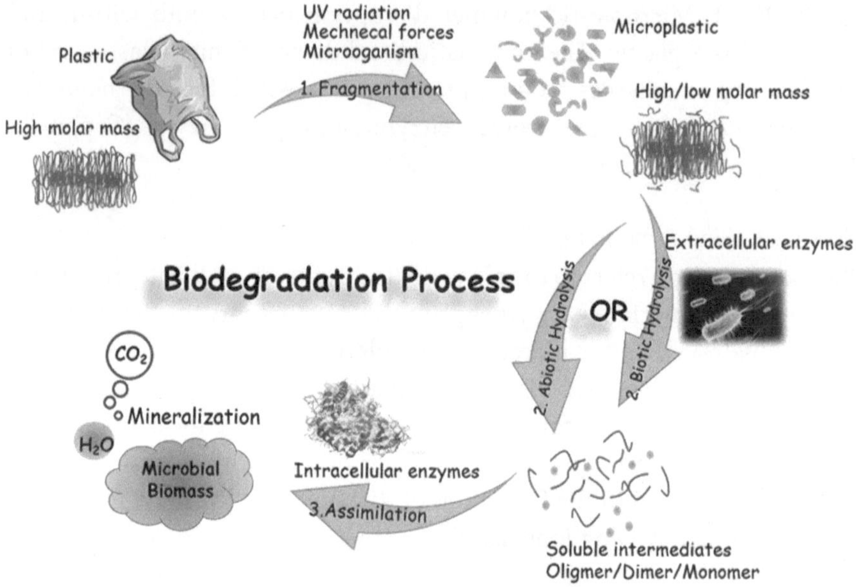

FIGURE 10.9 Schematic illustration of the biodegradation of microplastics (Wang, Ge-Xia and Huang, Dan et al., 2020).

substances (EPS) around the surface of the plastic. The development of EPS, which develops a biofilm surrounding the plastic particles, aids in adherence (Bano et al., 2017).

Enzyme Secretion: After attachment, these microbes secrete extracellular enzymes, such as cutinases, esterases, and lipases, which are specific to the type of plastic they encounter. These extracellular enzymes have the ability to break the chemical bonds present in plastic (Kaushal et al., 2021).

Enzymatic Hydrolysis: After the secretion of extracellular enzymes, this enzyme catalyzes the breakdown of large plastic polymers into small plastic monomers. Different enzymes catalyze different hydrolytic reactions, such as esterases, which catalyze amide and ester bonds, which help to degrade polyethylene and polypropylene plastic polymers, and lipases catalyze ester bonds, which help to degrade polyethylene terephthalate (El Menofy & Khattab, 2023).

Fragmentation: After enzymatic hydrolysis, the microplastic particles are split into tiny particles that are so small that they can still persist in soil (Kaushal et al., 2021).

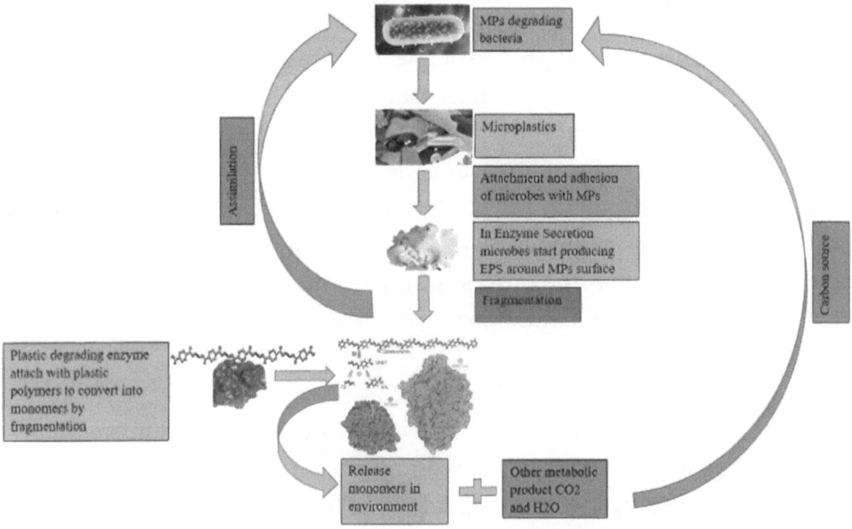

FIGURE 10.10 Schematic diagram of extracellular enzymatic degradation by microbes.

Uptake and Utilization: After fragmentation, some microbes ingest tiny fragments of plastic by phagocytosis to utilize tiny plastic particles as energy, carbon sources, and nutrient substrates (Amobonye et al., 2021).

Release of Monomers: Microplastics are broken down into small particles during the enzyme-mediated degradation process. When these polymers are discharged into the environment, they may provide another category of ecological problems than intact microplastics (Amobonye et al., 2021) (Figure 10.10).

It is critical to recognize that not all bacteria have the enzymatic ability to digest all forms of plastics. Extracellular enzymatic breakdown efficiency differs according to bacterial species, type of plastic, surrounding variables (e.g., pH level, temperature, availability of nutrients), and the years of age of plastic. Extracellular enzymatic degradation is merely one of the microbial routes involved in the total degradation of microplastics. Other mechanisms that contribute to the breakdown of microplastics in the environment include intracellular metabolic pathways and biofilm-mediated degradation. For successful microplastic removal and remediation in many ecosystems, including soil and aquatic habitats, the coordinated action of numerous microorganisms and degradation pathways is crucial. Microbial Degradation Pathways of Microplastics (Pometto 3rd, Lee, & Johnson, 1992).

10.12.4 Intracellular Enzymatic Degradation

Microorganisms ingest microplastic pieces or monomers via intracellular digestion. Following uptake, certain bacteria have the ability to metabolize plastic components intracellularly, turning them into intermediate products of metabolism or using them as a means of obtaining carbon (Kaushal et al., 2021). Depending on the kind of plastic and the metabolic capacity of the microorganisms involved, this process may involve multiple metabolic pathways, such as the process of glycolysis, the cycle of tricarboxylic acid, and the process of oxidation (Amobonye et al., 2021).

10.12.5 Trap and Release Mechanism

Currently, bioaggregation is considered an efficient way to remove microplastics from soil. Bioaggregation is a novel technique in which toxic substances such as heavy metal insecticides and tiny particles of microplastics accumulate in certain microorganisms. Bioaggregation is also a convenient way to recycle used plastics that have been wasted in incineration and landfills (Liu, Leung, Fang, & Chua, 2021). According to one study, it is expected that approximately 80% of plastics will end up in landfills, 12% will be burned, and only 9% will be recycled. As a consequence, recycling was chosen above incineration and landfills, indicating the need for increased recycling activities. Sylvia Yang Liu et al. engineered an efficient plastic degrader bacterium named Pseudomonas aeruginosa that has the ability to accumulate microplastics with the help of their adhesive matrix and then subsequently trigger the release of the entrapped microplastics for efficient subsequent retrieval and recovery of microplastics. There are numerous benefits to this 'trap-and-release' bioaggregation approach:

- Microplastics can accumulate regardless of their subject matter, shape, or structure, thus avoiding their particular concern.
- Following that, microplastics are localized in large amounts and may be quickly eliminated from the soil. A higher overall mass will facilitate elimination in containers by filtering or sedimentation (Liu et al., 2021).
- Treatment with an inducible stimulus might disperse biofilms and release microplastics for easy retrieval from the sticky exopolymeric matrix of biofilms (Urso & Pumera, 2022) (Figure 10.11).

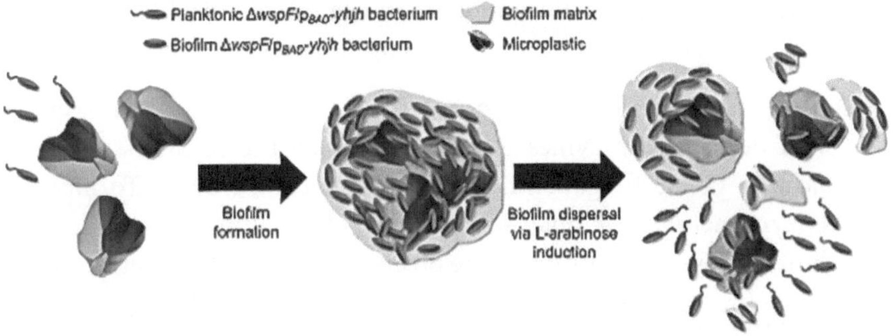

FIGURE 10.11 Trap and release mechanism for microplastic removal from soil. (Reference adapted from Liu et al., 2021.)

10.12.6 Case Study of Successful Genetically Modified Microbes for Microplastic Removal in Soil

Two case studies were conducted to evaluate the efficacy of microplastic degradation by using modified microorganisms such as bacteria and fungi. Some case studies are listed below.

10.12.6.1 Case Study 1: Genetic Engineering of Fungi for Microplastic Degradation

This case study aimed to develop a fungal strain to enhance the decomposition of microplastics, emphasizing polyethylene terephthalate (PET) polymers as well (Gaur et al., 2022).

Isolation and Screening of Fungal Strains: Experts acquired samples of sediment and soil from areas polluted with plastic, such as microplastics and beaches. The potential to break down PET plastics was discovered and tested in isolated fungi (Bugg, Williamson, & Alberti, 2021).

Key Enzyme Identification: By analyzing transcriptome and proteomic data, researchers were able to determine the enzymes responsible for PET breakdown in the designated fungi. Special attention was given to PETase and other enzymes involved in the breakdown of PET (Temporiti, Nicola, Nielsen, & Tosi, 2022).

Genetic Engineering Approach: The identified PET-degrading enzyme-encoding genes were extracted from the fungal strains by genetic engineering. The genes were subsequently transcribed and produced in greater amounts inside the identical fungal strain employing the methods of genetic engineering (Bugg et al., 2021).

Optimization and Conditions: The synthesis and release of PET-degrading enzymes were improved by the researchers' modification of the fungal strain's environment for growth. To increase enzyme activity, variables such as the pH level, humidity, and accessibility of essential nutrients were precisely modified (Gaur et al., 2022).

Biodegradation Assay: PET films were used as a substrate for biodegradation experiments, which were performed on the modified fungal strain and its wild-type equivalents. Between the modified strain and the wild-type strain, the progress of PET disintegration was assessed and evaluated (Temporiti et al., 2022).

Results: Compared to the wild-type fungal strain, the modified strain significantly increased the progress of PET biodegradation. PETase and other PET-degrading enzymes were elevated, which increased their ability to produce enzymes and improved their capacity for PET degradation. The modified fungal strain destroyed PET polymers well, slicing them into fragments, and demonstrated an increased rate of degradation compared with the wild-type strain (Temporiti et al., 2022).

10.12.6.2 Case Study 2: Genetic Engineering of Pseudomonas for Microplastic Degradation

The renowned bacterial genus *Pseudomonas sp.*, which has a variety of enzymatic abilities, has been effectively manipulated by scientists for effective microplastic breakdown. *Pseudomonas species* were identified under a range of habitats and have traits that render them viable candidates for microbial strain engineering (Kaur, Sharma, Shree, & Mehrotra, 2023).

To improve *Pseudomonas sp.*'s capacity to break down microplastics, this case study focused on two types of plastic that are frequently found in land and coastal surroundings, including polyethylene (PE) and polypropylene (PP) (Saravanan et al., 2021).

Isolation and Characterization of the Plastic-Degrading Bacterial Strain: *Pseudomonas sp.* strains have been isolated and characterized from plastic-contaminated habitats, such as microplastics and coastlines. The ability of isolated bacteria to degrade plastic was then described.

Identification of Important Genes: *Pseudomonas sp.* strains had specific genes that contribute to the decomposition of microplastic, which were discovered using whole-genome sequencing and computational biology. Such genes included PETase and cutinases, which are known for their ability to break down plastic (Bugg et al., 2021).

Advance Genetic Engineering: Modern genetic engineering techniques were employed to knock out the discovered genes in *Pseudomonas sp.* strains. The intention was to boost the strain's capacity to degrade PE and PP plastics by increasing the synthesis of PETase and cutinases.

Testing and Validation: Under carefully monitored laboratory conditions, the modified *Pseudomonas sp.* strains were put to the test. The effectiveness of the breakdown of both polyester and polypropylene plastics was measured in contrast to that of the normal-form *Pseudomonas sp.* strains (Saravanan et al., 2021).

Result: When compared to the wild-type strains, the modified *Pseudomonas sp.* strains showed a dramatically improved capacity for degrading microplastics. Degradation of both polyethylene and polypropylene plastics occurred more quickly and effectively as a result of increased enzymatic activity brought on by PETase and cutinase overexpression. The modified strains also showed the ability to persist in the soil environment, demonstrating their suitability for use in practical situations (Kaur et al., 2023).

Ethical Concern: Social and moral issues are still crucial although the effective development of *Pseudomonas sp.* strains for microplastic breakdown offers potential methods to combat pollution caused by plastic. To investigate prospective adverse environmental hazards associated with the inheritance of genes (Bugg et al., 2021) and any unintended adverse effects of unleashing the modified strains into the environment, scientists performed extensive risk evaluations. The ethical debates centered on the proper application of genetically modified microbes, the potential dangers to species that are not intended targets, and the significance of accessibility in studies, as well as involvement by the public (Saravanan et al., 2021).

Conclusion: The development of *Pseudomonas sp.* for effective microplastic breakdown is an important milestone in the discipline of microbial strain engineering. Experts have shown that modified microbes have the capability to play a key role in microplastic cleanup initiatives by employing the naturally occurring plastic-degrading ability of *Pseudomonas sp.* and increasing their ability to produce enzymes. To properly evaluate the detrimental effects on the environment and the durability of these modified strains in practical applications, more investigation, particularly field tests and ecological monitoring is needed.

10.13 COMPATIBILITY AND INTERACTION OF ENGINEERED MICROBES WITH NATIVE SOIL MICROBIOTA

The viability and conceivable interactions of designed microbes with the native soil fauna must be taken into account when using microbial strain engineering procedures, such as bioaugmentation or biostimulation. Microorganisms that have been genetically altered or introduced into the soil environment may have an intricate impact on the environment (X. Zhang et al., 2021).

10.13.1 Conflict over Resources

Engineered microorganisms and naturally occurring soil microbiota are going to battle for assets, such as food and surface area. The capacity of modified strains to integrate into and compete with the current microbial ecosystem may be a factor in their success.

10.13.2 Predatory Behavior and Grazing

Predators that eat other kinds of microbes, such as protozoa, are among the wide variety of microbes found in soil. Because engineered strains may be prey to predators, their ability to persist in soil environments may be hampered (Zurier & Goddard, 2021).

10.13.3 Overlap in Niche

Examine whether the synthetic strains have the same ecological niche as the natural bacteria. Crossing niches can result in immediate rivalry, which could reduce the efficacy of the modified strains.

10.13.4 Exchange of Genetic Information

In soil ecosystems, horizontal gene transfer occurs naturally. To change the features and activities of native microbes, engineered genes may be inserted into them. This change needs to be carefully monitored because it can have unexpected effects (Niu et al., 2020).

10.13.5 Impact on Soil Ecosystem

Take into account any possible adverse impacts that inserting modified strains may have on the broader soil habitat. Changes in microbial diversity and activity may have an adverse impact on the nitrogen cycle and processes in the soil (X. Zhang et al., 2021).

10.13.6 Pathogenicity and Biological Control

Make sure the inserted strains produce no detrimental impact on non-target organisms. This is known as biological control and pathogenicity. Analyze whether altered strains have the potential to be pathogenic or to possess other negative characteristics (Niu et al., 2020).

10.13.7 Persistence and Survival

Assess the longevity and persistence of engineered strains. Some bacteria could struggle to develop on soil or might be surpassed by native microbes.

10.13.8 Assessment of Ecological Risk

Prior to releasing modified strains into the environment, a complete evaluation of ecological risks should be performed (Romdhane et al., 2022).

10.13.9 Limits for Containment and Remediation

Take into account whether containment procedures are needed to avoid the unintentional spread of the engineered strains outside the designated restoration zone (Niu et al., 2020).

10.13.10 Assessment and Regulation

Periodic soil environment screening is employed to keep tabs on the activity of modified strains and gauge their influence on the natural soil fauna. Be sure to follow any applicable natural rules and laws regarding the use of GM organisms (Niu et al., 2020).

It is essential to carry out research and laboratory trials before the implementation of engineered microbes in soil to reduce risks and maximize the compatibility of engineered microbes with native soil microbiota. A collaborative strategy involving specialists in environmental microbiology, microbial ecology, and risk assessment can also help guarantee the responsible and effective application of microbial strain engineering for soil microplastic remediation.

10.14 FUTURE PROSPECTS AND ETHICAL CONCERNS OF MICROBIAL STRAIN ENGINEERING TECHNIQUES

Microorganisms such as fungal and bacterial strains are used in microbial strain engineering to decompose microplastics. This approach has tremendous potential because it may immediately target and break down microplastics into nontoxic metabolites (Feng et al., 2023). There are some important potential outcomes of microbial strain engineering for the remediation of soil.

Microbial strain enhancement through genetic engineering or other methods of selection involves researchers constantly attempting to improve microbial strains. These strains have the enzymes needed for effective microplastic breakdown and can be adapted to survive in particular soil environments. Future developments of strains optimized for diverse types of soil and components of microplastic can be anticipated (Tripathy et al., 2021).

Biodegradable plastic-degrading microorganism studies are being conducted to identify and cultivate native microbes that can devour and degrade microplastics. As our knowledge of microbial communities expands, we could find and manipulate microbes that are especially effective at breaking down microplastics, offering a long-lasting and environmentally benign option for the removal of microplastics in soil (Rodríguez-Narvaez, Goonetilleke, Perez, & Bandala, 2021).

10.14.1 Bioremediation and Bioaugmentation

Microbial strain engineering is a tool that can be employed in bioaugmentation and bioremediation techniques. To speed up microplastic degradation rates, targeted microbial communities can be added to contaminated soils. This strategy shows potential for extensive soil cleanup initiatives, especially in regions severely affected by microplastic pollution (Feng et al., 2023).

Biotechnological Application: Microbial strain engineering is also applicable to biotechnology processes. Wastewater treatment plants may be able to filter out microplastics from effluents before they enter aquatic and terrestrial ecosystems by using microbes designed to break down microplastics. The source of pollution may be halted by using this technique (Meera, Bhattacharyya, Nizam, & Kumar, 2022).

Environmental Sustainability: By using microbes to break down microplastics, researchers may mitigate the detrimental effects of conventional plastic removal methods such as dumping and burning. This assists in developing an eco-friendly and sustainable economy where plastic trash is converted by microorganisms into chemicals that are harmless (Xiang, Zhang, Wu, & Li, 2023).

Challenges and Ethical Issues: Although microbial strain engineering has considerable potential, it also brings up significant ethical and safety challenges. Realizing the full potential of this technology will depend on ensuring that genetically modified microbes do not have unexpected ecological effects and resolving social issues related to bioengineering in regard (Xiang et al., 2023) to the natural world (Rodríguez-Narvaez et al., 2021).

In conclusion, microbial strain engineering offers a promising future for resolving the critical issue of microplastic pollution in soil. We ought to anticipate the development of novel and persistent methods for the removal of microplastics as research progresses and our comprehension of microbial ecology increases. However, it is essential to approach this technology cautiously, considering both its potential advantages and the ethical issues it brings. In the end, the effective use of microbial strain engineering for microplastic removal has the ability to greatly improve the health of soil, conserve ecological systems, and ultimately lead to a healthier future for the global community.

REFERENCES

Amobonye, A., Bhagwat, P., Singh, S., & Pillai, S. (2021). Plastic biodegradation: Frontline microbes and their enzymes. *Science of the Total Environment, 759*, 143536.

Andrady, A. L., Barnes, P. W., Bornman, J. F., Gouin, T., Madronich, S., White, C. C., ... & Jansen, M. A. K. (2022). Oxidation and fragmentation of plastics in a changing environment; from UV-radiation to biological degradation. *Science of the Total Environment, 851*, 158022.

Bano, K., Kuddus, M., Zaheer, M. R., Zia, Q., Khan, M. F., Md Ashraf, G., … & Aliev, G. (2017). Microbial enzymatic degradation of biodegradable plastics. *Current Pharmaceutical Biotechnology, 18*(5), 429–440.

Baskaran, S., & Sathiavelu, M. (2022). Bioaugmentation and biostimulation of dumpsites for plastic degradation. In S. Kathi, S. Devipriya, K. Thamaraiselvi (Eds.), *Cost Effective Technologies for Solid Waste and Wastewater Treatment* (pp. 9–23): Elsevier.

Bharath, K. M., Muthulakshmi, A., & Usha, N. (2022). Microplastic contamination around the landfills: Distribution, characterization and threats-A review. *Current Opinion in Environmental Science & Health, 31*, 100422.

Bratovcic, A. (2019). Degradation of micro-and nano-plastics by photocatalytic methods. *Journal of Nanoscience and Nanotechnology, 3*, 206.

Braun, M., Mail, M., Krupp, A. E., & Amelung, W. (2023). Microplastic contamination of soil: Are input pathways by compost overridden by littering? *Science of the Total Environment, 855*, 158889.

Bugg, T. D., Williamson, J. J., & Alberti, F. (2021). Microbial hosts for metabolic engineering of lignin bioconversion to renewable chemicals. *Renewable and Sustainable Energy Reviews, 152*, 111674.

Cappitelli, F., Principi, P., & Sorlini, C. (2006). Biodeterioration of modern materials in contemporary collections: Can biotechnology help? *Trends in Biotechnology, 24*(8), 350–354.

Chang, X., Fang, Y., Wang, Y., Wang, F., Shang, L., & Zhong, R. (2022). Microplastic pollution in soils, plants, and animals: A review of distributions, effects and potential mechanisms. *Science of the Total Environment, 850*, 157857.

Chia, R. W., Lee, J.-Y., Jang, J., Kim, H., & Kwon, K. D. (2022). Soil health and microplastics: A review of the impacts of microplastic contamination on soil properties. *Journal of Soils and Sediments, 22*(10), 2690–2705.

Chia, R. W., Lee, J.-Y., Kim, H., & Jang, J. (2021). Microplastic pollution in soil and groundwater: A review. *Environmental Chemistry Letters, 19*(6), 4211–4224.

Chu, S., Zhang, B., Zhao, X., Soo, H. S., Wang, F., Xiao, R., & Zhang, H. (2022). Photocatalytic conversion of plastic waste: from photodegradation to photosynthesis. *Advanced Energy Materials, 12*(22), 2200435.

Cosgrove, L., McGeechan, P., Handley, P., & Robson, G. (2010). Effect of biostimulation and bioaugmentation on degradation of polyurethane buried in soil. *Applied and Environmental Microbiology, 76*(3), 810–819.

Crawford, C. B., & Quinn, B. (2016). *Microplastic Pollutants*: Elsevier.

De Gisi, S., Gadaleta, G., Gorrasi, G., La Mantia, F. P., Notarnicola, M., & Sorrentino, A. (2022). The role of (bio) degradability on the management of petrochemical and bio-based plastic waste. *Journal of Environmental Management, 310*, 114769.

de Souza Machado, A. A., Lau, C. W., Kloas, W., Bergmann, J., Bachelier, J. B., Faltin, E., … & Rillig, M. C. (2019). Microplastics can change soil properties and affect plant performance. *Environmental Science & Technology, 53*(10), 6044–6052.

de Souza Machado, A. A., Lau, C. W., Till, J., Kloas, W., Lehmann, A., Becker, R., & Rillig, M. C. (2018). Impacts of microplastics on the soil biophysical environment. *Environmental Science & Technology, 52*(17), 9656–9665.

Dioses-Salinas, D. C., Pizarro-Ortega, C. I., & De-la-Torre, G. E. (2020). A methodological approach of the current literature on microplastic contamination in terrestrial environments: Current knowledge and baseline considerations. *Science of the Total Environment, 730*, 139164.

Du, H., Xie, Y., & Wang, J. (2022). Environmental impacts of microplastics on fishery products: An overview. *Gondwana Research, 108*, 213–220.

Dussud, C., & Ghiglione, J.-F. (2014). Bacterial degradation of synthetic plastics. *Paper Presented at the CIESM Workshop Monogr*, France.

El Menofy, N. G., & Khattab, A. M. (2023). Plastics biodegradation and biofragmentation. In G. A. M. Ali, A. S. H. Makhlouf (Eds.), *Handbook of Biodegradable Materials* (pp. 571–600): Springer.

Feng, W., Deng, Y., Cao, Y., Liu, J., Han, Y., Liu, J., … & Giesy, J. P. (2023). Biotechnology remediation and environmental behavior of microplastics in soils: A review. *Reviews of Environmental Contamination and Toxicology, 261*(1), 1–15.

Gaur, V. K., Gupta, S., Sharma, P., Gupta, P., Varjani, S., Srivastava, J. K., … & Bui, X.-T. (2022). Metabolic cascade for remediation of plastic waste: A case study on microplastic degradation. *Current Pollution Reports, 8*(1), 30–50.

Haben Fesseha, M. (2019). Degradation of plastic materials using microorganisms: A review. *Public Health Open Journal, 4*(2), 57–63.

He, D., Luo, Y., Lu, S., Liu, M., Song, Y., & Lei, L. (2018). Microplastics in soils: Analytical methods, pollution characteristics and ecological risks. *TrAC Trends in Analytical Chemistry, 109*, 163–172.

Helmberger, M. S., Tiemann, L. K., & Grieshop, M. J. (2020). Towards an ecology of soil microplastics. *Functional Ecology, 34*(3), 550–560.

Huang, J., Chen, H., Zheng, Y., Yang, Y., Zhang, Y., & Gao, B. (2021). Microplastic pollution in soils and groundwater: Characteristics, analytical methods and impacts. *Chemical Engineering Journal, 425*, 131870.

Huang, Y., Liu, Q., Jia, W., Yan, C., & Wang, J. (2020). Agricultural plastic mulching as a source of microplastics in the terrestrial environment. *Environmental Pollution, 260*, 114096.

Jang, Y., Kim, M., Kim, Y., Yu, J., Kim, S.-K., Han, J., … & Min, J. (2023). Enhancing biodegradation of PBAT through bio-stimulation using Pseudozyma jejuensis for effective plastic waste reduction. *Chemosphere, 340*, 139867.

Kaur, K., Sharma, S., Shree, N., & Mehrotra, R. (2023). Recent advancements and mechanism of plastics biodegradation promoted by bacteria: A key for sustainable remediation for plastic wastes. *Biosciences Biotechnology Research Asia, 20*(1), 1–12.

Kaushal, J., Khatri, M., & Arya, S. K. (2021). Recent insight into enzymatic degradation of plastics prevalent in the environment: A mini-review. *Cleaner Engineering and Technology, 2*, 100083.

Liu, S. Y., Leung, M. M.-L., Fang, J. K.-H., & Chua, S. L. (2021) Engineering a microbial 'trap and release' mechanism for microplastics removal. *Chemical Engineering Journal, 404*, 127079. ISSN 1385-8947.

Lu, J., Hou, R., Wang, Y., Zhou, L., & Yuan, Y. (2022). Surfactant-sodium dodecyl sulfate enhanced degradation of polystyrene microplastics with an energy-saving electrochemical advanced oxidation process (EAOP) strategy. *Water Research, 226*, 119277.

Martic, S., Tabobondung, M., Gao, S., & Lewis, T. (2022). Emerging electrochemical tools for microplastics remediation and sensing. *Frontiers in Sensors, 3*, 958633.

Meera, S. P., Bhattacharyya, M., Nizam, A., & Kumar, A. (2022). A review on microplastic pollution in the mangrove wetlands and microbial strategies for its remediation. *Environmental Science and Pollution Research, 29*, 1–15.

Möller, J. N., Heisel, I., Satzger, A., Vizsolyi, E. C., Oster, S. J., Agarwal, S., … & Löder, M. G. (2022). Tackling the challenge of extracting microplastics from soils: A protocol to purify soil samples for spectroscopic analysis. *Environmental Toxicology and Chemistry, 41*(4), 844–857.

Monkul, M. M., & Özhan, H. O. (2021). Microplastic contamination in soils: A review from geotechnical engineering view. *Polymers, 13*(23), 4129.

Nguyen, P. Q., Huang, X., Collins, D. S., Collins, J. J., & Lu, T. (2023). Harnessing synthetic biology to enhance ocean health. *Trends in Biotechnology, 41*, 860–874.

Niu, B., Wang, W., Yuan, Z., Sederoff, R. R., Sederoff, H., Chiang, V. L., & Borriss, R. (2020). Microbial interactions within multiple-strain biological control agents impact soil-borne plant disease. *Frontiers in Microbiology, 11*, 585404.

Okoffo, E. D., O'Brien, S., Ribeiro, F., Burrows, S. D., Toapanta, T., Rauert, C., ... & Thomas, K. V. (2021). Plastic particles in soil: state of the knowledge on sources, occurrence and distribution, analytical methods and ecological impacts. *Environmental Science: Processes & Impacts, 23*(2), 240–274.

Pometto 3rd, A., Lee, B., & Johnson, K. E. (1992). Production of an extracellular polyethylene-degrading enzyme (s) by Streptomyces species. *Applied and Environmental Microbiology, 58*(2), 731–733.

Prasad, M. N., & Elchuri, S. V. (2022). Pharmaceuticals and personal care products in the environment with emphasis on horizontal transfer of antibiotic resistance genes. *Chemistry-Didactics-Ecology-Metrology,* 27, 35–51.

Qi, R., Jones, D. L., Li, Z., Liu, Q., & Yan, C. (2020). Behavior of microplastics and plastic film residues in the soil environment: A critical review. *Science of the Total Environment, 703*, 134722.

Rillig, M. C., Leifheit, E., & Lehmann, J. (2021). Microplastic effects on carbon cycling processes in soils. *PLoS Biology, 19*(3), e3001130.

Rodríguez-Narvaez, O. M., Goonetilleke, A., Perez, L., & Bandala, E. R. (2021). Engineered technologies for the separation and degradation of microplastics in water: A review. *Chemical Engineering Journal, 414*, 128692.

Romdhane, S., Spor, A., Aubert, J., Bru, D., Breuil, M.-C., Hallin, S., … & Philippot, L. (2022). Unraveling negative biotic interactions determining soil microbial community assembly and functioning. *The ISME Journal, 16*(1), 296–306.

Sa'adu, I., & Farsang, A. (2023). Plastic contamination in agricultural soils: A review. *Environmental Sciences Europe, 35*(1), 13.

Saravanan, A., Kumar, P. S., Vo, D.-V. N., Jeevanantham, S., Karishma, S., & Yaashikaa, P. (2021). A review on catalytic-enzyme degradation of toxic environmental pollutants: Microbial enzymes. *Journal of Hazardous Materials, 419*, 126451.

Setälä, O., Norkko, J., & Lehtiniemi, M. (2016). Feeding type affects microplastic ingestion in a coastal invertebrate community. *Marine Pollution Bulletin, 102*(1), 95–101.

Sharma, B., & Jain, P. (2020). Deciphering the advances in bioaugmentation of plastic wastes. *Journal of Cleaner Production, 275*, 123241.

Shi, J., Sun, Y., Wang, X., & Wang, J. (2022). Microplastics reduce soil microbial network complexity and ecological deterministic selection. *Environmental Microbiology, 24*(4), 2157–2169.

Shi, X., Chen, Z., Liu, X., Wei, W., & Ni, B.-J. (2022). The photochemical behaviors of microplastics through the lens of reactive oxygen species: Photolysis mechanisms and enhancing photo-transformation of pollutants. *Science of the Total Environment, 846*, 157498.

Shim, W. J., Hong, S. H., & Eo, S. E. (2017). Identification methods in microplastic analysis: A review. *Analytical Methods, 9*(9), 1384–1391.

Song, Y. K., Hong, S. H., Jang, M., Han, G. M., Jung, S. W., & Shim, W. J. (2017). Combined effects of UV exposure duration and mechanical abrasion on microplastic fragmentation by polymer type. *Environmental Science & Technology, 51*(8), 4368–4376.

Sun, Y., Li, X., Cao, N., Duan, C., Ding, C., Huang, Y., & Wang, J. (2022). Biodegradable microplastics enhance soil microbial network complexity and ecological stochasticity. *Journal of Hazardous Materials, 439*, 129610.

Tareen, A., Saeed, S., Iqbal, A., Batool, R., & Jamil, N. (2022). Biodeterioration of microplastics: A promising step towards plastics waste management. *Polymers, 14*(11), 2275.

Temporiti, M. E. E., Nicola, L., Nielsen, E., & Tosi, S. (2022). Fungal enzymes involved in plastics biodegradation. *Microorganisms, 10*(6), 1180.

Thakur, B., Singh, J., Singh, J., Angmo, D., & Vig, A. P. (2023). Biodegradation of different types of microplastics: Molecular mechanism and degradation efficiency. *Science of the Total Environment, 877*, 162912.

Thakur, S., Mathur, S., Patel, S., & Paital, B. (2022). Microplastic accumulation and degradation in environment via biotechnological approaches. *Water, 14*(24), 4053.

Tofa, T. S., Kunjali, K. L., Paul, S., & Dutta, J. (2019). Visible light photocatalytic degradation of microplastic residues with zinc oxide nanorods. *Environmental Chemistry Letters, 17*, 1341–1346.

Tripathy, A., More, R. D., Gupta, S., Samuel, J., Singh, J., & Prasad, R. (2021). Present and future prospect of algae: A potential candidate for sustainable pollution mitigation. *The Open Biotechnology Journal, 15*(1), 142–156.

Tudor, V. C., Mocuta, D. N., Teodorescu, R. F., & Smedescu, D. I. (2019). The issue of plastic and microplastic pollution in soil. *Materiale Plastice, 56*(3), 484–487.

Uheida, A., Mejía, H. G., Abdel-Rehim, M., Hamd, W., & Dutta, J. (2021). Visible light photocatalytic degradation of polypropylene microplastics in a continuous water flow system. *Journal of Hazardous Materials, 406*, 124299.

Urso, M., & Pumera, M. (2022). Nano/microplastics capture and degradation by autonomous nano/microrobots: A perspective. *Advanced Functional Materials, 32*(20), 2112120.

Vaksmaa, A., Polerecky, L., Dombrowski, N., Kienhuis, M. V., Posthuma, I., Gerritse, J., … & Niemann, H. (2023). Polyethylene degradation and assimilation by the marine yeast Rhodotorula mucilaginosa. *ISME Communications, 3*(1), 68.

Wang, J., Liu, X., Li, Y., Powell, T., Wang, X., Wang, G., & Zhang, P. (2019). Microplastics as contaminants in the soil environment: A mini-review. *Science of the Total Environment, 691*, 848–857.

Wang, W., Ge, J., Yu, X., & Li, H. (2020). Environmental fate and impacts of microplastics in soil ecosystems: Progress and perspective. *Science of the Total Environment, 708*, 134841.

Xiang, P., Zhang, T., Wu, Q., & Li, Q. (2023). Systematic review of degradation processes for microplastics: Progress and prospects. *Sustainability, 15*(17), 12698.

Xu, B., Liu, F., Cryder, Z., Huang, D., Lu, Z., He, Y., ... & Tang, C. (2020). Microplastics in the soil environment: Occurrence, risks, interactions and fate–A review. *Critical Reviews in Environmental Science and Technology, 50*(21), 2175–2222.

Ya, H., Jiang, B., Xing, Y., Zhang, T., Lv, M., & Wang, X. (2021). Recent advances on ecological effects of microplastics on soil environment. *Science of the Total Environment, 798*, 149338.

Yongming, L., Qian, Z., Haibo, Z., Xiangliang, P., Chen, T., Lianzhen, L., & Jie, Y. (2018). Pay attention to research on microplastic pollution in soil for prevention of ecological and food chain risks. *Bulletin of Chinese Academy of Sciences (Chinese Version), 33*(10), 1021–1030.

You, X., Wang, S., Li, G., Du, L., & Dong, X. (2022). Microplastics in the soil: A review of distribution, anthropogenic impact, and interaction with soil microorganisms based on meta-analysis. *Science of the Total Environment, 832*, 154975.

Yuan, J., Ma, J., Sun, Y., Zhou, T., Zhao, Y., & Yu, F. (2020). Microbial degradation and other environmental aspects of microplastics/plastics. *Science of the Total Environment, 715*, 136968.

Zhang, M., Tan, M., Ji, R., Ma, R., & Li, C. (2022). Current situation and ecological effects of microplastic pollution in soil. *Reviews of Environmental Contamination and Toxicology, 260*(1), 11.

Zhang, X., Li, Y., Ouyang, D., Lei, J., Tan, Q., Xie, L., ... & Farooq, T. H. (2021). Systematical review of interactions between microplastics and microorganisms in the soil environment. *Journal of Hazardous Materials, 418*, 126288.

Zhou, J., Gui, H., Banfield, C. C., Wen, Y., Zang, H., Dippold, M. A., ... & Jones, D. L. (2021). The microplastisphere: Biodegradable microplastics addition alters soil microbial community structure and function. *Soil Biology and Biochemistry, 156*, 108211.

Zurier, H. S., & Goddard, J. M. (2021). Biodegradation of microplastics in food and agriculture. *Current Opinion in Food Science, 37*, 37–44.

CHAPTER 11

Mechanisms and Influencing Factors Making Agricultural Soil as a Sink for Microplastics

Asif Farooq, Manzoor Ahmad Ahanger, and Khalid Muzamil Gani

11.1 INTRODUCTION

Worldwide plastic production reached a staggering 3.68 × 10^8 tons in 2019 (Su et al., 2022), with a significant portion of plastic waste making its way into landfills, accounting for 79% of total plastic waste (Geyer et al., 2017). Due to processes like photodegradation, mechanical abrasion, and bioturbation, plastics are prone to fragmentation, resulting in the formation of microplastics (MPs) (<5 mm in size) and has emerged as a pressing environmental concern in recent years (Sharma et al., 2023). According to Hartline et al. (2016), the escalation of MP pollution in terrestrial systems, expressing that aggregate quantity of microplastics present in terrestrial ecosystems could be 4–23 times greater than that found

DOI: 10.1201/9781003487579-11

in marine environments. While the recovery and recycling rates of plastics worldwide may be alarmingly low, approximately 32% of plastic waste has been identified in the soil ecosystem (De Souza Machado et al., 2018b; Jambeck et al., 2015). In a study conducted by Nizzetto et al. (2016), it has been found that terrestrial soils have been found to accumulate a higher concentration of MPs compared to aquatic ecosystems.

Recent studies delve into the dynamics and behavior of MPs within the soil ecosystem, including their tendency for bioaccumulation within food chains, their impact on soil microorganisms from a toxicological perspective, and the methodologies employed for their analysis (Qi et al., 2020; Guo et al., 2020; Li et al., 2020 a). Several factors play an important role in shaping the deposition, persistence, and movement of MPs within the soil environment, encompassing suboptimal waste management protocols, improper disposal, climatic conditions, topography as well as the physical characteristics of plastic particles (Oliviero et al., 2019; He et al., 2018b). The occurrence of MPs within soil can exert a substantial influence on soil-water interactions, thereby influencing parameters such as water retention capacity, bulk density, microbial functions, and the overall structural integrity of the soil (De Souza Machado et al., 2019, 2018a). The alteration in bulk density due to MP pollution may also accelerate evaporation, impacting plant growth (Wang et al., 2019).

Plastic films and mulch films used in agricultural practices are recognized as a potential origin of MP contamination in the agro ecosystem (Zhang and Liu, 2018). Additionally, the widespread application of sewage sludge from wastewater treatment plants (WWTPs) to agricultural lands represents a major input of MPs into agro ecosystems, with uncertain consequences for sustainability and food security (Nizzetto et al., 2016). MPs possess the capacity to undergo bioaccumulation within the food chain, influencing soil biota across multiple trophic levels and potentially impacting entire ecosystems (Rillig and Lehmann, 2020; He et al., 2018).

11.2 SOURCES OF MICROPLASTICS (MPS) IN AGRO ENVIRONMENT

Plastic contamination in agricultural soils is a growing concern, with multiple sources contributing to the presence of MPs in these environments. Plastic materials have found widespread application in diverse agricultural activities, encompassing areas such as cultivation, fertilization, and the implementation of plastic mulching techniques. As this plastic degrades in the soil, it releases a significant proportion of MPs, which can accumulate in crops (Zhang et al., 2020). One of the major sources of MPs in agro ecosystem is plastic mulch films, which are extensively used in cold and arid regions to maintain suitable temperatures and enhance crop yields (Liu et al., 2021). Some major studies especially conducted in China have revealed high concentrations of MPs in agricultural soils, with polyethylene (PE) being the primary composition of these MPs with mulch film concentration positively correlated (Chen et al., 2020; Li et al., 2020b; Liu et al., 2018; Huang et al., 2020).

Compost, employed for enhancing soil fertility, constitutes another noteworthy contributor to the presence of MPs within agricultural soils. During the composting process, the high temperature and microbial activity contribute to the breakdown of larger plastic particles into MPs through fragmentation, increasing their concentration in agricultural soils (Braun et al., 2021). Utilizing organic materials over an extended period in agricultural environments may result in substantial accretion of MPs within the soil. Common organic fertilizers, such as household bio-waste and mature compost, which we use for fertilizers, often serve as unrecognized sources of MPs (Yang et al., 2021; Weithmann et al., 2018).

Sewage sludge and wastewater play a crucial role in introducing MPs into agricultural soils. Inefficient removal of minute plastic particles in wastewater treatment facilities results in significant quantities of MP particles retained within their residuals (Nizzetto et al., 2016). These MPs are subsequently released into agricultural fields through irrigation, leading to elevated concentrations in soils (He et al., 2018). The impact of irrigation with wastewater on soil MP content is evident, with higher concentrations observed in vegetable soils compared to buffer soils (Van den Berg et al., 2020).

Apart from these sources, MPs can also enter into agricultural soils through atmospheric deposition (Mbachu et al., 2020). Various studies have reported the deposition of MPs in urban and suburban areas, with the composition primarily consisting of fibers (Wright et al., 2020; Dris et al., 2017). The density of the surrounding population in urban areas is associated with higher MP deposition, potentially posing a lower risk in rural regions.

The deposition of atmospheric MPs on soil surfaces can be influenced by weather conditions, including rainfall and snow (Allen et al., 2019). While the immediate influence of atmospheric MP particles and their potential to transport from urban areas to rural fields and deposit on crops and soil surfaces raises concerns about ecological risks to agricultural soils and human health (Liu et al., 2020).

11.3 MICROPLASTIC POLYMERS PRESENT IN AGRICULTURAL SOILS

The persistence of MPs in agricultural soils originates from a diverse range of sources that will contribute to specific types of polymer-based MPs. The most prominent sources are plastic mulching films and sludge from WWTPs which introduce various polymer types into the soil. These major sources consist of materials like PE, polypropylene (PP), and polyvinyl chloride (PVC) (Liu et al., 2022; Wang et al., 2022; Yu et al., 2021; Zhou et al., 2020). PE has been recognized as the predominant component of MPs originating from agricultural mulching films. This underscores its substantial role in soil contamination, encompassing various forms such as high-density PE, low-density PE, and linear low-density PE as per the research conducted by Blasing and Amelung (2016). In agricultural fields where greenhouses are utilized, the introduction of MPs is associated with greenhouse films, which may be composed of translucent or transparent plastics. White and transparent/translucent MPs, frequently detected in such environments, further highlight the polymer diversity, underlining the presence of various plastic materials.

In addition to the more common polymer types, MPs found in compost, another source of soil contamination within the agricultural context, can have a variety of polymer compositions. Compost often contains plastics like PE, PP, and polystyrene, reflecting the range of materials incorporated into organic waste streams (Corradini et al., 2019a, 2019b). Moreover, the utilization of diverse additives, including plasticizers, flame retardants, and antioxidants, within a multitude of plastic items, serves to augment the array of attributes exhibited by MPs originating from distinct origins. For example, plasticizers are primarily added to soften PVC, flame retardants are prevalent in polystyrene, and antioxidants are used to prevent aging in polymers like PE and PP. These additives, in combination with the polymer types, shape the properties, behavior, and ecological risks associated with MPs in agro ecosystems. Subsequent research endeavors should prioritize assessing the environmental implications of these distinct polymer types and additives, as the diversity of MPs from different sources remains a subject in need of further elucidation. Understanding the intricacies of polymer types and their additive compositions is crucial for effectively addressing the environmental and agricultural concerns associated with MPs in soils.

11.4 SHAPE AND COLOR FEATURES OF MPS IN AGRICULTURAL SOIL

The primary cause of MPs in agro environments has been conclusively attributed to the extensive utilization of plastic mulching films (Liu et al., 2022; Wang et al., 2022; Yu et al., 2021; Zhou et al., 2020). These plastic mulch films, which are frequently employed in agricultural practices, have been found to consistently contribute to the presence of both film and fragment MPs. The prevalence of MPs in agricultural lands subjected to mulching is discernible in various forms, including translucent/transparent, black, and white MPs (Zhang et al., 2022; Wang et al., 2022). White and transparent/translucent MPs exhibit a higher predominance in agricultural fields where greenhouses are employed, often constructed using greenhouse films. The deterioration of plastic mulches and greenhouse films has been associated with the generation of both fragment and film MPs (Yu et al., 2021; Liu et al., 2022; Wang et al., 2022). For the purpose of securing and tying down greenhouses, the use of plastic ropes has been found to contribute to the introduction of fiber MPs into the agro soil environment (Wang et al., 2022). While addressing and mitigating the potential environmental and agricultural impacts, understanding the origin and distribution of MPs in agricultural soils is important.

11.5 IMPACTS OF MICROPLASTICS (MPS) ON AGRO ENVIRONMENT

MPs in agricultural soils have gained significant attention as the research investigations indicate that the presence of MP particles has the potential to affect and alter the physical properties of soil, impacting parameters like water infiltration and soil structure (Liu et al., 2022). For example, the introduction of PE MPs at a concentration of 0.2% (w/w) was found to reduce soil infiltration by 11.7%. Certain types of MPs, such as PES (polyester) and PAA, have been observed to decrease the stability of soil aggregates, potentially making the soil more susceptible to wind and water erosion, with implications for soil fertility and productivity (Liang et al., 2021; Zhou et al., 2020).

The influence of MPs on soil nutrient composition and availability remains a sensitive matter of concern, which can create conflict in various studies. Some researchers suggest that MPs may lead to a reduction in soil nutrient content (Li et al., 2021), while others indicate potential benefits such as improved nutrient retention and reduced nitrate leaching

(Lozano et al., 2021a). MPs can also impact soil microbial communities, with different types of MPs potentially causing shifts in microbial composition and reductions in microbial diversity (Feng et al., 2022). However, the effect of MPs on soil respiration, a process connected to microbial decomposition of organic matter, varies among different studies. Zhao et al. (2021) in their findings reported that it can decrease while Lozano et al. (2021a) in their investigation found that it can increase.

Arbuscular mycorrhizal fungi (AMF) which is crucial for plant health and nutrition, are not immune to the effects of MPs. However, the outcomes of these interactions are mixed which do not clearly define the impact of MPs on AMF (Leifheit et al., 2021). Other soil organisms, such as nematodes, have been observed to suffer adverse effects due to the presence of MPs, which can disrupt the soil ecosystem. This disruption can result in reduced growth, reproduction, and overall health of these organisms, which in turn may impact soil material cycling, soil properties, and soil micro-structure. For instance, earthworm mortality has been associated with a decrease in macropores in the soil, which can have negative consequences for soil aeration, structure, and nutrient recycling (Schon et al., 2017).

The effects of MPs on crops are highly variable, contingent on factors like crop type, the specific type and shape of MPs, and the concentration of MPs. Some crops, such as wheat and Chinese cabbage, have shown diverse responses to MP exposure, affecting growth, nutrient uptake, and photosynthesis. For instance, PVC MPs have been found to limit nitrogen and phosphorus utilization by wheat, while PE MPs increased nitrogen uptake (Liu et al., 2022). Generally, the impacts on crops are often observed in the form of alterations to root microbial activities and photosynthesis (De Souza Machado et al., 2019), which can influence overall plant health and productivity.

11.6 GREEN FARMING PRACTICES

Biodegradable plastics (BDPs) in agricultural soil are of paramount importance. Initially designed to integrate into the soil matrix, BDP mulch is intended for consumption by native microorganisms to prevent long-term plastic residue formation. However, the actual degradation process in agricultural soil is intricate. Soil microbiota, chiefly consisting of bacteria and fungi, colonize the polymer surface, producing extracellular enzymes that depolymerize the plastics and consume the resulting BDP hydrolysis products. Soil microorganisms also interact with residual plastic fragments, potentially giving rise to microplastics and nanoplastics, while releasing chemical byproducts like monomers and additives, which could migrate into crops and soil-dwelling organisms. This complexity underscores the need for a comprehensive understanding of the consequences of BDP use within the agricultural ecosystem (Jin et al., 2022; Azeem et al., 2021, 2022; Adeel et al., 2021).

11.7 TECHNIQUES FOR ANALYZING MPS IN AGRICULTURAL SOIL

11.7.1 Sampling Methods

Sampling of MPs from agricultural soils is crucial to assess their presence and impact. Sampling depth and location should be determined considering the soil type and potential sources of MPs. Various techniques, such as coring, shovel sampling, and grid sampling, can be employed but the most useful is the transect field technique that emerged as the predominant method for MP sampling followed by the random sampling approach within agricultural soils (Beriot et al., 2021; Ding et al., 2020; Crossman et al., 2020). Taking cues from the methodology put forth by Yang et al. (2023), it is evident that the standardized practice for soil collection involves obtaining samples from (0 to 20 cm) soil stratum at all designated sampling points. These samples are diligently collected and preserved in clean glass jars. Subsequently, they undergo sieving to eliminate any stones, plant residue, and other impurities that may be present. The excess soil, not required for analysis, is methodically removed using the quadrat method. Following these procedures, the samples are securely sealed and stored, ensuring the integrity of the collected data for subsequent analysis. In another study, surface soil (0–25 cm) was obtained using a metallic soil auger. Samples were carefully transferred and preserved in 3 mm thick polypropylene bags to the laboratory where the samples were transferred into trays and stored in an air oven at 40°C ± 2°C and stored for further analysis (Corradini et al., 2019b).

11.7.2 Sample Processing and Extraction Techniques

There are many standard procedures for processing and extraction of MPs from agricultural soil. Hurley et al. (2018), Liu et al. (2018), and Tagg et al. (2017) in their studies used 200 g of sieved soil placed in clean beakers. Then saturated NaCl solution ($\rho = 1.2$ g/cm^3) was added to the soil in order to carry out density separation, allowing it to settle for 24 hours. Fenton's reagent was used to complete the process of organic matter digestion. Glass fiber filters were used to filter the resulting supernatant through vacuum filtration. On the other hand, following standard protocol, Jia et al. (2022) in his study took 5 g of soil, which had been sieved and air-dried, and subjected it to a digestion process in which 150 mL of a Fenton's reagent was utilized for a duration of 3 hours. Following this, the digestion vessel was transferred to an oven set at 50°C and allowed to remain there for a period spanning between 18 and 24 hours until complete desiccation was achieved. Subsequently, 200 mL of a saturated sodium chloride solution (ρ =1.2 g/cm^3) was introduced into the digestion vessel. The contents were subjected to agitation using a magnetic stirrer operating at 120 rpm for a duration of 30 minutes, ensuring the thorough dispersion of soil samples. After allowing the system to undergo static settlement for 24 hours at room temperature, approximately 100 mL of supernatant, containing MPs, were carefully collected in a separate beaker. This collected supernatant was then subjected to filtration using a vacuum filtration assembly. The resultant filter was stored within Petri dishes, earmarked for future analysis.

11.7.3 Quality Assurance in MP Analysis of Agricultural Soils

In order to maintain proper quality assurance (QA), nitrile rubber gloves and cotton lab coats should be worn as a precautionary measure to prevent the introduction of MPs. All experimental equipment should be performed in non-plastic consumable materials to ensure the avoidance of potential contamination. Prior to conducting experiments, all glassware and instruments should be meticulously washed three times, employing a combination of ethanol and Milli-Q water, followed by thorough drying (Ding et al., 2021). To maintain a pristine and uncontaminated environment, MPs will be meticulously isolated, extracted, and observed within a controlled environment in a biosafety cabinet (Wu et al., 2023). To validate the absence of any exogenous interference from MPs, a series of blank experiments will be undertaken in a triplicate manner. These blank experiment samples should consist of deionized water devoid of MPs, and they will be concurrently run alongside the laboratory samples to conclusively demonstrate the absence of any extraneous MPs influence (Ding et al., 2021).

11.8 CHARACTERIZATION AND QUANTIFICATION TECHNIQUES OF MICROPLASTICS (MPS)

μ-FTIR Spectroscopy: This is a versatile technique for investigating particles ranging from greater than 500 μm down to as small as 20 μm. This technique is used to determine the molecular structure and composition of the target substance which involves exposing the samples to a defined range of infrared radiation. *μ-FTIR* Spectroscopy is recognized for its emergent, rapid, and reliable nature, rendering it a nondestructive approach. Analyzing nontransparent particles and those smaller than 20 μm necessitates pre-treatment. Further, this spectroscopic technique can be relatively costly and is applicable and limited only to infra-red active samples (Li et al., 2018; He et al., 2018a).

μ-Raman Spectroscopy: Micro-Raman spectroscopy is a valuable analytical technique that is capable of examining MP particles (>1 μm), revealing substance-specific spectra through Raman scattering principles, enabling precise identification. This method provides high-resolution images of MPs, enhancing visualization and analysis. However, nontransparent or dark particles may require additional sample preparation and refinements. Nevertheless, it is a time-intensive procedure, demanding meticulous attention (Li et al., 2018; He et al., 2018a).

Scanning Electron-Energy Dispersive X-Ray Spectroscopy (SEM-EDS): This is an analytical technique used for chemical elemental analysis and is capable of examining high-resolution imaging of microscale particles. This method involves the generation of electrons that interact with the sample, enabling the measurement of secondary ions. The output of this process includes high-resolution images, facilitating detailed particle analysis. However, this technique typically necessitates the coating of samples and may provide relatively less informative data, as observed in studies conducted by Blasing and Amelung (2018), Zhou et al. (2018), and De Souza Machado et al. (2018a).

Gas Chromatography-Mass Spectrometry (GC-MS): This is a sensitive and reliable technique for the characterization of MPs. It is particularly effective for particles larger than 500 μm. This method involves using a GC column coupled to a quadrupole MS to generate spectra, which can be compared to existing plastic databases for identification. One of its advantages is its ability to minimize potential background contamination. However, due to a limited plastic database, this technique can only analyze one sample at a time within a specific weight range (Dümichen et al., 2017; Ivleva et al., 2017).

Liquid Chromatography (LC): This is a valuable technique, particularly known for handling large sample sizes, making it an ideal choice for studies that require comprehensive data collection. LC offers the advantage of selective sample preparation, enhancing its efficiency in isolating specific polymers and yielding better recovery rates when analyzing selected polymers. However, it is essential to acknowledge the limitations of LC, which primarily restrict its utility to certain polymer types, PE and PET, making it less suitable for the analysis of a broader spectrum of environmental samples (Elert et al., 2017; Hintersteiner et al., 2015).

Microscopy: This is a quantification technique used in the analysis of microplastics in various substrates. This method employs stereomicroscopes to identify particles as small as a micrometer. Additionally, digital microscopes can be employed to enhance precision. Visual microscopy is recognized for its cost-effectiveness, speed, and ease of application which involves quantification and identification of MP particles within samples based on shape and color (He et al., 2018a).

Tagging: This method enables the enumeration and visualization of microscale MPs in environmental samples. Tagging involves subjecting MP particles to irradiation with blue light, followed by the adsorption of a hydrophobic dye, rendering them fluorescent and easily distinguishable. This technique is characterized by its efficiency, cost-effectiveness, and relative ease of implementation. However, it is essential to note that tagging may have limitations in cases where impurities are present in the samples, as these impurities can potentially lead to overestimations of the MP quantities (He et al., 2018a; Shim et al., 2017).

11.9 CONCLUSION

MPs in agricultural soils exhibit a complex interaction with soil properties, biota, and crop health, with outcomes subject to multiple variables, including the type and concentration of MPs, soil characteristics, and crop varieties. MPs in agricultural soils have multiple sources, including plastic mulch films, compost, sewage sludge, wastewater, and atmospheric deposition. Understanding the contribution of each source and the characteristics of MPs from these sources is essential for addressing the ecological and human health risks associated with plastic contamination in agricultural soils. Application of biodegradable plastics in agriculture practices could be an alternative against plastic hazards but to determine the fate of BDPs in agricultural soil, understanding the mechanism of decomposition is a crucial process to determine its fate. There is need for continued research to gain a deeper understanding of the fate and effects of MPs in agricultural soils, as well as the development of strategies to mitigate their potential adverse impacts on soil health and crop production.

REFERENCES

Adeel, M., Shakoor, N., Shafiq, M., Pavlicek, A., Part, F., Zafiu, C., Raza, A., Ahmad, M.A., Jilani, G., White, J.C., Ehmoser, E.K., Lynch, I., Ming, X., Rui, Y.K., 2021. A critical review of the environmental impacts of manufactured nano-objects on earthworm species. *Environ. Pollut.* 290, 118041. https://doi.org/10.1016/j.envpol.2021.118041

Allen, S., Allen, D., Phoenix, V.R., Roux, G.L., Durántez Jiménez, P., Simonneau, A., Binet, S., Galop, D., 2019. Atmospheric transport and deposition of microplastics in a remote mountain catchment. *Nat. Geosci.* 12, 339–344.

Azeem, I., Adeel, M., Ahmad, M.A., Shakoor, N., Jiangcuo, G.D., Azeem, K., Ishfaq, M., Shakoor, A., Ayaz, M., Xu, M., Rui, Y., 2021. Uptake and accumulation of nano/microplastics in plants: a critical review. *Nanomaterials* 11, 2935. https://doi.org/10.3390/nano11112935

Azeem, I., Adeel, M., Ahmad, M.A., Shakoor, N., Zain, M., Yousef, N., Yinghai, Z., Azeem, K., Zhou, P.F., White, J.C., Ming, X., Rui, Y.K., 2022. Microplastic and nanoplastic interactions with plant species: trends, meta-analysis, and perspectives. *Environ. Sci. Technol. Lett.* 9, 482–492. https://doi.org/10.1021/acs.estlett.2c00107

Beriot, N., Peek, J., Zornoza, R., Geissen, V., Huerta Lwanga, E., 2021. Low density-microplastics detected in sheep faeces and soil: A case study from the intensive vegetable farming in Southeast Spain. *Sci. Total Environ.* 755. https://doi.org/10.1016/j.scitotenv.2020.142653

Blasing, M., Amelung, W., 2018. Plastics in soil: analytical methods and possible sources. *Sci. Total Environ.* 612, 422–435. https://doi.org/10.1016/j.scitotenv.2017.08.086

Braun, M., Mail, M., Heyse, R., Amelung, W., 2021. Plastic in compost: prevalence and potential input into agricultural and horticultural soils. *Sci. Total Environ.* 760. https://doi.org/10.1016/j.scitotenv.2020.143335

Chen, Y., Leng, Y., Liu, X., Wang, J., 2020. Microplastic pollution in vegetable farmlands of suburb Wuhan, central China. *Environ. Pollut.* 257. https://doi.org/10.1016/j.envpol.2019.113449

Corradini, F., Bartholomeus, H., Lwanga, E., Gertsen, H., Geissen, V., 2019a. Predicting soil microplastic concentration using vis-NIR spectroscopy. *Sci. Total Environ.* 650, 922–932.

Corradini, F., Meza, P., Eguiluz, R., Casado, F., Huerta-lwanga, E., Geissen, V., 2019b. Evidence of microplastic accumulation in agricultural soils from sewage sludge disposal. *Sci. Total Environ.* 671, 411–420. https://doi.org/10.1016/j.scitotenv.2019.03.368

Crossman, J., Hurley, R.R., Futter, M., Nizzetto, L., 2020. Transfer and transport of microplastics from biosolids to agricultural soils and the wider environment. *Sci. Total Environ.* 724. https://doi.org/10.1016/j.scitotenv.2020.138334

De Souza Machado, A.A., Kloas, W., Zarfl, C., Hempel, S., Rillig, M.C., 2018a. Microplastics as an emerging threat to terrestrial ecosystems. *Global Change Biol.* 24(4), 1405–1416. https://doi.org/10.1111/gcb.14020

De Souza Machado, A.A., Lau, C.W., Till, J., Kloas, W., Lehmann, A., Becker, R., Rillig, M.C., 2018b. Impacts of microplastics on the soil biophysical environment. *Environ. Sci. Technol.* 52(17), 9656–9665. https://doi.org/10.1021/acs.est.8b02212

De Souza Machado, A.A., Lau, C.W., Kloas, W., Bergmann, J., Bachelier, J.B., Faltin, E., Becker, R., Gorlich, A.S., Rillig, M.C., 2019. Microplastics can change soil prop- € erties and affect plant performance. *Environ. Sci. Technol.* 53(10), 6044–6052. https://doi.org/10.1021/acs.est.9b01339

Ding, L., Zhang, S., Wang, X., Yang, X., Zhang, C., Qi, Y., Guo, X., 2020. The occurrence and distribution characteristics of microplastics in the agricultural soils of Shaanxi Province, in north-western China. *Sci. Total Environ.* 720. https://doi.org/10.1016/j.scitotenv.2020.137525

Ding, L., Wang, X., Ouyang, Z., Chen, Y., Wang, X., Liu, D., Liu, S., Yang, X., Jia, H., Guo, X., 2021. The occurrence of microplastic in Mu Us Sand Land soils in northwest China: different soil types, vegetation cover and restoration years. *J. Hazard. Mater.* 403, 123982. https://doi.org/10.1016/j.jhazmat.2020.123982

Dris, R., Gasperi, J., Mirande, C., Mandin, C., Guerrouache, M., Langlois, V., Tassin, B., 2017. A first overview of textile fibers, including microplastics, in indoor and outdoor environments. *Environ. Pollut.* 221, 453–458. https://doi.org/10.1016/j.envpol.2016.12.013

Dümichen, E., Eisentraut, P., Bannick, C.G., Barthel, A.K., Senz, R., Braun, U., 2017. Fast identification of microplastics in complex environmental samples by a thermal degradation method. *Chemosphere* 174, 572–584. https://doi.org/10.1016/j.chemosphere.2017.02.010

Elert, A.M., Becker, R., Duemichen, E., Eisentraut, P., Falkenhagen, J., Sturm, H., Braun, U., 2017. Comparison of different methods for MP detection: what can we learn from them, and why asking the right question before measurements matters? *Environ. Pollut.* 231, 1256–1264. https://doi.org/10.1016/j.envpol.2017.08.074

Feng, X., Wang, Q., Sun, Y., Zhang, S., Wang, F., 2022. Microplastics change soil properties, heavy metal availability and bacterial community in a Pb-Zn contaminated soil. *J. Hazard Mater.* 424, 127364.

Geyer, R., Jambeck, J.R., Law, K.L., 2017. Production, use, and fate of all plastics ever made. *Sci. Adv.* 3, e1700782.

Guo, J.J., Huang, X.P., Xiang, L., Wang, Y.Z., Li, Y.W., Li, H., Cai, Q.Y., Mo, C.H., Wong, M.H., 2020. Source, migration and toxicology of microplastics in soil. *Environ. Int.* 137, 105263. https://doi.org/10.1016/j.envint.2019.105263

Hartline, N.L., Bruce, N.J., Karba, S.N., Ruff, E.O., Sonar, S.U., Holden, P.A., 2016. Microfiber masses recovered from conventional machine washing of new or aged garments. *Environ. Sci. Technol.* 50(21), 11532–11538. https://doi.org/10.1021/acs.est.6b03045

He, D., Luo, Y., Lu, S., Liu, M., Song, Y., Lei, L., 2018a. Microplastics in soils: analytical methods, pollution characteristics and ecological risks. *Trends Anal. Chem.* 109, 163–172. https://doi.org/10.1016/j.trac.2018.10.006

He, L., Wu, D., Rong, H., Li, M., Tong, M., Kim, H., 2018b. Influence of nano-and microplastic particles on the transport and deposition behaviors of bacteria in quartz sand. *Environ. Sci. Technol.* 52(20), 11555–11563. https://doi.org/10.1021/acs.est.8b01673

Hintersteiner, I., Himmelsbach, M., Buchberger, W.W., 2015. Characterization and quantitation of polyolefin microplastics in personal-care products using hightemperature gel-permeation chromatography. *Anal. Bioanal. Chem.* 407(4), 1253–1259. https://doi.org/10.1007/s00216-014-8318-2

Huang, Y., Liu, Q., Jia, W., Yan, C., Wang, J., 2020. Agricultural plastic mulching as a source of microplastics in the terrestrial environment. *Environ. Pollut.* 260. https://doi.org/10.1016/j.envpol.2020.114096

Hurley, R.R., Lusher, A.L., Olsen, M., Nizzetto, L., 2018. Validation of a method for extracting microplastics from complex, organic-rich, environmental matrices. *Environ. Sci. Technol.* 52(13), 7409–7417. https://doi.org/10.1021/acs.est.8b01517

Ivleva, N.P., Wiesheu, A.C., Niessner, R., 2017. Microplastic in aquatic ecosystems. *Angew. Chem. Int. Ed.* 56(7), 1720–1739. https://doi.org/10.1002/anie.201606957

Jambeck, J.R., Geyer, R., Wilcox, C., Siegler, T.R., Perryman, M., Andrady, A., Narayan, R., Law, K.L., 2015. Plastic waste inputs from land into the ocean. *Science* 347(6223), 768–771.

Jia, W., Karapetrova, A., Zhang, M., Xu, L., Li, K., Huang, M., Wang, J., Huang, Y., 2022. Automated identification and quantification of invisible microplastics in agricultural soils. *Sci. Total Environ.* 844. https://doi.org/10.1016/j.scitotenv.2022.156853

Jin, Y., Cai, F.F., Song, C., Liu, G.Q., Chen, C., 2022. Degradation of biodegradable plastics by anaerobic digestion: morphological, micro-structural changes and microbial community dynamics. *Sci. Total Environ.* 834, 155167. https://www.sciencedirect.com/science/article/pii/S0048969722022604?via%3Dihub

Leifheit, E.F., Lehmann, A., Rillig, M.C., 2021. Potential effects of microplastic on arbuscular mycorrhizal fungi. *Front. Plant Sci.* 12, 626709. https://doi.org/10.3389/fpls.2021.626709

Li, J., Liu, H., Chen, J.P., 2018. Microplastics in freshwater systems: a review on occurrence, environmental effects, and methods for microplastics detection. *Water Res.* 137, 362–374. https://doi.org/10.1016/j.watres.2017.12.056

Li, J., Song, Y., Cai, Y., 2020a. Focus topics on microplastics in soil: analytical methods, occurrence, transport, and ecological risks. *Environ. Pollut.* 257, 113570. https://doi.org/10.1016/j.envpol.2019.113570

Li, W., Wufuer, R., Duo, J., Wang, S., Luo, Y., Zhang, D., Pan, X., 2020b. Microplastics in agricultural soils: extraction and characterization after different periods of polythene film mulching in an arid region. *Sci. Total Environ.* 749. https://doi.org/10.1016/j.scitotenv.2020.141420

Li, M., Liu, Y., Xu, G., Wang, Y., Yu, Y., 2021. Impacts of polyethylene microplastics on bioavailability and toxicity of metals in soil. *Sci. Total Environ.* 760, 144037.

Liang, Y., Lehmann, A., Yang, G., Leifheit, E.F., Rillig, M.C., 2021. Effects of microplastic fibers on soil aggregation and enzyme activities are organic matter dependent. *Front. Environ. Sci.* 9, 650155. https://doi.org/10.3389/fenvs.2021.650155

Liu, M., Lu, S., Song, Y., Lei, L., Hu, J., Lv, W., Zhou, W., Cao, C., Shi, H., Yang, X., He, D., 2018. Microplastic and mesoplastic pollution in farmland soils in suburbs of Shanghai, China. *Environ. Pollut.* 242, 855–862. https://doi.org/10.1016/j.envpol.2018.07.051

Liu, K., Wang, X., Song, Z., Wei, N., Li, D., 2020. Terrestrial plants as a potential temporary sink of atmospheric microplastics during transport. *Sci. Total Environ.* 742. https://doi.org/10.1016/j.scitotenv.2020.140523

Liu, Y., Shao, H., Liu, J., Cao, R., Shang, E., Liu, S., Li, Y., 2021. Transport and transformation of microplastics and nanoplastics in the soil environment: a critical review. *Soil Use Manage.* 37(2), 224–242. Blackwell Publishing Ltd. https://doi.org/10.1111/sum.12709

Liu, Xi, Lin, H., Xu, S., Yan, Y., Yu, R., Hu, G., 2022. Occurrence, distribution, and characteristics of microplastics in agricultural soil around a solid waste treatment center in southeast China. *J. Soils Sediments* 23(2), 936–946. https://doi.org/10.1007/s11368-022-03341-6

Lozano, Y.M., Aguilar-Trigueros, C.A., Onandia, G., Maaß, S., Zhao, T., Rillig, M.C., 2021a. Effects of microplastics and drought on soil ecosystem functions and multifunctionality. *J. Appl. Ecol.* 58, 988–996.

Mbachu, O., Jenkins, G., Pratt, C., Kaparaju, P., 2020. A new contaminant superhighway? A review of sources, measurement techniques and fate of atmospheric microplastics. *Water Air Soil Pollut.* 231(2). Springer Science and Business Media Deutschland GmbH. https://doi.org/10.1007/s11270-020-4459-4

Nizzetto, L., Futter, M., Langaas, S., 2016. Are agricultural soils dumps for microplastics of urban origin? *Environ. Sci. Technol.* 50(20), 10777–10779. American Chemical Society. https://doi.org/10.1021/acs.est.6b04140

Oliviero, M., Tato, T., Schiavo, S., Fern andez, V., Manzo, S., Beiras, R., 2019. Leachates of micronized plastic toys provoke embryotoxic effects upon sea urchin Paracentrotuslividus. *Environ. Pollut.* 247, 706–715. https://doi.org/10.1016/j.envpol.2019.01.098

Qi, R., Jones, D.L., Li, Z., Liu, Q., Yan, C., 2020. Behavior of microplastics and plastic film residues in the soil environment: a critical review. *Sci. Total Environ.* 703, 134722. https://doi.org/10.1016/j.scitotenv.2019.134722

Rillig, M.C., Lehmann, A., 2020. Microplastics in terrestrial ecosystems. *Science* 368(6498), 1430–1431.

Schon, N.L., Mackay, A.D., Gray, R.A., van Koten, C., Dodd, M.B., 2017. Influence of earthworm abundance and diversity on soil structure and the implications for soil services throughout the season. *Pedobiologia* 62, 41–47.

Sharma, S., Bhardwaj, A., Thakur, M., Saini, A., 2023. Understanding microplastic pollution of marine ecosystem: a review. *Environ. Sci. Pollut. Res.* 31, 41402–41445. Springer Science and Business Media Deutschland GmbH. https://doi.org/10.1007/s11356-023-28314-1

Shim, W.J., Hong, S.H., Eo, S.E., 2017. Identification methods in microplastic analysis: a review. *Anal. Methods* 9(9), 1384–1391. https://doi.org/10.1039/C6AY02558G

Su, L., Xiong, X., Zhang, Y., Wu, C., Xu, X., Sun, C., Shi, H., 2022. Global transportation of plastics and microplastics: a critical review of pathways and influences. *Sci. Total Environ.* 831. https://doi.org/10.1016/j.scitotenv.2022.154884

Tagg, A.S., Harrison, J.P., Ju-Nam, Y., Sapp, M., Bradley, E.L., Sinclair, C.J., Ojeda, J.J., 2017. Fenton's reagent for the rapid and efficient isolation of microplastics from wastewater. *Chem. Commun.* 53(2), 372–375. https://doi.org/10.1039/c6cc08798a

Van den Berg, P., Huerta-Lwanga, E., Corradini, F., Geissen, V., 2020. Sewage sludge application as a vehicle for microplastics in eastern Spanish agricultural soils. *Environ. Pollut.* 261. https://doi.org/10.1016/j.envpol.2020.114198

Wang, J., Liu, X., Li, Y., Powell, T., Wang, X., Wang, G., Zhang, P., 2019. Microplastics as contaminants in the soil environment: a mini-review. *Sci. Total Environ.* 691, 848–857. https://doi.org/10.1016/j.scitotenv.2019.07.209

Wang, K., Chen, W., Tian, J., Niu, F., Xing, Y., Wu, Y., Zhang, R., Zheng, J., Xu, L., 2022. Accumulation of microplastics in greenhouse soil after long-term plastic film mulching in Beijing, China. *Sci. Total Environ.* 828, 154544.

Weithmann, N., Möller, J.N., Löder, M.G.J., Piehl, S., Laforsch, C., Freitag, R., 2018. Organic fertilizer as a vehicle for the entry of microplastic into the environment. *Sci. Adv.* 4. https://doi.org/10.1126/sciadv.aap8060

Wright, S.L., Ulke, J., Font, A., Chan, K.L.A., Kelly, F.J., 2020. Atmospheric microplastic deposition in an urban environment and an evaluation of transport. *Environ. Int.* 136. https://doi.org/10.1016/j.envint.2019.105411

Wu, X., Liu, H., Guo, X., Zhang, Z., Zhang, J., Huang, X., 2023. Microplastic distribution and migration in soil, water and sediments in Caohai Lake under the different hydrological periods, Southwest China. *Sci. Total Environ.* 865, 161292.

Yang, J., Li, R., Zhou, Q., Li, L., Li, Y., Tu, C., Zhao, X., Xiong, K., Christie, P., Luo, Y., 2021. Abundance and morphology of microplastics in an agricultural soil following long-term repeated application of pig manure. *Environ. Pollut.* 272. https://doi.org/10.1016/j.envpol.2020.116028

Yang, X., Zhang, Z., Guo, X., 2023. Impact of soil structure and texture on occurrence of microplastics in agricultural soils of karst areas. *Sci. Total Environ.* 902. https://doi.org/10.1016/j.scitotenv.2023.166189

Yu, L., Zhang, J., Liu, Y., Chen, L., Tao, S., Liu, W., 2021. Distribution characteristics of microplastics in agricultural soils from the largest vegetable production base in China. *Sci. Total Environ.* 756, 143860.

Zhang, G.S., Liu, Y.F., 2018. The distribution of microplastics in soil aggregate fractions in southwestern China. *Sci. Total Environ.* 642, 12–20.

Zhang, B., Yang, X., Chen, L., Chao, J., Teng, J., Wang, Q., 2020. Microplastics in soils: a review of possible sources, analytical methods and ecological impacts. *J. Chem. Technol. Biotechnol.* 95(8), 2052–2068. John Wiley and Sons Ltd. https://doi.org/10.1002/jctb.6334

Zhang, Z., Wu, X., Zhang, J., Huang, X., 2022. Distribution and migration characteristics of microplastics in farmland soils, surface water and sediments in Caohai Lake, southwestern plateau of China. *J. Clean. Prod.* 366, 132912.

Zhao, T., Lozano, Y.M., Rillig, M.C., 2021. Microplastics increase soil pH and decrease microbial activities as a function of microplastic shape, polymer type, and exposure time. *Front. Environ. Sci.* 9. https://doi.org/10.3389/fenvs.2021.675803

Zhou, Q., Zhang, H., Fu, C., Zhou, Y., Dai, Z., Li, Y., Xue, Y., Fu, C., Tu, C., Luo, Y., 2018. The distribution and morphology of microplastics in coastal soils adjacent to the Bohai Sea and the Yellow Sea. *Geoderma* 322, 201–208. https://doi.org/10.1016/j.geoderma.2018.02.015

Zhou, B., Wang, J., Zhang, H., Shi, H., Fei, Y., Huang, S., Tong, Y., Wen, D., Luo, Y., Barceló, D., 2020. Microplastics in agricultural soils on the coastal plain of Hangzhou Bay, east China: multiple sources other than plastic mulching film. *J. Hazard Mater.* 388, 121814.

CHAPTER 12

Microplastics in Agricultural Soil

Amal I. Hassan and Hosam M. Saleh

12.1 INTRODUCTION

Microplastics (MPs), which are often characterized as plastic particles measuring less than 5 mm, have emerged as a significant environmental concern impacting several ecosystems (Priya et al., 2022). MPs have been detected widely in marine and freshwater ecosystems around the globe due to the widespread utilization of plastics in our everyday activities. There is a growing body of research indicating that MPs can have detrimental impacts on both aquatic and terrestrial creatures (Fu et al. 2020; Pothiraj et al., 2023). These effects can be attributed to several factors, including the tiny size of MPs, the additives used during plastic manufacture, and the presence of poisonous substances that become adsorbed onto the surface of the particles. MPs have the potential to serve as carriers for several contaminants, including persistent organic pollutants, heavy metals, and antibiotics. Once organisms have internalized these substances, they can induce a decrease in eating behavior, oxidative stress, genotoxic effects, retardation of development, and potentially even mortality. Production of plastic goods throughout the world has been on the rise recently, reaching 359 million tons in 2018 (30% of which came from China; Plastics Europe 2019; Yu et al., 2021; Sakthipriya 2022). However, the rate of plastic recycling is poor, and some plastic trash ends up in landfills or on fields. MPs are defined as plastic particles less than 5 mm in size. Soil organisms'

DOI: 10.1201/9781003487579-12

development and reproduction are put at risk by the ingestion of MPs because of their size, abundance, and pervasive distribution (Chen et al., 2021a). Soil biota at various trophic levels and even entire ecosystems may be negatively impacted by MP accumulation in the food web (Rose et al. 2023). The roots of plants may take in MPs through a crack-entry route and then transfer them to the shoots (Li et al. 2020a). MPs have a large specific surface area and strong hydrophobicity and can adsorb, serving as carriers of persistent organic pollutants (such as phthalates, polychlorinated biphenyls, polybrominated diphenyl ethers, organochlorine pesticides, and polycyclic aromatic hydrocarbons), heavy metals, antibiotics, and other pollutants, which may threaten soil systems and exert a negative impact on human health (Zhang et al. 2022a).

The infiltration of MPs into agricultural soil represents a burgeoning environmental and agronomic concern that has gained increasing recognition in recent years. MPs, minuscule plastic particles measuring less than 5 mm, have been studied extensively in the context of marine ecosystems, yet their presence and implications in agricultural soils have received less attention (Rafique et al. 2020).

The prevalence of MPs in agricultural soil arises from multiple sources. Primary MPs are intentionally manufactured at the micro-scale for various applications, including agricultural products like plastic mulch films, and are used in other sectors, such as personal care products. On the other hand, secondary MPs result from the degradation and fragmentation of larger plastic items that have entered the environment (Raju et al. 2023).

Microplastics infiltrate agricultural soils through several routes, including the application of plastic mulch, the breakdown of plastic waste in landfills, and even atmospheric deposition. This introduction marks the pivotal entry point for MPs into the terrestrial environment, where their presence and persistence in the soil matrix pose multifaceted challenges (Wong et al. 2020).

The consequences of MPs in agricultural soil are intricate and manifold. These tiny particles have the potential to modify the physical and chemical properties of the soil, influencing factors such as water retention, compaction, and aeration. Such alterations can have detrimental effects on soil health and crop growth, ultimately affecting agricultural productivity (Raju et al. 2023).

Additionally, MPs may serve as carriers for contaminants, absorbing and transporting harmful substances into the soil, where they can affect both plant and soil health. Furthermore, there is growing concern about how MPs may disrupt the soil's delicate ecosystem, potentially impacting beneficial microorganisms that contribute to nutrient cycling and overall soil fertility (Zhu et al., 2022).

This introductory overview highlights the urgent need for further research, understanding, and practical solutions to address the pervasive issue of MPs in agricultural soil. As we delve into the subsequent sections of this review, we will explore emerging strategies, technologies, and interdisciplinary approaches aimed at mitigating the impact of MPs on soil health and food production.

Table 12.1 provides a concise summary of the key aspects related to the pervasive issue of MPs in agricultural soil, including its sources, impact, research methods, mitigation strategies, and the need for awareness and future challenges.

TABLE 12.1 Comprehensive Overview of the Pervasive Issue of Microplastics in Agricultural Soil

Aspect of Issue	**Description**
Nature of contaminants	Microscopic plastic particles, including primary microplastics (e.g., plastic mulch) and secondary microplastics (resulting from plastic degradation).
Sources of microplastics	• **Primary Sources:** agricultural practices, plastic waste, and various industries • **Secondary Sources:** weathering, fragmentation, and breakdown of larger plastic items
Infiltration routes	• Application of plastic mulch in farming • Microplastics from plastic waste in landfills • Atmospheric deposition of airborne microplastics
Impact on soil	• Alters physical properties (water retention, compaction) • Potential disruption of the soil microbiome • Adsorption and transport of contaminants
Crop productivity	Reduced crop yields due to soil health deterioration
Ecological implications	Microplastics entering the food chain through crops and soil organisms
Research methods	Innovative techniques include spectroscopy, DNA-based analysis, isotopic tracing, AI, biomarker analysis, and advanced soil sampling methods
Mitigation strategies	Sustainable farming practices, biodegradable alternatives, waste management solutions, and soil remediation technologies are being explored
Need for awareness	Raising awareness about microplastics in agricultural soil and promoting interdisciplinary research to address the issue
Future challenges	Developing practical, cost-effective methods for mitigating microplastics and restoring soil health

12.2 INNOVATIVE RESEARCH METHODS FOR ASSESSING MICROPLASTIC CONTAMINATION AND ITS IMPACTS

Microplastic have been discovered in a variety of marine and terrestrial environments (Booth and Sørensen 2022), (Zhang et al. 2021). MPs have the potential to negatively impact organisms in a variety of ways because to their tiny sizes, high specific surface areas, and significant propensity to load additional contaminants (Zhang and Chen 2020).

Effective techniques of extracting MPs from environmental samples are primarily responsible for the measurement of MPs, which is necessary for determining the contamination level of MPs (Yusuf et al. 2022). Microplastic contamination in various environments, including agricultural soil, has garnered significant attention due to its potential repercussions on ecosystems and human health (Ullah et al. 2021). To address this pervasive issue, researchers have been diligently working to develop innovative research methods that enable a more comprehensive assessment of microplastic contamination and its impacts (Figure 12.1).

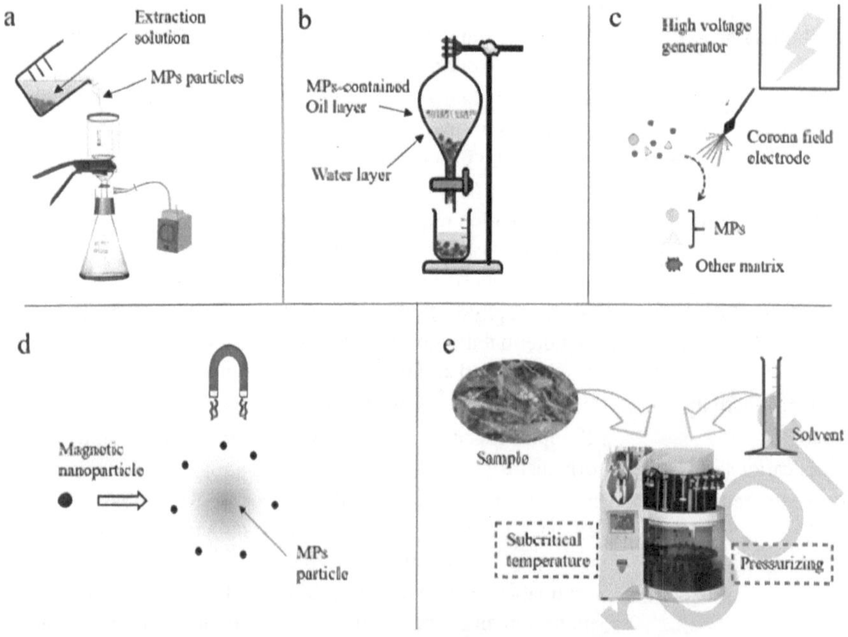

FIGURE 12.1 Schematic diagrams of five methods for separating microplastics (MPs) from complex environmental matrices. (a) density separation, (b) oil separation, (c) electrostatic separation, (d) magnetic separation, (e) solvent extraction separation. (b–d refer to Mani et al. (2019), Felsing et al. (2018), Grbic et al. (2019), and He et al. (2018), Elsevier, Permission NO. 5646380182068.)

Key techniques, their principles, advantages, limitations, and applications across soil types are discussed. Future research needs for standardized operating protocols and method optimizations are highlighted. The enormous environmental impact and possible health hazards associated with MPs need extensive research and development of appropriate technologies to identify and monitor MPs in diverse environmental media (Ivleva et al. 2017). To better regulate MPs, encourage the use of alternatives to plastics, and set up waste management policies that can be defended, it would be helpful if analytical methodologies could be standardized to offer more trustworthy and robust data.

With Asia accounting for 50% of global plastic output in 2021, the region was followed by North America and Europe. The three largest end-use markets were packaging (44%), construction (18%), and the automobile sector (8%), in that order (Degli Esposti et al. 2021; Johansen et al. 2023). Polypropylene (19.3%), low-density polyethylene (14.4%), polyvinyl-chloride (12.9%), high-density polyethylene (12.5%), polyethylene terephthalate (PET) (6.2%), polyurethane (PUR) (5.5%), and polystyrene (PS) (5.3%) were the most commonly produced polymers; 90.2% of the resins used were derived from fossil fuels. Roughly 9.8% of all plastics produced are circular plastics, which include recycled and biobased/bio-attributed plastics (Johansen et al. 2023). In 2015, it was estimated that 8.3 billion tons of plastics were manufactured worldwide, with 6.3 billion tons becoming waste. Of this total, only 9% was recycled, 12% was recovered for energy purposes, and the remaining 79% was either dumped in landfills or escaped from formal waste collection systems and accumulated in terrestrial or marine environments as single-use packaging (Zhang and Liu 2018). Separate garbage collection allows for recycling rates that are 13 times greater than mixed waste collection systems, per Plastics Europe (Horodytska et al. 2019).

Each polymer's environmental behavior may be affected by its unique combination of physical features. For instance, the density of a polymer is typically connected to the particle's propensity to float or settle in sediments in aquatic environments (Bellasi et al. 2020). Predicting the fate of polymers in the environment is difficult even when physical qualities are well known (Bellasi et al. 2020). This is because buoyant particles of PP and PE can sink and be held inside sediments. Plastics have been found in a wide variety of environments, including sediments in deep oceans, an underwater canyon, Arctic sea ice, and enormous accumulations at the surface waters, such as the Great Pacific Garbage Patch (Bellasi et al. 2020). In addition, plastiglomerates (Figure 12.1), which resemble solid rocks, may be formed when plastics fuse with volcanic rocks, sediments, and organic components (De-la-Torre et al. 2022). Chemical additives have also been demonstrated to seep out throughout the product's lifetime (He et al. 2018). In addition, plastics can pick up various chemical pollutants (including metals, PCBs, PAHs, and pesticides) from the environment thanks to their hydrophobic surface (He et al. 2018; Pozo et al. 2020). It is possible for their toxicity to live creatures to be amplified by synergistic interactions (Pozo et al. 2020; Zimmermann et al. 2020). Plastic particles, if ingested by biota, might behave as Trojan horses, transporting other toxins up and down the food chain (Zimmermann et al. 2020). Plastic debris has the potential to undergo degradation through various mechanisms, including physical, chemical, and biological processes. These drivers, such as ultraviolet radiation, wind, or water erosion, among others, can cause the plastic litter to break down into smaller fragments. It is important to note that the specific terminology used to describe these size categories may vary among different sources (Ter Halle and Ghiglione 2021). The issue of plastic debris was recognized as a significant environmental concern during the 1970s. However, it was not until 2004 that the word "microplastic" was officially coined and used to describe this particular kind of plastic waste (Tiller et al. 2022). At now, there is widespread awareness and concern over the environmental pollution resulting from MPs. This topic is recognized as a developing problem of international significance, as evidenced by its inclusion in public discourse and scholarly literature (Henderson and Green 2020). MPs may be categorized into two distinct groups: primary MPs, which are intentionally manufactured for industrial applications, such as the usage of polyethylene (PE) microbeads as exfoliating agents in personal care products; and secondary MPs, which are formed by the breakdown of bigger objects (Kawecki and Nowack 2019). MPs are representative of the predominant proportion of MPs that are typically seen in their native habitat (Hüffer et al. 2019).

Several recent studies have advanced novel analytical approaches for extracting and quantifying MPs from soil matrices. Zhang et al. (2020a) developed a density separation method using sodium polytungstate (SPT) solutions to isolate MPs from agricultural soil samples. Adjusting the density of SPT solutions enabled optimized microplastic recovery rates up to 95% for polyethylene particles spiked into soil. Fluorescent staining using Nile red coupled with fluorescence microscopy allowed visual confirmation of MPs separated from soil particles. A key advantage of this SPT density separation approach is the ability to process large soil volumes for more robust microplastic quantification.

Dehaut et al. (2019) applied a similar density separation technique to isolate MPs from marine sediments. They used sodium iodide solutions tuned to densities of 1.6 g/cm^3 to extract microplastic particles. Micro-Fourier transform infrared (μFTIR) spectroscopy analysis produced chemical "fingerprints" to identify microplastic polymer types. Their method yielded 82%–99% recovery rates for spiked samples. However, limitations included potential toxicity of NaI solutions and inability to detect smaller nanoplastic particles.

Fuller and Gautam (2016) developed an elutriation method to extract MPs from wastewater sludge samples. Their approach used agitation and rising air bubbles within an aqueous slurry to liberate low-density MPs from the sludge matrix. They reported 82%–92% recovery rates for spiked samples. A disadvantage was increased processing times required for the elutriation. The authors highlighted the need for a standard elutriation protocol optimized across sample types.

Felsing et al. (2018) developed a new thermal degradation method using thermogravimetric analysis coupled to mass spectrometry (TGA-MS). This detected MPs extracted from field soil samples based on their thermal decomposition patterns. Polymer types including PE, PP, and PET were identified. Hildebrandt et al. (2022) optimized a novel laser direct infrared (LDIR) chemical imaging technique. It enabled automated identification and quantification of MPs from soil using focal plane array detectors. Advantages included speed, cost, and minimal sample preparation.

Hurley et al. (2020) combined FTIR imaging with a machine learning model for automated counting and classification of MPs isolated from agricultural soils. The model was trained on FTIR data from reference polymers. This high-throughput approach could accelerate analysis.

Table 12.2 provides a comprehensive comparison of various methods for the analysis of MPs in soil, highlighting their principles, advantages, limitations, and specific applicability. Researchers can use this table as a reference to choose the most suitable method based on their research goals and the characteristics of the MPs and environmental samples being analyzed.

12.3 CUTTING-EDGE SUSTAINABLE FARMING PRACTICES TO PREVENT MICROPLASTIC ACCUMULATION

One of the innovative methods is biodegradable mulching materials to combat microplastic contamination involves replacing traditional plastic mulch films with biodegradable alternatives. Research, such as a study by Mason et al. (2018), has shown that biodegradable mulches made from

TABLE 12.2 Comparative Analysis of Microplastic Detection Methods in Soil

Method	Principle	Pros	Cons	Applicability	References
Density separation	Based on density differences between MPs and environmental materials	Widely applicable, effective in many scenarios	Less efficient for very small MPs with similar densities to natural materials	General MPs in soil	He et al. (2018), Shao et al. (2022)
Oil Separation	Exploits the oleophilic property of plastics, with plastics adhering to oil	Effective for hydrophobic plastics, straightforward	Less efficient for hydrophilic plastics, may not work for all samples	Hydrophobic MPs	Wolf et al. (2018), Rani et al. (2023)
Electrostatic separation	Separates MPs based on electrical charge differences	Efficient for MPs with distinct electrostatic properties	Requires specialized equipment, not universally effective	Specific MPs with unique electrostatic properties	Fu et al. (2022), Wang et al. (2023)
Magnetic separation	Separates magnetic or magnetizable MPs using magnets	Rapid and efficient for magnetic MPs	Limited to magnetic MPs	Magnetic or magnetizable MPs	Alves et al. (2019)
Solvent extraction	Dissolves the sample in a solvent, separating MPs through filtration	Effective for specific sample types	Applicability depends on the solubility characteristics of MPs	Specific MPs with known solubility properties	Elert et al. (2017)

(*Continued*)

TABLE 12.2 (*Continued*) Comparative Analysis of Microplastic Detection Methods in Soil

Method	Principle	Pros	Cons	Applicability	References
Spectroscopy and imaging	Utilizes spectroscopy (e.g., Raman, FTIR) and imaging (e.g., SEM, micro-CT) for identification	High precision in identifying and characterizing MPs	May require specialized equipment and expertise	General MPs, especially when their identification is essential	Querido et al. (2017)
DNA-based analysis	Uses DNA markers to detect and quantify MPs	Sensitive and reliable for identifying MPs	Limited by the presence of specific genetic markers in MPs	Specific MPs with known genetic markers	Melchionda et al. (2022)
Isotopic tracing	Distinguishes MPs based on isotope ratios (e.g., ^{13}C)	Useful for identifying the origin of carbon in MPs	Limited to the isotope characteristics of MPs	MPs with distinct isotope characteristics	Mason et al. (2018), Barrett et al. (2023)
Machine learning and AI	Applies AI algorithms to process large datasets of soil samples	Automates detection and quantification, identifies patterns	Requires training and data for AI models	Large-scale analysis of diverse MPs	Lambert and Wagner (2016), Feeny et al. (2020)
Biomarker analysis	Uses biomarkers in soil organisms as indicators of MP exposure	Offers insights into the presence and effects of MPs in soil ecosystems	Limited to biomarkers present in specific organisms	MPs in soil ecosystems with relevant organisms	Ziajahromi et al. (2017), Suman et al. (2021)

plant-based materials like cornstarch and cellulose are effective in weed control and moisture retention while avoiding the microplastic pollution associated with traditional plastics.

Biodegradable mulching materials are typically made from natural sources such as starch, cellulose, biopolymers, or other organic materials. These materials are selected for their ability to decompose into harmless byproducts, leaving no microplastic residues behind.

Just like traditional plastic mulches, biodegradable alternatives are used to control weeds, conserve soil moisture, and regulate soil temperature. They offer the same benefits while minimizing environmental impacts. Biodegradable mulching materials are designed to break down through natural processes, such as microbial activity, UV degradation, and hydrolysis. This ensures that they return to the environment as non-toxic substances without posing a threat to soil health or wildlife (Samir et al. 2022).

The primary advantage of biodegradable mulches is their ability to prevent the creation of MPs. Traditional plastic mulch films degrade into tiny plastic particles, while biodegradable mulches do not, reducing the risk of microplastic contamination in agricultural soil (Qi et al. 2020).

As biodegradable materials break down, they contribute organic matter to the soil, enhancing its structure and fertility. This, in turn, promotes healthier plant growth and minimizes the need for additional plastic-based agrochemicals (El-Beltagi et al. 2022).

By replacing traditional plastic mulches with biodegradable alternatives, the generation of plastic waste is significantly reduced. This aligns with the principles of sustainable and environmentally responsible farming.

The European standard EN 17033 outlines requirements and test methods for biodegradable mulch films. It provides guidelines for manufacturers to ensure the quality and biodegradability of these materials (Degli Innocenti and Breton 2020).

Researchers worldwide are conducting field trials to assess the performance and environmental impact of biodegradable mulching materials under various conditions. These trials help refine the materials and their applications, promoting their adoption in agriculture (Moshood et al. 2022). Biodegradable mulching materials are a prime example of how innovative farming practices are contributing to sustainable agriculture and environmental conservation. By choosing these materials, farmers not only enhance soil health but also play a vital role in reducing the spread of MPs, making agriculture more environmentally responsible.

Agroecology and regenerative farming practices prioritize the creation of holistic, self-sustaining ecosystems within farmlands. By focusing on natural nutrient cycling and diverse crop rotations, these methods reduce the reliance on synthetic inputs, including plastic-based agrochemical containers and plastic mulches (Schettini et al. 2022). The Rodale Institute's long-term trial on regenerative agriculture showcases its potential to restore soil health, sequester carbon, and mitigate the introduction of MPs. Agroecological and regenerative farming practices offer a holistic approach to create sustainable agroecosystems while eliminating reliance on plastic-based inputs that contribute to microplastic pollution. The core principles of agroecology include biodiversity, synergies, efficiency, resilience, recycling, and co-creation of knowledge (Vicente-Vicente et al. 2023). Practices that align with these principles, such as intercropping, crop rotations, organic soil amendments, and integration of livestock, aim to develop self-sufficient systems with closed nutrient loops.

Regenerative agriculture builds upon agroecology to regenerate degraded soils, maximize biodiversity, improve watersheds, and strengthen ecosystem services that industrial agriculture disrupts (Tittonell et al. 2022). A prominent example demonstrating the potential of regenerative agriculture is the Rodale Institute Farming Systems Trial, started in 1981 to compare conventional and organic crop production. As the trial does not rely on synthetic fertilizers or pesticides applied in plastic containers, it mitigates microplastic accumulation from these sources. The practice of cover cropping and complex rotations builds soil organic matter (SOM), further reducing the need for plastic mulching. After 40 years, the organic systems had 40% higher soil carbon levels and required 45% less energy (Reichelt et al., 2023). Global adoption of regenerative agriculture could therefore significantly reduce the microplastic burden while regenerating farmland ecosystems.

Cutting-edge soil management techniques, like biochar application and precision soil testing, improve soil health and nutrient retention. Biochar, a type of charcoal produced from organic waste, has been studied for its ability to enhance soil structure and organic matter content while reducing the introduction of MPs. Research by Ren et al. (2021) highlights biochar's potential to mitigate the risks associated with microplastic accumulation. Biochar is a form of charcoal produced through the pyrolysis of organic waste materials, such as wood chips, agricultural residues, or other biomass. It is a stable carbon-rich substance that, when added to soil, can enhance its physical, chemical, and biological properties. Biochar has been recognized for its ability to improve soil structure, increase water and nutrient retention, and enhance overall soil health (Ren et al. 2021).

Precision soil testing involves the precise analysis of soil properties, including nutrient content, pH levels, and texture. This technique allows farmers to tailor their soil management strategies to the specific needs of their fields, reducing the overuse of fertilizers and other agrochemicals (Mohamed et al. 2018).

Biochar's remarkable properties make it an innovative solution to prevent microplastic accumulation in agricultural soil. Biochar improves soil structure by increasing porosity and water-holding capacity. This aids in preventing MPs from becoming lodged in compacted soils, allowing them to be flushed away during rain events instead.

Biochar fosters beneficial microbial activity in the soil. These microbes can break down organic matter, which includes biochar itself, reducing the risk of microplastic persistence in the soil.

Biochar has a high surface area and is known for its adsorption properties. It can adsorb certain organic compounds and, potentially, MPs. This reduces the concentration of MPs in the soil, limiting their impact on soil health.

The study conducted by Zhu et al. (2019) investigated the efficacy of biochar in reducing the presence of MPs in soil. The researchers examined the impact of biochar on microplastic retention and distribution in the soil (Zhu et al. 2019). Their findings suggested that biochar has the potential to mitigate the risks associated with microplastic accumulation in agricultural soils.

The study demonstrated that biochar, when added to soil, altered the physical properties of the soil matrix, making it less conducive to microplastic retention. This resulted in reduced microplastic concentrations in the soil.

The use of biochar and precision soil testing offers environmentally friendly and sustainable solutions to soil management. By enhancing soil health and mitigating the introduction of MPs, these techniques promote agricultural sustainability and help protect the environment. Furthermore, they align with the principles of responsible farming, resource conservation, and the prevention of microplastic contamination, making them valuable tools for the future of agriculture. More recent studies have further investigated biochar for microplastic immobilization (Miao et al. 2023). Chahal et al. (2023) reported 70%–80% decreased leaching of PET MPs from soil columns with 5% biochar additions. The biochar's porous structure and charge effects were responsible for enhanced microplastic retention. Zhang et al. (2020b) found that wheat straw biochar incorporated at just 0.5% into green soil reduced uptake of microplastic polyethylene into Vicia faba plants by up to 42%. Biochar shows particular promise as a sustainable tool to restrict plant exposure and leaching losses of MPs from agricultural soils.

Precision nutrient management based on site-specific soil testing and variable rate fertilization also reduces overapplication of synthetic fertilizers in plastic containers. Hou et al. (2021) used on-the-go near-infrared sensors to rapidly characterize soil nutrient variability within fields. This guided variable rate nitrogen and phosphorus applications, achieving 10%–20% fertilizer reductions without impacting wheat yields (Trontelj Ml and Chambers 2021). Expanding such precision soil analysis and fertilization techniques can improve nutrient use efficiency and lower plastic packaging waste.

Urban agriculture is witnessing the growth of closed-loop systems, where waste materials are recycled and repurposed within urban farming operations. These systems reduce the generation of plastic waste and its entry into agricultural soil. The "Growing Underground" project in London, which cultivates crops in repurposed underground tunnels using hydroponics and light emitting diode (LED) lighting, is an example of sustainable urban agriculture minimizing plastic waste. Urban agriculture systems are adopting circular economy designs to eliminate plastic waste and its contamination of growing soils. Closed-loop systems recycle organic waste streams into valuable agricultural inputs to minimize reliance on plastic packaging and mulching films. For example, integrated aquaponic systems utilize fish waste nutrients to fertilize hydroponically grown crops, while livestock manure can be composted and applied to urban farms (Weidner et al. 2019).

A leading example of a closed-loop urban farm is Growing Underground in London, which grows salad crops hydroponically under LED lighting in repurposed underground tunnels (Oh and Lu 2023). This soilless setup eliminates plastic contamination risks while producing zero waste as nutrients are recirculated in the water system.

Vertical farming is another promising closed-loop approach. Yip et al. (2022) modeled a conceptual large-scale vertical farm that utilized waste-to-energy and renewable energy systems, nutrient recycling, and rainwater harvesting to operate independently of plastic imports. Though challenges exist for widespread adoption, such circular designs exemplify innovative farming without plastic waste.

Plasticulture, which involves the use of plastic materials for crop protection and irrigation, is a significant source of MPs in agriculture. Researchers are exploring alternatives like plant-based biofilms and microbial coatings that can provide the same protective benefits while being biodegradable. The work of scientists like Gregory et al. (2021) highlights the potential of such alternatives. Engaging citizens and farmers in microplastic research through citizen science initiatives, such as the "Plastic & Bay" project in the Chesapeake Bay region, helps gather data on microplastic contamination. These efforts raise awareness, promote responsible plastic use, and contribute to the development of sustainable farming practices (Bikker et al. 2020). Community scientists follow established protocols to quantify accumulation rates across watersheds. The project has helped identify hotspots of microplastic pollution, track sources, and educate the public.

Similar community-driven initiatives have expanded globally. The "Global Microplastics Count" brings together citizens worldwide to sample freshwater MPs using an app that guides and registers sampling (Camargo et al. 2022). From 2019 to 2021, volunteers across six continents provided over 1,000 water samples, greatly expanding the MPs dataset.

Such participatory research could be applied to assess MPs in agricultural soils through on-farm citizen monitoring. This raises awareness among farmers, provides data to guide interventions, and empowers citizens to engage with sustainable farming solutions. As Martyn and Galbraith (2022) conclude, participatory microplastic research is crucial for bridging knowledge gaps while encouraging environmentally responsible behaviors.

These innovative methods and studies collectively point to a future where agriculture embraces sustainability as a means to prevent microplastic accumulation in soil. By prioritizing eco-friendly practices, minimizing plastic use, and enhancing soil health, agriculture can play a crucial role in mitigating this global environmental challenge. As farmers, researchers, and policymakers continue to explore these approaches, we move closer to a future where resilient food systems thrive, and the detrimental impacts of MPs are mitigated.

12.4 TECHNOLOGICAL ADVANCEMENTS IN WASTE MANAGEMENT FOR EFFECTIVE MICROPLASTIC CONTROL

The effective management of plastic waste is critical to prevent further accumulation of MPs in the environment. Emerging technologies for improved waste sorting, recycling, and upcycling show promise in diverting plastic from landfills and waterways (Martyn and Galbraith 2022). For example, automated laser sorting can detect the chemical composition of plastics for separation by polymer types, while hyperspectral imaging can identify even small microplastic fragments in the recycling stream. Chemical recycling through processes like pyrolysis breaks down unrecyclable plastic into oils and chemical feedstocks. Plastic waste can also be upcycled into 3D printing filaments, an advancement pioneered by companies like Refil and Deltec. However, waste management requires systemic solutions beyond technological innovations. Reducing unnecessary plastic usage through bans on single-use items and plastic-free product designs is equally important. A multifaceted approach combining responsible production, consumption, and advanced waste technologies is key to curtailing microplastic pollution (Koelmans et al. 2017).

New technologies, such as compatibilization, have arisen to lessen the financial burden of upcycling plastic trash. Most compatibilizers are multicomponent polymers having specialized thermodynamic interactions with the immiscible plastics in a specific combination (Zhao et al. 2022b). These polymers may have a variety of topologies, including graft, block, and random copolymers. Adding a chemical (the compatibilizer) designed specifically for an incompatible mixture of plastics or materials may improve their compatibility and stability (Eagan et al. 2017). Recycling mixed plastics into goods of similar or better value is possible with the help of compatibilizers, which may also be employed to reduce costs.

Depolymerization is a cutting-edge technique for recycling plastic scraps. Developing selective catalysts to depolymerize plastics into monomers or building blocks for manufacturing new polymers (Clark & Shaver, 2024) may be a solution to plastic waste concerns since the bulk of monomers used for plastic synthesis are obtained from fossil fuels (Jia et al. 2016).

Since depolymerization has not yet been proved on a large, industrial scale (Jiao et al. 2020), it is important to think about the long-term economic sustainability of the approach before committing to it. Depolymerization is a cutting-edge technique for recycling plastic scraps. Developing selective catalysts to depolymerize plastics into monomers or building blocks for manufacturing new polymers (Hussain et al. 2022b) may be a solution to plastic waste concerns since the bulk of monomers used for plastic synthesis are obtained from fossil fuels.

Since depolymerization has not yet been proved on a large, industrial scale (Mohanty et al. 2022), it is important to think about the long-term economic sustainability of the approach before committing to it.

Using biodegradable polymers like poly(3-hydroxybutyrate) (PHB) to deal with the problem of where packaging plastics end up beyond their useful life (Hussain et al. 2022b) is one last option. Biodegradable polymers have seen some development, but they lack the mechanical and other physical qualities necessary to fully replace traditional plastics at this time. In comparison to common commercial polymers like polyethylene, PHB has worse mechanical qualities and chemical recyclability (recycling efficiency and selectivity). As a result, polycarbonates (PCs), polyurethanes (PURs), and polyesters—three nonbiodegradable, chemically recyclable polymers—are widely used.

However, there has been no systematic research on the impact of plastic waste streams with varying physicochemical qualities. Overall, the most effective methods for recycling and upcycling plastic trash have not yet been determined due to a lack of analysis.

Because of the variety of things that may be recycled or upcycled, there is no one-size-fits-all approach that will be successful. Plastic upcycling encompasses a wide range of plastic materials, including post-consumer plastics (such as bottles and packaging) and industrial plastic waste (like manufacturing scraps). These diverse feedstocks provide ample opportunities for innovation in converting plastics into valuable products. Various methods and technologies are employed for plastic upcycling. These include mechanical recycling, chemical recycling, and advanced processing methods. Each technique offers unique advantages and is tailored to the specific plastic waste stream. By upcycling plastic waste, there is a reduced likelihood of plastic degradation into MPs. Upcycled products are often designed to be more durable, with longer life spans, reducing the need for frequent replacement and, therefore, less plastic waste generation. Plastic waste can be transformed into building materials, such as construction blocks or roof tiles. These materials are not only eco-friendly but also lightweight and durable, offering a sustainable alternative to traditional construction materials.

One crucial kind of chemical recycling is catalytic depolymerization to monomers, often known as monomer chemical recycling (Chen et al. 2021b). Catalytic depolymerization of original monomers permits the creation of recovered plastic with qualities that are identical to virgin plastic if pure monomers can be acquired, setting it apart from mechanical recycling. To create new processes toward value-added chemicals, fuels, and materials, upcycling relies heavily on both monomers and partly deconstructed polymers as feedstocks (Thiounn and Smith 2020).

The existence of weak chemical bonds in the polymers' structural backbone is essential for their catalytic depolymerization to monomers. Due to the ease with which esters may be chemolyzed, polyesters, and in particular PET, are now the primary focus of catalytic depolymerization to monomers.

Many different approaches to catalytic depolymerization have been studied for their potential to break down PET into its component monomers. Glycolysis with excess glycols (such as ethylene glycol [EG], diethylene glycol [DEG], propylene glycol [PG], polyethylene glycol [PEG], 1,4-butanediol, and hexylene glycol) toward bis(hydroxylethylene) terephthalate (BHET or other corresponding esers) via a transesterification reaction; alcoholysis with alcohol (methanol, ethanol, etc.) to depolymerization of other polyester polymers like polybutylene terephthalate (PBT) is also possible using the same techniques (Ghosal and Nayak 2022). The most common technique for catalytically depolymerizing PURs is glycolysis. Commercialization of these processes is hindered by many factors, including their severe reaction conditions, the need for large amounts of solvents and depolymerizing agents, and the production of unwanted by-products.

Some fashion brands are incorporating upcycled plastics into their clothing lines. This includes recycled polyester made from plastic bottles or upcycled ocean plastics, which are turned into stylish and sustainable apparel.

Upcycled plastic materials are being used as 3D printing filament, enabling the creation of various objects and prototypes using recycled plastics.

Upcycled plastics can be utilized in the production of eco-friendly packaging materials, reducing the demand for new plastic packaging and potentially mitigating microplastic contamination.

While plastic upcycling holds great promise, it also faces challenges, such as the need for efficient collection and sorting of plastic waste, as well as the development of cost-effective upcycling processes. Additionally, upcycled products must meet quality and safety standards to gain consumer trust and market acceptance.

Plastic upcycling plays a vital role in microplastic control by preventing plastic waste from entering the environment and degrading into MPs. By upcycling plastics into higher-value products, the overall volume of plastic waste is reduced, contributing to cleaner ecosystems and healthier agricultural soil. This approach aligns with the principles of a circular economy, emphasizing resource conservation, sustainability, and environmental responsibility.

Plastic upcycling is a testament to human innovation and the potential for transforming a global waste challenge into an opportunity for positive change. It highlights the creative and practical solutions that can be harnessed to combat microplastic contamination and reduce the environmental footprint of plastic waste.

Polymers that can be broken down chemically have been developed in recent years to help with the problem of what to do with used plastics.

The quality of the polymer may decrease, and residual contaminants can be generated after numerous cycles of mechanical reprocessing, as previously documented (Eriksen et al. 2018). However, by a process of depolymerization, the precursor building blocks may be recovered through chemical recycling. On the other hand, polymers that can be quickly and selectively depolymerized back to monomers often need low-temperature (e.g., −40°C) polymerization routes and lack the mechanical/physical qualities necessary for practical usage.

Polymers that may be recycled chemically often have poor selectivity during depolymerization, indicating that a wide variety of products may be generated and need to be separated and purified. Due to its limited selectivity, polymer depolymerization and subsequent polymerization back to the original product is not easily achieved. The capacity to depolymerize a polymer typically comes at the expense of other desirable qualities. Because of this, many virgin (or chemically non-recyclable) polymers have superior performance and utility than chemically recyclable polymers. It's also not possible to extract monomers from complicated mixes without a significant investment of time, money, and energy. It is difficult to depolymerize plastics in a short reaction time at low temperatures and with a high tolerance for additives. But these difficulties have prompted scientists to seek for answers. Polymer systems based on γ-butyrolactone via a trans-ring fusion with repeatable chemical recyclability were designed by Zhu et al. (2018), and the resulting polymer can be synthesized at room temperature in the absence of solvents. Recovering the monomers from common additives, even in mixed plastic waste streams, was discovered by Christensen et al. (2019) using plastics polymerized through dynamic (reversible) covalent diketoenamine linkages. It is possible to employ recovered monomers to create both identical polymer formulations with the same performance and alternative polymer formulations with unique features. The value of poly(diketoenamine) plastic remains intact throughout its entire life cycle.

A sustainable society must have a system like a closed-loop polymer life cycle. These reactive covalent polymers may 1 day be used in lieu of non-recyclable thermosets (Christensen et al. 2019).

Circular economy models represent a paradigm shift in how we approach resource management and waste reduction (Pichlak and Szromek 2022). These models prioritize reducing, reusing, and recycling materials as core principles to minimize waste generation and the potential release of MPs into the environment. By encouraging sustainable product design, responsible consumption, and innovative recycling methods, circular economy models provide a holistic and effective approach to controlling MPs and promoting environmental sustainability (Pichlak and Szromek 2022). Circular economy models emphasize designing products that are built to last. This discourages single-use plastics, which are a significant source of MPs. Reducing excessive and non-recyclable packaging materials not only decreases waste but also limits the introduction of MPs into the environment.

Promoting the reuse of products and packaging, such as refillable containers and durable goods, reduces the need for disposable plastics and limits microplastic contamination.

Circular economy models advocate for efficient and high-quality recycling systems. These systems ensure that plastics are collected, sorted, and recycled effectively, reducing the risk of plastics breaking down into MPs. Research and investment in advanced recycling methods, including chemical recycling and upcycling, contribute to reducing plastic waste and the associated microplastic pollution (Nikiema and Asiedu 2022).

By reducing the use of single-use plastics, promoting durable goods, and efficient recycling, circular economy models minimize the creation of MPs at the source. Circular economy models promote the efficient use of resources and reduce the demand for new plastics, thereby decreasing the potential for microplastic pollution (Zhou et al. 2022). These models encourage consumers to make conscious choices, reduce waste, and favor products that can be reused or recycled, further mitigating microplastic contamination. Manufacturers are encouraged to design products with recyclability and reusability in mind. This shift in product design contributes to the reduction of waste and MPs. The implementation of closed-loop recycling systems ensures that plastic materials are continuously cycled back into production, limiting their dispersion into the environment.

While circular economy models offer a promising solution, they are not without challenges. These include the need for comprehensive infrastructure, consumer behavior change, and regulatory support. Additionally, the effectiveness of these models relies on the cooperation of various stakeholders, from producers and consumers to waste management authorities. Many countries have implemented Extended Producer Responsibility (EPR) programs, which hold manufacturers responsible for the entire life cycle of their products, from design to disposal (Diggle and Walker 2020). This encourages producers to design products with reusability and recyclability in mind. Companies are increasingly adopting plastic-free packaging solutions, opting for materials that are easily recyclable or biodegradable, reducing the risk of microplastic contamination. Circular economy models require consumer participation. Public awareness campaigns and education initiatives promote responsible consumption and recycling.

Waste sorting technologies are becoming more sophisticated. Automated systems, using artificial intelligence and robotics, can efficiently sort plastics from other waste streams. This ensures that plastic materials are properly managed and do not contribute to microplastic pollution (Diggle and Walker 2020).

Advanced filtration systems are being developed to capture MPs from wastewater and runoff. These systems employ technologies such as ultrafiltration and nanofiltration to remove MPs, preventing their discharge into water bodies that may eventually affect agricultural soil (Ngo et al. 2019).

Some waste-to-energy technologies can convert certain types of plastic waste into energy, reducing the need for landfill disposal. This approach decreases the potential for microplastic contamination in soil from decomposing plastics (Aragaw and Mekonnen 2021).

Innovative robots and autonomous vehicles are being deployed for the collection of plastics in various environments, including rivers and coastal areas. By efficiently removing plastic waste, Blockchain technology is being used to trace the life cycle of plastic products and packaging. This transparency enables better management of plastic waste, reducing the likelihood of plastic materials degrading into MPs in the environment.

Technological tools and platforms are aiding regulatory authorities in monitoring and enforcing waste management regulations. These technologies can track the movement of waste materials, ensuring compliance and reducing the potential for microplastic contamination (Mousavi et al. 2023).

Advanced monitoring systems, including remote sensing and data analytics, help identify areas with high microplastic pollution. This information guides targeted waste management efforts to prevent further contamination (Agarwala 2021).

Some technologies aim to convert MPs collected from waste streams into useful products. By repurposing MPs, these innovations not only prevent their spread but also promote recycling.

These technological advancements in waste management are integral to effective microplastic control. By reducing plastic waste, improving waste sorting and recycling, and deploying innovative solutions for waste collection, these technologies contribute to a cleaner and more sustainable environment, protecting agricultural soil from the detrimental impacts of MPs.

12.5 EXPLORING PRECISION AGRICULTURE TECHNIQUES FOR MINIMIZING MICROPLASTIC EXPOSURE IN SOIL

When MPs enter soils, they can change soils' physical and chemical properties, such as soil structure, porosity, pH, and nutrient availability (Wang et al. 2022), and destroy soil microbial community (Liu et al. 2019). In addition, MPs can adsorb pollutants through hydrogen bonding and electrostatic interactions (Wang et al. 2022), such as heavy metals, organic pollutants, pathogens, and resistance genes (Ren et al. 2021). The combination of two or more pollutants causes serious degradation of soil health. Moreover, MPs in soil can inhibit seed germination and be transported to stems and leaves from the roots, thereby interfering with plant growth (Luo et al. 2024) affecting photosynthesis and nutrient metabolism (Zhou et al. 2022), inducing oxidative damage (Li et al. 2020a) and altering genotoxicity (Zhou et al. 2022). It was reported that nanoplastics (50 nm) can enter plant cells, pass through the food chain, and cause harm to the human body (Sun et al. 2022).

Figure 12.2 illustrates the primary origins and routes by which MPs are introduced. Notably, agricultural film mulching and organic fertilizer (Zhang et al. 2022b) are identified as the main contributors to the presence of MPs in soil. The use of plastic agricultural film is prevalent in contemporary agricultural practices. In the year 2018, the global production of plastic goods, including but not limited to thermoplastics, polyurethane, thermosetting plastics, and elastomers, exceeded 425 million tons (Mt). The production of organic fertilizer involves the processing of biological materials, such as animal and plant residues, with the purpose of removing hazardous and detrimental components (Okoffo et al. 2020). The phenomenon of soil MPs being assimilated by plants, which are then consumed by animals, results in the presence of MPs inside organic fertilizer. Furthermore, it should be noted that the use of plastic materials is prevalent in the packaging of animal feed and the instruments employed for storing, processing, and transporting such feed. Consequently, these plastic goods have the potential to enter animal waste, including manure, and other related byproducts (Zhang et al. 2022b). Hence, the use of organic fertilizer is a significant contributor to the presence of soil MPs. The worldwide problem of plastic pollution arises from the use of organic fertilizers. The transportation of microplastic debris through the air is a phenomenon that should be considered, since the distribution of these particles may be influenced by airflow dynamics. According to a study conducted by Dris et al. (2017), the presence of MPs is seen to be higher in indoor environments, ranging from 1.00 to 60.0 particles/m^3, compared to outdoor environments where the range is between 0.300 and 1.50 particles/m^3. The transportation of MPs to distant regions through the atmosphere has been seen, including their presence in the deposits of Arctic ice and snow, as shown by Bergmann et al. (2019) with concentrations ranging from 0 to 14.4×10^3 particles/L. The deposition of atmospheric particles is a significant contributor to the presence of MPs in soil. Nevertheless, the present body of research pertaining to air MPs mostly emphasizes the identification and measurement aspects, while the implementation of effective mitigation strategies remains limited.

The identification of air MPs is a significant challenge in scientific research because to the inherent unpredictability in environmental circumstances, which subsequently introduces ambiguity in the detection process. Another problem that arises is the potential interference of molecules with comparable properties to MPs among air pollutants, such as those containing carbon, silica, and other organic components, which might hinder the accurate detection of MPs. It is important to build a more robust mathematical model in order to effectively address the situation at hand. Untreated landfill has been identified as an additional potential contributor of MPs in soil, as suggested by He et al. (2019). According to Geyer et al. (2017), the global production and use of plastics are projected to result in the generation of around 12.0 billion tons of plastic garbage by the year 2050. The prolonged presence of plastic debris in the soil environment leads to its degradation via weathering processes, resulting in the formation of MPs. This degradation process also causes the release of harmful compounds, such as phthalates, as shown by Xu et al. (2020). The intricate nature of landfills may result in significantly elevated levels of organic substances, and typical pretreatment techniques may not entirely eliminate contaminants. The digesting period might be extended, or a powerful acid or alkali solution can be used to facilitate the breakdown of an excessive amount of organic materials. Therefore, it is recommended to do further microscopic analysis and categorization of the samples. Moreover, the existing study mostly focuses on macroscopic MPs, whereas the investigation of microscopic MPs at the micrometer (μm) and nanometer (nm) scales remains unexplored. In order to mitigate the introduction of MPs into soil via untreated leachate, it is essential to enhance the effective management of landfills to reduce leachate leakage. Additional sources of soil MPs include illicit dumping and improper trash disposal, tidal movements in the ocean, and the degradation of plastic items and automobile tires. The act of unlawfully disposing of waste materials may result in the buildup of plastic items inside soil, subsequently causing the degradation of soil MPs (Zhang et al. 2022c). Furthermore, the phenomenon of tidal action has resulted in the extensive dispersion of MPs across the ocean and their subsequent deposition in coastal soils (Yin et al. 2022). The detection of MPs in soil necessitates rigorous sample handling procedures.

According to Yang et al. (2021), the first stage of soil microplastic extraction involves the process of screening. The density separation technique often constitutes the latter phase, with the need of chemical digestion both before to and subsequent to the density separation process (Zhang et al. 2022b). The therapy methods are succinctly shown in Figure 12.2. The analytical technique for soil MPs often involves pretreating samples using a screening approach in order to streamline the detection process. The screening procedure is conducted based on particle size, where the grid width is maintained below 5.0 mm (Perez et al. 2022). According to Van Cauwenberghe et al. (2015), the density of typical MPs falls within the range of 0.800–1.40 g/cm^3, while soil density varies from 2.60 to 2.70 g/cm^3. Consequently, the analysis of soil MPs may be conducted by considering disparities in density.

Table 12.3 provides a comprehensive overview of various techniques employed in the extraction of MPs from soil samples. Each extractant or method is meticulously described, along with its associated application in the realm of microplastic research. These extraction methods play a fundamental role in the accurate isolation of MPs from soil matrices, catering

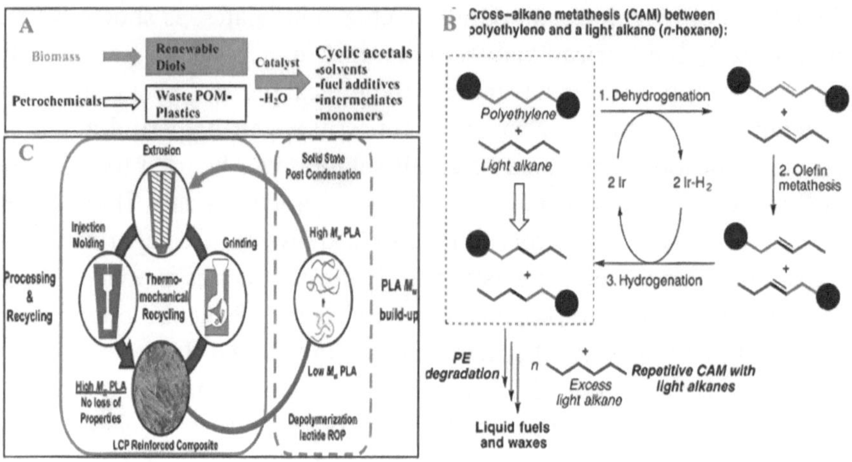

FIGURE 12.2 Some emerging technologies for plastic waste conversion. (a) An efficient pathway for upcycling the polyoxymethylene plastic in combination with biomass-derived diols into versatile cyclic acetals. (Reproduced with permission from ELSIEVER License Number 5647201036401.)

to a range of soil compositions and research objectives. Researchers rely on these techniques to ensure the reliability and precision of their findings in the ongoing endeavor to understand and mitigate the impact of MPs in agricultural soil. Plastics, including PE, PP, PS, PMMA, PC, and ABS, have been shown to be efficiently separated from the NaCl solution by the use of several extractions, extended ultrasonic treatment, and flotation techniques (Liu et al. 2018).

According to the report "Plastics - the Facts" (Europe 2020), PET and PVC accounted for 17.9% of the global plastic production by the year 2019. The process of separating high-density polymers, such as PET, PVC, and PMMA, continues to provide significant challenges.

It is anticipated that the use of high-activity salt solutions, such as $CaCl_2$ and $ZnBr_2$, would enhance the efficiency of the extraction process. The extraction effectiveness of $CaCl_2$ is much higher compared to that of NaCl solution. However, it should be noted that the presence of Ca^{2+} ions may result in the bridging of negative charges on organic molecules, causing the agglomeration of SOM. This agglomeration might potentially interfere with the accurate identification of MPs in later analyses (Scheurer and Bigalke 2018). According to Van Cauwenberghe et al. (2015), it was suggested that the optimal density for extracting MPs was between 1.60 and 1.80 g/cm^3. In the study conducted by

TABLE 12.3 Common Extractants and Methods for Soil Microplastic Extraction

Extractant	Extraction Method	Application	References
Sodium chloride (NaCl) solution	Density separation	Effective for separating microplastics based on density differences	Quinn et al. (2017), He et al. (2021)
Sucrose solution	Density separation	Used as an alternative to NaCl solution, particularly for sensitive analyses	Vortmann et al. (2019)
Hydrogen peroxide (H_2O_2)	Oxidation extraction	Oxidizes organic matter, leaving microplastics behind for subsequent analysis	Priya et al. (2023)
Enzymes (e.g., lipase)	Biological extraction	Organic matter degradation with the aid of enzymes to isolate microplastics	Mbachu et al. (2021)
Sodium hexametaphosphate (SHMP)	Dispersing agent	Used in combination with wet sieving to disperse soil particles and facilitate microplastic separation	Choudhary et al. (2022)
Hydrofluoric acid (HF)	Mineral digestion	Dissolves minerals in the soil, allowing for the recovery of microplastics	Müller et al. (2022)
Organic solvents (e.g., acetone, toluene)	Solvent extraction	Effective for extracting microplastics from organic-rich soils	Kadac-Czapska et al. (2023)
Sodium hydroxide (NaOH)	Alkaline digestion	Reduces organic matter content in the soil, enabling the isolation of microplastics	Li et al. (2020b)
Hydrochloric acid (HCl)	Acid digestion	Dissolves carbonates and organic matter, leaving microplastics for analysis	Budimir et al. (2018)

Nuelle et al. (2014), a solution containing a mixture of sodium chloride (NaCl) and sodium iodide (NaI) was used for the purpose of separating MPs. The separation efficiency achieved varied between 68% and 99% for different kinds of MPs. Based on the potential environmental risks and embryotoxic effects of NaI on aquatic creatures, it is suggested by Liu et al. (2019) that the use of NaBr solution is a preferable approach for the extraction of MPs in soil samples.

Coppock et al. (2017) created a sediment-microplastics-sediment separation unit using the concept of massively parallel signature sequencing. The unit was composed of a PVC tube, PVC ball valve, and $ZnCl_2$ solution, which was stirred by a magnetic stir bar. The gadget has a high extraction rate, affordability, and portability. Nevertheless, the examination of PVC should be precluded due to the potential abrasion of the PVC tube. Additional density solutions have been documented in the literature, such as $ZnBr_2$, $NaCl–Na_2WO_4$ (Frere et al. 2017), and a solution containing 50% KI (Phuong et al. 2018). The density separation approach has been shown to have a recovery rate over 90% for MPs with a size greater than 50.0 μm. Nevertheless, it is important to note that the method being discussed may not be appropriate for the analysis of tiny plastic particles measuring less than 50.0 μm in size. These smaller particles constitute a significant proportion, ranging from 35% to 90%, of the total plastic particles as reported by Okoffo et al. (2020). Furthermore, the task of distinguishing MPs with comparable densities, such as PVC (1.37–1.42 g/cm^3) and PET (1.35–1.40 g/cm^3), is a significant challenge. It is recommended to integrate the sample treatment with supplementary methodologies, such as electrostatic separation.

The extraction and analysis of MPs from soil samples necessitate a variety of methods to effectively remove organic matter and isolate these minuscule plastic particles (Shim et al. 2017). Among the techniques, electrostatic separation is founded on the disparity in conductivity between soil minerals and MPs. With soil minerals exhibiting higher conductivity, freeze-drying is commonly employed to eliminate the influence of moisture on electrical conductivity. This process has proven successful in adapting the electrostatic separator for microplastic separation, offering a valuable means of differentiating these particles based on their electrical properties (Caldwell et al. 2022).

Oil stripping, on the other hand, serves as an efficient method for extracting MPs that have infiltrated an oil layer within soil samples (Boyle and Örmeci 2020). A range of oils with differing densities and polarities can be selected as extractants, making this technique flexible and reliable. With recovery rates spanning from 90% to 97%, this method is not only effective but also simple, safe, and cost-efficient. However, it is essential to consider potential interference from fat residues and the method's inability to eliminate SOM.

Pressurized fluid extraction (PFE) is another approach for microplastic recovery that offers distinct advantages (He et al. 2021). It leverages the power of pressurized fluids to extract MPs from solid materials like soil. PFE is particularly valuable for capturing smaller MPs, but it does come with the caveat that it may alter the morphology of these particles.

The recovery efficiency is high, with recovery rates ranging from 84.5% to 94.0%. Notably, one-step extraction with PFE has been explored for various microplastic types and has yielded extraction efficiencies exceeding 80%. Although it can slightly affect the physical characteristics of MPs, it significantly reduces the time required for extraction.

Removing SOM is a critical step in enhancing the accuracy of microplastic recovery, especially when dealing with soil samples rich in SOM (Thomas et al. 2020). It is well-known that SOM, organic fibers, and MPs share comparable densities, posing challenges for visual and spectral examinations of MPs. To address this, several methods can be employed, including digestion with hydrogen peroxide or strong acids or bases. The elimination of SOM is paramount for ensuring the integrity of microplastic recovery and analysis.

Acid digestion and alkaline digestion are two chemical techniques used to eliminate organic matter from soil samples. Acid digestion, typically employing strong mineral acids such as HNO_3 and HCl, can achieve recovery rates of 93%–98%. However, the choice of acid can impact certain types of MPs. In contrast, alkaline digestion is a milder method and minimally affects the structure of MPs. Nevertheless, it necessitates a longer digestion period (Prata et al. 2019).

Finally, hydrogen peroxide and Fenton reagent oxidation represent methods for removing organic materials from soil samples (Hurley et al. 2018). These oxidants offer an efficient means of eliminating organic matter, with the Fenton process being particularly promising. It can effectively remove both organic and inorganic components and retain the original surface structure of certain MPs. However, further investigation is required to assess its impact on weathered MPs and its ability to remove all biogenic materials effectively.

A variety of analytical techniques have been applied for the detection and characterization of MPs in environmental samples (Table 12.4). Simple visual identification can provide preliminary screening, while spectroscopy methods like infrared (IR) and Raman spectroscopy allow definitive polymer identification. Electron microscopy techniques like SEM and TEM enable high-resolution imaging and detailed analysis of microplastic morphology and size. Chemical staining agents like Nile Red improve microplastic visualization. FTIR microspectroscopy combines IR spectroscopy with microscopy to both detect MPs and determine their spatial distribution. Thermal analysis provides differentiation based on polymer decomposition and ignition temperatures. Pyrolysis GC-MS identifies the

TABLE 12.4 Comparison of Analytical Techniques for Microplastic Detection and Characterization in Soils

Detection Technology	Principle	Advantages	Limitations	Applications	References
Visual identification	Visual inspection	Simple and cost-effective method	Limited to qualitative assessment	Preliminary screening	Horton et al. (2017)
Infrared spectroscopy (IR)	Measures infrared light	Identifies chemical composition	Requires specialized equipment	Chemical analysis, polymer identification	Löder et al. (2015)
Raman spectroscopy	Measures light scattering	Provides molecular structure information	Sensitive to sample impurities	Polymer identification, chemical analysis	Scheurer and Bigalke (2018)
Chemical staining (e.g., Nile Red)	Fluorescence dyes	Enhances visualization of microplastics	May require specific lighting conditions	Enhancing visibility of microplastics	Francischini and Arruda (2021)
Scanning electron microscopy (SEM)	Electron microscopy	High-resolution imaging of microplastics	Time-consuming and expensive	Detailed characterization of microplastics	Gniadek and Dąbrowska (2019)
Transmission electron microscopy (TEM)	Electron microscopy	High-resolution imaging and analysis	Requires specialized expertise and equipment	Detailed characterization of microplastics	Caputo et al. (2021)
FTIR microspectroscopy	Combines microscopy and IR	Identifies and locates microplastics	Requires sophisticated instrumentation	Characterization and spatial analysis	Primpke et al. (2020)
X-ray microcomputed tomography (μCT)	X-ray imaging	Non-destructive 3D visualization	Limited to larger microplastics	Spatial distribution analysis, volume measurement	Kontiza et al. (2022)

(*Continued*)

TABLE 12.4 (*Continued*) Comparison of Analytical Techniques for Microplastic Detection and Characterization in Soils

Detection Technology	Principle	Advantages	Limitations	Applications	References
DNA barcoding	Genetic analysis	Identifies microplastic sources and types	Requires DNA extraction and sequencing	Source tracking, ecological impact assessment	Parrish and Fahrenfeld (2019)
Microplastic extraction methods	Various techniques	Isolates microplastics from complex matrices	Specific extraction method may be required	Sample preparation for analysis	Thomas et al. (2020)
Remote sensing	Hyperspectral imaging	Detects microplastics on a larger scale	Limited to surface detection, not within soil	Identifying microplastic pollution in large areas	Vidal and Pasquini (2021)
Thermal analysis	Uses differential decomposition, melting, and ignition temperatures	Simple instrumentation, useful for polymer differentiation	Limited by polymer thermal stability range	Quality control, polymer typing	Francischini and Arruda (2021)
Pyrolysis GC-MS	Identifies monomers and additives from pyrolytic breakdown	Provides chemical fingerprints for polymer identification	Extensive sample preparation, expensive equipment	Detailed polymer characterization and degradation studies	Fries et al. (2013)

monomers and chemical additives in plastic samples based on pyrolytic breakdown products. While powerful, many advanced techniques require expensive instrumentation, extensive sample preparation, and technical expertise. Ongoing improvements aim to optimize analytical protocols for reliable microplastic quantification across environmental matrices.

12.6 INNOVATIVE SOIL REMEDIATION TECHNOLOGIES USING ADVANCED MATERIALS SCIENCE AND NANOTECHNOLOGY

Soil and groundwater treatment are only two of the numerous applications of nanotechnology. Soil and groundwater that have been polluted by petroleum pollutants and heavy metals may now benefit from nanoremediation, a fast and efficient process (Hussain et al. 2022a).

The primary goal of soil and groundwater remediation is to reduce contamination to an acceptable and safe level. This objective for soil and groundwater remediation has been accomplished by the use of physical, chemical, and biological approaches (Li et al. 2019). Soil features and pollutants, as well as the type of the remediation method that is chosen and planned, play important roles in determining the best course of action for soil and groundwater cleanup (Hussain et al. 2022a). By contrast, ex situ procedures like carbon adsorption, air stripping, chemical oxidation/precipitation, and biological reactors are used to remove pollutants from groundwater that has been pumped from wells using conventional methods like pump-and-treat. High operational expenses and waste contamination are typical of these approaches (Panagopoulos et al. 2019). Emerging remediation technology, such as surfactant-enhanced remediation (SER), has been found to be successful for cleaning up groundwater and soil polluted with organic pollutants in the form of dense nonaqueous phase liquids (DNAPLs). However, there are dangers connected with these technologies; when DNAPLs' interfacial force decreases, they may undergo uncontrolled vertical migration (Zhao et al. 2022a). Nanotechnology has gained attention in many different areas in recent years. Due to the many applications that nanoparticles (NPs) permit, they exhibit a wide range of desirable qualities (Kim et al. 2018; Shah et al. 2021). Multidisciplinary methods, including concepts of molecular manufacturing and engineering, are used to create NPs. Nanotechnology is a broad term for a set of techniques that build particles between 1 and 100 nm in size, study the physics of those particles, and put them to use in various contexts. Industries as diverse as the chemical, electrical, medicinal, and biotechnological make use of nanotechnology. Water and wastewater treatment, pollution control, and the treatment/remediation of polluted soil and groundwater are only some of the environmental protection tasks that have seen efforts to use nanotechnology (Alazaiza et al. 2021).

Nanoremediation technologies have been utilized to clean up polluted environments since 2009. Few field-scale applications have been studied in order to assess nanoremediation technology (Corsi et al. 2018). Reduced cost and cleanup time, complete degradation of some contaminants without the need to dispose of polluted soil, and no need to transfer the soil or pump groundwater are some of the main advantages of using nanoremediation for soil and groundwater remediation, especially for extensive site cleaning (Wang et al. 2022).

Nanoremediation techniques use reactive NPs for pollutant conversion and elimination. Catalysis and chemical reduction are the primary remedial methods by NPs (Hussain et al. 2022a). Given that NPs have large surface-area-to-mass ratios and a distinct distribution of active sites, they are also effective removal mechanisms (Varier et al. 2019). Many engineering NPs, because of their novel surface coating and minute size, offer very practical features for in situ remediation applications. Also, unlike microparticles, NPs may possibly travel far distances and achieve broader geographic dispersion because to their ability to diffuse and enter the microscopic crevices in the subsurface and remain suspended in groundwater for a long period (Huang et al. 2022).

Due to their nanoscale features, NPs' physical movement and/or transport in groundwater is dominated by random motion or Brownian movement rather than the wall effect (Alazaiza et al. 2021). Since NPs are so much smaller and lighter than microscale particles, they are not affected by gravity sedimentation and may float freely in groundwater throughout the restoration process without settling. Consequently, NPs provide a practical treatment strategy by enabling direct injection into the subsurface in polluted areas (Alazaiza et al. 2021).

Soil and groundwater nanoremediation has been the subject of several investigations (Boregowda et al. 2022). However, the environmental impacts of those NPs are still unknown, and additional research is needed to understand the environmental fate and toxicity of NPs, concerns that are vital for environmental protection practice but have so far received little attention.

Many toxins have been examined in the lab to see whether they can be removed using nanomaterials for use in soil and groundwater remediation, and the findings have been encouraging (Mukhopadhyay et al. 2022). However, nanomaterials should be considered in a case-by-case setting because of the potential for both beneficial and negative effects on living creatures, the environment, society, and the economy (Gottardo et al. 2021). Popularity and acceptance of nanoremediation technologies may be boosted by proper risk documentation, field-scale confirmation of remediation outcomes, discussions at the science-policy interface, and appropriate market development activities (Gottardo et al. 2021).

Nanoremediation methods have been the subject of a lot of study and development for cleaning up polluted soil and water. It is widely agreed that nanoremediation is a green technique. Since this is a traditional site remediation approach, it is a viable option (Boregowda et al. 2022).

Site cleanup may be made more efficient and inexpensive by the use of nanotechnology. NPs, such as, metal oxides, nanoscale zeolites, nanometals, carbon nanotubes, and titanium dioxide, have all been employed for nanoremediation.

Twenty years ago, it was discovered that NPs may stay active in injected soil for up to 56 days and migrate with groundwater for more than 20 m (Zhang 2003). Since then, NPs have been widely used in large-scale field applications for soil and groundwater remediation. More than 99% of trichloroethene (TCE) may be removed from polluted locations within a few days, as reported by Zhang (2003).

Engineered nanomaterials (ENMs) are gaining prominence as innovative tools for in situ remediation of contaminated soils, particularly in cases where contaminants span a wide spectrum, including heavy metals, metalloids, and organic compounds (Mukherjee et al. 2022). This emerging field of environmental nanotechnology offers promising solutions to address the challenges associated with soil contamination and environmental pollution. Let's delve into the applications and potential of ENMs in remediating soils contaminated with these diverse pollutants.

ENMs, such as iron-based NPs (e.g., zero-valent iron), are utilized to immobilize heavy metals and metalloids in the soil. These NPs create insoluble precipitates through redox reactions, rendering contaminants less bioavailable and mobile (Ahmed et al. 2021).

Some ENMs, like carbon nanotubes and graphene oxide, possess high surface area and reactivity (Ahmed et al. 2021). They are employed to enhance the sorption of heavy metals and metalloids, effectively capturing and sequestering these contaminants in the soil matrix.

ENMs can be incorporated into phytostabilization approaches, where plants are used to immobilize and accumulate contaminants. NPs can enhance plant uptake and reduce the translocation of heavy metals, thus improving soil quality.

Nanoremediation with Zero-Valent Iron (nZVI) has demonstrated effectiveness in degrading a wide range of organic compounds, including chlorinated solvents and petroleum hydrocarbons. nZVI promotes reductive dechlorination and facilitates the breakdown of organic pollutants (Ebrahimbabaie and Pichtel 2021).

ENMs, especially those with catalytic properties, like palladium NPs, can serve as catalysts in advanced oxidation processes (Peramune et al. 2022). These catalysts facilitate the degradation of organic contaminants through processes such as Fenton-like reactions.

Nanoscale materials can enhance the activity of microorganisms involved in bioremediation. Nano-sized carriers loaded with beneficial microorganisms can improve the degradation of organic pollutants in contaminated soils. Also on the rise in recent years is the practice of combining nanotechnology with bioremediation (such as phytoremediation and micro-remediation) (Liu et al., 2023; Deng et al., 2024). Adsorption has been crucial to the treatment of heavy metal- and metalloid-contaminated soils, whereas catalytic degradation and reduction reactions have been used mostly for the removal of organic pollutants. Importantly, in real-world circumstances, ENMs that could remove numerous contaminants at once via combined or sequential treatments tended to exhibit significant value in soil remediation (Gao et al. 2024). The presence of ENMs might aid in the adsorption and redox degradation of micropollutants in a simultaneous adsorption and oxidation scenario (Aliyari Rad et al. 2023). The need for these versatile materials to reduce refractory and co-exposed pollution is on the rise.

Understanding the fate, transport, and potential ecological impacts of ENMs in the environment is crucial. Their persistence and interactions with native soil components must be carefully assessed. The use of ENMs in soil remediation is subject to regulatory scrutiny and ethical considerations, especially regarding their long-term effects and unintended consequences (Carboni et al. 2021). Effective remediation often requires site-specific engineering of ENMs, considering the contaminant types and soil properties unique to each location. A thorough risk-benefit analysis is essential to weigh the advantages of ENM-based remediation against potential risks and uncertainties.

Table 12.5 provides an overview of how ENMs can be applied for the in situ remediation of contaminated soils containing heavy metals,

TABLE 12.5 Engineered Nanomaterial Applications in In Situ Remediation of Contaminated Soils

Contaminant Type	Engineered Nanomaterials (ENMs) and Applications	Advantages	Challenges	Examples of ENMs	References
Heavy Metals and Metalloids					
Immobilization	Zero-valent iron nanoparticles for in situ immobilization	Reduces bioavailability and mobility	Environmental fate of ENMs; long-term efficacy	Zero-valent iron (nZVI)	Maciel et al. (2019), Song et al. (2022)
Enhanced sorption	Carbon nanotubes and graphene oxide for sorption	Increases contaminant capture in soil	ENM stability and potential leaching	Carbon nanotubes, graphene oxide	Vu et al. (2017), Tshangana et al. (2021)
Phytostabilization	ENMs integrated with plants for phytostabilization	Enhances plant uptake, reduces translocation	ENM-plant interactions; ecological impacts	Iron oxide nanoparticles, silica	Duncan and Owens (2019), Bajaj et al. (2023)
Organic Compounds					
Nanoremediation	Zero-valent iron nanoparticles for organic degradation	Promotes reductive dechlorination and breakdown	Fate of degradation byproducts; site-specific needs	Zero-valent iron (nZVI)	Machado and Pacheco (2016), Galdames et al. (2020)
Nanomaterial-supported catalysts	Palladium nanoparticles for catalytic degradation	Accelerates advanced oxidation processes	Cost of catalyst; potential toxicity of catalyst	Palladium nanoparticles	Nayna and Tareq (2019)
Nano-enhanced bioremediation	Nano-sized carriers with microorganisms for enhanced bioremediation	Improves microbial activity	Compatibility with indigenous soil microorganisms	Nano-sized bacterial carriers	Chinni et al. (2023)

metalloids, and organic compounds. Each application is accompanied by its advantages and associated challenges, as well as examples of relevant ENMs.

The bioremediation of soil polluted with Pb and Cd by a Gram-negative bacterium, *Halomonas* sp., was enhanced by the addition of NPs of FeO coated with polyvinylpyrrolidone (PVP). Compared to removal by bacteria or just NPs, this method has considerably eliminated almost 100% of Pb after 24 hours and Cd after 48 hours (Alabresm et al. 2018). Adsorptive removal of 795, 1,355, and 985 μmol/g for Cu, Pb, and Ni, respectively, was accomplished using a biosorbent comprising magnetic Fe_3O_4 NPs treated with *S. aureus*, with a surface encapsulated with phthalic acid (as an *n*-Fe_3O_4-Phth-*S* complex). Recoveries of Cu^{2+} were from 83.1% to 89.5%, Pb^{2+} from 99.4% to 100%, and Ni^{2+} from 92.6% to 7.5%. Based on the results of a study comparing *n*-Fe_3O_4-Phth-*S* to dried *S. aureus* for heavy metals (HMs) removal, researchers conclud ed that both the NPs' core of *n*-Fe_3O_4-Phth-*S* and the functional groups present on the microbial surface were important in the removal of HMs (Mahmoud et al. 2016). According to the results of this study, the removal of pollutants is significantly influenced by the NPs' core as well as functional groups present on the microbial surface.

A recent study on the removal of Cu, Cd, Cr, and Pb using HM-resistant bacteria like *B. cereus* (PMBL-3) and *L. macroides* (PMBL-7) evidently confirmed that ZnO NPs at 5 mg/L synergistically removed the Cr by 60%, the Cu by 70%, and the Pb by 85%. This is in comparison to *B. cereus* (80% and 60%) and *L. macroides* (55% and 50%) at neutral pH. Surface negative charges on ZnO NPs at neutral pH facilitate electrostatic interactions with metal cations; however, when the pH is lowered, the HMs precipitate as hydroxides, and the hydrogen ions compete for binding with adsorbent materials. The *B. cereus* strain XMCr-6 has also been found to use an enzyme-mediated mechanism to decrease Cr^{6+} levels. By forming coordination bonds with the surface functional group of the bacterial cell wall, the reduced Cr^{3+} was shown to have a binding affinity to the cells. Cr_2O_3 NPs were discovered to have formed on the surface of the cells as a byproduct (Laslo et al. 2022).

Metalloids are a group of elements that possess a diverse range of qualities, namely exhibiting both metallic and nonmetallic physicochemical characteristics. Naturally occurring sources of these substances have been identified; nevertheless, the advent of industrialization has led to the emergence of concentrations that are beyond the permissible threshold

(Fordyce et al. 2017). The six elements often referred to as metalloids in the periodic table are antimony, arsenic, boron, germanium, tellurium, and silicon (Yazdi et al. 2016). The worldwide problem of metalloid contamination in soil arises from the deposition of these chemical elements in the environment, resulting in adverse impacts on human health, plant life, and animal populations. Moreover, it is important to note that these substances pose a significant risk to human health regardless of their concentration, including even at very low levels (Martínez-Alcalá and Bernal 2020). Moreover, the presence of heavy metals/metalloids may have a significant influence on the abundance, composition, and variety of soil microbial communities, potentially resulting in alterations to ecosystem functioning (Huang et al. 2021). Concerns remain concerning nanotechnology's presence in environmental realms, its destiny, and the toxicological repercussions that might result from its use. Because of their expanded uses, particularly in remediation procedures, NPs are a primary focus of this section, in which we want to present a clear overview of their destiny and the rising environmental issues associated with their intentional emission. Some recent publications have shown how widespread use of NPs in farming has negative consequences. Soil qualities may be altered, for instance, if NPs are intentionally applied and accumulate or the concentration of their components increases in the soil (Sheteiwy et al. 2021). Soil pH is known to be altered by NPs, which has major implications for nutrient availability, microbial dynamics, soil health, and plant growth and development (Chaudhary et al. 2022).

12.7 CONCLUSION

In conclusion, the issue of MPs in agricultural soil is a multifaceted challenge with wide-ranging implications. This review has explored innovative research methods, sustainable farming practices, and advanced soil management techniques to address this complex problem.

Sustainable farming practices like biodegradable mulching materials and agroecology offer promising ways to reduce microplastic accumulation in soil while promoting its health and fertility.

Technological advancements in waste management, including plastic upcycling and circular economy models, offer solutions to reduce microplastic pollution.

Innovative sorting technologies and microplastic filtration systems help efficiently capture and prevent the discharge of MPs into agricultural soils.

Engineered nanomaterials show promise for remediating soil contaminated with heavy metals, metalloids, and organic compounds, though environmental safety must be a primary consideration.

In essence, addressing MPs in agricultural soil requires a comprehensive approach involving research, sustainable practices, responsible waste management, and innovative remediation techniques. Collaboration among various stakeholders is key to mitigating the impact of MPs and ensuring the health of our soils and food systems.

REFERENCES

Agarwala N (2021) Managing marine environmental pollution using Artificial Intelligence. *Marit Technol Res* 3:120–136.

Ahmed T, Noman M, Ijaz M, et al. (2021) Current trends and future prospective in nanoremediation of heavy metals contaminated soils: A way forward towards sustainable agriculture. *Ecotoxicol Environ Saf* 227:112888.

Alabresm A, Chen YP, Decho AW, Lead J (2018) A novel method for the synergistic remediation of oil-water mixtures using nanoparticles and oil-degrading bacteria. *Sci Total Environ* 630:1292–1297.

Alazaiza MYD, Albahnasawi A, Ali GAM, et al. (2021) Recent advances of nanoremediation technologies for soil and groundwater remediation: A review. *Water* 13:2186.

Aliyari Rad S, Nobaharan K, Pashapoor N, et al. (2023) Nano-microbial remediation of polluted soil: A brief insight. Sustainability 15(1):876. https://doi.org/10.3390/su15010876

Alves MN, Miró M, Breadmore MC, Macka M (2019) Trends in analytical separations of magnetic (nano)particles. *TrAC Trends Anal Chem* 114:89–97. https://doi.org/10.1016/j.trac.2019.02.026

Aragaw TA, Mekonnen BA (2021) Current plastics pollution threats due to COVID-19 and its possible mitigation techniques: A waste-to-energy conversion via pyrolysis. *Environ Syst Res* 10:1–11.

Bajaj T, Alim H, Ali A, Patel N (2023) Phytotoxicity responses and defence mechanisms of heavy metal and metal-based nanoparticles. In: Husen A, (ed.), *Nanomaterials and Nanocomposites Exposures to Plants: Response, Interaction, Phytotoxicity and Defense Mechanisms*. Springer, pp. 59–96.

Barrett R, Adebowale M, Birch H, et al. (2023) Planktic foraminiferal resilience to environmental change associated with the PETM. *Paleoceanogr Paleoclimatol* 38:1–16. https://doi.org/10.1029/2022PA004534

Bellasi A, Binda G, Pozzi A, et al. (2020) Microplastic contamination in freshwater environments: A review, focusing on interactions with sediments and benthic organisms. *Environments* 7:30.

Bergmann M, Mützel S, Primpke S, et al. (2019) White and wonderful? Microplastics prevail in snow from the Alps to the Arctic. *Sci Adv* 5:eaax1157.

Bikker J, Lawson J, Wilson S, Rochman CM (2020) Microplastics and other anthropogenic particles in the surface waters of the Chesapeake Bay. *Mar Pollut Bull* 156:111257. https://doi.org/10.1016/j.marpolbul.2020.111257

Booth AM, Sørensen L (2022) Microplastic fate and impacts in the environment. In: Rocha-Santos T, Costa MF, Mouneyrac C, (eds.), *Handbook of Microplastics in the Environment*. Springer, pp. 757–779.

Boregowda N, Jogigowda SC, Bhavya G, et al. (2022) Recent advances in nanoremediation: Carving sustainable solution to clean-up polluted agriculture soils. *Environ Pollut* 297:118728.

Boyle K, Örmeci B (2020) Microplastics and nanoplastics in the freshwater and terrestrial environment: A review. *Water* 12:2633.

Budimir S, Setälä O, Lehtiniemi M (2018) Effective and easy to use extraction method shows low numbers of microplastics in offshore planktivorous fish from the northern Baltic Sea. *Mar Pollut Bull* 127:586–592.

Caldwell J, Taladriz-Blanco P, Lehner R, et al. (2022) The micro-, submicron-, and nanoplastic hunt: A review of detection methods for plastic particles. *Chemosphere* 293:133514.

Camargo ALG, Girard P, Sanz-Lazaro C, et al. (2022) Microplastics in sediments of the Pantanal Wetlands, Brazil. *Front Environ Sci* 10:1–12. https://doi.org/10.3389/fenvs.2022.1017480

Caputo F, Vogel R, Savage J, et al. (2021) Measuring particle size distribution and mass concentration of nanoplastics and microplastics: Addressing some analytical challenges in the sub-micron size range. *J Colloid Interface Sci* 588:401–417.

Carboni A, Slomberg DL, Nassar M, et al. (2021) Aquatic mesocosm strategies for the environmental fate and risk assessment of engineered nanomaterials. *Environ Sci Technol* 55:16270–16282.

Chahal S, Wang P, Bueno V, et al. (2023) Effect of emerging contaminants on soil microbial community composition, soil enzyme activity, and strawberry plant growth in polyethylene microplastic-containing soils. *Environ Sci Adv* 2:629–644. https://doi.org/10.1039/d2va00233g

Chaudhary P, Chaudhary A, Bhatt P, et al. (2022) Assessment of soil health indicators under the influence of nanocompounds and Bacillus spp. in field condition. *Front Environ Sci* 9:769871.

Chen G, Li Y, Wang J (2021a) Occurrence and ecological impact of microplastics in aquaculture ecosystems. *Chemosphere* 274:129989.

Chen H, Wan K, Zhang Y, Wang Y (2021b) Waste to wealth: Chemical recycling and chemical upcycling of waste plastics for a great future. *ChemSusChem* 14:4123–4136. https://doi.org/10.1002/cssc.202100652

Chinni SV, Sanniraj A, Koduru JR, Lebaka VR (2023) Nanobioremediation—An emerging eco-friendly approach for a sustainable environment. In: Koduru JR, Karri RR, Mujawar MN, (eds.), *Hybrid Nanomaterials for Sustainable Applications*. Elsevier, pp. 199–219.

Choudhary S, Neelavanan K, Saalim SM (2022) Microplastics in the surface sediments of Krossfjord-Kongsfjord system, Svalbard, Arctic. *Mar Pollut Bull* 176:113452.

Christensen PR, Scheuermann AM, Loeffler KE, Helms BA (2019) Closed-loop recycling of plastics enabled by dynamic covalent diketoenamine bonds. *Nat Chem* 11:442–448. https://doi.org/10.1038/s41557-019-0249-2

Clark R, Shaver MP (2024) Depolymerization within a circular plastics system. *Chemical Reviews* 124(5):2617–2650.

Coppock RL, Cole M, Lindeque PK, et al. (2017) A small-scale, portable method for extracting microplastics from marine sediments. *Environ Pollut* 230:829–837.

Corsi I, Winther-Nielsen M, Sethi R, et al. (2018) Ecofriendly nanotechnologies and nanomaterials for environmental applications: Key issue and consensus recommendations for sustainable and ecosafe nanoremediation. *Ecotoxicol Environ Saf* 154:237–244.

De-la-Torre GE, Pizarro-Ortega CI, Dioses-Salinas DC, et al. (2022) First record of plastiglomerates, pyroplastics, and plasticrusts in South America. *Sci Total Environ* 833:155179.

Degli Esposti M, Morselli D, Fava F, et al. (2021) The role of biotechnology in the transition from plastics to bioplastics: An opportunity to reconnect global growth with sustainability. *FEBS Open Bio* 11:967–983.

Degli Innocenti F, Breton T (2020) Intrinsic biodegradability of plastics and ecological risk in the case of leakage. *ACS Sustain Chem Eng* 8:9239–9249. https://doi.org/10.1021/acssuschemeng.0c01230

Dehaut A, Hermabessiere L, Duflos G (2019) Current frontiers and recommendations for the study of microplastics in seafood. *TrAC Trends Anal Chem* 116:346–359. https://doi.org/10.1016/j.trac.2018.11.011

Deng S, Zhang X, Zhu Y, Zhuo R (2024). Recent advances in phyto-combined remediation of heavy metal pollution in soil. *Biotechnology Advances* 72:108337.

Diggle A, Walker TR (2020) Implementation of harmonized extended producer responsibility strategies to incentivize recovery of single-use plastic packaging waste in Canada. *Waste Manag* 110:20–23.

Dris R, Gasperi J, Mirande C, et al. (2017) A first overview of textile fibers, including microplastics, in indoor and outdoor environments. *Environ Pollut* 221:453–458.

Duncan E, Owens G (2019) Metal oxide nanomaterials used to remediate heavy metal contaminated soils have strong effects on nutrient and trace element phytoavailability. *Sci Total Environ* 678:430–437.

Eagan JM, Xu J, Di Girolamo R, et al. (2017) Combining polyethylene and polypropylene: Enhanced performance with PE/iPP multiblock polymers. *Science* 355:814–816. https://doi.org/10.1126/science.aah5744

Ebrahimbabaie P, Pichtel J (2021) Biotechnology and nanotechnology for remediation of chlorinated volatile organic compounds: Current perspectives. *Environ Sci Pollut Res* 28:7710–7741.

El-Beltagi HS, Basit A, Mohamed HI, et al. (2022) Mulching as a sustainable water and soil saving practice in agriculture: A review. *Agronomy* 12:1881.

Elert AM, Becker R, Duemichen E, et al. (2017) Comparison of different methods for MP detection: What can we learn from them, and why asking the right question before measurements matters? *Environ Pollut* 231:1256–1264. https://doi.org/10.1016/j.envpol.2017.08.074

Eriksen MK, Pivnenko K, Olsson ME, Astrup TF (2018) Contamination in plastic recycling: Influence of metals on the quality of reprocessed plastic. *Waste Manag* 79:595–606.

Feeny AK, Chung MK, Madabhushi A, et al. (2020) Artificial intelligence and machine learning in arrhythmias and cardiac electrophysiology. *Circ Arrhythm Electrophysiol* 13:e007952. https://doi.org/10.1161/CIRCEP.119.007952

Felsing S, Kochleus C, Buchinger S, et al. (2018) A new approach in separating microplastics from environmental samples based on their electrostatic behavior. *Environ Pollut* 234:20–28. https://doi.org/10.1016/j.envpol.2017.11.013

Fordyce FM, Everett PA, Bearcock JM, Lister TR (2017) Soil metal/metalloid concentrations in the Clyde Basin, Scotland, UK: Implications for land quality. *Earth Environ Sci Trans R Soc Edinburgh* 108:191–216.

Francischini DS, Arruda MAZ (2021) When a picture is worth a thousand words: Molecular and elemental imaging applied to environmental analysis–A review. *Microchem J* 169:106526.

Frere L, Paul-Pont I, Rinnert E, et al. (2017) Influence of environmental and anthropogenic factors on the composition, concentration and spatial distribution of microplastics: A case study of the Bay of Brest (Brittany, France). *Environ Pollut* 225:211–222.

Fries E, Dekiff JH, Willmeyer J, et al. (2013) Identification of polymer types and additives in marine microplastic particles using pyrolysis-GC/MS and scanning electron microscopy. *Environ Sci Process Impacts* 15:1949–1956.

Fu W, Min J, Jiang W, et al. (2020) Separation, characterization and identification of microplastics and nanoplastics in the environment. *Sci Total Environ* 721:137561.

Fu Y, Wei Z, Wan Z, et al. (2022) Recent process of multimode stimuli-responsive flexible composites based on magnetic particles filled polymers: Characteristics, mechanism and applications. *Compos Part A Appl Sci Manuf* 163:107215. https://doi.org/10.1016/j.compositesa.2022.107215

Fuller S, Gautam A (2016) A procedure for measuring microplastics using pressurized fluid extraction. *Environ Sci Technol* 50:5774–5780. https://doi.org/10.1021/acs.est.6b00816

Galdames A, Ruiz-Rubio L, Orueta M, et al. (2020) Zero-valent iron nanoparticles for soil and groundwater remediation. *Int J Environ Res Public Health* 17:5817.

Gao Y, Zhou L, Ouyang S, Sun J, Zhou Q (2024) Environmental applications and risks of engineered nanomaterials in removing petroleum oil in soil. *Science of the Total Environment* 946:174165. https://doi.org/10.1016/j.scitotenv.2024.174165

Geyer R, Jambeck JR, Law KL (2017) Production, use, and fate of all plastics ever made. *Sci Adv* 3:e1700782.

Ghosal K, Nayak C (2022) Recent advances in chemical recycling of polyethylene terephthalate waste into value added products for sustainable coating solutions-hope vs. hype. *Mater Adv* 3:1974–1992. https://doi.org/10.1039/d1ma01112j

Gniadek M, Dąbrowska A (2019) The marine nano-and microplastics characterisation by SEM-EDX: The potential of the method in comparison with various physical and chemical approaches. *Mar Pollut Bull* 148:210–216.

Gottardo S, Mech A, Drbohlavová J, et al. (2021) Towards safe and sustainable innovation in nanotechnology: State-of-play for smart nanomaterials. *NanoImpact* 21:100297.

Grbic J, Nguyen B, Guo E, et al. (2019) Magnetic extraction of microplastics from environmental samples. *Environ Sci Technol Lett* 6:68–72. https://doi.org/10.1021/acs.estlett.8b00671

Gregory DA, Tripathi L, Fricker AT, et al. (2021) Bacterial cellulose: A smart biomaterial with diverse applications. *Materials Science and Engineering: R: Reports* 145:100623. https://doi.org/10.1016/j.mser.2021.100623

He D, Luo Y, Lu S, et al. (2018) Microplastics in soils: Analytical methods, pollution characteristics and ecological risks. *TrAC Trends Anal Chem* 109:163–172.

He D, Zhang X, Hu J (2021) Methods for separating microplastics from complex solid matrices: Comparative analysis. *J Hazard Mater* 409:124640.

He P, Chen L, Shao L, et al. (2019) Municipal solid waste (MSW) landfill: A source of microplastics? Evidence of microplastics in landfill leachate. *Water Res* 159:38–45.

Henderson L, Green C (2020) Making sense of microplastics? Public understandings of plastic pollution. *Mar Pollut Bull* 152:110908.

Hildebrandt L, El Gareb F, Zimmermann T, et al. (2022) Spatial distribution of microplastics in the tropical Indian Ocean based on laser direct infrared imaging and microwave-assisted matrix digestion. *Environ Pollut* 307:119547.

Horodytska O, Cabanes A, Fullana A (2019) Plastic waste management: Current status and weaknesses. In: Stock F, Reifferscheid G, Brennholt N, Kostianaia E, (eds.), *Plastics in the Aquatic Environment-Part I: Current Status and Challenges*. Springer, pp. 289–306.

Horton AA, Walton A, Spurgeon DJ, et al. (2017) Microplastics in freshwater and terrestrial environments: Evaluating the current understanding to identify the knowledge gaps and future research priorities. *Sci Total Environ* 586:127–141.

Huang C-C, Liang C-M, Yang T-I, et al. (2021) Shift of bacterial communities in heavy metal-contaminated agricultural land during a remediation process. *PLoS One* 16:e0255137.

Huang D, Chen H, Shen M, et al. (2022) Recent advances on the transport of microplastics/nanoplastics in abiotic and biotic compartments. *J Hazard Mater* 438:129515.

Hüffer T, Metzelder F, Sigmund G, et al. (2019) Polyethylene microplastics influence the transport of organic contaminants in soil. *Sci Total Environ* 657:242–247.

Hou Y, Liang H, Tang A, Du X, Mei Z (2021) Hot-electron photocurrent detection of near-infrared light based on ZnO. *Applied Physics Letters* 118(6):063501. https://doi.org/10.1063/5.0031719

Hurley RR, Lusher AL, Olsen M, Nizzetto L (2018) Validation of a method for extracting microplastics from complex, organic-rich, environmental matrices. *Environ Sci Technol* 52:7409–7417.

Hussain A, Rehman F, Rafeeq H, et al. (2022a) In-situ, ex-situ, and nano-remediation strategies to treat polluted soil, water, and air–A review. *Chemosphere* 289:133252.

Hussain I, Ganiyu SA, Alasiri H, Alhooshani K (2022b) A state-of-the-art review on waste plastics-derived aviation fuel: Unveiling the heterogeneous catalytic systems and techno-economy feasibility of catalytic pyrolysis. *Energy Convers Manag* 274:116433.

Ivleva NP, Wiesheu AC, Niessner R (2017) Microplastic in aquatic ecosystems. *Angew Chemie Int Ed* 56:1720–1739.

Jia X, Qin C, Friedberger T, et al. (2016) Efficient and selective degradation of polyethylenes into liquid fuels and waxes under mild conditions. *Sci Adv* 2:1–8. https://doi.org/10.1126/sciadv.1501591

Jiao X, Zheng K, Chen Q, et al. (2020) Photocatalytic conversion of waste plastics into C_2 fuels under simulated natural environment conditions. *Angew Chemie Int Ed* 59:15497–15501. https://doi.org/10.1002/anie.201915766

Johansen MB, Donslund BS, Larsen E, et al. (2023) Closed-loop recycling of polyols from thermoset polyurethanes by tert-amyl alcohol-mediated depolymerization of flexible foams. *ACS Sustain Chem Eng* 11:10737–10745.

Kadac-Czapska K, Trzebiatowska PJ, Knez E, et al. (2023) Microplastics in food-a critical approach to definition, sample preparation, and characterisation. *Food Chem* 418:135985.

Kawecki D, Nowack B (2019) Polymer-specific modeling of the environmental emissions of seven commodity plastics as macro-and microplastics. *Environ Sci Technol* 53:9664–9676.

Kim D, Kadam A, Shinde S, et al. (2018) Recent developments in nanotechnology transforming the agricultural sector: A transition replete with opportunities. *J Sci Food Agric* 98:849–864.

Koelmans AA, Besseling E, Foekema E, et al. (2017) Risks of plastic debris: Unravelling fact, opinion, perception, and belief. *Environ Sci Technol* 51:11513–11519. https://doi.org/10.1021/acs.est.7b02219

Kontiza A, Semitekolos D, Milickovic TK, et al. (2022) Double cantilever beam test and micro-computed tomography as evaluation tools for self-healing of CFRPs loaded with DCPD microcapsules. *Compos Struct* 279:114780.

Lambert S, Wagner M (2016) Formation of microscopic particles during the degradation of different polymers. *Chemosphere* 161:510–517. https://doi.org/10.1016/j.chemosphere.2016.07.042

Laslo V, Pinzaru SC, Zaguła G, et al. (2022) Synergic effect of selenium nanoparticles and lactic acid bacteria in reduction cadmium toxicity. *J Mol Struct* 1247:131325.

Li C, Zhou K, Qin W, et al. (2019) A review on heavy metals contamination in soil: Effects, sources, and remediation techniques. *Soil Sediment Contam Int J* 28:380–394.

Li L, Luo Y, Li R, et al. (2020a) Effective uptake of submicrometre plastics by crop plants via a crack-entry mode. *Nat Sustain* 3:929–937.

Li X, Chen L, Ji Y, et al. (2020b) Effects of chemical pretreatments on microplastic extraction in sewage sludge and their physicochemical characteristics. *Water Res* 171:115379.

Liu K, Wang X, Fang T, et al. (2019) Source and potential risk assessment of suspended atmospheric microplastics in Shanghai. *Sci Total Environ* 675:462–471.

Liu M, Lu S, Song Y, et al. (2018) Microplastic and mesoplastic pollution in farmland soils in suburbs of Shanghai, China. *Environ Pollut* 242:855–862.

Liu C J, Deng S G, Hu C Y, et al (2023). Applications of bioremediation and phytoremediation in contaminated soils and waters: CREST publications during 20182022. *Critical Reviews in Environmental Science and Technology* 53(6): 723–732.

Löder MGJ, Kuczera M, Mintenig S, et al. (2015) Focal plane array detector-based micro-Fourier-transform infrared imaging for the analysis of microplastics in environmental samples. *Environ Chem* 12:563–581.

Luo C, Zhang L, Ali MM, Xu Y, Liu, Z (2024) Environmental risk substances in soil on seed germination: chemical species, inhibition performance, and mechanisms. *Journal of Hazardous Materials* 471:134518. https://doi.org/10.1016/j.jhazmat.2024.134518

Machado S, Pacheco JG (2016) Nanoremediation with zero-valent iron nanoparticles. In: Albergaria, JT, Nouws, HPA, (eds.), *From Soil Remediation,* 1st ed. CRC Press, pp. 108–120.

Maciel EVS, de Toffoli AL, Neto ES, et al. (2019) New materials in sample preparation: Recent advances and future trends. *TrAc Trends Anal Chem* 119:115633.

Mahmoud ME, Abdou AEH, Mohamed SMS, Osman MM (2016) Engineered staphylococcus aureus via immobilization on magnetic Fe_3O_4-phthalate nanoparticles for biosorption of divalent ions from aqueous solutions. *J Environ Chem Eng* 4:3810–3824.

Mani T, Frehland S, Kalberer A, Burkhardt-Holm P (2019) Using castor oil to separate microplastics from four different environmental matrices. *Anal Methods* 11:1788–1794. https://doi.org/10.1039/c8ay02559b

Martínez-Alcalá I, Bernal MP (2020) Environmental impact of metals, metalloids, and their toxicity. In: Deshmukh R, Tripathi DK, Guerriero G, (eds.), *Metalloids in Plants: Advances and Future Prospects*. John Wiley & Sons, Incorporated, pp. 451–488.

Martyn R, Galbraith H (2022) Re-materialising: Considering dominant understandings of value and systems of production within industrial plastics and the plastic arts. *Swamphen J Cult Ecol (ASLEC-ANZ)* 8:1–25.

Mason SA, Welch VG, Neratko J (2018) Synthetic polymer contamination in bottled water. *Front Chem* 6:407. https://doi.org/10.3389/fchem.2018.00407

Mbachu O, Jenkins G, Pratt C, Kaparaju P (2021) Enzymatic purification of microplastics in soil. *MethodsX* 8:101254.

Melchionda F, Silvestrini B, Robino C, et al. (2022) Development and validation of MPS-based system for human appearance prediction in challenging forensic samples. *Genes (Basel)*. 13:1688.

Miao J, Chen Y, Zhang E, et al. (2023) Effects of microplastics and biochar on soil cadmium availability and wheat plant performance. *GCB Bioenergy* 15:1046–1057. https://doi.org/10.1111/gcbb.13083

Reichelt F, Holder D, Maier T (2023). Interdisciplinary Practice in Industrial Design, Vol. 100, 2023, 109-118 AHFE. Interdisciplinary Practice in Industrial Design, 109.

Mohamed ES, Saleh AM, Belal AB, Gad A (2018) Application of near-infrared reflectance for quantitative assessment of soil properties. *Egypt J Remote Sens Sp Sci* 21:1–14. https://doi.org/10.1016/j.ejrs.2017.02.001

Mohanty AK, Wu F, Mincheva R, et al. (2022) Sustainable polymers. *Nat Rev Methods Prim* 2:46.

Moshood TD, Nawanir G, Mahmud F, et al. (2022) Sustainability of biodegradable plastics: New problem or solution to solve the global plastic pollution? *Curr Res Green Sustain Chem* 5:100273. https://doi.org/10.1016/j.crgsc.2022.100273

Mousavi SH, Kavianpour MR, Alcaraz JLG, Yamini OA (2023) System dynamics modeling for effective strategies in water pollution control: Insights and applications. *Appl Sci* 13:9024.

Nuelle MT, Dekiff JH, Remy D, Fries E (2014) A new analytical approach for monitoring microplastics in marine sediments. *Environmental Pollution* 184:161–169.

Mukherjee S, Shantha Kumar J, Nagar A, Pradeep T (2022) Concepts of sustainability in clean water technologies. In: Boul PJ (ed.), *Energy Transition: Climate Action and Circularity*. ACS Publications, pp. 625–657.

Mukhopadhyay A, Duttagupta S, Mukherjee A (2022) Emerging organic contaminants in global community drinking water sources and supply: A review of occurrence, processes and remediation. *J Environ Chem Eng* 10:107560.

Müller A, Kocher B, Altmann K, Braun U (2022) Determination of tire wear markers in soil samples and their distribution in a roadside soil. *Chemosphere* 294:133653.

Nayna OK, Tareq SM (2019) Application of semiconductor nanoparticles for removal of organic pollutants or dyes from wastewater. In: Ismail AF, Ahsan A (eds.), *Nanotechnology in Water and Wastewater Treatment*. Elsevier, pp. 267–290.

Ngo PL, Pramanik BK, Shah K, Roychand R (2019) Pathway, classification and removal efficiency of microplastics in wastewater treatment plants. *Environ Pollut* 255:113326.

Nikiema J, Asiedu Z (2022) A review of the cost and effectiveness of solutions to address plastic pollution. *Environ Sci Pollut Res* 29:24547–24573.

Oh S, Lu C (2023) Vertical farming - Smart urban agriculture for enhancing resilience and sustainability in food security. *J Hortic Sci Biotechnol* 98:133–140. https://doi.org/10.1080/14620316.2022.2141666

Okoffo ED, Ribeiro F, O'Brien JW, et al. (2020) Identification and quantification of selected plastics in biosolids by pressurized liquid extraction combined with double-shot pyrolysis gas chromatography–mass spectrometry. *Sci Total Environ* 715:136924.

Panagopoulos A, Haralambous K-J, Loizidou M (2019) Desalination brine disposal methods and treatment technologies-A review. *Sci Total Environ* 693:133545.

Parrish K, Fahrenfeld NL (2019) Microplastic biofilm in fresh-and wastewater as a function of microparticle type and size class. *Environ Sci Water Res Technol* 5:495–505.

Peramune D, Manatunga DC, Dassanayake RS, et al. (2022) Recent advances in biopolymer-based advanced oxidation processes for dye removal applications: A review. *Environ Res* 215:114242.

Perez CN, Carré F, Hoarau-Belkhiri A, et al. (2022) Innovations in analytical methods to assess the occurrence of microplastics in soil. *J Environ Chem Eng* 10:107421.

Phuong NN, Poirier L, Lagarde F, et al. (2018) Microplastic abundance and characteristics in French Atlantic coastal sediments using a new extraction method. *Environ Pollut* 243:228–237.

Pichlak M, Szromek AR (2022) Linking eco-innovation and circular economy—A conceptual approach. *J Open Innov Technol Mark Complex* 8:121.

Plastics Europe (2020) Plastics–the facts 2020. *PlasticEurope* 1:1–64.

Pothiraj C, Gokul TA, Kumar KR, et al. (2023) Vulnerability of microplastics on marine environment: A review. *Ecol Indic* 155:111058.

Pozo K, Urbina W, Gómez V, et al. (2020) Persistent organic pollutants sorbed in plastic resin pellet—"Nurdles" from coastal areas of Central Chile. *Mar Pollut Bull* 151:110786.

Prata JC, da Costa JP, Duarte AC, Rocha-Santos T (2019) Methods for sampling and detection of microplastics in water and sediment: A critical review. *TrAC Trends Anal Chem* 110:150–159.

Primpke S, Godejohann M, Gerdts G (2020) Rapid identification and quantification of microplastics in the environment by quantum cascade laser-based hyperspectral infrared chemical imaging. *Environ Sci Technol* 54:15893–15903.

Priya AK, Jalil AA, Dutta K, et al. (2022) Microplastics in the environment: Recent developments in characteristic, occurrence, identification and ecological risk. *Chemosphere* 298:134161.

Priya AK, Muruganandam M, Imran M, et al. (2023) A study on managing plastic waste to tackle the worldwide plastic contamination and environmental remediation. *Chemosphere* 341:139979.

Qi R, Jones DL, Li Z, et al. (2020) Behavior of microplastics and plastic film residues in the soil environment: A critical review. *Sci Total Environ* 703:134722. https://doi.org/10.1016/j.scitotenv.2019.134722

Querido W, Falcon JM, Kandel S, Pleshko N (2017) Vibrational spectroscopy and imaging: Applications for tissue engineering. *Analyst* 142:4005–4017. https://doi.org/10.1039/c7an01055a

Quinn B, Murphy F, Ewins C (2017) Validation of density separation for the rapid recovery of microplastics from sediment. *Anal Methods* 9:1491–1498.

Rafique A, Irfan M, Mumtaz M, Qadir A (2020) Spatial distribution of microplastics in soil with context to human activities: A case study from the urban center. *Environ Monit Assess* 192:1–13.

Raju M, Gandhimathi R, Nidheesh PV (2023) The cause, fate and effect of microplastics in freshwater ecosystem: Ways to overcome the challenge. *J Water Process Eng* 55:104199.

Rani M, Ducoli S, Depero LE, et al. (2023) A complete guide to extraction methods of microplastics from complex environmental matrices. *Molecules* 28:5710.

Ren X, Tang J, Wang L, Sun H (2021) Combined effects of microplastics and biochar on the removal of polycyclic aromatic hydrocarbons and phthalate esters and its potential microbial ecological mechanism. *Front Microbiol* 12:1–12. https://doi.org/10.3389/fmicb.2021.647766

Rose PK, Yadav S, Kataria N, Khoo KS (2023) Microplastics and nanoplastics in the terrestrial food chain: Uptake, translocation, trophic transfer, ecotoxicology, and human health risk. *TrAC Trends Anal Chem* 167:117249.

Sakthipriya N (2022) Plastic waste management: A road map to achieve circular economy and recent innovations in pyrolysis. *Sci Total Environ* 809:151160.

Samir A, Ashour FH, Hakim AAA, Bassyouni M (2022) Recent advances in biodegradable polymers for sustainable applications. *NPJ Mater Degrad* 6:68. https://doi.org/10.1038/s41529-022-00277-7

Schettini E, Scarascia-Mugnozza G, Blanco I, et al. (2022) Agricultural plastic waste. In: Colangelo F, Cioffi R, Farina IBT-H of SC and IWM (eds.), *Woodhead Publishing Series in Civil and Structural Engineering*. Woodhead Publishing, pp. 255–268.

Scheurer M, Bigalke M (2018) Microplastics in Swiss floodplain soils. *Environ Sci Technol* 52:3591–3598.

Shah SS, Shaikh MN, Khan MY, et al. (2021) Present status and future prospects of jute in nanotechnology: A review. *Chem Rec* 21:1631–1665.

Shao L, Li Y, Jones T, et al. (2022) Airborne microplastics: A review of current perspectives and environmental implications. *J Clean Prod* 347:131048. https://doi.org/10.1016/j.jclepro.2022.131048

Sheteiwy MS, Shaghaleh H, Hamoud YA, et al. (2021) Zinc oxide nanoparticles: Potential effects on soil properties, crop production, food processing, and food quality. *Environ Sci Pollut Res* 28:36942–36966.

Shim WJ, Hong SH, Eo SE (2017) Identification methods in microplastic analysis: A review. *Anal Methods* 9:1384–1391.

Song Q, Zhu J, Niu X, et al. (2022) Interfacial assembly of micro/nanoscale nanotube/silica achieves superhydrophobic melamine sponge for water/oil separation. *Sep Purif Technol* 280:119920.

Suman KH, Haque MN, Uddin MJ, et al. (2021) Toxicity and biomarkers of micro-plastic in aquatic environment: A review. *Biomarkers* 26:13–25. https://doi.org/10.1080/1354750X.2020.1863470

Sun H, Lei C, Yuan Y, Xu J, Han M (2022) Nanoplastic impacts on the foliar uptake, metabolism and phytotoxicity of phthalate esters in corn (Zea mays L.) plants. *Chemosphere* 304:135309. https://doi.org/10.1016/j.chemosphere.2022.135309

Tavşanoğlu ÜN, Koraltan İ, Kankılıç GB, et al. (2024) Assessing microplastic pollution in a river basin: A multidisciplinary study on circularity, sustainability, and socio-economic impacts. *Environmental Research* 262:119819. https://doi.org/10.1016/j.envres.2024.119819

Ter Halle A, Ghiglione JF (2021) Nanoplastics: A complex, polluting terra incognita. *Environ Sci Technol* 55:14466–14469.

Thiounn T, Smith RC (2020) Advances and approaches for chemical recycling of plastic waste. *J Polym Sci* 58:1347–1364. https://doi.org/10.1002/pol.20190261

Thomas D, Schütze B, Heinze WM, Steinmetz Z (2020) Sample preparation techniques for the analysis of microplastics in soil—A review. *Sustainability* 12:9074.

Tiller R, Booth AM, Cowan E (2022) Risk perception and risk realities in forming legally binding agreements: The governance of plastics. *Environ Sci Policy* 134:67–74.

Tittonell P, El Mujtar V, Felix G, et al. (2022) Regenerative agriculture—Agroecology without politics? *Front Sustain Food Syst* 6:844261. https://doi.org/10.3389/fsufs.2022.844261

Trontelj MJ, Chambers O (2021) Machine learning strategy for soil nutrients prediction using spectroscopic method. *Sensors (Basel)* 21:4208. https://doi.org/10.3390/s21124208

Tshangana CS, Muleja AA, Kuvarega AT, et al. (2021) The applications of graphene oxide quantum dots in the removal of emerging pollutants in water: An overview. *J Water Process Eng* 43:102249.

Ullah R, Tsui MT, Chen H, et al. (2021) Microplastics interaction with terrestrial plants and their impacts on agriculture. *J Environ Qual* 50:1024–1041.

Van Cauwenberghe L, Devriese L, Galgani F, et al. (2015) Microplastics in sediments: A review of techniques, occurrence and effects. *Mar Environ Res* 111:5–17.

Varier KM, Gudeppu M, Chinnasamy A, et al. (2019) Nanoparticles: Antimicrobial applications and its prospects. *Adv Nanostructured Mater Environ Remediat* 25:321–355.

Vicente-Vicente JL, Borderieux J, Martens K, et al. (2023) Scaling agroecology for food system transformation in metropolitan areas: Agroecological characterization and role of knowledge in community-supported agriculture farms connected to a food hub in Berlin, Germany. *Agroecol Sustain Food Syst* 47:857–889. https://doi.org/10.1080/21683565.2023.2187003

Vidal C, Pasquini C (2021) A comprehensive and fast microplastics identification based on near-infrared hyperspectral imaging (HSI-NIR) and chemometrics. *Environ Pollut* 285:117251

Vortmann S, Remy D, Klasmeier J, Beeken M (2019) Student experiments on the topic of micro-plastics in the environment for secondary education level II: The coffee pad machine experiment and sediment analysis. *World J Chem Educ* 7:96–101.

Vu HC, Dwivedi AD, Le TT, et al. (2017) Magnetite graphene oxide encapsulated in alginate beads for enhanced adsorption of Cr (VI) and As (V) from aqueous solutions: Role of crosslinking metal cations in pH control. *Chem Eng J* 307:220–229.

Wang Q, Guo S, Ali M, et al. (2022) Thermally enhanced bioremediation: A review of the fundamentals and applications in soil and groundwater remediation. *J Hazard Mater* 433:128749.

Wang Y, Xiang L, Amelung W, et al. (2023) Micro- and nanoplastics in soil ecosystems: Analytical methods, fate, and effects. *TrAC Trends Anal Chem* 169:117309. https://doi.org/10.1016/j.trac.2023.117309

Weidner T, Yang A, Hamm MW (2019) Consolidating the current knowledge on urban agriculture in productive urban food systems: Learnings, gaps and outlook. *J Clean Prod* 209:1637–1655. https://doi.org/10.1016/j.jclepro.2018.11.004

Wolf MP, Salieb-Beugelaar GB, Hunziker P (2018) PDMS with designer functionalities—Properties, modifications strategies, and applications. *Prog Polym Sci* 83:97–134. https://doi.org/10.1016/j.progpolymsci.2018.06.001

Wong JKH, Lee KK, Tang KHD, Yap P-S (2020) Microplastics in the freshwater and terrestrial environments: Prevalence, fates, impacts and sustainable solutions. *Sci Total Environ* 719:137512.

Xu Z, Xiong X, Zhao Y, et al. (2020) Pollutants delivered every day: Phthalates in plastic express packaging bags and their leaching potential. *J Hazard Mater* 384:121282.

Yang L, Zhang Y, Kang S, et al. (2021) Microplastics in soil: A review on methods, occurrence, sources, and potential risk. *Sci Total Environ* 780:146546.

Yazdi MH, Sepehrizadeh Z, Mahdavi M, et al. (2016) Metal, metalloid, and oxide nanoparticles for therapeutic and diagnostic oncology. *Nano Biomed Eng* 8:246–267.

Yin W, Zhang B, Zhang H, et al. (2022) Vertically co-distributed vanadium and microplastics drive distinct microbial community composition and assembly in soil. *J Hazard Mater* 440:129700.

Yip A, Smith-Roberge J, Khorasani SH, et al. (2022) Calibrating spatiotemporal models of microbial communities to microscopy data: A review. *PLOS Comput Biol* 18:e1010533.

Yu L, Zhang J, Liu Y, et al. (2021) Distribution characteristics of microplastics in agricultural soils from the largest vegetable production base in China. *Sci Total Environ* 756:143860.

Yusuf A, Sodiq A, Giwa A, et al. (2022) Updated review on microplastics in water, their occurrence, detection, measurement, environmental pollution, and the need for regulatory standards. *Environ Pollut* 292:118421.

Zhang B, Yang X, Chen L, et al. (2020a) Microplastics in soils: A review of possible sources, analytical methods and ecological impacts. *J Chem Technol Biotechnol* 95:2052–2068.

Zhang GS, Liu YF (2018) The distribution of microplastics in soil aggregate fractions in southwestern China. *Sci Total Environ* 642:12–20.

Zhang Q, He Y, Cheng R, et al. (2022a) Recent advances in toxicological research and potential health impact of microplastics and nanoplastics in vivo. *Environ Sci Pollut Res* 29:40415–40448.

Zhang Q, Song Y, Wu Z, et al. (2020b) Effects of six-year biochar amendment on soil aggregation, crop growth, and nitrogen and phosphorus use efficiencies in a rice-wheat rotation. *J Clean Product* 242:118435. https://doi.org/10.1016/j.jclepro.2019.118435

Zhang S, Li Y, Chen X, et al. (2022b) Occurrence and distribution of microplastics in organic fertilizers in China. *Sci Total Environ* 844:157061.

Zhang W (2003) Nanoscale iron particles for environmental remediation: An overview. *J Nanoparticle Res* 5:323–332.

Zhang Y, Gao T, Kang S, et al. (2021) Microplastics in glaciers of the Tibetan Plateau: Evidence for the long-range transport of microplastics. *Sci Total Environ* 758:143634.

Zhang Z, Chen Y (2020) Effects of microplastics on wastewater and sewage sludge treatment and their removal: A review. *Chem Eng J* 382:122955.

Zhang Z, Zhao S, Chen L, et al. (2022c) A review of microplastics in soil: Occurrence, analytical methods, combined contamination and risks. *Environ Pollut* 306:119374.

Zhao S, Zhang J, Feng S-J (2022a) The era of low-permeability sites remediation and corresponding technologies: A review. *Chemosphere* 313:137264.

Zhao X, Korey M, Li K, et al. (2022b) Plastic waste upcycling toward a circular economy. *Chem Eng J* 428:131928. https://doi.org/10.1016/j.cej.2021.131928

Zhou J, Wen Y, Cheng H, et al. (2022) Simazine degradation in agroecosystems: Will it be affected by the type and amount of microplastic pollution? *L Degrad Dev* 33:1128–1136.

Zhu F, Yan Y, Doyle E, et al. (2022) Microplastics altered soil microbiome and nitrogen cycling: The role of phthalate plasticizer. *J Hazard Mater* 427:127944.

Zhu F, Zhu C, Wang C, Gu C (2019) Occurrence and ecological impacts of microplastics in soil systems: A review. *Bull Environ Contam Toxicol* 102:741–749. https://doi.org/10.1007/s00128-019-02623-z

Zhu J-B, Watson EM, Tang J, Chen EY- X (2018) A synthetic polymer system with repeatable chemical recyclability. *Science* 360:398–403. https://doi.org/10.1126/science.aar5498

Ziajahromi S, Neale PA, Rintoul L, Leusch FDL (2017) Wastewater treatment plants as a pathway for microplastics: Development of a new approach to sample wastewater-based microplastics. *Water Res* 112:93–99. https://doi.org/10.1016/j.watres.2017.01.042

Zimmermann L, Göttlich S, Oehlmann J, et al. (2020) What are the drivers of microplastic toxicity? Comparing the toxicity of plastic chemicals and particles to Daphnia magna. *Environ Pollut* 267:115392.

CHAPTER 13

Agricultural Soils Containing Micro/ Nanoplastics and Related Risks

Naveen Chand, Manish Chaudhary,
Vineet Singh, and Surindra Suthar

ABBREVIATIONS

DOC Dissolved organic carbon
DOM Dissolved organic matter
HDPE high-density polyethylene
LDPE low-density polyethylene
MFs microplastic fibre(s)
MPs microplastic(s)
NPs nanoplastic(s)
PES Polyethersulfone
PLA polylactic acid
PP polypropylene
PS polystyrene

DOI: 10.1201/9781003487579-13

PTFE polytetrafluorethylene
PU polyurethane
PVC polyvinyl chloride

13.1 INTRODUCTION

Over the past decade, the high abundance of microplastics (MPs) in environmental frameworks has gained a significant amount of attention throughout the globe. Their high persistence and classification as emerging pollutants urge further investigation. These MPs are defined as plastic particles <5 mm (Bethanis and Golia, 2023), apart from that particles from 1 μm and 1 nm are defined as nanoplastics (NPs) (Frias and Nash, 2019). Further, their classification is made based on their origin namely, primary and secondary. Primary MPs are those which are intentionally manufactured for industrial, cosmetic, and cleaning products, etc. Secondary plastics are those which are developed by the weathering of the larger plastic particles (Tang et al., 2019). Consistent with that 367 million tonnes of worldwide plastic output has been reported in the year 2020 and it is estimated to reach 3.3 billion tonnes by 2030 (Hua et al., 2023). All these plastic particles are affecting the aquatic ecosystem and the terrestrial ecosystem. Simultaneously studies have documented that MPs have higher soil contamination as compared to the aquatic ecosystem and are approximately 4–23 times greater (de Souza Machado et al., 2018a; Sridharan et al., 2021). Thus it becomes more important to unveil the impact of MPs in soil systems mainly in agricultural soil.

Notably, it has been found that MPs-related hazards and their distribution in the terrestrial ecosystem have not been explored as much as in the aquatic environment. The reason behind this is the lack of a common procedure (Zantis et al., 2023). Subsequently, it has been indicated in the literature that among distinct terrestrial ecosystems, agricultural soils tend to have significantly higher MPs in the system (Yu et al., 2021a). However, the higher abundance in agricultural soil could be primarily linked to the common farming techniques used in the agricultural frameworks for cultivation, which may include untreated wastewater used for irrigation (Li et al., 2018), ground cover fabrics (Huang et al., 2020), granular fertilisers (Bian et al., 2022), different soil improvers, and sewage sludge (Weithmann et al., 2018; Vithanage et al., 2021). Subsequently, the source of MPs in the agricultural frameworks includes airborne deposition (Dris et al., 2016), surface runoff (Nizzetto et al., 2016), and direct littering.

It is noteworthy to understand the phenomena of how these MPs are distributed in agricultural soil. When these MPs reach the agricultural soils, they are carried forward by ploughing into the subsurface layers. Further, they may persist in soil via pores formed by microorganisms for a longer time frame because of their strong resistance to degradation (Fei et al., 2020; Guo et al., 2021). Research advances have unveiled that MPs have the potential to alter distinct biological and physicochemical characteristics of the soil. This alteration is likely to include porosity, nutrient availability, acidity, evaporation rate, and water-holding capacity (Qi et al., 2020a; Colzi et al., 2022; Wang et al., 2022a). However, whatever the case MPs have a serious impact on soil functionality and the health of crops and the organisms exposed to it. Simultaneously, it has the risk of trophic transmission via the food chain (Jin et al., 2022; Iqbal et al., 2023). Since soil is indicated as the repository of MPs, it is critical to disclose recent developments in agricultural soil research. Therefore, the purpose of this chapter is to discuss the advancements in the research conducted in agricultural soil and the hazards possessed by MPs. The source, distribution, biological and ecological hazards, impacts on agricultural production, and other elements including trophic transmission are summarised in this chapter.

13.2 MICROPLASTICS SOURCE IN SOIL

Human activity is primarily linked to the source of MPs in the soil environment (Figure 13.1). The primary contributing factors to MP pollution in soil environments include air deposition, composting, plastic mulching,

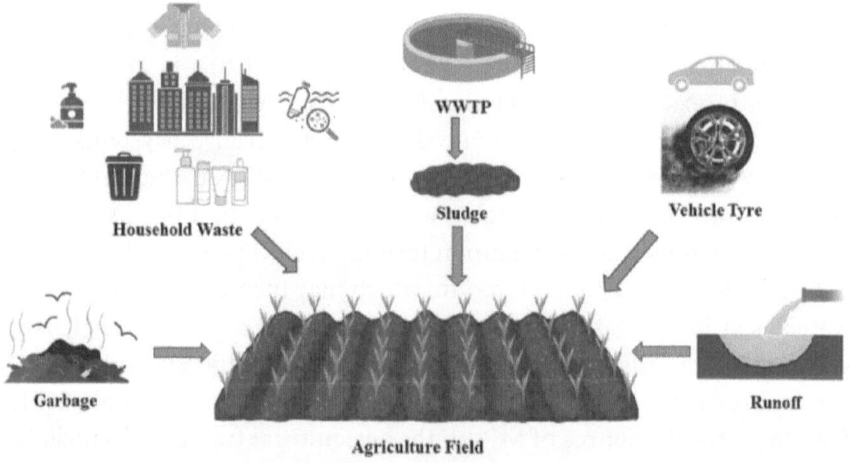

FIGURE 13.1 Sources of microplastics in agricultural soil.

littering, and sewage sludge. Mulch is mostly made of polyethylene (PE) and polyvinyl chloride (PVC), and it is a technology that is frequently globally employed in agricultural production. In the case of mulch, it is widely employed because of its substantial economic advantages, which include increased yield, improved fruit quality, and better water utilisation efficiency. The agricultural plastic films market was valued at 4 million tonnes globally in 2016 and is projected to reach 5.6 million tonnes annually by 2030 (Huang et al., 2020). Plastic mulch covers almost 20 million hectares of cropland worldwide. Since it takes a lot of work and time to remove all of the mulch from the field, plastic films—or portions of them—are frequently left in agricultural soils, where they can decompose into tiny particles and become brittle (Astner et al., 2019). Therefore, improper disposal of agricultural plastic films may cause MPs to build up in agricultural soils, which would be extremely dangerous for terrestrial species and food security (Ramos et al., 2015) hence, mulch becomes a substantial source of MPs in agricultural soil.

On the other hand, utilising organic wastes, such as compost, as soil amendments and fertilisers is a crucial practice in the circular bio-economy for enhancing or maintaining soil fertility parameters, such as soil organic carbon stocks, water-holding capacity, and soil nutrient contents (Hargreaves et al., 2008; Diacono and Montemurro, 2011). As a result, composts made from a range of source materials, such as home bio-waste, garden and green cuttings from commercial and park areas, and waste from food processing and agriculture, have been created for use in agriculture (Braun et al., 2021). Although plastic in compost has already been identified as a hazard, larger pieces from cosmetics have received most of the focus. Compost made from bio-waste is primarily contaminated with plastic due to inappropriate waste disposal, littering, and the use of regular plastic bags for collecting bio-waste (Bläsing and Amelung, 2018; Braun et al., 2021). As a result, compost plants employ several techniques, such as manual sorting and sifting both before and after composting, to lower the proportion of plastic in compost. However, plastic is still included in the finished product and is, for the most part, allowed by fertiliser laws (Braun et al., 2021). Consequently, using compost that contains MPs increases the amount of plastics in agricultural soil. Similarly, improper disposal of plastic waste contributes significantly to MP contamination in soil. Due to the over development of plastics and haphazard or insufficient management practices, a diverse spectrum of MP waste is present in the soil environment (Akdogan and Guven, 2019; Kumar et al., 2022).

Like aquatic habitats, MPs may originate from substantial plastic debris (Bläsing and Amelung, 2018; Wong et al., 2020). An estimated 6.3 billion tonnes of plastic waste were produced worldwide between 1950 and 2015; 4.97 billion of those were accumulated in landfills and other natural areas (Geyer et al., 2017). Unmanaged plastic waste is typically found in the dirt, next to roadways, or at unapproved disposal locations (Zhang and Liu, 2018). Many researchers discovered during fieldwork that plastic bottles and bags used for fertiliser and insecticides are dispersed across the cropland. This theory is supported by the morphological features of MPs, which show that the majority of MPs in agriculture have heavily weathered surfaces (Li et al., 2020). Nevertheless, no research has yet measured the quantity of MPs that end up in the soil as a result of illegal dumping or littering. These MPs may seep into the ground through flooding, street runoff, and wind. However, no quantitative research has yet been done on the MPs that enter into the soil as a result of unlawful rubbish disposal. In the sphere of agriculture, atmospheric deposition is another source of MPs. Through atmospheric transportation and deposition, smaller plastic particles from unprotected or improperly managed landfill sites and litter from urban areas can disperse into suspended particles in the air and then penetrate or deposit everywhere, including agricultural soils. Dris et al. (2016) documented MP deposition between 2 and 355 particles/m^2/day in the urban areas of Paris, France. As indicated above there are a variety of sources via which the MPs can be deposited to farmlands. Subsequently, these MPs get distributed within the soil environment causing various hazards and impacts.

13.3 DISTRIBUTION OF MICROPLASTICS IN AGRICULTURAL SOIL

Plastic items are widely used in modern agriculture because they enhance the safety, affordability, efficiency, and convenience of routine activities. The distribution is mostly driven by crops grown under foil tunnels, the use of fertiliser, pests, and resources like water. Additionally, synthetic fibres emitted from washing machines may be present in sewage sludge, and compost made from domestic garbage may contain packaging residues which may result in the contamination of croplands (Table 13.1). Similarly, crops shielded from pests and unpredictable weather with nets and mats tend to yield higher. In addition, practically all agricultural products are transported and stored in plastic containers (Zubris and Richards, 2005; Weithmann et al., 2018).

TABLE 13.1 Abundance of Microplastics in Agricultural Soil

Province	Source	Abundance (particle/kg)	MP Polymer Types	Land Use	References
Schleswig Holstein, Germany	Wastewater irrigation, organic fertiliser	0–217.8	PR, PP, PA, others	Winter-rapeseed-winter barley-winter wheat	Harms et al. (2021)
China	—	5,386±835 5,124±632	—	Mulching land, greenhouse land	Huang et al. (2020)
Qinghai – Tibet plateau, China	Mulching film, wastewater irrigation, littering	53.2±29.7 43.9±22.3	PP, PE, PS, PA and others	Farmland. mulching land, greenhouse land	Feng et al. (2022)
Shouguang, China	Mulching film, organic fertiliser, littering, runoff	275–5,411 179–7,175 307–4,507	PP, EPC, PE, PS, PES, PU, ABS, PMMA and others	Vegetable land	Yu et al. (2021a)
Ontario, Canada	Sludge application	4–541	PS, PP, PE, PES, Arc, PU, PMMA, PA and others	—	Crossman et al. (2020)
Yeoju, Korea	The mulching film, car tyres, atmospheric deposition	664–3,440	PE, PP, PS, PVC	Orchard field, greenhouse and paddy field	Choi et al. (2021)
Yong-In, Korea	Littering, mulching film	20–325 10–265 75–7,630	PE, PP, PET	Paddy land, mulching and greenhouse land	Kim et al. (2021)
Mellipilla, Chile	Sludge application	600–10,400	—	Cornland	Corradini et al. (2019)
Southeast Germany	—	4–541	PE, PS, PP	—	Piehl et al. (2018)
Netherland	—	664	PE, PP, SBR	Agriculture land	Choi et al. (2021)
Pakistan	Littering, runoff	675 mg	PE, PP, PES, PS, and PVC	Agriculture land	Sajjad et al. (2022)
Argentina	Sludge, littering	125 mg	PE, PP, and PS	Agriculture land	Rodriguez-Seijo et al. (2017)

Studies on this subject are still lacking, even though a variety of parameters, including planting age, soil texture, irrigation techniques, and others, may have an impact on the distribution of MPs in soil. MPs in soils can move by a variety of methods, such as mechanical disturbance, leaching, and bioturbation (Li et al., 2020), though the vertical MP distribution is not always consistent. For instance, in the open-field croplands of China (Shanghai), the abundance of MPs within soil layer 0–3 cm was much higher than that in soil layer 3–6 cm (Liu et al., 2018). However, in the facility croplands and lakeside wetlands converted from croplands near Dianchi Lake (Southwest China), there was no noticeable difference in soil layers for MPs abundance i.e. between 0–5 and 5–10 cm (Liu, 2018). Likewise, similar patterns from family gardens in Mexico were discovered for the 0–10 and 10–20 cm soil layers (Huerta et al., 2017). Since the distribution of MPs depends on several variables, it is crucial to take all of these aspects into account to comprehend their mobility pores (Rahmatpour et al., 2018). For instance, research has shown that soil texture, which is related to the distribution and continuity of soil pores, may have an impact on the accumulation and mobility of MPs in soil frameworks. Planting age is therefore an additional factor that could affect the distribution because MPs might be incorporated into the soil during planting operations (Bläsing and Amelung, 2018). Moreover, another source of MPs in the soil frameworks is irrigation water frameworks (Bläsing and Amelung, 2018). Literature has shown that adding MPs to natural soils with irrigation water may go as (i.e. surface irrigation > sprinkler irrigation > drip irrigation) (Yu et al., 2021b). Some MPs persist in the soil instantly following their release from human production activities, while others are continually enhanced in the food chain (Wu et al., 2017). The research indicates that surface soil contains MPs up to 80–100 pieces/kg of soil (Liu et al., 2018). Furthermore, some MPs travel horizontally through the action of wind and water to other areas of the farmland (Scheurer and Bigalke, 2018). Notably, the residual MPs in the soil may be moved vertically and end up deep down in the soil (Lwanga et al., 2017). When these MPs act as the vector for additional contaminants, their migration may become even more dangerous. For instance, they can adsorb heavy metals, persistent organic pollutants (Guo et al., 2020, 2018; Hüffer et al., 2018) and some emerging contaminants (such as pesticides, and antibiotics) (Avio et al., 2015). The transport of these pollutants could lead to various adverse effects on the soil organisms and soil characteristics. Once these MPs enter the cropland they are subjected to distinct effects including decreased soil fertility, impact on food production and altered ecological functioning (Scheurer and Bigalke, 2018). Additionally, the presence of MPs in crops may promote their trophic transmission across the food chain, which may affect human health (Wu et al., 2017; Zhou et al., 2018).

13.4 POTENTIAL HAZARDS OF MICROPLASTICS IN AGRICULTURAL SOIL

13.4.1 Fertility and Nutrients

Nutrients are important for the cultivation of crops and these nutrients in the soil originate from the breakdown of minerals and organic matter as indicated in the literature (Wang et al., 2022a). Subsequently, enzymes play a vital role in regulating nutrient cycles, which serve as a measure of fertility (Liu et al., 2017; Guo et al., 2020). But with the presence of the MPs in the soil the physicochemical parameters of the soil are impacted. For instance, a study by Fei et al. (2020) indicated how, in acidic soil, PE and PVC influence microbial activity and promote urease and acid phosphatase. Likewise, an important nitrogen cycle enzyme *N*-(leucine aminopeptidase) is also less active when exposed to PS NPs (Bandopadhyay et al., 2019; Awet et al., 2018). Additionally, advances in research have indicated that MPs affect the cycling of carbon and nutrients via DOM, as the addition of PP MPs to soil significantly raises the amounts of DOC, N, and P in the soil (Liu et al., 2017). Similar research demonstrated that plastic film mulching has a significant impact on the nitrogen cycle and lowers the level of inorganic nitrogen (Ya et al., 2021), while PP and PE MPs drastically increase DOM by 72%–324% (Yu et al., 2021a). Another research documented that the content of accessible P was reduced by approximately 30%, while the content of available N in a rice paddy soil polluted with 1% (w/w)PVC MPs was reduced by 10%–13%(Yan et al., 2021). Similarly, available P in the soil was reduced by adding PE MPs (Yu et al., 2020) and soil-accessible N and P content was decreased by PS MP and PTFE MPs (Dong et al., 2021). According to Lozano et al. (2021a) MFs (microplastic fibers) decreased $NO_3{}^-{}_{-N}$ leaching by about 70%. This is probably because MPs have a beneficial effect on soil aggregation, which increases the soil's ability to hold onto nutrients. In contradiction in a different study, 0.2% (w/w) MPs did not affect the availability of nutrients in the soil supply (Li et al., 2021b). However according to Chen et al. (2020), PLA MPs considerably raised the NO_3^- and NO_2-N content of soil under low carbon conditions, while decreasing the NH_4^+-N content. This suggests that PLA MPs might help with the nitrification of NH_4^+-N.

To summarise based on the aforementioned data we can suggest that MPs do affect soil nutrients directly and indirectly via a variety of mechanisms which may vary depending on the characteristics of MPs such as the type, dose, form, and size. These effects may be interconnected and interact with one another. Furthermore, the evidence now available is derived from limited studies. Nothing about the impact of MPs on soil micronutrients, including Fe, Mn, Zn, and Cu, is known. Therefore, studies related to how MPs induce changes in soil, nutrient cycling, and plants should be the primary focus (Wang et al., 2022a).

13.4.2 pH and Soil Physical Properties

pH of the soil is one of the most important parameters that determine soil qualities including soil organic carbon, bioavailability, binding capacity of minerals, soil organic carbon, microbial composition and adsorption of nutrients and pollutants (Rousk et al., 2010). Therefore, when MPs are introduced in the soil they have the potential to create alteration in soil pH. These MPs when interacting with distinct organic and inorganic components such as protons and cations (Boots et al., 2019) may lead to alteration in soil pH. In an investigation, Qi et al. (2020b) reported the influence of LDPE and biodegradable mulch on soil chemistry and discovered that MPs raise soil pH. Similarly, a study reported that PE MPs lowered the pH in acidic soils and boosted the pH in alkaline soils (Dissanayake et al., 2022). Likewise, few other studies documented a rise in soil pH by the addition of MPs such as polylactic acid (PLA) MPs and HDPE (Yang et al., 2021) as well as LDPE and bio-plastic mulch films (Qi et al., 2020a). This demonstrates how MPs modify pH in various soil types. However, it is unclear what mechanisms are driving these changes, thus more research is required to be conducted in the field as well as the laboratory scale. The investigation is required to address a variety of soil and changes in the soil characteristics with different types of MPs.

The primary element governing soil stability, porosity, and structure is the soil aggregates. Approximately 72% of MPs in soil considerably diminish soil water aggregates (Zhang and Liu, 2018), which makes it problematic that MPs may reduce biological activity and function (Rillig and Lehmann, 2020; Huang et al., 2022). The dry-wet cycle causes MPs to disperse into the soil matrix. Eventually, this dispersion of MPs in soil affects physical qualities such as water-holding capacity, aggregate stability and soil bulk density (Ya et al., 2021; Guo et al., 2020). The impacts vary according to the type of MPs, such as polyester, which affects soil bulk density (Xu et al., 2020). Moreover, MPs may alter soil porosity, increasing evaporation rates and causing soil cracking (Wan et al., 2019; de Souza et al., 2018). Apart from that varying MPs have varying influences on water and water-holding capacity for instance polyester fibres have the highest effect when compared to PE and polyacrylic water-holding capacity (de Souza Machado et al., 2018b). Further studies have reported no significant impact on the bulk density of soil hence more research to work out how distinct forms of MPs affect the bulk density of soil is required. In line with that Zhang et al. (2019) revealed that PMF did not influence soil bulk density. Hence there is a wide gap in understanding the fate of the MPs and their impact on the soil.

13.4.3 Soil Microorganisms

Microbial consortia and their roles are dependent on soil factors and can differ significantly amongst soil compartments (Garbeva et al., 2008). It is commonly known that a range of soil parameters, including soil type, fertility, organic matter concentration pH, and pollution status, has a significant influence on soil microbial composition within the soil matrix. However, different soil microbial communities react differently to disturbances depending on their resistance and resilience (Griffiths and Philippot, 2013; Shade et al., 2012). In one instance, it was discovered that the bacterial populations in paddy and red soil showed different contrasting effects with the addition of PVC MPs (Yan et al., 2021). Food production and the functions provided by the soil ecosystem depend heavily on soil microbes. Regrettably, the presence of MPs in soil harms its microbial and fungal populations, which are crucial to the health of the soil (Frontiers Production, 2021; Azeem et al., 2021). According to de Souza Machado et al. (2018b), the activity and variety of microbial communities are reduced by polyester and polyacrylic MPs, particularly in acidic soils.. Additionally, the variety of soil microbial populations is increased by membranous PE and fibrous PP (Sridharan et al., 2022; Lei et al., 2018b; Besseling et al., 2013).

There is a scarcity of information on microbial biota and MPs in soil frameworks. Subsequently, research on what extent these MPs have an impact on how MPs affect soil microbial society and functions is currently lacking. This makes it difficult to determine the relationship between microbial community functionality and MP exposure. Such studies ought to cover neurological impacts, immune system regulation, and reproductive disturbance in earthworms that impact, among other things, oxidative stress, sperm viability, and coelomocyte viability (Kwak and An, 2021). Additionally, MPs have an impact on the growth, reproduction, and energy metabolism of *Caenorhabditis elegans*, snails, and springtails (Lei et al., 2018a and 2018b). Nevertheless, the majority of experiments are performed within a controlled environment (laboratory), and further, in-situ-empirical data from field research is required to comprehend the impact on the soil ecosystem.

13.4.4 Effects on Plant Growth and Physiology

Plant roots and MPs interact in a variety of ways. For example, plastic litter may hinder the growth of root hairs, which could compromise the plant's ability to absorb nutrients and water (Bosker et al., 2019). Early crop stages may be directly harmed by soil MPs if the pores in the seed capsule or roots are physically blocked (Khalid et al., 2020). In fact, with MP exposure, very short-term detrimental effects on the growth of edible plants may manifest as early as 6 days after sowing (Pignattelli et al., 2020). Subsequently, limited photosynthetic rate and leaf expansion may further impair root system development, which is another indicator of N deficiency since root growth is dependent on the plant's carbohydrate content (Postma et al., 2014). The way MPs affect plant growth varies according to the species. Research findings have indicated the accumulation of MPs in the cortex has an impact on the length and growth of the roots, which reduces the absorption of nutrients and water (Bosker et al., 2019; Lian et al., 2020). It has also been indicated that MPs impact the number of fruits, height, and seedlings of plants (Pignattelli et al., 2020). Literature reveals that LDPE and biodegradable plastic films had a significant impact on wheat growth, while the treatment with PS-NPs (200 nm) significantly reduced fresh weight and plant height in *Arabidopsis* seedlings (Qi et al., 2018). Conversely, root biomass is increased by polyester and PS MPs, indicating the need for additional research on the effects of MP exposure on plant physiology and growth (de Souza Machado et al., 2019).

The physiology of plants is also impacted by MPs in addition to growth. Plant growth depends heavily on photosynthesis, which is impacted by several MPs. These MPs reduce the amount of chlorophyll (chl-a and chl-b) in leaves while preventing protein synthesis and altering energy and amino acid metabolism (Wang et al., 2022b; Gao et al., 2021). Furthermore, MPs impact physiology and growth by raising the reactive oxygen species (ROS) content in leaves and roots, which raises the levels of the antioxidant enzymes SOD and CAT (Gao et al., 2019b). Furthermore,

Zong et al.'s (2021) hydroponic wheat seedling experiment showed that PS MPs improve photosynthesis and lower reactive oxygen species while reducing the accumulation of cuprum and cadmium in wheat seedlings, suggesting a decrease in the harmful impacts of heavy metals. According to Dong et al. (2020), PS and PTFE may also shield as lesions from chlorophyll. These findings collectively demonstrate the intricate consequences on agricultural ecosystems of concurrent exposure to MPs and heavy metals.

It is well known that the growth of agriculturally interesting plants may be impacted by the direct uptake of MPs from soils via apoplastic (uptake via passive diffusion) and symplastic (uptake via osmosis) pathways and that these MPs may then be transported to the entire plant via the vascular system (Kumar et al., 2022). However, the persistence of MPs in soils and their long-term effects are difficult to assess (Kumari et al., 2022). Furthermore, stomatal foliar absorption may allow even the smallest plastic particles to reach the plants (Azeem et al., 2021).

13.5 CONCLUSION AND FUTURE PERSPECTIVES

Agricultural land undergoes various consequences due to the presence of the MPs such as reduced permeability, disruption of the structure, water retention capacity, and nutrient cycling. Subsequently, these MPs also disturb the composition of the soil microbiota which leads to an ecological imbalance in the soil frameworks and eventually it leads to reduced biodiversity in the soil matrix. Similarly, these MPs can also infiltrate into plant tissues which increases the risk of reduced growth rate and trophic transmission of the MPs. As a result, both humans and other forms of life are subjected to the potential threat of MPs. Therefore, developing MP mitigation strategies and focusing on sustainable agricultural growth becomes crucial.

It is essential to comprehend the characteristics and environmental behaviours of MPs to promote sustainable agricultural production and implement the required methods into practice. For instance, the following strategies might be taken into account: encouraging cropland management techniques, cutting back on the excessive use of plastic films, looking into substitute materials, refining the processes used to treat sludge and wastewater, raising environmental awareness, and encouraging collaboration in scientific research. It is necessary to conduct more research to determine the effects of MP interactions with other contaminants in soil frameworks. Future studies should take into account the simultaneous long-term effects of environmental ageing on MPs in agricultural soil systems from many sources, with an emphasis on human health and food security.

REFERENCES

Akdogan, Z., & Guven, B. (2019). Microplastics in the environment: A critical review of current understanding and identification of future research needs. *Environmental Pollution* 254, 113011.

Astner, A. F., Hayes, D. G., O'Neill, H., Evans, B. R., Pingali, S. V., Urban, V. S., & Young, T. M. (2019). Mechanical formation of micro-and nano-plastic materials for environmental studies in agricultural ecosystems. *Science of the Total Environment, 685*, 1097–1106.

Avio, C. G., Gorbi, S., Milan, M., Benedetti, M., Fattorini, D., d'Errico, G., ... & Regoli, F. (2015). Pollutants bioavailability and toxicological risk from microplastics to marine mussels. *Environmental Pollution, 198*, 211–222.

Awet, T. T., Kohl, Y., Meier, F., Straskraba, S., Grün, A. L., Ruf, T., ... & Emmerling, C. (2018). Effects of polystyrene nanoparticles on the microbiota and functional diversity of enzymes in soil. *Environmental Sciences Europe, 30*(1), 1–10.

Azeem, I., Adeel, M., Ahmad, M. A., Shakoor, N., Jiangcuo, G. D., Azeem, K., ... & Rui, Y. (2021). Uptake and accumulation of nano/microplastics in plants: a critical review. *Nanomaterials, 11*(11), 2935.

Bandopadhyay, S., Sintim, H. Y., & DeBruyn, J. M. (2019). Structural and functional responses of soil microbial communities to biodegradable plastic film mulching in two agroecosystems. *BioRxiv*, 650317.

Besseling, E., Wegner, A., Foekema, E. M., Van Den Heuvel-Greve, M. J., & Koelmans, A. A. (2013). Effects of microplastic on fitness and PCB bioaccumulation by the lugworm Arenicola marina (L.). *Environmental Science & Technology, 47*(1), 593–600.

Bethanis, J., & Golia, E. E. (2023). Revealing the combined effects of microplastics, Zn, and Cd on soil properties and metal accumulation by leafy vegetables: a preliminary investigation by a laboratory experiment. *Soil Systems, 7*(3), 65.

Bian, W., An, L., Zhang, S., Feng, J., Sun, D., Yao, Y., ... & Zhang, M. (2022). The long-term effects of microplastics on soil organomineral complexes and bacterial communities from controlled-release fertilizer residual coating. *Journal of Environmental Management, 304*, 114193.

Bläsing, M., & Amelung, W. (2018). Plastics in soil: analytical methods and possible sources. *Science of the Total Environment, 612*, 422–435.

Boots, B., Russell, C. W., & Green, D. S. (2019). Effects of microplastics in soil ecosystems: above and below ground. *Environmental Science & Technology, 53*(19), 11496–11506.

Bosker, T., Bouwman, L. J., Brun, N. R., Behrens, P., & Vijver, M. G. (2019). Microplastics accumulate on pores in seed capsule and delay germination and root growth of the terrestrial vascular plant Lepidium sativum. *Chemosphere, 226*, 774–781.

Braun, M., Mail, M., Heyse, R., & Amelung, W. (2021). Plastic in compost: prevalence and potential input into agricultural and horticultural soils. *Science of the Total Environment, 760*, 143335.

Chen, H., Wang, Y., Sun, X., Peng, Y., & Xiao, L. (2020). Mixing effect of polylactic acid microplastic and straw residue on soil property and ecological function. *Chemosphere, 243*, 125271.

Choi, Y. R., Kim, Y.-N., Yoon, J.-H., Dickinson, N., & Kim, K.-H. (2021). Plastic contamination of forest, urban, and agricultural soils: a case study of Yeoju City in the Republic of Korea. *Journal of Soils and Sediments, 21*, 1962–1973. https://doi.org/10.1007/s11368-020-02759-0

Colzi, I., Renna, L., Bianchi, E., Castellani, M. B., Coppi, A., Pignattelli, S., … & Gonnelli, C. (2022). Impact of microplastics on growth, photosynthesis and essential elements in Cucurbita pepo L. *Journal of Hazardous Materials, 423*, 127238.

Corradini, F., Meza, P., Eguiluz, R., Casado, F., Huerta-Lwanga, E., & Geissen, V. (2019). Evidence of microplastic accumulation in agricultural soils from sewage sludge disposal. *Science of the Total Environment, 671*, 411–420. https://doi.org/10.1016/J.SCITOTENV.2019.03.368

Crossman, J., Hurley, R. R., Futter, M., & Nizzetto, L. (2020). Transfer and transport of microplastics from biosolids to agricultural soils and the wider environment. *Science of the Total Environment, 724*, 138334. https://doi.org/10.1016/J.SCITOTENV.2020.138334

de Souza Machado, A. A., Kloas, W., Zarfl, C., Hempel, S., & Rillig, M. C. (2018a). Microplastics as an emerging threat to terrestrial ecosystems. *Global Change Biology, 24*(4), 1405–1416.

de Souza Machado, A. A., Lau, C. W., Kloas, W., Bergmann, J., Bachelier, J. B., Faltin, E., … & Rillig, M. C. (2019). Microplastics can change soil properties and affect plant performance. *Environmental Science & Technology, 53*(10), 6044–6052.

de Souza Machado, A. A., Lau, C. W., Till, J., Kloas, W., Lehmann, A., Becker, R., & Rillig, M. C. (2018b). Impacts of microplastics on the soil biophysical environment. *Environmental Science & Technology, 52*(17), 9656–9665.

Diacono, M., & Montemurro, F. (2011). Long-term effects of organic amendments on soil fertility. *Sustainable Agriculture, 2*, 761–786.

Dissanayake, P. D., Kim, S., Sarkar, B., Oleszczuk, P., Sang, M. K., Haque, M. N., … & Ok, Y. S. (2022). Effects of microplastics on the terrestrial environment: a critical review. *Environmental Research, 209*, 112734.

Dong, Y., Gao, M., Qiu, W., & Song, Z. (2021). Effect of microplastics and arsenic on nutrients and microorganisms in rice rhizosphere soil. *Ecotoxicology and Environmental Safety, 211*, 111899.

Dong, Y., Gao, M., Song, Z., & Qiu, W. (2020). Microplastic particles increase arsenic toxicity to rice seedlings. *Environmental Pollution, 259*, 113892.

Dris, R., Gasperi, J., Saad, M., Mirande, C., & Tassin, B. (2016). Synthetic fibers in atmospheric fallout: a source of microplastics in the environment? *Marine Pollution Bulletin, 104*(1–2), 290–293.

Fei, Y., Huang, S., Zhang, H., Tong, Y., Wen, D., Xia, X., … & Barceló, D. (2020). Response of soil enzyme activities and bacterial communities to the accumulation of microplastics in an acid cropped soil. *Science of the Total Environment, 707*, 135634.

Feng, X., Wang, Q., Sun, Y., Zhang, S., & Wang, F. (2022). Microplastics change soil properties, heavy metal availability and bacterial community in a Pb-Zn-contaminated soil. *Journal of Hazardous Materials, 424*, 127364. https://doi.org/10.1016/J.JHAZMAT.2021.127364

Frias, J. P., & Nash, R. (2019). Microplastics: finding a consensus on the definition. *Marine Pollution Bulletin, 138*, 145–147.

Gao, G., Zhao, X., Jin, P., Gao, K., & Beardall, J. (2021). Current understanding and challenges for aquatic primary producers in a world with rising micro-and nano-plastic levels. *Journal of Hazardous Materials, 406*, 124685.

Gao, M., Liu, Y., & Song, Z. (2019b). Effects of polyethylene microplastic on the phytotoxicity of di-n-butyl phthalate in lettuce (Lactuca sativa L. var. ramosa Hort). *Chemosphere, 237*, 124482.

Garbeva, P., Van Elsas, J. D., & Van Veen, J. A. (2008). Rhizosphere microbial community and its response to plant species and soil history. *Plant and Soil, 302*, 19–32.

Griffiths, B. S., & Philippot, L. (2013). Insights into the resistance and resilience of the soil microbial community. *FEMS Microbiology Reviews, 37*(2), 112–129.

Guo, J. J., Huang, X. P., Xiang, L., Wang, Y. Z., Li, Y. W., Li, H., … & Wong, M. H. (2020). Source, migration and toxicology of microplastics in soil. *Environment International, 137*, 105263.

Guo, Q. Q., Xiao, M. R., & Zhang, G. S. (2021). The persistent impacts of polyester microfibers on soil bio-physical properties following thermal treatment. *Journal of Hazardous Materials, 420*, 126671.

Guo, X., Hu, G., Fan, X., & Jia, H. (2020). Sorption properties of cadmium on microplastics: the common practice experiment and a two-dimensional correlation spectroscopic study. *Ecotoxicology and Environmental Safety, 190*, 110118.

Guo, X., Pang, J., Chen, S., & Jia, H. (2018). Sorption properties of tylosin on four different microplastics. *Chemosphere, 209*, 240–245.

Hargreaves, J. C., Adl, M. S., & Warman, P. R. (2008). A review of the use of composted municipal solid waste in agriculture. *Agriculture, Ecosystems & Environment, 123*(1–3), 1–14.

Harms, I. K., Diekötter, T., Troegel, S., & Lenz, M. (2021). Amount, distribution and composition of large microplastics in typical agricultural soils in Northern Germany. *Science of the Total Environment, 758*, 143615. https://doi.org/10.1016/J.SCITOTENV.2020.143615

Hua, Z., Ma, S., Ouyang, Z., Liu, P., Qiang, H., & Guo, X. (2023). The review of nanoplastics in plants: detection, analysis, uptake, migration and risk. *TrAC Trends in Analytical Chemistry, 158*, 116889.

Huang, D., Wang, X., Yin, L., Chen, S., Tao, J., Zhou, W., … & Xiao, R. (2022). Research progress of microplastics in soil-plant system: ecological effects and potential risks. *Science of the Total Environment, 812*, 151487.

Huang, Y., Liu, Q., Jia, W., Yan, C., & Wang, J. (2020). Agricultural plastic mulching as a source of microplastics in the terrestrial environment. *Environmental Pollution, 260*, 114096.

Huerta, L. E., Mendoza, V. J., Ku Quej, V., Chi, J. L. A., Sanchez Del Cid, L., Chi, C., … & Geissen, V. (2017). Field evidence for transfer of plastic debris along a terrestrial food chain. *Scientific Reports, 7*, 14071.

Geyer, R., Jambeck, J. R., & Law, K. L. (2017). Production, use, and fate of all plastics ever made. *Science Advances*, 3(7), e1700782.

Hüffer, T., Weniger, A. K., & Hofmann, T. (2018). Sorption of organic compounds by aged polystyrene microplastic particles. *Environmental Pollution, 236*, 218–225.

Iqbal, B., Zhao, T., Yin, W., Zhao, X., Xie, Q., Khan, K. Y., … & Du, D. (2023). Impacts of soil microplastics on crops: a review. *Applied Soil Ecology, 181*, 104680.

Jin, T., Tang, J., Lyu, H., Wang, L., Gillmore, A. B., & Schaeffer, S. M. (2022). Activities of microplastics (MPs) in agricultural soil: a review of MPs pollution from the perspective of agricultural ecosystems. *Journal of Agricultural and Food Chemistry, 70*(14), 4182–4201.

Khalid, N., Aqeel, M., & Noman, A. (2020). Microplastics could be a threat to plants in terrestrial systems directly or indirectly. *Environmental Pollution, 267*, 115653.

Kim, S. K., Kim, J. S., Lee, H., & Lee, H. J. (2021). Abundance and characteristics of microplastics in soils with different agricultural practices: importance of sources with internal origin and environmental fate. *Journal of Hazardous Materials, 403*, 123997. https://doi.org/10.1016/J.JHAZMAT.2020.123997

Kumar, R., Ivy, N., Bhattacharya, S., Dey, A., & Sharma, P. (2022). Coupled effects of microplastics and heavy metals on plants: uptake, bioaccumulation, and environmental health perspectives. *Science of the Total Environment, 836*, 155619.

Kumari, A., Rajput, V. D., Mandzhieva, S. S., Rajput, S., Minkina, T., Kaur, R., … & Glinushkin, A. P. (2022). Microplastic pollution: an emerging threat to terrestrial plants and insights into its remediation strategies. *Plants, 11*(3), 340.

Kwak, J. I., & An, Y. J. (2021). Microplastic digestion generates fragmented nanoplastics in soils and damages earthworm spermatogenesis and coelomocyte viability. *Journal of Hazardous Materials, 402*, 124034.

Lei, L., Liu, M., Song, Y., Lu, S., Hu, J., Cao, C., … & He, D. (2018a). Polystyrene (nano) microplastics cause size-dependent neurotoxicity, oxidative damage and other adverse effects in Caenorhabditis elegans. *Environmental Science: Nano, 5*(8), 2009–2020.

Lei, L., Wu, S., Lu, S., Liu, M., Song, Y., Fu, Z., … & He, D. (2018b). Microplastic particles cause intestinal damage and other adverse effects in zebrafish Danio rerio and nematode Caenorhabditis elegans. *Science of the Total Environment, 619*, 1–8.

Li, H. Z., Zhu, D., Lindhardt, J. H., Lin, S. M., Ke, X., & Cui, L. (2021b). Long-term fertilization history alters effects of microplastics on soil properties, microbial communities, and functions in diverse farmland ecosystem. *Environmental Science & Technology, 55*(8), 4658–4668.

Li, J., Song, Y., & Cai, Y. (2020). Focus topics on microplastics in soil: analytical methods, occurrence, transport, and ecological risks. *Environmental Pollution*, 257, 113570.

Li, X., Chen, L., Mei, Q., Dong, B., Dai, X., Ding, G., & Zeng, E. Y. (2018). Microplastics in sewage sludge from the wastewater treatment plants in China. *Water Research, 142*, 75–85.

Lian, J., Wu, J., Xiong, H., Zeb, A., Yang, T., Su, X., … & Liu, W. (2020). Impact of polystyrene nanoplastics (PSNPs) on seed germination and seedling growth of wheat (Triticum aestivum L.). *Journal of Hazardous Materials, 385*, 121620.

Liu, H., Yang, X., Liu, G., Liang, C., Xue, S., Chen, H., ... & Geissen, V. (2017). Response of soil dissolved organic matter to microplastic addition in Chinese loess soil. *Chemosphere*, *185*, 907–917.

Liu, M., Lu, S., Song, Y., Lei, L., Hu, J., Lv, W., ... & He, D. (2018). Microplastic and mesoplastic pollution in farmland soils in suburbs of Shanghai, China. *Environmental Pollution*, *242*, 855–862.

Liu, Y. F. (2018). Quantity and distribution of microplastics in farmland soil of lake shore in Dianchi Lake. Yunnan University, Kunmig.

Lozano, Y. M., Aguilar-Trigueros, C. A., Onandia, G., Maaß, S., Zhao, T., & Rillig, M. C. (2021a). Effects of microplastics and drought on soil ecosystem functions and multi-functionality. *Journal of Applied Ecology*, *58*(5), 988–996.

Lwanga, E. H., Gertsen, H., Gooren, H., Peters, P., Salánki, T., van der Ploeg, M., ... & Geissen, V. (2017). Incorporation of microplastics from litter into burrows of Lumbricus terrestris. *Environmental Pollution*, *220*, 523–531.

Nizzetto, L., Bussi, G., Futter, M. N., Butterfield, D., & Whitehead, P. G. (2016). A theoretical assessment of microplastic transport in river catchments and their retention by soils and river sediments. *Environmental Science: Processes & Impacts*, *18*(8), 1050–1059.

Office, F. P. (2021). Erratum: microplastic shape, polymer type, and concentration affect soil properties and plant biomass. *Frontiers in Plant Science*, *12*, 714541.

Piehl, S., Leibner, A., Löder, M. G. J., Dris, R., Bogner, C., & Laforsch, C. (2018). Identification and quantification of macro- and microplastics on an agricultural farmland. *Scientific Reports*, *8*(1), 17950. https://doi.org/10.1038/s41598-018-36172-y

Pignattelli, S., Broccoli, A., & Renzi, M. (2020). Physiological responses of garden cress (L. sativum) to different types of microplastics. *Science of the Total Environment*, *727*, 138609.

Postma, J. A., Dathe, A., & Lynch, J. P. (2014). The optimal lateral root branching density for maize depends on nitrogen and phosphorus availability. *Plant Physiology*, *166*(2), 590–602. https://doi.org/10.1104/pp.113.233916

Qi, Y., Beriot, N., Gort, G., Lwanga, E. H., Gooren, H., Yang, X., & Geissen, V. (2020a). Impact of plastic mulch film debris on soil physicochemical and hydrological properties. *Environmental Pollution*, *266*, 115097.

Qi, Y., Ossowicki, A., Yang, X., Lwanga, E. H., Dini-Andreote, F., Geissen, V., & Garbeva, P. (2020b). Effects of plastic mulch film residues on wheat rhizosphere and soil properties. *Journal of Hazardous Materials*, *387*, 121711.

Qi, Y., Yang, X., Pelaez, A. M., Lwanga, E. H., Beriot, N., Gertsen, H., ... & Geissen, V. (2018). Macro-and micro-plastics in soil-plant system: effects of plastic mulch film residues on wheat (Triticum aestivum) growth. *Science of the Total Environment*, *645*, 1048–1056.

Rahmatpour, S., Mosaddeghi, M. R., Shirvani, M., & Šimůnek, J. (2018). Transport of silver nanoparticles in intact columns of calcareous soils: the role of flow conditions and soil texture. *Geoderma*, *322*, 89–100.

Ramos, L., Berenstein, G., Hughes, E. A., Zalts, A., & Montserrat, J. M. (2015). Polyethylene film incorporation into the horticultural soil of small periurban production units in Argentina. *Science of the Total Environment*, *523*, 74–81.

Rillig, M. C., & Lehmann, A. (2020). Microplastic in terrestrial ecosystems. *Science, 368*(6498), 1430–1431.

Rodriguez -Seijo, A., Lourenço, J., Rocha-Santos, T. A. P., da Costa, J., Duarte, A. C., Vala, H., & Pereira, R. (2017). Histopathological and molecular effects of microplastics in Eisenia andrei Bouché. *Environmental Pollution, 220*, 495–503. https://doi.org/10.1016/J.ENVPOL.2016.09.092

Rousk, J., Bååth, E., Brookes, P. C., Lauber, C. L., Lozupone, C., Caporaso, J. G., … & Fierer, N. (2010). Soil bacterial and fungal communities across a pH gradient in an arable soil. *The ISME Journal, 4*(10), 1340–1351.

Sajjad, M., Huang, Q., Khan, S., Khan, M. A., Liu, Y., Wang, J., Lian, F., Wang, Q., & Guo, G. (2022). Microplastics in the soil environment: a critical review. *Environmental Technology & Innovation, 27*, 102408. https://doi.org/10.1016/J.ETI.2022.102408

Scheurer, M., & Bigalke, M. (2018). Microplastics in Swiss floodplain soils. *Environmental Science & Technology, 52*(6), 3591–3598.

Shade, A., Peter, H., Allison, S. D., Baho, D. L., Berga, M., Bürgmann, H., … & Handelsman, J. (2012). Fundamentals of microbial community resistance and resilience. *Frontiers in Microbiology, 3*, 417.

Sridharan, S., Kumar, M., Bolan, N. S., Singh, L., Kumar, S., Kumar, R., & You, S. (2021). Are microplastics destabilizing the global network of terrestrial and aquatic ecosystem services?. *Environmental Research, 198*, 111243.

Sridharan, S., Kumar, M., Saha, M., Kirkham, M. B., Singh, L., & Bolan, N. S. (2022). The polymers and their additives in particulate plastics: what makes them hazardous to the fauna? *Science of the Total Environment, 824*, 153828.

Tang, C. C., Chen, H. I., Brimblecombe, P., & Lee, C. L. (2019). Morphology and chemical properties of polypropylene pellets degraded in simulated terrestrial and marine environments. *Marine Pollution Bulletin, 149*, 110626.

Vithanage, M., Ramanayaka, S., Hasinthara, S., & Navaratne, A. (2021). Compost as a carrier for microplastics and plastic-bound toxic metals into agroecosystems. *Current Opinion in Environmental Science & Health, 24*, 100297.

Wan, Y., Wu, C., Xue, Q., & Hui, X. (2019). Effects of plastic contamination on water evaporation and desiccation cracking in soil. *Science of the Total Environment, 654*, 576–582.

Wang, F., Wang, Q., Adams, C. A., Sun, Y., & Zhang, S. (2022a). Effects of microplastics on soil properties: current knowledge and future perspectives. *Journal of Hazardous Materials, 424*, 127531.

Wang, W., Yuan, W., Xu, E. G., Li, L., Zhang, H., & Yang, Y. (2022b). Uptake, translocation, and biological impacts of micro (nano) plastics in terrestrial plants: progress and prospects. *Environmental Research, 203*, 111867.

Weithmann, N., Möller, J. N., Löder, M. G., Piehl, S., Laforsch, C., & Freitag, R. (2018). Organic fertilizer as a vehicle for the entry of microplastic into the environment. *Science Advances, 4*(4), eaap8060.

Wong, K. H. J., Lee, K. K., Tang, K. H. D., & Yap, P. S. (2020). Microplastics in the freshwater and terrestrial environments: Prevalence, fates, impacts and sustainable solutions. *Science of the Total Environment*, 137512.

Wu, W. M., Yang, J., & Criddle, C. S. (2017). Microplastics pollution and reduction strategies. *Frontiers of Environmental Science & Engineering, 11*, 1–4.

Xu, B., Liu, F., Cryder, Z., Huang, D., Lu, Z., He, Y., … & Xu, J. (2020). Microplastics in the soil environment: occurrence, risks, interactions and fate–a review. *Critical Reviews in Environmental Science and Technology, 50*(21), 2175–2222.

Ya, H., Jiang, B., Xing, Y., Zhang, T., Lv, M., & Wang, X. (2021). Recent advances on ecological effects of microplastics on soil environment. *Science of the Total Environment, 798*, 149338.

Yan, Y., Chen, Z., Zhu, F., Zhu, C., Wang, C., & Gu, C. (2021). Effect of polyvinyl chloride microplastics on bacterial community and nutrient status in two agricultural soils. *Bulletin of Environmental Contamination and Toxicology, 107*, 602–609.

Yang, W., Cheng, P., Adams, C. A., Zhang, S., Sun, Y., Yu, H., & Wang, F. (2021). Effects of microplastics on plant growth and arbuscular mycorrhizal fungal communities in a soil spiked with ZnO nanoparticles. *Soil Biology and Biochemistry, 155*, 108179.

Yu, H., Hou, J., Dang, Q., Cui, D., Xi, B., & Tan, W. (2020). Decrease in bioavailability of soil heavy metals caused by the presence of microplastics varies across aggregate levels. *Journal of Hazardous Materials, 395*, 122690.

Yu, L., Zhang, J. D., Liu, Y., Chen, L. Y., Tao, S., & Liu, W. X. (2021a). Distribution characteristics of microplastics in agricultural soils from the largest vegetable production base in China. *Science of the Total Environment, 756*, 143860. https://doi.org/10.1016/J.SCITOTENV.2020.143860

Yu, Z. F., Song, S., Xu, X. L., Ma, Q., & Lu, Y. (2021b). Sources, migration, accumulation and influence of microplastics in terrestrial plant communities. *Environmental and Experimental Botany, 192*, 104635.

Zantis, L. J., Borchi, C., Vijver, M. G., Peijnenburg, W., Di Lonardo, S., & Bosker, T. (2023). Nano-and microplastics commonly cause adverse impacts on plants at environmentally relevant levels: a systematic review. *Science of the Total Environment, 867*, 161211.

Zhang, G. S., & Liu, Y. F. (2018). The distribution of microplastics in soil aggregate fractions in southwestern China. *Science of the Total Environment, 642*, 12–20.

Zhang, G. S., Zhang, F. X., & Li, X. T. (2019). Effects of polyester microfibers on soil physical properties: perception from a field and a pot experiment. *Science of the Total Environment, 670*, 1–7.

Zhou, Q., Zhang, H., Fu, C., Zhou, Y., Dai, Z., Li, Y., … & Luo, Y. (2018). The distribution and morphology of microplastics in coastal soils adjacent to the Bohai Sea and the Yellow Sea. *Geoderma, 322*, 201–20

Zong, X., Zhang, J., Zhu, J., Zhang, L., Jiang, L., Yin, Y., & Guo, H. (2021). Effects of polystyrene microplastic on uptake and toxicity of copper and cadmium in hydroponic wheat seedlings (Triticum aestivum L.). *Ecotoxicology and Environmental Safety*, 217, 112217.

Zubris, K. A. V., & Richards, B. K. (2005). Synthetic fibers as an indicator of land application of sludge. *Environmental Pollution, 138*(2), 201–211.

CHAPTER 14

Impact of Microplastics on Soil's Biodiversity and Public Health

Fábio Carvalho Nunes, Lander de Jesus Alves,
Cláudia Csekö Nolasco de Carvalho,
Majeti Narasimha Vara Prasad, and
José Rodrigues de Souza Filho

14.1 INTRODUCTION

We live in the Plasticene [1], as plastic materials are increasingly present in our daily lives. They have become ubiquitous on the Earth's surface and can be found in the air, water, soil, sediments, rocks, flora, fauna and in the human body, including the lungs, and heart [2,3]. So, could we say that plastic has become so important that it also makes us pulse? Yes, but metaphorically! Pulses of reflections! Pulses that direct us to investigate the impacts of the plastic era on ecosystems and our lives. Pulses that alert us to the need to look for alternatives.

The plastic inputs that permeate our days undergo a series of degradation mechanisms, including photodecomposition and hydrolysis [4], resulting in fragmentation into smaller pieces known as microplastics

DOI: 10.1201/9781003487579-14

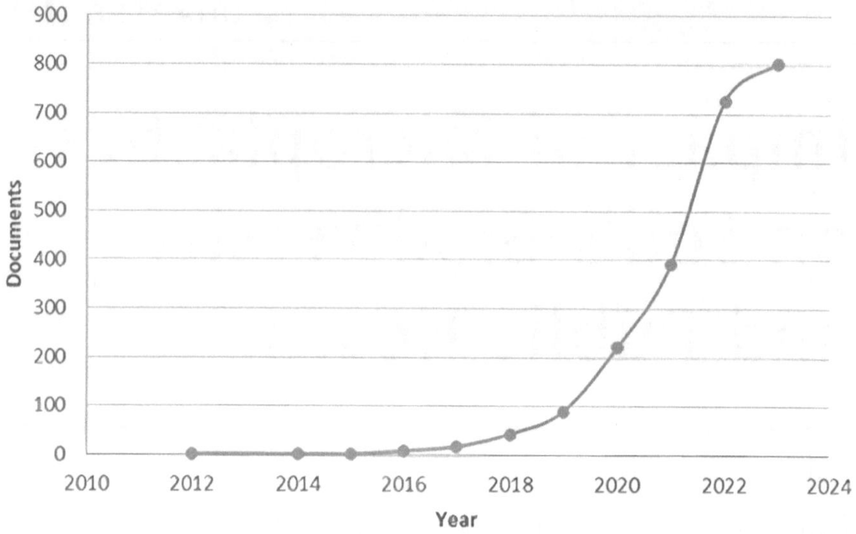

FIGURE 14.1 Number of documents by year on "microplastics" and "soil" (*Source:* Scopus database until October 19, 2023).

(MP),[1] particles smaller than 5 μm. Studies on MP in soils are still incipient [5], however, this reality began to change in 2017 (Figure 14.1), when studies began to expand. The research is promising, the methods are becoming more refined, and the results are showing the complex relationships between MP, soils, and the environment.

Currently, the countries that develop the most work on this topic are in Asia, Europe, and North America (Figure 14.2), highlighting China, USA, India, and Germany, according to data collected in Scopus using the descriptors "microplastics" and "soil". Due to the importance of the topic, the number of articles on the subject is still low. However, today it is possible to state that MP permeates soil ecosystems around the world, being introduced through various pathways, such as effluents, waste, atmospheric deposition, and degradation of macroplastic debris [6].

Rilling [7] was the first author to warn in an English-language journal with wide international circulation about the problem of MP in the soil, through the article "Microplastic in Terrestrial Ecosystems and the Soil?" published in the Journal Environ. Sci. Technol. In the article, the author emphasizes that MP are small enough to be absorbed by biota, adsorb pollutants, and accumulate in the food chain, probably causing harm to human health [8].

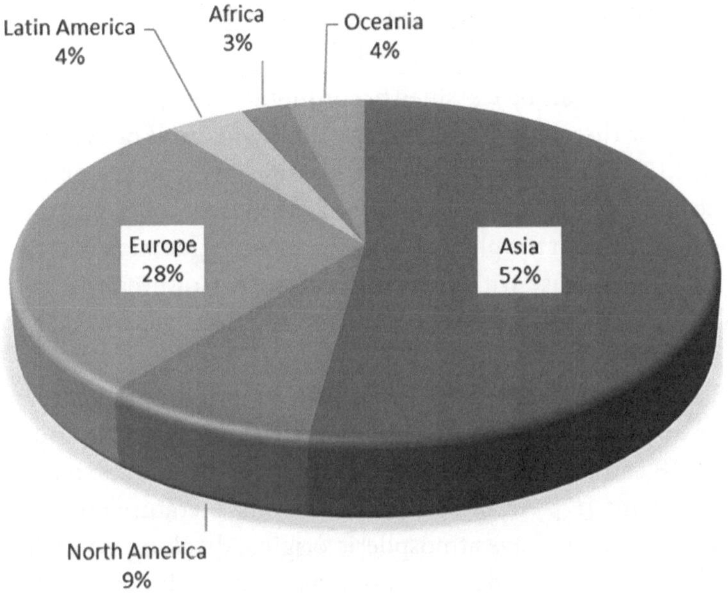

FIGURE 14.2 Percentage of documents by continent on "microplastics" and "soil" (*Source:* Scopus database until October 19, 2023).

MP have emerged as a significant pollutant as they affect organisms of different habitats, trophic levels, sizes, feeding mechanisms, and behaviors [9].

Part of the lack of research on MP in soils is related to methodological difficulties, mainly sampling, quantification, and extraction [10]. This is due to the complexity of the soil matrix and the electrostatic attraction forces between colloids and other solid, liquid, and gaseous fractions [5]. The soil has many components, mineral and organic fractions (living and non-living), ions, and gases that interact with each other and with the environment, often in a non-linear way [11], which makes sampling, quantification, and extraction of microplastics difficult.

The soil is a complex ecological system where materials and energies permeate that affect the atmosphere, hydrosphere, lithosphere, and biosphere. The interactions of soil components are responsible for its ability to self-regulate, which is critical not only for its own dynamic balance but for the balance of other terrestrial systems [11]. Therefore, studying MP in soils and the effects on their biodiversity is very important, which is what we do in this chapter. Furthermore, we discuss its sources, transfers, and health risks to organisms and humans.

14.2 SOURCES AND TRANSFER OF MICROPLASTICS IN SOILS

Plastic materials can be classified according to their processability (thermoplastics or thermosets) or biodegradability (natural or synthetic polymers) [12,13]. Plastic (P) can also be classified as primary or secondary [14], the first refers to raw materials used in industry [15], while the second is produced through the physical and chemical degradation of larger parts [16], which is the case with most microplastics. However, MP in the soil can also have a primary origin, such as microspheres generated by the cosmetics industry.

MP classification can help investigate its origin and reflect on the impacts on the soil-plant system, terrestrial geosystems, and human beings. MP can be added to the soil from different sources, mainly from agroecosystems [17]. According to the relevant literature on the subject, MP in soils can also have atmospheric origins, drinking water, food, rising groundwater, wetlands, rivers, oceans, rocks, sediments, organic and inorganic wastes, and the influence of fauna and flora (Table 14.1).

Agroecosystems are the main sources of MP in soil, which can come mainly from the use of plastic mulch [18–20], but also from fertilization, such as sewage sludge, composting, and irrigation [21–24]. Plastic mulch (PM) improves agricultural productivity by allowing crops to utilize more sunlight, heat, water, and nutrients, resulting in earlier maturity, higher yield, and harvest quality [18,57,58]. As studies have shown concern about the contribution of MP to soils due to the use of PM, researchers have recently dedicated themselves to studying more ecological alternatives, such as the use of biodegradable PM.

TABLE 14.1 Direct and Indirect Sources of PM in Soils

Sources	References
Agroecossistems	[17–27]
Atmosphere (air and rain)	[28–41]
Ocean	[42–45]
Rivers and lakes	[46–51]
Rocks and sediments	[28,52]
Potable water and foods	[53–55]
Biota (fauna and flora)	[56]
Landfills (waste)	[10]

Source: Adapted from Nunes et al. [5].

According to Bandopadhyay et al. [59], biodegradable PM offers a more sustainable alternative than conventional PM, as it does not need to be removed after use. However, the authors point out that uncertainties remain regarding the long-term impacts of incorporating biodegradable PM into soils. Prolonged use of PM can cause the accumulation of PM in the soil, increase its bioavailability [18] and favor mechanical transport by organisms and in suspension in water and air [5]. Improving PM removal methods after harvest and using biodegradable plastics could mitigate the impact on soil and the environment [18]. However, complete removal of plastic is not possible, and the impacts of biodegradables need to be better studied [59].

The use of sewage sludge and various organic compounds in agriculture are alternatives to provide nutrients for plants [21], but studies show that they can be rich in MP, heavy metals (Cd, Pb, Ni, Zn, Cr, and Cu), and pathogens [23, 60–63]. The use of urban wastewater in agriculture is also a source of MP for the soil [23,64,65], even with the improvement of effluent treatment techniques [66]. MP that enters the soil through agroecosystems are certainly transferred to the air, surface and groundwater, oceans, fauna, and flora [5,67].

MP in soils can also be sourced from the atmosphere, capillary rise, input from rivers, oceans, flora, and fauna,[2] rocks, sediments, effluents, and degradation of solid waste [10,27,28,30,38,46,52,68], such as face masks and medical equipment used in the COVID-19 pandemic, which released fibrous materials and polypropylene, heavy metals, and organic contaminants such as acetophenone, benzothiazole, and environmentally persistent free radicals (EPFR) [69].

Winds that blow over the surface of land [10] and oceans disperse MP into the atmosphere, which then return to the land by dry [38] or wet deposition [36,37], through the formation of condensation nuclei [5]. MP dispersed in the atmosphere can be transported over long distances, which is why they are found in remote areas such as poles, remote ocean areas, and mountain [35,42,70–72].

As can be seen, there is already information available on the sources and transport of MP in terrestrial environments and soils [5,73], but further investigation is needed, mainly on (1) direct and indirect sources of primary and secondary MP; (2) biotic and abiotic transport in and out of the soil. The scientific community has a long way to go, but the flanks are open, pointing out promising research paths.

14.3 IMPACT OF MICROPLASTICS ON SOIL'S BIODIVERSITY

The decline in soil biodiversity is the reduction of life forms living in soils, both in terms of quantity and variety [74]. The reduction in soil biodiversity can mean the loss of environmental functions that are fundamental to the balance of life on the planet, such as providing food, habitats, and nurseries for many species. Many threats to soil biodiversity have been identified over the past few years, including MP.

To obtain information on the current state of knowledge on the impacts of MP on soil biodiversity, a systematic review on the subject was carried out, using the Scopus main collection database, focusing on articles in English published up to October 2023. Initially, we extracted a total of 2,302 documents using the keywords "microplastics" and "soil". To refine the search, we incorporated the word "biodiversity", resulting in 73 articles (Figure 14.3), 35 of which dealt specifically with the impacts of microplastics on soil biota and biodiversity.

Analysis of the articles showed that most studies on MP impacts on biodiversity do not focus on biodiversity *stricto-sensu*, but on examining changes in the abundance and/or diversity of individual groups of soil organisms, as well as impacts negative in each organism or group of organisms. Similar results were found by Tibbett et al. [74], when they reviewed the main threats to soil biodiversity, highlighting plastic as one of the threats.

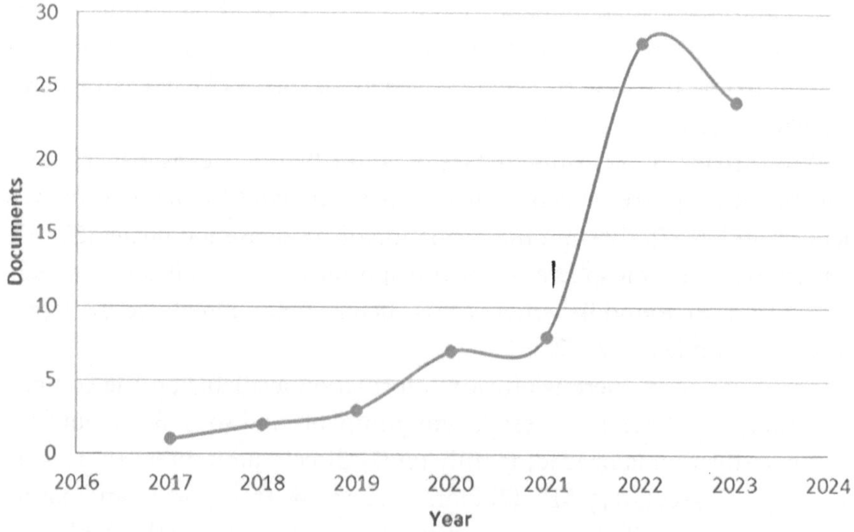

FIGURE 14.3 Number of documents by year on "microplastics" and "soil" and "biodiversity" (*Source:* Scopus database until October 19, 2023).

The articles show that the presence and accumulation of MP in the soil can affect several physical, chemical, and biological processes, all extremely important for its functioning [75,76] (Table 14.2). PM from agrosystems, for example, have significant effects on the soil microbiota and its processes, regardless of their composition, as they modify the pedoclimate, reducing evaporation, gas exchange, and light transmissivity, increasing temperature, soil humidity, and root exudation, favoring the accumulation of CO_2 and affecting the availability of nutrients [77–81].

MP can be lethal to many soil organisms or inflict a range of ecotoxicity effects, including impacts on their development, reproduction, and bioaccumulation of pollutants [6,82,83]. Furthermore, they affect nutrient cycling, plant growth, metabolism, enzymatic activities, and general soil health [8,12,84–86]. However, many of the harmful effects may not be directly linked to MP but to the release of contaminants used as additives in its manufacture or adsorbed on them [87].

MP can accumulate in the food chain because they are easily ingested by organisms at different trophic levels [6,88,106], or through trophic cascades [113,114], that is, resulting from indirect interactions among numerous alternatives. MP accumulation can cause organismal lethality [105],

TABLE 14.2 Probable Impacts of MP on Soil Fauna and Flora

Impacts	References
Bioaccumulation of contaminants	[6,88]
Toxicity	[6,82,83,89–91]
Genotoxicity	[92,93]
Alters plant growth, concentration of carbon, nitrogen, biomass, and leaf metabolism	[8,12,84,94–98]
Nutrient cycling	[12,84–86,94,96]
Changes metabolism	[12,83,94,99–101]
Increase in proteobacteria, actinobacteria, and mycotoxigenic fungi	[102–104]
Changes diet, causes intestinal damage, affects development, mobility, reproductive behavior, and can be lethal to soil fauna	[83,86,105–108]
Endocrine and energy metabolism disruption	[92,109]
Oxidative stress	[92,93,98,106,108–110]
Immune and neurotransmission dysfunction	[92,109]
Reduced bacterial diversity and activity	[111]
Threatens soil biodiversity	[96]
Reduced interspecific competition	[96]
Impacts photosynthesis	[97,98,112]

reduced feeding activity, growth inhibition [105], endocrine and energy metabolism disruption [109], oxidative stress [109], immunological [109] and neurotransmission dysfunction, and genotoxicity [92], can reduce biodiversity and bacterial activity [111] and increase the presence of pathogens [102,103].

Specifically regarding soil flora, MP alters physiological and biochemical properties, such as: (1) root length, carbon, nitrogen content, biomass, and leaf metabolism [95]; (2) cause genotoxicity and oxidative damage [93]; (3) inhibit plant growth and favor pathogen proliferation [104]; (4) reduce the proportion of roots, plant growth and reduce interspecific competition, interfering with nutrient cycling and threatening soil biodiversity [96]; (5) can concentrate in the roots and affect nutrient transport, blocking cellular connections or cell wall pores [93] and; (6) alter photosynthesis [97].

All the effects of MP on physical, chemical, and biological processes in the soil impact its biodiversity and can affect public health, as pathogens and contaminants can reach humans via accumulation in the food chain. Furthermore, they can compromise soil functions or ecosystem services of support and/or regulation, provision, culture, leisure, and religion [5,6,11]. In Table 14.3 can see examples of possible impacts of plastic coverings and MP on soil functions, signaling paths for future research.

Methods used to understand the impacts of MP on soil biodiversity are evolving, but most still focus on discrete, specific, non-integrated aspects. Li et al. [6] observing this lack and based on the management tool for Life Cycle Assessment (LCA),[3] carried out a conceptual proposal to link the different effects of MP on soil organisms to damage in the loss of soil species, loss of functional diversity, and loss of ecosystem services.

The proposal by Li et al. [6] includes: (1) Threshold Indicator Rate Analysis, seeking to translate the ecotoxicological effects of MP on soil biodiversity into quantitative measures; (2) seeks to link the loss of soil species to the loss of functional diversity; (3) seeks to measure the loss of functional diversity and damage to ecosystem services by linking ecological networks and Bayesian beliefs. These authors present an interesting proposal, which will serve as a conceptual instrument to characterize MP damage to soil biodiversity and ecosystem services.

14.4 IMPACT OF MICROPLASTICS ON PUBLIC HEALTH

MP can accumulate in the soil, change physically and chemically, pass through the food chain, and cause damage to plant, animal, and human health. Accumulation of these plastics has been reported in tissues of

TABLE 14.3 Probable Impacts of Plastic and Microplastic Materials on Soil Functions

Soil Functions	Impacts (Synthetic Description)
	Regulation and/or Support Functions
Water purification, assimilation, and recycling of pollutants	MP and inorganic or organic materials adsorbed on them can compromise the ability to purify water and recycle pollutants
Provide refuge, nurseries, and habitats for organisms	MP and adsorbed inorganic or organic materials can compromise the use of soil as a nursery, refuge, and habitat for organisms
Support bacterial culture for antibiotics production	The use of soil as a bacterial culture medium can be compromised, as some bacteria are sensitive to environmental changes
Nutrient cycling	Physical, chemical, and biological changes caused by plastic materials and MP can affect nutrient cycling
Climate regulation	Plastic mulch reduces evaporation, gas exchange, light transmissivity and increases soil temperature, locally modifying soil-atmosphere interactions
	Provision Functions
Natural production of foods, fibers, and medicines	MP can be transported to remote and even uninhabited areas of the planet through winds, rivers, groundwater, oceans, and organisms, potentially compromising the natural supply of some foods, fibers, and natural pharmaceuticals
Genetic resources	Physical, chemical, and/or biological modifications caused by plastic materials and MP can affect the supply of genetic resources
Materials for pharmaceuticals and cosmetics	Physical, chemical, and/or biological changes caused by plastic materials and MP can affect the supply of resources used in the manufacture of pharmaceuticals and cosmetics
	Functions of Information, Culture, Leisure, and Religion
Plaeoenvironmental information and cultural heritage	Environments degraded by plastic materials and MP can compromise paleoenvironmental and cultural records that are very susceptible to physical, chemical and/or biological changes
Recreation, leisure, and educational activities	Environments highly degraded by plastic materials may not be suitable for recreational and educational activities
Religious rituals	Environments that are degraded or highly degraded by plastic materials can compromise rituals or religious activities that require a more balanced environment and materials for purification

several species, such as liver, intestine, lungs, and kidneys [115–118]. Studies have also reported MP in human organs [2,3].

Lwanga et al. [119], for example, evaluated micro and macroplastics in the soil, in earthworms, and chicken organs used for human consumption, concluding that MP increased in the organs studied. The data shows that

MP can enter terrestrial food chains and reaching humans. Therefore, the effects and risks of MP on soil need to be better studied [120], since the damage can affect the entire food chain and our health [8,121].

Humans are exposed to MP every day through ingestion, inhalation, or skin contact. Most ingested plastics are excreted by the body, but inhaled MP can accumulate in the lungs and even reach the brain [122]. According to several authors, the smaller the plastic fragment, the greater its capacity to harm human health. These small fragments are being called nanoplastic (NP) [123,124].

Arriving in humans, MP can affect your health. Silver [31] suggests that MPs can cause pneumonia and cancer; D'Angelo and Meccariello [14] report that they are threats to male fertility because the study showed deleterious effects on reproduction and sperm quality in mammals; Jin et al. [125] also confirmed the relationship between MP and male reproductive dysfunction in mice.

The studies mentioned raise the alarm about the impacts of MP on human health, but the concern does not stop there. MP are also transport vectors for several dangerous chemicals [126], that is, they carry with them co-contaminants that: (1) were intentionally added to the plastic during production to color, provide transparency, and improve resistance to degradation [127]; (2) were adsorbed from the environment [28].

Additives are chemicals added to plastic that can be toxic, such as heavy metals, and include inert or reinforcing fillers, plasticizers, antioxidants, ultraviolet light stabilizers, lubricants, dyes, and flame retardants [127]. Table 14.4 adapted from Nunes et al. [5] below presents the most toxic heavy metals contained in plastic materials and their likely effects on human health.

14.5 FINAL CONSIDERATIONS

The threats from human activities to soil biodiversity, including MP, are a worrying fact and require more attention from the scientific community [74]. Scientists have only begun to unravel the threats of MP to soil biodiversity and public health, but the results already signal that we should be concerned, as they are ubiquitous pollutants in the pedosphere, and the adverse effects can be serious.

The soil has a wide variety of macro, meso, and microorganisms, which have specific ecological niches and perform important environmental functions, such as biomass production and soil protection (Table 14.5). The death of soil biota would represent the death of the soil itself and,

TABLE 14.4 Main Use of Heavy Metals as Additives in Polymer Products and Their Effects on Human Health

Heavy Metals	Sources	Probable Effects on Human Health	References
Aluminum (Al)	Additives: stabilizers, inorganic pigments, and flame retardants; Adsorbed: rock weathering and degradation of household items, packaging, vehicles, etc.	Metal–estrogen; breast cancer	[28,127,128]
Antimony (Sb)	Additives: flame retardants and biocides; Adsorbed: degradation of metal alloys and various materials	Metal–estrogen; breast cancer	[127,129,130]
Arsenic (As)	Additives: biocides; Adsorbed: degradation of treated wood and leather, degradation of metal, poisons, etc.	Congenital disabilities; Carcinogen: lung, skin, liver, bladder, kidneys; gastrointestinal damage, breast cancer, death	[28,127,131]
Barium (Ba)	Additives: inorganic pigments and ultraviolet light stabilizers	Metal–estrogen, breast cancer, cardiovascular and kidney disease, metabolic, neurological, and mental disorders	[28,127]
Bromine (Br)	Additives: flame retardants	Apoptosis and genotoxicity	[28,127]
Cadmium (Cd)	Additives: heat and ultraviolet light stabilizers, and inorganic pigments	Changes in calcium, phosphorus and bone metabolism, osteomalacia and bone fractures in postmenopausal women, lipid peroxidation and in the promotion of carcinogenesis; cell apoptosis and ADN methylation, breast cancer	[28,127,132,133]
Chrome (Cr)	Additives: inorganic pigments	Allergic reactions in the body, septonasal ulcer, cardiovascular, respiratory, hematologic, gastrointestinal, renal, hepatic, and neurologic disorders and death, breast câncer	[28,130,134]

(Continued)

TABLE 14.4 (*Continued*) Main Use of Heavy Metals as Additives in Polymer Products and Their Effects on Human Health

Heavy Metals	Sources	Probable Effects on Human Health	References
Cobalt (Co)	Additives: inorganic pigments	Formation of more reactive oxygen species, neurological, cardiovascular and endocrine problems, breast cancer	[28,130,134]
Copper (Cu)	Additives: biocides	Formation of more reactive oxygen species, inducing ADN strand breaks and oxidation	[28,127]
Lead (Pb)	Additives: heat and ultraviolet light stabilizers, and inorganic pigments	Anemia, hypertension, miscarriages, nervous system disruption, brain and cell damage, infertility and oxidative stress, breast câncer	[28,127,133]
Manganese (Mn)	Additives: inorganic pigments	Neurodegenerative disorders	[28,127]
Mercury (Hg)	Additives: biocides	Carcinogen and brain damage	[28]
Tin (Sn)	Additives: UV light stabilizers and biocides	Metal–estrogen, breast cancer; skin rashes, intestinal and stomach problems, nausea, vomiting, diarrhea; headaches and palpitations	[28,127,129,134]
Titanium (Ti)	Additives: UV light stabilizers and inorganic pigments	Cytotoxicity of lung and colon epithelial cells	[28,127]

Source: Adapted from Nunes et al. [5].

TABLE 14.5 Some Environmental Functions Performed by Soil Organisms

Soil Organisms	Some Environmental Functions
Bacteria	They transform contaminants, such as heavy metals and pathogens, retain water, help fix nitrogen, and produce biomass
Fungi	Transformation or immobilization of contaminants, stimulates the use of nutrients and biomass production
Protozoa	Transform contaminants, consume pathogens, and support biomass production
Flora	Soil protection against erosion, revolves, cools, favors percolation of water and air, produces biomass, fixes nutrients, in addition to transforming, immobilizing, extracting, and volatilizing contaminants
Fauna (in general)	It revolves and aerates the soil, produces biomass, transforms waste, and contaminants

therefore, suppression of the ecosystem services it offers [11]. Therefore, there is a need to in-depth study the impacts of MP on soil biodiversity and public health.

Soil ecosystems are crucial for the provision of ecosystem services and are increasingly pressured by human activities. MP associated with various human activities have become prevalent in various soil ecosystems [6] and may pose a threat to soil security. Studying soil security is a multifaceted framework that considers soils as integral to society's concerns regarding global environmental challenges [135], which includes food and nutrition security, energy and water security, mitigating ecosystem degradation, protecting biodiversity, and maintenance of ecosystem services [136].

Maintaining soil security is only possible as long as its vital ecological functions function fully [11]. Therefore, it is also important to understand how the presence of MP can compromise the physical and chemical properties of the soil, as they directly and indirectly interfere with soil biodiversity. Several studies have been developed in this direction and highlight the importance of considering microplastic diversity and its impacts on morphological, physical, chemical, and ecological attributes of soils [75, 137–142].

Studies on the effects of MP on morphological, physical, and chemical properties of the soil are still beginning, but the high heterogeneity of the soil in time and space determined by its formation factors [11], signals the importance of studying the impacts in different soil classes, considering climatic and landscape units [5].

It is also important to highlight, as indicated by Ren et al. [12], the need to advance the following studies on the effects of MP: (1) soil structure, which controls water retention and infiltration, gas exchange, soil nutrients, and erodibility; (2) in biogeochemical processes, such as the carbon and nitrogen cycles; (3) in microbial processes, such as mutualism, microbial competition, and nutrient cycling; (4) in plant growth, physiology, and behavior, such as biomass production, nutrient distribution, and contaminant mediation, such as extraction, immobilization, transformation, and volatilization.

Finally, it is worth highlighting that for a better assessment of MP in the soil, it is necessary to monitor its variation in space and time, considering environmental variables such as climate, relief, and land use from a systemic perspective [5]. This can allow researchers to model and consider interrelationships more effectively when recommending sampling intensity or when evaluating changes in evaluation results [135].

NOTES

1 In this work, the term microplastic includes nanoplastics. It is important to make this reservation, as in the literature the terms can be found separately.

2 Examples: (1) earthworms, springtails, mites, and small mammals fragment, ingest, and digest macroparticles and then release microparticles into the environment [10]; (2) fungi, bacteria, and protozoa biochemically degrade larger plastic particles [68].

3 It is a management tool used to evaluate the environmental performance of the life cycles of products and technologies (ISO 14040:2006). Access: https://www.iso.org/standard/37456.html

REFERENCES

1. Reed, C. (2015). Dawn of the Plasticene age. *New Scientist*, 225, 28–32. https://doi.org/10.1016/S0262-4079(15)60215-9
2. Yang, Y. et al. (2023). Detection of various microplastics in patients undergoing cardiac surgery. *Environ. Sci. Technol.*, 57(30), 10911–10918. https://doi.org/10.1021/acs.est.2c07179
3. Pauly, J.L. et al. (1998). Inhaled cellulosic and plastic fibers found in human lung tissue. *Cancer Epidemiol. Biomark. Prev.*, 7(5), 419–428. https://doi.org/10.1023/A:1008833422577
4. Wong, J.K.H. et al. (2020). Microplastics in the freshwater and terrestrial environments: prevalence, fates, impacts and sustainable solutions. *Sci. Total Environ.*, 719, Article 137512.
5. Nunes, F.C. et al. (2023). Critical review of microplastics in soil. In: Vithanage, M.P., Vara, M.N., (Eds.), *Microplastics in the Ecosphere: Air, Water, Soil, and Food.* 1 ed. London: John Wiley & Sons, Incorporated, Vol. 1, pp. 249–269.

6. Li, T. et al. (2023). Micro-and nanoplastics in soil: linking sources to damage on soil ecosystem services in life cycle assessment. *Sci. Total Environ.*, 904, 166925. https://doi.org/10.1016/j.scitotenv.2023.166925
7. Rilling, M.C. (2012). Microplastic in terrestrial ecosystems and the soil? *Environ. Sci. Technol.*, 46, 6453–6454. https://doi.org/10.1021/es302011r
8. Ivy, N. et al. (2023). Effects of microplastics and arsenic on plants: interactions, toxicity and environmental implications. *Chemosphere,* 338, 139542. https://doi.org/10.1016/j.chemosphere.2023.139542
9. Kukkola, A. et al. (2021). Nano and microplastic interactions with freshwater biota-current knowledge, challenges, and future solutions. *Environ. Int.*, 152, Article 106504. https://doi.org/10.1016/j.envint.2021.106504
10. Browne, M.A. et al. (2008). Ingested microscopic plastic translocates to the circulatory system of the mussel, Mytilus edulis (L.). *Environ. Sci. Technol.*, 42, 5026–5031.
11. Nunes, F.C. et al. (2020). Soil as a complex ecological system for meeting food and nutritional security. In: Prasad, M.J.V., Pietrzykowski, M., (Eds.), *Climate Change and Soil Interactions.* Amsterdam: Elsevier, pp. 229–269. https://doi.org/10.1016/B978-0-12-818032-7.00009-6
12. Ren, X. et al. (2022). Microplastics in plant-microbes-soil system: a review on recent studies. *Sci. Total Environ.*, 816 151523. https://doi.org/10.1016/j.scitotenv.2021.151523
13. Sander, M. (2019). Biodegradation of polymeric mulch films in agricultural soils: concepts, knowledge gaps, and future research directions. *Environ. Sci. Technol.*, 53, 2304–2315. https://doi.org/10.1021/acs.est.8b05208
14. D'Angelo, S. and Meccariello, R. (2021). Microplastics: a threat for male fertility. *Int. J. Environ. Res. Public Health,* 18(5), 2392. https://doi.org/10.3390/ijerph18052392
15. Browne, M.A. et al. (2011). Accumulation of microplastic on shorelines worldwide: sources and sinks. *Environ. Sci. Technol.*, 45, 9175–9179.
16. Cole, M. et al. (2011). Microplastics as contaminants in the marine environment: a review. *Marine Pollut. Bull.*, 62 (12), 2588–2597.
17. Nizzetto, L. et al. (2016). A theoretical assessment of microplastic transport in river catchments and their retention by soils and river sediments. *Environ. Sci. Process. Impacts*, 18, 1050–1059. https://doi.org/10.1039/C6EM00206D
18. Li, S. et al. (2022). Macro- and microplastic accumulation in soil after 32 years of plastic film mulching. *Environ. Pollut.,* 300, 118945. https://doi.org/10.1016/j.envpol.2022.118945
19. Xu, L. et al. (2022). Is mulch film itself the primary source of meso- and microplastics in the mulching cultivated soil? A preliminary field study with econometric methods. *Environ. Pollut.*, 299, 118915. https://doi.org/10.1016/j.envpol.2022.118915
20. Tian, X. et al. (2022). Plastic mulch film induced soil microplastic enrichment and its impact on wind-blown sand and dust. *Sci. Total Environ.*, 813, 152490. https://doi.org/10.1016/j.scitotenv.2021.152490
21. Gao, D. et al. (2020). Source, occurrence, migration and potential environmental risk of microplastics in sewage sludge and during sludge amendment to soil. *Sci. Total Environ.,* 742, 140355. https://doi.org/10.1016/j.scitotenv.2020.140355

22. Corradini, F. et al. (2019). Evidence of microplastic accumulation in agricultural soils from sewage sludge disposal. *Sci. Total Environ.*, 671, 411–420. https://doi.org/10.1016/j.scitotenv.2019.03.368
23. Pérez-Reverón, R. et al. (2022). Recycled wastewater as a potential source of microplastics in irrigated soils from an arid-insular territory (Fuerteventura, Spain). *Sci. Total Environ.*, 817, 152830. https://doi.org/10.1016/j.scitotenv.2021.152830
24. Zhou, B. et al. (2020). Microplastics in agricultural soils on the coastal plain of Hangzhou Bay, East China: multiple sources other than plastic mulching film. *J. Hazard. Mater.*, 388, 121814. https://doi.org/10.1016/j.jhazmat.2019.121814
25. Alves, L.J. et al. (2020). Microbial approach for alleviation of potentially toxic elements in agricultural soils. In: Prasad, M.J.V., Pietrzykowski, M., (Eds.), *Climate Change and Soil Interactions*. Amsterdam: Elsevier, pp. 271–303. https://doi.org/10.1016/B978-0-12-818032-7.00010-2
26. Zhou, B. et al. (2020). Microplastics in agricultural soils on the coastal plain of Hangzhou Bay, East China: multiple sources other than plastic mulching film. *J. Hazard. Mater.*, 388, 121814. https://doi.org/10.1016/j.jhazmat.2019.121814
27. Rillig, M.C., Ingra, R., and de Souza Machado, A. (2017). Microplastic incorporation into soil in agroecosystems. *Front. Plant Sci.*, 8, 1805.
28. Campanale, C. et al. (2020). A detailed review study on potential effects of microplastics and additives of concern on human health. *Int. J. Environ. Res. Saúde Pública*, *17*, 1212. https://doi.org/10.3390/ijerph17041212
29. Alfonso, M.B. et al. (2021). Continental microplastics: presence, features, and environmental transport pathways. *Sci. Total Environ.*, 799, 149447. https://doi.org/10.1016/j.scitotenv.2021.149447
30. Dris, R. et al. (2015). Microplastic contamination in an urban area: a case study in Greater Paris. *Environ. Chem.*, 12(5), 592–599. https://doi.org/10.1071/EN14167
31. Prata, J.C. (2017). Airborne microplastics: consequences to human health. *Environ. Pollut.*, 234, 115–126. https://doi.org/10.1016/j.envpol.2017.11.043
32. Liu, Z. et al. (2022). Distribution and possible sources of atmospheric microplastic deposition in a valley basin city (Lanzhou, China). *Ecotoxicol. Environ. Safety*, 233, 113353. https://doi.org/10.1016/j.ecoenv.2022.113353
33. Abbadi, S. et al. (2022). Atmospheric transport of microplastics during a dust storm. *Chemosphere*, 292, 133456. https://doi.org/10.1016/j.chemosphere.2021.133456
34. Abbasi, S., Turner, A. (2021). Dry and wet deposition of microplastics in a semi-arid region (Shiraz, Iran). *Sci. Total Environ.*, 786, 147358.
35. Zhang, Y. et al. (2019). Importance of atmospheric transport for microplastics deposited in remote areas. *Environ. Pollut.*, 254. https://doi.org/10.1016/j.envpol.2019.07.121
36. Zhang, Y. et al. (2020). Atmospheric microplastics: a review on current status and perspectives. *Earth Sci. Rev.*, 203, 103118. https://doi.org/10.1016/j.earscirev.2020.103118

37. Liu, K. et al. (2019). Consistent transport of terrestrial microplastics to the ocean through atmosphere. *Environ. Sci. Technol.*, 53(18), 10612–10619. https://doi.org/10.1021/acs.est.9b03427
38. Szewc, K., Graca, B., and Dolega, A., (2021). Atmospheric deposition of microplastics in the coastal zone: characteristics and relationship with meteorological factors. *Sci. Total Environ.*, 761. https://doi.org/10.1016/j.scitotenv.2020.143272
39. Koutnik, V.S. et al. (2022). Microplastics retained in stormwater control measures: where do they come from and where do they go? *Water Res.*, 210, 118008. https://doi.org/10.1016/j.watres.2021.118008
40. Boni, W., Arbuckle-Keil, G., and Fahrenfeld, N.L. (2021). Inter-storm variation in microplastic concentration and polymer type at stormwater outfalls and a bioretention basin. *Sci. Total Environ.*, 809, 151104. https://doi.org/10.1016/j.scitotenv.2021.151104
41. Liu, Z. et al. (2022). Distribution and possible sources of atmospheric microplastic deposition in a valley basin city (Lanzhou, China). *Ecotoxicol. Environ. Safety*, 233, 113353. https://doi.org/10.1016/j.ecoenv.2022.113353
42. Ding, J. et al. (2022). Atmospheric microplastics in the Northwestern Pacific Ocean: distribution, source, and deposition. *Sci. Total Environ.*, 829, 154337. https://doi.org/10.1016/j.scitotenv.2022.154337
43. Pushan, Z.A. et al. (2022). A critical review of the emerging research on the detection and assessment of microplastics pollution in the coastal, marine, and urban Bangladesh. *Front. Environ. Sci. Eng.*, 16(10), 128. https://doi.org/10.1007/s11783-022-1563-2
44. Ferrero, L. et al. (2022). Airborne and marine microplastics from an oceanographic survey at the Baltic Sea: an emerging role of air-sea interaction? *Sci. Total Environ.*, 824, 153709. https://doi.org/10.1016/j.scitotenv.2022.153709
45. Alfaro-Núñez, A. et al. (2022). Author Correction: Microplastic pollution in seawater and marine organisms across the tropical Eastern Pacific and Galápagos. *Sci. Rep.*, 12, 3502. https://doi.org/10.1038/s41598-022-07504-w
46. Napper, I.E. et al. (2021). The abundance and characteristics of microplastics in surface water in the transboundary Ganges River. *Environ. Pollut.*, 274, 116348. https://doi.org/10.1016/j.envpol.2020.116348
47. Alfonso, M.B. et al. (2020). First evidence ofmicroplastics in nine lakes across Patagonia (South America). *Sci. Total Environ.*, 733. https://doi.org/10.1016/j.scitotenv.2020.139385
48. Zhao, S. et al. (2019). Analysis of suspended microplastics in the Changjiang Estuary: implications for riverine plastic load to the ocean. *Water Res.*, 161, 560–569. Elsevier Ltd. https://doi.org/10.1016/j.watres.2019.06.019
49. Carvalho, A.R. et al. (2021). Urbanization and hydrological conditions drive the spatial and temporal variability of microplastic pollution in the Garonne River. *Sci. Total Environ.*, 769, 144479. https://doi.org/10.1016/j.scitotenv.2020.144479
50. Uurasjärvi, E. et al. (2020). Microplastic concentrations, size distribution, and polymer types in the surface waters of a northern European lake. *Water Environ. Res.*, 92, 149–156. https://doi.org/10.1002/wer.1229

51. Maheswaran, B. et al. (2022). Assessment, characterization, and quantification of microplastics from river sediments. *Chemosphere*, 298, 134268. https://doi.org/10.1016/j.chemosphere.2022.134268
52. Fernandino, G. et al. (2020). Anthropoquinas: first description of plastics and other man-made materials in recently formed coastal sedimentary rocks in the southern hemisphere. *Marine P. Bull.*, 154, 111044. https://doi.org/10.1016/j.marpolbul.2020.111044
53. Souza, V.G.L., and Fernando, A.L. (2016). Nanoparticles in food packaging: Biodegradability and potential migration to food—a review. *Food Packag. Shelf Life*, 8, 63–70.
54. Waring, R.H., Harrisa, R.M., and Mitchell, S.C. (2018). Plastic contamination of the food chain: a threat to human health? *Maturitas*, 115, 64–68.
55. Pivokonsky, M. et al. (2018). Occurrence of microplastics in raw and treated drinking water. *Sci. Total Environ.*, 643, 1644–1651.
56. Rezania, S. et al. (2018). Microplastics pollution in different aquatic environments and biota: a review of recent studies. *Mar. Pollut. Bull.*, 133, 191–208.
57. Wang, Z. et al. (2021). Agronomic performance of polyethylene and biodegradable plastic film mulches in a maize cropping system in a humid continental climate. *Sci. Total Environ.*, 786, 147460.
58. Gao, H.H. et al. (2019). Exploring optimal soil mulching to enhance yield and water use efficiency in maize cropping in China: a meta-analysis. *Agric. Water Manag.*, 225, 105741.
59. Bandopadhyay, S. et al. (2018). Biodegradable plastic mulch films: impacts on soil microbial communities and ecosystem functions. *Front. Microbiol.*, 9, Article 819, 1–7. https://doi.org/10.3389/fmicb.2018.00819
60. Li, X.-Y. et al. (2022). Effects of typical sludge treatment on microplastics in China—characteristics, abundance and micro-morphological evidence. *Sci. Total Environ.*, 826, 154206. https://doi.org/10.1016/j.scitotenv.2022.154206
61. Harley-Nyang, D. et al. (2022). Investigation and analysis of microplastics in sewage sludge and biosolids: a case study from one wastewater treatment works in the UK. *Sci. Total Environ.*, 823, 153735. https://doi.org/10.1016/j.scitotenv.2022.153735
62. Chilian, A. et al. 2022. Extraction of heavy metals and phosphorus from sewage sludge with elimination of antibiotics and biological risks. *Chem. Eng. J.*, 437, 135298. https://doi.org/10.1016/j.cej.2022.135298
63. Wu, R.-T. et al. (2021). Occurrence of microplastic in livestock and poultry manure in South China. *Environ. Pollut.*, 277, 116790. https://doi.org/10.1016/j.envpol.2021.116790
64. Cheng, Y.L. et al. (2021). Occurrence and removal of microplastics in wastewater treatment plants and drinking water purification facilities: a review. *Chem. Eng. J.*, 410, 128381. https://doi.org/10.1016/j.cej.2020.128381.
65. Xu, Z. et al. (2021). Removal and generation of microplastics in wastewater treatment plants: a review. *J. Clean. Prod.*, 291, 125982. https://doi.org/10.1016/j.jclepro.2021.125982

66. Bayo, J., López-Castellanos, J., and Olmos, S. (2020). Membrane bioreactor and rapid sand filtration for the removal of microplastics in an urban wastewater treatment plant. *Mar. Pollut. Bull.*, 156, 111211. https://doi.org/10.1016/j.marpolbul.2020.111211
67. Nizzetto, L. et al. (2016). Are agricultural soils dumps for microplastics of urban origin? *Environ. Sci. Technol.*, 50, 10777–10779. https://doi.org/10.1021/acs.est.6b04140
68. So, M.W.K. et al. (2022). Fate of plastic in the environment: from macro to nano by macrofauna. *Env. Pollut.*, 300, 118920. https://doi.org/10.1016/j.envpol.2022.118920
69. Liu, Z. et al. (2022). Generation of environmental persistent free radicals (EPFRs) enhances ecotoxicological effects of the disposable face mask waste with the COVID-19 pandemic. *Environ. Pollut.*, 301, 119019. https://doi.org/10.1016/j.envpol.2022.119019
70. Zhang, Y. et al. (2021). Microplastics in glaciers of the Tibetan plateau: evidence for the long-range transport of microplastics. *Sci. Total Environ.*, 758, 143634. https://doi.org/10.1016/j.scitotenv.2020.143634
71. Ambrosini, R. et al. (2019). First evidence of microplastic contamination in the supraglacial debris of an alpine glacier. *Environ. Pollut.*, 253, 297–301. https://doi.org/10.1016/j.envpol.2019.07.005
72. Evangeliou, N. et al. (2020). Atmospheric transport is a major pathway of microplastics to remote regions. *Nat. Commun.*, 11(1). https://doi.org/10.1038/s41467-020-17201-9
73. Lwanga, E.H. et al. (2022). Review of microplastic sources, transport pathways and correlations with other soil stressors: a journey from agricultural sites into the environment. *Chem. Biol. Technol. Agric.*, 9, 20. https://doi.org/10.1186/s40538-021-00278-9
74. Tibbett, M. et al. (2020). Identifying potential threats to soil biodiversity. *Peer J.*, 8, e9271. https://doi.org/10.7717/peerj.9271
75. Souza Machado, A.A. et al., (2018). Impacts of microplastics on the soil biophysical environment. *Environ. Sci. Technol.*, 52, 9656–9665. https://doi.org/10.1021/acs.est.8b02212
76. Wan, Y. et al., (2019). Effects of plastic contamination on water evaporation and desiccation cracking in soil. *Sci. Total Environ.*, 654, 576–582. https://doi.org/10.1016/j.scitotenv.2018.11.123
77. Kasirajan, S., and Ngouajio, M. (2012). Polyethylene and biodegradable mulches for agricultural applications: a review. *Agron. Sustain. Dev.*, 32, 501–529. https://doi.org/10.1007/s13593-011-0068-3
78. Qin, W. et al. (2015). Soil mulching significantly enhances yields and water and nitrogen use efficiencies of maize and wheat: a meta-analysis. *Sci. Rep.*, 5, 16210. https://doi.org/10.1038/srep16210
79. Liu, J. et al. (2015). Response of nitrogen use efficiency and soil nitrate dynamics to soil mulching in dryland maize (Zea mays L.) fields. *Nutr. Cycling Agroecosyst.*, 101, 271–283. https://doi.org/10.1007/s10705-015-9678-5

80. Yu, Y. et al. (2016). Plastic mulching increased soil CO_2 concentration and emissions from an oasis cotton field in Central Asia. *Soil Use Manag.*, 32, 230–239. https://doi.org/10.1111/sum.12266
81. Wang, Y. P. et al. (2016). Multi-site assessment of the effects of plastic-film mulch on the soil organic carbon balance in semiarid areas of China. *Agric. For. Meteorol.*, 228, 42–51. https://doi.org/10.1016/j.agrformet.2016.06.016
82. Palacios-Mateo, C. et al. (2023). Enzymes for microplastic-free agricultural soils. *Ecotoxicol. Environ. Safety*, 258, 114982. https://doi.org/10.1016/j.ecoenv.2023.114982
83. Tang, N. et al. (2023). Microplastic pollution destabilized the osmoregulatory metabolism but did not affect intestinal microbial biodiversity of earthworms in soil. *Environ. Pollut.*, 320, 121020. https://doi.org/10.1016/j.envpol.2023.121020
84. Khan, A. et al. (2023). Ecological risks of microplastics contamination with green solutions and future perspectives. *Sci. Total Environ.*, 899, 165688. https://doi.org/10.1016/j.scitotenv.2023.165688
85. Huang, F. et al. (2023). Are biodegradable mulch films a sustainable solution to microplastic mulch film pollution? A biogeochemical perspective. *J. Hazard. Mater.*, 459, 132024. https://doi.org/10.1016/j.jhazmat.2023.132024
86. Bouaicha, O. et al. (2022). Microplastics make their way into the soil and rhizosphere: a review of the ecological consequences. *Rhizosphere*, 22, 100542. https://doi.org/10.1016/j.rhisph.2022.100542
87. Jiménez-Skrzypek, G. et al. (2021). Microplastic-adsorbed organic contaminants: analytical methods and occurrence. *Trends Anal. Chem.*, 136, 116186. https://doi.org/10.1016/j.trac.2021.116186
88. Alvarez-Lopeztello, J. et al. (2021). Microplastic pollution in neotropical rainforest, savanna, pine plantations, and pasture soils in lowland areas of Oaxaca, Mexico: preliminary results. *Ecol. Indic.*, 121, 107084. https://doi.org/10.1016/j.ecolind.2020.107084
89. Möhrke, A.C.F. et al. (2022). Underestimated and ignored? The impacts of microplastic on soil invertebrates-current scientific knowledge and research needs. *Front. Environ. Sci.*, 10, 975904. https://doi.org/10.3389/fenvs.2022.975904
90. Maity, S. and Pramanick, K. (2020). Perspectives and challenges of micro/nanoplastics-induced toxicity with special reference to phytotoxicity. *Glob. Change Biol.*, 26, 3241–3250. https://doi.org/10.1111/gcb.15074
91. Wang, F. et al. (2020). Interactions of microplastics and cadmium on plant growth and arbuscular mycorrhizal fungal communities in an agricultural soil. *Chemosphere*, 254, 126791. https://doi.org/10.1016/j.chemosphere.2020.126791
92. Xu, B. et al. (2020). Microplastics in the soil environment: occurrence, risks, interactions and fate–a review. *Crit. Rev. Environ. Sci.*, 50(21), 2175–2222.
93. Jiang, X. et al. (2019). Ecotoxicity and genotoxicity of polystyrene microplastics on higher plant Vicia faba. *Environ. Pollut.*, 250, 831–838. https://doi.org/10.1016/j.envpol.2019.04.055

94. Liu, M. et al. (2023). Microplastics effects on soil biota are dependent on their properties: a meta-analysis. *Soil Biol. Biochem.*, 178, Article 108940.
95. Lian, J. et al. (2020). Impact of polystyrene nanoplastics (PSNPs) on seed germination and seedling growth of wheat (Triticum aestivum L.). *J. Hazard. Mater.*, 385, 121620. https://doi.org/10.1016/j.jhazmat.2019.121620
96. Kleunen, M. et al. (2019). A microplastic used as infillmaterial in artificial sport turfs reduces plant growth. *Plants People Planet*, 2, 157–166. https://doi.org/10.1002/ppp3.10071
97. Ren, X. et al. (2021). Microplastics in soil-plant system: effects of nano/microplastics on plant photosynthesis, rhizosphere microbes and soil properties in soil with different residues. *Plant Soil*, 462, 561–576. https://doi.org/10.1007/s11104-021-04869-1
98. Campanale, C. et al. (2022). Microplastics pollution in the terrestrial environments: poorly known diffuse sources and implications for plants. *Sci. Total Environ.*, 805, 150431. https://doi.org/10.1016/j.scitotenv.2021.150431
99. Hu, C. et al. (2023). Black carbon derived from pyrolysis of maize straw and polystyrene microplastics affects soil biodiversity. *Sci. Total Environ.*, 881, 163398. https://doi.org/10.1016/j.scitotenv.2023.163398
100. Temporiti, M.E.E. et al. (2022). The analysis of the mycobiota in plastic polluted soil reveals a reduction in metabolic ability. *J. Fungi*, 8, 1247. https://doi.org/10.3390/jof8121247
101. Yang, B. et al. (2022). Concentration-dependent impacts of microplastics on soil nematode community in bulk soils of maize: evidence from a pot experiment. *Front. Environ. Sci.*, 10, 872898. https://doi.org/10.3389/fenvs.2022.872898
102. Yongguan, Z. et al. (2019). Impacts of (micro)plastics on soil ecosystem: progress and perspective. *J. Agro-Environ. Sci.*, 38, 1–6.
103. Zhang, M. et al. (2019). Microplastics from mulching film is a distinct habitat for bacteria in farmland soil. *Sci. Total Environ.*, 688, 470–478. https://doi.org/10.1016/j.scitotenv.2019.06.108
104. Qi, Y. et al. (2018). Macro- and micro- plastics in soil-plant system: effects of plastic mulch film residues on wheat (Triticum aestivum) growth. *Sci. Total Environ.*, 645, 1048–1056. https://doi.org/10.1016/j.scitotenv.2018.07.229
105. Lwanga, H. et al. (2016). Microplastics in the terrestrial ecosystem: implications for Lumbricus terrestris (Oligochaeta, Lumbricidae). *Environ. Sci. Technol.*, acs.est.5b05478. https://doi.org/10.1021/acs.est.5b05478
106. Daghighi, E. et al. (2023). The forgotten impacts of plastic contamination on terrestrial micro-and mesofauna: a call for research. *Environ. Res.*, 231, 116227. https://doi.org/10.1016/j.envres.2023.116227
107. Holzinger, A. et al. (2022). Comparison of fitness effects in the earthworm Eisenia fetida after exposure to single or multiple anthropogenic pollutants. *Sci. Total Environ.*, 838, 156387. https://doi.org/10.1016/j.scitotenv.2022.156387
108. Zhang, Y. et al. (2022). Interaction of microplastics and soil animals in agricultural ecosystems. *Environ. Sci. Health*, 26, 100327.

109. Chen, K. et al. (2022). Transcriptomic and metabolic responses of earthworms to contaminated soil with polypropylene and polyethylene microplastics at environmentally relevant concentrations. *J. Hazard. Mater.*, 427, 128176. https://doi.org/10.1016/j.jhazmat.2021.128176
110. Zhang, Y. et al. (2023). A new quantitative insight: interaction of polyethylene microplastics with soil - microbiome – crop. *J. Hazard. Mater.*, 460, 132302. https://doi.org/10.1016/j.jhazmat.2023.132302
111. Fei, Y. et al. (2020). Response of soil enzyme activities and bacterial communities to the accumulation of microplastics in an acid cropped soil. *Sci. Total Environ.*, 707, 135634. https://doi.org/10.1016/j.scitotenv.2019.135634
112. Gentili, R. et al. (2022). Toxic impact of soil microplastics (PVC) on two weeds: changes in growth, phenology and photosynthesis efficiency. *Agronomy*, 12, 1219. https://doi.org/10.3390/agronomy12051219
113. He, D.F. et al. (2021). Micro(nano)plastic contaminations from soils to plants: human food risks. *Curr. Opin. Food Sci.*, 41, 116–121.
114. Liang, Y.Q. et al. (2023). Micro(nano)plastics in plant-derived food: source, contamination pathways and human exposure risks. *TrAC-Trend. Anal. Chem.*, 165, Article 117138.
115. Lu, Y. et al. (2016). Uptake and accumulation of polystyrene microplastics in zebrafish (Danio rerio) and toxic effects in liver. *Environ. Sci. Technol.*, 50, 4054–4060.
116. Li, B. et al. (2020). Polyethylene microplastics affect the distribution of gut microbiota and inflammation development in mice. *Chemosphere*, 244, 125492.
117. Lim, D. et al. (2021). Inhalation toxicity of polystyrene micro(nano)plastics using modified OECD TG 412. *Chemosphere*, 262, 128330.
118. Deng, Y., Zhang, Y., Lemos, B., and Ren, H. (2017). Tissue accumulation of microplastics in mice and biomarker responses suggest widespread health risks of exposure. *Sci. Rep.*, 7, 46687.
119. Lwanga, E.H. et al. (2017). Field evidence for transfer of plastic debris along a terrestrial food chain. *Sci. Rep.*, 7, 14071. https://doi.org/10.1038/s41598-017-14588-2
120. Xue, M. et al. (2019). Migration of organic contaminants into dry powdered food in paper packaging materials and the influencing factors. *J. Food Eng.*, 262, 75–82. https://doi.org/10.1016/j.jfoodeng.2019.05.018
121. Yang, L. et al. (2021). Microplastics in soil: a review on methods, occurrence, sources, and potential risk. *Sci. Total Environ.*, 780, 146546. https://doi.org/10.1016/j.scitotenv.2021.146546
122. Seung-Woo, H. et al. (2024). Recent progress and future directions of the research on nanoplastic-induced neurotoxicity. *Neural Regen. Res.*, 19(2), 331–335. https://doi.org/10.4103/1673–5374.379016
123. Banerjee, A. and Shelver, W.L. (2021). Micro- and nanoplastic induced cellular toxicity in mammals: a review. *Sci. Total Environ.*, 755(Pt 2), 142518.
124. Yong, C.Q.Y. et al. (2020). Toxicity of microplastics and nanoplastics in mammalian systems. *Int. J. Environ. Res. Public Health*, 17, 1509.

125. Jin, H. et al. (2022). Chronic exposure to polystyrene microplastics induced male reproductive toxicity and decreased testosterone levels via the LH-mediated LHR/cAMP/PKA/StAR pathway. *Part Fibre Toxicol.,* 19, 13. https://doi.org/10.1186/s12989-022-00453-2
126. Lambert, S. and Wagner, M. (2018). Microplastics are contaminants of emerging concern in freshwater environments: an overview. In: Martin Wagner, Scott Lambert (Ed.), Freshwater microplastics. Emerging environmental contaminants? Switzerland: Springer Open. 1-23p.
127. Hahladakis, J.N. et al. (2018). Closing the loop on plastic packaging materials: what is quality and how does it affect their circularity? *Sci. Total Environ.,* 630, 1394–1400. https://doi.org/10.1016/j.scitotenv.2018.02.330
128. Linhart, C. et al. (2017). Use of underarm cosmetic products in relation to risk of breast cancer: a case-control study. *EBioMedicine*, 21, 79–85, https://doi.org/10.1016/j.ebiom.2017.06.005
129. Niehoff, N.M. et al. (2021). Metals and breast cancer risk: a prospective study using toenail metal biomarkers. *Am. J. Epidemiol.*, 190, 2360–2373, https://doi.org/10.1093/aje/kwab204
130. White, A.J. et al. (2019). Metallic air pollutants and breast cancer risk in a nationwide cohort study. *Epidemiology*, 30, 20–28, https://doi.org/10.1097/EDE.0000000000000917
131. Pullella, K. and Kotsopoulos, J. (2020). Arsenic exposure and breast cancer risk: a re-evaluation of the literature. *Nutrients*, 12, 1–17, https://doi.org/10.3390/nu12113305
132. Filippini, T. et al. (2020). Cadmium exposure and risk of breast cancer: a dose-response meta-analysis of cohort studies. *Environ. Int.*, 142, 1–9, https://doi.org/10.1016/j.envint.2020.105879
133. Gaudet, M.M. et al. (2019). Blood levels of cadmium and lead in relation to breast cancer risk in three prospective cohorts. *Int. J. Cancer*, 144, 1010–1016, https://doi.org/10.1002/ijc.31805
134. O'Brien, K.M. (2019). Toenail-based metal concentrations and young-onset breast cancer. *Am. J. Epidemiol.*, 188, 646–655, https://doi.org/10.1093/aje/kwy283
135. Bradley, S.C. et al. (2021). Soil health spatial-temporal variation influence soil security on Midwestern, U.S. farms. *Soil Security*, 3, 100005. https://doi.org/10.1016/j.soisec.2021.100005
136. Amundson, R. et al. (2015). Soil and human security in the 21st century. *Science*, 80, 348. https://doi.org/10.1126/science.1261071.
137. Qi, R. et al. (2020). Behavior of microplastics and plastic film residues in the soil environment: a critical review. *Sci. Total Environ.*, 703, 134722. https://doi.org/10.1016/j.scitotenv.2019.134722
138. Qi, Y. et al. (2020). Effects of plastic mulch film residues on wheat rhizosphere and soil properties. *J. Hazard. Mater.*, 387, 121711. https://doi.org/10.1016/j.jhazmat.2019.121711
139. Boots, B., Russell, C.W., and Green, D.S. (2019). Effects ofmicroplastics in soil ecosystems: above and below ground. *Environ. Sci. Technol.*, 53, 11496–11506. https://doi.org/10.1021/acs.est.9b03304

140. Machado, A.A. et al. (2018). Impacts of microplastics on the soil biophysical environment. *Environ. Sci. Technol.*, 52, 9656–9665. https://doi.org/10.1021/acs.est.8b02212
141. He, L.-X. et al. (2023). Increasing soil microplastic diversity decreases community biomass via its impact on the most dominant species. *Ecol. Indic.*, 155, 111010. https://doi.org/10.1016/j.ecolind.2023.111010
142. Machado, A.A.S. et al. (2019). Microplastics can change soil properties and affect plant performance. *Environ. Sci. Technol.*, 53, 6044–6052. https://doi.org/10.1021/acs.est.9b01339

CHAPTER 15

Minimizing Microplastics

Impacts on Restoring Soil

Nimra Qureshi, Khadija Amjad, Syeda Gul Nisa, Mahnoor Malik, Fatima Malik, Naeema Khanum, Zeeshan Khan, and Tariq Shah

15.1 INTRODUCTION

15.1.1 Background

Globally, the utility of plastic has escalated during the past decades because of its low cost, outstanding properties, and versatility [1–3]. Plastic is a synthetic polymer made up of identical or similar units [4,5]. The voluntary as well as involuntary disposal of plastic in environment results in the production of MPs over time [6]. In the year 2004, Thompson Richard Charles coined the word "MP (MP)" for plastic particles smaller than 5 mm in diameter [7]. MPs contain different types of polymers and additives that can be combined to create a wide range of goods. Typically, MPs can be identified by using various hues and a minimum of seven polymers [8]. The non-decomposable nature of MPs makes them a threat to human health as they remain undegraded in the environment for hundreds of years. Although plastic has been used extensively for a long time, its by-product MPs have only recently gained international attention due to their possible effects on the aboveground and subsurface environments.

DOI: 10.1201/9781003487579-15

15.1.1.1 Historical Evolution of MPs Research

The field of MPs research officially began after the anecdotal reports of entanglement and ingestion of plastics by marine animals in the 1960s [9]. The development of the first commercial thermoplastics, including polyvinyl chloride (PVC), low-density polyethylene, and polymethyl methacrylate, began with the onset of World War II in 1939. Although researchers have occasionally noted the existence of tiny plastic particles in water as early as the 1970s, a groundbreaking study headed by marine ecologist Richard Thompson in 2004 marked the official start of the investigation into the distribution and effects of these particles [10]. The fact that MPs are widespread and have a unique impact on the environment urged the scientist to consider the fate, contamination, and consequences of MPs on Earth's ecosystems, natural cycles, and living organisms in their studies. Global research on MPs over the past 14 years has produced significant advancements in the understanding of the origins, fate, and impacts of MPs and the substances they are connected with [11].

15.1.1.2 Emergence of Soil MPs as a Critical Concern

Possibly one of the most pervasive and persistent human-caused alterations to the earth's surface is MP contamination [12]. The global conservation of biodiversity has found MP pollution to be one of the most significant issues [13]. Compared to their aquatic counterparts, terrestrial systems have gotten significantly less scientific attention in terms of MP pollution. Although about 80% of plastic debris that finds its way into the ocean is created, used, and frequently disposed of on land, not much is known about the possible impact of MPs on terrestrial ecosystems [11]. Compared to marine basins, soil may hold more MPs [14]. There are several ways that MPs can get into the soil ecosystem. The most frequent additions to soil are the usage of plastic mulch covers and the addition of organic wastes (such as composts and biosolids) that may include plastic contaminants [15]. Given that plastics decompose at a remarkably slow rate compared to their rate of entrance into the system, which causes a gradual accumulation in soil, it is currently anticipated that the detrimental effects of plastics on ecosystem health would grow over time [16].

15.2 SIGNIFICANCE OF MPs IN SOIL

MPs are now considered a globally recognized soil pollutant of concern as they are widespread in prevalence and have the potential to harm the biological system. The significance of MPs in soil lies in their persistent nature, which could endanger biodiversity, the quality of soil, and possibly human health. These tiny particles that originate from diverse sources, seep into the soil where they negatively affect plant interactions, microbial communities, and also serve as carriers for other pollutants [17]. Understanding the effects of MPs in soil systems is critical to study and manage the environment for the continuous efforts to reduce the adverse effects of MPS on terrestrial ecosystems as well as human health.

15.2.1 Ecological Consequences

Due to MP contamination in the soil, various ecological repercussions have been identified. Various studies have indicated that MPs can have an adverse effect on the chemical as well as physical characteristics of soil. MPs can also diminish the soil fertility, modify the biogeochemical processes of soil, and influence the functions and activities of soil microbiota [18]. Furthermore, MPs have the ability to alter the developmental process of plants, soil flora and fauna, and germination of seeds. Alteration of all these processes can impact the soil biodiversity and ecosystem function [19,20]. Changes in the rates at which soil water evaporates have been associated with the presence of MPs in the soil; these changes may have an impact on soil water dynamics and plant growth [21].

15.2.1.1 MPs and Soil Fauna Interactions

The understanding of how MPs affect the health of land animals is significantly less advanced than that of aquatic creatures. Only a small number of soil invertebrates, such as nematodes, oligochaeta (earthworms), and isopods have been studied in this regard. Upon exposure of *Caenorhabditis elegans* to polystyrene (PS) particles of various sizes ranging from 0.1 to 5 μm for 3 days, Lei et al. [22] discovered that the group treated with the particle size of 0.1 μm showed the lowest survival rate. In addition to the notable decline in the body length. Given that 1.0 μm PS particles accumulated more than other sizes, it is possible that the moderately sized plastic particles were easier to be ingested by nematodes, which would account for the highest toxicity of 1.0 μm PS particles. Research on MP toxicity on oligochaeta demonstrates that the degree of exposure greatly influences the impact of MPs.

15.2.1.2 Impact on Soil Microbial Communities

There is still much to explore about the interaction between soil microbiota and MPs. Only a small portion of research has been done on the effects of MPs in soil systems, namely on bacterial translocation, antibiotic resistance genes (ARGs), and overall microbial activity. According to reports, the activity of soil microorganisms is positively impacted by polypropylene (PP) particles (7% and 28%), whereas polyester (0.05%–0.4%) and polyacrylic (0.05%–0.4%) have the opposite effect [23]. He, Wu [24] showed the effect of MPs on the deposition and transport of soil microorganisms. It was found that PS particles, under conditions of low ionic strength showed negligible effect on the transport of Escherichia coli in quartz sand whereas the transport of bacteria was stimulated by the plastic particles under high ionic strength conditions. Plastics at the nanoscale (20 nm) were found to be the primary mechanisms responsible for the increased cell transport; at the microscale (2 μm), plastics primarily increased cell transport by competing for deposition sites on sand. Both processes involved the adsorption of plastic particles onto the cell surface and the repel effect. The proliferation of pathogenic bacteria and the likelihood of microbial interaction with other adsorbed contaminants are increased when specific microbes are absorbed by MPs and form a single, distinct habitat.

15.2.2 Agricultural and Ecological Implications

Humans have deliberately manipulated the agroecosystems for assured well-being. However, agroecosystems are negatively impacted by certain human activities. Certain agricultural inputs, such as plant growth regulators, pesticides, fertilizers, and most recently, MPs have become cause of environmental pollution [25]. The environment is highly impacted by MP variables such as ubiquity, sheer volume, tiny sizes, composite chemicals, and growing anthropogenic activities. MP pollution has also been reported to affect terrestrial plants. A strong negative effect of polyethylene MP residues has been reported on wheat plants in both reproductive as well as vegetative growth periods [26]. MPs may build up in cucumber plant roots and transported to the flowers, leaves, and fruits of the plant. Similarly, MPs may build up in the roots of lettuce and wheat and then move to the shoots and leaves of the plant [27]. Additionally interfering with the microbial populations found in the soil, MPs eventually have an adverse influence on plant health. MPs may have a detrimental impact on plants if they have an impact on the soil's nitrogen-fixing bacteria and root symbionts [28].

15.2.2.1 Crop Productivity and Soil Health

MPs, due to their unique characteristics, can affect soil qualities by altering the physical structures of the soil. MPs' binding to the organic components and soil minerals of the soil can change how the soil aggregates [29]. The binding varies with varying size, composition, shape, and type of the plastic particles. For example, it has been reported that water-soluble aggregates increase in the presence of polyester fibers [20], whereas in the presence of polyethylene films, the proportion of macroaggregates decreases [30]. As microbial metabolites, or exo-biopolymers like polysaccharides, act as glue agents and enhance soil stability [31], desorption of hazardous chemicals, like phthalate esters, used in plastic production, might influence microbial activity and hence soil aggregation. It has been established that MPs in soil decrease the abundance of *Actinobacteria*, one of the major bacterial groups that contributes to soil aggregation [32]. Second, MPs can lower the bulk density of soil since plastics are frequently less dense than many of the soil minerals [20]. This increases soil aeration, which in turn facilitates the penetration of roots. Early crop stages may be directly harmed by MPs if the pores in the roots or seed capsule are physically blocked [33]. In fact, with MP exposure, very short-term detrimental effects on the growth of food plants may manifest as early as 6 days after sowing [34]. MPs have the ability to damage the biophysical environment of farmed soil, which could result in financial losses. They could also enter the trophic food chain and have an impact on human nutrition and health.

15.2.2.2 Ecological Disruption in Terrestrial Ecosystems

Many aquatic and terrestrial vulnerable species, populations, and habitats have already come to be considered at risk by MPs. Based on existing research, it appears that ecosystems contaminated with MPs exhibit decreased productivity, storage of carbon in flowers, biomass, and fauna. Considering that MPs may prevent oxygen-producing bacteria from photosynthezising, it is important to determine whether MPs have an effect on the ecosystems that are contaminated in terms of overall oxygen generation [35]. MPs present on the surface of soil can be transported to nearby waterbodies via water runoff and thus cause contamination [36]. Although MPs are seen as a threat to the world's ecosystems, little research has been done on how they might affect the temporary, regulatory, and socioeconomic ecosystem services [35].

15.3 PURPOSE OF SOIL REMEDIATION

Recently, various parts of global society have begun deal with MP contamination in the terrestrial environment. The careless use and mishandling of the many plastic products is contributing to the rise of MPs pollution [37]. The increasing amount of plastic waste in the environment & its threat to human health and ecosystems can no longer be ignored. Thus, it is imperative to address the issue of plastic contamination in the soil ecosystem as a critical concern and find possible remediation strategies [38].

15.3.1 Ecological Restoration

The health and functionality of the soil ecosystem have suffered greatly as a result of MP contamination. This widespread harm to the soil ecosystem emphasizes the need for ecological restoration. As mentioned above, MPs have damaged the health of the soil ecosystem by disrupting the nutrient cycles, soil structure, and the balance of naturally occurring biodiversity. The goals of ecological restoration must be customized to meet these problems with an emphasis on the restoration of natural habitats, enhancement of biodiversity, as well as sustainable management of soil and water.

15.3.1.1 Rehabilitating Biodiversity Hotspots

The rehabilitation of biodiversity hotspots refers to the restoration and conservation initiatives that are meant to improve the biodiversity in the regions that are rich in diverse species and are of ecological significance. Biodiversity hotspots, which are regions with high concentrations of rare, threatened, or rich species, are being identified more often to assist in determining conservation priorities [39]. These hotspots are facing significant threat due to MP contamination. Restoring the general health and functionality of these vital ecosystems is the aim of rehabilitation, which also aims to address species threats and environmental deterioration. Key elements included in the rehabilitation of hotspots biodiversity are the implementation of policies to improve and restore the natural ecosystems in these hotspots so that they can sustain a wide variety of animals and plants and simultaneously address the negative effects of MPs on soil, flora, and fauna [40].

15.3.1.2 Soil Habitat Preservation

The cycling of nutrients, filtration of water, and sustenance of a wide variety of plant and animal life are all made possible by soil ecosystems, which are essential for the continuation of life on Earth. MPs are a danger to these vital activities because they have the ability to alter the soil structure and hence the biodiversity [25]. A comprehensive approach is necessary to preserve soil ecosystem in the midst of MP pollution. In order to preserve the soil habitats, minimization of MP input into the environment should be done via proper waste disposal as well as awareness programs [41]. The restoration of MP-impacted soils is greatly aided by initiatives that improve the structure of soil and encourage biodiversity. Maintaining soil habitats is also essential in order to ensure that agricultural systems are productive and to reduce the possibility of MPs making their way up the food chain.

15.3.2 Human Health Protection

Consumer products that can be a source of MPs include toothpaste, cosmetics, hand soaps, scrubs etc. These chemicals have the potential to cause exposure to phthalates or PCBs that have passed into them during production operations, may result in a number of detrimental repercussions on human health. In fact, not only do they carry these chemicals but also other pollutants which adhere themselves to their surfaces as well [42]. A prohibition on the use of plastic bags is gaining popularity as a way to reduce MP pollution globally. To minimize MP contamination and their adverse effects on humans, countries like Sweden, Belgium, Australia, Luxembourg, Germany, and the Netherlands have imposed bans on the use of microbeads in products falling under the category of personal care products [43].

15.3.2.1 Safeguarding Food Supply Chains

Concerns are growing over the presence of MPs in the food chain, for example. Since the risk assessment of MPs in food is still in its infancy, further investigation and observation of MPs in food items are necessary [44]. The European Food Safety Authority has concluded that more knowledge on the occurrence of MPs in food is needed to solve this issue. While it is now believed that there is little chance of acute health consequences from plastic contamination of the food chain, the long-term effects are still unclear.

15.3.2.2 Minimizing Human Exposure Risks

Reducing the dangers of human exposure to MPs is crucial due to their potential health impacts. MPs can enter the body through various means, including absorption, inhalation, or physical contact, raising

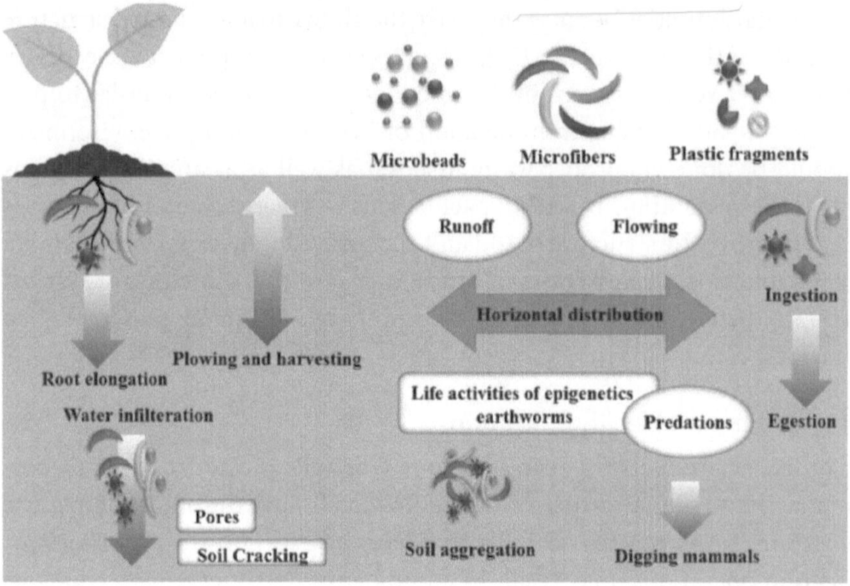

FIGURE 15.1 Various factors that influence MP migration in soil. Indicated by vertical arrows is the vertical migration of MPs through soil.

concerns about their persistent effects over time [45]. Ensuring the well-being of people, both physically and emotionally, is paramount. Therefore, it is imperative to tackle this issue head-on. Numerous global and local efforts, as well as policies, are striving towards reducing MP pollution's effects. The United Nations has launched the Clean Seas initiative, which aims to eliminate major sources of oceanic plastic and MP pollution in an effort to combat MP contamination on a global scale [46]. The campaign aims to boost public awareness, enhance waste management procedures, and advocate for the decrease in usage of disposable plastic products [46,47]. Although it is widely known that MPs can enter the human body through inhalation, ingestion, and skin contact, still not much is known about the health hazards connected to each of these pathways (Figure 15.1).

15.4 IMPACTS OF MPs ON SOIL

15.4.1 Ecological Consequences

Exposure to MPs (MP) has complicated the ecological impacts on the functions and compositions of soil ecosystems. MPs can alter the nitrogen cycle and soil structure. MPs pose a hazard to the survival, growth, and development of soil animals and plants and can also alter soil enzyme activity and microbial populations [48,49].

15.4.2 Effects on Soil Microorganisms

Limited studies are available on the interaction between MPs and soil microorganisms. MPs affect the soil microbiota by reducing the soil carbon and nitrogen content and releasing toxic chemicals that inhibit microbe growth [50,51]. On the other hand, many soil microorganisms are reported to degrade plastic polymers and are considered one of the effective solutions to degrade MPs [52].

15.4.2.1 Microbial Diversity and Functional Shifts

Different studies reported variable effects on soil microbial diversity in the presence of MP, so it is difficult to draw any conclusion. The study conducted by Judy et al. [53] reveals that the presence of 1% (w/w) polyethylene and polyethylene terephthalate MP has no significance on soil microbial communities. Whereas Fei et al. [32] reported that higher concentrations of polyvinyl chloride MPs significantly altered the soil microbiome and stimulated soil nitrogen fixation.

The addition of MPs to the soil caused a considerable change in the bacterial communities, and the variation grew throughout the incubation period, indicating that MPs may be able to support the diversity of the soil bacterial community.

15.4.2.2 Soil Enzyme Activities

Soil enzymes are essential for decomposing organic matter and encouraging the cycling of elements like C, N, and P. Fluorescein diacetate hydrolase (FDAse) is a crucial indicator of soil microbial metabolic activity that can promptly reveal alterations in soil quality [54]. While PVC MPs had a greater impact on soil acid phosphatase and FDAse activities, and PE MPs had a greater impact on soil urease activity, the addition of PE and PVC to soil decreased the activity of FDAse and increased the activity of urease and acid phosphatase [32].

15.4.3 Impact on Plant Growth

There is very little evidence available regarding the possible impacts of MP contamination on terrestrial plants. A study was conducted by Qi et al. [26] on the effects of MPs on the growth of wheat (*Triticum aestivum*).

Results showed that macro and MPs altered the wheat growth both at the vegetative and reproductive stages. de Souza Machado et al. [55] examined how different MPs affected the performance of spring onions (*Allium fistulosum*). The results showed that exposure to MPs could change the plant's overall biomass, the elemental makeup of its tissue, and its root characteristics, though the precise effects varied depending on the type of particle.

15.4.3.1 Root Morphology and Architecture

Plants utilize various strategies to adapt to stressful conditions. One such strategy involves root system expansion, which raises the root/shoot ratio. This is because larger, deeper root systems facilitate greater water and nutrient uptake, enabling vigorous root growth to withstand stress [56]. MP contamination is known to effect crop growth, particularly seed germination and root development [57,58]. The experiment conducted by Li et al. [59] revealed that the presence of *Solidago canadensis L* and MP caused a substantial decrease in root width, root height, and root area of rice plants. A study reported that when *Lolium perenne* (perennial ryegrass) HDPE MPs, a significant increase in root biomass was observed as compared to the control [19].

15.4.3.2 Nutrient Uptake Dynamics

The variation in the elemental composition of the leaves, or the changes in the N and C contents and their ratios, was another intriguing result observed in PES-exposed plants. Variations in the plant's physiological condition or nutritional availability are frequently linked to intraspecific variations in this particular feature [60]. It's probable that PES treatment can change plant physiology since PES-treated soils had far larger water-holding capacities, sustained higher water saturation for extended periods of time, and raised water levels in aboveground plant tissues [55]. Plant nutrient content and nutrient availability can both be impacted by mycorrhizal association. This symbiotic relationship frequently promotes healthy plant growth, which may have helped PES-exposed plants produce more biomass [61].

15.4.4 Soil Invertebrates and MPs

MPs have mechanical or physical impacts that are exhibited externally to the organism or following ingestion [62]. It has been demonstrated that the effect's potency inversely correlates with the size of the plastic particles, smaller particles are more likely to be consumed [63].

Among the soil invertebrates, earthworms are the major organisms that are affected by MPs [64]. It has been reported that 2% of polystyrene has a lethal effect on earthworms and is involved in damaging the self-defense system of the organism [65]. Soil collembolan species are also reported to be sensitive to MPs. When exposed to 0.1% PVC MPs for 56 days, the growth and reproduction of *Folsomia candida* were significantly inhibited [66]. In addition to that, the gut bacterial community of *F. candida* was also altered [67]. According to these findings, collembolan has the potential to be a useful bioindicator of soil MP disturbance [68].

15.4.4.1 Terrestrial Invertebrates Response

Different termite species have been demonstrated to degrade PE but avoid alternative plastic cable sheathings. This implies that termites excrete ground MP particles, yet it is unclear what effects this has on metabolism [69]. Observable tissue damage in the colon and stomach was noted in terrestrial snails exposed to MP/NP [70].

15.4.4.2 Soil Mesofauna Interactions

Although it is commonly known that meso- and macrofauna, such as mites, springtails, and earthworms, are essential to preserve soil quality, but the impact of MP on these vital organisms could jeopardize the health of agroecosystems [71,72]. Exposure to plastic by collembola or enchytraeid generally resulted in decreased motility, altered feeding and growth patterns, avoided of MP-contaminated areas, and decreased reproduction [73,74]. Enchytraeids play a crucial function in the mineralization of soil and can consume soil and MP as fibers, which has a significant impact on body size and number as well as reproduction [74,75]. Maaß et al. [76] noted that when exposed in a Petri plate, two collembola species, *F. candida* and *P. minuta*, could transport polymers up to 4 cm. Due to their small size, plastics bigger than >100 μm are not predicted to be consumed by soil organisms. Collembolans can readily attach plastics from the earth to their bodies or push them with their heads and legs. On the other hand, because they are smaller than 1 μm, nanoplastic trash can be consumed by collembolans and enter the food chain [77].

15.5 HUMAN HEALTH CONCERNS

MPs have been discovered by research in seafood, bottled water, tap water, and the air [78]. When MPs invade our systems, they promote inflammation, which may lead to a variety of health issues [79]. Furthermore, they possess the capacity to discharge toxic substances into the body, which can cause a variety of health concerns, including cancer, issues with reproduction, and developmental abnormalities [80].

15.5.1 Soil-Plant-Human Transfer

Multiple paths of exposure allow MPs to enter the human body. These pathways include inhaling PM2.5, consuming MP-polluted food or salt, drinking MP-polluted water directly, using plastic tea bags, and through skin contact [81]. According to studies, there may be a chance that MNPs will move up to the stem, fruit, and other edible portions of crops after first entering plant roots in the soil [82]. Thus, after crops are harvested, MNPs have the potential to contaminate food. Consequently, plastic can re-enter human bodies through MNPs that contain fruit and vegetable food [83].

15.5.1.1 Uptake Pathways in Edible Crops

Crop plants are one of the main sources of food calories, and agricultural soils supply 98.8% of the calories consumed worldwide by humans. However, there is a significant possibility that MPs will enter the food chain because of their extensive distribution in soils, primarily through crop contamination. MPs from the nearby soil and water can enter lettuce and wheat crops through fissures where new lateral roots are beginning to emerge [57]. MPs are unable to directly penetrate plant tissues as would be predicted due to the large size. Despite the fact that NP can penetrate plant cell walls directly, tobacco studies revealed that plant cells absorbed 20–40 nm nanopolystyrene beads [84]. Plastic particle absorption and translocation by plants are significantly influenced by transpiration pull. Plastic particles penetrate the primary and secondary roots of wheat's epidermal tissue, are activated by the pericycle, and then migrate into the xylem. These particles can get to the plant's aerial portion via the xylem inside the central cylinder [27].

15.5.1.2 Contaminant Accumulation Factors

The physical-chemical characteristics of contaminants and MPs determine the processes by which MPs absorb and release chemicals [85]. Particles of plastic float on the ocean's surface and readily absorb pollutants. The adsorption of these pollutants to MPs at concentrations much higher than those found in seawater and possible accumulation in organisms may be facilitated by the combination of increased surface area from weathering, extended exposure times in the marine environment, and the hydrophobicity of organic xenobiotics [86]. Exposure of animals to high concentrations of MPs carrying adsorbed chemicals may increase their vulnerability because of the likelihood that the kinetics may favor the desorption of contaminants from the ingested MPs into the tissues [87,88].

15.5.2 Health Risks

According to studies, ingestion of MPs damages the intestinal wall, which can result in oxidative stress, inflammation, and changes in the microbiota of the gut [89,90]. Unwanted health effects include cancer, neurological illnesses, and autoimmune disorders could result from this. Breathing in MPs can irritate and harm the respiratory system and might contribute to the emergence of respiratory diseases, including chronic obstructive pulmonary disease (COPD) and asthma [91].

15.5.2.1 Toxicological Impacts

MP cause toxicity at different trophic levels. MP cause adverse effects to phytoplankton, including impaired growth due to reduced chlorophyll content resulting in poor photosynthetic efficiency [92]. MPs are also reported in the bodies of marine mammals and birds, including their fecal remains that act as a major carrier of MPs in marine environment [93–95]. Polypropylene (PP) MPs' cause toxicity to several human cell lines. The toxicity of PP particles on cell lines was mostly dependent on their size and concentration. Via ROS-mediated oxidative stress, higher doses and small size PP caused toxicity to human dermal fibroblast cells, RAW264.7 murine macrophage cells, and peripheral blood mononuclear cells [96].

15.5.2.2 Epidemiological Correlations

An effective way to show the link between exposure to MPs and unfavorable health effects is through epidemiological research. On the other hand, epidemiological research on MPs is limited [97]. Workers in the Netherlands' polymer industries are more prone to develop chronic respiratory conditions as a result of occupational exposure. Interstitial lung illness linked to the job is identified in workers at nylon flocking factories in the US and Canada [98]. Yan et al. [99] find that patients with inflammatory bowel disease (IBD) have fecal MP concentrations that are much greater than those of healthy individuals and that these concentrations are positively connected with the severity of IBD. Furthermore, assessment of the mice exposed to MPs revealed that their fasting insulin and blood glucose levels were noticeably higher. These findings highlight the critical need for a large-scale cohort assessment on the prevalence and future prognosis of insulin resistance following exposure to MPs [100].

15.6 ASSESSMENT AND DETECTION

15.6.1 Methods for Detecting MPs in soil

Since the occurrence of MPs has become an issue of concern, various quantitative, qualitative, and combined methods have been developed to analyze MPs in soil samples.

15.6.1.1 Microscopy and Imaging

Visual identification of MPs under a light microscope is the oldest and simplest method [101]. Visual identification is a crucial step as it reveals surface texture and other features of possible MPs [102]. Morphological characteristics such as size, color, and shape are the initial basis for scrutinizing MPs. Stereomicroscopes are also used effectively in determining the physical properties of MPs [103]. Visual inspection of MPs is doubtful as it may exhibit errors ranging from 20% to 70%. Therefore, visual identification should be validated by other physical or chemical technologies [104].

15.6.1.1.1 Scanning Electron Microscopy (SEM) MP less than 1 mm in size can most probably be miscounted or missed when using a light microscope. Therefore, suspected MPs are further analyzed by Scanning Electron Microscopy. SEM is utilized to identify MPs. It provides a clearer structural image and greater magnification of MPs [105]. SEM is a costly and laborious process. Preparation of SEM requires other coatings which may lead to erroneous detection of the color and surface texture of MPs [102]. SEM provides high-resolution images of the particles which helps in discriminating MPs from organic particles.

15.6.1.1.2 Fluorescence Microscopy Fluorescence microscopy is a convenient strategy for MP detection, particularly transparent and white MPs as they innately fluoresce. This technique, coupled with imaging diminishes the chance of MPs detection failure and reduces the size limit of detected MPs [106]. Fluorescence microscopy with Nile red staining provides a relatively accurate technique for MPs detection in soil. After Nile red staining, MPs show strong fluorescent signals under a fluorescent microscope, whereas natural particles show weak signals and do not glow. This method for MPs detection is still under debate as chemical additives are utilized in its synthesis process [107].

15.6.1.2 Spectroscopy Techniques

Visual identification can sometimes be misleading, as microscopes cannot delineate the chemical composition of MPs accurately. Hence, there is a need to pick suspected MPs and analyze them by vibrational spectroscopy such as Raman or Fourier Transform Infrared (FTIR) spectroscopy. Their characteristics absorption spectrum identifies the functional group possessed by MPs [108]. These analytical methods are considered prestigious in MPs research as they accomplish precise identification of polymer shape, size, abundance, and types.

15.6.1.2.1 Fourier Transform Infrared Spectroscopy (FTIR) FTIR spectroscopy is an authentic identification technique because it documents distinct chemical bonds present between chemicals. FTIR optimization technologies such as attenuated total reflectance FTIR (ATR-FTIR), focal plane array FTIR (FPA-FTIR), and micro FTIR (μ-FTIR) possess updated MPs characterization potential [109]. These spectroscopic devices can also be coupled with a microscope for the detection of micro-scaled plastic particles. MPs detection limit of μ-FTIR is as low as 10 μm. FPA-FTIR can automatically scan the tentative sample for the presence of MPs in size ranging from 10 to 500 μm [110]. Due to the strong signal-to-noise ratio, ATR-FTIR can detect MPs >500 μm [111].

15.6.1.2.2 Raman Spectroscopy Another propitious approach for MP detection is Raman spectroscopy. It can identify MPs smaller than 1 μm when it is attached to a microscope. Raman microspectroscopy chemical imaging can resolve pixels as fine as 500 nm [112]. Raman spectroscopy has the significant benefit of processing wet materials and simultaneously identifying pigments or fillers [113]. Furthermore, Raman spectroscopy is not affected by ambient H_2O or CO_2. However, in certain situations, MPs cannot be identified due to interference from the spectrum caused by organic materials or background fluorescence of pigments contained in the polymer. Furthermore, there is a notable difference in the run duration between FTIR and Raman imaging [114].

15.6.1.2.3 Energy-Dispersive X-Ray Spectroscopy (EDS) SEM can be paired with Energy-Dispersive X-ray Spectroscopy (EDS), which along with producing high-resolution images, also provides the elemental composition of MP particles. The elemental composition of the particle is determined to differentiate inorganic particles from carbon-dominating plastics [102].

In SEM-EDX, an electron beam scans the particle surface. When an electron beam encounters a sample, it results in the emission of secondary electrons. As a result of these electrons, an image of the particle is generated, and its elemental composition can also be determined [115]. A limited number of samples can be processed as SEM-EDX is laborious, time-consuming, and expensive.

15.6.1.3 Chemical Analysis

Although FTIR and Raman spectroscopy can provide authentic MP identification information, this process is time-consuming. Mass spectroscopy can overcome the flaw of time consumption and can rapidly provide information about the mass concentration of MPs. Mass spectroscopy has emerged as a key tool for analyzing MPs, offering specificity and sensitivity to identify various polymer types [116].

15.6.1.3.1 Gas Chromatography-Mass Spectroscopy Three mass spectroscopic techniques are being employed for the detection of MPs, which include thermal extraction desorption-gas chromatography-mass spectroscopy (TED-GC-MS), thermo-gravimetric analysis-mass spectroscopy (TGA-MS), and pyrolysis-gas chromatography-mass spectroscopy (Pyr-GC-MS) [109]. TED-GC-MS is regarded as the first single-step technique to quantify polyethylene (PE) MPs in soil. TGA-MS is appropriate for quantifying polyethylene terephthalate (PET) MPs in standard loamy sand [108]. Pyr-GC-MS is a well-established technique for the mass quantification of polymers and their organic additives in environmental samples [117].

15.6.1.3.2 Liquid Chromatography-Mass Spectroscopy Liquid chromatography with tandem mass spectroscopy (LC-MS-MS) is an analytical technique that cannot detect gas samples directly. To analyze MPs using LC-MS-MS, the MPs should be disintegrated into monomers or small molecules using appropriate solvents. For example, PET can be broken down into its monomer i.e. terephthalic acid, using a combination of pentanol and potassium hydroxide as a solvent. This terephthalic acid can be identified and quantified using LC-MS-MS [118]. This method can be utilized to analyze municipal water, pet food and feces, human feces, sludge, sediments, and indoor dust [116].

15.6.2 Challenges in Quantifying MP Mass

At present, no readily applicable method can determine the abundance of MPs in the soil, as soil is such a complex and organic content-rich substrate. Current methods used to ascertain soil MPs are not consistent including, the patterns of spatial and temporal variability, the impact of environmental factors, reporting units, contamination control, and so on. These inconsistencies lead to incomparable results among different studies, as well as doubts about whether these results represent the true levels of MP contamination. Scientists must standardize protocols for effective comparison and monitoring of MPs in soil [104].

15.6.3 Emerging Technologies for MP Detection

The establishment of modern analytical instruments and combinations of new technologies with existing ones can resolve the current difficulties of MPs identification. The detectable size limit of MPs can be enhanced using these advanced technologies. Currently available methods can detect MPs up to a few micrometers. Smaller MP particles exhibit greater toxic effect, therefore, there is a crucial need to determine their fate, distribution, existence, and toxicity [119]. These methods include Atomic Force Microscopy (AFM)-IR, AFM-Raman [120], Nile Red Staining of plastics [121], Near IR Spectra Analysis method [122], Vis-NIR spectroscopy [123], and Nano Thermal Analysis (nano-TA) [124] (Figure 15.2).

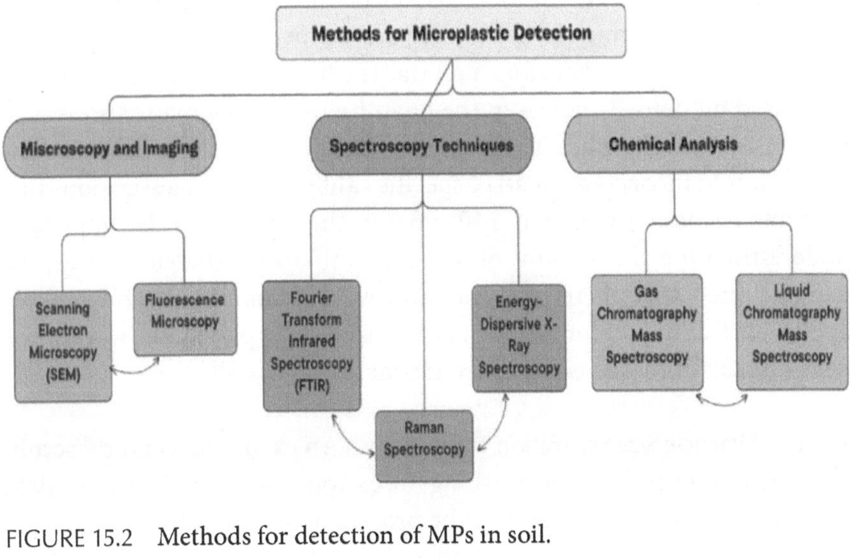

FIGURE 15.2 Methods for detection of MPs in soil.

15.7 SOIL REMEDIATION TECHNIQUES

Concern over contamination caused by MPs is growing among the scientific community as well as the general population [125]. Briefly, techniques like plastic filtering, pyrolysis, switching to biodegradable polymers, and subsequent chemical or biological breakdown might decrease the MP pollution in soil and groundwater [8].

15.7.1 Physical Methods

Physical and chemical treatment procedures are more assertive than biological processes, but they may treat soils with more persistent toxins or at greater contamination concentrations. The methods tend to reach remedial aims faster than their natural counterparts. https://www.fao.org/3/cb4894en/online/src/html/chapter-13-4.html

15.7.1.1 Soil Washing

In situ soil washing is a method for decontamination, the contaminants are removed from the soil by applying a washing solution by overflowing or sprinkling it over the unexcavated contaminated zone [126]. In order to remove a wide range of organic, inorganic, and radioactive pollutants from soil, the soil washing system to be discussed is a water-based physical separation method that depends on conventional physical and chemical extraction and separation procedures [127]. Fenton's reagent can be in conjunction with density separation to remove MPs [128]. For organic pollutants, the technique based on hydrogen peroxide is more effective. Overall, it was found that canola oil was the best technique for separating soil based on density; however, the quantity of organic matter present in the soil affected the efficacy of the procedure. This result emphasizes how crucial it is to incorporate matrix-specific calibration in investigations that take into account a variety of MP types in the future in order to prevent underestimating the amount of MP contamination. Therefore, it can be demonstrated that there are straightforward, economical, and broadly applicable techniques for removing MPs from soils, paving the way for the development of MP research in terrestrial settings [129].

15.7.1.1.1 Particle Size Separation Techniques An ex-situ mechanical scrubbing technique called "soil washing" uses water to get rid of unwanted pollutants from soils. Either way, the process uses basic particle size separation techniques (like those in sand and gravel operations) to concentrate the contaminants into a smaller volume of soil, or it dissolves or suspends them in the wash solution (which is then treated by conventional wastewater treatment methods).

The concept behind using particle size separation to reduce soil contamination stems from the observation that most organic and inorganic pollutants have a tendency to bond to organic soil particles, silt, and clay, either chemically or physically. The physical mechanisms of compaction and adhesion are in charge of silt and clay binding to sand and gravel particles. The process of washing, which separates the fine (tiny) clay and silt particles from the coarser sand and gravel soil particles, effectively separates and concentrates contaminants into a smaller volume of soil for further treatment or disposal. Particles with high or low specific gravity, including those containing heavy metals (radium oxide, lead, etc.), can be effectively removed via gravity separation. Coarser particles lose adhering contamination coatings when they are scrubbed by attrition. Attrition washing, however, may result in higher particles in the treated soils. The site can be used again with the cleaned, greater fraction (https://www.frtr.gov/matrix2/section4/4-19.html).

15.7.1.1.2 Centrifugation and Sedimentation Processes The centrifuges can be used to separate two immiscible liquids, break up emulsions of two immiscible liquids, produce a clear solution from a liquid-solid combination, and clarify nonsettling solutions containing finely split particles. Properties including particle size, density, and solubility are involved in separation processes, which deal with the separation of liquids from liquids, solids from gas, and solids from solids [130]. Centrifugation is a superior method to the method to fractional distillation for extracting polypropylene MPs from swine manure [131]. Centrifugation with a $ZnCl_2$ solution is one method of removing inorganic particles, such as sand, glass, etc. Consequently, using silicone tubes in centrifuges has the benefit of being able to be pinched off above the heavy fraction threshold at the bottom. As a result, while pouring out of the tube, there is no chance of the light and inorganic fractions coming into contact. MP that floated were extracted from the supernatant [132].

15.7.1.2 Electrokinetic Soil Remediation

Via the application of electric currents, heavy metals, radionuclides, mixed inorganic species, specific organic compounds, and some organic wastes can be extracted from soils and slurries using electrokinetic remediation, also known as electrochemical soil processing, electrokinetic decontamination, electromigration, or electroreclamation [133]. Pollutant removal in the electrokinetic process is dependent on a number of variables, including the electrodes' composition and placement within the electrokinetic system [134]. By applying a DC current across electrodes implanted into a soil mass, contaminants can be desorbened, transported, captured, and removed from fine-grained soils by electrokinetic soil processing [135]. When applied with a 100 V DC current, electrokinetic soil remediation can successfully remove 137Cs and 90Sr from soil, depending on the sorption and diffusion properties of the soil [136]. Gnanasundar and Raj [137] contend that by applying an electric field to the specimen and decontaminating the soil using the processes of electro-osmotic, electrophoresis, and electromigration, electrokinetic technology can be used to remove infused contaminants, particularly in low hydraulic conductivity soils.

15.7.1.2.1 Electro Osmosis and Electrophoresis Mechanisms The three primary mechanisms via which contaminants are transported in soil under a direct current electric field are electro-osmosis, electrophoresis, and electromigration. Unlike pressure-driven flushing, which is ineffective in fine-grained soils, they are especially successful in them [138]. When an electric field is provided and an ionic liquid flows past a charged surface, such as the interior surfaces of a porous material, this phenomenon is known as electroosmosis [139]. It may be seen that saturated clay functions like an electrochemical cell. Induced currents are the source of chemical reactions. The relationship between soil porosity and electroosmotic conductivity is being proportional to each other [140]. A charged ion in solution is simply transported by electromigration. A liquid containing ions travels in electroosmosis in relation to a charged surface that is immobile, such as in a capillary pore. The reverse of this procedure, known as electrophoresis, involves the movement of a charged particle with respect to a stationary liquid. In compacted soil systems, the role of electrophoretic transport of colloidal particles is insignificant [141]. One emerging remediation technique for restoring soils, sludge, and sediments polluted with inorganic and/or organic substances is electrokinetic remediation [142].

15.7.1.2.2 Electrochemical Enhancement Strategies One of the most promising methods for decontaminating soil is electrokinetic remediation, which works well in low-permeability soils like clay and has a high removal efficiency and time-effectiveness. Combining electromigration, electroosmosis, diffusion, and electrolysis of water with other processes, electrokinetic soil processing can eliminate ionic pollutants and non-polar organics [143]. Heavy metal and organic pollutants are efficiently removed from contaminated soils using ultrasonically enhanced electrokinetic remediation, which increases permeability, outflow, and the rate at which contaminants are removed [144]. Utilizing solubilizing chemicals that desorb the pollutants and promote their dissolution/solubilization in the interstitial fluid is crucial for the electro-osmosis removal of organic contaminants. The elimination of such hydrophobic organics from soils has been attempted with varying degrees of success using co-solvents, cyclodextrins, biosurfactants, and surfactants. But sustained high electro-osmotic flow is also necessary for the procedure to be effective. This may be accomplished by employing pulsed electric fields and controlling the pH of the soil with the appropriate washing solution in electrodes [142]. When ionic and non-ionic surfactants are used in electrokinetic soil remediation, the removal of hydrophobic organic pollutants is enhanced; SDS is most efficient in removing n-hexadecane and anthracene [145]. Chlorobenzene and trichloroethylene may be significantly removed from soggy podzolic soil by using a mixed reagent that contains a chelating agent and surfactant in the electrokinetic approach for the removal of harmful hydrophobic organic chemicals [146].

15.7.2 Chemical Methods

Chemical approaches seek to stabilize contaminants and transform them into less hazardous forms that are safe for aquatic life, plants, and human beings by adding chemicals or solvents to the contaminated soil. Usually, biomaterials, metallic oxides, or clays are the materials employed for such purposes. The kind of metal pollutant, soil organic matter content, soil texture, and other factors all affect how effective these products are at remediating soil [147].

15.7.2.1 Soil Amendments

A variety of composted agricultural byproducts, wood ash, animal manures and litter, sugar beetroot lime, municipal biosolids, coal combustion products like log yard waste, fly ash, composted biosolids, neutralizing lime products, and conventional agricultural fertilizers are among the frequently used amendments. Correct application of soil amendments reduces exposure to contaminants by impeding several paths of exposure and immobilizing pollutants to decrease their bioavailability. By adding organic matter, bringing the pH back into balance, increasing the soil's ability to store water, reestablishing microbial populations, and lessening soil compaction, amendments enhance the quality of the soil [148].

15.7.2.1.1 Organic Amendments It has been demonstrated that adding organic waste-based amendments to deteriorated and contaminated soils is an efficient and affordable way to improve soil quality. Fertility levels and the physical and biological qualities of the soil are indicators of this change in soil quality, which permits and supports the vegetative covering of formerly degraded lands. Nowadays, one of the most promising recovery methods for contaminated soils is the inclusion of organic amendments [149].

15.7.2.1.1.1 Biochar and Organic Matter Additions Biochar is used as a substitution to activated carbon for the adsorption-based elimination of organic micropollutants from wastewater and water bodies [150]. Electrostatic interactions, pore-filling, hydrophobic interactions, H-bonding, and cation/p/π–π interactions are key interactions in the removal of organic contaminants [151]. Although the effects of biochar as a soil supplement on MP-contaminated soil vary depending on temperature and the kind of pyrolysis feedstock, biochar can be utilized to improve the overall quality of the soil. In order to improve the soil quality in MP-contaminated soils, choosing an appropriate biochar is crucial [152].

15.7.2.1.1.2 Advanced Organic Amendments Biochar filters can remove and immobilize of 10 μm diameter MP spheres (above 95%) via three morphologically controlled mechanisms called "Stuck," "Trapped" and "Entangled" whilst the MP spheres being "Stuck" in the sand filter [153].

One effective defense against the detrimental impact of polystyrene microplastics (MPs) on soil chemistry has been demonstrated to be chitosan. When applied, it has positive outcomes, such as improving organic matter (OM) and reducing nutrient depletion in soil. Shifting focus to crop growth, As a growth inhibitor, polystyrene significantly lowers the biomass and chlorophyll content of maize plants. On the other hand, chitosan acts as a growth promoter, increasing plant height, leaf area, and the dry weights of the roots and shoots [154].

Following bioremediation by vermicomposting, sewage sludge polluted with MPs of various kinds and sizes may be safely applied to agricultural soils again. Making vermicompost using earthworms (Eudrilus eugeniae) [155].

15.7.2.1.2 Inorganic Amendments These amendments generally use minerals associated with soil fertility [156]. Paddy straw, farmyard manure, and green manure applied in combination with inorganic fertilizer in the rice-wheat cropping system have been shown to significantly improve the soil's quality and nutrient availability [157].

15.7.2.1.2.1 Zeolites and Soil Structure Enhancements One relatively abundant mineral resource is zeolite. Its unique free structural water storage, surface adsorption characteristics, and cation exchange capacity (CEC) offer a wide range of potential uses in soil amelioration [158]. They have a vast interior surface area that may be used for a range of responses due to their unique structure of cavities, channels, and cages. This surface area can also be used to protect, enhance, and remediate the quality of the soil [159]. The term "aluminosilicates" refers to engineered nano-porous zeolites that possess micro- (less than 2 nm), meso- (between 2 and 50 nm), and macropores (more than 50 nm) pores. These pores have the ability to adsorb and exchange ions, making them suitable to be utilized as fertilizer carriers [160]. The nano-Fe@ZIF-8 composite can eliminate upto 94% of human endocrine-disrupting phenols (1 mg/L), such as bisphenol A and 4-tert-butylphenol, as well as upto 98% of polystyrene microsphere trash (25 mg/L) within 5 minutes of treatment [161]. Thus, the application of zeolite may be beneficial for agricultural practices by improving nutrient use efficiency (NUE) and water use efficiency (WUE), which can lower the risk of groundwater and surface water contamination [162]. As a result, zeolites are widely used worldwide for remediating both organic and inorganic contaminants [163].

15.7.2.1.2.2 Metal Oxide Nanoparticles for MPs Binding Given of the enhanced reactive nature of nanoparticles and their potential for in situ treatment, remediation using engineered nanomaterials (ENMs) they offers more cost-effective and efficient procedures than traditional methods. Three examples of how ENMs are utilized in soil remediation include nanoscale metal oxides for the adsorption of metals, nanoscale calcium peroxide for the breakdown of organics (like petrol), and nanoscale zero-valent iron for the breakdown of halogenated organic compounds [164]. By combining surface adsorption, adsorptive bubble separation, and catalytic degradation, the low-cost metal oxides Fe_2O_3-MnO_2-based micromotor effectively removes both suspended MPs and aqueous organics. Over 10% of the suspended MPs were removed from the contaminated water in just two due to the adsorptive separation [165]. Through hydrogen bonding, bare Fe_3O_4 nanoaggregates effectively adsorb to a range of common MPs, such as polypropylene, high-density polyethylene, polystyrene, PVC, and PET. Due to its nanozyme activity at high temperatures, this permits total MP breakdown and complete magnetic removal at 1% of the plastic weight [166]. Metal and metal oxide nanomaterials have high adsorption capacities and fast kinetics, making them very efficient adsorbents. Nanoparticles are widely employed for environmental cleanup because they may be used both in situ and ex situ in aquatic settings [167].

15.7.2.2 Chemical Degradation

15.7.2.2.1 Oxidation Processes Introducing chemical oxidants into underground soil and groundwater to eliminate organic pollutants is known as "in situ chemical oxidation," a rapidly developing technique. When contaminants can be mineralized into carbon dioxide and water or oxidized into a comparatively less toxic state that may be used for further treatment, this method shows potential. The method of oxidation and the oxidant to be used depend on the kind and quantity of contamination, the degree of cleanup needed, the feasibility of delivering the oxidant, the site's soil characteristics, and its hydrogeology [168].

This entails pumping oxidants down one well and groundwater that has been combined with oxidants out of another well. More oxidant is added, the mixture is pumped out, and then it is pumped back and recirculated through the first well. Recirculation speeds up the treatment of a larger region. Another alternative for clayey soil that prevents groundwater and oxidants from flowing is to use mechanical augers or excavation equipment to mix and inject oxidants [169].

Through chemical interactions with oxidants, such as sodium persulfate, hydrogen peroxide (also known as CHP or modified Fenton's reagent), sodium and potassium permanganate, a broad variety of organic compounds can be destroyed through oxidative degradation [170]. For the ex-situ restoration of the problematic sediments, chemical was oxidation established to be a successful remediation technique that was easily adaptable. The combination of hydrogen peroxide, potassium permanganate, and modified Fenton's reagent yields the greatest remediation results when the oxidant dose is around 100 mmol/30 g sediment sample [171]. When it comes to many of the often-targeted organic pollutants in soil and groundwater systems, persulfate can be useful. These consist of MTBE, BTEX, PAHs, chlorinated benzenes, and chlorinated solvents [172]. Such a proven remediation technique for the in situ treatment of a variety of organic environmental pollutants is chemical oxidation [173].

15.7.2.2.1.1 Advanced Oxidation Technologies (AOTs) For the purpose of treating soils contaminated by organic substances that are challenging to eliminate, advanced oxidation processes represent a promising approach [174]. To break down organic contaminants that are not biodegradable in industrial effluents, advanced oxidation procedures (O_3, O_3/UV, H_2O_2/UV, O_3/H_2O_2/UV, Fe_2^+/H_2O_2, O_3/H_2O_2) present an alluring alternative to traditional treatment techniques. Because of its strong oxidant capacity, this technique, involving the generation of highly reactive and oxidizing free radicals, is becoming popular. These radicals can be produced by combining hydrogen peroxide with ferrous ions in a process known as Fenton's reaction or by utilizing single oxidants, ozone, hydrogen peroxide, and UV rays in combination [175]. These species are in charge of initiating the breakdown of plastic by introducing changed C-H moieties and newly generated functional moieties like hydroxyl and carbonyl into the framework. MP deterioration is facilitated by the framework's modifications. The oxidation kinetics are accelerated during these procedures, which results in the full mineralization of pollutants into carbon dioxide or other safe byproducts [176]. It is possible to eliminate around 80% of MPs using sophisticated treatment plants. Although very effective in removing and degrading MPs, advanced oxidation techniques (photocatalytic degradation, ozone degradation, electrochemical degradation, Fenton/Fenton-based AOPs, and photodegradation) are difficult to deploy on a wide scale [175].

15.7.2.2.1.2 Photocatalysis for MPs Degradation Photocatalysis is the process involved in light-induced redox reactions that can remove contaminants from the environment. When photon energy ($h\omega$) is equal to or greater than band gap energy (Ebg), electrons are moved from the valence band into the conduction band on the photocatalyst surface. This causes the creation of electron–hole pairs in both bands, respectively. The positive hole that is made in the valence band oxidizes the hydroxyl radical that the pollutant or water molecules produce, and the excited electrons on the photocatalyst reduce the amount of oxygen that has been adsorbed. The excited electrons within the conduction band have a large reduction capacity when they interact with dissolved molecular oxygen in water or adsorbed oxygen on the outermost layer of the photocatalytic material [177]. The method in question is based on the standard electrical structure of several metal oxide semiconductors that, when exposed to radiation, are stimulated to produce pairs of electrons and holes [178].

Due to their numerous good qualities—such as chemical stability, broad availability, low cost, and stable structural and electrical properties—TiO_2-based photocatalysts are widely regarded as the most widely used materials for POP destruction. Basically, when photocatalytic materials (like TiO_2) are irradiated to light, electrons (e^-) and holes (h^+) may be produced on their surface (*hv* higher than or equal to *Eg*). Consequently, water (H_2O) and oxygen (O_2) may be combined by (h^+) and (e^-) to produce diverse reactive radicals, such as active oxygen species (O2·$^-$) and hydroxyl radicals (OH^-). As a result, POPs and (OH^-) hydroxyl radicals will continually react to dehalogenate and shorten the chain (from Cn to Cn−1), producing the anticipated end products of H_2O and carbon dioxide (CO_2) [179].

Regular MPs, such as polystyrene (PS) microspheres and polyethylene (PE), can be degraded by photocatalysis when exposed to UV light on TiO_2 nanoparticle coatings. Within 12 hours, a TiO_2 nanoparticle coating prepared with Triton X-100 showed full mineralization (98.40%) of 400 nm PS; deterioration for other PS sizes was also examined. After 36 hours, a strong photodegradation rate was shown in the PE degradation experiment. The primary final product discovered was CO_2 [180].

15.7.2.2.2 Hydrolysis Techniques MPs can be degraded via a process called hydrolysis, in which an interaction with water breaks down the polymer chains. Depending on the kind of polymer, either basic or acidic conditions can be used for the hydrolysis process [181]. In water, the presence of a small amount of catalysts accelerates hydrolysis. The ions that are present in the water might act as catalysts. The quantity of water within the material affects the pace of hydrolysis in addition to the polymer's sensitive chemical linkages. Water in the polymer is necessary for both hydrolysis by pure water and catalysis by the presence of salts [182]. With about 70 million tonnes produced globally each year for usage in textiles and packaging, PET is the most common polyester plastic worldwide [183].

In a manufacturing facility, ester bond cleavage can depolymerize PET to its constituent parts [184]. When it comes to the establishment of facilities for the direct synthesis of PET from TPA and EG, hydrolysis—a technique for recovering PET waste chemicals—is gaining popularity [185].

Sulfuric acid is the most often utilized acid, while concentrated acids like phosphoric or nitric can also be employed for acidic hydrolysis of PET. The main disadvantage of this approach is the separation of the ethylene glycol from the highly acidic solution, despite the method's normally excellent yields. The quantity of acid required to industrialize this process also presents challenges for the environment, the economy, and the process. Typically, 4–20 weight percent NaOH is used in aqueous solutions for alkaline hydrolysis. With this method, sodium terephthalate and EG are produced in comparatively high quantities, with up to 100% PET conversion [186].

Complete hydrolysis of PVC-coated PET woven fibers in the presence of a 1 M NaOH solution can occur between 120°C and 180°C, producing water-soluble Na2TP and terephthalic acid quantitatively following PVC filtrate. This allows for the simultaneous recycling of PVC and PEC [187]. Using a 5% or 10% w/v NaOH solution, complete polycarbonate breakdown may be achieved at 160°C after 40 or 10 minutes of microwave irradiation [188].

PET can also be broken down by alkaline hydrolysis, which involves heating an aqueous NaOH solution to 120°C–200°C and a nonaqueous KOH solution in methyl Cellosolve to 110°C–120°C. When sulfuric acid is applied to the disodium or dipotassium terephthalate, it produces very pure terephthalic acid (TPA) [189].

15.7.3 Biological Methods

Since biodegradation technology is very efficient, inexpensive, environmentally friendly, and promotes sustainable development, it is believed to have the most chance of being applied in the future to mitigate MP pollution [190]. Putting in succinctly, abiotic and biotic factors collaborate to trigger microbial MP breakdown. MPs are formed from the fragmentation of plastic waste, and they have the ability to further break down into nanoscale (or tiny plastic particle) forms. Polymers experience bio-fragmentation and bio-deterioration processes due to the combined action of extracellular enzymes and abiotic stimuli, which releases oligomers and monomers [191]. Several factors, such as functional group, molecular weight, chemical structure, additives, and crystallinity influence, biological degradation [192].

15.7.3.1 Microbial Degradation

Scientists are now interested in using microorganisms, including fungi, bacteria, and algae, as a cure for MPs. The enzymatic processes that the microorganisms produce are strongly linked to the decomposition of MPs [193]. As the foundation of enzymatic degradation, microbial colonization on plastic surfaces via adhesion or contact is the first step in the microbial destruction of plastics. The coupling of an enzyme and polymer matrix is called hydrolysis, and it is followed by catalytic hydrolysis and cracking in the second stage. Organic matter hydrolysis is catalyzed by the bound enzyme, a hydrolase. Microorganisms' external and intracellular enzymes are involved in two crucial activities. Hydrolytic cleavage of the polymer chain is induced by extracellular enzymes such as hydrolases and depolymerases. Small monomers or oligomers that bacteria can integrate into the cell are produced by the polymer chain when it is attacked by enzyme enhancement. In addition, the cell's enzymes can absorb tiny oligomers or monomers following their metabolism [194]. A biofilm of microorganisms develops on the outer region of pollutants, resulting in a plastisphere—a zone where these microorganisms interact and produce acid and other enzymes for the breakdown of MPs [195]. It has been established that the microbial species on the polymer debris, which forms the Plastisphere, vary from typical representatives of their surroundings in terms of both kind and quantity. Along with exhibiting improved horizontal gene transfer, varied metabolic pathways, and antibiotic resistance, they also produce islands of pathogenicity. For example, the hydrocarbon-degrading bacteria Arenibacter, Alcanivorax, and Marinobacter are abundant in seawater-filled polymer debris (PE) or bottles made of PET [196].

High-density polythene may be broken down by the bacterial species Pseudomonas and Arthrobacter by 15% and 12%, respectively. About 8% of low-density polythene was broken down by Pseudomonas organisms. Polythene may also be effectively broken down by *Aspergillus glaucus* up to 28.80%. Low-density polythene is broken down by *Aspergillus niger* and Aspergillus japonicas at rates of 8% and 12%, respectively. The blue-green algae Anabaena spiroides may break down polythene by 8.18% [197]. Apart from bacteria, fungi may also attach themselves on and make use of MP. In order to lessen the hydrophobic nature of MPs, fungi can encourage the formation of several chemical linkages, including ester, carboxyl, and carbonyl functional groups. Numerous types of fungus exist, and these microorganisms have robust reproductive capabilities and a broad distribution. Moreover, fungus may facilitate the alterations and movement of many chemicals [198]. Through the use of universal enzymes, fungi are able to both penetrate and detoxify many contaminants. Mycelium may penetrate three-dimensional surfaces, and fungi create proteins that are hydrophobic which can adhere mycelium to hydrophobic-natured substrates [199]. It has been discovered that petroleum-based polymers may be broken down primarily by ascomycetes, then by basidiomycetes and zygomycetes [200].

15.7.3.2 Phytoremediation

Since MPs are big molecules and are unable to pass through plant cell walls, it is typically not expected that they would be carried into plant tissues [201]. Particles of nanoplastics could be more able to get past plant barriers. In fact, a significant amount of nanoplastics may penetrate plant roots and travel to the aerial portions of the plant [202]. NPs have a higher propensity than MPs to penetrate plant cell walls. Moreover, transpiration pull is important in the absorption and transport of plastic particles by plants. MPs barely affect physiological and biochemical markers in plants in a negative way [203]. Plant bioavailability of micro/nanoplastics may be altered by their partitioning to various soil matrices, such as soil particles, naturally existing soil organic materials, pore fluids, and soil fauna. The apoplastic and symplastic routes allow roots to absorb the minute micro/nanoplastic particles. Cellular endocytosis and aquaporin may be important in this approach. Additionally, the form of the polymers can control how quickly they are absorbed by roots; roots are more likely to absorb spherical polymers than other shapes [204]. Plastic beads smaller than 0.2 micrometres (0.2 μm) and bigger than 2 μm might penetrate the root at the sites of lateral root initiation where 0.2 μm extracellularly caught in the highly hydrated polysaccharide known as the root cap mucilage of the lettuce plant. The majority of polystyrene microbeads were found in the vascular system and on the cortical tissue of the roots' cell walls, which suggested that the apoplastic transport mechanism was used by the beads to move through the intercellular space. The vascular system followed the transpiration stream to move the 0.2 μm polystyrene microbeads from the roots towards the stems and finally leaves once they were within the central cylinder. The PS beads were also found to stick to one another and methodically self-assemble into clusters that resembled grapes and strings in the vascular tissue of the lettuce plant's root and stem. PS beads were distributed throughout the leaf tissue, as opposed to the root and stem [205]. Polystyrene microplastics (1 μm) can enter and gather the inside intercellular layer of the carrot roots but are unable to enter the cells as compared to 0.2 μm particles which can be transported to leaves. Larger MPs can enter the roots and leaves in the presence of As (III) [206].

Particles of polymethylmethacrylate and polystyrene, both micrometre and submicrometer in size, can penetrate the stele of wheat (*Triticum aestivum*) and lettuce (*Lactuca sativa*) through the crack-entry process at the locations of lateral root emergence. Submicrometre plastic is efficiently absorbed due to the crack-entry channel and characteristics of the polymeric particles. Plastic particles then moved from the roots up towards the shoots. Elevated rates of transpiration facilitated the absorption of plastic particles, indicating that transpirational pull was the primary catalyst for motion [27].

15.7.3.2.1 Plant Species Selection for MPs Uptake MP adsorption and accumulation are facilitated by a number of processes, the most significant of which are the presence of periphyton, leaf shape, and electrostatic forces. Herbivores may readily consume absorbed MPs from plant tissues. Therefore, plants may be a potential entry point for MPs into aquatic food webs. However, MPs' potent interactions with plants may be used to their advantage for phytostabilization and eventual environmental clearance [207]. When soil accumulates contaminants as a result of organic fertilizer, surface runoff, plastic mulching, wastewater irrigation, sludge, plastic littering, or atmospheric deposition of airborne MP, land edible plants become exposed to microplastics directly [208]. Through root accumulation and translocation to aboveground tissues, rice seedlings may move accumulated nano/microplastics to higher trophic levels farther up the food chain. Confocal laser scanning microscopy confirmed the presence of both nano- and micro-sized PS microspheres in the roots, stems, and leaves of rice seedlings 14 and 40 days after exposure. Equally, the 1 μm and 80 nm polystyrene microspheres get accumulated in the vascular systems of plant tissues, especially vascular bundles of stems, veins of leaves, root stele, and most of them accumulating on cell walls and intercellular regions [209]. Also, when compared to the control in maize (*Zea mays*) seedlings, confocal laser scanning micrographs verified the presence of strong fluorescence signals from tiny polystyrene beads (0.2 μm) in all tissues. After treating root tissues with 5.0 μm polystyrene beads, no polystyrene beads were found with 2.0 μm polystyrene beads mostly dispersed throughout the xylem. The root tip zone was less likely to absorb polystyrene beads than the root maturation zone was [210]. When polystyrene

MPs are absorbed by a lettuce root, they can pass through the vascular tissue and enter the vascular string. Small MPs (SMPs, 100 nm) are more capable of translocating than large MPs (LMPs, 500 nm) [211].

Numerous pollutants, including endocrine disruptors, persistent organic pollutants, heavy metals, medications, and pharmaceutical and personal care items, are absorbed by vascular plants and the ecosystems surrounding them [212]. Many factors, including particle size, shape, content, plant species, and soil conditions, render it more difficult to accurately assess the presence of plastic contamination in plant tissue [213]. Plants can become a source of MPs as a result of weak adsorption and environmental conditions that allow the entrapped MPs to desorb, resuspend, or escape. Due to the possibility of ingestion by higher trophic levels of the food chain, the MPs that have been trapped and accumulated in plants give rise to concerns over potential ecotoxicity [214].

15.7.3.2.2 Rhizosphere Interactions and Enhanced Phytoremediation The possible effects of MP pollution on soil characteristics, which are essential for plant development and soil fertility, make soil contamination a reason for uncertainty [215]. The community structure, diversity, microbial richness, and function between MP surfaces and the surrounding "rhizosphere-like" soil are significantly diverse, resulting in a "rhizosphere-like neighbour avoidance effect" between MP surfaces and nearby regions. MPs lower the microbial richness and diversity on their surface than those in the surrounding "rhizosphere-like" soil [216]. Rhizosphere soil of wheat shows lower alpha diversities (richness, evenness, and diversity) of microbiota due to the presence of polyethylene MPs [217]. MP treatment was observed to dramatically lower the bacterial richness and diversity in the rhizosphere soil of highland barley [218]. Low-density polyethylene and biodegradable MP at high doses (2.5%) decrease α-diversity in the rhizosphere microbial communities of common beans (*Phaseolus vulgaris* L.) when compared to doses of 0.5% and 1.0%. This suggests that a 2.5% dose of MPs may have a selective effect on rhizosphere bacterial populations [219].

15.8 CHALLENGES IN MP SOIL REMEDIATION

The soil chemistry, metabolic activities, and general functionality of the soil microbiota can all be affected by MPs in the soil. This could therefore have a major impact on plant productivity. MPs are able to enter the soil ecosystem in a number of ways. The two most common additions to soil are organic wastes (composts and biosolids), which may contain plastic pollutants, and plastic mulch covers. A lot of agricultural systems use plastic mulch films (4–8 μm thick) to increase plant growth and water efficiency. Despite their great effectiveness, they are rarely recycled and are infamously hard to collect from soil following crop harvest. As a result, plastic films are frequently mixed with the soil in areas where tillage and UV radiation promote soil fragmentation [220]. MPs may improve soil structure, aeration, and C storage in agricultural settings, but the possible drawbacks are expected to exceed any advantages by a wide margin. Due to the difficulties in catching these tiny particles and the variety of sources from which they come, conventional waste management systems typically fail to effectively mitigate the problem of MP contamination. Diverse cutting-edge techniques and technologies have been created to collect and remove environmental MPs. These solutions cover a range of techniques, such as biodegradation techniques and filtration systems [221]. However, there are a number of obstacles to overcome before implementing MP remediation solutions, such as scalability, cost-effectiveness, and the possibility of unintentionally harming organisms that are not the intended targets. Moreover, the removal of MPs from terrestrial settings presents a significant difficulty because of the vastness and restricted accessibility of some places [222]. Researchers are looking into ways to recycle MPs so that their environmental impact can be minimized and waste management concerns can be addressed, as opposed to just trying to eradicate the particles entirely. Several disciplines, including materials science, engineering, and even the arts, are involved in this idea of using MPs for good [223].

By raising public knowledge of MP pollution's negative effects and encouraging responsible consumption, we can significantly reduce the amount of MP pollution entering the ecosystem. In order to encourage the adoption of sustainable practices, it is imperative that information on the harmful impacts of MPs be disseminated through numerous platforms to consumers, industries, and politicians [224]. Regulatory frameworks and international agreements that place limitations on the use of MPs in particular products and industries have the potential to significantly reduce the amount of MP pollution. Enacting regulations that support responsible plastic production, use, and disposal requires industry, government, and non-governmental organizations to work together [225].

15.9 CONCLUSION

The complex challenges posed by MP pollution in soil ecosystems and stresses the immediate need for aggressive actions to tackle this critical environmental problem. Although progress has been made in comprehending the origins, varieties, and environmental consequences of soil MPs, there are still considerable gaps in our understanding, especially concerning their extended durability and their effects on land-dwelling organisms and human well-being. In the future, it is crucial to make joint efforts to progress research, create new remediation methods, and enhance regulatory frameworks to efficiently reduce soil MP contamination. Encouraging cooperation among academics, policymakers, industrial stakeholders, and the public may help develop sustainable solutions to conserve soil ecosystems, biodiversity, and global environmental health and resilience.

REFERENCES

1. Nielsen TD, Hasselbalch J, Holmberg K, Stripple J. Politics and the plastic crisis: a review throughout the plastic life cycle. *Wiley Interdisciplinary Reviews: Energy and Environment.* 2020;9(1):e360.
2. Shen M, Huang W, Chen M, Song B, Zeng G, Zhang Y. (Micro)plastic crisis: un-ignorable contribution to global greenhouse gas emissions and climate change. *Journal of Cleaner Production.* 2020;254:120138.
3. Yadav V, Sherly M, Ranjan P, Tinoco RO, Boldrin A, Damgaard A, et al. Framework for quantifying environmental losses of plastics from landfills. *Resources, Conservation and Recycling.* 2020;161:104914.
4. Laskar N, Kumar U. Plastics and MPs: a threat to environment. *Environmental Technology & Innovation.* 2019;14:100352.
5. Mishra UN, Das S, Kandali R. Bioremediation of synthetic polymers: present and future prospects of plastic biodegradation. *International Journal of Current Microbiology and Applied Sciences.* 2020;9(12):1234–47.
6. Bajt O. From plastics to MPs and organisms. *FEBS Open Bio.* 2021;11(4):954–66.
7. Picó Y, Barceló D. Analysis and prevention of MPs pollution in water: current perspectives and future directions. *ACS Omega.* 2019;4(4):6709–19.
8. Chia RW, Lee J-Y, Kim H, Jang J. Microplastic pollution in soil and groundwater: a review. *Environmental Chemistry Letters.* 2021;19(6):4211–24.
9. Ryan PG. A brief history of marine litter research. *Marine Anthropogenic Litter.* 2015:1–25.
10. Thompson RC, Olsen Y, Mitchell RP, Davis A, Rowland SJ, John AW, et al. Lost at sea: where is all the plastic? *Science.* 2004;304(5672):838.
11. Rochman CM. MPs research—from sink to source. *Science.* 2018;360(6384):28–9.

12. Barnes DK, Galgani F, Thompson RC, Barlaz M. Accumulation and fragmentation of plastic debris in global environments. *Philosophical Transactions of the Royal Society B: Biological Sciences.* 2009;364(1526):1985–98.
13. Sutherland WJ, Clout M, Depledge M, Dicks LV, Dinsdale J, Entwistle AC, et al. A horizon scan of global conservation issues for 2015. *Trends in Ecology & Evolution.* 2015;30(1):17–24.
14. Nizzetto L, Bussi G, Futter MN, Butterfield D, Whitehead PG. A theoretical assessment of microplastic transport in river catchments and their retention by soils and river sediments. *Environmental Science: Processes & Impacts.* 2016;18(8):1050–9.
15. Mahon AM, O'Connell B, Healy MG, O'Connor I, Officer R, Nash R, et al. MPs in sewage sludge: effects of treatment. *Environmental Science & Technology.* 2017;51(2):810–8.
16. Rillig MC, Ziersch L, Hempel S. Microplastic transport in soil by earthworms. *Scientific Reports.* 2017;7(1):1362.
17. Bläsing M, Amelung W. Plastics in soil: analytical methods and possible sources. *Science of the Total Environment.* 2018;612:422–35.
18. Yu H, Zhang Y, Tan W, Zhang Z. MPs as an emerging environmental pollutant in agricultural soils: effects on ecosystems and human health. *Frontiers in Environmental Science.* 2022;10:217.
19. Boots B, Russell CW, Green DS. Effects of MPs in soil ecosystems: above and below ground. *Environmental Science & Technology.* 2019;53(19):11496–506.
20. de Souza Machado AA, Lau CW, Till J, Kloas W, Lehmann A, Becker R, et al. Impacts of MPs on the soil biophysical environment. *Environmental Science & Technology.* 2018;52(17):9656–65.
21. Ding L, Huang D, Ouyang Z, Guo X. The effects of MPs on soil ecosystem: a review. *Current Opinion in Environmental Science & Health.* 2022;26:100344.
22. Lei L, Liu M, Song Y, Lu S, Hu J, Cao C, et al. Polystyrene (nano) MPs cause size-dependent neurotoxicity, oxidative damage and other adverse effects in Caenorhabditis elegans. *Environmental Science: Nano.* 2018;5(8):2009–20.
23. Awet T, Kohl Y, Meier F, Straskraba S, Grün A-L, Ruf T, et al. Effects of polystyrene nanoparticles on the microbiota and functional diversity of enzymes in soil. *Environmental Sciences Europe.* 2018;30(1):1–10.
24. He L, Wu D, Rong H, Li M, Tong M, Kim H. Influence of nano-and microplastic particles on the transport and deposition behaviors of bacteria in quartz sand. *Environmental Science & Technology.* 2018;52(20):11555–63.
25. Okeke ES, Okoye CO, Atakpa EO, Ita RE, Nyaruaba R, Mgbechidinma CL, et al. MPs in agroecosystems-impacts on ecosystem functions and food chain. *Resources, Conservation and Recycling.* 2022;177:105961.
26. Qi Y, Yang X, Pelaez AM, Lwanga EH, Beriot N, Gertsen H, et al. Macro-and micro-plastics in soil-plant system: effects of plastic mulch film residues on wheat (Triticum aestivum) growth. *Science of the Total Environment.* 2018;645:1048–56.
27. Li L, Luo Y, Li R, Zhou Q, Peijnenburg WJ, Yin N, et al. Effective uptake of submicrometre plastics by crop plants via a crack-entry mode. *Nature Sustainability.* 2020;3(11):929–37.

28. Tang KHD. Effects of MPs on agriculture: a mini-review. *Asian Journal of Environment & Ecology*. 2020;13(1):1–9.
29. Rillig MC, Ingraffia R, de Souza Machado AA. Microplastic incorporation into soil in agroecosystems. *Frontiers in Plant Science*. 2017;8:1805.
30. Zhang G, Zhang F, Li X. Effects of polyester microfibers on soil physical properties: perception from a field and a pot experiment. *Science of the Total Environment*. 2019;670:1–7.
31. Lehmann A, Zheng W, Rillig MC. Soil biota contributions to soil aggregation. *Nature Ecology & Evolution*. 2017;1(12):1828–35.
32. Fei Y, Huang S, Zhang H, Tong Y, Wen D, Xia X, et al. Response of soil enzyme activities and bacterial communities to the accumulation of MPs in an acid cropped soil. *Science of the Total Environment*. 2020;707:135634.
33. Khalid N, Aqeel M, Noman A. MPs could be a threat to plants in terrestrial systems directly or indirectly. *Environmental Pollution*. 2020;267:115653.
34. Pignattelli S, Broccoli A, Renzi M. Physiological responses of garden cress (L. sativum) to different types of MPs. *Science of the Total Environment*. 2020;727:138609.
35. Sridharan S, Kumar M, Bolan NS, Singh L, Kumar S, Kumar R, et al. Are MPs destabilizing the global network of terrestrial and aquatic ecosystem services? *Environmental Research*. 2021;198:111243.
36. Schell T, Hurley R, Buenaventura NT, Mauri PV, Nizzetto L, Rico A, et al. Fate of MPs in agricultural soils amended with sewage sludge: is surface water runoff a relevant environmental pathway? *Environmental Pollution*. 2022;293:118520.
37. Anik AH, Hossain S, Alam M, Sultan MB, Hasnine MT, Rahman MM. MPs pollution: a comprehensive review on the sources, fates, effects, and potential remediation. *Environmental Nanotechnology, Monitoring & Management*. 2021;16:100530.
38. Chae Y, An Y-J. Current research trends on plastic pollution and ecological impacts on the soil ecosystem: a review. *Environmental Pollution*. 2018;240:387–95.
39. Reid WV. Biodiversity hotspots. *Trends in Ecology & Evolution*. 1998;13(7):275–80.
40. Adla K, Dejan K, Neira D, Dragana Š. Degradation of ecosystems and loss of ecosystem services. In Prata JC, Ribeiro AI, Rocha-Santos T (Eds.), *One Health*: Elsevier; 2022. pp. 281–327.
41. Schuhen K, Sturm MT. Microplastic pollution and reduction strategies. In Rocha-Santos T, Costa M, Mouneyrac C (Eds.), *Handbook of MPs in the Environment*: Springer; 2020. pp. 1–33.
42. Lassen C, Hansen SF, Magnusson K, Hartmann NB, Jensen PR, Nielsen TG, et al. *MPs: Occurrence, Effects and Sources of Releases to the Environment in Denmark*. Danish Environmental Protection Agency; 2015.
43. Perschbacher E. Microbeads-legislative update. 2016.
44. Rainieri S, Barranco A. MPs, a food safety issue? *Trends in Food Science & Technology*. 2019;84:55–7.
45. Sharma S, Chatterjee S. Microplastic pollution, a threat to marine ecosystem and human health: a short review. *Environmental Science and Pollution Research*. 2017;24:21530–47.

46. Usman S, Abdull Razis AF, Shaari K, Azmai MNA, Saad MZ, Mat Isa N, et al. The burden of MPs pollution and contending policies and regulations. *International Journal of Environmental Research and Public Health.* 2022;19(11):6773.
47. Nikiema J, Asiedu Z. A review of the cost and effectiveness of solutions to address plastic pollution. *Environmental Science and Pollution Research.* 2022;29(17):24547–73.
48. Guo J-J, Huang X-P, Xiang L, Wang Y-Z, Li Y-W, Li H, et al. Source, migration and toxicology of MPs in soil. *Environment International.* 2020;137:105263.
49. Zhao S, Zhang Z, Chen L, Cui Q, Cui Y, Song D, et al. Review on migration, transformation and ecological impacts of MPs in soil. *Applied Soil Ecology.* 2022;176:104486.
50. Kong X, Jin D, Jin S, Wang Z, Yin H, Xu M, et al. Responses of bacterial community to dibutyl phthalate pollution in a soil-vegetable ecosystem. *Journal of Hazardous Materials.* 2018;353:142–50.
51. Ng EL, Lin SY, Dungan AM, Colwell JM, Ede S, Lwanga EH, et al. Microplastic pollution alters forest soil microbiome. *Journal of Hazardous Materials.* 2021;409:124606.
52. Xuan C, Jiaen Z, Hui W. Research progress and prospect on transportation, transformation and ecotoxicology of MPs in environment. *Asian Journal of Ecotoxicology.* 2021;2021(6):70–86.
53. Judy JD, Williams M, Gregg A, Oliver D, Kumar A, Kookana R, et al. MPs in municipal mixed-waste organic outputs induce minimal short to long-term toxicity in key terrestrial biota. *Environmental Pollution.* 2019;252:522–31.
54. Muscolo A, Settineri G, Attinà E. Early warning indicators of changes in soil ecosystem functioning. *Ecological Indicators.* 2015;48:542–9.
55. de Souza Machado AA, Lau CW, Kloas W, Bergmann J, Bachelier JB, Faltin E, et al. MPs can change soil properties and affect plant performance. *Environmental Science & Technology.* 2019;53(10):6044–52.
56. Mašková T, Herben T. Root: shoot ratio in developing seedlings: how seedlings change their allocation in response to seed mass and ambient nutrient supply. *Ecology and Evolution.* 2018;8(14):7143–50.
57. Sun X-D, Yuan X-Z, Jia Y, Feng L-J, Zhu F-P, Dong S-S, et al. Differentially charged nanoplastics demonstrate distinct accumulation in Arabidopsis thaliana. *Nature Nanotechnology.* 2020;15(9):755–60.
58. Wang J, Liu X, Li Y, Powell T, Wang X, Wang G, et al. MPs as contaminants in the soil environment: a mini-review. *Science of the Total Environment.* 2019;691:848–57.
59. Li G, Zhao X, Iqbal B, Zhao X, Liu J, Javed Q, et al. The effect of soil MPs on Oryza sativa L. root growth traits under alien plant invasion. *Frontiers in Ecology and Evolution.* 2023;11:1172093.
60. Faucon M-P, Houben D, Lambers H. Plant functional traits: soil and ecosystem services. *Trends in Plant Science.* 2017;22(5):385–94.
61. Bender SF, Wagg C, van der Heijden MG. An underground revolution: biodiversity and soil ecological engineering for agricultural sustainability. *Trends in Ecology & Evolution.* 2016;31(6):440–52.

62. Padervand M, Lichtfouse E, Robert D, Wang C. Removal of MPs from the environment. a review. *Environmental Chemistry Letters*. 2020;18:807–28.
63. da Costa JP, Santos PS, Duarte AC, Rocha-Santos T. (Nano) plastics in the environment–sources, fates and effects. *Science of the Total Environment*. 2016;566:15–26.
64. Wang H-T, Ding J, Xiong C, Zhu D, Li G, Jia X-Y, et al. Exposure to MPs lowers arsenic accumulation and alters gut bacterial communities of earthworm Metaphire californica. *Environmental Pollution*. 2019;251:110–6.
65. Cao D, Wang X, Luo X, Liu G, Zheng H. Effects of polystyrene MPs on the fitness of earthworms in an agricultural soil. *IOP Conference Series: Earth and Environmental Science*. 2017;61:012148. IOP Publishing.
66. Zhu D, Chen Q-L, An X-L, Yang X-R, Christie P, Ke X, et al. Exposure of soil collembolans to MPs perturbs their gut microbiota and alters their isotopic composition. *Soil Biology and Biochemistry*. 2018;116:302–10.
67. Ju H, Zhu D, Qiao M. Effects of polyethylene MPs on the gut microbial community, reproduction and avoidance behaviors of the soil springtail, Folsomia candida. *Environmental Pollution*. 2019;247:890–7.
68. Zhu F, Zhu C, Wang C, Gu C. Occurrence and ecological impacts of MPs in soil systems: a review. *Bulletin of Environmental Contamination and Toxicology*. 2019;102:741–9.
69. Lenz M, Creffield JW, Evans TA, Kard B, Vongkaluang C, Sornnuwat Y, et al. Resistance of polyamide and polyethylene cable sheathings to termites in Australia, Thailand, USA, Malaysia and Japan: a comparison of four field assessment methods. *International Biodeterioration & Biodegradation*. 2012;66(1):53–62.
70. Song Y, Cao C, Qiu R, Hu J, Liu M, Lu S, et al. Uptake and adverse effects of polyethylene terephthalate MPs fibers on terrestrial snails (Achatina fulica) after soil exposure. *Environmental Pollution*. 2019;250:447–55.
71. Dissanayake PD, Kim S, Sarkar B, Oleszczuk P, Sang MK, Haque MN, et al. Effects of MPs on the terrestrial environment: a critical review. *Environmental Research*. 2022;209:112734.
72. George PB, Keith AM, Creer S, Barrett GL, Lebron I, Emmett BA, et al. Evaluation of mesofauna communities as soil quality indicators in a national-level monitoring programme. *Soil Biology and Biochemistry*. 2017;115:537–46.
73. Ding J, Zhu D, Wang H-T, Lassen SB, Chen Q-L, Li G, et al. Dysbiosis in the gut microbiota of soil fauna explains the toxicity of tire tread particles. *Environmental Science & Technology*. 2020;54(12):7450–60.
74. Pflugmacher S, Huttunen JH, von Wolff M-A, Penttinen O-P, Kim YJ, Kim S, et al. Enchytraeus crypticus avoid soil spiked with microplastic. *Toxics*. 2020;8(1):10.
75. Selonen S, Dolar A, Kokalj AJ, Skalar T, Dolcet LP, Hurley R, et al. Exploring the impacts of plastics in soil–the effects of polyester textile fibers on soil invertebrates. *Science of the Total Environment*. 2020;700:134451.
76. Maaß S, Daphi D, Lehmann A, Rillig MC. Transport of MPs by two collembolan species. *Environmental Pollution*. 2017;225:456–9.

77. Bergami E, Rota E, Caruso T, Birarda G, Vaccari L, Corsi I. Plastics everywhere: first evidence of polystyrene fragments inside the common Antarctic collembolan Cryptopygus antarcticus. *Biology Letters*. 2020;16(6):20200093.
78. Yee MS-L, Hii L-W, Looi CK, Lim W-M, Wong S-F, Kok Y-Y, et al. Impact of MPs and nanoplastics on human health. *Nanomaterials*. 2021;11(2):496.
79. Landrigan PJ, Stegeman JJ, Fleming LE, Allemand D, Anderson DM, Backer LC, et al. Human health and ocean pollution. *Annals of Global Health*. 2020;86(1):151.
80. Jaishankar M, Tseten T, Anbalagan N, Mathew BB, Beeregowda KN. Toxicity, mechanism and health effects of some heavy metals. *Interdisciplinary Toxicology*. 2014;7(2):60.
81. Hodson ME, Duffus-Hodson CA, Clark A, Prendergast-Miller MT, Thorpe KL. Plastic bag derived-MPs as a vector for metal exposure in terrestrial invertebrates. *Environmental Science & Technology*. 2017;51(8):4714–21.
82. He D, Zhang Y, Gao W. Micro (nano) plastic contaminations from soils to plants: human food risks. *Current Opinion in Food Science*. 2021;41:116–21.
83. Toussaint B, Raffael B, Angers-Loustau A, Gilliland D, Kestens V, Petrillo M, et al. Review of micro-and nanoplastic contamination in the food chain. *Food Additives & Contaminants: Part A*. 2019;36(5):639–73.
84. Bandmann V, Müller JD, Köhler T, Homann U. Uptake of fluorescent nano beads into BY2-cells involves clathrin-dependent and clathrin-independent endocytosis. *FEBS Letters*. 2012;586(20):3626–32.
85. Mosca Angelucci D, Tomei MC. Uptake/release of organic contaminants by MPs: a critical review of influencing factors, mechanistic modeling, and thermodynamic prediction methods. *Critical Reviews in Environmental Science and Technology*. 2022;52(8):1356–400.
86. Ogata Y, Takada H, Mizukawa K, Hirai H, Iwasa S, Endo S, et al. International Pellet Watch: global monitoring of persistent organic pollutants (POPs) in coastal waters. 1. Initial phase data on PCBs, DDTs, and HCHs. *Marine Pollution Bulletin*. 2009;58(10):1437–46.
87. Avio CG, Gorbi S, Milan M, Benedetti M, Fattorini D, d'Errico G, et al. Pollutants bioavailability and toxicological risk from MPs to marine mussels. *Environmental Pollution*. 2015;198:211–22.
88. Ribeiro F, O'Brien JW, Galloway T, Thomas KV. Accumulation and fate of nano-and micro-plastics and associated contaminants in organisms. *TrAC Trends in Analytical Chemistry*. 2019;111:139–47.
89. Osman AI, Hosny M, Eltaweil AS, Omar S, Elgarahy AM, Farghali M, et al. Microplastic sources, formation, toxicity and remediation: a review. *Environmental Chemistry Letters*. 2023;21(4):2129–69.
90. Qiao R, Sheng C, Lu Y, Zhang Y, Ren H, Lemos B. MPs induce intestinal inflammation, oxidative stress, and disorders of metabolome and microbiome in zebrafish. *Science of the Total Environment*. 2019;662:246–53.
91. Amato-Lourenço LF, dos Santos Galvão L, de Weger LA, Hiemstra PS, Vijver MG, Mauad T. An emerging class of air pollutants: potential effects of MPs to respiratory human health? *Science of the Total Environment*. 2020;749:141676.

92. Prata JC, da Costa JP, Lopes I, Duarte AC, Rocha-Santos T. Effects of MPs on microalgae populations: a critical review. *Science of the Total Environment*. 2019;665:400–5.
93. Burkhardt-Holm P, N'Guyen A. Ingestion of MPs by fish and other prey organisms of cetaceans, exemplified for two large baleen whale species. *Marine Pollution Bulletin*. 2019;144:224–34.
94. Hammer S, Nager R, Johnson P, Furness R, Provencher J. Plastic debris in great skua (Stercorarius skua) pellets corresponds to seabird prey species. *Marine Pollution Bulletin*. 2016;103(1–2):206–10.
95. Provencher JF, Vermaire JC, Avery-Gomm S, Braune BM, Mallory ML. Garbage in guano? Microplastic debris found in faecal precursors of seabirds known to ingest plastics. *Science of the Total Environment*. 2018;644:1477–84.
96. Hwang J, Choi D, Han S, Choi J, Hong J. An assessment of the toxicity of polypropylene MPs in human derived cells. *Science of the Total Environment*. 2019;684:657–69.
97. Li Y, Tao L, Wang Q, Wang F, Li G, Song M. Potential health impact of MPs: a review of environmental distribution, human exposure, and toxic effects. *Environment & Health*. 2023;1(4):249–57.
98. Kremer A, Pal T, Boleij J, Schouten J, Rijcken B. Airway hyperresponsiveness, prevalence of chronic respiratory symptoms, and lung function in workers exposed to irritants. *Occupational and Environmental Medicine*. 1994;51(1):3.
99. Yan Z, Liu Y, Zhang T, Zhang F, Ren H, Zhang Y. Analysis of MPs in human feces reveals a correlation between fecal MPs and inflammatory bowel disease status. *Environmental Science & Technology*. 2021;56(1):414–21.
100. Shi C, Han X, Guo W, Wu Q, Yang X, Wang Y, et al. Disturbed Gut-Liver axis indicating oral exposure to polystyrene microplastic potentially increases the risk of insulin resistance. *Environment International*. 2022;164:107273.
101. Möller JN, Löder MG, Laforsch C. Finding MPs in soils: a review of analytical methods. *Environmental Science & Technology*. 2020;54(4):2078–90.
102. Shim WJ, Hong SH, Eo SE. Identification methods in microplastic analysis: a review. *Analytical Methods*. 2017;9(9):1384–91.
103. Wang W, Wang J. Investigation of MPs in aquatic environments: an overview of the methods used, from field sampling to laboratory analysis. *TrAC Trends in Analytical Chemistry*. 2018;108:195–202.
104. He D, Luo Y, Lu S, Liu M, Song Y, Lei L. MPs in soils: analytical methods, pollution characteristics and ecological risks. *TrAC Trends in Analytical Chemistry*. 2018;109:163–72.
105. Zhou Q, Zhang H, Fu C, Zhou Y, Dai Z, Li Y, et al. The distribution and morphology of MPs in coastal soils adjacent to the Bohai Sea and the Yellow Sea. *Geoderma*. 2018;322:201–8.
106. Mariano S, Tacconi S, Fidaleo M, Rossi M, Dini L. Micro and nanoplastics identification: classic methods and innovative detection techniques. *Frontiers in Toxicology*. 2021;3:636640.

107. Shruti V, Pérez-Guevara F, Roy PD, Kutralam-Muniasamy G. Analyzing MPs with Nile Red: emerging trends, challenges, and prospects. *Journal of Hazardous Materials*. 2022;423:127171.
108. Zhang B, Yang X, Chen L, Chao J, Teng J, Wang Q. MPs in soils: a review of possible sources, analytical methods and ecological impacts. *Journal of Chemical Technology & Biotechnology*. 2020;95(8):2052–68.
109. Wang W, Ge J, Yu X, Li H. Environmental fate and impacts of MPs in soil ecosystems: progress and perspective. *Science of the Total Environment*. 2020;708:134841.
110. Löder MGJ, Kuczera M, Mintenig S, Lorenz C, Gerdts G. Focal plane array detector-based micro-Fourier-transform infrared imaging for the analysis of MPs in environmental samples. *Environmental Chemistry*. 2015;12(5):563–81.
111. Renner G, Schmidt TC, Schram J. Characterization and quantification of MPs by infrared spectroscopy. *Comprehensive Analytical Chemistry*. 2017; 75:67–118. Elsevier.
112. Käppler A, Windrich F, Löder MG, Malanin M, Fischer D, Labrenz M, et al. Identification of MPs by FTIR and Raman microscopy: a novel silicon filter substrate opens the important spectral range below 1300 cm^{-1} for FTIR transmission measurements. *Analytical and Bioanalytical Chemistry*. 2015;407:6791–801.
113. Zhao S, Zhu L, Gao L, Li D. Limitations for microplastic quantification in the ocean and recommendations for improvement and standardization. In Zeng EY (Ed.), *Microplastic Contamination in Aquatic Environments*: Elsevier; 2018. pp. 27–49.
114. Käppler A, Fischer D, Oberbeckmann S, Schernewski G, Labrenz M, Eichhorn K-J, et al. Analysis of environmental MPs by vibrational microspectroscopy: FTIR, Raman or both? *Analytical and Bioanalytical Chemistry*. 2016;408:8377–91.
115. Eriksen M, Mason S, Wilson S, Box C, Zellers A, Edwards W, et al. Microplastic pollution in the surface waters of the Laurentian Great Lakes. *Marine Pollution Bulletin*. 2013;77(1–2):177–82.
116. Zhang J, Fu D, Feng H, Li Y, Zhang S, Peng C, et al. Mass spectrometry detection of environmental MPs: advances and challenges. *TrAC Trends in Analytical Chemistry*. 2023:117472.
117. Nuelle M-T, Dekiff JH, Remy D, Fries E. A new analytical approach for monitoring MPs in marine sediments. *Environmental Pollution*. 2014;184:161–9.
118. Wang L, Zhang J, Hou S, Sun H. A simple method for quantifying polycarbonate and polyethylene terephthalate MPs in environmental samples by liquid chromatography–tandem mass spectrometry. *Environmental Science & Technology Letters*. 2017;4(12):530–4.
119. Woo H, Seo K, Choi Y, Kim J, Tanaka M, Lee K, et al. Methods of analyzing microsized plastics in the environment. *Applied Sciences*. 2021;11(22):10640.
120. Luo H, Xiang Y, Li Y, Zhao Y, Pan X. Photocatalytic aging process of Nano-TiO_2 coated polypropylene MPs: combining atomic force microscopy and infrared spectroscopy (AFM-IR) for nanoscale chemical characterization. *Journal of Hazardous Materials*. 2021;404:124159.

121. Erni-Cassola G, Gibson MI, Thompson RC, Christie-Oleza JA. Lost, but found with Nile red: a novel method for detecting and quantifying small MPs (1 mm to 20 µm) in environmental samples. *Environmental Science & Technology*. 2017;51(23):13641–8.
122. Paul A, Wander L, Becker R, Goedecke C, Braun U. High-throughput NIR spectroscopic (NIRS) detection of MPs in soil. *Environmental Science and Pollution Research*. 2019;26:7364–74.
123. Corradini F, Bartholomeus H, Lwanga EH, Gertsen H, Geissen V. Predicting soil microplastic concentration using vis-NIR spectroscopy. *Science of the Total Environment*. 2019;650:922–32.
124. Luo H, Zeng Y, Zhao Y, Xiang Y, Li Y, Pan X. Effects of advanced oxidation processes on leachates and properties of MPs. *Journal of Hazardous Materials*. 2021;413:125342.
125. da Costa JP, Paço A, Santos PS, Duarte AC, Rocha-Santos T. MPs in soils: assessment, analytics and risks. *Environmental Chemistry*. 2018;16(1):18–30.
126. Abumaizar RJ, Smith EH. Heavy metal contaminants removal by soil washing. *Journal of Hazardous Materials*. 1999;70(1–2):71–86.
127. Mann MJ. Full-scale and pilot-scale soil washing. *Journal of Hazardous Materials*. 1999;66(1–2):119–36.
128. Hurley RR, Lusher AL, Olsen M, Nizzetto L. Validation of a method for extracting MPs from complex, organic-rich, environmental matrices. *Environmental Science & Technology*. 2018;52(13):7409–17.
129. Radford F, Zapata-Restrepo LM, Horton AA, Hudson MD, Shaw PJ, Williams ID. Developing a systematic method for extraction of MPs in soils. *Analytical Methods*. 2021;13(14):1695–705.
130. Varzakas T. Centrifugation–filtration. In Varzakas T, Tzia C (Eds.), *Food Engineering Handbook: Food Process Engineering*: CRC Press; 2014. p. 61.
131. Wu R-T, Cai Y-F, Xing S-C, Yang Y-W, Mi J-D, Liao X-D. A novel method for extraction of polypropylene MPs in swine manure. *Environmental Science and Pollution Research*. 2021;28:13021–30.
132. Wolff S, Kerpen J, Prediger J, Barkmann L, Müller L. Determination of the MPs emission in the effluent of a municipal waste water treatment plant using Raman microspectroscopy. *Water Research X*. 2019;2:100014.
133. Acar YB, Gale RJ, Alshawabkeh AN, Marks RE, Puppala S, Bricka M, et al. Electrokinetic remediation: basics and technology status. *Journal of Hazardous Materials*. 1995;40(2):117–37.
134. Méndez E, Pérez M, Romero O, Beltrán E, Castro S, Corona J, et al. Effects of electrode material on the efficiency of hydrocarbon removal by an electrokinetic remediation process. *Electrochimica Acta*. 2012;86:148–56.
135. Acar YB, Alshawabkeh AN, Gale RJ. Fundamentals of extracting species from soils by electrokinetics. *Waste Management*. 1993;13(2):141–51.
136. Yu J-W, Neretnieks I. Theoretical evaluation of a technique for electrokinetic decontamination of soils. *Journal of Contaminant Hydrology*. 1997;26(1–4):291–9.
137. Gnanasundar V, Raj RA. Remediation of inorganic contaminants in soil using electrokinetics, phytoremediation techniques. *Materials Today: Proceedings*. 2021;45:950–6.

138. Paillat T, Moreau E, Grimaud P, Touchard G. Electrokinetic phenomena in porous media applied to soil decontamination. *IEEE Transactions on Dielectrics and Electrical Insulation*. 2000;7(5):693–704.
139. Shapiro AP, Probstein RF. Removal of contaminants from saturated clay by electroosmosis. *Environmental Science & Technology*. 1993;27(2):283–91.
140. Segall BA, Bruell CJ. Electroosmotic contaminant-removal processes. *Journal of Environmental Engineering*. 1992;118(1):84–100.
141. Probstein RF, Hicks RE. Removal of contaminants from soils by electric fields. *Science*. 1993;260(5107):498–503.
142. Cameselle C, Reddy KR. Development and enhancement of electro-osmotic flow for the removal of contaminants from soils. *Electrochimica Acta*. 2012;86:10–22.
143. Kim S-O, Moon S-H, Kim K-W. Removal of heavy metals from soils using enhanced electrokinetic soil processing. *Water, Air, and Soil Pollution*. 2001;125:259–72.
144. Chung HI, Kamon M. Ultrasonically enhanced electrokinetic remediation for removal of Pb and phenanthrene in contaminated soils. *Engineering Geology*. 2005;77(3–4):233–42.
145. Boulakradeche MO, Akretche DE, Cameselle C, Hamidi N. Enhanced electrokinetic remediation of hydrophobic organics contaminated soils by the combination of non-ionic and ionic surfactants. *Electrochimica Acta*. 2015;174:1057–66.
146. Kolosov AY, Popov K, Shabanova N, Artem'eva A, Kogut B, Frid A, et al. Electrokinetic removal of hydrophobic organic compounds from soil. *Russian Journal of Applied Chemistry*. 2001;74(4):631–5.
147. Koul B, Taak P, Koul B, Taak P. Chemical methods of soil remediation. In Koul B, Taak P (Eds.), *Biotechnological Strategies for Effective Remediation of Polluted Soils*: Springer; 2018. pp. 77–84.
148. (EPA) USEPA. *The Use of Soil Amendments for Remediation, Revitalization and Reuse*: Solid Waste and Emergency Response; 2007.
149. Forján R, Asensio V, Guedes R, Rodríguez-Vila A, Covelo E, Marcet P. Remediation of soils polluted with inorganic contaminants: role of organic amendments. In Anjum NA, Gill SS, Tuteja N (Eds.), *Enhancing Cleanup of Environmental Pollutants: Volume 2: Non-Biological Approaches*: Springer; 2017. pp. 313–37.
150. Tong Y, McNamara PJ, Mayer BK. Adsorption of organic micropollutants onto biochar: a review of relevant kinetics, mechanisms and equilibrium. *Environmental Science: Water Research & Technology*. 2019;5(5):821–38.
151. Qiu M, Liu L, Ling Q, Cai Y, Yu S, Wang S, et al. Biochar for the removal of contaminants from soil and water: a review. *Biochar*. 2022;4(1):19.
152. Palansooriya KN, Sang MK, Igalavithana AD, Zhang M, Hou D, Oleszczuk P, et al. Biochar alters chemical and microbial properties of microplastic-contaminated soil. *Environmental Research*. 2022;209:112807.
153. Wang Z, Sedighi M, Lea-Langton A. Filtration of microplastic spheres by biochar: removal efficiency and immobilisation mechanisms. *Water Research*. 2020;184:116165.

154. Rassaei F. Chitosan as an organic amendment to improve soil properties and plant growth in the presence of polystyrene MPs. *Environmental Progress & Sustainable Energy*. 2023:e14301.
155. Ragoobur D, Huerta-Lwanga E, Somaroo GD. Reduction of MPs in sewage sludge by vermicomposting. *Chemical Engineering Journal*. 2022;450: 138231.
156. Maiti SK, Ahirwal J. Ecological restoration of coal mine degraded lands: topsoil management, pedogenesis, carbon sequestration, and mine pit limnology. In Pandey VC (Ed.), *Phytomanagement of Polluted Sites*: Elsevier; 2019. pp. 83–111.
157. Ghosh S, Wilson B, Ghoshal S, Senapati N, Mandal B. Organic amendments influence soil quality and carbon sequestration in the Indo-Gangetic plains of India. *Agriculture, Ecosystems & Environment*. 2012;156:134–41.
158. Xiubin H, Zhanbin H. Zeolite application for enhancing water infiltration and retention in loess soil. *Resources, Conservation and Recycling*. 2001;34(1):45–52.
159. Doula MK, Kavvadias VA, Elaiopoulos K. Zeolites in soil remediation processes. In Inglezakis VJ, Zorpas AA (Eds.), *Handbook of Natural Zeolites: Bentham Science Publishers*; 2012. pp. 519–68.
160. Sharma V, Javed B, Byrne H, Curtin J, Tian F. Zeolites as carriers of nano-fertilizers: from structures and principles to prospects and challenges. *Applied Nano*. 2022;3(3):163–86.
161. Pasanen F, Fuller RO, Maya F. Fast and simultaneous removal of MPs and plastic-derived endocrine disruptors using a magnetic ZIF-8 nanocomposite. *Chemical Engineering Journal*. 2023;455:140405.
162. Nakhli SAA, Delkash M, Bakhshayesh BE, Kazemian H. Application of zeolites for sustainable agriculture: a review on water and nutrient retention. *Water, Air, & Soil Pollution*. 2017;228:1–34.
163. Williams CD. Application of zeolites to environmental remediation. *Urban Pollution: Science and Management*. 2018;1:249–58.
164. Mueller NC, Nowack B. Nanoparticles for remediation: solving big problems with little particles. *Elements*. 2010;6(6):395–400.
165. Ye H, Wang Y, Liu X, Xu D, Yuan H, Sun H, et al. Magnetically steerable iron oxides-manganese dioxide core–shell micromotors for organic and microplastic removals. *Journal of Colloid and Interface Science*. 2021;588:510–21.
166. Zandieh M, Liu J. Removal and degradation of MPs using the magnetic and nanozyme activities of bare iron oxide nanoaggregates. *Angewandte Chemie International Edition*. 2022;61(47):e202212013.
167. Guerra FD, Attia MF, Whitehead DC, Alexis F. Nanotechnology for environmental remediation: materials and applications. *Molecules*. 2018;23(7):1760.
168. Kluck C, Achari G. Chemical oxidation techniques for in situ remediation of hydrocarbon impacted soils. *Environmental Engineering*. 2004;2004:1–8.
169. United States Environmental Protection Agency. *A Citizen's Guide to In Situ Chemical Oxidation*: United States Environmental Protection Agency, Office of Solid Waste and Emergency Response; 2012.

170. Siegrist RL, Crimi M, Brown RA. In situ chemical oxidation: technology description and status. In Siegrist RL, Crimi M, Simpkin TJ, Borden RC (Eds.), *In* Situ *Chemical Oxidation for Groundwater Remediation*: Springer; 2011. pp. 1–32.
171. Ferrarese E, Andreottola G, Oprea IA. Remediation of PAH-contaminated sediments by chemical oxidation. *Journal of Hazardous Materials*. 2008;152(1):128–39.
172. Tsitonaki A, Petri B, Crimi M, Mosbaek H, Siegrist RL, Bjerg PL. In situ chemical oxidation of contaminated soil and groundwater using persulfate: a review. *Critical Reviews in Environmental Science and Technology*. 2010;40(1):55–91.
173. Dombrowski PM, Kakarla P, Caldicott W, Chin Y, Sadeghi V, Bogdan D, et al. Technology review and evaluation of different chemical oxidation conditions on treatability of PFAS. *Remediation Journal*. 2018;28(2):135–50.
174. Cheng M, Zeng G, Huang D, Lai C, Xu P, Zhang C, et al. Hydroxyl radicals based advanced oxidation processes (AOPs) for remediation of soils contaminated with organic compounds: a review. *Chemical Engineering Journal*. 2016;284:582–98.
175. Azbar N, Yonar T, Kestioglu K. Comparison of various advanced oxidation processes and chemical treatment methods for COD and color removal from a polyester and acetate fiber dyeing effluent. *Chemosphere*. 2004;55(1):35–43.
176. Rizwan K, Bilal M. Developments in advanced oxidation processes for removal of MPs from aqueous matrices. *Environmental Science and Pollution Research*. 2022;29(58):86933–53.
177. Saravanan A, Kumar PS, Vo D-VN, Yaashikaa PR, Karishma S, Jeevanantham S, et al. Photocatalysis for removal of environmental pollutants and fuel production: a review. *Environmental Chemistry Letters*. 2021;19:441–63.
178. Ahmad K, Ghatak HR, Ahuja S. A review on photocatalytic remediation of environmental pollutants and H2 production through water splitting: a sustainable approach. *Environmental Technology & Innovation*. 2020;19:100893.
179. Nguyen V-H, Smith SM, Wantala K, Kajitvichyanukul P. Photocatalytic remediation of persistent organic pollutants (POPs): a review. *Arabian Journal of Chemistry*. 2020;13(11):8309–37.
180. Nabi I, Li K, Cheng H, Wang T, Liu Y, Ajmal S, et al. Complete photocatalytic mineralization of microplastic on TiO_2 nanoparticle film. *iScience*. 2020;23(7):101326.
181. Zoppas FM, Sacco N, Soffietti J, Devard A, Akhter F, Marchesini FA. Catalytic approaches for the removal of MPs from water: recent advances and future opportunities. *Chemical Engineering Journal Advances*. 2023;16:100529.
182. Padsalgikar AD. Biological properties of plastics. In Padsalgikar AD (Ed.), *Plastics in Medical Devices for Cardiovascular Applications*: William Andrew Publishing; 2017. pp. 83–102.
183. Tournier V, Topham C, Gilles A, David B, Folgoas C, Moya-Leclair E, et al. An engineered PET depolymerase to break down and recycle plastic bottles. *Nature*. 2020;580(7802):216–9.

184. Austin HP, Allen MD, Donohoe BS, Rorrer NA, Kearns FL, Silveira RL, et al. Characterization and engineering of a plastic-degrading aromatic polyesterase. *Proceedings of the National Academy of Sciences.* 2018;115(19):E4350–E7.
185. Spaseska D, Civkaroska M. Alkaline hydrolysis of poly (ethylene terephthalate) recycled from the postconsumer soft-drink bottles. *Journal of the University of Chemical Technology and Metallurgy.* 2010;45(4):379–84.
186. Thiounn T, Smith RC. Advances and approaches for chemical recycling of plastic waste. *Journal of Polymer Science.* 2020;58(10):1347–64.
187. Kumagai S, Hirahashi S, Grause G, Kameda T, Toyoda H, Yoshioka T. Alkaline hydrolysis of PVC-coated PET fibers for simultaneous recycling of PET and PVC. *Journal of Material Cycles and Waste Management.* 2018;20:439–49.
188. Tsintzou GP, Achilias DS. Chemical recycling of polycarbonate based wastes using alkaline hydrolysis under microwave irradiation. *Waste and Biomass Valorization.* 2013;4:3–7.
189. Karayannidis G, Chatziavgoustis A, Achilias D. Poly (ethylene terephthalate) recycling and recovery of pure terephthalic acid by alkaline hydrolysis. *Advances in Polymer Technology: Journal of the Polymer Processing Institute.* 2002;21(4):250–9.
190. Xiang P, Zhang T, Wu Q, Li Q. Systematic review of degradation processes for MPs: progress and prospects. *Sustainability.* 2023;15(17):12698.
191. Anand U, Dey S, Bontempi E, Ducoli S, Vethaak AD, Dey A, et al. Biotechnological methods to remove MPs: a review. *Environmental Chemistry Letters.* 2023;21:1–24.
192. Thakur B, Singh J, Singh J, Angmo D, Vig AP. Biodegradation of different types of MPs: molecular mechanism and degradation efficiency. *Science of the Total Environment.* 2023;877:162912.
193. Othman AR, Hasan HA, Muhamad MH, Ismail NI, Abdullah SRS. Microbial degradation of MPs by enzymatic processes: a review. *Environmental Chemistry Letters.* 2021;19:3057–73.
194. Cai Z, Li M, Zhu Z, Wang X, Huang Y, Li T, et al. Biological degradation of plastics and MPs: a recent perspective on associated mechanisms and influencing factors. *Microorganisms.* 2023;11(7):1661.
195. Mishra S, Swain S, Sahoo M, Mishra S, Das AP. Microbial colonization and degradation of MPs in aquatic ecosystem: a review. *Geomicrobiology Journal.* 2022;39(3–5):259–69.
196. Dąbrowska A. Microbial degradation of MPs. In Inamuddin, Ahamed MI, Prasad R (Eds.), *Recent Advances in Microbial Degradation*: Springer; 2021. pp. 373–87.
197. Abbasi HZ, Amjad U, Zeb BS, Abbasi MHA. Critical evaluation of plastic waste remediation through microbial degradation. *2021 Third International Sustainability and Resilience Conference:* Climate Change, Sakheer, Bahrain; 2021: IEEE.
198. Yuan J, Ma J, Sun Y, Zhou T, Zhao Y, Yu F. Microbial degradation and other environmental aspects of MPs/plastics. *Science of the Total Environment.* 2020;715:136968.

199. Liu L, Xu M, Ye Y, Zhang B. On the degradation of (micro) plastics: degradation methods, influencing factors, environmental impacts. *Science of the Total Environment*. 2022;806:151312.
200. Sánchez C. Fungal potential for the degradation of petroleum-based polymers: an overview of macro-and MPs biodegradation. *Biotechnology Advances*. 2020;40:107501.
201. Li L, Yang J, Zhou Q, Peijnenburg WJ, Luo Y. Uptake of MPs and their effects on plants. In He D, Luo Y (Eds.), *MPs in Terrestrial Environments: Emerging Contaminants and Major Challenges*: Springer; 2020. pp. 279–98.
202. Del Real AEP, Mitrano DM, Castillo-Michel H, Wazne M, Reyes-Herrera J, Bortel E, et al. Assessing implications of nanoplastics exposure to plants with advanced nanometrology techniques. *Journal of Hazardous Materials*. 2022;430:128356.
203. Azeem I, Adeel M, Ahmad MA, Shakoor N, Jiangcuo GD, Azeem K, et al. Uptake and accumulation of nano/MPs in plants: a critical review. *Nanomaterials*. 2021;11(11):2935.
204. Maity S, Guchhait R, Sarkar MB, Pramanick K. Occurrence and distribution of micro/nanoplastics in soils and their phytotoxic effects: a review. *Plant, Cell & Environment*. 2022;45(4):1011–28.
205. Li L, Zhou Q, Yin N, Tu C, Luo Y. Uptake and accumulation of MPs in an edible plant. *Chinese Science Bulletin*. 2019;64(9):928–34.
206. Dong Y, Gao M, Qiu W, Song Z. Uptake of MPs by carrots in presence of As (III): combined toxic effects. *Journal of Hazardous Materials*. 2021;411:125055.
207. Kalčíková G. Aquatic vascular plants–a forgotten piece of nature in microplastic research. *Environmental Pollution*. 2020;262:114354.
208. Ebere EC, Wirnkor VA, Ngozi VE. Uptake of MPs by plant: a reason to worry or to be happy? *World Scientific News*. 2019;(131):256–67.
209. Liu Y, Guo R, Zhang S, Sun Y, Wang F. Uptake and translocation of nano/MPs by rice seedlings: evidence from a hydroponic experiment. *Journal of Hazardous Materials*. 2022;421:126700.
210. Li H, Chang X, Zhang J, Wang Y, Zhong R, Wang L, et al. Uptake and distribution of MPs of different particle sizes in maize (Zea mays) seedling roots. *Chemosphere*. 2023;313:137491.
211. Hua Z, Zhang T, Luo J, Bai H, Ma S, Qiang H, et al. Internalization, physiological responses and molecular mechanisms of lettuce to polystyrene MPs of different sizes: validation of simulated soilless culture. *Journal of Hazardous Materials*. 2024;462:132710.
212. Yin L, Wen X, Huang D, Du C, Deng R, Zhou Z, et al. Interactions between MPs/nanoplastics and vascular plants. *Environmental Pollution*. 2021;290:117999.
213. Dehghanian Z, Lajayer BA, Atigh ZBQ, Nayeri S, Ahmadabadi M, Taghipour L, et al. Micro (nano) plastics uptake, toxicity and detoxification in plants: challenges and prospects. *Ecotoxicology and Environmental Safety*. 2023;268:115676.

214. Tang KHD. Phytoremediation of MPs: a perspective on its practicality. *Industrial and Domestic Waste Management*. 2022;3(2):90–102.
215. Mbachu O, Jenkins G, Kaparaju P, Pratt C. The rise of artificial soil carbon inputs: reviewing microplastic pollution effects in the soil environment. *Science of the Total Environment*. 2021;780:146569.
216. Yu H, Zhang Y, Tan W. The "neighbor avoidance effect" of MPs on bacterial and fungal diversity and communities in different soil horizons. *Environmental Science and Ecotechnology*. 2021;8:100121.
217. Zhu J, Liu S, Wang H, Wang D, Zhu Y, Wang J, et al. Microplastic particles alter wheat rhizosphere soil microbial community composition and function. *Journal of Hazardous Materials*. 2022;436:129176.
218. Xiang P, Liao W, Xiong Z, Xiao W, Luo Y, Peng L, et al. Effects of polystyrene MPs on the agronomic traits and rhizosphere soil microbial community of highland barley. *Science of the Total Environment*. 2024;907:167986.
219. Meng F, Harkes P, van Steenbrugge JJ, Geissen V. Effects of MPs on common bean rhizosphere bacterial communities. *Applied Soil Ecology*. 2023;181:104649.
220. Picó Y, Barceló DJ. MPs and other emerging contaminants in the environment after COVID-19 pandemic: the need of global reconnaissance studies. *Current Opinion in Environmental Science & Health*. 2023;33:100468.
221. Thacharodi A, Hassan S, Meenatchi R, Bhat MA, Hussain N, Arockiaraj J, et al. Mitigating microplastic pollution: a critical review on the effects, remediation, and utilization strategies of MPs. *Materials (Basel)*. 2024;351:119988.
222. Robert D, Alle PH, Keller N, Dzuila M-A, Garcia-Muñoz P. Challenges and opportunities for microplastic and nanoplastic removal from industrial wastewater. In Tyagi RD, Pandey A, Drogui P, Yadav B, Pilli S (Eds.), *Current Developments in Biotechnology and Bioengineering*: Elsevier; 2023. pp. 425–46.
223. Chia RW, Lee J-Y, Cha J, Çelen-Erdem İ. A comparative study of soil microplastic pollution sources: a review. *Environmental Pollutants and Bioavailability*. 2023;35(1):2280526.
224. Sadia M, Mahmood A, Ibrahim M, Irshad MK, Quddusi AHA, Bokhari A, et al. MPs pollution from wastewater treatment plants: a critical review on challenges, detection, sustainable removal techniques and circular economy. *Environmental Technology & Innovation*. 2022;28:102946.
225. Ge J, Wang M, Liu P, Zhang Z, Peng J, Guo XJTTi AC. A systematic review on the aging of MPs and the effects of typical factors in various environmental media. *TrAC Trends in Analytical Chemistry*. 2023;162:117025.

CHAPTER 16

Microplastic Pollution in the Environment

Effects of Microplastic Particles on Geotechnical and Other Properties of Soil

Adeyemi Babayemi Fajobi, Moruf Afolabi Asani, Olanrewaju Rotimi Bodede, Ufuoma Phoebe Williams Bello, Dele Adedapo Daramola, Damilola Gbadebo Bobola, and Isaiah Adesola Oke

SYMBOLS AND ABBREVIATIONS

A-PE-MPs	tetracycline and microplastic particles exposed to air
C	CK + 10 mg/kg CIP
CE	CK + 1% PE + 10 mg/kg CIP
CIP	ciprofloxacin
CK	original and raw soil
DOC	dissolved organic carbon
E	CK + 1% stands for PE

DOI: 10.1201/9781003487579-16

FPP	fibrous polypropylene
GCLs	geosynthetic clay liners
GHGs	greenhouse gases
HDPE	high-density polyethylene
LDA	linear discriminant analysis
LDPE	low-density polyethylene
LMWOAs	low-molecular-weight organic acids
MPE soil	membranous polyethylene-amended soil
MPE	membranous polyethylene
MPP soil	microsphere polypropylene-amended soil
PAN	polyacrylonitrile
PE	polyethylene
PET	polyethylene terephthalate
PP	polypropylene
PVC	polyvinylchloride
S-PE-MPs	tetracycline and microplastic particles exposed to soil
TC	tetracycline
UV	ultraviolet
w/w	weight by weight
W-PE-MPs	tetracycline and microplastic particles exposed to water

16.1 INTRODUCTION

Plastic solid wastes and products are produced regularly at an increasing rate and in vast quantities on a large scale. It has been estimated and calculated that 400×10^6 tonnes of plastic solid waste and production are made yearly, and these quantities and amounts are expected to multiply and more than double by the year 2050 (Schöpfer et al., 2020; Monkul and Özhan, 2021; Lim, 2021; Hellevik and Cyvin, 2023). Global plastics manufacturing and production have risen exponentially during the last four decades. Less than 10% of plastic solid waste and products are recycled, and the remaining balance is frequently unconfined into the environment. The main classes of non-fibre plastics production are polypropylene (PP), polyethylene (PE), polyethylene terephthalate (PET) and polyvinyl chloride (PVC). Polyester, largely PET, accounts for 70% of all fibre classes of plastic production. Once they enter the environment, plastics can break down into smaller pieces through various mechanical, chemical and biological processes (Van Gestel and Selonen, 2018; Wang et al., 2020a and b; Song et al., 2019; Schöpfer et al., 2020; Hellevik and Cyvin, 2023).

Microplastic materials are fine and small plastic solid wastes and debris of a size less than 5 mm. These materials frequently originate from larger plastic solid wastes' oxidation, degradation or deformation. These materials comprise a heterogeneous size, composition and shape (Wang et al., 2014; Cole, 2016; Chen et al., 2017; Song et al., 2019; Schöpfer et al., 2020; Wu et al., 2020a and b; Hellevik and Cyvin, 2023). In the last three decades and in recent years, microplastic particles have been identified and detected end-to-end around the world, particularly in aquatic environments such as lakes, oceans and rivers. Microplastic particle pollution is raising concerns. It has been enlisted as the second most vital scientific issue in the field of environmental science, engineering and ecology (Song et al., 2019). More on microplastic particles and environmental pollution can be established through the literature such as Daniel and Wu (1993); Estebenet (1995); Bernet et al. (1999); Ghiselli et al. (2000); Gaylor et al. (2013); Goldstein et al. (2014); Jia et al. (2014); Cole and Galloway (2015); Rochman et al. (2015); Yang et al. (2015a and b); Cole et al. (2016); Duis and Coors (2016); Gajendiran et al. (2016); Green (2016); Huerta Lwanga et al. (2016); Imhof et al. (2016); Ioakeimidis et al. (2016); Lambert and Wagner (2016); Zhang et al. (2016); Dris et al. (2017); Geyer et al. (2017); Horton et al. (2017); Alimi et al. (2018); Beiras et al. (2018); Brooks et al. (2018); De et al. (2018); Gigault et al. (2018); He et al. (2018); Jabeen et al. (2018); Jovanovic et al. (2018); Mai et al. (2018); Qu et al. (2018); Yin et al. (2018); Yu et al. (2018); Zhang et al. (2018a and b); Fuhr and Franklin (2019); Schöpfer et al. (2020); Song et al. (2019); Xu et al. (2020a and b); Da Costa et al. (2020); Zhao et al. (2021); and Hellevik and Cyvin (2023).

It has been documented that most plastic solid wastes and rubbish in receiving water bodies originated from terrestrial human activities. Microplastic particles are regarded as emerging pollutants and hazards to terrestrial ecosystems, in which soils constitute a larger tank for plastics and solid wastes than seas (Song et al., 2019). There are many possible routes through which microplastics can enter the soil, including soil amendment with compost and sewage sludge, plastic mulching, irrigation and flooding, littering and atmospheric deposition (Blasing and Amelung, 2018). Compost materials and sewage sludges are commonly utilized as fertilizers in North America and Europe and may be an important source of soil and plastic pollution in these mentioned regions. The key sources of soil microplastic particles stem from industrial actions through plastic manufacturing and production, atmospheric deposition, garbage, soil sedimentation and road dust. Microplastic particles in the soil are accumulated, stored, eroded, degraded, translocated, leached or percolated into the water level or groundwater, and these actions endanger organisms throughout the terrestrial ecosystem. More on the effects of microplastic particles on aquatic animals, humans and soil can be found in literature such as Thompson et al. (2004); Koerner (2005a and b); Rosenkranz et al. (2009); Ward and Kach (2009); von Moos et al. (2012); Larba and Soltani (2014); Mathalon and Hill (2014); UNEP (2014); Watts et al. (2014); Lange and Mwinzi (2015); Li et al. (2015); Van Cauwenberghe et al. (2015); Hu et al. (2016); Lu et al. (2016); Ped et al. (2016); Blasing and Amelung (2017); Cao et al. (2017); Horton et al. (2017); Huerta Lwanga et al. (2017); Pereira et al. (2017); Rillig et al. (2017); Rist et al. (2017); Anbumani and Kakkar (2018); Chae and An (2018); Dawson et al. (2018); Liu et al. (2018); Mai et al. (2018); Ng et al. (2018); Scheurer and Bigalke (2018); Zhou et al. (2018); Song et al. (2019); Wang et al. (2019); Ziajahromi et al. (2019); PlasticsEurope (2020); Cheng et al. (2021); Narevski et al. (2021); Snowdon (2021); and Zhang et al. (2023).

Monkul and Özhan (2021) documented that the top ten disciplines that are working and conducting research on microplastic particles and soils between 2016 and 2022 are Environmental Sciences, Environmental Engineering, Multidisciplinary Sciences, Water Resources, Toxicology, Chemical Engineering, Chemistry, Soil Science and Green Sustainable Science Technology. In addition, Figure 16.1 presents two engineering disciplines, which are chemical and environmental that show critical interest in plastic pollution in the environment. This figure shows the awareness of microplastic particles and soils, but there is little or no data from Civil engineering. This outcome shows that Civil Engineers need to be involved and state the present global status and awareness of the Civil Engineering communities on microplastic particles, which is extremely restricted and limited. The main objective of the current research is to provide geotechnical properties of soils with respect to microplastic particle pollution in the soil. The geotechnical engineering option is one of the key options of the civil engineering profession, which specializes in soils, soil properties, soil distribution, engineering behaviour and soil properties. Some of these soils and engineering behaviours are permeability, particle size distribution, shear strength, compressibility, settlement, porosity and densities of the soil, which are vital in all civil engineering developments and projects. The research will enhance the discipline to advance engineers' understanding and thought of microplastic particle contamination in soils from several perspectives.

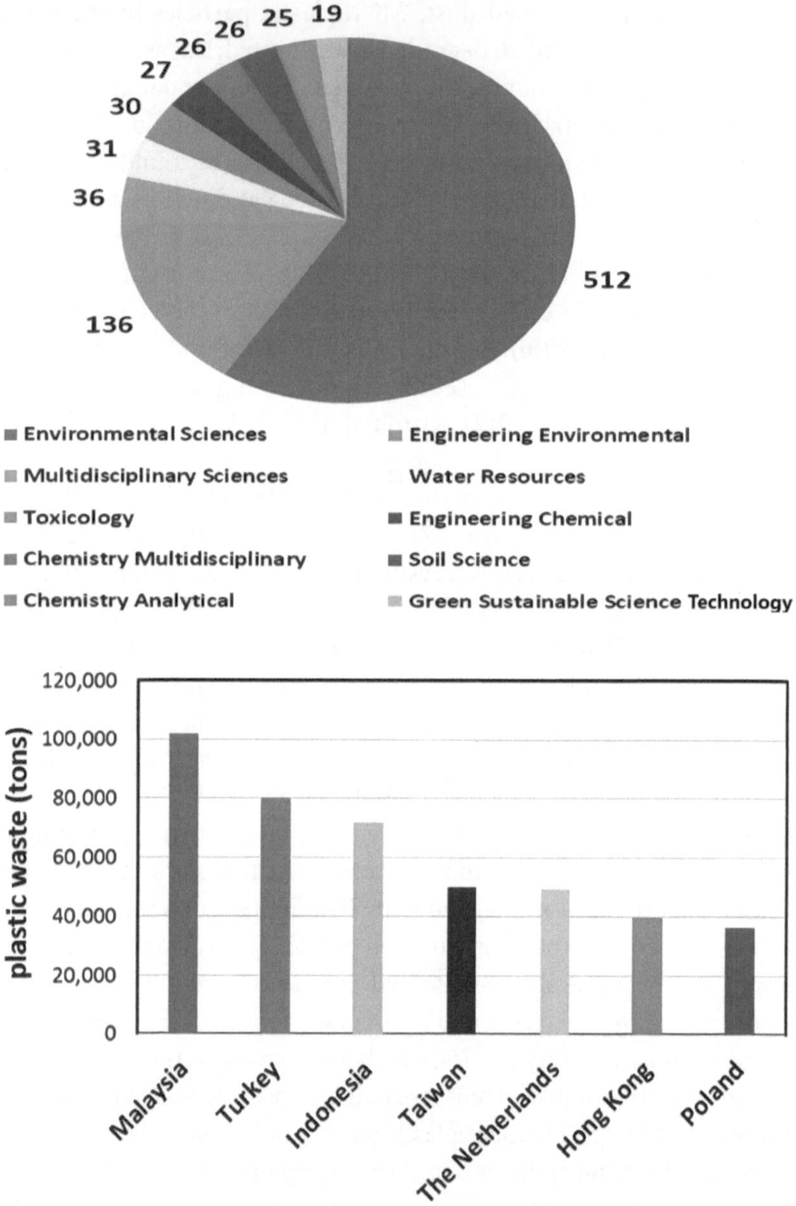

FIGURE 16.1 (a) Engineering disciplines with critical publication on plastic pollution in the environment. (b) The list of seven countries that received plastic solid wastes from the United Kingdom in the year 2018. (Monkul and Özhan, 2021.)

16.2 RESEARCH STUDY DESIGN

This dynamic and essential study on plastic and microplastic substances treatment and removal from water and wastewater streams includes the appraisal of the literature on the basis of accomplished information on microplastic particle presence, the sampling, examination, evaluation and laboratory analysis of microplastic in freshwater samples, and the environment using various techniques. The data and information were collected from published databanks such as PubMed link (https://www.pubmed.ncbi.nlm.nih.gov), Taylor and Francis Online link (http://www.tandfonline.com/), ScienceDirect link (http://www.sciencedirect.com/), Google Scholar link (https://scholar.google.com/), Scopus link (https://www.scopus.com/), Web of Science link (https://webofknowledge.com/), Elsevier link (https://www.elsevier.com/), Wiley Online Library link (http://onlinelibrary.wiley.com/), SpringerLink (http://link.springer.com/), American Chemical Society (ACS) Publications link (http://pubs.acs.org), Magiran link (https://www.magiran.com) and Royal Society of Chemistry (RSC) Publishing link (http://pubs.rsc.org) using these keywords microplastic, detection of microplastics, analysis of microplastics, computation of microplastic in treated and drinking water, microplastic substances in freshwater, treatment of microplastic and microplastic substances, microplastic substances in marine and aquatic environment.

16.3 RESEARCH FINDINGS

Figure 16.2a presents the outcome of the findings and search. The results of the search provided 8,562 conference proceedings, journal articles and edited books. The clarifications based on the year of publication revealed that 7,321 of these materials are recent materials between 2002 and 2023 (21 years) at the rate of 349 new data and information per year, while 1,241 of these materials were older materials between 1980 and 2001 (21 years) at the rate of 59 materials per year. This result shows an increment in the information and data on plastic materials and microplastic particles. The differences (6,080 materials and 290 materials/year) revealed more understanding, research, experiments and studies on the subject of plastic pollution of the environment.

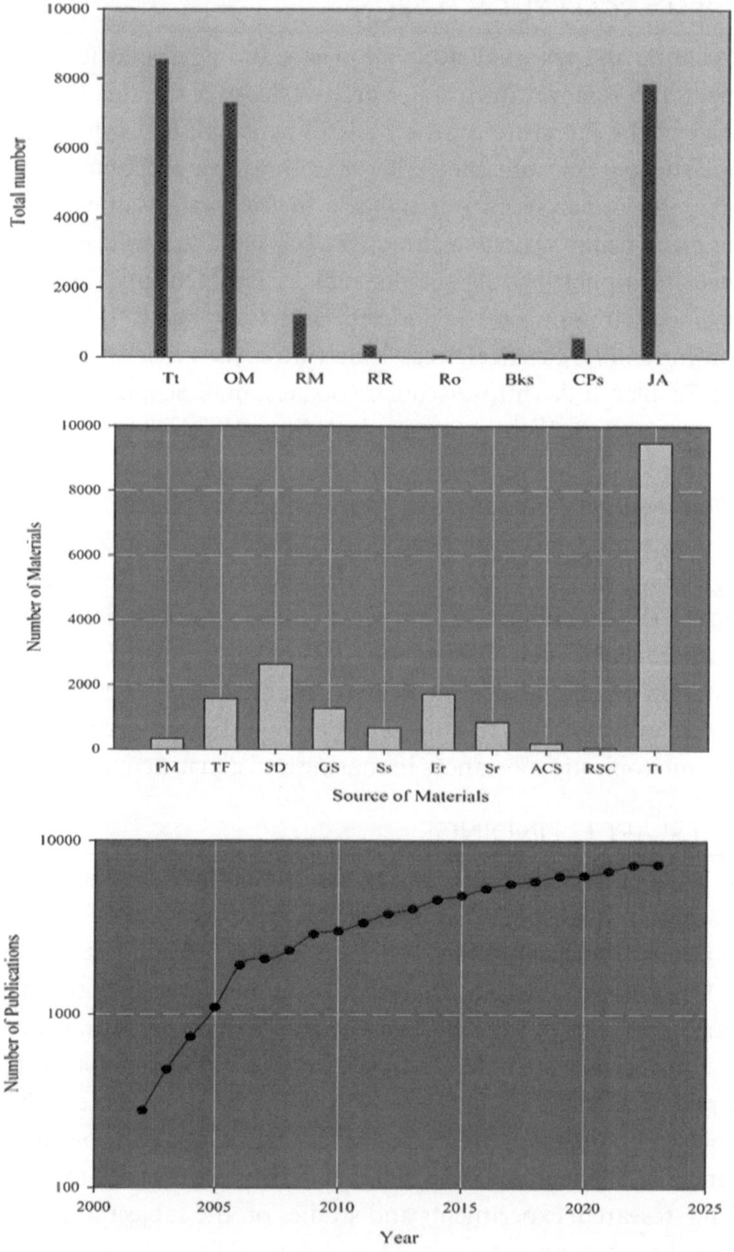

FIGURE 16.2 (a) The outcome of the findings and search (b) the outcome of the findings and search (c) the growth in the data by the year for both old and recent articles and materials (d) conceptual and simplified drawings of the pull-and-push dynamic between the ultimate development of international agreements, researchers and policymakers. (Hellevik and Cyvin, 2023.)

(Continued)

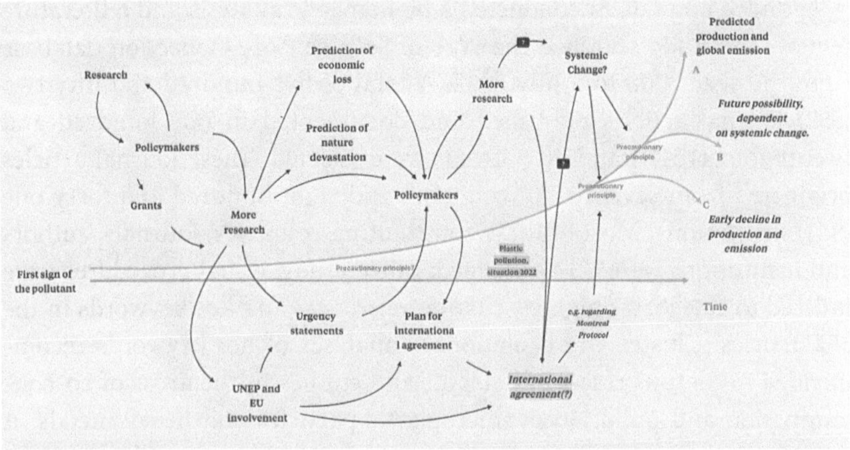

FIGURE 16.2 (*Continued*) (a) The outcome of the findings and search (b) the outcome of the findings and search (c) the growth in the data by the year for both old and recent articles and materials (d) conceptual and simplified drawings of the pull-and-push dynamic between the ultimate development of international agreements, researchers and policymakers. (Hellevik and Cyvin, 2023.)

A breakdown of the materials into categories of publications shows that 130 were books, 653 were conference proceedings and 7,869 were journal articles. A further breakdown identified that the materials were from PubMed (362), Taylor and Francis (594), ScienceDirect (2643), Google Scholar (1299), Scopus (696), Elsevier (1732), Springer (875), ACS (94), Magiran (222) and RSC (75) (Figure 16.2b). Out of a total of 7,321 recent materials, materials that explained or identified microplastic in the urban areas and environment (in the air, water, plants and materials) were 5,132 (70.10%) and the rest were on other related topics on microplastic. The materials on microplastic particles in freshwater and related topics were made up from the referred journals. Figure 16.2c shows the growth in the data by the year for old and recent articles and materials. The figure revealed that the growth and publishing rate were logarithms. The materials on microplastic, emerging and microplastic substances treatments, reduction and removal were made up of 1,458 articles from the following journals from various sources. Figure 16.2d provides full information and data on the simplified and conceptual drawing of the push-and-pull dynamic between researchers, policymakers and the ultimate development of international agreements.

Li and Wang (2023) conducted a bibliometric analysis and a literature review. The study searched the Web of Science Core Collection database from the year 2008 to 5 July 2022. A total of five hundred and fifty-two (552) journal articles published and documented in one hundred and twenty-four (124) standard journals were selected. These journal articles originated from seventy (70) countries and eight hundred and forty-one (841) institutions. Most of these contributing countries, journals, authors and institutions were acknowledged. In the study, visual procedures were utilized to categorize high co-citation references and hot keywords in the 552 articles. Cluster and evolutionary analyses of hot keywords recommended numerous research hotspots and studies in the areas of co-contamination and pollution of microplastic particles and heavy metals. It was stated that research work on the adsorption behaviour, bioaccumulation, environmental pollution, desorption behaviours, risk assessment, toxicity and detection and characterization techniques is needed. It was concluded that based on the current studies and research status, several research directions of priority are recommended to comprehend the interactions between microplastic particles, the environment and heavy metals.

16.4 PREVIOUS AND RECENT STUDIES ON ANALYSIS, MONITORING AND DETECTION OF MICROPLASTIC

Wu et al.'s (2020a and b, 2022) study on cytotoxicity performance examined desorption behaviours of contaminated and polluted microplastic particles in aquatic and aquatic intestinal environments after ingestion of microplastic particles by either cold- or warm-blooded organisms. The research helped in the explanation of the significant fate and possible cytotoxicity of polluted microplastic particles and the probable implications of the microplastic particles as vital vectors for bisphenols to the increment of possible health risks. Table 16.1 presents the first-order kinetic model correlation results for bisphenol release and desorption from bisphenol-attached PVC microplastic particles in water, cold and warm situations. Xu et al. (2020) investigated the adsorption of polar diazepam and nonpolar phenanthrene into two soils (Oxisol and Inceptisol). The research utilized batch adsorption experiments to evaluate the impacts of PE, PS and PP, microplastic particles on the properties of the soils. It was concluded that microplastic particles can change the adsorption of diazepam and phenanthrene onto soil and influence environmental risk and their mobility in the soil ecosystems. Figure 16.3 presents the mechanism of adsorption. Wang et al. (2020a and b) examined the degradation

TABLE 16.1 The first-order kinetic model correlation results for bisphenol release and desorption from bisphenol-attached PVC microplastic particles in water, cold, and warm situations (Source: Wu et al., 2020) and Wu et al., 2022).

Adsorbate	Solution	k_{rap} (10^{-2})*	Ratio⁋	k_{slow} (10^{-4})*	k_{vs} (10^{-5})*	F_{rap}^{+}	F_{slow}^{+}	F_{vs}^{+}
BPA	Water#	0.746±0.206		5.979±0.555	6.997±0.788	0.640±0.003	0.284±0.004	0.076±0.004
	Cold-I#	0.809±0.302	1.08	10.90±0.407	9.442±0.716	0.561±0.003	0.395±0.003	0.043±0.005
	Warm-I#	2.163±0.456	2.90	11.40±0.898	9.069±0.647	0.677±0.004	0.303±0.046	0.019±0.003
BPAF	Water	0.794±0.230		5.983±0.498	6.314±0.606	0.660±0.006	0.272±0.010	0.068±0.002
	Cold-I	1.007±0.318	1.26	9.957±0.755	<0.01	0.588±0.005	0.409±0.002	0.003±0.001
	Warm-I	2.609±0.521	3.28	13.80±2.684	<0.01	0.684±0.004	0.315±0.007	<0.001
BPB	Water	0.777±0.285		6.751±0.274	6.992±0.534	0.660±0.004	0.293±0.006	0.046±0.002
	Cold-I	0.864±0.209	1.11	10.80±0.737	7.718±0.488	0.484±0.004	0.456±0.005	0.059±0.001
	Warm-I	2.479±0.452	3.19	18.90±2.818	<0.01	0.682±0.003	0.308±0.003	0.010±0.001
BPF	Water	0.619±0.179		6.659±0.441	<0.01	0.578±0.003	0.371±0.004	0.051±0.002
	Cold-I	0.662±0.241	1.07	6.906±0.579	2.367±0.112	0.548±0.014	0.335±0.002	0.017±0.004
	Warm-I	1.744±0.373	2.82	9.877±1.415	5.778±0.325	0.669±0.005	0.308±0.003	0.005±0.001
BPS	Water	0.580±0.174		6.789±0.198	<0.01	0.561±0.003	0.410±0.007	0.028±0.002
	Cold-I	0.587±0.228	1.01	7.612±0.336	12.03±0.685	0.532±0.002	0.383±0.001	0.085±0.002
	Warm-I	1.497±0.391	2.58	10.00±1.191	11.338±0.322	0.663±0.005	0.325±0.005	<0.001

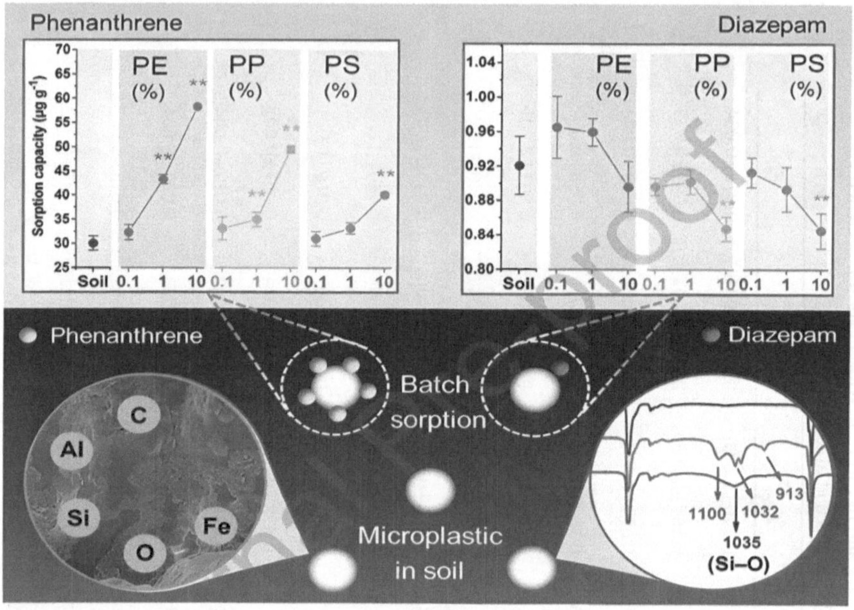

FIGURE 16.3 The mechanism of adsorption of microplastic particles in the soil. (Xu et al., 2020a and b.)

behaviour of ciprofloxacin (CIP) and PE applied to soil microcosm and evaluated the combined impacts of these materials on soil microbial communities. The study revealed that the specific modification of CIP, the co-structure of CIP and microplastic particles declined the CIP decomposition and degradation effectiveness throughout the 35 days of cultivation time. In the case of the microbial colony and community structure, the microbial configurations at the phylum concentration level were steady among all treatment techniques utilized, and the common and frequent phyla were proteobacteria, Chloroflexi and Actinobacteria. Arthrobacter was the only genus that had an unusual change in the CIP-amended soil equated to the change in the blank or control, but the four genera were significantly changed in the microplastic particles and CIP co-amended soil. Figure 16.4 is the schematic presentation of the structure shift of bacterial colonies and communities (dominant and common with phyla, significant and important genera) and variations in the physicochemical properties within blank sample or control, CIP-loaded soil, and CIP-MPs amended soil. Figure 16.5 presents ciprofloxacin decomposition and degradation curves comparison between C and CE treatments. Kim et al. (2020)

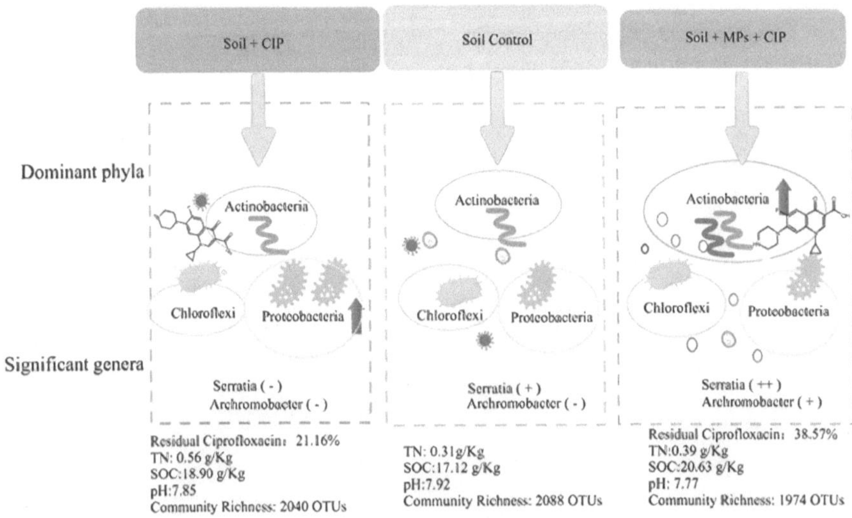

FIGURE 16.4 The schematic presentation of the structure shift of bacterial colonies and communities. (Wang et al., 2020a and b.)

established that poly-ethylene terephthalate fragments known as PET and polyacrylonitrile fibres also known as PAN showed the highest toxicity, while polypropylene termed PP, polystyrene fragments confirmed as PS and high-density polyethylene known as HDPE induced relatively less adverse influences on nematodes. In addition, low-density polyethylene which stands for LDPE induced little or no toxicity.

Wang et al. (2021) examined the adsorption behaviours of tetracycline (TC) and Cu(II) on PE microplastic particles that were exposed to soil (S-PE-MPs), air (A-PE-MPs), and water (W-PE-MPs) and environments for a long time. The study established that the contribution of microplastic particles to antibiotic and heavy metal movement and migration could be influenced by the exposed situations and environments. Yu et al. (2020) explored the effect of LDPE known as low-density polyethylene, and networks of earthworm burrows on the saturated water flow and soil hydraulic properties. The study concluded that the microplastic particles did not reveal or have any significant impacts or effects on the saturated wastewater and water flow, which may be attributed to the low-level concentrations of microplastic particles utilized. Ren et al. (2019) investigated the impacts of two particle sizes of microplastic particles on dissolved organic carbon (DOC) and its relative functional groups, fluxes of greenhouse gases (GHGs) and the fungal and bacterial communities in

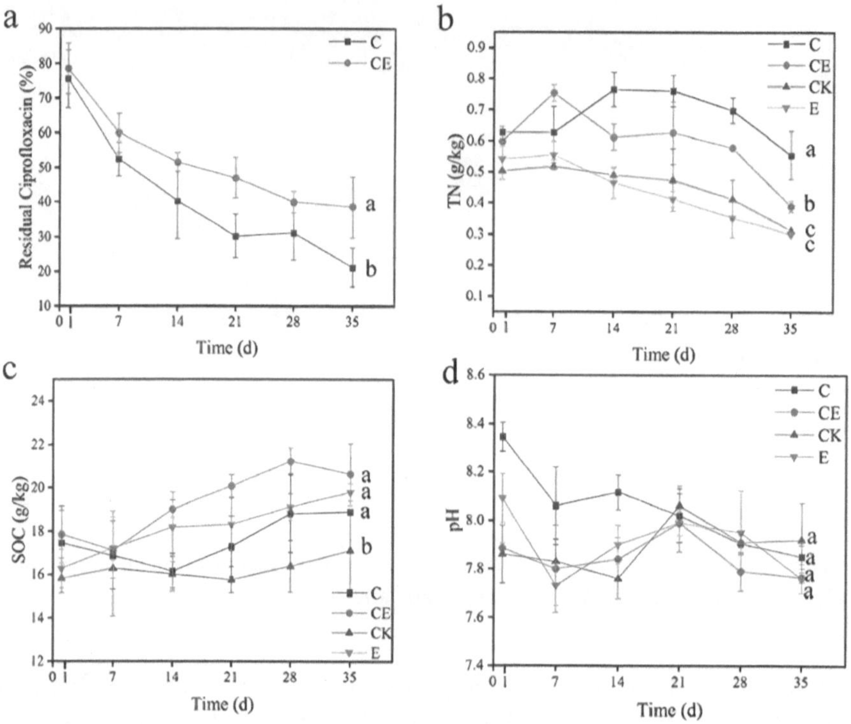

FIGURE 16.5 Ciprofioxacin decomposition and degradation curves comparison between selected treatment techniques. (a) In the figure stands for residual CIP in control; (b) in the figure stands for Concentration of TN; (c) in the figure stands for concentration of SOC; and (d) in the figure stands for pH. (Wang et al., 2020a and b.)

fertilized soil. The research revealed that microplastic particles decreased the global warming probable of soil. The particle size of the material influenced *Actinobacteria* to replace *Proteobacteria* and alpha diversity as the dominant phylum in the soil with microplastic particles. It was concluded that microplastic particles have selective impacts on microbes, which can possibly have a greater influence on terrestrial and biogeochemical cycles. Ju et al. (2019) exposed the soil springtail *Folsomia candida* to artificial soils contaminated with PE microplastic particles of size less than 500 mm for 28 days to explore the impacts of microplastic particles on reproduction, gut microbiota and avoidance of *Folsomia candida*. The research revealed that the relative quantities and presence of *Wolbachia* meaningfully reduced while the quantities and presence of the *Bradyrhizobiaceae*, *Stenotrophomonas* and *Ensifer* meaningfully increased. It was concluded

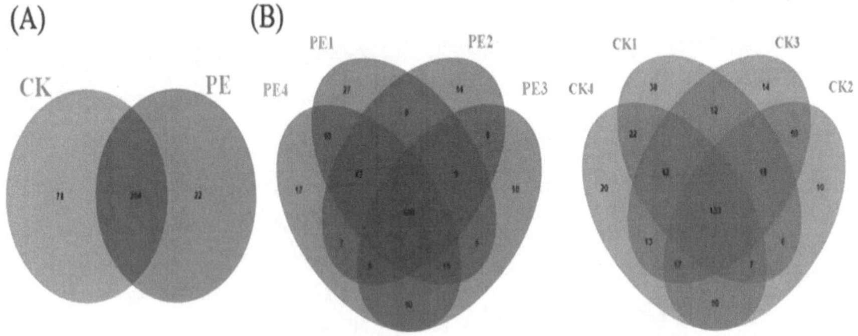

FIGURE 16.6 The Venn mappings and diagrams of the number and quantity of bacterial operational taxonomic units shared within and between groups of samples. Label (a) stands for the number of operational taxonomic units shared between the control (0% microplastic particles in the soil) and microplastic particle exposure groups. Label (b) stands for the amount of operational taxonomic units shared between the soil samples for each of the two groups, control and microplastic particle exposure. (CK, control; PE, polyethylene microplastic particles). (Ju et al., 2019.)

that microplastic particles exerted a significant toxic influence on springtails and changed the gut microbial community, which is known to provide clearer and more useful data and information for risk assessment of microplastic particles in terrestrial ecosystems. Figure 16.6 presents the Venn mappings and diagrams of the number of bacterial operational taxonomic units shared within and between groups of samples. Figure 16.7 shows linear discriminant analysis (LDA) scores calculated for taxa differentially abundant between the control and microplastic particles treated groups.

Yang et al. (2019) investigated the Cu^{2+} adsorption capacities of six kinds of microplastic particles namely polyethylene (PE), polyamide-6(PA), polyethylene terephthalate (PET), polystyrene (PS), polymethyl methacrylate (PMMA) and polyvinyl chloride (PVC) and the impacts of different soil environmental and operational factors, which include the presence of cations, pH, low-molecular-weight organic acids (LMWOAs) and ultraviolet (UV) ageing on the adsorption. The research established that the soil environmental factors can change the ability of different microplastic particles to adsorb or absorb Cu^{2+} and influence the movement and transport of plastic pollutants as the key carriers.

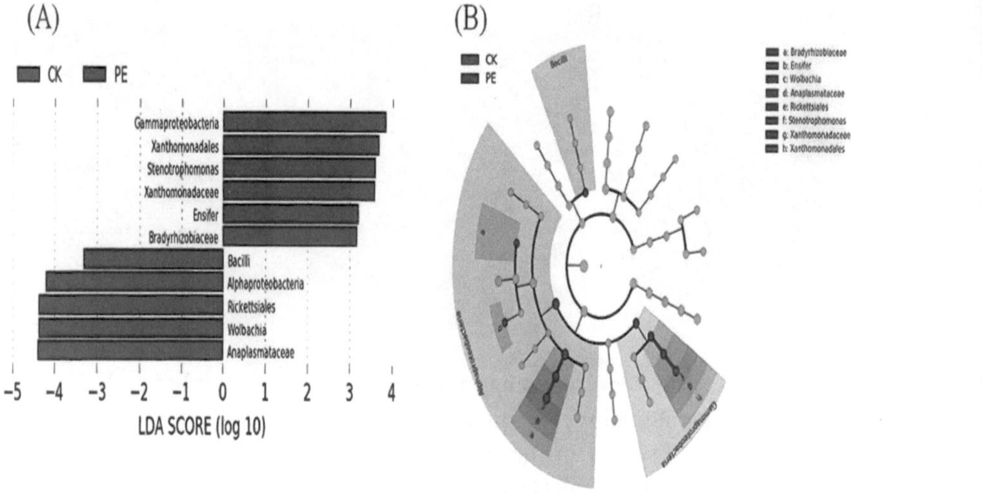

FIGURE 16.7 Linear discriminant analysis (LDA) scores computed for taxa differentially abundant between the control and microplastic particles treated groups (a) stands for the linear discriminant analysis (LDA) scores computed for taxa differentially abundant between the control (0% microplastic particles in soil) and microplastic particles treated groups. Six structures were differentially abundant in the microplastic particles treated group and five features were in the CK group. (b) Stands for the Cladograms showing differences in taxa between the control and microplastic particles treated groups. The circled figure from inside to outside represents the phylum, type, order, genus, and family, respectively. The diameter of each circle in the figure is proportional to the quantity of taxon. Green and red colours in the figure indicate taxon are significantly more abundant in the control and microplastic particles treated group. (CK, control; PE, polyethylene MP). (Ju et al., 2019.)

Yi et al. (2020) investigated the effects of three microplastic particles, namely, membranous polyethylene (MPP), microsphere polypropylene and fibrous polypropylene (FPP), on microbial community structure and enzyme activities in both loamy soil and sandy soil. The research showed that the effects of three microplastic particles on urease, alkaline phosphatase activities and dehydrogenase were in the order of FPP greater than MPP greater than microsphere polypropylene, MPP greater than microsphere polypropylene greater than FPP and FPP greater than microsphere polypropylene greater than MPP, respectively. It was concluded that microplastic particles significantly changed the microbial community structure, especially for the enrichment of *Bacteroidetes* and *Acidobacteria*, the depletion of *Chloroflexi. Aeromicrobium, Streptomyces, Mycobacterium,*

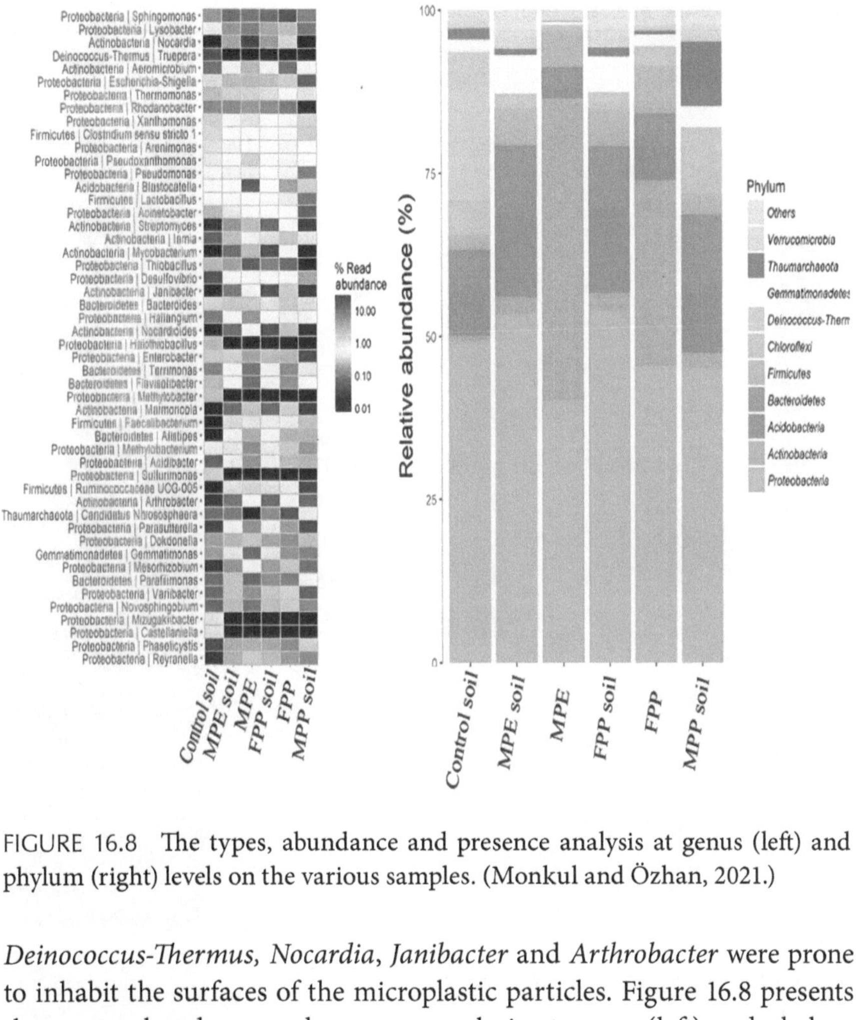

FIGURE 16.8 The types, abundance and presence analysis at genus (left) and phylum (right) levels on the various samples. (Monkul and Özhan, 2021.)

Deinococcus-Thermus, Nocardia, *Janibacter* and *Arthrobacter* were prone to inhabit the surfaces of the microplastic particles. Figure 16.8 presents the types, abundance and presence analysis at genus (left) and phylum (right) levels on the various samples (Tables 16.2 and 16.3).

16.5 PREVIOUS STUDIES ON EXISTENCE AND FORMATION OF MICROPLASTIC PARTICLES IN SOILS

Monkul and Özhan (2021) reported that microplastic particle pollution affects various ecosystems, which include marine (Jahnke et al., 2017) and freshwater (surface and groundwater) environments (Poleć et al., 2018), and soils (Zhou et al., 2020, Zhao et al., 2021), terrestrial soils are the fairly less researched among these three ecosystems. Literature such as Nizzetto et al. (2016), Zhang et al. (2018a and b), Da Costa et al. (2019, 2020), Zhu et al. (2019), Wang et al. (2020a) and Möller et al. (2020) have all

TABLE 16.2 Name and Structural Formula of Selected Plastics

Name	Structural Formula	Average Diameter (μm)	Zeta Potential (mV)	BET (m^2/g)	Morphology
PA	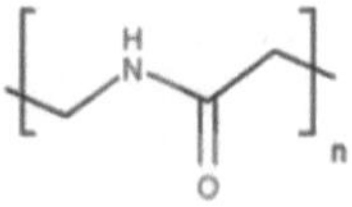	69.86	−7.59	8.710	
PE	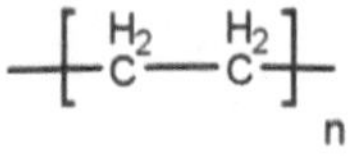	204.7	−15.2	2.110	
PS	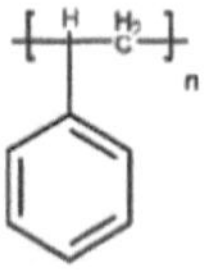	192.0	−16.1	<0.001	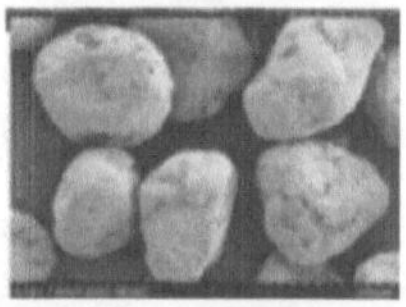
PET	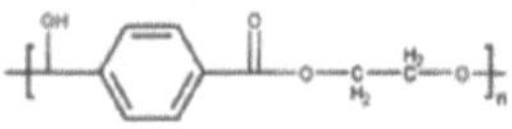	350.8	−16.7	0.475	
PVC	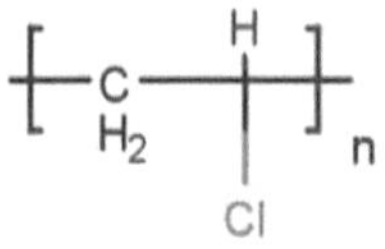	137.8	−8.70	0.556	
PMMA	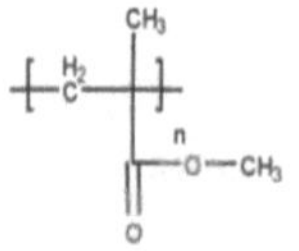	74.76	−15.9	0.110	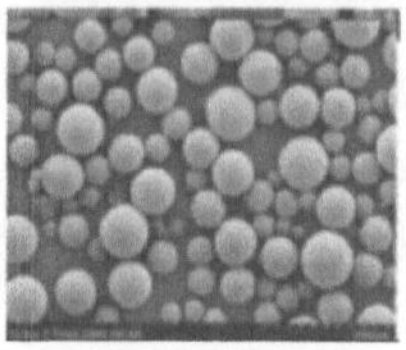

Source: Yang et al. (2019).

explicitly stated that research works and studies on microplastic particles in the soil environment are still comparatively limited in quantity and content related to other environments. Researchers have listed several factors and reasons for such observations which include the following:

TABLE 16.3 Type of Soils and Information on Plastics Present

Soil	Sampling Depth	Extraction Method	Identification Method	Number Concentration (particles/kg)[a]	Mass Concentration (mg/kg)[b]	Size	Shape[c]	Composition[c]	Location	References
					Microplastic (<5 mm)					
Floodplain soils	5 cm	Density separation (NaCl)[d]	Raman and FTIR spectroscopy	0–593	5 (0–55.5)[e]	125–500 mm (88%)		PE (88%) > PS > PP	Switzerland	Scheurer and Bigalke (2018)
Farmland soils	5 cm	Size fractioning (water)	Microscope and FTIR spectroscopy	0.34 ± 0.36 g (0–1.25)[e]	35 g/ha	1–5 mm	Film (44%), fragment (44%), fibre (12%)	PE (62.5%), PP (25%), PS (12.5%)	Franconia, Germany	Piehl et al. (2018)
Farmland soils	3 cm	Density separation (NaCl)	Microscope and FTIR spectroscopy	78.0 ± 12.9	—	1.91 ± 0.13 mm[g]	Fibre (53%), fragment (38%), film (7%)	PP (51%), PE (43%), PES (6%)	Shanghai, China	Liu et al. (2018)
Farmland soils	3–6 cm	Density separation (NaCl)	Microscope and FTIR spectroscopy	62.5 ± 13.0	—	1.48 ± 0.11 mm	Fibre (38%), film (34%), fragment (28%)	PP (51%), PE (43%), PES (6%)	Shanghai, China	Liu et al. (2018)
Coastal soils	2 cm	Density separation (NaCl, NaI)	Microscope and FTIR spectroscopy	740 (1.3–14,713)	—	<5 mm (~60% < 1 mm)	Flakes (69%), foams (28%), fragment (1%)	PE, PP, PS	Shandong, China	Zhou et al. (2018)
Coastal soils	2 cm	Density separation (NaCl, NaI)	Microscope	634	—	1.56 ± 0.63 mm (50% <1 mm)	Granules (76%), fragment (20%), fibre (2%)	—	Hebei, China	Zhou et al. (2016)
Home garden soils	20 cm	Density separation (water)	Microscope	870 ± 1,900	—	10–20 mm (59%), 20–50 mm (34%), >50 mm (5%)	—	—	Campeche, Mexico	Lwanga et al. (2017)
Agricultural soils	30 cm	Density separation (water)	Microscope	40 ± 126–320 ± 329	0.008–0.54	>100 mm (domain)	—	—	Loess plateau, China	Yang, et al. (2018)
Agricultural soils	25 cm	Density separation (NaCl, $ZnCl_2$)	Microscope	600–10,400	0.57–12.9	20 mm (width) × 0.97 mm (length)	Fibre (97%)	—	Mellipilla, Chile	Corradini et al. (2019)

(Continued)

TABLE 16.3 (*Continued*) Type of Soils and Information on Plastics Present

Soil	Sampling Depth	Extraction Method	Identification Method	Number Concentration (particles/kg)[a]	Mass Concentration (mg/kg)[b]	Size	Shape[c]	Composition[c]	Location	References
Agricultural soils	10 cm	Density separation (NaI)	Microscope	18,760 (7,100–42,960)	—	0.05–1 mm (95%)	Fibre (92%), film and fragment (8%)	—	Kunming, China	Zhang and Liu (2018)
Agricultural soils	15 cm	Density separation ($ZnCl_2$)	FTIR spectroscopy	82,000 (0–165,000)	6.26 (0–16.5)	10–130 mm	—	PP (56%), PE (39%), Nylon (5%)	Northern Jutland Denmark	Vollertsen and Hansen (2017)
Agricultural soils	15 cm	Density separation ($ZnCl_2$)	FTIR spectroscopy	236,000 (53,000–528,000)	51.1 (0.1–224.3)	20–245 mm	—	PE (89%), Nylon (12%), PP (1%)	Northern Jutland Denmark	Vollertsen and Hansen (2017)
Industrial soils	—	Pressurized fluid extraction	Microscope and FTIR spectroscopy	—	300–67,500	—	—	PVC (>80%), PE, PS	Sydney, Australia	Fuller and Gautam (2016)
Paddy soils	10 cm	Density separation (NaCl)	Microscope and FTIR spectroscopy	10.3 ± 2.2	—	0.02–1 mm (domain)	Fibre (domain), fragment, film	PE (61%), PP (35%), PVC (4%)	Shanghai, China	Lv et al. (2019)
Mesoplastic (5–25 mm)										
Floodplain soils	5 cm	Density separation (NaCl)	Raman and FTIR spectroscopy	0–89[h]	0–295.3[h]	5 mm to 2.5 cm	—	PE (35%)> PA > latex	Switzerland	Scheurer and Bigalke (2018)
Farmland soils	3 cm	Density separation (NaCl)	Microscope and FTIR spectroscopy	6.75 ± 1.51	—	5.1–16 mm	Fibre (53%), fragment (38%), film (7%)	PP (51%), PE (43%), PES (6%)	Shanghai, China	Lu et al. (2016)
Farmland soils	3–6 cm	Density separation (NaCl)	Microscope and FTIR spectroscopy	3.25 ± 1.04	—	5.1–16 mm	Fibre (38%), film (34%), fragment (28%)	PP (51%), PE (43%), PES (6%)	Shanghai, China	Lu et al. (2016)
Farmland soils	5 cm	Size fractioning (water)	Microscope and FTIR spectroscopy	206 pieces/ha	66 g/ha	5–50 mm (73%)	Film (65%), fragment	PE (68%), PS (14%), PP (10%)	Franconia Germany	Piehl et al. (2018)

(*Continued*)

TABLE 16.3 (*Continued*) Type of Soils and Information on Plastics Present

Soil	Sampling Depth	Extraction Method	Identification Method	Number Concentration (particles/kg)[a]	Mass Concentration (mg/kg)[b]	Size	Shape[c]	Composition[c]	Location	References
					Macroplastic (>25 mm)					
Horticultural soils	10 cm	Manual collection	Visual observation	—	30±19 kg/ha	28±13 cm^2 (25%)	Film	PE	Buenos Aires, Argentina	Ramos et al. (2015)
Agricultural soils	30 cm	Manual collection	Visual observation	—	0–503 kg/ha	—	Film	—	Xinjiang, China	Liu et al. (2016)
Agricultural soils	30 cm	Manual collection	Visual observation	5,670,000 pieces/ha	44.7 kg/ha	<4 cm^2 (53%), 4–25 cm^2 (40%)	Film	—	Yunnan, China	Kang et al. (2013)
Agricultural soils	30 cm	Manual collection	Visual observation	3,930,000–9,650,000	59.1–103.4 kg/ha	<4 cm^2 (65%), 4–25 cm^2 (30%)	Film	—	Hebei, China	Ma et al. (2008)
Agricultural soils	20 cm	Manual collection	Visual observation	—	26.8 kg/ha	—	Film	—	Hebei, China	Hu et al. (2016)
Agricultural soils	40 cm	Manual collection	Visual observation	—	121.82–352.38 kg/ha	—	Film	—	Xinjiang, China	Wang et al. (2018)

Source: Xu et al. (2019).

PA, polyamide; PE, polyethylene; PES, polyether sulfones; PP, polypropylene; PS, polystyrene; PVC, polyvinyl chloride.

[a] The number of plastic per kg or otherwise stated as number per ha.

[b] The weight of plastic per kg or otherwise stated as mass per ha.

[c] Only the first three types listed.

[d] Extracted by distilled/de-ionized water, NaCl, NaI or $ZnCl_2$ in the bracket.

[e] The mean value (the whole range of measured concentration or size in the bracket).

[f] Not provided in the literature.

[g] The mean value±standard error.

[h] Data requested from the authors of the literature

a. relative postponements in the global consciousness between different disciplines (such as Civil Engineering and Environmental Sciences; Environmental Science and Engineering against Chemistry and Geology) on microplastic particle pollution and contamination;

b. nonexistence of multidisciplinary research work and study on microplastic particles in soils between soil-water related disciplines and others;

c. historical and regular progress of microplastic particle consciousness, which initially originated in the marine environment (Monkul and Özhan, 2021).

However, research work and relevant and related studies on microplastic particles in soils have gained an impetus over the last few decades or years. The current areas of research work and studies include the analytical and determination methods for microplastic particles sampling, collections, determinations and evaluations in soils (Zhu et al., 2019; Da Costa et al., 2019, 2020; Möller et al., 2020, Wang et al., 2020b); their movement and migration, environmental and ecological risks (Zhao et al., 2021;Zhou et al., 2020; Li et al., 2020, Guo et al., 2020); their accumulation and transfer in agricultural soils (Corradini et al., 2019; Crossman et al., 2020; Zhang and Liu, 2018); their impacts and effects on the soil functions and qualities (de Souza MacHado et al., 2018; Steinmetz et al., 2016). The stated studies above on microplastic particles and soil interaction inexplicably focused mostly on health aspects and environmental and agricultural sections. These studies provided little or no information and data on the engineering perspective of soil and microplastic particle relation or interaction. de Souza MacHado et al. (2018) reported that microplastic particles influenced the water-holding capacity and bulk density of soils. The study explained the outcome with respect to an agronomical perspective without engineering or geotechnical properties of the soil applications.

The study established that there is a good relationship between the rootability of soil and the bulk density of the soil. It was reported that a reduction in the bulk density of the soil would naturally imply a better rootability of the plat. This can be attributed to increased porosity of the soil which will help root growth (de Souza MacHado et al., 2018; Steinmetz et al., 2016). de Souza MacHado et al. (2018) documented that a decrease or decline in the bulk density of the soil due to the presence of microplastic particles could be misleading in respect of rootability of the soil. The issue of soil porosity was not critically addressed as the densest microplastic particle polymer used in the research work (polyester) was still considerably lighter in density, less than 1,370 kg/m^3, compared to the density of the control soil, which is not less than 1,439 kg/m^3.

It was stated that the observed reduction in bulk density of soil was due to microplastic particle accumulation in the soil throughout the study, which could have been more influenced by the addition of a lighter microplastic material than by an increase in soil porosity. With reference to typical geotechnical properties of the soil and from a soil engineering perspective, porosity and bulk density of the soil naturally relate to the settlement, shear strength and compressibility characteristics of the soil (Dexter, 2004; Monkul et al., 2020; Monkul and Özden, 2021). Both compressibility and shear strength are two very significant perceptions related to many sections, options and aspects in geotechnical and civil engineering, which include slope stability (Duncan et al., 2005), foundation design in structural engineering (Coduto, 2001), reclamation and dredging works (Von'f Hoff and Van der Kolff, 2012) and sanitary landfill design (Qian et al., 2001). The void ratio of the soil is defined as the ratio of the volume of voids to the volume of solids in a typical soil matrix. It is a popular geotechnical parameter of the soil, more so than another similar parameter, the porosity of the soil. The porosity of soil is the ratio of the volume of voids of the soil to the total volume in a typical soil matrix. In a typical geotechnical engineering research work and practice these two parameters are vital. Furthermore, these changes in the soil densities due to the presence of microplastic particle contamination or pollution are not expected to influence the overall geotechnical engineering properties and behaviour of the soil in terms of settlement, compaction or shear strength, except if extreme concentration or levels of pollution or contamination exist in the soil.

16.6 PREVIOUS STUDIES ON DEGRADATION PROCESSES

It has been reported that there are diverse degradation processes for microplastic particle formation, these degradation processes are known to have nonrapid rates depending on the ambient situations and conditions. The key mechanism of degradation involves physical, chemical, biological or a combination of these physical, chemical or biological processes. It has been documented these polymeric degradations play a major role in microplastic particle formation. These polymeric degradations can be in the form of photo-oxidative, mechanochemical, thermal, ozone-induced, catalytic types and biological processes (Lambert and Wagner, 2016; Singh and Sharma, 2008). These degradation processes influence the toxicity of microplastic particles (Guo et al., 2020; Zou et al., 2020). O'Kelly et al. (2021)

specified that ozone-induced degradation, photo-oxidative degradation and biodegradation degradation are the key degradation categories contributing to the breaking down of macrosized and mesosized plastics, which are commonly formed as plastic particles in soils.

Singh and Sharma (2008) discussed in detail the mechanisms of ozone-induced degradation, photo-oxidative degradation and biodegradation degradation processes. Photo-oxidative degradation was defined as a decomposition process of plastic that occurs by the absorption of UV and visible light by the plastic or polymers. The oxidation and degradation reactions are the two reactions after the adsorption of UV and visible light. The functional groups and impurities attached to the polymer determine and control these reactions. Ozone-induced degradation is cauterized by the ozone present in the air. Ozone quickens the ageing process of the polymeric materials. This phenomenon of ozonation is followed by the concentrated formation of oxygen-containing compounds, which change the mechanical properties, the molecular weight and the electrical properties of the plastic materials. Ozone-induced degradations of plastic materials occur by the attachment of the ozone molecule to the unsaturated polymer in unsaturated polymers. The reaction propagates in three key steps as follows; the formation of ozone olefin adducts, the decomposition of the primary ozonide to carbonyl compounds and a carbonyl oxide and the fate of the carbonyl oxide (Singh and Sharma, 2008).

The biodegradation process of polymers is typically cauterized by the presence of enzymes and microorganisms of a particular habit in the soils. These microorganisms can be different bacteria and fungi, while the enzymes can be either polyurethanase or lipase. These enzymes and microorganisms can contribute to the biodegradation process (Singh and Sharma, 2008; Kliem et al., 2020; Zheng et al., 2005). Crabbe et al. (1994) examined four fungus categories isolated from soil. The fungi are *Cladosporium sp, Curvularia sene galensis, Aureobasidium pullulans* and *Fusarium solani*. The study concluded that these fungi can biodegrade ester-based plastic such as polyurethane. Yoshida et al. (2016) revealed a novel bacterium namely *Ideonella sakaiensis*, which can utilize PET known as polyethylene terephthalate as a source of energy. The bacterium utilizes a carbon source for living, which can be summarized by the fact that the organism perfectly biodegrades PET. Feng et al. (2021) studied the biodegradation mechanism of PET with the molecular quantum and dynamics mechanisms and molecular mechanics methods. The study focused on the two key enzyme systems namely *IsMHETase and IsPETase* in the presence of *Ideonella sakaiensis*. Orhan et al. (2004) evaluated biodegradation of plastic compost bags under regulated soil situations and conditions. In the study, Orhan et al. (2004) examined the response of non-degradable and degradable LDPE and HDPE in the soil mixed with 50% weight-to-weight mature municipal solid waste compost originating from municipal refuse. The study established that the biodegradation rate of these polymers is influenced by some environmental factors, which are biological activity, moisture and temperature. Similar situations and environmental factors were confirmed by Kliem et al. (2020) for the biodegradation of diverse polymers. Kyrikou and Briassoulis (2007) explained in detail that biodegradable polymers are expected to leave no harmful and toxic substances in the environment, which indicates that these polymers should be converted to water, biomass, mineral and carbon dioxide, without any ecotoxicity. Kyrikou and Briassoulis (2007) explained that various polymers that are considered and established as biodegradable materials are indeed degradable depending on operational and environmental situations and conditions. These biodegradable materials and polymers can be degradable under hydro-degradable, photo-degradable or oxo-degradable operational conditions. In addition, photo-oxidative degradation and ozone-induced degradations can encourage the biodegradation possible of polymers due to the breakdown of the polymer chain, which can increase the surface area for colonization of microorganisms and decrease the molecular weight (O'Kelly et al., 2021; Singh and Sharma, 2008; Palmisano and Pettigrew, 1992; Tokiwa et al., 2009).

16.7 PREVIOUS STUDIES ON THE EXISTENCE OF MICROPLASTIC PARTICLES IN SOILS

The presence of microplastic particles in soil conditions and the environment is a worldwide emerging problem, which should not be the concern of only the scientific and engineering communities but must be the policymaker and public's problem too. Several research studies have reported microplastic particle concentrations present and availability in different soils worldwide. Lwanga et al. (2017) examined surficial samples of soils between 0.0 and 20.0 cm depth of soils from ten different home garden soils in Mexico. These sites were selected based on economic demographics and ethnicity. It was reported microplastic particle concentrations in the soils were between 2,770 and 930 particles/kg of soil. Zhang and Liu (2018) examined the distributions of plastic particles over four agricultural locations, which involved cropped soils and another site in a riparian forest used as a buffer zone in China. In all samples collected an average of 18,760 particles/kg of soil were reported, in which 95% of the particles were in the microplastic particle range 0.05–1.0 mm based on the classification. Zhang and Liu (2018) noted that microplastic particle concentrations were higher in the agricultural soil locations in the study compared to the forest used as a buffer zone. Crossman et al. (2020) evaluated microplastic particle concentrations and levels in soil depths of between 0.0 and 15.0 cm at three agricultural locations and sites in Ontario, Canada. The average microplastic particle concentrations and levels for these three sites vary significantly.

Van den Berg et al. (2020) reviewed soils from 16 agricultural locations in Spain. The study established that the average microplastic particle concentration ranged from 230 to 1,670 particles/kg of soil for light and low-density microplastic particles (density less than 1.0 g/cc). For high-density microplastic particles (density greater than 1.0 g/cc), the concentration ranged between 560 and 1,680 particles/kg of soil, based on surficial samples with depths between 0.0 and 30.0 cm. Dahl et al. (2021) examined the contamination and pollution of seagrass soils, which are known as marine soils at three locations along the Spanish Mediterranean coast. The study concluded that microplastic particle contamination and pollution were negligible until the year 1975 when microplastic particle pollution dramatically increased. The soil samples were collected from shallow depths of the soil surface between 0.0 and 15.0 cm. The concentrations and levels changed between 68 particles/kg of soil and 3,819 particles/kg of soil depending on the location. Dahl et al. (2021) specified that there is a strong correlation between the strengthening of the agricultural industry in a particular region and the microplastic particle concentrations and levels in the soils. Dahl et al. (2021) revealed that concentrations, levels, existence and presence of microplastic particles in soils vary with country and location and characteristics of the location, and may be any of these factors such as agricultural soil, garden soil, marine soil and landfill soil, regional factors and industrial practices (such as agricultural applications). Other interlinked sub-factors may be the usage of plastic in products, water treatment plants, the level and quality of solid waste, regional population intensity and economic level. It was observed some of the studies focused on the microplastic particles' presence within the first 20.0 or 30.0 cm below the ground surface, which is within the topsoil. It was established that soil life is more active within these depths, typical rooting and ploughing depths for agricultural purposes do not exceed 30.0 cm. In the evaluation for engineering purposes, this depth of range between 0.0 and 30.0 cm is very shallow from the perspective of foundation engineering and geotechnical engineering. It was then concluded that more studies and research works are needed to quantify the presence of microplastic particles at greater depths below depths the ground surface. Hydraulic conductivity or permeability is one of the main parameters in geotechnical and foundation engineering as well as in hydrogeology and geology, which indicates the tendency of water and other fluids to flow through the voids between the soil grains (Coduto et al., 2010). Hydraulic conductivity is a function of several different factors, which include viscosity of the pore fluid, void ratio and density of the soil, type of the soil (clay, sand and silt) and effective pore size between the soil grains (Ma et al., 2008; Coduto et al., 2010, Holtz et al., 2010; Kang et al., 2013; Ramos et al., 2015; Fuller and Gautam, 2016; Zhou et al., 2016; Vollertsen and Hansen, 2017; Yang et al., 2018; Piehl et al., 2018; Lv et al., 2019). Researchers assumed that hydraulic conductivity can be among the main parameters for the spatial distribution of microplastic particles in soils both in vertical and horizontal directions, which merits attention in future research on microplastic particles in soils.

Several studies and research works are available in the literature. This literature presents data and information on the migration and mechanism of microplastic particles in soils. O'Connor et al. (2019) reported an experimental work on the vertical migration of microplastic particles in sandy soils from a wetting and drying cycles perspective. The study evaluated the mobility of five different microplastic particles that have different particle densities and sizes. It was established that the samples consisted of PE and PP particles. The study established that the peak penetration depth of microplastic particles through sandy soils had a nearly linear correlation with the number of wetting and drying cycles. The microplastic particle size transformed into smaller sizes, mobility of these particles in the sandy soil increased. O'Connor et al. (2019) stated that microplastic particle concentration or levels at the volume and surface of infiltration liquid had negligible impacts or weak effects on depth of migration. The study projected the long-term penetration depths of the microplastic particles based on weather information and data from 347 Chinese urban, and semi-urban and the experimental findings and model for the work. The research work computed an average penetration depth of these microplastic particles to be 5.24 m/100 years. This study (O'Connor et al., 2019) is critical in terms of the influence of wetting and drying cycles on microplastic particles and penetration in sandy soils. With reference to geotechnical and foundation design engineering assessments, sandy soils are one of the numerous soil categories in engineering classification (clays, gravels, silts and mixtures of clays, gravels and silts). Therefore, the penetration depth of microplastic particles can be expected to change with several other factors explained in the literature such as void ratio, soil type and hydraulic conductivity.

de Souza MacHado et al. (2018) established experimental studies on loamy sand soils. The research work evaluated the hydraulic conductivity. The study stated that the presence of microplastic particles in loamy sand soils did not significantly change the hydraulic conductivity of the loamy sand soil. In line with the geotechnical and foundation design engineering point of view, sandy soils and loamy sand typically have high hydraulic conductivities, which range between 1.0 and 1.0×10^{-2} cm/s for clean coarse sandy soils, and between 1.0×10^{-2} and $1.0 \times 10^{-4} cm^2/s$ for clayey sand soils (Coduto et al., 2010; Terzaghi et al., 1996). de Souza MacHado et al. (2018) established that the hydraulic conductivity of the sandy soil that was used in the experimental study and program can be high enough to be unimpacted and unaffected by the presence of microplastic particle concentrations and adopted hydraulic conductivity measurements in the study.

Wu et al. (2020b) investigated the vertical transfer and migration response of polystyrene nanoplastics in three normal soils in China with different physicochemical properties. The varied physicochemical properties are salt composition, zeta potentials and ionic strength. The research work reported that pH and soil mineralogy influence the transfer and migration of nanoplastic particles in soil medium. The migration and transfer of nanoplastic particles were reported and documented to be sensitive to cation type and ionic strength. It was reported that several valuable and critical studies evaluated the presence, existence, transfer and migration of microplastic particles in soils from diverse aspects.

In line with the geotechnical and foundation design engineering perspective, researchers need to emphasize and remind the importance of the following studies, factors and research gaps for future research studies:

a. distribution, presence and migration of microplastic particles in non-shallow soil depths of 30.0 cm

b. influence and effects of hydraulic conductivity on spatial distribution, presence and migration of microplastic particles in soils both in vertical and horizontal directions;

c. consideration of the influence and effects of different soil categories utilized in engineering classifications on microplastic particle distribution and transfer in the soils.

All these three research aspects highlighted above deserve further methodical multi-disciplinary research involving geotechnical and foundation design engineering.

16.8 PREVIOUS STUDIES ON THE PERSISTENCE OF MICROPLASTIC PARTICLES IN SOILS

It is acknowledged that microplastic particles are relatively persistent in the soils. The degradation process of microplastic particles involves a very long period of time. The molecular structures of some of these frequently encountered polymers in the soils are presented in Figure 16.9. Cooper and Corcoran (2010) examined plastic particles from five beach soils on Kauai

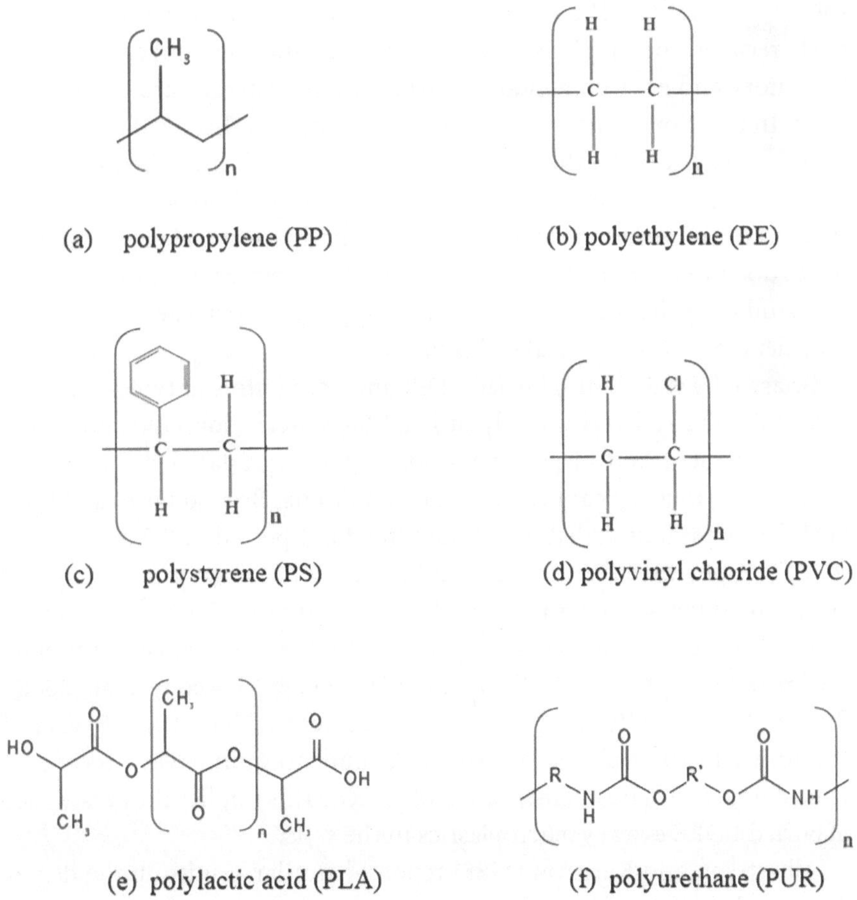

FIGURE 16.9 Molecular structures of some of the frequently confronted polymers in the soils. (Monkul and Özhan, 2021.)

Island, Hawaii. The study cautioned that microplastic particles formed by the disintegration of macroplastic particles remain in the location and environment indefinitely. It was stated that this act cannot be resolved simply by utilizing more rapidly degrading polymer categories, particularly with the fast-tracking trend of plastic utilization. A similar observation was made in the literature by Krueger et al. (2015). It was stated that the present-day synthetic polymers are quite persistent against biodegradation. Some of these polymers have a degradation time of decades or even centuries. This period of time cannot counteract the devastating pollution of the environment and soils with plastics. Krueger et al. (2015) assembled some laboratory and research studies on the biodegradation of synthetic polymers in different environments, which included marine conditions, soil and freshwater. The study stated that most plastic materials are relatively recalcitrant, and have low reaction rates and even under optimized laboratory and environmental conditions these materials are still recalcitrant. In addition, Krueger et al. (2015) documented that available laboratory and experimental studies can be strongly biased to successes attained under optimized environmental and laboratory conditions. This observation has limited the transferability of these findings to real environments. It was documented that there are few relatively long-term field and laboratory studies in the literature, which investigated the persistence of several polymers in real and typical soil conditions.

Briassoulis et al. (2015) buried LDPE mulching films in typical agricultural soil. The soil was initially utilized for watermelon cultivation. The study simulated and observed the long-term degradation behaviour of the material in a typical field and soil conditions. Briassoulis et al. (2015) buried pro-oxidant-added mulching films for a period of 8.5 years in the soil. These pro-oxidants are special additives, which involve primarily carbonyl functional groups and metals blended with some ingredients such as magnesium stearate or cobalt acetylacetonate. These ingredients accelerate the breakdown process of PE (Briassoulis et al., 2015; Reddy et al., 2008). Briassoulis et al. (2015) reported and documented that after 8.5 years of remaining in the soil, the buried LDPE mulching films were recovered almost intact with no disintegration observed. This implies the persistence of buried LDPE even as macroplastics in the soils.

Albertsson and Karlsson (1988) reported similar results on the degradation behaviours of buried LDPE film in laboratory and experimental conditions for a period of 10 years of cultivation period with soil. Otake et al. (1995) evaluated the biodegradation of some polymers when these polymers were buried in the soil for a period of 32 years. The study determined the biodegradation of different synthetic polymer categories in Japanese garden soil from depths 10.0 and 50.0 cm from the ground surface, which included LDPE, polystyrene, PVC and urea-formaldehyde resin buried in soil between 32 and 37 years. The study reported that for PVC, urea formaldehyde and polystyrene resin, no biodegradation was observed after 32 years of study. It was stated that LDPE samples showed some signs of degradation. Otake et al. (1995) stated that the signs of degradation of polymers were more visible for samples collected from shallow soil depths less than 10.0 cm than those collected from soils with relatively deeper levels greater than 50.0 cm. This can be attributed to having more aerobic activities at shallow depths. The higher persistence of microplastic particles in deeper soil layers can be attributed to the smaller microbial community and population available, compared to the microbial community and population available in the shallow depths of the soil. These actions reduced the degradation potential of these materials.

Tabone et al. (2010) examined life cycle, mass from renewable sources, atom economy, biodegradability, the distance to the furthest feedstock, percentage recycled, price, life cycle energy use and health hazards which are the sustainability metrics of the selected (12) polymers. It was stated that biodegradable polymers include polylactic acid and polyhydroxyalkanoate. More data and information on biodegradable polymers are itemized as the green design in the literature. These biodegradable polymers exhibit comparatively large environmental effects and impacts due to the production processes. It was documented that PLA products (in Figure 16.9e) are being progressively utilized in short shelf-life products such as in bags and compostable food-packaging films. It was highlighted that solid wastes that involve PLA products are increasing in the environment. Karamanlioglu and Robson (2013) studied the degradation behaviours of PLA products such as packages through commercial packaging and buried them in soil and compost at a temperature range of between 25°C and 55°C. It was observed that there was no change in the molecular weight and tensile strength after a research period of 1 year at a relatively low-temperature range of between 25°C and 37°C. This observation implies a critical problem of PLA product persistence in the soils. The study observed that at elevated or higher temperatures of 50°C microbes improved the biodegradation process of PLA products, which indicated the importance and significance of soil temperature as an operational factor in the biodegradation process of PLA products.

16.9 DESIGN OF FACILITIES THAT LIMIT THE MOVEMENT OF MICROPLASTIC PARTICLES

Speedy population progress and industrialization are the two crucial factors that contribute to and support environmental and plastic pollution. In addition, invention diversity and variations in consumption characters and habits affect the growth in the volume of solid and liquid waste materials. These solid and liquid waste products are mostly composed of municipal solid wastes. The solid waste materials are used by human beings and are thrown away daily. Some of these solid wastes are partially reused, recycled or burnt (as open burning), with the critical aim of reducing waste materials, but it is not possible to abolish all the particles that will emanate from these wastes (Ghosh and Agamuthu, 2019). It is, therefore, aimed to keep and store these municipal solid wastes in specific and special solid waste disposal facilities that are called municipal solid waste landfills or sanitary landfills with a perfect aim of minimizing the harmful and toxic effects of these wastes on the environment. In these facilities, the wastes are isolated from the groundwater and the subsoil or topsoil environment (Monkul and Özden, 2021).

In order to collect and discharge the surface fluid or flow, a drainage facility or system has to be in place and the sanitary landfill has to be constructed and developed in accordance with standards. These standards are with natural and typical environmental situations and conditions such as growing or planting specific plants on top of sanitary fills or by making the place a social area for public guarding (Hughes et al., 2005). Separation or sorting and classification of these solid wastes is a vital step for making some of these solid wastes reused and recycled, reducing the usage of natural resources. These systems of reuse and recycling prevent environmental and plastic pollution. The separated or sorted solid wastes that are known to have harmful and toxic effects on human health and the environment have to be kept in sanitary landfills that are designed and developed for various purposes with a critical determination of the geotechnical engineering properties of the soil profile, the groundwater and water table situations and conditions, evaluation of the piezometric heads in the confined aquifers, the hydraulic conductivity and permeability of the soil. These engineering properties and classifications of the geochemical situations and conditions play significant roles in the selection of the location or site of a sanitary landfill (Qian et al., 2004). The base or the bottom and the sides of a sanitary landfill have to be protected or covered with a semi-permeable barrier or liner materials with the vital role of preventing or controlling soil and groundwater pollution and contamination.

In an attempt to protect groundwater and soil contamination, compacted soil known as clay liners (CCLs) or geosynthetics are commonly prepared and utilized as pollution barrier materials in sanitary landfills (Qian et al., 2004). The focal aim of putting in place such liner material in a sanitary landfill is to control or prevent the infiltration of the leachate through the barrier materials to the groundwater and subsoil. These compacted soil or clay liners are prepared and composed of natural clay soil or deposits. The hydraulic conductivity or permeability of the compacted soil or the clay depends on and is a function of the properties of the clay such as the method of compaction, clay mineralogy, water content and void ratio of the clay throughout the compaction process. CCLs are frequently compacted at water contents greater than the optimum water content of the clay attained in laboratory conditions with the key objective of achieving lower permeability or hydraulic conductivity in the clay liners. In line with the demand for vital hydraulic conductivity, the hydraulic conductivity of the CCL is correlated with the void ratio of the clay soil. The CCL with the best properties is selected for usage in the sanitary landfill and satisfies several geotechnical standards and criteria. In some of the standards it is known that the hydraulic conductivity of CCL must be 1.0×10^{-9} m/s; the dry weight percentage of the fine soil particles that pass through sieve size of 0.075 mm (sieve number 200) must be 50% with the plasticity index between 7% and 10% and the dry weight percentage of the soil particles that remain on sieve dimension of 4.75 mm (sieve number 4) must be 20% (Sharma and Lewis, 1994).

Geotextiles and geomembranes are the two main geosynthetic materials used for numerous purposes in sanitary landfills. Geotextile materials are flexible textile substances and materials with synthetic fibres. These materials are frequently provided for filtration in a sanitary landfill. The geomembranes materials are thin or small sheets of water-resistant plastic materials. Geomembrane layers are frequently placed over the CCL to provide inflexibility and water resistance against leachates. Geomembranes are the additional lining layers over a CCL that are employed to enhance the barrier material capacity of the clay liner in a solid waste disposal facility. The geomembranes can simply be punctured or damaged by sharp particles that are frequently accessible in solid waste. In addition, installation defects or damages on geomembranes are not infrequent when the construction excellence assurance and quality control measures are not firmly applied (Ben Othmen and Bouassida, 2013). In a situation of such damage, the leachate that involves microplastic particles can flow or seep through into the underlying CCL. Geosynthetic clay liners (GCLs) are lining materials and barrier materials that consist of a thin or smaller bentonite layer sandwiched within two geotextiles (Koermer, 2004). In the last 20 years or decades, sanitary landfill design engineers, including environmental engineers, have chosen to utilize GCLs as an alternative or substitute to CCLs as the barrier materials in solid waste disposal facilities or sanitary landfills (Bouazza, 2002). Furthermore, the GCLs materials have some merits over the CCLs materials such as a lower hydraulic conductivity of less than 10^{-10} m/s, lower and very small thickness of between 4.0 and 10.0 mm, lower initial cost, less labour work and faster and easier installation (Benson and Meer, 2009; Güler et al., 2018). The most significant component of a GCL material that determines the hydraulic and permeability performance of the barrier material is the bentonite layer (Liu et al., 2013; Özhan and Güler, 2013, 2016). A typical solid waste disposal zone and area in Kütahya, Turkey with and without a barrier system, which was composed of a geomembrane-laminated material is presented in Figure 16.10.

The GCLs can be classified and grouped into three main groups based on the production or manufacturing process. The classifications or grouping are as follows:

a. Adhesive-bonded GCLs are composed of bentonite layers attached to the lower and upper geotextiles with water-soluble adhesive materials without any reinforcement materials.

FIGURE 16.10 Solid waste disposal location, site or area in Kütahya, Turkey Label (a) in the figure stands for sanitary landfill without a barrier material for water table protection; Label (b) in the figure stands for sanitary landfill with the barrier material that consisted of a geomembrane-laminated GCL for water table protection. (Monkul and Özhan, 2021.)

b. Needle-punched GCLs are manufactured or produced by punching the needle-like fibre particles from the upper part geotextile through the bentonite layers to the lower part geotextile. This action is due to the reinforcement supply provided by the needle-punching process, the movement and migration of the bentonite from the GCLs are frequently prevented in this category.

c. Stitch-bonded GCLs are another reinforced GCL type. In the stitch-bonded GCLs, the lower part and upper part geotextiles are stitched together with parallel-oriented yarns and materials, which keep the bentonite layers inside the GCLs (Monkul and Özden, 2021)

A typical cross-section of these three different GCL categories and types is presented in Figure 16.11. In typical sanitary landfill liner systems, first, the GCL materials which are utilized as the barrier and lining materials are placed on the subsoil or the natural soil. Geomembrane layers are utilized within the drainage layers and the GCL materials with the aim of protecting the lining materials against probable sharp gravel or solid waste particles that can damage, puncture or break the GCL and act as an impermeable edge between the drainage layer and GCL materials used. The gravel layers with a minimum thickness of 30.0 cm are placed on the geomembrane, which acts as the drainage layers of the sanitary landfill. The effectiveness of the operation of a sanitary landfill depends or is a function of the installation of proper and adequate drainage systems that collect, drain and remove the leachate from the sanitary landfill properly (Monkul and Özden, 2021). The key factor of selection of the transmission pipes

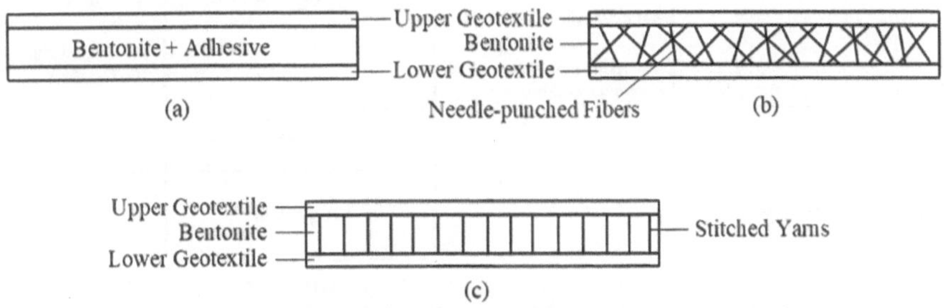

FIGURE 16.11 Cross-sectional views of GCL categories Label (a) in the figure stands for adhesive-bonded GCL; Label (b) in the figure stands for needle-punched GCL; Label (c) in the figure stands for stitch-bonded GCL. (Monkul and Özhan, 2021.)

and the barrier materials for the leachate removal and drainage are the critical steps in the design, development and construction of the sanitary landfill drainage systems. In line with a total collection of the leachate, the drainage systems have to be designed, developed and constructed with an inclination or bed slope of between 0.5% and 1.0% (Giroud and Soderman, 2000; Orsini and Rowe, 2001; Monkul and Özhan, 2021). The core reason for this bed slope or inclination is to provide or support the collection of the leachate by producing the needed hydraulic energy or head on the barrier materials. The hydraulic and volumetric capacities of the drainage systems play a critical and significant role in the determination of the exact inclination and length of the drainage pipes or the transmission pipes. The perforated pipes or collectors, which are designed and installed in the leachate collection systems are used to remove the leachate from the sanitary landfill. The outcome of the design and development of the sanitary system is that the leachate is sent to the surface of the sanitary landfill for purification in the wastewater treatment plant (Monkul and Özhan, 2021). The geotextile materials that are placed on the gravel drainage layer serve as filtration layers detailed is as presented in Figure 16.12. The geotextile materials are utilized to prevent the clogging of the drainage layers with the particles from the solid waste, which includes plastic particles. The solid wastes are dumped into the sanitary landfill and can be kept in a single-liner system as presented in Figure 16.12a. The geotechnical and environmental engineering designs of the sanitary landfill determine the

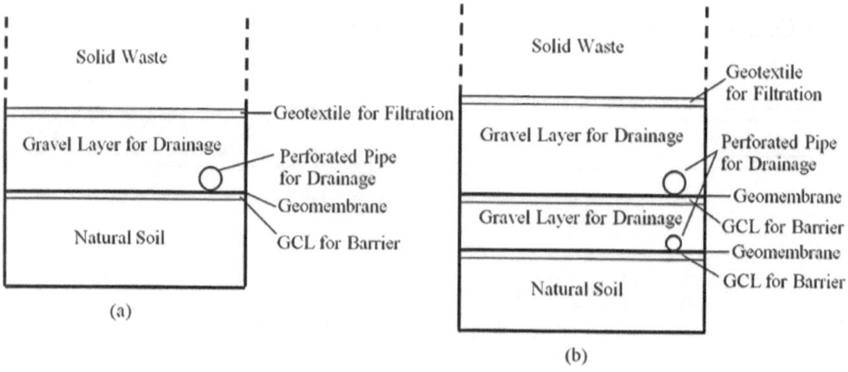

FIGURE 16.12 Cross-sectional view of sanitary landfill liner systems: Label (a) in the figure stands for single-liner system; Label (b) in the figure stands for double-liner system (upper layer for leachate collection and lower layer for leak detection). (Monkul and Özhan, 2021.)

performance, efficacy and durability of the liner system. The demolition debris and construction are typically buried in sanitary landfills that have a single-liner system.

In addition, single-liners are not a favourite to be utilized to store municipal solid wastes. A double-liner system composed of two single-liners is shown in Figure 16.11b. In these double-liner systems, the upper liners are utilized in the collection of the leachate and the lower liners serve as leak-detection systems. The gravel layers in the leachate-collection systems from the sanitary landfill have a height of at least 30.0 cm and the height of the gravel layer in the leak-detection system is equivalent to half of the height of the upper gravel layer that functions for drainage for the sanitary landfill (Monkul and Özhan, 2021). These double-liner systems are commonly utilized in both hazardous waste and municipal solid waste sanitary landfills. It is known that double-liner systems have more lining and barrier components than single-liner systems (Figure 16.12). The critical issue is that there is no assurance and guarantee that the leachate from the sanitary landfill will totally remain in the natural subsoil environment. This indicates that there is still the risk of pollution and contamination of the groundwater and the soil. This leachate and the risk are potential threats for different concentrations and levels of microplastic particle movement and migration into the surrounding soil environment (Monkul and Özhan, 2021).

16.10 PREVIOUS STUDIES ON THE IMPACTS, INFLUENCE AND EFFECTS OF MICROPLASTIC AND NANOPLASTIC PARTICLES ON THE SOIL AND SOIL TERRESTRIAL ANIMALS AND PLANTS

Yue et al. (2023) stated that nanoplastic and microplastic particles are emerging as potential and possible threats to human health and worldwide ecosystems. In the last two decades, the individual and combined effects of nanoplastic and microplastic particles on soil–terrestrial plants and soil systems have attracted increasing consideration. Yue et al. (2023) summarized the origins and sources of nanoplastic and microplastic particles (Figure 16.13) and the interaction between nanoplastic and microplastic particles and coexisting pollutants in the soil environment. The steadily induced multilevel impacts of nanoplastic and microplastic particles on soil properties and terrestrial plants are shown in Figure 16.14. Agroecosystems and soil are key long-term sinks of secondary and primary nanoplastic and microplastic particles with extensive origins and

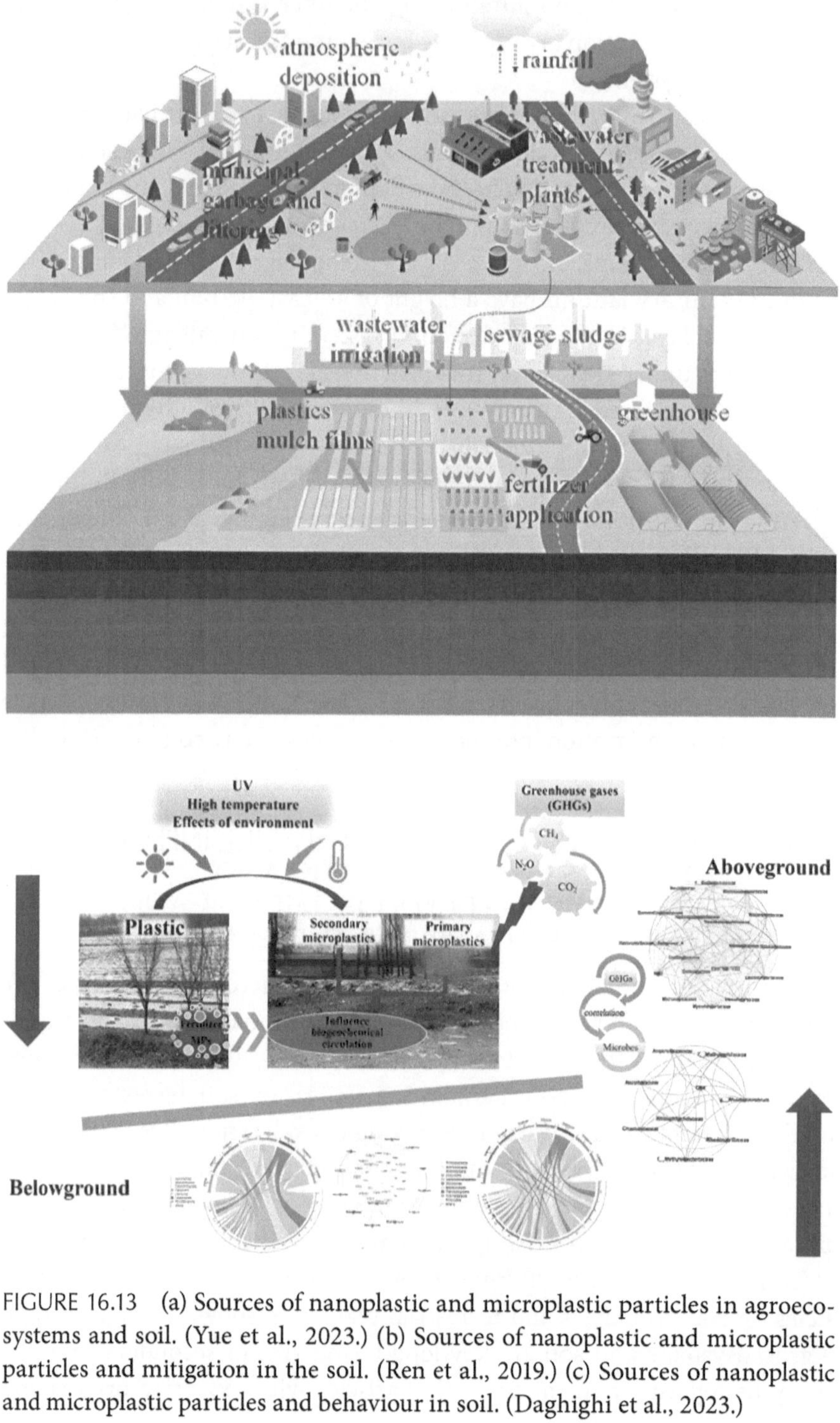

FIGURE 16.13 (a) Sources of nanoplastic and microplastic particles in agroecosystems and soil. (Yue et al., 2023.) (b) Sources of nanoplastic and microplastic particles and mitigation in the soil. (Ren et al., 2019.) (c) Sources of nanoplastic and microplastic particles and behaviour in soil. (Daghighi et al., 2023.)

(Continued)

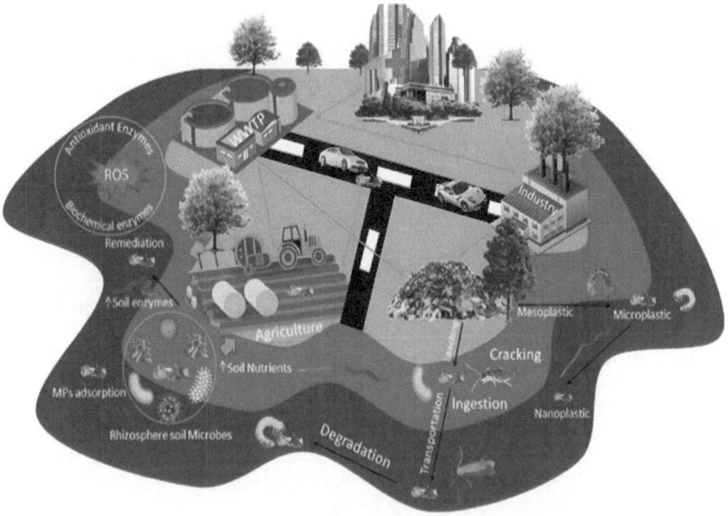

FIGURE 16.13 (*Continued*) (a) Sources of nanoplastic and microplastic particles in agroecosystems and soil. (Yue et al., 2023.) (b) Sources of nanoplastic and microplastic particles and mitigation in the soil. (Ren et al., 2019.) (c) Sources of nanoplastic and microplastic particles and behaviour in soil. (Daghighi et al., 2023.)

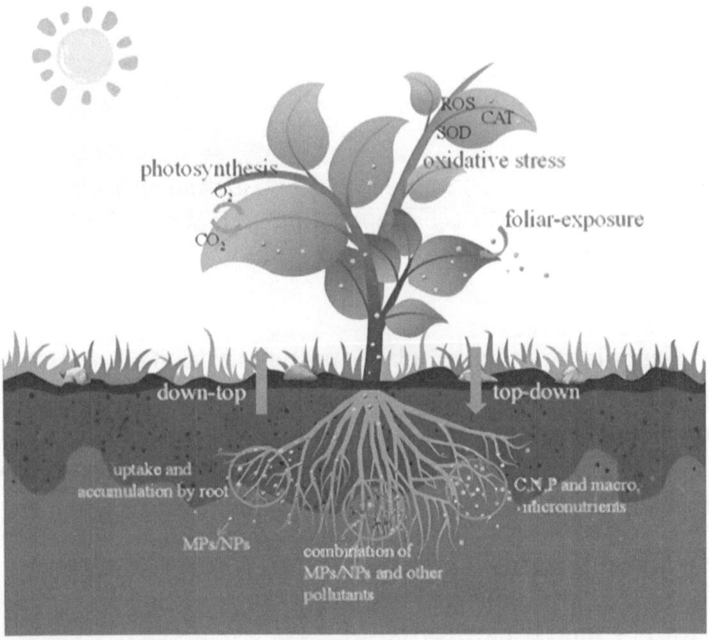

FIGURE 16.14 (a) Sources of nanoplastic and microplastic particles and mitigation in the soil. (Red et al., 2018.) (b) Effects of nanoplastic and microplastic particles on soil properties. (Yue et al., 2023.)

(*Continued*)

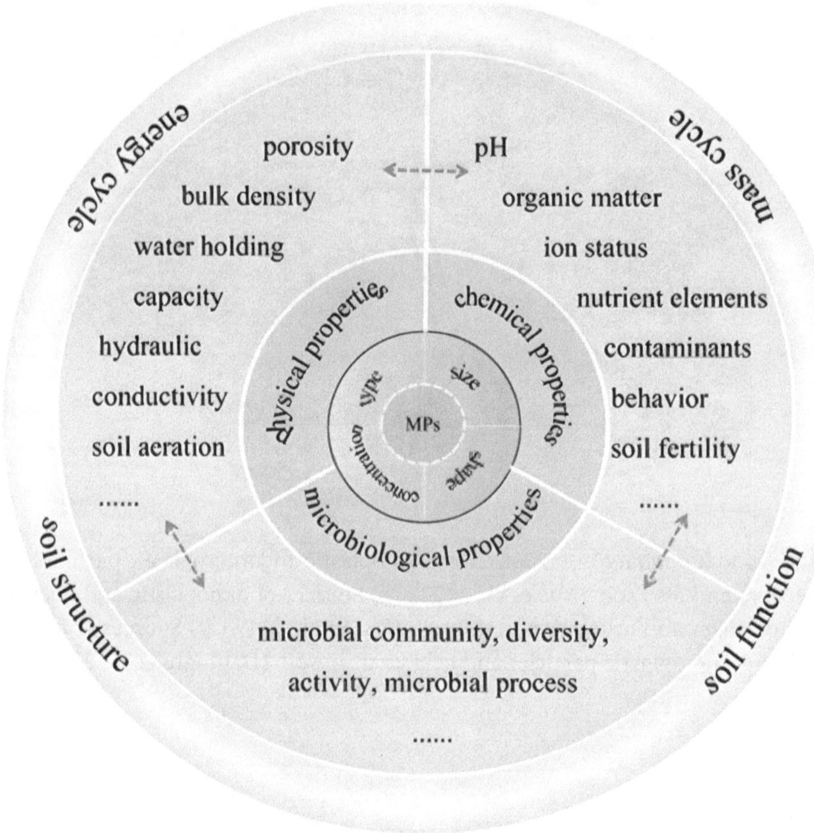

FIGURE 16.14 (*Continued*) (a) Sources of nanoplastic and microplastic particles and mitigation in the soil. (Red et al., 2018.) (b) Effects of nanoplastic and microplastic particles on soil properties. (Yue et al., 2023.)

sources. Nanoplastic and microplastic particles exhibit adsorption capacities. These nanoplastic and microplastic particles can serve as vectors for varied biological contaminants, organic matters and heavy metals. Tables 16.4–16.11 present data and information on the effects of nanoplastic and microplastic particles on various organisms and plants.

16.11 THE RISK OF MICROPLASTIC PARTICLES MIGRATION FROM LANDFILLS TO SOIL AND OTHER ENVIRONMENTS

Plastics and plastic solid wastes constitute an important volume of solid waste. Literature reported that 79% of plastic solid wastes are either kept or stored in sanitary landfills or are unconfined in natural and typical

TABLE 16.4 A Summary of the Effects of Polymers on Soil Protist and Tardigrades Communities

Species Testing	Polymer Type	Size	Concentration	Exposure Conditions	Observed Endpoints	References
Protists free-living ciliated protists	Green, fluorescent amino formaldehyde microspheres	1–5 μm diameter	0%, 0.1%, 1% w/w	Microcosm experiment (petri dish) for 14 days	Keep MP within their food vacuoles at low concentrations (0.1% w/w), and ingestion causes a decline in protist abundance with increasing concentrations (1% w/w)	Kanold et al. (2021)
Protists free-living ciliated protists	PE+As	200 nm (NP) and 200 μm (MP)	0, As (40 mg/kg), MP (2,000 mg/kg), As+MPs (40, 2,000 mg/kg, respectively), NP (200 mg/kg), As+NPs (40, 200 mg/kg, respectively)	Pot experiment (Lactuca sativa L.) for 75 days	The composition and structure of the community and the relative abundance were affected. Plastic presence can increase the adverse effects of As	Zhu et al. (2021)
Soil plastisphere community (bacteria, fungi, and protists)	PE	±1.2 mm	1%	Microcosm experiment during 60 days with *collembola*, potworms and soil plastisphere	Protist time-decay curves (total and abundant taxa) exhibited a bigger reduction without soil mesofauna than with mesofauna, whereas bacterial and fungi showed the opposite trend	Xiang et al. (2022)

(*Continued*)

TABLE 16.4 (*Continued*) A Summary of the Effects of Polymers on Soil Protist and Tardigrades Communities

Species Testing	Polymer Type	Size	Concentration	Exposure Conditions	Observed Endpoints	References
D. discoideum	PE+PS	Nano- (20 and 100 nm) and microplastics (1,000 nm)	25, 250, and 2,500 mg/kg on w/w of PS/water	Agar plates and soil incubation	Adverse impact on fitness and development, including nutrient and energy metabolism. Nano- and microplastic ingestion through phagocytosis but packed and excreted during slug migration—possible PS biodegradation	Zhang et al. (2022)
Tardigrada not mentioned	PE, PP, PVC, PET	20–1,000 µm	1,000 particles/L	Microcosm experiment. Beach sediments	Increase the ecological impacts on meiofaunal assemblages	Corinaldesi et al. (2022)

Source: Daghighi et al. (2023).

NP, nanoplastic; PE, polyethylene; PET, polyethylene terephthalate; PP, polypropylene; PS, polystyrene; PVC, polyvinyl chloride.

TABLE 16.5 A Summary of the Impacts of Microplastics and Nanoplastics on Nematodes

Species Testing	Polymer Type	Size	Concentration	Exposure Conditions	Observed Endpoints	References
Nematodes group of nematodes bacterivores	PP	<250 μm	0%, 0.5%, 1%, 2%, w/w	A pot experiment with corn	Negatively affected soil nematode diversity and ecological functioning	Yang et al. (2022a)
Caenorhabditis elegans	Fluorescently labelled PS	100–1,000 nm	100 μL suspension of $1.0\times109\ mL^{-1}$ of 0.5 μm and $1.0\times108\ mL^{-1}$ of 1.0 μm microspheres in *S*-basal buffer	Agar plates. 30 minutes	Uptake and accumulation in the intestine and pharynx	Kiyama et al. (2012)
	PA, PE, PP, PVC Fluorescently labelled PS	<0.70, 0.1, 1.0 and 5.0 μm	MP suspension in K-medium (32 mM/L KCl 51 mM/L) was added to the nematode growth medium at different concentrations (0.5, 1.0, 5.0 and 10.0 mg/m^2)	Agar-padded slides 2-day	Inhibition of growth, survival and reproduction, decreased intestinal calcium levels, microplastic accumulation in the intestine, and oxidative stress (increased expression of Glutathione *S*-transferase 4)	Lei et al. (2018a)
	PS	100, 500 nm (NanoPS) and 1, 2 and 5 μm (MP)	1 mg/L	3 days	The lowest survival rate, decreased body length and shortest lifespan for those exposed to the 1.0 μm group reproduction	Lei et al. (2018b)

(Continued)

TABLE 16.5 (*Continued*) A Summary of the Impacts of Microplastics and Nanoplastics on Nematodes

Species Testing	Polymer Type	Size	Concentration	Exposure Conditions	Observed Endpoints	References
	Nano-PS	50 and 200 nm	17.3 and 86.8 mg/L	Agar-padded slide. 1 day.	Induction of oxidative stress (ROS production). Changes in energy metabolism (reduction of TCA cycle intermediates, lactic acid and glucose). Uptake and accumulation of nanoplastics in the intestine	Kim et al. (2020)
	HDPE, PET, PP, PS, LDPE, PAN	<250–1,000 µm	0.001–1% w/w	Acute toxicity test (24 hours)	Decrease Nematode offspring. PET and PAN show the highest toxicity—less toxicity under HDPE, PP and PS treatments. LDPE treatment was not toxic. Acute toxicity was mainly attributed to the extractable additive	Kim et al. (2020)
	TiO_2-Nano-PS	108.2±4.5 nm	0.01, 0.1, or 1 µg/L by 1% solid suspension in water	Agar-padded slide. 1 day	Inhibition of locomotion and induction of intestinal ROS production	Dong et al. (2018)
Nematodes community	LDPE	0.3–400 µm	+5, +10 and +15 g/m	Field experiment for 287 days	Nematode abundance and Changes in the community composition are significantly affected by microplastic additions	Lin et al. (2020)

(*Continued*)

TABLE 16.5 (*Continued*) A Summary of the Impacts of Microplastics and Nanoplastics on Nematodes

Species Testing	Polymer Type	Size	Concentration	Exposure Conditions	Observed Endpoints	References
C. elegans	LDPE, PLA, PBAT	LDPE: $57 \pm 40\,\mu m$. PLA and PBAT: $40 \pm 31\,\mu m$	0, 1, 10, and 100 mg MP/L	Agar plates (~6 days)	Nematodes had fewer offspring. The decline was independent of the plastic type	Schöpfer et al. (2020)
	PS and PS-COOH	1.018 (PS) and 1.021 (PS–OOH) μm	0.1–100 μg/L	Exposure for 24 hours	PS-COOH is highly neurotoxic, including damage to neurones	Yu et al. (2022)
	PE microbeads and different organic contaminants (ibuprofen, sertraline, amoxicillin, and simazine)	212–300 μm	Dose of 0.1% w/w of MPs	Exposure for 4 and 30 days	Did not affect growth, reproduction, or survival	Fajardo et al. (2022)
C. elegans, Acrobeloides nanus, Plectus acuminatus	PS microspheres	1 μm	Suspensions 10^{-7} PS beads/mL	Nematode growth gelrite. 49 days	Ingestion of 1.0-μm PS beads by three nematode species. Significant differences in population growth	Müller et al. (2012)
C. elegans, A. nanus, Panagrolaimus thienemanni, P. acuminatus, Poikilolaimus regenfussi	PS beads	>0.5, 1, 3 and 6 μm	PS suspension. 10^7 and 3×10^6 PS beads/mL	4, 24 and 72 hours	Ingestion by nematodes with a buccal cavity, and species-specific. Influence of concentration and exposure on ingestion	Fueser et al. (2019)

(*Continued*)

TABLE 16.5 (*Continued*) A Summary of the Impacts of Microplastics and Nanoplastics on Nematodes

Species Testing	Polymer Type	Size	Concentration	Exposure Conditions	Observed Endpoints	References
C. elegans	NanoPS SiO_2-NPs, Al_2O_3-NPs and TiO_2-NPs	30 nm	0, 0.1, 1, 10 and 100 µg/L	Not mentioned	NP is more toxic than contaminated by metal oxides	Li et al. (2020)
	NanoPS + chlordane and hexachlorocyclo-hexane	100 nm	NanoPS: 1.0 mg/L, contaminants (0.1–10 mg/L)	Exposure for 72 hours	Increased toxicity under combined exposure (lifespan, increased oxidative stress)	Li et al. (2020)
	NanoPS	20 and 100 nm	0.1–100 µg/L	Exposure for 6.5 days	Smaller size have more toxicity (transgenerational toxicity) than the bigger size	Liu et al. (2021)
	NanoPS	100 nm	1–100 µg/L	Exposure for 6.5 days	Activation of mitochondrial unfolded protein response	Liu and Wang (2021)
	NanoPS, amino-modified NanoPS	35 nm	1, 10, 100, and 1,000 µg/L	Exposure for 2 days	Damage on gonad development and reproductive capacity. amino-modified NanoPS showed higher toxicity than pristine PS	Qu et al. (2018)
	NanoPS	100 nm	0, 1, 10, 100, 1,000, and 10,000 µg/L	Exposure for 4.5 days	Transgenerational toxicity on concentrations >100 µg/L. Probable translocation into the gonad	Zhao et al. (2017)
	PS	1 µm	0, 0.1, 1, 10, and 100 µg/L	Exposure for 72 hours in liquid solutions	Adverse physiological effects. Oxidative stress response on higher concentrations. Intestine accumulation	Yu et al. (2020)

Source: Daghighi et al. (2023).

LDPE, low-density polyethylene; NanoPS, nanopolystyrene; PBAT, poly(butylene adipate-co-terephthalate; PE, polyethylene; PLA, biodegradable polymers polylactide; PS, polystyrene; PS–COOH, carboxyl-modified polystyrene microplastics.

TABLE 16.6 A Summary of the Impacts and Influences of Microplastics and Nanoplastics on Potworms

Species Testing	Polymer Type	Size	Concentration	Exposure Conditions	Observed Endpoints	References
				Enchytraeid E. crypticus		
	PA	30 µm	1,000 mg/kg (dry weight)	Agriculture soil. 21 days	No mortality. Increased reproduction	Ma et al. (2020a)
	PVC			Agriculture soil. 21 days	No mortality. Significantly reduced reproduction	
	PA+TC		1,000 mg/kg (PA) and 20 mg/kg soil (TC) (dry weight)	Agriculture soil. 21 days	No mortality. Significantly reduced reproduction. Tetracycline and microplastics disturbed the microbial community in *E. crypticus*. Increased TC bioaccumulation	
	PVC+TC		1,000 mg/kg (PVC) and 20 mg/kg soil (TC) (dry weight)	Agriculture soil. 21 days	No mortality. Significantly reduced reproduction. Tetracycline and microplastics disturbed the microbial community in *E. crypticus*. Increased TC bioaccumulation	
	PES	12 µm–2.87 mm to 4–24 mm	0%, 0.02%, 0.06%, 0.17%, 0.5% and 1.5% w/w	LUFA soil type no. 2.2.21 days	Survival was slightly decreased only at moderate fibre concentrations (0.17%–0.5%) Long fibres in soil negatively affected the reproduction at all concentrations except for 0.06%	Selonen et al. (2020)
	Tyre particles	80–110 µm	0%, 0.02%, 0.06%, 0.17%, 0.5% and 1.5% w/w	LUFA soil type no. 2.2.21 days	Slight decrease in the reproduction of *E. crypticus* was not dose-dependent	Selonen et al. (2021)
	PS	50–100 nm	Eq. 1,000 mg/kg	Petri dish+Oats	No mortality	

(*Continued*)

TABLE 16.6 (*Continued*) A Summary of the Impacts and Influences of Microplastics and Nanoplastics on Potworms

Species Testing	Polymer Type	Size	Concentration	Exposure Conditions	Observed Endpoints	References
	PS+TC	50–100 nm	(TC 10 mg/kg, 4 Ml nanoscale PS solution (2.5% w/v))	Contamination. 14 days.	No mortality. Loss weight. PS and TC together reversibly perturb the microbial community of *E. crypticus*. TC bioaccumulation	Ma et al. (2020b)
				E. crypticus		
	PA	13–18 and 90–150 μm	20, 50, 90 and 120 g/kg	LUFA soil type no. 2.2.21 days	No effects on survival. Reproduction significantly reduced. Ingestion of PA particle	Lahive et al. (2019)
	PA	63–90 μm	90 g/kg		No effects on survival. Reproduction significantly reduced. Ingestion of PA particle	
	PVC	106–150 μm	90 g/kg		No effects on survival. Reproduction significantly reduced	
	Tyre particles	13–1,400 μm	0%, 0.0048%, 0.024%, 0.12%, 0.6%, and 3% of dry soil weight	Fluvo-aquic soil. 21 days	Adverse effects on survival, reproduction and disturbed the microbiota	Ding et al. (2020)
	HDPE	4 mm	0%, 2%, 4% or 8% w/w	Turf-free soil (20% lingo fibres, 35% cocopeat washed, 10% spelt fermented, and 35% substrate compost). 2	Avoided plastic-spiked soil. Slight mortality at higher concentrations (8%) oxidative stress response (an increase of catalase and glutathione *S*-transferase activities)	Pflugmacher et al. (2020)

(*Continued*)

TABLE 16.6 (*Continued*) A Summary of the Impacts and Influences of Microplastics and Nanoplastics on Potworms

Species Testing	Polymer Type	Size	Concentration	Exposure Conditions	Observed Endpoints	References
	PS fibres + Ag nanoparticles	50–3,000 μm	PS fibres (0.01% dry weight) + Ag nanoparticles (32, 100, 320, 1,000, 3,200 mg Ag/kg)	LUFA soil type no. 2.2.21 days	No significative effects on reproduction. Microplastic fibre influences on Ag toxicity (slight uptake and bioaccumulation of Ag at higher levels with PS)	Tourinho et al. (2021)
	PES	12–2,870 μm	0.02%, 0.06%, 0.17%, 0.5% and 1.5% w/w via soil and food		Ingestion and excretion	Mendes (2021)
	Nano-PS	100 nm	Nanoscale PS (eq. 1,000 mg/kg)	Agricultural soils. 21 days	No effects on reproduction or body weight. NanoPS disturbed the gut microbiome	Yang et al. (2022b)
	Nano-PS + TC	100 nm	(TC 20 mg/kg, L nanoscale PS (eq. 1,000 mg/kg))	Agricultural soils. 21 days	Highest body weight. Significative effects for reproduction. NanoPS disturbed the gut microbiome, enhanced the toxicity of TC and promoted the enrichment of antibiotic genes resistance	

Source: Daghighi et al. (2023).

PA, Polyamides; PES, Polyester; PVC, Polyvinyl chloride; PS, Polystyrene; TC, Tetracycline. *Tyre particles (usually generic reference, but they're a mixture of synthetic rubbers including styrene-butadiene rubber (SBR), butadiene rubber (BR), and butyl rubber (IIR)) (Ding et al., 2020; Selonen et al., 2021, 2023).

HDPE; high-density polyethylene; NanoPS, nanopolystyrene; PA, polyamides; PES, polyester; PVC, polyvinyl chloride; PS, polystyrene; TC, tetracycline. *Tyre particles (usually generic reference, but they're a mixture of synthetic rubbers including styrene-butadiene rubber (SBR), butadiene rubber (BR), and butyl rubber (IIR)).

TABLE 16.7 A Summary of the Impacts and Influences of Microplastics and Nanoplastics on Springtails

Species Testing	Polymer Type	Size	Concentration	Exposure Conditions	Observed Endpoints	References
			Springtails F. candida			
	UF, PET	<100 and 100–200 μm	2.5 mg of the <100 μm 5 mg of the 100–200 μm fraction	Direct exposure to Petri dishes. Without substrate. 7 days	Ability to transport and distribute MP	Maaß et al. (2017)
	PVC	80–250 μm	0 and 1 gr MP/kg soild	OECD artificial soil. 56 days	Inhibition of growth and reproduction. Changes in gut microbiota and in the carbon and nitrogen elemental absorption	Zhu et al. (2018a)
	PVC	80–250 μm	5,000 items/dish	Direct exposure to Petri dishes. Without substrate. 7 days	Transport and distribution of MP up to 9 cm.	Zhu et al. (2018b)
	PE	<50–500 μm	0.5%–1% w/w	Artificial soil. 28 days	Avoidance of plastic-spiked soils. Reproduction significantly inhibited gut microbiota alteration	Ju et al. (2019)
	PS	2–2.9 μm	0%, 1% w/w	90 mm × 15 mm glass. Petri dish microorganism	Altered gut microbiome and antibiotic resistance in exposed to mixed contamination	Xiang et al. (2019)

(Continued)

TABLE 16.7 (*Continued*) A Summary of the Impacts and Influences of Microplastics and Nanoplastics on Springtails

Species Testing	Polymer Type	Size	Concentration	Exposure Conditions	Observed Endpoints	References
	PS+Sulfamethoxazole		0% and 1% w/w (PS)+0% and 1% mg/L		Altered gut microbiome and antibiotic resistance in exposed to mixed contamination	
	PES	12 μm–2.87 mm to 4–24 mm	0%, 0.02%, 0.06%, 0.17%, 0.5% and 1.5% w/w	LUFA soil type no. 2.2.28 days	There were no differences in springtail survival or reproduction	Selonen et al. (2020)
	PP (face mask)	<300 μm	1,000 mg/kg dry soil	LUFA soil type no. 2.2.28 days	No adverse effects on survival, esterase activity or oxidative stress. Significative reduction on reproduction and juvenile growth	Kwak and An (2021)
			F. candida			
	Tyre particles[a]	80–110 μm	0%, 0.02%, 0.06%, 0.17%, 0.5% and 1.5% w/w	LUFA soil type no. 2.2.28 days	Tyre particles slightly decreased reproduction and survival	Selonen et al. (2021)
	PES+chlorpyrifos	220±200 μm	0.05% w/w and chlorpyrifos: nominal concentrations of 0.01, 0.03, 0.11, 0.33 and 1.0 mg/kg	LUFA soil type no. 2.2.28 days	Microplastics did not affect chlorpyrifos toxicity to springtail reproduction. Tyre particles significantly decreased the chlorpyrifos-induced mortality of springtails, while polyester fibres did not	Selonen et al. (2023)

(*Continued*)

TABLE 16.7 (*Continued*) A Summary of the Impacts and Influences of Microplastics and Nanoplastics on Springtails

Species Testing	Polymer Type	Size	Concentration	Exposure Conditions	Observed Endpoints	References
	Tyre particles[a] + chlorpyrifos	80–110 μm	0.05% w/w and chlorpyrifos: nominal concentrations of 0.01, 0.03, 0.11, 0.33 and 1.0 mg/kg	LUFA soil type no. 2.2.28 days	Microplastics did not affect chlorpyrifos toxicity to springtail reproduction. Tyre particles significantly decreased the chlorpyrifos-induced mortality of springtails, while polyester fibres did not	Selonen et al. (2023)
	PP + NH_4NO_{3+} $CdCl_2$ + E. coli	<100 and 100–200 μm	50 mg	Soil (origin not indicated)	Ability to transport MP on soils, including contaminated MP with nutrients (N), contaminants (Cd) and microorganisms (E. coli)	Luo et al. (2023)
Lobella sokamensis (Deharveng and Weiner, 1984)	PE and PS	0.47–0.53, 27–32, and 250–300 μm	Several concentrations (4–1000 mg/kg)	LUFA soil type no. 2.2 <1 day	Decreased locomotion.	Kim and An (2019)
Proisotoma minuta	UF, PET	<100 and 100–200 μm	2.5 mg of the <100 μm 5 mg of the 100–200 μm fraction	Direct exposure to Petri dishes. Without substrate. 7 days	Ability to transport and distribute MP	Maaß et al. (2017)
Cryptopygus antarcticus	PS	<100 μm		Field collection	Plastic ingestion	Bergami et al. (2020)

PE, polyethylene; PES, polyester; PET, polyethylene terephthalate; PP, polypropylene; PS, polystyrene; PVC, polyvinyl chloride; UF, urea-formaldehyde.

[a] Tyre particles (usually generic reference, but they're a mixture of synthetic rubbers including styrene-butadiene rubber (SBR), butadiene rubber (BR), and butyl rubber (IIR)).

TABLE 16.8 A Summary of the Impacts and Influences of Microplastics and Nanoplastics on Mites

Specie Testing	Polymer Type	Size	Concentration	Exposure Conditions	Observed Endpoints	References
Mites						
Diverse mites' species	PE (plastic bottles)	<4,800	0%, 5%, 30%, 60% and 90% w/w (manure)	Field experiment. Forest/agricultura area. Microcosm experiment	Detrimental effects on the abundance of mites (>60%)	Stamatiadis and Dindal (1990)
	PS (packing material)	>2,000	0%, 5%, 30%, 60% and 90% w/w (manure)	Field experiment. Forest/agricultura area. Microcosm experiment	Detrimental effects on the abundance of mites (>60%)	Stamatiadis and Dindal (1990)
Hypoaspis aculeifer	PVC	80–250 μm	5,000 items/dish	Direct exposure to Petri dishes	Passive transport	Zhu et al. (2018b)
			Damaeus exspinosus			
Diverse mites: Tectocepheus velatus	PP, PES	Microfibres (length) 2–3 mm or 5–6 mm; diameter of fibres 22.92 ± 0.17 μm for PE and 33.33 ± 0.07 μm for PP fibres	0.4% w/w	MP addition to microcosm experiment (550 mL volume)	Oribatid species richness was not affected by MP addition. No effects on reproduction (new juveniles)	Barreto et al. (2020)

(*Continued*)

TABLE 16.8 (*Continued*) A Summary of the Impacts and Influences of Microplastics and Nanoplastics on Mites

Specie Testing	Polymer Type	Size	Concentration	Exposure Conditions	Observed Endpoints	References
Oppiella nova *Moritzoppia unicarinata*	PP, PES	Microfibres (length 2–3 mm or 5–6 mm; diameter of fibres 22.92 ± 0.17 µm for PE and 33.33 ± 0.07 µm for PP fibres	0.4% w/w	MP addition to microcosm experiment (550 mL volume)	Oribatid species richness was not affected by MP addition. No effects on reproduction (new juveniles)	Barreto et al. (2020)
Suctobelbella sp. *Scheloribates sp.* *Nr. Laevigatus, Minuthozetes semirufus*	PP, PES	Microfibres (length 2–3 mm or 5–6 mm; diameter of fibres 22.92 ± 0.17 µm for PE and 33.33 ± 0.07 µm for PP fibres	0.4% w/w	MP addition to microcosm experiment (550 mL volume)	Oribatid species richness was not affected by MP addition. No effects on reproduction (new juveniles)	Barreto et al. (2020)
Trichoribates novus, Eupelops curtipilus	PP, PES	Microfibres (length 2–3 mm or 5–6 mm; diameter of fibres 22.92 ± 0.17 µm for PE and 33.33 ± 0.07 µm for PP fibres	0.4% w/w	MP addition to microcosm experiment (550 mL volume)	Oribatid species richness was not affected by MP addition. No effects on reproduction (new juveniles)	Barreto et al. (2020)
O. nitens	PES	12 µm–2.87 mm to 4–24 mm	0.5 w/w	LUFA soil type no. 2.2.28 days	There were no differences in survival and reproduction.	Selonen et al. (2020)

PE, Polyethylene; PES, Polyester; PP, Polypropylene; PVC, Polyvinyl chloride.

TABLES 16.9 A Summary of the Impacts and Influences of Microplastics and Nanoplastics on Terrestrial Plants

	MPs/NPs			
Plant Species	**Type**	**Size**	**Concentration**	**Main Effects**
Cress (Lepidium sativum L.)	Green fluorescent plastic	50, 500, 4,800 nm	10^3–10^7 particles/mL	MPs exposure resulted in short-term and transient effects on germination rate and root growth
Cucurbita pepo L	PE, PVC, PP, PET	40–50 µm	0.02%, 0.10%, 0.20%	MPs impaired root and shoot growth and influenced leaf size, chlorophyll content, photosynthetic efficiency and micro- and macro elemental profile; PVC was the most toxic and PE was less toxic
Vicia faba L.	PS	100 nm, 5 µm	10, 50, 100 mg/L	5 µm PS decreased biomass and CAT enzyme activity, increased SOD and POD enzyme activity; 100 nm PS (100 mg/L) decreased growth; 100 nm PS induced higher genotoxic and oxidative damage than 5 µm PS; 100 nm PS accumulated in root
Allium cepa L.	PES fibres	1.70 µm	0.4% (w:w)	PES fibre ES fiberrss increased aboveground biomass
Lettuce (Lactuca sativa L.)	PVC	a: 100 nm–18 µm b: 18–150 µm	0.5%, 1%, 2%	0.5% a and 1% a increased the total length, surface area, volume, and diameter of roots; 1% a increased the SOD activity; PVC-a was related to photosynthesis, PVC-b was correlated with root morphology
Soybean (Glycine max (L.) Merr.)	PE, (Bio) mulch film	2×2 cm, 1×1 cm, 0.5×0.5 cm debris	0%, 0.1%, 0.5%, 1%	PE reduced plant height, culm diameter, leaf area and root/shoot ratio while Bio debris showed adverse effects on germination viability and root biomass
Cucumber (Cucumis sativus L.)	PS	100, 300, 500, 700 nm	50 mg/L	300 nm PS significantly increased root activity MDA and root proline content; PS significantly increased soluble protein in cucumber fruits; decreased the levels of Mg, Ca and Fe, and the effect depends on PS particle sizes

(Continued)

TABLES 16.9 (*Continued*) A Summary of the Impacts and Influences of Microplastics and Nanoplastics on Terrestrial Plants

	MPs/NPs			
Plant Species	**Type**	**Size**	**Concentration**	**Main Effects**
Wheat (Triticum aestivum L.)	PS	100 nm	0.01–10 mg/L	Enhanced wheat seedling growth, growth parameters and chlorophyll content reduced the shoot-to-root biomass ratio and micronutrient contents, altered metabolic profiles
Lettuces (L. sativa L.)	PS	93.6 nm	0, 0.1, 1 mg/L	Decreased the dry weight, height and leaf area, plant pigment content and nutritional quality, produced oxidative stress
Spring onion (Allium fistulosum L.)	PA beads: 15–20 μm PES fibres: 5,000 μm length, 8 μm diameter PEHD, PP: 2–3 mm spheres PS, PET: 2–3 mm cylinders		PES: 0.2% Others: 2.0%	Have different effects on plant performance including plant biomass, tissue elemental composition, root traits, and soil microbial activities depending on particle types
Common bean (Phaseolus vulgaris L.)	LDPE, Bio	250–500 μm, 500–1,000 μm	0.5%, 1.0%, 1.5%, 2.0%, 2.5% (w/w)	Effect of LDPE-MP depends on its concentration: ≥1.0% showed significantly higher specific root nodules, 2.5% showed significantly higher specific root length, 1.0% caused higher leaf area and 0.5% caused lower leaf relative chlorophyll content Bio-MP treatments showed significantly higher specific root length and specific root nodules, lower shoot, root and fruit biomass
Garden cress (L. sativum L.)	PP, PE, PVC, PE+PVC	<0.125 mm	184 mg/kg	Caused negative effect on biometric traits, depending on MPs' types and exposure time
Wheat (T. aestivum L.)	MacroLDPE: 6.92×6.10 mm MacroBio: 6.98 mm×6.01 mm Micro: 50 μm^{-1} mm	1% (w/w)		Affected above-ground and below-ground parts, biodegradable plastic residues showed stronger negative effects than PE
Flowering Chinese cabbage (Brassica rapa L.)	PS	70 nm, 5 μm	10 mg/kg	PS influenced the plant photosynthesis and growth depending on MP size

(*Continued*)

TABLES 16.9 (*Continued*) A Summary of the Impacts and Influences of Microplastics and Nanoplastics on Terrestrial Plants

	MPs/NPs			
Plant Species	**Type**	**Size**	**Concentration**	**Main Effects**
Tomato (Lycopersicon esculentum L.)	PS, PP, PE	52–368 μm	10, 100, 500, 1,000 mg/L	MPs (≤500 mg/L) had inhibitory effects on seed germination, and then alleviated under 1,000 mg/L conditions; PE was more toxic to seedling growth than PS and PP
Arabidopsis (Arabidopsis thaliana L.) Wheat (T. aestivum L.)	PS	40 nm, 1 μm	8.3×1,011 n/mL 5.3×107 n/mL	PS accumulated at Arabidopsis and wheat root cap cells
Maize (Zea mays L. var. Jubilee)	PE microbeads	3 μm	0.0125 mg/L 100 mg/L	PE bioaccumulation in the rhizosphere decreased transpiration, nitrogen content, and growth; PE may accumulate in the rhizosphere, impairing water and nutrient uptake, and eventually reaching root eaters
Arabidopsis (A. thaliana L.)	PS-SO_3H: 55 nm PS-NH_2: 71 nm	0.3, 1.0 g/kg 10, 50, 100 μg/mL		Decreased above-ground biomass, seedling growth, root elongation Arabidopsis can take up and transport PS
Rice (Oryza sativa L.)	BM and PE mulch film	50 μm	1% (w/w)	Reduced the height and dry weight of rice plants, induced oxidative stress; caused negative effects on the growth of rice plants via nitrogen metabolism and photosynthesis
Wheat (T. aestivum L.)	PVC, PE	125 μm	1%, 5%, 10%, 20%	MPs had a negative, dose-dependent impact on plant growth affecting both above- and below-ground productivity

Source: Yue et al. (2023).

Bio: biodegradable plastic; (Bio) mulch film, biodegradable plastic mulch film; BM, PBAT based biodegradable mulch film; CAT, catalase; LDPE, low-density polyethylene; MDA, malondialdehyde PA, polyamide; PBAT, butyleneadipate-co-terephthalate; PE, polyethylene; PEHD, polyethylene high density; PES, polyester; PET, polyethylene terephthalate; POD, peroxidase; PP, polypropylene; PS, polystyrene; PS-NH_2, amino-modified polystyrene nanoparticles; PS-SO_3H, sulfonic-acid-modified polystyrene nanoparticles; PVC, polyvinyl chloride; SOD, superoxide dismutase.

TABLE 16.10 A Summary of the Translocation and Uptake of Microplastics and Nanoplastics by Terrestrial Plants

	MPs/NPs			
Plant Species	**Type**	**Size**	**Concentration**	**Main Effects**
Cress (Lepidium sativum L.)	Green fluorescent plastic	50, 500, 4,800 nm	10^3–10^7 particles/mL	PS caused physical blockage of the pores in the seed capsule
Mung bean (Vigna radiata (L.) Wilczek)	Carboxylate-modified polystyrene	28 nm	0, 10, 100 mg/kg dry soil	NPs were detected in leaves
Carrots (Daucus carota var. sativa Hoffm.)	PS	0.1–1 μm 5 μm	10 and 20 mg/L	1 μm PS can enter carrot roots and accumulate in the intercellular layer but are unable to enter the cells; 0.2 μm PS can migrate to the leaves.
Vicia faba L.	PS	50 nm	0.01, 0.1 1 g/L	PS could internalize into different external compartments
Italian lettuce (Lactuca sativa L.), radish (Raphanus sativus L.), wheat (Triticum aestivum L.) and corn (Zea mays L.)	PS	100 nm, 5 μm	1, 10 mg/L	Fluorescent nano-PS in the roots or germs of the tested crops suggests that nanoplastics can be taken up by plants even at a very early growth stage
V. faba L.	PS	5 mm, 100 nm	10, 50, 100 mg/L	100 nm PS accumulated in root and most probably blocked cell connections or cell wall pores
Pea (Pisum sativum L.)	PS	20 nm	20, 40 mg/kg	MP translocation in cell wall of vascular bundle
Wheat (T. aestivum L.) Lettuce (L. sativa L.)	PS, PMMA	0.2 μm, 2 μm	50 mg/L	PS and PMMA penetrated the stele via the crack-entry mode, transpirational pull was the main driving force
Wheat (T. aestivum L.)	PS	0.2 μm	0.5 mg/g	Uptake of PS into the root outer cortical exosome space and vascular tissue, transported to the aboveground stem vascular bundle and leaf vascular tissue
Cucumber (Cucumis sativus L.)	PS	100, 300, 500, 700 nm	50 mg/L	PS initially accumulated in the root system, and then was transported to the aboveground parts. PS was distributed in the leaves, flowers, and fruits, through the stems

(Continued)

TABLE 16.10 (*Continued*) A Summary of the Translocation and Uptake of Microplastics and Nanoplastics by Terrestrial Plants

	MPs/NPs			
Plant Species	**Type**	**Size**	**Concentration**	**Main Effects**
Wheat (T. aestivum L.)	PS	100 nm	0.01–10 mg/L	PS was taken up by wheat roots and subsequently down-top transported to leaves via xylem pathways
Lettuce (L. sativa L.)	PS	93.6 nm	0, 0.1, 1 mg/L	Possible absorption of PSNPs through leaves stoma and the translocation downwards to plant roots
Rice (Oryza sativa L.)	PS	80 nm, 1 μm	40 mg/L	Both nano- and micro-sized PS could be absorbed by rice roots and translocated to aerial parts, apoplastic transport may be the main pathway
Wheat (T. aestivum L.) *Lettuce (L. sativa L.)*	PS, PS-Eu	200 nm	0–5,000 μg/L	MPs accumulated mainly in the roots, while transport to the shoots was limited
Arabidopsis (A. thaliana L.)	PS-SO_3HPS-NH_2	55 nm 71 nm	0.3, 1.0 g/kg 10, 50, 100 μg/mL	PS can accumulate in Arabidopsis thaliana, depending on surface charge
Maize (Z. mays L.)	PS-COOH PS-NH_2	22.0 ± 1.5 nm 24.0 ± 2.2 nm	0, 10, 50, 100, 200, 400, 500 ng/spot	PS could accumulate on maize leaves PS in the leaves would transfer to the vasculature mainly through stomatal opening and move down to the roots through vascular bundle
Wheat (T. aestivum L.)	PS-NH_2PS-NH_2PS-COOHPS-COOH	38.3 nm 191.2 nm 34.4 nm 101.2 nm	20 mg/L	PS could be taken up by wheat root and cells, and the translocation is dependent on particle size and surface characterization

Source: Yue et al. (2023).

PMMA, polymethylmethacrylate; PS, polystyrene; PS-Eu, polystyrene (PS) particles doped with the europium chelate Eu–β-diketonate; PS-COOH, carboxy-modified polystyrene nanoparticles; PS-NH_2, amino-modified polystyrene nanoparticles; PS-SO_3H, sulfonic-acid-modified polystyrene nanoparticles.

TABLE 16.11 A Summary of the Combined Impacts and Influences of Microplastics and Nanoplastics and Coexisting Pollutants on Terrestrial Plants

	MPs/NPs				
Plant Species	**Type**	**Size**	**Concentration**	**Coexisting Pollutants**	**Combined Effect**
Carrots (Daucus carota var. sativa Hoffm.)	PS	0.1–1 μm 5 μm	10, 20 mg/L	As: 1, 2, 4 mg/L	As increases the negatively charged area of PS and causes a greater amount of microplastics to enter the carrot; As exacerbates the effect of PS on carrots
Rice (Oryza sativa L.)	PS, PTFE	10 μm	0.04, 0.1, 0.2 g/L	As: 1.6, 3.2, 4.0 mg/L	PS and PTFE reduced As uptake, and absorbed As decreased with the increasing concentration of microparticles
Red lettuce (Lactuca sativa L. cv. Red Sails)	PS	100–1,000 nm, >10,000 nm	0.25, 0.50, 1.00 g/L	DBP: 5 mg/L	PS reduced the DBP bioavailability, caused decrease in photosynthetic, and serious oxidative damage, and reduced the quality of DBP-treated-red lettuce
Rapeseed (Brassia campestris L.)	PMMA	<100 nm	0, 0.05, 0.5, 5 g/L	As: 0, 10, 20, 40, 60 mg/L	Caused synergistic effect on rapeseed germination, promoted As uptake in rapeseed under high concentration
Lettuce (L. sativa L. var. ramosa Hort)	PE	~23 μm	0.25, 0.50, 1.00 mg/mL	DBP: 5 mg/L	MP can inhibit growth, hinder photosynthesis and interfere with the antioxidant defence system in lettuce; exposure to MP exacerbated the damage to lettuce by DBP
Lettuce (L. sativa L. var. ramosa Hort)	PS	100~1,000 nm, >10,000 nm	0.25, 0.50, 1.00 mg/mL	DBP: 5 mg/L	PS reduced lettuce biomass and DBP enrichment in roots and leaves, exacerbated oxidative stress and subcellular damage
Rape (B. napus L.)	PE	293 μm	0.001%, 0.01%, 0.1%	Cu: 50, 100 mg/kg Pb: 25, 50 mg/kg	PE increased accumulation and toxicity of heavy metals to rape
Wheat (Triticum aestivum L.)	PS	87 nm	0, 10 mg/L	Cd: 0, 20 μM	PS partially alleviated Cd-induced toxicity in wheat
Strawberry (Fragaria × ananassa Duch.)	HDPE	20 μm thick, 2~5 mm	0.2 g/kg	Cd: 3 mg/L	HDPE increased Cd bioavailability and accumulation in roots, decreased the total number of fruits and total biomass per plant
Maize (Zea mays L. var. Wannuoyihao)	HDPE, PS	100–154 μm	0.1%, 1%, 10%	Cd: 5 mg/kg	high-dose of HDPE (10%) amplified Cd phytotoxicity PS negatively affected maize growth and phytotoxicity further increased in the presence of Cd

(Continued)

TABLE 16.11 (*Continued*) A Summary of the Combined Impacts and Influences of Microplastics and Nanoplastics and Coexisting Pollutants on Terrestrial Plants

	MPs/NPs				
Plant Species	**Type**	**Size**	**Concentration**	**Coexisting Pollutants**	**Combined Effect**
Maize (Z. mays L. var. Wannuoyihao)	PE, PLA	100~154 μm	0.1%, 1%, 10%	Cd: 0, 5 mg/kg	PLA caused higher Cd bioavailability than PE, but no alterations in plant Cd content. MPs and Cd drove shifts in maize performance and root symbiosis
Lettuce (L. sativa L.)	PE	<0.5 mm	0.1%, 1%, 10%	Cd: 0.49, 1.75, 4.38 mg/kg	Co-exposure of PE increased the toxicity, uptake, accumulation and bioavailability of Cd
Soybean (Glycine max (L.) Merr.)	PS	100 nm, 1, 10, 100 μm	10 mg/kg	Phenanthrene, 1 mg/kg	PS decreased the uptake of Phe in soybean roots and leaves but caused combined toxicity to soybean plants
Lettuce (L.sativa L.)	PS	100 μm, 100 nm	100, 1,000 mg/kg	Cu: 82.00 mg/kg Zn: 174.84 mg/kg Pb: 42.08 mg/kg Cd: 0.20 mg/kg	MPs increased the uptake of heavy metals in lettuce
Maize (Z. mays L. var. Wannuoyihao)	HDPE, PLA	100–154 μm	0.1%, 1%, 10%	ZnO: 30±10 nm 0, 50, 500 mg/kg soil	HDPE and PLA increased Zn accumulation in roots, decreased Zn translocation to aerial parts
Brassica chinensis L	PS	75 μm	0.5%, 1.0%, 1.5%, 2.0%	Cd: 10 mg/kg	PS-Cd co-pollution produced higher phytotoxicity than PS alone, PS mitigated the phytotoxicity of Cd alone and reduce Cd uptake
Wheat (T. aestivum L.)	PS	0.5 μm	100 mg/L	Cu: 2 mg/L, Cd: 1 mg/L	Mitigated Cu and Cd bioavailability and toxicity

Source: Yue et al. (2023).

DBP, dibutyl phthal; HDPE, high–density polyethylene; PE, polyethylene; Phe, phenanthrene; PLA, polylactic acid; PMMA, polymethylmethacrylate; PS, polystyrene; PTFE, polytetrafluoroethylene.

environments (Geyer et al., 2017). Literature reported that between 21% and 42% of the worldwide plastic solid wastes that are neither burnt nor recycled are deposited in sanitary landfills (He et al., 2019). It was stated that nearly 12.0×10^9 tons of plastic solid waste are projected to be stored and kept in sanitary landfills or unconfined to the environment by the year 2050 (Geyer et al., 2017; He et al., 2019). Separating the sides and bottom of a sanitary landfill with a barrier technique and system can prevent or at least reduce and limit the seepage of the leachate from the solid wastes and the movement of the microplastic particles to the groundwater and the subsoil. These plastic wastes that are kept and stored in sanitary landfills can undergo both biochemical and physical transformation due to many environmental, biological and physical factors.

The stated factors include biodegradation, severe changes in temperature and acidity (such as pH fluctuations between 4.5 and 9), generation of gases such as CH_4 and CO_2, high pressure and compaction as well as high salinity (Hanson et al., 2005; Mahon et al., 2017; Sundt et al., 2014; Zettler et al., 2013). As an outcome of these transformations, these plastic wastes can be fragmented or segregated into smaller dimensions or pieces that are categorized or classified as megaplastic, microplastic and nanoplastic particles. The groundwater and the soil can be polluted or contaminated because of the probable transfer and movement of these stated particles from the sanitary landfill to the subsoil environment. The lack of adequate liner techniques and systems for abandoned solid waste dumps or defects and liner damages in sanitary landfills at different locations can cause environmental problems due to the effluent or discharge of the leachate that transmits the stated or transformed particles produced from the plastic solid wastes into the soil. It was highlighted that 52.0×10^6 tons of municipal waste materials were stored and confined in sanitary landfills in the European Union countries in 2018 (Xu et al., 2020a and b). Xu et al. (2020a and b) documented that this quantity corresponds to $10.40 \times 10^6 m^3$ of leachate formations and 3.03×10^9 of plastic particles released into the environment. Establishing these quantities and values of plastic particles in the European Union countries alone, the volume of the plastic particles released worldwide would be abundant. Yang et al. (2015a and b) and Xu et al. (2020a and b) stated the leachate formation average from $1.0 \times 10^3 kg$ of solid waste that was dumped in sanitary landfills in 30 different urban and semi-urban areas of China was between 1.30 and $3.20 m^3$. Figure 16.15 provides a schematic view and diagram of the environmental and non-environmental processes of plastic particles in soil. He et al. (2019) detected plastic particles in all 12 leachate samples collected from

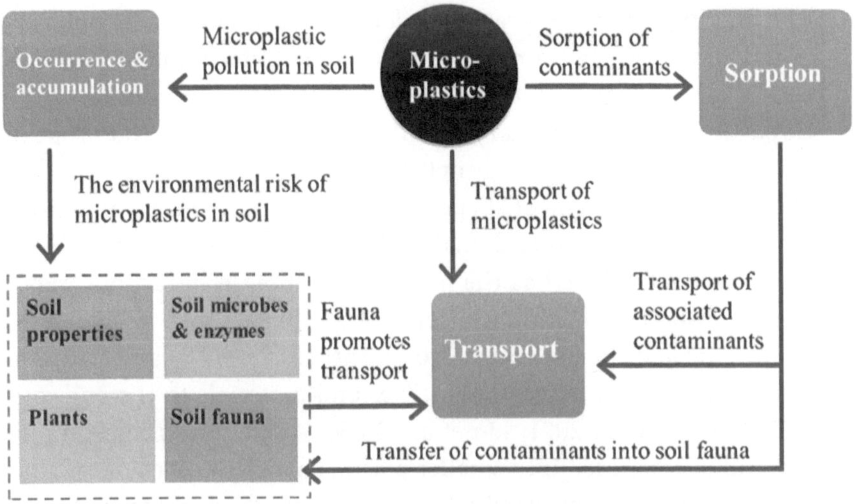

FIGURE 16.15 Schematic diagram of the environmental risks and processes of microplastic particles in soil. (Xu et al., 2020a and b.)

six different sanitary landfills in China. The occurrence, presence and accumulation of microplastic particles cause probable adverse influences on soil systems. These influences include the transformation of microbes and enzymes, soil properties, plants and fauna. These plastic particles can transport or move horizontally and vertically in soil, which is enhanced by soil fauna such as *collembolans* and earthworms. The associated pollutants and contaminants such as pathogens, heavy metals and organic contaminants on these plastic particles are adsorbed and transported by microplastic particles and later transferred into soil fauna.

It has been reported that PP and PE are the established two major polymer categories among the plastic particles in the studies of soil. It is established that nearly 77.5% of the plastic particles in the soil had a particle size and dimension between 0.1 and 1.0 mm. The concentration and quantity of the plastic particles in the sanitary landfill leachates were between the range of 0.42 and 24.58 particles per litre (ptcl./L). However, the concentration and the number of plastic particles in the leachate of sanitary landfills from Northern Europe were less than 4.51 ptcl./L (Van Praagh et al., 2019). These differences in concentrations and quantities can be attributed to the strict environmental laws and regulations, environmental awareness and lower population of the Northern European countries when compared with the population of China. He et al. (2019), and Su et al. (2019) reported that the most commonly encountered polymer categories of the plastic particles in sanitary landfill leachate from China were PP and PE, while Van Praagh et al. (2019) have shown that polyurethane and PE were the two foremost polymers in sanitary landfill leachates in the North European countries. The particle size and dimension of the plastic particles from the sanitary landfill leachates in the North European countries were between the range of 0.05 and 5.0 mm.

Additionally, the foremost shape of the plastic particles in numerous leachate samples collected and analyzed was angular with sharp edges. Sanitary landfill leachate has the potential to contain hazardous materials and substances including bisphenol A (BPA). With reference to physicochemical, diffusion and hydrolysis processes, BPA can easily and quickly be released from plastic particles and materials to the environment (Ficociello et al., 2021; Liu et al., 2019). As the plastics are transformed and decomposed into smaller plastic particles and these smaller plastic particles are transferred with the leachate fluid, these BPA substances can be mixed with the groundwater and the soil. The concentrations and quantities of BPA and smaller plastic particles when analyzed were measured as the same in the leachate samples that were collected from three sanitary landfills in Southeastern Europe. These results indicated that the concentration and volume of the smaller plastic particles were between the range of 0.64 and 2.16 milligrams per litre (mg/L) while the BPA concentration and quantity were between 0.70 and 2.72 mg/L. It was reported that with reference to the high absorption capacity of smaller plastic particles, the smaller plastic particles may simply and easily absorb toxic organic substances (Beckingham and Ghosh, 2017) and thus increase the concentration of heavy metal in the soils (Hodson et al., 2017). Soil particles with these smaller plastic particles carried by the leachates seeping into the soil in the vicinity of sanitary landfills are sometimes used in agricultural practices and the toxic materials and substances that are absorbed by these smaller plastic particles might easily and quickly deteriorate quality and health of the soil and alter the plant growth (Monkul and Özhan, 2021). There is a comparatively limited amount of research work and studies that specify the occurrence and presence of microplastic particles in groundwater and soil layers beneath sanitary landfills. It has been documented that the wrinkles in the geomembrane liner systems could cause the leachate with microplastic particles to move and migrate through the barrier substance and material to the subsoil (Foose et al., 2001; Monkul and Özhan, 2021). Manikanda et al. (2021) detected and established microplastic particles in the groundwater from the subsoil beneath municipal solid waste sanitary landfills in South India. The microplastic particle concentrations and quantity in the groundwater were determined and measured, and the result showed that the quantity was between 2.0 and 80.0 ptcl./L. The results revealed that the main polymer categories of the microplastic particles were grouped as polypropylene (Figure 16.9) and polystyrene (Figure 16.9). In another research study, microplastic particles were established in some soil samples from a sanitary landfill in Bangladesh (Monkul and Özhan, 2021). These polymer categories of the microplastic particles were expressed as cellulose acetate and PE. The particle diameter of the microplastic particles detected and established in the soil underneath the sanitary landfill was between the range of 1.0×10^{-3} and 2.0×10^{-3} mm (Afrin et al., 2020; Monkul and Özhan, 2021). Both soil and leachate samples were collected from several sanitary landfills in Thailand (Puthcharoen and Leungprasert, 2019; Monkul and Özhan, 2021). The soil samples collected were from the upper layer of the subsoil under the sanitary landfill bottom at a depth of between 10.0×10^{-3} and 20.0×10^{-3} m using a standard facility (a hand auger). The microplastic particle concentrations were established to be higher in soil from the sanitary landfills than in leachate from the same. Puthcharoen and Leungprasert (2019) reported that the soil samples collected from these sites had plastic particles of nearly 1,500 ptcl./kg and there were 20 ptcl./kg plastic particles in the leachate samples analyzed. The foremost polymer categories of the plastic particles attained from both the leachate and the soil samples were PET, PP and PE. The main microplastic particle patterns were films and granules for both the leachate and the soil samples. The other plastic particle patterns detected in the leachate and the soil samples were irregular fibres and sphere fibres. The sphere, irregular and granule-shaped plastic particles were established to be largely decomposed from water bottles, plastic food and storage containers.

In addition, bags and plastic packages were the foremost source or origin of the film-shaped microplastic particles. The fibre-shaped microplastic particles were largely formed by the transformation and decomposition of synthetic fibres from synthetic clothes (Monkul and Özhan, 2021; Browne et al., 2015). Monkul and Özhan (2021) reported that the solid waste samples collected from different sanitary landfills in Shangai, China were examined and evaluated. The results revealed that the older sanitary landfills contained a greater and wider range of microplastic particles in terms of shape. It was documented the contribution of the sanitary landfill age and retention time to the microplastic particles shape in leachates from the sanitary landfills was not as significant and critical as the factor of solid wastes in the sanitary landfills (Monkul and Özhan, 2021).

The foremost agents and factors that contribute to or may be attributed to the horizontal transportation of the microplastic particles were surface runoff, flooding, wind erosion, and animal and human scavenging (Monkul and Özhan, 2021). It was documented that lighter and smaller microplastic particles can be easily and quickly transported by the wind from the surface of the soils and sanitary landfills to the surface of water resources and other lands, while the denser and larger microplastic particles are more likely to migrate and move vertically into the soil (Horton et al., 2017; Zylstra, 2013; Monkul and Özhan, 2021). Some research studies have indicated and established that these microplastic particles can be transferred from sanitary landfills to other land or surface water resources or other lands. It was stated that microplastic particles were confirmed and detected in a river (surface water resources) that was close to a sanitary landfill located around a coastal area (Monkul and Özhan, 2021; Kilponen, 2016). These investigations established and showed that the sanitary landfill leachate had moved the microplastic particles to the river. In some of these research studies, it was documented that the sanitary landfills that are built on beaches or coastal areas might fail or collapse due to the geotechnical properties of the soil and erosion. Global and worldwide warming that originated or led to elevated or increased sea levels is the foremost reason for the erosion which will trigger failures. The failure and collapse of sanitary landfills might originate and cause the scattering or spreading of the microplastic particles into the sea, thus polluting and contaminating freshwater (Monkul and Özhan, 2021; Hurley et al., 2018; Kazour et al., 2019). With reference to the atmospheric and environmental transport of the smaller and lighter microplastic particles from sanitary landfills to either a lake, any sea or river horizontally or to other terrestrial regions, there are risks for human beings to inhale these microplastic particles (Monkul and Özhan, 2021; Rezaei et al., 2019). These inhaled microplastic particles may probably reach the stomach and lungs, and thus threaten the digestive system and human respiratory health (Prata et al., 2020; Wright and Kelly, 2017). The microplastic particles transfer or movement from sanitary landfills to soils, the atmosphere and both surface water and groundwater resources have been examined and evaluated in several studies. These studies reported that the leachates from sanitary landfills are the foremost agents or sources that move or transport microplastic particles from sanitary landfills to the soil deposits vertically. It was stated that the size of the plastic particles that were established in the samples collected from both the soil and leachates in sanitary landfills samples underneath sanitary landfill bases varied in a wide range, the main polymer categories of the microplastic particles were PE and PP. It was reported that the principal origin and sources for the transformation of plastic particles and the formation of microplastic particles in sanitary landfills were plastic packages, plastic bottles and plastic food storage containers. These microplastic particles could move from sanitary landfills to surface water resources (which include rivers, ponds, lakes, streams and seas) to terrestrial regions or top soils far away from the sanitary landfills or to the atmosphere which includes the troposphere through wind erosion and surface runoff. These stated plastic particles in leachates from sanitary landfills or soil under sanitary landfills can contain hazardous materials or toxic substances, which can be transferred to the soil from a sanitary landfill without a liner, barrier material or an adequate drainage system and barrier material that is damaged (Monkul and Özhan, 2021).

16.12 SELECTED GEOTECHNICAL AND FOUNDATION ENGINEERING FACILITY AS POTENTIAL SOURCE FOR MICROPLASTIC PARTICLES IN THE ENVIRONMENT

Sanitary landfills are measured by scientists and researchers as one of the significant origins and sources of microplastic particle pollution and contamination in soils, which merits further studies and attention. It was reported that there are other geotechnical and foundation engineering applications, in which polymer-based substances and materials (such as tyre chips) or already polluted and contaminated soils (such as dredged soils) are utilized. These applications fundamentally have the possibility to generate and create microplastic particles as well. It was documented that the scale and level of contamination and pollution are expected to be minimal and smaller than the ones from the main source (sanitary landfills). Some of these selected applications are clarified as follows (Monkul and Özhan, 2021):

a. **Geotechnical Engineering Properties of Soil Enhancement with Tyre Chips**
   Tyre chips are shredded solid waste tyres that originate from the abrasion of road vehicle and aeroplane tyres. Tyre solid waste is one of the most important sources of microplastic particles contributing to their migration and movement into the soil and surface water sources (Monkul and Özhan, 2021). With reference to the breakdown of the shredded solid waste tyres, these microplastic particle substances can enter the soil quickly and easily. The worldwide emission

of microplastic particles with respect to the abrasion of road vehicle tyres is nearly 0.81 kg/year/capita (Kole et al., 2017). The most frequent source of yearly microplastic particle quantity emission per capita for the European Union countries was established to be tyre wear with an average of 1,784.8 g/capita (Meixner et al., 2020). These microplastic particles transformed or formed by the breakdown of the solid waste tyres were detected in some sanitary landfill leachate samples. In addition, solid waste tyres are utilized as geomaterials or substances to enhance the shear strength of soils in civil and construction engineering projects. In an attempt to minimize the negative influences of shredded solid waste tyres on the environment, and establish economical engineering solutions, soil deposits are mixed with tyre chips and utilized in selected geotechnical and foundation engineering applications. These applications include backfills of retaining structures, road embankments and pavement subgrades (Daud, 2018; Promputthangkoona and Karnchanachetaneea, 2013). Singh and Sonthwal (2017) and Solanki et al. (2017) reported that the addition of tyre chips to clayey soils increased the shear strength of the clay soil. Daud (2018) documented that the shear strength of a sandy deposit and the internal friction angle of the soil were improved by the addition of tyre chips. Al-Neami (2018) examined the California bearing ratio (CBR) of soil by the addition of tyre chips to sand mixtures and the outcomes indicated that tyre chips addition with a fixed composition of 8% by dry weight increased the load-bearing capacity of the soil by 1.6 times higher than the sand without any addition of tyre chips. It was stated that unconfined compression and CBR tests were evaluated on tyre chips-cement-soil mixtures established that both the load-bearing capacity and the shear strength of these selected soils improved due to the presence of tyre chips (Reddy et al., 2016; Monkul and Özhan, 2021).

b. **Utilization of Dredge Sediments Originating from Water Resources Facilities as Filling Materials**

Dredge sludge or sediments are typically emptied from the bottom of harbours, lakes, rivers or any other surface water resources to specific locations, dumped and deposited to fill lowlands. The dredge sludge or sediments are mostly utilized for urban or semi-urban soil reconstruction, coastal land reclamation, shoreline stabilization, upland placement for agriculture, as covers for sanitary landfills as fertilizers in farmlands and as a cover for mine tailings dams (Ji et al., 2021; Monkul and Yükselen-Aksoy, 2016; Islam et al., 2013). It is well-known that these microplastic particles can easily and quickly be dispersed into aquatic and freshwater environments, thus dredge sludge or sediments can accumulate significant quantities of microplastic particles (Monkul and Özhan, 2021). It has been reported that greater than $200.0 \times 10^6 m^3$ of dredge sludge or sediments have been emptied per annum from the bottom of 400 ports in the United States of America. These same quantities have been used mainly for urban and semi-urban soil reconstruction and plateau placement for agriculture (Monkul and Özhan, 2021). In some research works or studies, the dredge sludge or sediment that primarily was composed of fine sand was emptied from riverbanks and used as filling materials for land reclamations (Islam et al., 2013). It was documented that the total sediment quantity that has been emptied or dredged per annum in China is greater than $5.0 \times 10^9 m^3$ (Zhejiang et al., 2017; Monkul and Özhan, 2021). It was stated that between 136.0×10^6 and $116.0 \times 10^6 m^3$ of dredge sludge or sediments were emptied from the riverbanks in the Zhejiang Province, China for reconstruction facilities between 2016 and 2017 (Zhejiang et al., 2017; Monkul and Özhan, 2021). Ji et al. (2021) documented that 3.0×10^6 out of $10.0 \times 10^6 m^3$ of the emptied dredge sludge or sediments from the riverbanks in Yueqing, China were transported or moved, dumped and deposited into sanitary landfills to serve as the cover soil in coastal reclamation projects. The balance of $7.0 \times 10^6 m^3$ of the dredge sludge or sediment was placed in storage piles to be used as agricultural fertilizers in selected farmlands. The microplastic particle concentrations or levels in the soil within and around the storage piles were confirmed to be higher in the dry season than in the wet season. These increments can be because of wind dispersion, while the river surrounded by the storage piles had a higher quantity of microplastic particles in the wet season which can be attributed to the surface runoff (Monkul and Özhan, 2021).

c. **Soil Improvement with Synthetic Polymer-Based Fibres**

Polymers have been used in the fibre form for several geotechnical and foundation engineering applications over the last 20 years. These stated materials have been used for an improvement in the geotechnical and foundation engineering properties of the ground. These polymer-based fibres are characteristically added to the soil with the aim of enhancing the strength properties or parameters of the soil. These fibre materials serve as the coating substances or agents that are capable of filling or reducing the voids within the soil grains, thus acting as reinforcement substances in soil fabric. In addition, these fibre materials can be mixed with other binding materials namely cement or lime in the soil matrix for an improvement in the compressive and shear strengths of the stated soil (Tingle et al., 2007; Monkul and Özhan, 2021). In various research works and studies, synthetic polymer-based fibres were added to sandy soils, clayey or soil-cement mixtures. The results of the geotechnical properties tests (triaxial compression, unconfined compression and direct shear) conducted revealed that fibre addition increased both the compressive and shear strengths of the soil. These results indicated an improvement in the geotechnical properties of the soil for foundation or structural engineering design (Abdel Hadi, 2016; Chen et al., 2015; Hamidi and Hooresfand, 2013; Tang et al., 2007; Ayeldeen and Kitazume, 2017). Ayeldeen and Kitazume (2017) added a portion of polypropylene which is a synthetic polymer-based fibre to a clayey soil mixed with cement (15% by dry weight) and fibre (0.5%). The results revealed that there was an improvement of 240% in the shear strength of the mixture (clay-cement mixture). In addition, a further addition or increase in the fibre content of the soil mixed resulted in a reduction in the shear strength of the soil mixture. It was reported that these synthetic polymer-based fibres can be used as secondary additives to improve the ductility of stabilized soils that are primarily used as barriers in sanitary landfills or subgrade soils underneath roadways (Öncü and Bilsel, 2017). Yilmaz and Sevencan (1994) introduced polypropylene fibres to a soil mixture in addition to fly ash. The addition of the fibres to a soil mixture led to an improvement in the unconfined compressive strength of the soil. Furthermore, the addition of waste ashes and polypropylene fibres to a sulphate-rich expansive soil increased the shrinkage strain and the swell index of the expansive soil (Punthutaecha et al., 2006). In the case by Öncü and Bilsel (2017), 2% of the dry weight of a polymeric fibre was added to an expansive soil-sand mixture. The results revealed that the compressive strength and split tensile strength values of the soil mixture improved (Monkul and Özhan, 2021).

In addition, it is well-known that polymer-based fibre materials that are introduced to soils for the purpose of reinforcement can easily be fragmented into smaller and lighter particles (Wei et al., 2018; Monkul and Özhan, 2021). In much of the literature, polypropylene has been reported to be among the most abundant plastic particle and polymer categories in both soil and wastewater environments (Ahmed et al., 2013; Monkul and Özhan, 2021). In summary, it should be noted that there is a significant risk of megaplastic, microplastic and nanoplastic particle contamination and pollution when soils are stabilized through the addition of synthetic polymer-based fibres for various civil, geotechnical and foundation engineering projects (Monkul and Özhan, 2021).

d. **Geotechnical and Foundation Applications with Expanded Polystyrene as Lightweight Fill Materials**

Engineering principles of sanitary landfills are rationalized by scientists, engineers and researchers as one of the substantial origins and sources of microplastic particle pollution and contamination in groundwater and soils, which merits further studies and attention. It was reported that there are several geotechnical and foundation engineering applications, in which polymer-based substances and materials or already polluted and contaminated soils are utilized. These applications fundamentally have the possibility to generate and create microplastic particles as well. It was documented that the scale and level of contamination and pollution are expected to be minimal, reduced and smaller than the ones from the main source (sanitary landfills). Some of these selected applications are clarified as follows (Monkul and Özhan, 2021):

i. **Utilization of the Polystyrene Based Microplastic Particles in Soils**

Polystyrene (PS) is one of the polymer types that is classically used for the manufacture of hard plastics, which include beverage cups, food packages, beverage plates, laboratory wares and food containers (Monkul and Özhan, 2021; Hwang et al., 2020; Schellenberg, 2009). It was documented that as soon as polystyrene-based plastic solid wastes are unrestricted to the environment, these materials can be easily fragmented into smaller and lighter pieces. These fragmented materials are carried to the surface water resources, terrestrial lands and atmosphere and then move farther away from the locations where the solid wastes were originally dumped, deposited or alternatively can infiltrate or penetrate into the soil vertically and pollute groundwater (Kukharchyk and Chernyuk, 2020; Monkul and Özhan, 2021). It is well-established that plastic particles with a particle dimension or size of 1.0×10^{-6}m formed by the fragmentation process or the disintegration of polystyrene-based plastic wastes are considered a serious threat to animals and human beings when inhaled. It is documented that once the polystyrene-based plastic particles are inhaled or the foods that comprise these particles are eaten, these polystyrene plastic particles may destabilize red blood cells and finally could lead to the disease known as haemolysis (Monkul and Özhan, 2021).

ii. **Geotechnical and Foundation Engineering Applications with Polystyrene-Based Lightweight Fills**
Expanded polystyrene is one of the most favoured lightweight fill materials that is used in geotechnical and foundation engineering applications. These applications include embankment constructions for bridges and roads, backfills of retaining walls and slope stabilization (Abbasimaedeh et al., 2021; Özer and Akay, 2016; Monkul and Yükselen-aksoy 2016; Stark et al., 2004; Tsuchida and Egashira, 2004). In bridge and road constructions on soft soils, lightweight fill embankment materials with expanded polystyrene are used with the key objective of reducing the bearing pressures on the soft soil (Arellano et al., 2013; Gan and Tan, 2003; Monkul and Özhan, 2021). Utilizations of these expanded polystyrenes as lightweight fill materials are preferred to protect buried pipelines and culverts with the critical aim of decreasing the soil pressures acting as another structure on the soil (Moorsel et al., 2014; Monkul and Özhan, 2021). In literature, prefabricated vertical expanded polystyrene drainage systems were used in the reduction of the swelling capability of expansive soil underneath the foundations of some buildings in India (Daigavane et al., 2010). Literature established that prefabricated vertical expanded polystyrene drainage systems stabilized the expansive soil through a reduction of the rate of settlements and consolidation. Though these utilizations of these expanded polystyrenes as lightweight fill materials in many geotechnical and

foundation engineering applications have several vital benefits from the engineering points of view, the probable propagation of the plastic particles (mega, micro and nanoparticles) into the soil may lead to at least soil, surface water and groundwater pollutions in the long term. The level or weight of plastic particle pollution and contamination in soils due to these applications is perhaps relatively small compared to the contamination and pollution levels caused by massive quantities of global plastic waste generated or originated systematically. In summary, it can be concluded that the goal or objective of researchers on the pollution of the environment by plastic particles is not to disapprove or discourage these geotechnical and foundation engineering applications, but rather to make these engineers aware of the possible risks of plastic particle pollution and contamination (Monkul and Özhan, 2021).

e. **Summary and Conclusions**

In this chapter and reviewed literature, plastic particle contamination and pollution in soils were assessed and evaluated from the geotechnical and foundation engineering perspective. It was highlighted that there is a vital delay in worldwide awareness of plastic particle pollution and contamination in the soils between different disciplines. A basic statistical analysis of the published literature for the last six years revealed that there are only two engineering disciplines namely Chemical Engineering and Environmental Engineering, which are among the top ten disciplines conducting research and studies on plastic particles and soils. However, in the case of civil engineers, geotechnical and foundation engineers as researchers on plastic particles, it can be concluded that the worldwide awareness of the Civil Engineering communities on plastic particles is extremely limited. It is well-known that microplastic particle pollution and contamination in soils is a worldwide problem that requires to be extremely measured.

It would be quite favourable and good to adopt a geotechnical and foundation engineering scene in a way to provide a better and clearer understanding of the interaction between soil, microplastic particles, air and water. The focus of the geotechnical design of sanitary landfills is on CCL and GCL. The lack of sufficient liner systems for abandoned waste dumps or defects in the liner or damages in landfills might also significantly contribute to the microplastic particle contamination in soils. Furthermore, it would be of interest and a critical research topic to investigate the performance and efficacies of GCLS for filtering or holding microplastic particles, which is presently unknown or not established. It is possible that GCLs could act as mitigation tools or facilities in reducing the microplastic particle contamination in soils. In fact, the researchers of this paper have been working on a research project proposal on the topic. Civil, foundation and Geotechnical Engineers should be aware of the potential risks for additional microplastic particle contamination due to engineering applications.

REFERENCES

Abbasimaedeh, P., Ghanbari, A., O'Kelly, B., Tavanafar, M., Irdmoosa, K. Geomechanical behaviour of uncemented expanded polystyrene (EPS) beads–clayey soil mixtures as lightweight fill. *Geotechnics* 2021, 1, 38–58.

Abdel Hadi, N.A.R. Utilization of polymer fibres and crushed limestone sand for stabilization of expansive clays in Amman area. *Int. J. Geotech. Eng.* 2016, 10, 428–434.

Afrin, S., Uddin, K., Rahman, M. Microplastics contamination in the soil from urban landfill site, Dhaka, Bangladesh. *Heliyon* 2020, 6, e05572.

Ahmed, M.R., Meguid, M., Whalen, J., Eng, P. Laboratory measurement of the load reduction on buried structures overlain by EPS geofoam. In *Proceedings of the 66th Canadian Geotechnical Conference*, Montreal, QC, Canada, 29 September–3 October 2013.

Albertsson, A.C., Karlsson, S. The three stages in degradation of polymers—polyethylene as a model substance. *J. Appl. Polym. Sci.* 1988, 35, 1289–1302.

Alimi, O.S., Farner Budarz, J., Hernandez, L.M., Tufenkji, N. Microplastics and nanoplastics in aquatic environments: aggregation, deposition, and enhancedcontaminant transport. *Environ. Sci. Technol.* 2018, 52, 1704–1724.

Al-Neami, M.A. Stabilization of sandy soil using recycle waste tire chips. *Int. J.* 2018, 15, 175–180.

Alomar, C., Sureda, A., Capo, X., Guijarro, B., Tejada, S., Deudero, S. Microplastic ingestion by Mullus surmuletus Linnaeus, 1758 fish and its potential for causing oxidative stress. *Environ. Res.* 2017, 159, 135–142.

Anbumani, S., Kakkar, P. Ecotoxicological effects of microplastics on biota: a review. *Environ. Sci. Pollut. Res.* 2018, 25, 14373–14396.

Arellano, D., Stark, T.D., Horvath, J.S., Leshchinsky, D. Guidelines for geofoam applications in slope stability projects. In Research Results Digest 380; National Cooperative Highway Research Program, The National Academies Press: Washington, WA, USA, 2013.

Ayeldeen, M., Kitazume, M. Using fiber and liquid polymer to improve the behaviour of cement-stabilized soft clay. *Geotext. Geomembranes* 2017, 45, 592–602.

Baile Xu, Fei Liu, Zachary Cryder, Dan Huang, Zhijiang Lu, Yan He, Haizhen Wang, Zhenmei Lu, Philip C. Brookes, Caixian Tang, Jay Gan & Jianming Xu (2019): Microplastics in the soil environment: Occurrence, risks, interactions and fate – A review, Critical Reviews in Environmental Science and Technology, DOI: 10.1080/10643389.2019.1694822

Barreto, C., Rillig, M., Lindo, Z. Addition of polyester in soil affects litter decomposition rates but not microarthropod communities. *Soil Org.* 2020, 92(2), 109–119. https://doi.org/10.25674/so92iss2pp109

Beckingham, B., Ghosh, U. Differential bioavailability of polychlorinated biphenyls associated with environmental particles: microplastic in comparison to wood, coal and biochar. *Environ. Pollut.* 2017, 220, 150–158.

Beiras, R., Bellas, J., Cachot, J., Cormier, B., Cousin, X., Engwall, M., Gambardella, C., Garaventa, F., Keiter, S., Le Bihanic, F., Lopez-Ibanez, S., Piazza, V., Rial, D., Tato, T., Vidal-Linán, L. Ingestion and contact with polyethylene microplastics does not cause acute toxicity on marine zooplankton. *J. Hazard Mater.* 2018, 360, 452–460.

Ben Othmen, A., Bouassida, M. Detecting defects in geomembranes of landfill liner systems: durable electrical method. *Int. J. Geotech. Eng.* 2013, 7, 130–135.

Benson, C.H., Meer, S.R. Relative abundance of monovalent and divalent cations and the impact of desiccation on geosynthetic clay liners. *J. Geotech. Geoenviron. Eng.* 2009, 135, 349–358.

Bergami, E., Rota, E., Caruso, T., Birarda, G., Vaccari, L., Corsi, I. Plastics everywhere: first evidence of polystyrene fragments inside the common Antarctic collembolan Cryptopygus antarcticus. *Biol. Lett.* 2020, 16(6), 20200093. https://doi.org/10.1098/rsbl.2020.0093

Bernet, D., Schmidt, H., Meier, W., Burkhardt-Holm, P., Wahli, T. Histopathology in fish: proposal for a protocol to assess aquatic pollution. *J. Fish Dis.* 1999, 22, 25–34.

Blasing, M., Amelung, W. Plastics in soil: analytical methods and possible sources. *Sci. Total Environ.* 2018, 612, 422–435. https://doi.org/10.1016/j.scitotenv.2017.08.086

Blasing, M., Amelung, W. Plastics in soil: analytical methods and possible sources. *Sci. Total Environ.* 2017, 612, 422–435.

Bouazza, A. Geosynthetic clay liners. *Geotext. Geomembr.* 2002, 20, 3–17.

Briassoulis, D., Babou, E., Hiskakis, M., Kyrikou, I. Analysis of long-term degradation behaviour of polyethylene mulching films with pro-oxidants under real cultivation and soil burial conditions. *Environ. Sci. Pollut. Res.* 2015, 22, 2584–2598.

Brooks, A.L., Wang, S., Jambeck, J.R. The Chinese import ban and its impact on global plastic waste trade. *Sci. Adv.* 2018, 4, 52.

Browne, M.A., Underwood, A.J., Chapman, M.G., Williams, R., Thompson, R., Van Franeker, J.A. Linking effects of anthropogenic debris to ecological impacts. *Proc. R. Soc. B Biol. Sci.* 2015, 282, 20142929.

Cao, D., Wang, X., Luo, X., Liu, G., Zheng, H. Effects of polystyrene microplastics on the fitness of earthworms in an agricultural soil. *IOP Conf. Series Earth Environ. Sci.* 2017, 61, 012148. https://doi.org/10.1088/1755-1315/61/1/012148

Chae, Y., An, Y.J. Current research trends on plastic pollution and ecological impacts on the soil ecosystem: a review. *Environ. Pollut.* 2018, 240, 387–395.

Chen, M., Shen, S.-L., Arulrajah, A., Wu, H.-N., Hou, D.-W., Xu, Y.-S. Laboratory evaluation on the effectiveness of polypropylene fibres on the strength of fiber-reinforced and cement-stabilized Shanghai soft clay. *Geotext. Geomembr.* 2015, 43, 515–523.

Chen, Y., Si, Y., Zhou, D., Dang, F., Differential bioaccumulation patterns of nano-sized and dissolved silver in a land snail Achatina fulica. *Environ. Pollut.* 2017, 222, 50–57.

Cheng, Y., Song, W., Tian, H., Zhang, K., Li, B., Du, Z., Zhang W., Wang, J., Wang, J., Zhu, L. The effects of high-density polyethylene and polypropylene microplastics on the soil and earthworm Metaphire guillelmi gut microbiota. *Chemosphere* 2021, 267, 129219. doi: 10.1016/j.chemosphere.2020.129219

Coduto, D.P. *Foundation Design: Principles and Practices*, 2nd ed.; Pearson: London, UK, 2001.

Coduto, D.P., Yeung, M.R., Kitch, W.A. *Geotechnical Engineering: Principles and Practices*, 2nd ed.; Pearson: London, UK, 2010; ISBN 0131354256.

Cole, M. A novel method for preparing microplastic fibres. *Sci. Rep.* 2016, 6, 34519.

Cole, M., Galloway, T.S. Ingestion of nanoplastics and microplastics by Pacific oyster Larvae. *Environ. Sci. Technol.* 2015, 49, 14625.

Cole, M., Lindeque, P.K., Fileman, E., Clark, J., Lewis, C., Halsband, C., Galloway, T.S. Microplastics alter the properties and sinking rates of zooplankton faecal pellets. *Environ. Sci. Technol.* 2016, 50, 3239–3246.

Cooper, D.A., Corcoran, P.L. Effects of mechanical and chemical processes on the degradation of plastic beach debris on the island of Kauai, Hawaii. *Mar. Pollut. Bull.* 2010, 60, 650–654.

Corinaldesi, C., Canensi, S., Carugati, L., Lo Martire, M., Marcellini, F., Nepote, E., Sabbatini, S., Danovaro, R. Organic enrichment can increase the impact of microplastics on meiofaunal assemblages in tropical beach systems. *Environ. Pollut.* 2022, 292, 118415. https://doi.org/10.1016/j.envpol.2021.118415

Corradini, F., Meza, P., Eguiluz, R., Casado, F., Huerta-Lwanga, E., Geissen, V. Evidence of microplastic accumulation in agricultural soils from sewage sludge disposal. *Sci. Total Environ.* 2019, 671, 411–420.

Crabbe, J.R., Campbell, J.R., Thompson, L., Walz, S.L., Schultz, W.W. Biodegradation of a colloidal ester-based polyurethane by soil fungi. *Int. Biodeterior. Biodegrad.* 1994, 33, 103–113.

Crossman, J., Hurley, R.R., Futter, M., Nizzetto, L. Transfer and transport of microplastics from biosolids to agricultural soils and the wider environment. *Sci. Total Environ.* 2020, 724, 138334.

da Costa, J.P., Paço, A., Santos, P.S.M., Duarte, A.C., Rocha-Santos, T. Microplastics in soils: assessment, analytics and risks. *Environ. Chem.* 2019, 16, 18–30.

Da Costa, J.P., Rocha Santos, T., Duarte, A. The environmental impacts of plastics and micro-plastics use, waste and pollution: EU and national measures. *EU and National Measures*, EPRS: European Parliamentary Research Service 2020, 76. Belgium. Retrieved from https://coilink.org/20.500.12592/k9rb4j on 31 Dec 2024, COI: 20.500.12592/k9rb4j.

Daghighi, E., Shah, T., Chia, R.W., Jin-Yong, Shang, L.J., Rodríguez-Seijo, A. The forgotten impacts of plastic contamination on terrestrial micro- and meso-fauna: a call for research *Environ. Res.* 2023, 231, 116227.

Dahl, M., Bergman, S., Björk, M., Diaz-Almela, E., Granberg, M., Gullström, M., Leiva-Dueñas, C., Magnusson, K., Marco-Méndez, C., Piñeiro-Juncal, N. A temporal record of microplastic pollution in mediterranean seagrass soils. *Environ. Pollut.* 2021, 273, 116451.

Daigavane, P.B., Dawande, G.M., Gulhane, S.W., Chaudhari, N.D. Improvement of B. C. soil using geofoam prefabricated vertical drain. In *Proceedings of the Indian Geotechnical Conference GEOtrendz*, Mumbai, India, 16–18 December 2010.

Daniel, D.E., Wu, Y. Compacted clay liners and covers for arid sites. *J. Geotech. Eng.* 1993, 119, 223–237.

Daud, K.A. Soil improvement using waste tire chips. *Int. J. Civ. Eng. Technol.* 2018, 9, 1338–1345.

Dawson, A., Huston, W., Kawaguchi, S., King, C., Cropp, R., Wild, S., Eisenmann, P., Townsed, K., Bengtson Nash, S. Uptake and depuration kinetics infiuence microplastic bioaccumulation and toxicity in Antarctic Krill (Euphausia superba). *Environ. Sci. Technol.* 2018, 52, 3195–3201.

de Souza MacHado, A.A., Lau, C.W., Till, J., Kloas, W., Lehmann, A., Becker, R., Rillig, M.C. Impacts of microplastics on the soil biophysical environment. *Environ. Sci. Technol.* 2018, 52, 9656–9665.

De, A.S.M., Kloas, W., Zarfi, C., Hempel, S., Rillig, M.C. Microplastics as an emerging threat to terrestrial ecosystems. *Glob. Chang. Biol.* 2018, 24, 1405–1416.

Dexter, A.R. Soil Physical quality: part, I. Theory, effects of soil texture, density, and organic matter, and effects on root growth. *Geoderma* 2004, 120, 201–214.

Ding, J., Zhu, D., Wang, H.T., Lassen, S.B., Chen, Q.L., Li, G. Dysbiosis in the gut microbiota of soil fauna explains the toxicity of tire tread particles. *Environ. Sci. Technol.* 2020, 54(12), 7450–7460. https://doi.org/10.1021/acs.est.0c00917

Dong, S., Qu, M., Rui, Q., Wang, D. Combinational effect of titanium dioxide nanoparticles and nanopolystyrene particles at environmentally relevant concentrations on nematode Caenorhabditis elegans. Ecotoxicol. *Environ. Saf.* 2018, 161, 444–450. https://doi.org/10.1016/j.ecoenv.2018.06.021

Dris, R., Gasperi, J., Mirande, C., Mandin, C., Guerrouache, M., Langlois, V., Tassin, B. A first overview of textile fibres, including microplastics, in indoor and outdoor environments. *Environ. Pollut.* 2017, 221, 453–458.

Duis, K., Coors, A. Microplastics in the aquatic and terrestrial environment: sources (with a specific focus on personal care products), fate and effects. *Environ. Sci. Eur.* 2016, 28, 2.

Duncan, J.M., Wright, S.G. *Soil Strength Slope Stab*, 1st ed.; Wiley: Hoboken, NJ, USA, 2005; ISBN 9780471691631.

Estebenet, A.L. Food and feeding in pomacea canaliculata (gastropoda: ampullariidae). *Veliger* 1995, 38, 277–283.

Fajardo, C., Martín, C., Costa, G., S anchez-Fortún, S., Rodríguez, C., de Lucas Burneo, J.J.. Assessing the role of polyethylene microplastics as a vector for organic pollutants in soil: ecotoxicological and molecular approaches. *Chemosphere* 2022, 288, 132460. https://doi.org/10.1016/j.chemosphere.2021.132460

Feng, S., Yue, Y., Zheng, M., Li, Y., Zhang, Q., Wang, W. IsPETase - and IsMHETase-catalyzed cascade degradation mechanism toward polyethylene terephthalate. *ACS Sustain. Chem. Eng.* 2021, 9, 9823–9832.

Ficociello, G., Gerardi, V., Uccelletti, D., Setini, A. Molecular and cellular responses to short exposure to bisphenols A, F, and S and eluates of microplastics in C. elegans. *Environ. Sci. Pollut. Res.* 2021, 28, 805–818.

Foose, G.J., Benson, C.H., Edil, T.B. Predicting leakage through composite landfill liners. *J. Geotech. Geoenviron. Eng.* 2001, 127, 510–520.

Fueser, H., Mueller, M., Weiss, L., Hoss, S., Traunspurger, W. Ingestion of microplastics by nematodes depends on feeding strategy and buccal cavity size. *Environ. Pollut.* 2019, 255, 113227. https://doi.org/10.1016/j.envpol.2019.113227

Fuhr, L., Franklin, M. *The PLASTIC ATLAS 2019*; Heinrich Böll Foundation: Berlin, Germany, 2019; ISBN 9783869282114.

Fuller, S., & Gautam, A. (2016). A procedure for measuring microplastics using pressurized fluid extraction. Environmental Science & Technology, 50(11), 5774–5780. doi:10.1021/acs.est.6b00816

Gajendiran, A., Krishnamoorthy, S., Abraham, J. Microbial degradation of low-density polyethylene (LDPE) by Aspergillus clavatus strain JASK1 isolated from landfill soil. 3 Biotech 2016, 6, 52.

Gan, C.H., Tan, S.M. Some construction experiences on soft soil using light weight materials. In *Proceedings of the 2nd International Conference on Advances in Soft Soil Engineering and Technology*, Putrajaya, Malaysia, 2–4 July 2003.

Gaylor, M.O., Harvey, E., Hale, R.C. Polybrominated diphenyl ether (PBDE) accumulation by earthworms (Eisenia fetida) exposed to biosolids-, polyurethane foam microparticle-, and Penta-BDE-amended soils. *Environ. Sci. Technol.* 2013, 47, 13831–13839.

Geyer, R., Jambeck, J.R., Law, K.L. Production, use, and fate of all plastics ever made. *Sci. Adv.* 2017, 3, 25–29.

Ghiselli, A., Serafini, M., Natella, F., Scaccini, C. Total antioxidant capacity as a tool to assess redox status: critical view and experimental data. *Free Radic. Biol. Med.* 2000, 29, 1106–1114.

Ghosh, S.K., Agamuthu, P. Plastics in municipal solid waste: what, where, how and when? *Waste Manag. Res.* 2019, 37, 1061–1062.

Gigault, J., Ter Halle, A., Baudrimont, M., Pascal, P.Y., Gauffre, F., Phi, T.L., el Hadri, H., Grassl, B., Reynaud, S. Current opinion: what is a nanoplastic? *Environ. Pollut.* 2018, 235, 1030–1034.

Giroud, J., Soderman, K. Criterion for acceptable bentonite loss from a GCL incorporated into a liner system. *Geosynth. Int.* 2000, 7, 529–581.

Goldstein, M.C., Carson, H.S., Eriksen, M. Relationship of diversity and habitat area in North Pacific plastic-associated rafting communities. *Mar. Biol.* 2014, 161, 1441–1453.

Green, D.S. Effects of microplastics on European fiat oysters, Ostrea edulis and their associated benthic communities. *Environ. Pollut.* 2016, 216, 95–103.

Güler, E., Özhan, H.O., Karaoglu, S. Hydraulic performance of anionic polymer-treated bentonite-granular soil mixtures. *Appl. Clay Sci.* 2018, 157, 139–147.

Guo, J.J., Huang, X.P., Xiang, L., Wang, Y.Z., Li, Y.W., Li, H., Cai, Q.Y., Mo, C.H., Wong, M.H. Source, migration and toxicology of microplastics in soil. *Environ. Int.* 2020, 137, 105263.

Hamidi, A., Hooresfand, M. Effect of fiber reinforcement on triaxial shear behavior of cement treated sand. *Geotext. Geomembr.* 2013, 36, 1–9.

Hanson, J.L., Yesiller, N., Kendall, L.A. Integrated temperature and gas analysis at a municipal solid waste landfill. In *Proceedings of the 16th International Conference on Soil Mechanics and Geotechnical Engineering: Geotechnology in Harmony with the Global Environment*, Osaka, Japan, 12 September 2005.

He, D., Luo, Y., Lu, S., Liu, M., Song, Y., Lei, L. Microplastics in soils: analytical methods, pollution characteristics and ecological risks. *Trends Analyt.* Chem. 2018, 109, 163–172.

He, P., Chen, L., Shao, L., Zhang, H., Lü, F. Municipal solid waste (MSW) landfill: a source of microplasticsfi -evidence of microplastics in landfill leachate. *Water Res.* 2019, 159, 38–45.

Hellevik, C.C., Cyvin, J.B. Plastic pollution: about time to unify research methods and demand systemic changes. *Front. Environ. Sci.* 2023, 11, 1232974. https://doi.org/10.3389/fenvs.2023.1232974

Hodson, M.E., DuffuseHodson, C.A., Clark, A., PrendergasteMiller, M.T., Thorpe, K.L. Plastic bag derived-microplastics as a vector for metal exposure in terrestrial invertebrates. *Environ. Sci. Technol.* 2017, 51, 4714–4721.

Holtz, R.D., Kovacs, W.D., Sheahan, T.C. *An Introduction to Geotechnical Engineering*, 2nd ed.; Pearson: London, UK, 2010.

Horton, A.A., Walton, A., Spurgeon, D.J., Lahive, E., Svendsen, C. Microplastics in freshwater and terrestrial environments: evaluating the current understanding to identify the knowledge gaps and future research priorities. *Sci. Total Environ.* 2017, 586, 127–141.

Hu, L., Su, L., Xue, Y., Mu, J., Zhu, J., Xu, J., Shi, H. Uptake, accumulation and elimination of polystyrene microspheres in tadpoles of Xenopus tropicalis. *Chemosphere* 2016, 164, 611–617.

Huerta Lwanga, E., Gertsen, H., Gooren, H., Peters, P., Salanki, T., van der Ploeg, M., Besseling, E., Koelmans, A.A., Geissen, V. Microplastics in the terrestrial ecosystem: implications for Lumbricus terrestris (Oligochaeta, lumbricidae). *Environ. Sci. Technol.* 2016, 50, 2685–2691.

Huerta Lwanga, E., Gertsen, H., Gooren, H., Peters, P., Salanki, T., van der Ploeg, M., Besseling, E., Koelmans, A.A., Geissen, V. Incorporation of microplastics from litter into burrows of Lumbricus terrestris. *Environ. Pollut.* 2017, 220, 523–531.

Hughes, K.L., Christy, A.D., Heimlich, J.E. *Landfill Types and Liner Systems. Extension Fact Sheet*; The Ohio State University: Columbus, OH, USA, 2005.

Hurley, R.R., Lusher, A.L., Olsen, M., Nizzetto, L. Validation of a method for extracting microplastics from complex, organic-rich, environmental matrices. *Environ. Sci. Technol.* 2018, 52, 7409–7417.

Hwang, J., Choi, D., Han, S., Jung, S.Y., Choi, J., Hong, J. Potential toxicity of polystyrene microplastic particles. *Sci. Rep.* 2020, 10, 7391.

Imhof, H.K., Laforsch, C., Wiesheu, A.C., Schmid, J., Anger, P.M., Niessner, R., Ivleva, N.P. Pigments and plastic in limnetic ecosystems: a qualitative and quantitative study on microparticles of different size classes. *Water Res.* 2016, 98, 64–74.

Ioakeimidis, C., Fotopoulou, K.N., Karapanagioti, H.K., Geraga, M., Zeri, C., Papathanassiou, E., Galgani, F., Papatheodorou, G. The degradation potential of PET bottles in the marine environment: an ATR-FTIR based approach. *Sci. Rep.* 2016, 6, 23501.

Islam, M., Nasrin, M., Khan, A. Foundation alternatives in dredge fill soils overlaying organic clay. *Lowl. Technol. Int.* 2013, 15, 1–14.

Jabeen, K., Li, B., Chen, Q., Su, L., Wu, C., Hollert, H., Shi, H. Effects of virgin microplastics on goldfish (Carassius auratus). *Chemosphere* 2018, 213, 323–332.

Jahnke, A., Arp, H.P.H., Escher, B.I., Gewert, B., Gorokhova, E., Kühnel, D., Ogonowski, M., Potthoff, A., Rummel, C., Schmitt-Jansen, M. Reducing uncertainty and confronting ignorance about the possible impacts of weathering plastic in the marine environment. *Environ. Sci. Technol. Lett.* 2017, 4, 85–90.

Ji, X., Ma, Y., Zeng, G., Xu, X., Mei, K., Wang, Z., Chen, Z., Dahlgren, R., Zhang, M., Shang, X. Transport and fate of microplastics from riverine sediment dredge piles: implications for disposal. *J. Hazard. Mater.* 2021, 404, 124132.

Jia, R., Cao, L.P., Du, J.L., Wang, J.H., Liu, Y.J., Jeney, G., Xu, P., Yin, G.J. Effects of carbon tetrachloride on oxidative stress, infiammatory response and hepatocyte apoptosis in common carp (Cyprinus carpio). *Aquat. Toxicol.* 2014, 152, 11–19.

Jovanovic, B., Gokdag, K., Güven, O., Emre, Y., Whitley, E.M., Kideys, A.E. Virgin microplastics are not causing imminent harm to fish after dietary exposure. *Mar. Pollut. Bull.* 2018, 130, 123–131.

Ju, H., Zhu, D., Qiao, M. Effects of polyethylene microplastics on the gut microbial community, reproduction and avoidance behaviors of the soil springtail, Folsomia candida. *Environ. Pollut.* 2019, 247, 890–897. https://doi.org/10.1016/j.envpol.2019.01.097

Kang, P., Hu, Q., Lu, Y., Lei, B., He, G., & Wang, C. (2013). Residual plastic film in typical maize planting areas in Lijiang City of Yunnan Province. Hunan Agricultural Sciences, 3, 56–58. doi:10.16498/j.cnki.hnnykx.2013.03.034

Kanold, E., Rillig, M., Antunes, P.M. Microplastics and phagotrophic soil protists: evidence of ingestion: see video as supplementary material. *Soil Org.* 2021, 93(2), 133–140. https://doi.org/10.25674/so93iss2id160

Karamanlioglu, M., Robson, G.D. The infiuence of biotic and abiotic factors on the rate of degradation of poly(lactic) acid (PLA) coupons buried in compost and soil. *Polym. Degrad. Stab.* 2013, 98, 2063–2071.

Kaur, R., Kaur, J., Mahajan, J., Kumar, R., Arora, S., Oxidative stress–implications, source and its prevention. *Environ. Sci. Pollut. Res.* 2014, 21, 1599–1613.

Kazour, M., Terki, S., Rabhi, K., Jemaa, S., Khalaf, G., Amara, R. Sources of microplastics pollution in the marine environment: importance of wastewater treatment plant and coastal landfill. *Mar. Pollut. Bull.* 2019, 146, 608–618.

Kiyama, Y., Miyahara, K., Ohshima, Y. Active uptake of artificial particles in the nematode Caenorhabditis elegans. *J. Exp. Biol.* 2012, 215, 1178–1183. https://doi:10.1242/jeb.067199

Kilponen, J. Microplastics and harmful substances in urban runoffs and landfill leachates: possible emission sources to marine environment. Bachelor's Thesis in Environmental Engineering Summer 2016, Faculty of Technology, Degree programme in Environmental Technology, Volume Environmental Technology; Lahti University of Applied Sciences: Lahti, Finland, 2016.

Kim, S.W., An, Y.J. Soil microplastics inhibit the movement of springtail species. *Environ. Int.* 2019, 126, 699–706. https://doi.org/10.1016/j.envint.2019.02.067

Kim, S.W., Waldman, W.R., Kim, T.-Y., Rillig, M.C. Effects of different microplastics on nematodes in the soil environment: tracking the extractable additives using an ecotoxicological approach. *Environ. Sci. Technol.* 2020, 54, 13868–13878. https://doi.org/10.1021/acs.est.0c04641

Kliem, S., Kreutzbruck, M., Bonten, C. Review on the biological degradation of polymers in various environments. *Materials* 2020, 13, 4586.

Koerner, G.R. Assessing potential geomembrane damage from direct construction equipment contact. *Geo Front. Congr.* 2005a, 1, 1–3.

Koerner, R.M. *Designing with Geosynthetics*; Pearson: London, UK, 2005b.

Kole, P.J., Löhr, A.J., Van Belleghem, F.G.A.J., Ragas, A.M.J. Wear and tear of tyres: a stealthy source of microplastics in the environment. *Int. J. Environ. Res. Public Health* 2017, 14, 1265.

Krueger, M.C., Harms, H., Schlosser, D. Prospects for microbiological solutions to environmental pollution with plastics. *Appl. Microbiol. Biotechnol.* 2015, 99, 8857–8874.

Kukharchyk, T., Chernyuk, V. Microplastic of polystyrene in soil and water: fluxes study from industrial site. In Proceedings of the EGU General Assembly 2020, Vienna, Austria, 4–8 May 2020.

Kyrikou, I., Briassoulis, D. Biodegradation of agricultural plastic films: a critical review. *J. Polym. Environ.* 2007, 15, 125–150.

Kwak, J.I., An, Y.J. Post COVID-19 pandemic: biofragmentation and soil ecotoxicological effects of microplastics derived from face masks. *J. Hazard Mater.* 2021, 416, 126169. https://doi.org/10.1016/j.jhazmat.2021.126169

Lahive, E., Walton, A., Horton, A.A., Spurgeon, D.J., Svendsen, C. Microplastic particles reduce reproduction in the terrestrial worm Enchytraeus crypticus in a soil exposure. Environ. Pollut. 2019, 255, 113174. https://doi.org/10.1016/j.envpol.2019.113174

Lambert, S., Wagner, M. Formation of microscopic particles during the degradation of different polymers. *Chemosphere* 2016, 161, 510–517.

Lange, C.N., Mwinzi, M. Snail diversity, abundance and distribution in arabuko sokoke forest, Kenya. *Afr. J. Ecol.* 2015, 41, 61–67.

Larba, R., Soltani, N. Use of the land snail Helix aspersa for monitoring heavy metal soil contamination in Northeast Algeria. *Environ. Monit. Assess.* 2014, 186, 4987–4995.

Lei, L., Liu, M., Song, Y., Lu, S., Hu, J., Cao, C., Xie, B., Shi, H., He, D. Polystyrene (nano)microplastics cause size-dependent neurotoxicity, oxidative damages and other adverse effects in Caenorhabditis elegans. *Environ. Sci. Nano.* 2018a, 5, 2009–2020.

Lei, L., Wu, S., Lu, S., Liu, M., Song, Y., Fu, Z., Shi, H., Raley-Susman, K.M., He, D. Microplastic particles cause intestinal damage and other adverse effects in zebrafish Danio rerio and nematode Caenorhabditis elegans. *Sci. Total Environ.* 2018b, 619e620, 1e8.

Li, K., Wang, F. Global hotspots and trends in interactions of microplastics and heavy metals: a bibliometric analysis and literature review. *Environ. Sci. Pollut. Res.* 2023, 30, 93309–93322. https://doi.org/10.1007/s11356-023-29091-7

Li, C.C., Dang, F., Cang, L., Zhou, D.M., Peijnenburg, W.J. Internal distribution of Cd in lettuce and resulting effects on Cd trophic transfer to the snail: Achatina fulica. *Chemosphere* 2015, 135, 123–128.

Li, W., Wufuer, R., Duo, J., Wang, S., Luo, Y., Zhang, D., Pan, X. Microplastics in agricultural soils: extraction and characterization after different periods of polythene film mulching in an arid region. *Sci. Total Environ.* 2020, 749, 141420.

Lim, X.Z. Microplastics are everywhere—but are they harmful? *Nature* 2021, 593, 22–25.

Lin, D., Yang, G., Dou, P., Qian, S., Zhao, L., Yang, Y. Microplastics negatively affect soil fauna but stimulate microbial activity: insights from a field- based microplastic addition experiment. *Proc. Royal Soc. B.* 2020, 287 (1934), 20201268. https://doi.org/10.1098/rspb.2020.1268

Liu, H., Wang, D. Intestinal mitochondrial unfolded protein response induced by nanoplastic particles in Caenorhabditis elegans. *Chemosphere* 2021, 267, 128917. https://doi.org/10.1016/j.chemosphere.2020.128917

Liu, M., Lu, S., Song, Y., Lei, L., Hu, J., Lv, W., Zhou, W., Cao, C., Shi, H., Yang, X., He, D. Microplastic and mesoplastic pollution in farmland soils in suburbs of Shanghai, China. *Environ. Pollut.* 2018, 242, 855–862.

Liu, X., Shi, H., Xie, B., Dionysiou, D.D., Zhao, Y. Microplastics as both a sink and a source of bisphenol A in the marine environment. *Environ. Sci. Technol.* 2019, 53, 10188–10196.

Liu, Y., Gates, W.P., Bouazza, A. Acid induced degradation of the bentonite component utilised in geosynthetic clay liners. *Geotext. Geomembr.* 2013, 36, 71–80.

Liu, Y., Shao, H., Liu, J., Cao, R., Shang, E., Liu, S., Li, Y. Transport and transformation of microplastics and nanoplastics in the soil environment: a critical review. *Soil Use Manag.* 2021, 37, 224–242.

Lu, Y., Zhang, Y., Deng, Y., Jiang, W., Zhao, Y., Geng, J., Ding, L., Ren, H. Uptake and accumulation of polystyrene microplastics in Zebrafish (Danio rerio) and toxic effects in liver. *Environ. Sci. Technol.* 2016, 50, 4054–4060.

Luo, Y., Wang, L., Cao, T., Chen, J., Lv, M., Wei, S. Microplastics are transferred by soil fauna and regulate soil function as material carriers. *Sci. Total Environ.* 2023, 857, 159690. https://doi.org/10.1016/j.scitotenv.2022.159690

Lv, W., Zhou, W., Lu, S., Huang, W., Yuan, Q., Tian, M., He, D. (2019). Microplastic pollution in rice-fish co-culture system: A report of three farmland stations in Shanghai, China. Science of the Total Environment, 652, 1209–1218. doi:10.1016/j.scitotenv.2018.10. 321

Lwanga, E.H., Vega, J.M., Quej, V.K., de los Angeles Chi, J., Del Cid, L.S., Chi, C., Segura, G.E., Gertsen, H., Salánki, T., van der Ploeg, M. Field evidence for transfer of plastic debris along a terrestrial food chain. *Sci. Rep.* 2017, 7, 1–7.

Ma, H., Mei, X., Yan, C., He, W., & Li, K. (2008). The residue of mulching plastic film of cotton field in north. China. Journal of Agro-Environment Science, 02(2), 570–573. Retrieved from http://en.cnki.com.cn/Article_en/CJFDTOTAL-NHBH200802034.htm

Ma, J., Sheng, G.D., Chen, Q.L., O'Connor, P. Do combined nanoscale polystyrene and tetracycline impact on the incidence of resistance genes and microbial community disturbance in Enchytraeus crypticus? J. *Hazard Mater.* 2020b, 387, 122012. https://doi.org/10.1016/j.jhazmat.2019.122012

Ma, J., Sheng, G.D., O'Connor, P. Microplastics combined with tetracycline in soils facilitate the formation of antibiotic resistance in the Enchytraeus crypticus microbiome. *Environ. Pollut.* 2020a, 264, 114689. https://doi.org/10.1016/j.envpol.2020.114689

Maaß, S., Daphi, D., Lehmann, A., Rillig, M.C. Transport of microplastics by two collembolan species. *Environ. Pollut.* 2017, 225, 456–459. https://doi.org/10.1016/j.envpol.2017.03.009

Mahon, A.M., O'Connell, B., Healy, M., O'Connor, I., Officer, R., Nash, R., Morrison, L. Microplastics in sewage sludge: effects of treatment. *Environ. Sci. Technol.* 2017, 51, 810–818.

Mai, L., Bao, L.J., Wong, C.S., Zeng, E.Y. Microplastics in the terrestrial environment. In Zeng, E.Y. (Ed.), Microplastic Contamination in Aquatic Environments -An Emerging Matter of Environmental Urgency. Elsevier Inc.: Amsterdam, 2018; pp. 365–378.

Manikanda, B.K., Usha, N., Vaikunth, R., Praveen, K.R., Ruthra, R., Srinivasalu, S. Spatial distribution of microplastic concentration around landfill sites and its potential risk on groundwater. *Chemosphere* 2021, 277, 130263.

Mathalon, A., Hill, P. Microplastic fibres in the intertidal ecosystem surrounding Halifax harbor, Nova Scotia. *Mar. Pollut. Bull.* 2014, 81, 69–79.

Meixner, K., Kubiczek, M., Fritz, I. Microplastic in soil—current status in Europe with special focus on method tests with Austrian samples. *AIMS Environ. Sci.* 2020, 7, 174–191.

Mendes, L.A. Microplastics effects in the terrestrial environment. In: Rocha-Santos, T., Costa, M., Mouneyrac, C. (Eds.), Handbook of Microplastics in the Environment. 2021, Springer, Cham, pp. 1–30. https://doi.org/10.1007/978-3-030-10618-8_46-1

Möller, J.N., Löder, M.G.J., Laforsch, C. Finding microplastics in soils: a review of analytical methods. *Environ. Sci. Technol.* 2020, 54, 2078–2090.

Monkul, M.M., Aydin, N.G., Demirhan, B., Sahin, M. Undrained shear strength and monotonic behavior of different nonplastic silts: sand-like or clay-like? *Geotech. Test. J.* 2020, 43, 20180147.

Monkul, M.M., Ozden, G. Compressional behavior of clayey sand and transition fines content. *Eng. Geol.* 2007, 89, 195–205.

Monkul, M.M., Özhan, H.O. Microplastic contamination in soils: a review from geotechnical engineering view. *Polymers* 2021, 13, 4129. https://doi.org/10.3390/polym13234129

Monkul, M.M., Yükselen-Aksoy, Y. Sustainable usage of dredge materials at engineering fills and investigation of their performance with different additives. In Geotechnical Special Publication No 273: Sustainable Waste Management and Remediation. ASCE: Reston, VI, USA, 2016; pp. 471–480. ISBN 978-0-7844–8016-8. DOI: 10.1061/9780784480168.047

Moorsel, D., Kilpeläinen, T., Meuwissen, E., Neirinckx, L., Tepper, H., Thompsett, D., Zipp, K. *EPS White Book,* EUMEPS Background Information on Standardization of EPS; EUMEPS Construction: Maaseik, Belgium, 2014.

Müller, C., Townsend, K., Matschullat, J. Experimental degradation of polymer shopping bags (standard and degradable plastic, and biodegradable) in the gastrointestinal fiuids of sea turtles. *Sci. Total Environ.* 2012, 416, 464–467.

Narevski, A.C., Novaković M.I., Petrović M.Z., Mihajlović I.J., Maoduš, N.B., Vujić G.V. Occurrence of bisphenol A and microplastics in landfill leachate: lessons from South East Europe. *Environ. Sci. Pollut. Res.* 2021, 28, 42196–42203.

Ng, E.L., Huerta Lwanga, E., Eldridge, S.M., Johnston, P., Hu, H.W., Geissen, V., Chen, D. An overview of microplastic and nanoplastic pollution in agroecosystems. *Sci. Total. Environ.* 2018, 627, 1377–1388.

Nizzetto, L., Bussi, G., Futter, M.N., Butterfield, D., Whitehead, P.G. A theoretical assessment of microplastic transport in river catchments and their retention by soils and river sediments. *Environ. Sci. Process. Impacts* 2016, 18, 1050–1059.

O'Connor, D., Pan, S., Shen, Z., Song, Y., Jin, Y., Wu, W.M., Hou, D. Microplastics undergo accelerated vertical migration in sand soil due to small size and wet-dry cycles. *Environ. Pollut.* 2019, 249, 527–534.

O'Kelly, B.C., El-Zein, A., Liu, X., Patel, A., Fei, X., Sharma, S., Mohammad, A., Goli, V.S.N.S., Wang, J.J., Li, D., et al. Microplastics in soils: an environmental geotechnics perspective. *Environ. Geotech.* 2021, 2000179, 1–30.

Öncü, Ş., Bilsel, H. Infiuence of polymeric fiber reinforcement on strength properties of sand-stabilized expansive soil. *Polym. Technol. Eng.* 2017, 56, 391–399.

Orhan, Y., Hrenović J., Büyükgüngör, H. Biodegradation of plastic compost bags under controlled soil conditions. *Acta Chim. Slov.* 2004, 51, 579–588.

Orsini, C., Rowe, R.K. Testing procedure and results for the study of internal erosion of geosynthetic clay liners. In *Proceedings of the Geosynthetics*, Portland, OR, USA, 12–14 February 2001; pp. 189–201.

Otake, Y., Kobayashi, T., Asabe, H., Murakami, N., Ono, K. Biodegradation of low-density polyethylene, polystyrene, polyvinyl chloride, and urea formaldehyde resin buried under soil for over 32 years. *J. Appl. Polym. Sci.* 1995, 56, 1789–1796.

Özer, A.T., Akay, O. Interface shear strength characteristics of interlocked EPS-block geofoam. *J. Mater. Civ. Eng.* 2016, 28, 04015156.

Özhan, H.O., Güler, E. Factors affecting failure by internal erosion of geosynthetic clay liners utilised in fresh water reservoirs. *Environ. Eng. Geosci.* 2016, 22, 157–169.

Özhan, H.O., Güler, E. Use of perforated base pedestal to simulate the gravel subbase in evaluating the internal erosion of geosynthetic clay liners. *Geotech. Test. J.* 2013, 36, 418–428.

Palmisano, A.C., Pettigrew, C.A. Biodegradability of plastics. *BioScience* 1992, 42, 680–685.

Ped, C., Caccamo, L., Fossi, M.C., Gai, F., Andaloro, F., Genovese, L., Perdichizzi, A., Romeo, T., Maricchiolo, G. Intestinal alterations in European sea bass Dicentrarchus labrax (Linnaeus, 1758) exposed to microplastics: preliminary results. *Environ. Pollut.* 2016, 212, 251–256.

Pereira, A., Carvalho, A.P., Cruz, C., Saraiva, A. Histopathological changes and zootechnical performance in juvenile zebrafish (Danio rerio) under chronic exposure to nitrate. *Aquaculture* 2017, 473, 197–205.

Pflugmacher, S., Huttunen, J.H., Wolff, M.V., Penttinen, O.P., Kim, Y.J., Kim, S. Enchytraeus crypticus avoid soil spiked with microplastic. *Toxics* 2020, 8(1), 10. https://doi.org/10.3390/toxics8010010

Piehl, S., Leibner, A., L oder, M. G. J., Dris, R., Bogner, C., & Laforsch, C. (2018).Identification and quantification of macro- and microplastics on an agricultural farm-land. Scientific Reports, 8(1), 17950. doi:10.1038/s41598-018-36172-y

Plastics Europe—Association of Plastics Manufactures Plastics—The Facts 2020. PlasticsEurope 2020, 1–64. Available online: https://plasticseurope.org/knowledge-hub/plastics-the-facts-2020/ (accessed on 8 September 2023).

Połeć, M., Aleksander-Kwaterczak, U., Wator, K., Kmiecik, E. The occurrence of microplastics in freshwater systems—preliminary results from Krakow (Poland). *Geol. Geophys. Environ.* 2018, 44, 391.

Prata, J.C., da Costa, J.P., Lopes, I., Duarte, A.C., Rocha-Santos, T. Environmental exposure to microplastics: an overview on possible human health effects. *Sci. Total Environ.* 2020, 702, 134455.

Promputthangkoona, P., Karnchanachetaneea, B. Geomaterial prepared from waste tyres, soil and cement. *Procedia Soc. Behav. Sci.* 2013, 91, 421–428.

Punthutaecha, K., Puppala, A.J., Vanapalli, S.K., Inyang, H. Volume change behaviors of expansive soils stabilized with recycled ashes and fibres. *J. Mater. Civ. Eng.* 2006, 18, 295–306.

Puthcharoen, A., Leungprasert, S. Determination of microplastics in soil and leachate from the landfills. *Thai Environ. Eng. J.* 2019, 33, 39–46.

Qian, G., Koerner, R.M., Gray, D. *Geotechnical Aspects of Landfill Design and Construction*, Pearson: London, UK, 2001.

Qian, X., Gray, D.H., Koerner, R.M. Estimation of maximum liquid head over landfill barriers. *J. Geotech. Geoenviron. Eng.* 2004, 130, 488–497.

Qu, H., Ma, R., Wang, B., Zhang, Y., Yin, L., Yu, G., Deng, S., Huang, J., Wang, Y. Effects of microplastics on the uptake, distribution and biotransformation of chiral antidepressant venlafaxine in aquatic ecosystem. *J. Hazard Mater.* 2018, 359, 104–112.

Ramos, L., Berenstein, G., Hughes, E. A., Zalts, A., & Montserrat, J. M. (2015). Polyethylene film incorporation into the horticultural soil of small periurban production units in Argentina. Science of the Total Environment, 523, 74–81. doi:10.1016/j.scitotenv.2015.03.142

Reddy, M.M., Gupta, R.K., Gupta, R.K., Bhattacharya, S.N., Parthasarathy, R. Abiotic Oxidation studies of oxo-biodegradable polyethylene. *J. Polym. Environ.* 2008, 16, 27–34.

Reddy, V.R., Reddy, I.S., Prasad, D.S.V. Improvement of soil characteristics using shredded rubber. *IOSR J. Mech. Civ. Eng.* 2016, 13, 44–48.

Ren, X., Tang, J., Liu, X., Liu, Q. Effects of microplastics on greenhouse gas emissions and the microbial community in fertilized soil. *Environ. Pollut.* 2019, 113347. doi: 10.1016/j.envpol.2019.113347

Rezaei, M., Riksen, M.J., Sirjani, E., Sameni, A., Geissen, V. Wind erosion as a driver for transport of light density microplastics. *Sci. Total Environ.* 2019, 669, 273–281.

Rillig, M.C., Ziersch, L., Hempel, S. Microplastic transport in soil by earthworms. *Sci. Rep.* 2017, 7, 1362.

Rist, S., Baun, A., Hartmann, N.B. Ingestion of micro- and nanoplastics in Daphnia magna -quantification of body burdens and assessment of feeding rates and reproduction. *Environ. Pollut.* 2017, 228, 398–407.

Rochman, C.M., Tahir, A., Williams, S.L., Baxa, D.V., Lam, R., Miller, J.T., Teh, F.C., Werorilangi, S., Teh, S.J. Anthropogenic debris in seafood: plastic debris and fibres from textiles in fish and bivalves sold for human consumption. *Sci. Rep.* 2015, 5, 14340.

Rosenkranz, P., Chaudhry, Q., Stone, V., Fernandes, T.F.A. Comparison of nanoparticle and fine particle uptake by Daphnia magna. *Environ. Toxicol Chem.* 2009, 28, 2142–2149.

Schellenberg, J. *Syndiotactic Polystyrene: Synthesis, Characterization, Processing, and Applications*; John Wiley & Sons: Hoboken, NJ, USA, 2009.

Scheurer, M., Bigalke, M. Microplastics in Swiss fioodplain soils. *Environ. Sci. Technol.* 2018, 52, 3591–3598.

Schöpfer, L., Menzel, R., Schnepf, U., Ruess, L., Marhan, S., Brümmer, F., Pagel, H., Kandeler, E. Microplastics effects on reproduction and body length of the soil-dwelling nematode Caenorhabditis elegans. *Front. Environ. Sci.* 2020, 8, 41. https://doi.org/10.3389/fenvs.2020.00041

Selonen, S., Dolar, A., Jemec Kokalj, A., Skalar, T., Parramon Dolcet, L.I., Hurley, R.R. Exploring the impacts of plastics in soil – the effects of polyester textile fibers on soil invertebrates. *Sci. Total Environ.* 2020, 700, 134451. https://doi.org/10.1016/j.scitotenv.2019.134451

Selonen, S., Dolar, A., Kokalj, A.J., Sackey, L.N., Skalar, T., Fernandes, V.C. Exploring the impacts of microplastics and associated chemicals in the terrestrial environment – exposure of soil invertebrates to tire particles. *Environ. Res.* 2021, 201, 111495. https://doi.org/10.1016/j.envres.2021.111495

Selonen, S., Jemec Kokalj, A., Benguedouar, H., Alavian Petroody, S.S., Dolar, A., Drobne, D., van Gestel, C.A. Modulation of chlorpyrifos toxicity to soil arthropods by simultaneous exposure to polyester microfibers or tire particle microplastics. *Appl. Soil Ecol.* 2023, 181, 104657. https://doi.org/10.1016/j.apsoil.2022.104657

Sharma, D.H., Lewis, P.S. *Waste Containment Systems, Waste Stabilization and Landfills, Design and Evaluation, Waste Characterization and Solid-Waste Interaction*; John Wiley & Sons: Hoboken, NJ, USA, 1994.

Singh, B., Sharma, N. Mechanistic implications of plastic degradation. *Polym. Degrad. Stab.* 2008, 93, 561–584.

Singh, J., Sonthwal, V.K. Improvement of engineering properties of clayey soil using shredded rubber tyre. *Int. J. Theor. Appl. Sci.* 2017, 9, 1–6.

Snowdon, K. UK Plastic Waste Being Dumped and Burned in Turkey, Says Greenpeace. Available online: https://www.bbc.com/ news/uk-57139474 (accessed on 15 November 2021).

Solanki, D., Dave, M., Purohit, D.G.M. Stabilization of clay soil mixed with rubber tyre chips for design in road construction. *Int. J. Eng. Sci. Invent.* 2017, 6, 88–91.

Song, Y., Cao, C., Qiu, R., Hu, J., Liu, M., Lu, S., Shi, H., Raley-Susman, K.M., He, D. Uptake and adverse effects of polyethylene terephthalate microplastics fibres on terrestrial snails (Achatina fulica) after soil exposure. *Environ. Pollut.* 2019, 250, 447–455. doi: 10.1016/j.envpol.2019.04.066

Stamatiadis, S., Dindal, D.L. Coprophilous mite communities as affected by concentration of plastic and glass particles. *Exp. Appl. Acarol.* 1990, 8, 1–12. https://doi.org/10.1007/BF01193377

Stark, T.D., Arellano, D., Horvath, J.S., Leshchinsky, D. *Geofoam Applications in the Design and Construction of Highway Embankments*; Transportation Research Board: Washington, WA, USA, 2004.

Steinmetz, Z., Wollmann, C., Schaefer, M., Buchmann, C., David, J., Tröger, J., Muñoz, K., Frör, O., Schaumann, G.E. Plastic mulching in agriculture. Trading short-term agronomic benefits for long-term soil degradation? *Sci. Total Environ.* 2016, 550, 690–705.

Su, Y., Zhang, Z., Wu, D., Zhan, L., Shi, H., Xie, B. Occurrence of microplastics in landfill systems and their fate with landfill age. *Water Res.* 2019, 164, 114968.

Sundt, P., Schulze, P.E., Syversen, F. Sources of microplastic pollution to the marine environment. *Mepex Rep. Nor. Environ. Agency* 2014, 49–50.

Tabone, M.D., Cregg, J.J., Beckman, E.J., Landis, A.E. Sustainability metrics: life cycle assessment and green design in polymers. *Environ. Sci. Technol.* 2010, 45, 8264–8269.

Tang, C., Shi, B., Gao, W., Chen, F., Cai, Y. Strength and mechanical behavior of short polypropylene fiber reinforced and cement stabilized clayey soil. *Geotext. Geomembr.* 2007, 25, 194–202.

Terzaghi, K., Peck, R.B., Mesri, G. *Soil Mechanics in Engineering Practice*, 3rd ed.; John Wiley & Sons: Hoboken, NJ, USA, 1996.

Thompson, R.C., Olsen, Y., Mitchell, R.P., Davis, A., Rowland, S.J., John, A.W.G., Mcgonigle, D., Russell, A.E. Lost at sea: where is all the plastic? *Science* 2004, 304, 838.

Tingle, J.S., Newman, J.K., Larson, S.L., Weiss, C.A., Rushing, J.F. Stabilization mechanisms of nontraditional additives. *Transp. Res. Rec. J. Transp. Res. Board* 2007, 1989, 59–67.

Tokiwa, Y., Calabia, B.P., Ugwu, C.U., Aiba, S. Biodegradability of plastics. *Int. J. Mol. Sci.* 2009, 10, 3722–3742.

Tourinho, P.S., Loureiro, S., Talluri, V.S.S.L.P., Dolar, A., Verwejj, R., Chvojla, J. Microplastic fibers influence Ag toxicity and bioaccumulation in Eisenia andrei but not in Enchytraeus crypticus. *Ecotoxicology* 2021, 30, 1216–1226. https://doi.org/10.1007/s10646-021-02424-3

Tsuchida, T., Egashira, K. *The Lightweight Treated Soil Method-New Geomaterials for Soft Ground Engineering in Coastal Areas*; Balkema: Liden, The Netherlands, 2004.

UNEP (UN Environmental Programme). *UNEP Year Book 2014: Emerging Issues in Our Global Environment*; UNEP, Division of Early Warning and Assessment: Nairobi, Kenya, 2014; p. 71.

Van Cauwenberghe, L., Claessens, M., Vandegehuchte, M.B., Janssen, C.R., Microplastics are taken up by mussels (Mytilus edulis) and lugworms (Arenicola marina) living in natural habitats. *Environ. Pollut.* 2015, 199, 10–17.

van den Berg, P., Huerta-Lwanga, E., Corradini, F., Geissen, V. Sewage sludge application as a vehicle for microplastics in Eastern Spanish agricultural soils. *Environ. Pollut.* 2020, 261, 114198.

Van Gestel, C.A.M., Selonen, S. Ecotoxicological effects of microplastics in soil: comments on the paper by Zhu et al. (2018) "Exposure of soil collembolans to microplastics perturbs their gut microbiota and alters their isotopic composition." *Soil Biol. Biochem.* 116, 302–310. *Soil Biol. Biochem.* 124, 116–117.

Van Praagh, M., Hartman, C., Brandmyr, E. Microplastics in Landfill Leachates in the Nordic Countries. Tema.Nord. 2019. Nordic Council of Ministers, Copenhagen, TN2018–557. Available online: https://www.diva-portal.org/smash/get/diva2:1277395/FULLTEXT01.pdf (accessed on 8 September 2023).

Van't Hoff, J., van der Kolff, A.N. *Hydraulic Fill Manual: For Dredging and Reclamation Works*; CRC Press/Balkema Taylor & Francis Group: London, UK, 2012.

Vollertsen, J., & Hansen, A. A. (2017). Microplastic in Danish wastewater: Sources, occur-rences and fate. Retrieved from https://www.researchgate.net/publication/316966942. doi:10.3390/w11091935

von Moos, N., Burkhardt-Holm, P., Kohler, A. Uptake and effects of microplastics on cells and tissue of the blue mussel Mytilus edulis L. after an experimental exposure. *Environ. Sci. Technol.* 2012, 46, 11327–11335.

Wang, J., Coffin, S., Sun, C., Schlenk, D., Gan, J. Negligible effects of microplastics on animal fitness and HOC bioaccumulation in earthworm Eisenia fetida in soil. *Environ. Pollut.* 2019, 249, 776–784.

Wang, J., Liu, X., Dai, Y., Ren, J., Li, Y., Wang, X., Zhang, P., Peng, C. Effects of co-loading of polyethylene microplastics and ciprofloxacin on the antibiotic degradation efficiency and microbial community structure in soil. *Sci. Total Environ.* 2020a, 741, 140463. doi: 10.1016/j.scitotenv.2020.140463

Wang, W., Ge, J., Yu, X., Li, H. Environmental fate and impacts of microplastics in soil ecosystems: progress and perspective. *Sci. Total Environ.* 2020b, 708, 134841.

Wang, X., Liu, Z., Wang, W., Yan, Z., Zhang, C., Wang, W., Chen, L. Assessment of toxic effects of triclosan on the terrestrial snail (Achatina fulica). *Chemosphere* 2014, 108, 225–230.

Wang, Y., Wang, X., Li, Y., Li, J., Liu, Y., Xia, S., Zhao, J. Effects of exposure of polyethylene microplastics to air, water and soil on their adsorption behaviors for copper and tetracycline. *Chem. Eng. J.* 2021, 404, 126412. doi:10.1016/j.cej.2020.126412

Ward, J.E., Kach, D.J. Marine aggregates facilitate ingestion of nanoparticles by suspension-feeding bivalves. *Mar. Environ. Res.* 2009, 68, 137–142.

Watts, A.J.R., Lewis, C., Goodhead, R.M., Beckett, S.J., Moger, J., Tyler, C.R., Galloway, T.S. Uptake and retention of microplastics by the shore crab Carcinus maenas. *Environ. Sci. Technol.* 2014, 48, 8823–8830.

Wei, L., Chai, S.X., Zhang, H.Y., Shi, Q. Mechanical properties of soil reinforced with both lime and four kinds of fiber. *Constr. Build. Mater.* 2018, 172, 300–308.

Wright, S.L., Kelly, F.J. Plastic and human health: a micro issue? *Environ. Sci. Technol.* 2017, 51, 6634–6647.

Wu, B., Li, L.-W., Zu, Y.-X., Nan, J., Chen, X.-Q., Sun, K., Li, Z.-L. Microplastics contamination in groundwater of a drinking-water source area, northern China. *Environ. Res.* 2022, 214, 114048. doi: 10.1016/j.envres.2022.114048

Wu, P., Tang, Y., Jin, H., Song, Y., Liu, Y., Cai, Z. Consequential fate of bisphenol-attached PVC microplastics in water and simulated intestinal fluids. *Environmental Science and Ecotechnology* 2020a, 2, 100027. doi: 10.1016/j.ese.2020.100027

Wu, X., Lyu, X., Li, Z., Gao, B., Zeng, X., Wu, J., Sun, Y. Transport of polystyrene nanoplastics in natural soils: effect of soil properties, ionic strength and cation type. *Sci. Total Environ.* 2020b, 707, 136065.

Xiang, Q., Chen, Q., Yang, X., Li, G., Zhu, D. Soil mesofauna alter the balance between stochastic and deterministic processes in the plastisphere during microbial succession. *Sci. Total Environ.* 2022, 849, 157820. https://doi.org/10.1016/j.scitotenv.2022.157820

Xiang, Q., Zhu, D., Chen, Q.L., O'Connor, P., Yang, X.R., Qiao, M. Adsorbed sulfamethoxazole exacerbates the effects of polystyrene (~ 2 μm) on gut microbiota and the antibiotic resistome of a soil collembolan. *Environ. Sci. Technol.* 2019, 53(21), 12823–12834. https://doi.org/10.1021/acs.est.9b04795

Xu, B., Huang, D., Liu, F., Alfaro, D., Lu, Z., Tang, C., Gan, J., Xu, J. Contrasting effects of microplastics on sorption of diazepam and phenanthrene in soil. *J. Hazard. Mater.* 2020a, 406, 124312. DOI: 10.1016/j.jhazmat.2020.124312

Xu, B., Liu, F., Cryder, Z., Huang, D., Lu, Z., He, Y., Wang, H., Lu, Z., Brookes, P.C., Tang, C., Gan, J., Xu, J. Microplastics in the soil environment: occurrence, risks, interactions and fate – a review. *Crit. Rev. Environ. Sci. Technol.* 2019, 50, 1–48. https://doi.org/10.1080/10643389.2019.1694822

Xu, Z., Sui, Q., Li, A., Sun, M., Zhang, L., Lyu, S., Zhao, W. How to detect small microplastics (20–100 μm) in freshwater, municipal wastewaters and landfill leachatesfi A trial from sampling to identification. *Sci. Total Environ.* 2020b, 733, 139218.

Yang, B., Li, P., Entemake, W., Guo, Z., Xue, S. Concentration dependent impacts of microplastics on soil nematode community in bulk soils of maize: evidence from a pot experiment. *Front. Environ. Sci.* 2022a, 10, 872898. https://doi.org/10.3389/fenvs.2022.872898

Yang, J., Cang, L., Sun, Q., Dong, G., Ata-Ul-Karim, S.T., Zhou, D. Effects of soil environmental factors and UV aging on Cu^{2+} adsorption on microplastics. *Environ. Sci. Pollut. Res.* 2019, 26, 23027–23036. https://doi.org/10.1007/s11356-019-05643-8

Yang, L., Wang, X., Ma, J., Li, G., Wei, L.S., Sheng, G.D. Nanoscale polystyrene intensified the microbiome perturbation and antibiotic resistance genes enrichment in soil and Enchytraeus crypticus caused by tetracycline. *Appl. Soil Ecol.* 2022b, 174, 104426. DOI: 10.1016/j.apsoil.2022.104426

Yang, N., Damgaard, A., Kjeldsen, P., Shao, L.-M., He, P.-J. Quantification of regional leachate variance from municipal solid waste landfills in China. *Waste Manag.* 2015a, 46, 362–372.

Yang, Y., Yang, J., Wu, W.M., Zhao, J., Song, Y., Gao, L., Yang, R., Jiang, L. Biodegradation and mineralization of polystyrene by plastic-eating meal-worms: part 1. Chemical and physical characterization and isotopic tests. *Environ. Sci. Technol.* 2015b, 49, 12080–12086.

Yang, X., Bento, C. P. M., Chen, H., Zhang, H., Xue, S., Lwanga, E. H., Geissen, V. (2018). Influence of microplastic addition on glyphosate decay and soil microbial activities in Chinese loess soil. Environmental Pollution, 242, 338–347. doi:10.1016/j.envpol.2018.07.006

Yi, M., Zhou, S., Zhang, L., Ding, S. The effects of three different microplastics on enzyme activities and microbial communities in soil. *Water Environ. Res.* 2020, 93, 24–32. https://doi.org/10.1002/wer.1327

Yilmaz, Y., Sevencan, Ü. Investigation of some geotechnical properties of polypropylene fiber and fiy ash amended Ankara clay. In *Proceedings of the 13th National Conference on Soil Mechanics and Foundation Engineering*, Istanbul, Turkey, 5–10 January 1994; pp. 133–142.

Yin, L., Chen, B., Xia, B., Shi, X., Qu, K. Polystyrene microplastics alter the behavior, energy reserve and nutritional composition of marine jacopever (Sebastes schlegelii). *J. Hazard Mater.* 2018, 360, 97–105.

Yoshida, S., Hiraga, K., Takehana, T., Taniguchi, I., Yamaji, H., Maeda, Y., Toyohara, K., Miyamoto, K., Kimura, Y., Oda, K. A bacterium that degrades and assimilates poly (ethylene terephthalate). *Science* 2016, 351, 1196–1199.

Yu, M., der Ploeg, M., Ma, X., Ritsema, C.J., Geissen, V. Effects of microplastics and earthworm burrows on soil macropore water flow within a laboratory soil column setup. *Vadose Zone J.* 2020, 19(1). https://doi.org/10.1002/vzj2.20059

Yu, P., Liu, Z., Wu, D., Chen, M., Lv, W., Zhao, Y. Accumulation of polystyrene microplastics in juvenile Eriocheir sinensis and oxidative stress effects in the liver. *Aquat. Toxicol.* 2018, 200, 28–36.

Yu, Y., Xie, D., Yang, Y., Tan, S., Li, H., Dang, Y., Xiang, M., Chen, H. Carboxyl-modified polystyrene microplastics induces neurotoxicity by affecting dopamine, glutamate, serotonin, and GABA neurotransmission in Caenorhabditis elegans. *J. Hazard Mater.* 2022, 445, 130543. https://doi.org/10.1016/j.jhazmat.2022.130543

Yue, Y., Li, X., Wei, Z., Zhang, T., Wang, H., Huang, X., Tang, S. Recent advances on multilevel effects of micro(nano)plastics and coexisting pollutants on terrestrial soil-plants system. *Sustainability* 2023, 15, 4504. https://doi.org/10.3390/su15054504

Zettler, E., Mincer, T., Amaral-Zettler, L.A. Life in the "plastisphere": microbial communities on plastic marine debris. *Environ. Sci. Technol.* 2013, 47, 7137–7146.

Zhang, Y., Niu, J., Wei, Z., Zhou, X., Wu, L., Li, X., Ma, S., Shi, G. Spatial distribution characteristics, source analysis and risk assessment of polycyclic aromatic hydrocarbons in topsoil of a typical chemical industry park. *Front. Environ. Sci.* 2023, 11, 1209137. https://doi.org/10.3389/fenvs.2023.1209137

Zhang, G.S., Liu, Y.F. The distribution of microplastics in soil aggregate fractions in southwestern China. *Sci. Total Environ.* 2018, 642, 12–20.

Zhang, K., Shi, H., Peng, J., Wang, Y., Xiong, X., Wu, C., Lam, P.K.S. Microplastic pollution in China's inland water systems: a review of findings, methods, characteristics, effects, and management. *Sci. Total Environ.* 2018a, 630, 1641–1653.

Zhang, K., Su, J., Xiong, X., Wu, X., Wu, C., Liu, J. Microplastic pollution of lakeshore sediments from remote lakes in Tibet plateau, China. *Environ.* Pollut. 2016, 219, 450–455.

Zhang, S., He, Z., Wu, C., Wang, Z., Mai, Y., Hu, R. Complex bilateral interactions determine the fate of polystyrene micro- and nanoplastics and soil protists: implications from a soil amoeba. *Environ. Sci. Technol.* 2022, 56(8), 4936–4949. https://doi.org/10.1021/acs.est.1c06178

Zhang, S., Yang, X., Gertsen, H., Peters, P., Salánki, T., Geissen, V. A simple method for the extraction and identification of light density microplastics from soil. *Sci. Total Environ.* 2018b, 616–617, 1056–1065.

Zhao, C., Liu, M., Du, H., Gong, Y. The evolutionary trend and impact of global plastic waste trade network. Sustainability 2021, 13, 3662.

Zhao, L., Qu, M., Wong, G., Wang, D. Transgenerational toxicity of nanopolystyrene particles in the range of μg L–1 in the nematode Caenorhabditis elegans. *Environ. Sci.* 2017, Nano 4, 2356–2366. https://doi.org/10.1039/C7EN00707H

Zhejiang Provincial Bureau of Statistics. Power Conversion Structure Optimization and Quality Improvement-Zhejiang's Economic Performance in 2017. Available online: (accessed on 19 August 2023).

Zheng, Y., Yanful, E.K., Bassi, A.S. A review of plastic waste biodegradation. *Crit. Rev. Biotechnol.* 2005, 25, 243–250.

Zhou, Q., Zhang, H., Zhou, Y., Li, Y., Xue, Y., Fu, C., Luo, Y. (2016). Separation of microplastics from a coastal soil and their surface microscopic features. Chinese Science Bulletin, 61(14), 1604–1611. doi:10.1360/N972015-01098

Zhou, Q., Zhang, H., Fu, C., Zhou, Y., Dai, Z., Li, Y., Tu, C., Luo, Y. The distribution and morphology of microplastics in coastal soils adjacent to the Bohai Sea and the Yellow Sea. *Geoderma* 2018, 322, 201–208.

Zhou, Y., Wang, J., Zou, M., Jia, Z., Zhou, S., Li, Y. Microplastics in soils: a review of methods, occurrence, fate, transport, ecological and environmental risks. *Sci. Total Environ.* 2020, 748, 141368.

Zhu, D., Bi, Q., Xiang, Q., Chen, Q., Christie, P., Ke, X., et al. Trophic predator-prey relationships promote transport of microplastics compared with the single Hypoaspis aculeifer and Folsomia candida. *Environ. Pollut.* 2018b, 235, 150–154. https://doi.org/10.1016/j.envpol.2017.12.058

Zhu, D., Chen, Q., An, X., Yang, X., Christie, P., Ke, X. Exposure of soil collembolans to microplastics perturbs their gut microbiota and alters their isotopic composition. *Soil Biol. Biochem.* 2018a, 116, 302–310. https://doi.org/10.1016/j. soilbio.2017.10.027

Zhu, D., Li, G., Wang, H., Duan, G.. Effects of nano- or microplastic exposure combined with arsenic on soil bacterial, fungal, and protistan communities. *Chemosphere* 2021, 281, 130998. https://doi.org/10.1016/j.chemosphere.2021.130998

Zhu, F., Zhu, C., Wang, C., Gu, C. Occurrence and ecological impacts of microplastics in soil systems: a review. *Bull. Environ. Contam. Toxicol.* 2019, 102, 741–749.

Ziajahromi, S., Kumar, A., Neale, P.A., Leusch, F.D.L. Effects of polyethylene microplastics on the acute toxicity of a synthetic pyrethroid to midge larvae (Chironomus tepperi) in synthetic and river water. *Sci. Total Environ.* 2019, 671, 971–975. DOI: 10.1016/j.scitotenv.2019.03.425

Zou, W., Xia, M., Jiang, K., Cao, Z., Zhang, X., Hu, X. Photo-oxidative degradation mitigated the developmental toxicity of polyamide microplastics to zebrafish larvae by modulating macrophage-triggered proinfiammatory responses and apoptosis. *Environ. Sci. Technol.* 2020, 54, 13888–13898.

Zylstra, E. Accumulation of wind-dispersed trash in desert environments. *J. Arid. Environ.* 2013, 89, 13–15.

CHAPTER 17

Fate and Transport Pathways of Microplastics in Agricultural Soil and their Interaction with Agrochemicals

Megha R. Raj, Priya K.L., Baiju V., and Selvam S.

17.1 INTRODUCTION

Plastics are synthetic polymers designed for durability, but this very characteristic makes them resistant to natural degradation processes. As a result, plastic waste accumulates in terrestrial and aquatic environments, leading to a range of environmental problems. Plastics can be found in oceans, rivers, soils, and the atmosphere. Plastic pollution has become a prominent environmental issue recently, due to its widespread presence in various ecosystems, its persistence, and its potential adverse effects on the environment and living organisms. These macro-sized plastic particles undergo fragmentation through various mechanisms like physical, chemical, and microbial decomposition resulting in the generation of microplastics

DOI: 10.1201/9781003487579-17

(MPs) and nanoplastics (NPs). The pervasive presence of MPs and NPs in our environment has become a pressing concern with far-reaching consequences. These minuscule plastic fragments, having a size of less than 5 mm, have become ubiquitous in various ecosystems, including terrestrial and aquatic environments, posing complex challenges for the enhancement of knowledge in their fate, transport, degradation, and removal. It is therefore classified under the emerging contaminant category.

The escalating prevalence of MPs and NPs demands an in-depth exploration of the processes that govern their behavior, as well as the development of innovative strategies to address this issue. The MPs and NPs in the environment arise from various sources, including the breakdown of larger plastic debris, industrial processes, and even the release of consumer products containing MPs. Figure 17.1a represents the release of MPs/NPs from daily use products and Figure 17.1b shows the plastic consumed in different sectors. Figure 17.2 illustrates the proportion of MPs and NPs released from common sources.

Given the ubiquitous presence of plastics across various aspects of daily life, owing to their virtually limitless applications, and the potential challenges they pose for the environment, it is imperative to conduct additional research to identify regions contaminated with MPs. This is essential for preventing their inadvertent release into the environment. These plastic residues are known to exhibit toxicity however, the effects of which are not yet fully understood, and their susceptibility to ingestion by living organisms and their impacts are poorly known (Castro et al., 2016; Morais et al., 2020; Nan et al., 2020). These tiny plastic particles have pervaded oceans, rivers, lakes, soils, and the atmosphere, raising profound concerns about their potential repercussions on ecosystems and living organisms. The presence of MPs in aquatic environments, along with their ability to attract organic pollutants, is raising significant concerns about the increased uptake of contaminants by aquatic organisms (Nguyen et al., 2023; Akhila et al., 2023).

Understanding how MPs and NPs move through the soil environment is critical in assessing their environmental impact. These particles can be transported over vast distances via water and air currents, leading to their widespread distribution. According to Vivekanand et al. (2021), the atmosphere plays a crucial role in facilitating the transport of various suspended materials to terrestrial and aquatic environments. This transportation occurs through processes including wind speed and direction, upward and downward air currents, convection, and turbulence. The fate of MPs/NPs is determined by a complex interplay of physical, chemical, and biological factors, influencing their accumulation in different environmental

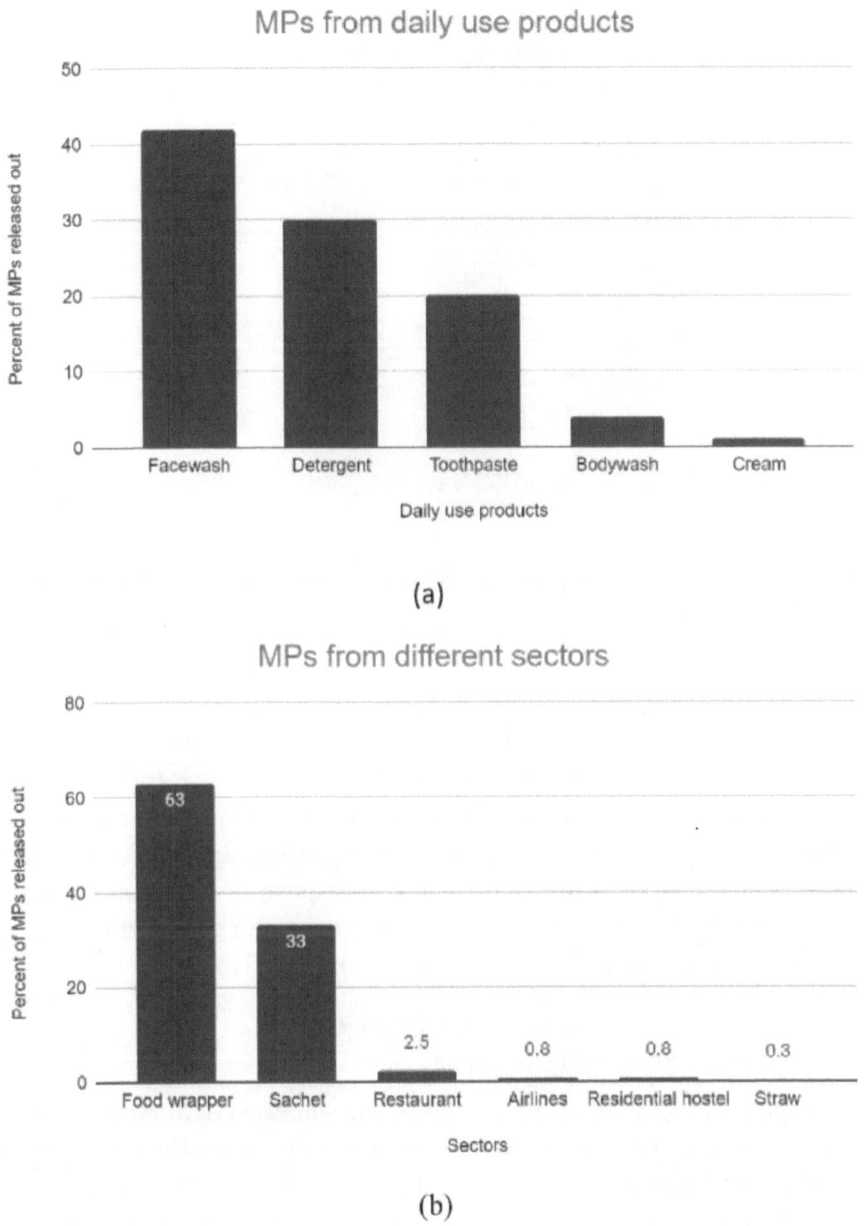

FIGURE 17.1 Microplastics released from (a) daily use products and (b) different sectors. (Mahmud et al., 2022.)

compartments. Despite their reputation for durability, plastics are not immune to degradation. Environmental factors such as ultraviolet radiation (UVR), mechanical wear and tear, and microbial activity can break

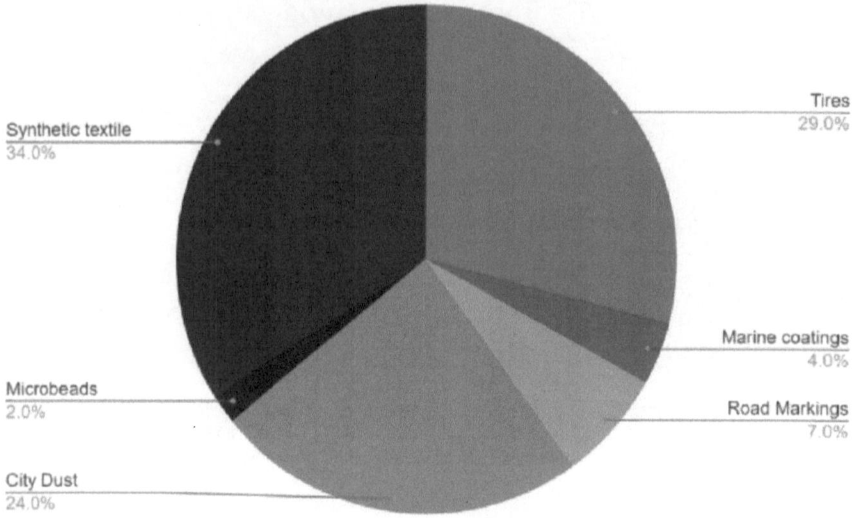

FIGURE 17.2 Percentage of microplastics emitted from common sources. (Birch et al., 2020.)

down MPs and NPs into smaller fragments in the soil environment, exacerbating the challenges they pose. Investigating the mechanisms of degradation in soil is essential to understanding the persistence and potential toxicity of these plastic particles. This chapter seeks to provide a comprehensive overview of our current understanding of MPs and NPs within the soil ecosystem. By examining their fate, transport, degradation mechanisms, and their influence on agrochemicals, we aim to contribute to a deeper comprehension of this critical environmental issue.

17.2 MICROPLASTICS AND NANOPLASTICS WITHIN SOIL ECOSYSTEM

The exploration of MPs and NPs within the soil ecosystem remains less investigated compared to other environmental domains. The abundance of MPs in soil ecosystems is estimated to be larger than their concentration within the oceans (Horton et al., 2017). Su et al. (2022) reported that the quantity of MPs in soil is almost double the quantity in river sediments and suggested that soil can act as a sink for MPs. The larger surface area and hydrophobic nature of MPs make them a good adsorbent for many contaminants present in soil and hence can serve as vectors to the pollutants (Dissanayake et al., 2022). The structure and metabolic characteristics of soil microbes can undergo significant changes due to the presence of MPs, which has implications for the organic pollutants in soil

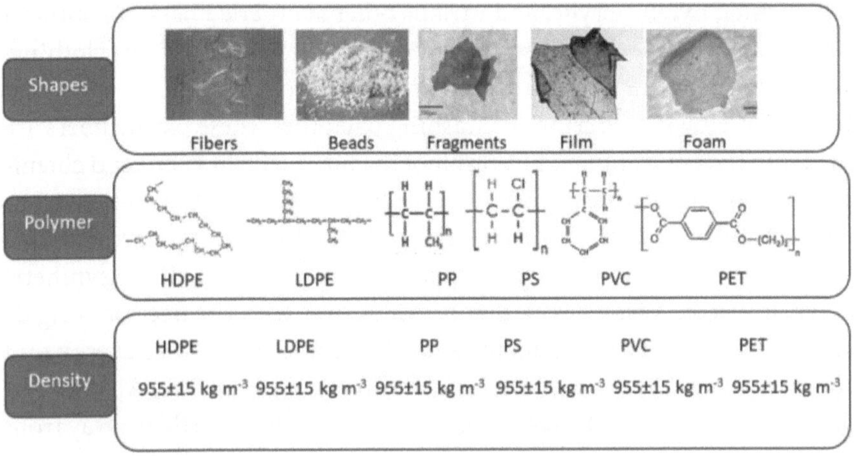

FIGURE 17.3 Various forms of existence of MPs and NPs (Singh et al., 2022; Priya et al., 2022). HDPE, high-density polyethylene; LDPE, low-density polyethylene; PET, polyethylene terephthalate; PP, polypropylene; PS, polystyrene; PVC, polyvinyl chloride.

(Zhou et al., 2022). These particles, despite their diverse shapes, sizes, and compositions, (Figure 17.3) share common attributes like low material density, small dimensions, and a high surface area, which make them adept at entering and traveling deep into the soil strata. Human activities on land, including littering, improper waste management, industrial accidents, and emissions from landfill sites, are all potential sources of MPs released into the soil environment.

17.2.1 Sources of MPs and NPs within the Soil Ecosystem

The deposition of plastic polymers in the soil ecosystem initiates a mechanism wherein this synthetic material undergoes fragmentation, resulting in the formation of MPs, which can further degrade into even smaller particles known as NPs. MP pollution arises from both the intentional release of engineered MPs and the breakdown of larger plastic debris. This fragmentation can occur due to mechanical stress, exposure to UV radiation, and the action of biological agents. Additionally, MPs are introduced into the soil environment through wastewater discharge, especially when treated effluent is used for irrigation. One significant source of soil MP is the shedding of fibers from synthetic textiles. These clothing fibers break down and are released into the terrestrial environment as fine particles, a process often driven by photo-oxidation (Priya et al., 2022).

Polyester, nylon, acrylic, and various other synthetic fibers, constitute a significant portion, approximately 60%, of the fabrics used in our clothing. These synthetic MP fibers are popular due to their cost-effectiveness and versatility. However, during the washing of clothes, these plastic fibers are shed into the environment as a result of the physical (abrasion) and chemical stresses that textiles undergo and ultimately reach the soil environment. Another source of MPs in soil is tires, which contain around 24% man-made rubber and 19% gum rubber (natural rubber). These synthetic polymers create a framework that provides tires with the necessary rigidity and traction. Tires gradually wear down due to thermal energy and abrasion generated during interaction with the road surface. Airflow and precipitation disperse the resulting tire particles and rinse them away from the pavement, leading to their deposition in the soil and eventual entry into water bodies. Dust from cities includes contributions from the wearing off of items like synthetic shoe soles, culinary utensils, wear and tear of structures like household surfaces, synthetic turf, and surface coatings on docks, harbors, and marine buildings. Additionally, it contains particles generated during abrasive blasting, the natural weathering of plastic polymers, and the application of detergents.

Road markings, which include polymer tape and paint, can contribute to the release of MPs through weathering and abrasion caused by vehicles. Approximately 7% of primary MPs found in the ocean come from debris originating from road markings (Birch et al., 2020). Marine coatings used by developers often contain plastics like polyurethane, epoxy coatings, vinyl, and lacquers. Primary MPs are discharged from these coatings as a consequence of weathering and inadvertent spills during their application, upkeep, and disposal. Engineered MPs, notably polyethylene-based microbeads, are commonly found in personal care and cosmetic products. These minute particles can effortlessly penetrate soil systems and ultimately enter the groundwater table. Plastic manufacturers produce a range of plastics typically in the form of pellets and powders, constituting the principal form of plastic. During manufacturing, processing, transportation, and recycling, pellets may inadvertently spill into the soil environment, thereby adding to the accumulation of plastic pellets.

The extensive use of face masks during the COVID-19 pandemic has become a notable origin of NPs within the terrestrial surroundings. N95 masks are typically made of synthetic polymers such as polypropylene, polyethylene (PE), or polyester. These materials can shed MP particles through processes like abrasion, degradation, or fragmentation, especially with extended use or improper disposal. A study estimated that each N95 or surgical mask released more than one billion NPs and MPs (Jie et al., 2021). Daily use-cosmetics products are a primary contributor to these NPs, as they contain fine bead particles composed of predominantly PE beads. Once plastic waste enters sewerage systems, it undergoes physical (agitation, turbulence, and abrasion) and chemical (hydrolysis, oxidation, and photodegradation) degradation. The increasing accumulation of plastic waste in sewage systems exacerbates the degradation of plastics into smaller particles that eventually break down into NPs, making them challenging to remove from the sewage system.

In recent years, there has been a growing focus among researchers on the presence of plastic contamination within agricultural soil. It is evident that various types of agricultural soil, such as arable lands, paddy fields, and uplands, as well as those used for irrigation and greenhouse cultivation, are being affected by plastic pollutants. These contaminants exhibit diverse characteristics in terms of size, shape, and distribution, manifesting spatial and temporal variations across the surface, subsurface, and soil profiles. Key contributors to plastic pollution in agricultural soil include practices like plastic mulching, the disposal of sludge and composite materials, as well as the abandonment of greenhouses. Moreover, the use of controlled-release fertilizers has emerged as a significant source, accounting for approximately 3% of plastic input into agricultural soil. Specifically, MPs originating from fertilization processes, such as those associated with NPK encapsulated in polymer materials and compost manure, contribute to this contamination. Consequently, addressing these sources of plastic pollution in agricultural soil is imperative for mitigating environmental and agricultural risks associated with such contamination.

17.2.2 Detection of MP in Soil

MP detection is a critical aspect of environmental monitoring aimed at understanding the pervasive presence and potential impacts of tiny plastic particles in various ecosystems. Detection methods encompass a range of techniques tailored to identify and quantify MPs in different environmental matrices. Larger and smaller MPs can be detected through visual analysis using the naked eye and forceps (Heo et al., 2013). Visual identification, often conducted under a microscope, relies on discerning physical characteristics such as size, shape, color, and surface

TABLE 17.1 Abundance of MPs in Agricultural Soil

Continent	Country	Polymer and Concentration	Agricultural Practice	References
Asia	China	PE, PP, PET 6,796 pieces/kg	Mulch	Crossman et al. (2020)
Asia	Japan	PE 144 mg/kg	Coated fertilizer	Katsumi et al. (2021)
Asia	Thailand	PE, LDPE, PP, PS 12–117 items/m^2	Mulch, plastic gauze, fertilizer	Fakour et al. (2021)
Europe	Germany	PES, PA, PVC, PAN 14.6 MPs/g	Sludge	Tagg et al. (2022)
Europe	Greece	PE, BMF 301 items/kg	Film	Isari et al. (2021)
South America	Chile	PE, PP, PS 306 pieces/kg	Mulch, plastic gauze, fertilizer	Kumar and Sheela (2021)
North America	Canada	PE, PP, PS	Biosolid	Crossman et al. (2020)

Source: Sa'adu et al. (2023).

texture (Dey et al., 2021). While manual observation is foundational, advanced microscopy techniques like scanning electron microscopy (SEM Analysis) and Fourier-transform infrared (FTIR) spectroscopy offer higher resolution and chemical composition analysis, aiding in accurate identification.

In addition to microscopy, spectroscopic methods such as FTIR and Raman spectroscopy unveil valuable details regarding the chemical constitution of MPs, enabling differentiation between various polymer types (Dey et al., 2021). Flow cytometry emerges as a rapid analysis tool, sorting particles based on optical properties and size, while fluorescence imaging enhances detection using fluorescent markers (Kaile et al., 2020). Density separation techniques, like centrifugation and floatation, isolate MPs from environmental samples based on their density relative to surrounding materials. Filtration methods capture particles on filters with specific pore sizes, facilitating subsequent analysis. Chemical digestion and passive sampling offer alternative approaches for isolating MPs from complex environmental matrices.

Furthermore, remote sensing techniques, including hyperspectral imaging and satellite observations, hold promise for large-scale monitoring and mapping of MP pollution in aquatic environments (Dey et al., 2021).

Combining these diverse detection methods ensures a comprehensive understanding of MP distribution and abundance, crucial for effective environmental management and mitigation strategies. Standardization of sampling and analysis protocols is paramount to ensure consistency and comparability of results across studies. Continued research and innovation in MP detection techniques are essential to address emerging challenges and refine our understanding of MP pollution dynamics.

17.3 MECHANISM OF FORMATION OF MPS

17.3.1 Photo-Oxidation of Plastics and Fragmentation in Soil Strata

Secondary MPs arise from the gradual breakdown of plastic debris, primarily due to the effects of weathering and exposure to solar UV radiation. The process of photo-oxidation plays a crucial role in weakening and making the plastic more brittle, ultimately leading to fragmentation when subjected to environmental mechanical stresses. The chemical degradation mechanism is depicted in Figure 17.4. Exposure of plastics to UV or sunlight irradiation can cause their polymer macromolecules to absorb photons directly, inducing excited states that trigger chain breakage, branching, cross-linking, and oxidation reactions (Priya et al., 2022).

The oxidation process primarily affects the amorphous portion of semi-crystalline PE and polypropylene (PP) polymers, which are the most

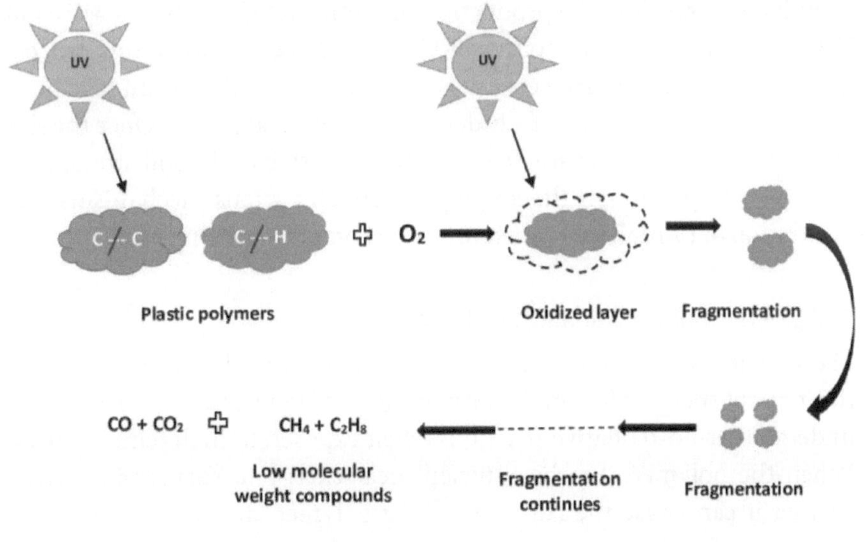

FIGURE 17.4 Mechanism of photo-oxidation of plastics. (Priya et al., 2022.)

common types of plastic found in the environment. As a result of this oxidation, surface pits or cracks may develop, and with further exposure to UVR, these flaws can extend further into the interior of the material, leading to the fragmentation of the plastic into numerous smaller pieces. It's essential to highlight that although photo-oxidation weakens the polymer material, the creation of smaller fragments or individual MPs usually requires additional mechanical pressures. Interestingly, even minimal force can lead to the fragmentation of highly photo-oxidized and weakened plastics, just by handling such materials can result in their breakdown. Fragmentation of virgin or slightly oxidized plastics demands higher levels of force but might still occur within beach slush zones.

In soil, the mechanical breakdown of pristine plastic debris may occur through abrasion in the slush zone, where plastics interact with rocks and sand. This process has been replicated in laboratory simulations (Andrady et al., 2022). Moreover, the fragmentation of relatively unoxidized plastics is observable in the creation of microfiber fragments during fabric laundering and the disintegration of virgin plastics when consumed by crustaceans. However, the true environmental impact of the biological fragmentation process remains uncertain (Andrady et al., 2022). Fragmentation results in an increase in the plastic material's specific surface area, which, in turn, provides more surface area for subsequent photo-oxidation and potential biodegradation processes. Photo-degradation can persist, and theoretically, it could result in the production of significant quantities of MPs and NPs, given that there is a source of UVR available to initiate this process (Andrady et al., 2022). However, it is essential to note that it can take hundreds of years for plastics to fully degrade (Wenxi et al., 2023). Once the size of the particles becomes smaller than the pore size of the soil strata, they penetrate deep through the soil layer through various mechanisms; the infiltration of rainfall and transport by soil organisms being predominant.

17.3.2 Thermal Degradation of Plastics

Thermal degradation refers to the process of breaking down plastics due to the input of energy, often in the form of elevated temperatures. Plastics can undergo thermo-oxidative reactions when exposed to high temperatures. When the polymer absorbs sufficient heat energy to surpass its energy barrier, it can cause the rupture of long polymer chains, leading to the formation of radicals. These radicals can then react with oxygen, creating hydroperoxide, which is a process similar to what occurs in the photo-degradation of plastics, where hydroxyl free radicals and alkoxy radicals

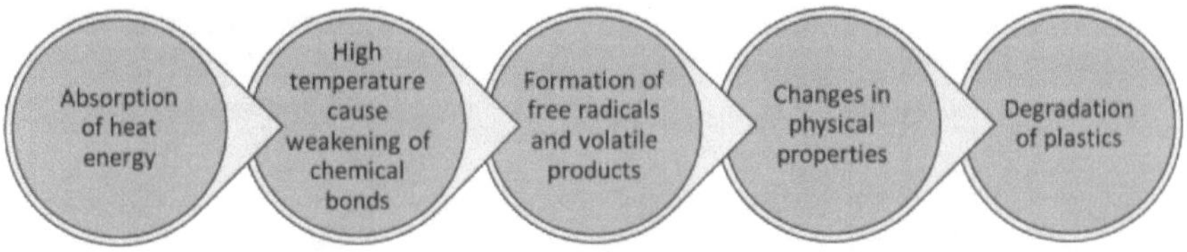

FIGURE 17.5 Mechanism of thermal degradation of plastics. (Zhang et al., 2021.)

are formed. This reaction can self-propagate until either the energy input stops or inert products are produced through the collision of two radicals. During thermal degradation, both molecular reduction (making the molecules smaller) and enlargement (making the molecules larger) can occur due to chain scission (breaking of polymer chains) and cross-linking (joining of polymer chains) processes (as described by Zhang et al., 2021). The mechanism of degradation is sequentially illustrated in Figure 17.5.

The temperature at which thermal degradation occurs depends on the thermal properties of plastics and the availability of oxygen. Two key thermal properties of plastics are the glass transition temperature (T_g) and its melting point (T_m). T_g represents the temperature at which a polymer shifts from a glassy state to an elastic state. Plastics are flexible above their T_g but become rigid below it due to changes in molecular mobility. The activation energy of a polymer determines the temperature at which thermal degradation begins. In the presence of oxygen, most plastics containing hydrogen as a component undergo an exothermic oxidation process at temperatures near 200°C. Plastics with a higher T_m typically require even higher temperatures for thermal degradation to occur (Zhang et al., 2021).

17.3.3 Ozone Degradation

Ozone, which is naturally present in the upper atmosphere, plays a significant role in the degradation of polymeric materials. Polymers exhibit extended durability in the absence of active oxidative processes. Despite its minimal presence in the atmosphere, ozone significantly affects polymers. Ozone deteriorates polymers by generating reactive oxygen species (ROS), which lead to alterations in the polymers' molecular weight, as well as their electrical and mechanical properties (Zeenat et al., 2021). When polymers are exposed to ozone, they generate diverse carbonyl and unsaturated carbonyl products, such as ketones, lactones, esters, and aromatic carbonyl compounds. These products are commonly linked with a phase referred to as the styrene phase. The degradation of large polymer molecules can be contributed to by the generation of unstable aromatic ring intermediates

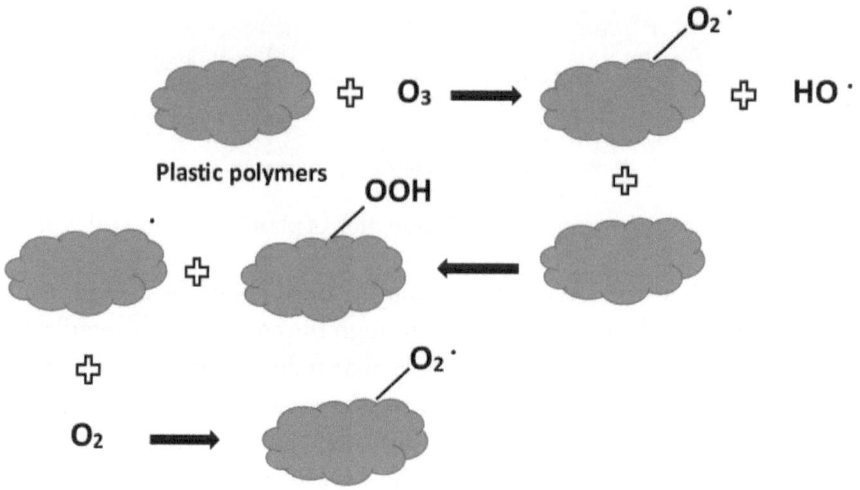

FIGURE 17.6 Ozone degradation of plastics. (Kefeli et al., 1971.)

(bipolar ions/peroxy radicals) resulting from ozone reactions with polymers containing C–C bonds and other saturated hydrocarbon links (as discussed in Zeenat et al., 2021). The ozone degradation mechanism of plastics is shown in Figure 17.6. The degraded products reach the soil strata through precipitation, deposition, wind, and other physical mechanisms.

17.3.4 Mechano-Chemical Degradation

The breakdown of polymer materials through mechano-chemical degradation involves the influence of both intense chemical and mechanical pressure. The role of mechanical force in this phenomenon leads to the fragmentation of molecular chains, followed by ensuing chemical reactions. This combined action of mechanical stress and chemical reactions gives rise to the term "Mechano-chemical degradation." For instance, when rubber is masticated (subjected to mechanical shearing or kneading), it can lead to the breaking of polymer chains and the development of plasticity. This phenomenon arises as rubber molecules undergo radical formation in the presence of oxygen. As oxygen serves as a scavenger for radicals, it reacts with them, leading to irreversible chain breakage (Zeenat et al., 2021).

17.3.5 Biodegradation

Biodegradation refers to the process in which microorganisms, such as *actinomycetes*, bacteria, and fungi, present in the soil bring about physical and chemical changes in both natural and synthetic plastics. This process involves the secretion of extracellular enzymes by the microorganisms, which then attach to the plastic's surface. These enzymes catalyze the hydrolysis of the plastic, breaking it down into shorter polymer intermediates. Subsequently, these intermediates are assimilated by the microbial cells as a carbon source, ultimately leading to the release of carbon dioxide (CO_2). Microbial polymer degradation encompasses various stages, including biodeterioration, biofragmentation, mineralization, and assimilation (Zeenat et al., 2021). Figure 17.7 depicts the overall degradation process by microbes.

Biodeterioration is a phenomenon that primarily affects the surface of plastics, causing significant alterations in their chemical, physical, and mechanical characteristics. The specific changes that occur depend on the composition and structure of the polymers involved, and environmental

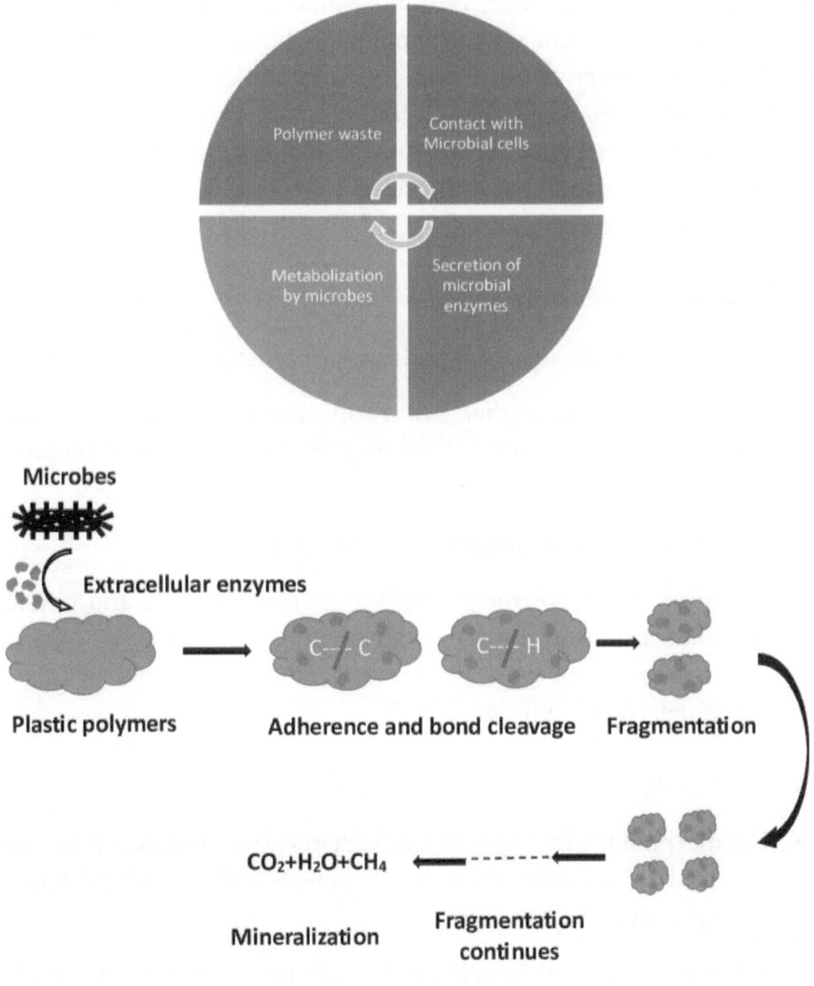

FIGURE 17.7 Biodegradation of plastics by microbes. (Zeenat et al., 2021.)

conditions play a significant role in influencing these alterations. During biodeterioration, two crucial processes take place: substrate formation within the plastic and the formation of biofilms. These processes set the stage for the subsequent step, known as biofragmentation, which involves the enzymatic breakdown of plastic polymers. Bacteria often employ oxygenases, enzymes capable of introducing oxygen molecules into the polymer's carbon chains, resulting in the formation of less harmful alcohol and peroxyl products. Additionally, lipases, esterases, and endopeptidases catalyze the transformation of carboxylic and amide groups within the polymers. The smaller monomers generated during biofragmentation can enter microbial cells, where they are oxidized to produce energy, subsequently used for biomass production. During the assimilation process, atoms from these monomers become integrated into microbial cells, facilitating complete degradation. Secondary metabolites might either be expelled from the cells or transferred to other microbes for additional breakdown, ultimately leading to the release of oxidized products such as CO_2, N_2, H_2O, and CH_4 during metabolite degradation (Zeenat et al., 2021).

Plastic biodegradability is influenced by the physical and chemical properties mentioned below (Zeenat et al., 2021):

- Availability of functional groups that enhance hydrophobicity.
- Structural complexity, such as linear or branched configurations.
- The type of bonds that are easily breakable, such as amide bonds and ester bonds. Coupling within the chain hierarchy (ester > ether > amide > urethane).
- Composition and arrangement of molecules
- The form of the polymer, including its characteristic and physical appearance.

Example: pellets, films, powder

- The density of the polymer and its molecular weight.
- The morphology of TM (thermomechanical) materials can be characterized by the proportion of amorphous and crystalline regions present within the material.
- Softer polymers with toughness degrade more quickly compared to harder or tougher ones.

In aerobic degradation, microorganisms break down large organic compounds into smaller ones by using oxygen as an electron acceptor. This process produces carbon dioxide and water as by-products as shown in Equation 17.1.

$$\text{Carbon (plastic)} + \text{Oxygen} \rightarrow \text{carbon dioxide} + \text{water} + \text{Carbon residual} \tag{17.1}$$

In anaerobic biodegradation, microorganisms can break down compounds even in the absence of oxygen. Instead of oxygen, anaerobic bacteria utilize alternative electron acceptors such as nitrate, iron, sulfate, manganese, and carbon dioxide to facilitate the decomposition of larger organic compounds into smaller ones. This process results in the production of byproducts such as methane (CH_4), carbon dioxide (CO_2), water (H_2O), and residual carbon. Anaerobic biodegradation as shown in Equation 17.2 is particularly important in environments where oxygen is limited or absent, such as in certain sediments, deep underground, or within the digestive systems of animals like cows and termites. It plays a crucial role in natural attenuation processes at sites with hazardous waste and in the decomposition of organic matter under anaerobic conditions.

$$\text{Carbon (plastic)} \rightarrow \text{methane} + \text{carbon dioxide} + \text{water} + \text{Carbon residual} \tag{17.2}$$

Exoenzymes, enzymes secreted externally by cells, degrade large polymers and generate small, water-soluble molecules. The molecules generated through the action of enzymes can traverse semi-permeable bacterial membranes and can be harnessed as an energy source. Depolymerization describes the breakdown of large polymers, whereas the transformation of resulting substances into inorganic compounds like H_2O, CH_4, and CO_2 is termed mineralization. In aerobic settings, only H_2O, CO_2, and microbial biomass were noted as final products. However, under anaerobic/methanogenic and sulfidogenic conditions, CH_4 and H_2S were identified as additional byproducts in addition to the three primary components.

17.4 TRANSPORT OF MICROPLASTICS AND NANOPLASTICS

17.4.1 Transport of MPs/NPs in Soil

MPs can be introduced to soil when plastic-based materials, such as mulch, films, or compost containing MPs, are applied to the surface. Atmospheric deposition can lead to the settling of MPs onto soil surfaces. These MPs

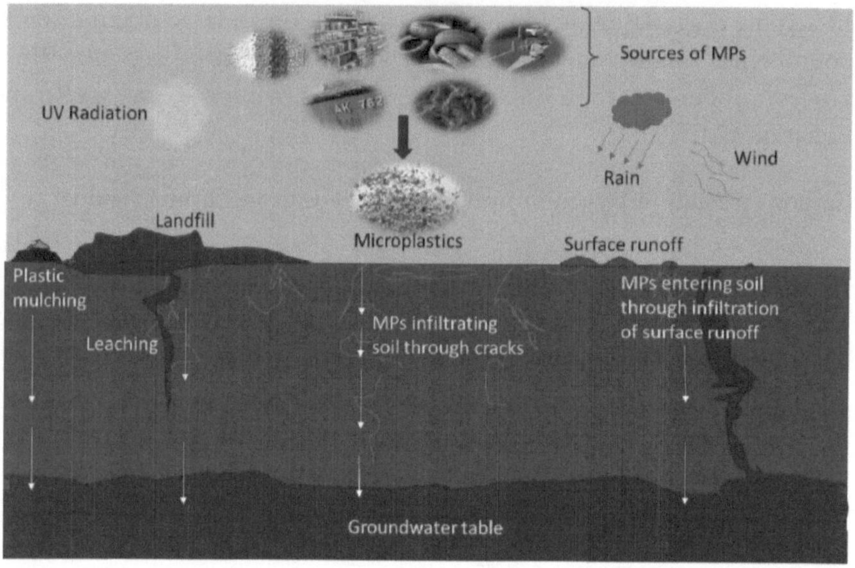

FIGURE 17.8 Various modes through which MPs can enter the soil strata.

may originate from the degradation of larger plastic items or be carried by the wind. Rainwater and irrigation can transport MPs from surfaces, such as roads, into the soil through runoff (Figure 17.8). The use of plastic-based agricultural materials, such as plastic mulch and irrigation tubing, can release MPs into the soil.

MPs can be transported downward in soil through a process called leaching. This occurs when water percolates through the soil, carrying MPs with it. Soil erosion, caused by factors like heavy rainfall, wind, and human activities, can also transport MPs from one location to another within a landscape. Soil-dwelling organisms, such as earthworms and burrowing insects, can ingest MPs and transport them vertically and horizontally within the soil profile. Plants can take up MPs from the soil through their roots. This can potentially lead to the transfer of MPs into the plant biomass as stated by Bank and Hansson (2021).

The texture of the soil (e.g., sandy, loamy, clayey) can influence the transport of MPs. Coarser soils with larger pore spaces may allow for faster movement of MPs, while finer soils may retain MPs more effectively. Soil with high organic matter content can bind MPs, potentially reducing their mobility. Organic matter can also influence the adsorption of contaminants onto MPs. The moisture content of soil also affects the mobility of MPs. Saturated or waterlogged soils may facilitate the downward transport of MPs through leaching. Furthermore, microorganisms in soil can interact with these MPs, potentially leading to their aggregation or degradation. MPs can adsorb contaminants from the surrounding environment, and these contaminants may affect their transport behavior in soil.

Understanding the transport of MPs in soil is essential for assessing their environmental impact and potential risks to ecosystems and human health. Research in this area is ongoing, and scientists are working to improve our knowledge of the fate and behavior of MPs in terrestrial environments. Efforts are also being made to develop sustainable practices that reduce the introduction of MPs into soil.

17.4.2 Transport of MPs/NPs in Groundwater

The transport pathway and dispersion of MPs in groundwater are still areas of active research, but some mechanisms and factors that influence their movement in groundwater systems have been identified. Groundwater is the water that resides beneath the Earth's surface within subsurface rock and soil formations, and the movement of MPs in this environment can differ from that in surface waters. MPs can enter groundwater when rainwater percolates through landfills containing plastic waste. The leachate from landfills may contain MPs and was identified in groundwater (Priya et al., 2023). If wastewater treatment plants do not effectively remove MPs, effluents from these facilities can introduce MPs into groundwater through discharge. MPs may be carried into groundwater through agricultural runoff, particularly in areas where plastic mulch or films are used in farming. MPs from personal care products or other sources can potentially enter groundwater through septic systems and leach fields.

MPs can interact with the porous soil or sediment matrix, with some particles adhering to soil particles or organic matter. This interaction can influence their movement through groundwater. The hydraulic conductivity of the subsurface materials (soil or sediment) affects how easily water and MPs can move through the ground. Materials with high hydraulic conductivity allow for more rapid movement. The presence of aquifers (permeable geological formations that can store and transmit groundwater) can facilitate the transport of MPs through the subsurface. The flow of groundwater is typically slow compared to surface water, and the direction of flow is determined by the topography and geological features of the

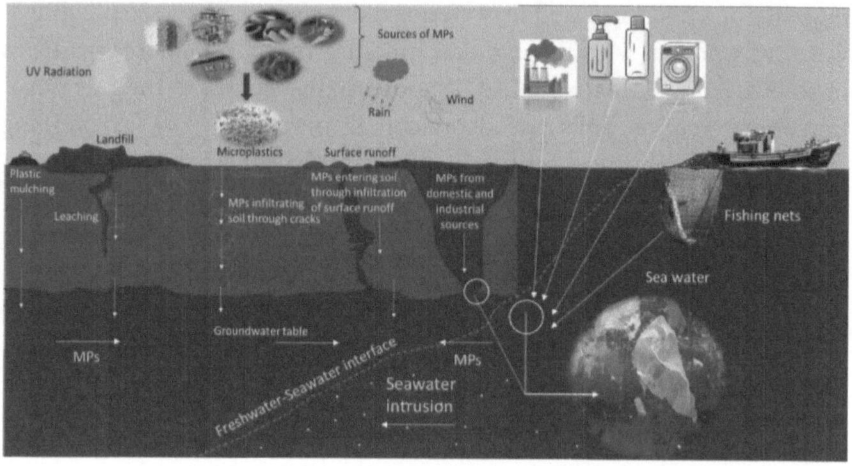

FIGURE 17.9 Transport of MPs to groundwater. (Singh et al., 2022.)

area. MPs can be transported with the flow of groundwater. Areas where groundwater is recharged by surface water, such as rivers and lakes, may have a higher potential for MP transport due to their proximity to surface water sources. The various pathways through which MPs can reach groundwater are shown in Figure 17.9.

17.5 INFLUENCE OF MICROPLASTICS ON AGROCHEMICALS

Primary contributors to plastic pollution in agricultural soil encompass various practices, including plastic mulching, the disposal of sludge and composite materials, and the abandonment of greenhouses. Additionally, the utilization of controlled-release fertilizers has surfaced as a notable source, constituting roughly 3% of plastic input into agricultural soil (Sa'adu and Farsang, 2023). Notably, MPs resulting from fertilization procedures, such as those linked with NPK encapsulated in polymer materials and compost manure, play a significant role in this contamination. MPs scattered across soil layers are recognized as effective adsorbents for various organic pollutants, like pesticides, due to their extensive specific surface area and hydrophobic nature. The presence of MPs in soil can modify the behavior of pesticides, affecting their adsorption, mobility, and degradation and adversely affecting living organisms (Figure 17.10). Smaller plastic particles typically exhibit greater pesticide adsorption due to the increased available adsorbent surface area and the presence of more effective sorption sites. Agrochemicals can interact with MPs through various

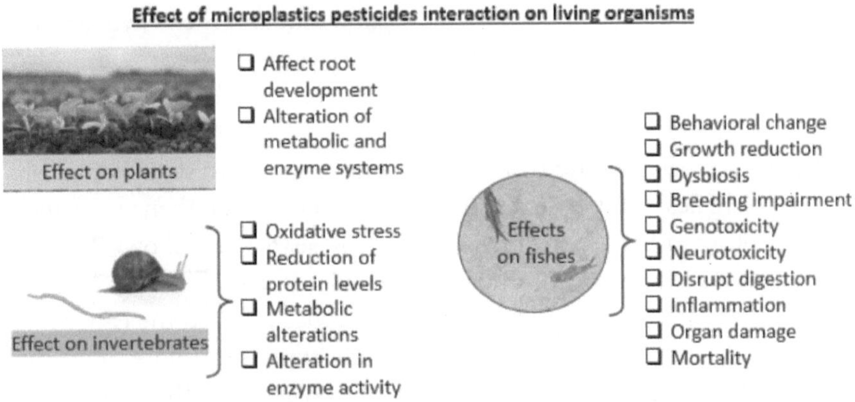

FIGURE 17.10 Effect of MP-pesticide interaction on living organisms. (Pena et al., 2023.)

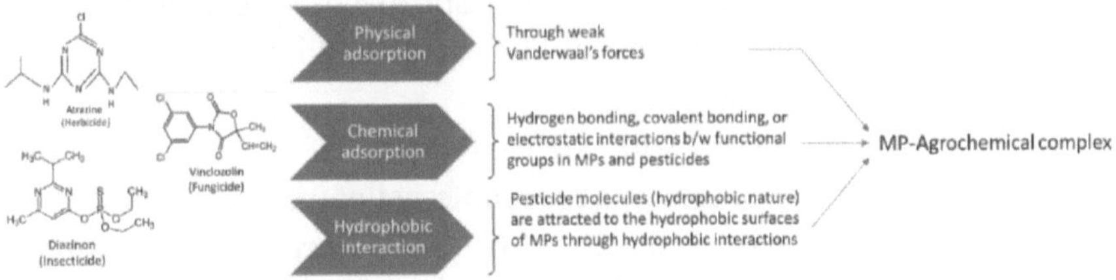

FIGURE 17.11 Various mechanisms of interaction between MPs and agrochemicals. (Yang et al., 2021.)

means as indicated by Figure 17.11. Interaction mainly happens through physical adsorption, chemical adsorption, or hydrophobic interaction.

Physical adsorption, also known as physisorption, occurs due to weak VanderWaals forces between the surface of MPs and the pesticide molecules. This type of adsorption does not involve any chemical bonding but relies on the proximity of the pesticide molecules to the MP surface. Pesticide molecules with nonpolar regions are attracted to the nonpolar surfaces of MP. Chemical adsorption involves the formation of chemical bonds between the pesticide molecules and the surface functional groups of MPs. MPs often possess functional groups like hydroxyl (–OH), carboxyl (–COOH), amino (–NH_2), and carbonyl (–C=O) groups, which can undergo chemical interactions with pesticides. As far as hydrophobic interaction is concerned, many pesticides are hydrophobic, meaning they tend to repel water and preferentially interact with nonpolar surfaces. MPs, especially those derived from PE, PP, or polystyrene (PS), possess hydrophobic surfaces. Pesticide molecules, due to their hydrophobic nature, are attracted to the hydrophobic surfaces of MPs through hydrophobic interactions. This phenomenon facilitates the adsorption of hydrophobic pesticides onto MPs.

Recent studies propose that MPs present in soil ecosystems may function as vehicles for pesticides, potentially increasing their movement within soil or other environmental areas. Research has shown that the existence of PE particles diminishes the soil adsorption of atrazine and 2,4-DB. Polyester fibers and PP particles likewise reduced the capacity of alluvial soil to retain different pesticides. This impact was more significant for hydrophobic pesticides and minimal for polar ones. The glass transition temperature (T_g) of MPs, indicating their flexibility, was identified as a significant factor influencing pesticide adsorption. PE showed higher pesticide adsorption compared to glassy state MPs like PS or polyvinyl chloride (PVC). As PE is the most prevalent plastic in agriculture, agricultural fields frequently contain residues of PE, which could potentially impact the fate of pesticides in soils.

The existence of MPs in soil further reduces the efficiency of agrochemicals. They alter the toxicity of organic pollutants by synergistic or antagonistic effects (Chang et al., 2022). In soil polluted with PS remnants and polyvinyl chloride beads, the dissipation of the pesticide thiacloprid was not significantly affected, indicating similar adsorption before and after plastic co-presence (Pena et al., 2023). Glyphosate, a herbicide with variable charges, showed inconsistent effects in the presence of different plastics (Pena et al., 2023). Notably, two studies revealed a reciprocal influence, where pesticides, specifically propiconazole and prothioconazole, accelerated the fragmentation of PE and biodegradable polybutylene adipate terephthalate (PBAT), resulting in irregularities and plastic oxidation. This suggests a dynamic interaction between pesticides and MPs in soil ecosystems, influencing the degradation of both pollutants and plastics. Pesticides adsorbed on MPs may impede their degradation, while pesticide residues in the soil may enhance microbial activity and induce the degradation of plastics, emphasizing the complex interplay between these environmental factors (Pena et al., 2023).

From the research reviews, it could be concluded that plastic materials show the capability to retain pesticides, particularly those that are hydrophobic, thereby decreasing their prompt accessibility for leaching, decomposition, absorption by plants, or consumption by soil organisms. This phenomenon raises concerns as it may result in inadequate pest control, ingestion of pollutant-carrying plastics by non-target organisms, or the gradual release of pesticides from polluted plastics with time, becoming a temporal source of agrochemicals in ecosystems. The coexistence of MPs and agrochemicals within soils could lead to disruptions with varying effects on soil-dwelling organisms over medium and long-term periods, depending on the specific MP-agrochemical system. Aged MPs typically exhibit more significant impacts compared to pristine ones on the fate and toxicity of pesticides. This is due to the emergence of new functional groups and microcracks during aging, which increase pesticide retention. Environmental factors like pH, dissolved organic matter, and salinity seem to have lesser significance in the interactions between MPs and pesticides, depending on the characteristics of both pollutants. MPs could act as conduits for pesticides to enter terrestrial animals, with the resulting harmful impacts contingent upon the pesticide's capacity to engage with the organism's biological system. The harmful effects of MP-pesticide interaction are depicted in Figure 17.10. Further research is highly recommended to thoroughly comprehend the mechanisms governing the uptake and movement of pesticides by crops cultivated in soils polluted with MPs.

17.6 CONCLUSIONS

MPs and NPs in soil, though less researched than in other environments, are of increasing concern, with concentrations potentially higher than in oceans. Human activities such as littering, tire wear, and the shedding of synthetic fibers contribute to plastic contamination. Mechanisms like photo-oxidation, thermal degradation, and biodegradation fragment plastics into smaller particles, which can penetrate soil strata. Soil acts as a sink for MPs, altering microbial characteristics and potentially affecting organic pollutant fate. Understanding these sources and mechanisms is vital for mitigating environmental and agricultural risks associated with plastic pollution in soil. The transport of MPs and NPs in soil involves various pathways, including deposition from plastic-based materials, atmospheric settling, runoff from rainwater and irrigation, and leaching downward through soil layers. Soil erosion, soil-dwelling organisms, and plant uptake further contribute to the movement of MPs within soil profiles. The texture and organic matter content of soil influence MP transport, with coarser soils allowing for faster movement and organic matter potentially binding MPs. MPs can adsorb contaminants and interact with microorganisms in the soil, affecting their mobility and fate. In groundwater, MPs can enter through landfills, wastewater treatment plants, agricultural runoff, and septic systems. The porous soil or sediment matrix, hydraulic conductivity, and presence of aquifers influence MP movement in groundwater. MPs can interact with pesticides in soil, altering their adsorption, mobility, and degradation, with hydrophobic pesticides showing higher adsorption onto plastic surfaces. The presence of MPs may reduce the efficiency of agrochemicals and influence their degradation, while pesticides can also affect the degradation of MPs. A dynamic interaction between MPs and pesticides occurs in soil ecosystems, with implications for environmental and human health. Further research is needed to understand the mechanisms and long-term effects of MP-pesticide interactions in soil.

REFERENCES

Akhila G., Meera G., Chellappan S., Sajithkumar K. J., Chingakham C., Priya K. L., & Indu S. (2023). Techniques for removal and degradation of microplastics. In Wang C., Babel S., Lichtfouse E. (Eds.), *Microplastic Occurrence, Fate, Impact, and Remediation* (pp. 127–153). Springer.

Andrady A. L., Barnes P. W., Bornman J. F., Gouin T., Madronich S., White C. C., Zepp R. G., & Jansen M. A. K. (2022). Oxidation and fragmentation of plastics in a changing environment: From UV-radiation to biological degradation. *Science of the Total Environment*, 851(Part 2), 158022.

Bank M. S., & Hansson S. V. (2021). The microplastic cycle: An introduction to a complex issue. In Bank M. S. (Ed.), *Microplastics in the Environment: Pattern and Process* (pp. 1–16). Springer.

Birch Q. P., Potter P., Pinto P., Dionysiou D., & Al-Abed S. (2020). Sources, transport, measurement, and impact of nano and microplastics in urban watersheds. *Reviews in Environmental Science and Bio/Technology*, 19, 275.

Castro R. O., Silva M. L., Marques M. R. C., & de Araújo F. V. (2016). Evaluation of microplastics in Jurujuba Cove, Niterói, RJ, Brazil, an area of mussels farming. *Marine Pollution Bulletin*, 110, 555–558.

Chang J., Fang W., Liang J., Zhang P., Zhang G., Zhang H., Zhang Y., & Wang Q. (2022). A critical review on interaction of microplastics with organic contaminants in soil and their ecological risks on soil organisms. *Chemosphere*, 306, 135573.

Choudhury A., Simnani F. Z., Singh D., Patel P., Sinha A., Nandi A., Ghosh A., Saha U., Kumari K., Jaganathan S. K., Kaushik N. K., Panda P. K., Suar M., & Verma S. K. (2023). Atmospheric microplastic and nanoplastic: The toxicological paradigm on the cellular system. *Ecotoxicology and Environmental Safety*, 259, 115018.

Crossman J., Hurley R. R., Futter M., & Nizzetto L. (2020). Transfer and transport of microplastics from biosolids to agricultural soils and the wider environment. *Science of the Total Environment*, 724, 138334.

Dey T. K., Uddin M. E., & Jamal M. (2021). Detection and removal of microplastics in wastewater: Evolution and impact. *Environmental Science and Pollution Research,* 28(14), 16925–16947.

Dissanayake P. D., Kim S., Sarkar B., Oleszczuk P., Sang M. K., Haque M. N., Ahn J. H., Bank M. S., & Ok Y. S. (2022). Effects of microplastics on the terrestrial environment: A critical review. *Environmental Research*, 209, 112734.

Fakour H., Lo S.-L., Yoashi N. T., Massao A. M., Lema N. N., Mkhontfo F. B., Jomalema P. C., Jumanne N. S., Mbuya B. H., & Mtweve J. T. (2021). Quantification and analysis of microplastics in farmland soils: Characterization, sources, and pathways. *Agriculture*, 11, 330.

Heo N. W., Hong S. H., Han G. M., Hong S., Lee J., Song Y. K., Jang M., & Shim W. J. (2013). Distribution of small plastic debris in cross-section and high strandline on Heungnam beach, South Korea. *Ocean Science Journal*, 48(2), 225–233.

Horton A. A., Walton A., Spurgeon D. J., Lahive E., & Svendsen C. (2017). Microplastics in freshwater and terrestrial environments: Evaluating the current understanding to identify the knowledge gaps and future research priorities. *Science of the Total Environment*, 586, 127–141.

Isari E. A., Papaioanno D., Kalavrouziotis, I. K., & Karapanagioti H. K. (2021). Microplastics in agricultural soils: A case study in cultivation of watermelons and canning tomatoes. *Water*, 13, 2168.

Kaile N., Lindivat M., Elio J., Thuestad G., Crowley Q. G., & Hoell I. A. (2020). Preliminary results from detection of microplastics in liquid samples using flow cytometry. *Frontiers in Marine Science*, 7. https://doi.org/10.3389/fmars.2020.552688

Katsumi N., Kusube T., Nagao S., & Okochi H. (2021). Accumulation of microcapsules derived from coated fertilizer in paddy fields. *Chemosphere*, 267, 129185.

Kefeli A. A., Razumovskii S. D., Zaikov G. Y. (1971). Interaction of polyethylene with ozone, *Polymer Science U.S.S.R*, 13(4), 904–911. https://doi.org/10.1016/0032-3950(71)90288-7

Kumar M. V., & Sheela A. M. (2021). Effect of plastic film mulching on the distribution of plastic residues in agricultural fields. *Chemosphere*, 273, 128590.

Li W., Zhao W., Zhu H., Li Z.-J., & Wang W. (2023). State of the art in the photochemical degradation of (micro)plastics: From fundamental principles to catalysts and applications. *Journal of Material Chemistry*, 11, 2503–2527.

Ma J., Chen F., Xu H., Jiang H., Liu J., Li P., Chen C. C., & Pan K. (2021). Face masks as a source of nanoplastics and microplastics in the environment: Quantification, characterization, and potential for bioaccumulation. *Environmental Pollution*, 288, 117748.

Mahmud A., Wasif M., Roy H., Mehnaz F., Ahmed T., Pervez M. N., Naddeo V., & Islam M. (2022). Aquatic microplastic pollution control strategies: Sustainable degradation techniques, resource recovery, and recommendations for Bangladesh. *Water*, 14, 3968.

Morais L. M. S. S., Sarti F., Chelazzi D., Cincinelli A., Giarrizzo T., & Martinelli Filho J. E. (2020). The sea anemone Bunodosoma cangicum as a potential biomonitor for microplastics contamination on the Brazilian Amazon coast. *Environmental Pollution,* 265, 114817.

Nabi I., Bacha A.-U.-R., Ahmad F., & Zhang L. (2021). Application of titanium dioxide for the photocatalytic degradation of macro- and micro-plastics: A review. *Journal of Environmental Chemical Engineering,* 9(5), 105964.

Nan B., Su L., Kellar C., Craig N. J., Keough M. J., & Pettigrove V. (2020). Identification of microplastics in surface water and Australian freshwater shrimp Paratya australiensis in Victoria, Australia. *Environmental Pollution,* 259, 113865.

Nguyen M.-K., Lin C., Nguyen H.-L., Le V.-R., Priya K. L., Singh J., Soon W. C., Um M.-J., & Duc Nguyen D. (2023). Emergence of microplastics in the aquatic ecosystem and their potential effects on health risks: The insights into Vietnam. *Journal of Environmental Management,* 344. https://doi.org/10.1016/j.jenvman.2023.118499

Pena, A., Rodriguez-Liebana J. A., & Delgado-Moreno L. (2023). Interactions of microplastics with pesticides in soils and their ecotoxicological implications. *Agronomy,* 13(4), 701.

Priya K. L., Renjith K. R., Joseph C. J., Indu M. S., Srinivas R., & Haddout S. (2022). Fate, transport and degradation pathway of microplastics in aquatic environment – A critical review. *Regional Studies in Marine Science,* 56, 102647.

Priya K. L., Shabana I., Gopika B., Michi M., Haddout S., & Athul M. M. (2023). Implications of solid waste dumps on the microplastic abundance in groundwater in Kollam, India. *Journal of Environmental management,* 348, 119224.

Sa'adu I., & Farsang A. (2023). Plastic contamination in agricultural soils: A review. *Environmental Science and Europe,* 35(13), 13.

Singh R., Kumar R., & Sharma P. (2022). Microplastic in the subsurface system: Extraction and characterization from sediments of River Ganga near Patna, Bihar. In Gupta P. K., Yadav B., Himanshu S. K. (Eds.), *Advances in Remediation Techniques for Polluted Soils and Groundwater* (pp. 191–217). Elsevier.

Su P., Wang J., Zhang D., Chu K., Yao Y., Sun Q., Luo Y., Zhang R., Su X., Wang Z., Bu N., & Li Z. (2022). Hierarchical and cascading changes in the functional traits of soil animals induced by microplastics: A meta-analysis. *Journal of Hazardous Materials,* 440, 129854.

Tagg A. S., Brandes E., Fischer F., Fischer D., Brandt J., & Labrenz M. (2022). Agricultural application of microplastic-rich sewage sludge leads to further uncontrolled contamination. *Science of the Total Environment,* 806(Part 4), 150611.

Vivekanand A. C., Mohapatra S., & Tyagi V. K. (2021). Microplastics in aquatic environment: Challenges and perspectives. *Chemosphere,* 282, 131151.

Yang C., Lim W., & Song G. (2021). Mechanisms of deleterious effects of some pesticide exposure on pigs. *Pesticide Biochemistry and Physiology,* 175, 104850.

Zeenat, Elahi A., Bukhari D. A., Shamim S., & Rehman A. (2021). Plastics degradation by microbes: A sustainable approach. *Journal of King Saud University – Science*, 33, 101538.
Zhang K., Hamidian A. H., Tubi A., Zhang Y., Fang J. K. H., W C., & Lam P. K. S. (2021). Understanding plastic degradation and microplastic formation in the environment: A review. *Environmental Pollution*, 274, 116554.
Zhou J., Wen Y., Cheng H. G., Zang H. D., & Jones D. L. (2022). Simazine degradation in agroecosystems: Will it be affected by the type and amount of microplastic pollution? *Land Degradation & Development*, 33, 1128–1136.

CHAPTER 18

Microplastics in Freshwater and Soil

Policy Implementation and Management

Zahwa Zaffar, Noor Fatima, Fatima Malik, Mahnoor Malik, Khadija Amjad, Zeeshan Khan, and Tariq Shah

18.1 INTRODUCTION

The abundance of microplastics (MPs) in freshwater and soil has emerged as a critical global issue, necessitating a deep understanding and effective management strategies. At present, nano-plastics (NPs), which are smaller in size than MPs, have been explored in a very limited amount. The initial findings indicate that NPs may be a particular concern. Experimental investigations in laboratory settings have demonstrated their capability to infiltrate tissues and cross the blood-brain barrier [1]. This chapter comprehensively explores the evolution of MPs as a pervasive environmental challenge. From delving into the historical context of MPs research to elucidating the alarming escalation of contamination in freshwater and soil, it unfolds the complexities of this alarming issue. The imperative for policy development becomes increasingly apparent, supported by a growing

DOI: 10.1201/9781003487579-18

awareness of environmental concerns and the interconnectedness of MPs pollution across ecosystems. This chapter extends to the examination of global initiatives by scrutinizing the integration of MPs goals within the UN Sustainable Development Goals (SDGs) and delve into international collaboration efforts. The national policy landscape takes center stage, unraveling legislative frameworks, regulatory mechanisms, and the legal provisions designed to safeguard freshwater and soil. Within this mosaic, it frameworks policy variances among nations, engaging in a comparative analysis of national approaches and navigating the challenges in uniform implementation. To address the current policies and regulations, a dissection of prominent directives such as the European Union (EU) Single-Use Plastics Directive and the United Nations (UN) Environment Programme's Global Partnership on Marine Litter will be done. This involves a detailed overview of these directives, their implementation challenges, and notable achievements. With a lens on policy implementation strategies, it explores monitoring and assessment protocols, regulatory measures, and stakeholder engagement strategies. The challenges in policy implementation are also addressed. These include standardization issues, resource constraints, and transboundary challenges. As the intricate landscape of MPs policy is traversed, the chapter extends towards future policy directions. Integration with the circular economy, technological advancements, and strengthening international cooperation become focal points, offering a glimpse into the evolving trajectory of MPs management policies. In conclusion, this chapter endeavors to provide a comprehensive understanding of the multifaceted nature of MPs pollution, policy responses, and the future directions necessary for sustainable environmental stewardship.

18.1.1 Evolution of Microplastics as a Global Issue

18.1.1.1 Historical Context of Micropalstics Research

From the 1950s to the 1970s, the production of plastic was minimal, resulting in relatively manageable plastic waste. However, the period between the 1970s and the 1990s witnessed a more than threefold increase in plastic waste generation, mirroring the surge in plastic production. In the early 2000s, the quantity of plastic waste generated experienced a sharper escalation in a single decade than it had in the preceding 40 years. The production of plastic has been consistently increasing year by year, reaching a staggering 368 million metric tons by 2019. However, the rate of increase experienced a slowdown in 2020, primarily attributed to reduced demand resulting from the COVID-19 pandemic [2], illustrated in Figure 18.1.

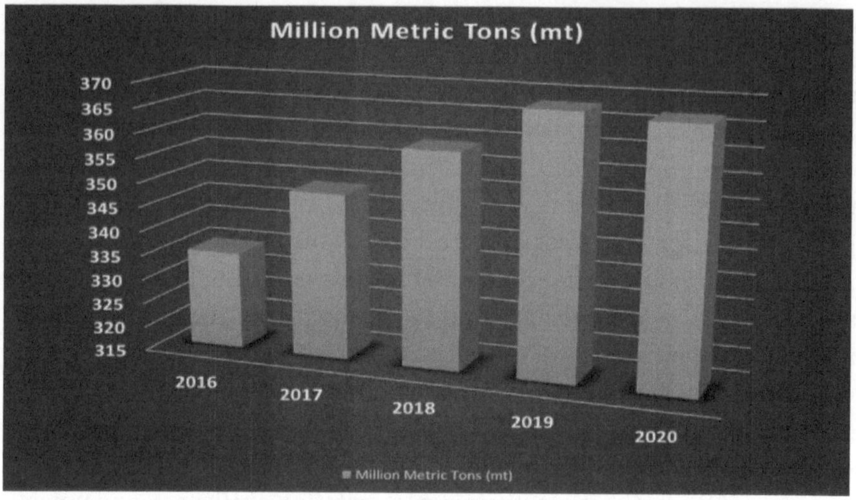

FIGURE 18.1 Year-on-year statistics of plastic production from 2016 to 2020 [2].

Presently, our annual production of plastic waste stands at approximately 400 million tonnes. If historical growth patterns persist, it is projected that the global production of primary plastic will reach 1,100 million tonnes by the year 2050. Additionally, there has been a concerning trend towards the use of single-use plastic products—items designed to be discarded after a brief, one-time use. Roughly 36% of the total plastic production is dedicated to packaging, which includes single-use plastic items for food and beverage containers. Alarmingly, about 85% of this packaging material finds its way into landfills or becomes unregulated waste [3].

18.1.1.2 Escalation of MPs Pollution in Freshwater and Soil

Undoubtedly, the surge in plastic production comes with consequences, as extensive amounts of the utilized plastics find their way into the environment, polluting both terrestrial and aquatic ecosystems. After their initial use, plastics are predominantly discarded on land [4]. Following their disposal, they are subsequently transferred to the aquatic environment. It is estimated that coastal countries generate approximately 275 million metric tons of plastic, of which 4.8–12.7 million metric tons enter the ocean [5]. A United Nations (UN) estimate disclosed the existence of approximately 51 trillion MPs in the seas, a quantity 500 times greater than the number of stars in the entire galaxy [6]. As depicted in Figure 18.2, MPs are widely classified into two categories depending on their originating sources.

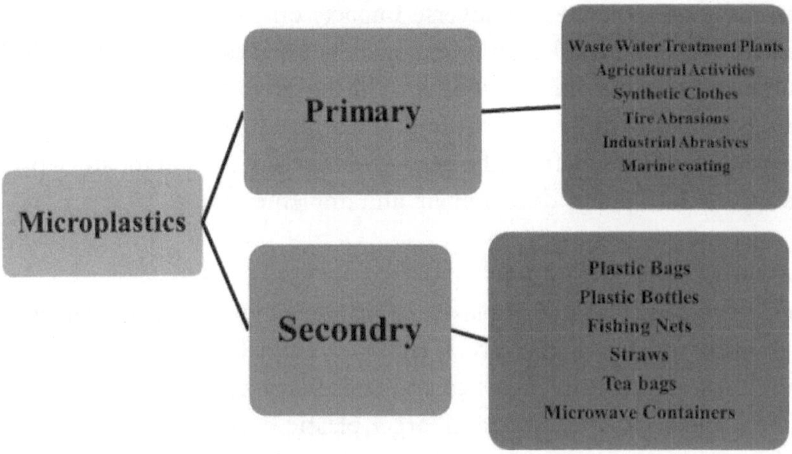

FIGURE 18.2 Types and sources of MPs pollution in the ocean.

Primary MPs are directly released into the environment as small particulate materials, constituting between 15% and 35% of the MPs found in oceans. Sources of primary MPs include synthetic clothes (35%), tire abrasions (28%), and synthesized MPs used as additives in personal care products like facial scrubs (2%). The vast majority, around 98%, originate from land-based activities, with only 2% stemming from sea-based activities. Major pathways for their introduction into the ocean include road runoff (66%), wastewater treatment systems (25%), and wind transfer (7%) [7]. On the other hand, secondary MPs contribute between 69% and 81% of the MPs in the ocean, resulting from the degradation of larger plastics, such as bags, bottles, and fishing nets, into smaller forms [6].

18.1.2 Necessity for Policy Development

The escalating issue of MPs pollution in freshwater and soil demands urgent attention and concerted efforts to formulate and implement effective policies. The historical context of MPs research reveals a substantial increase in plastic waste generation, particularly since the 1970s. With annual plastic waste production reaching a staggering 400 million tons, the adverse environmental impacts are pervasive. The interconnectedness of freshwater and soil MPs contamination amplifies the severity of the problem, necessitating the development of comprehensive policies. In a recent investigation, it was discovered that the existence of MPs in the soil resulted in diminished seed germination and inhibited plant growth. These effects could potentially be attributed to adverse impacts on earthworms. To mitigate these multifaceted challenges posed by MPs, policies must be meticulously crafted, considering global initiatives, collaborative efforts, and the legislative landscapes of individual nations. The urgency to safeguard ecosystems and human well-being from the pervasive effects of MPs mandates a proactive approach in policy development and implementation.

18.1.2.1 Growing Environmental Concerns

The predominant source of plastic in marine environments is terrestrial, originating from land and rivers. However, there has been comparatively limited research on the presence and consequences of MPs in freshwater systems and soil. After MPs and larger plastic particles enter the natural environment, monitoring their distribution and concentration becomes challenging, particularly when these particles become too small for easy visibility under a microscope. The movement of plastics through the environment is extensive, interacting with various ecosystems and animals, yet the specific patterns are not well understood at present. The development of markers capable of tracing plastics and MPs through the environment, along with identifying their source, would offer valuable insights into these complex dynamics.

18.1.2.2 Interconnectedness of Freshwater and Soil MPs Contamination

The interconnectedness of freshwater and soil MPs contamination underscores the complexity of plastic pollution dynamics within terrestrial ecosystems. MPs introduced into freshwater systems can undergo various pathways leading to soil contamination, and vice versa. Surface runoff from urban areas and agricultural lands, as well as atmospheric deposition, can transport MPs from freshwater bodies to soil environments. Similarly, soil erosion and runoff can carry MPs from terrestrial environments to freshwater systems. Once deposited in soil or water bodies, MPs can persist and interact with biota, affecting ecological processes and potentially entering food webs.

18.2 POLICY DEVELOPMENT AND FRAMEWORKS

18.2.1 Global Initiatives

Marine debris, primarily caused by production, consumption, and waste management, poses a significant threat to human health, marine life, and the environment [8]. To address this issue, global, regional, and national programs, guidelines, action plans, regulations, strategies, agreements, and conventions exist [9]. The IMO Marine Environment Protection Committee approved a six-category MARPOL port reception facility action plan, while the UNEP/IOC guidelines standardize beach, benthic, and surface water marine debris assessment. Market-based schemes and plastic bans have been implemented, and regional projects like MARELITT and GGGI physically remove marine debris. The Ad Hoc Open-ended Expert Group on Marine Litter and MPs emphasizes the need for global, regional, and national action [10].

18.2.1.1 UN Sustainable Development Goals (SDGs)

The SDGs aim to promote marine debris management, including clean water and sanitation, sustainable cities and communities, responsible consumption and production, and waste generation reduction [11]. These goals aim to reduce pollution, minimize hazardous chemicals and materials release, and ensure responsible consumption and production of chemicals and waste throughout their life cycle [12].

18.2.1.1.1 Integration of MPs Goals

With the National Sword policy, China banned global waste imports in 2018. The policy raised global waste management awareness [13]. The EU gave small and medium-sized waste-conversion technology firms 2.3 billion euros under Circular Economy Promotion laws [14]. The EU aims to reduce plastic bag use by 80% by 2019 [15]. UK government allocates £61 million to eliminate avoidable plastic waste by 2042 [16]. France, Kenya, and Peru ban single-use plastics. Zimbabwe bans plastic food packaging. Australians ban plastic shopping and pet waste bags. Canada would ban single-use plastics by 2021 [17]. Plastics were banned in Bangladesh first. Cambodia, Malaysia, Taiwan, Hong Kong, and Indonesia ban or tax plastics [18]. Norwegian single-use plastic bottlers pay an "environmental levy." The Philippines is a major ocean plastic waste producer and has studied plastic typologies and regulations [19].

18.2.1.1.2 Progress and Challenges

Over the past 70 years, industrialized nations have prioritized environmental law due to the global agreement on industrialization's environmental impacts [20]. Environmental controls are required by law nationwide to reduce and prevent pollution-related deaths [21]. The US Environmental Protection Agency and state agencies have transformed technological innovation to meet environmental quality standards. Environmental legal authority to reduce trash plastic has grown worldwide in the past decade. Legislation requiring recycling, collection, product limits, producer responsibility, and waste reduction must address all plastic trash [22]. MP production and use may need regulation to reduce environmental pollution. Furthermore, comprehensive monitoring systems are needed to track the implementation and efficacy of these regulations. Public awareness campaigns and education programs should also encourage responsible plastic waste management among individuals and businesses [23].

18.2.1.2 International Collaboration Efforts

International initiatives and partnerships are working to combat MPs, including the Global Partnership on Plastic Pollution and Marine Litter (GPML), which unites stakeholders in the fight against plastic pollution and marine litter. The EPA is participating in initiatives like the Basel Convention's Plastic Waste Partnership and the Asia-Pacific Economic Cooperation Virtual Working Group on Marine Debris. The IAEA and Argentina are conducting a research mission to Antarctica to gather evidence for describing and evaluating MP pollution in the ocean [24].

18.2.1.2.1 Joint Research Initiatives

To combat the effects of MPs, several groups have banded together to conduct research and undertake joint projects. MPs and their environmental impacts were the subject of a joint action by six funded projects called the Joint Programming Initiative Oceans. Teams of researchers from six different funded projects—ANDROMEDA, HOTMIC, FACTS, MicroplastiX, i-plastic, and RESPONSE—took part in this endeavor. With an emphasis on the need for ongoing research and global collaboration, the initiative brought attention to the causes, pathways, and consequences of MP contamination in aquatic ecosystems [25].

18.2.1.2.2 Cross Border Policy Harmonization

Harmonization of cross-border policies is necessary to address the global issue of MPs. The European Chemicals Agency (ECHA) has proposed limitations for MP particles that are intentionally added, demonstrating Europe's leadership in developing standardized regulatory standards. Combating marine debris is an issue that the US is actively pursuing. Organizations dedicated to standardization, such as the International Standard Organization (ISO), are working on creating testing criteria for MPs that are standardized worldwide [26]. A treaty to eliminate plastic pollution is currently being negotiated, which has led to an increase in international cooperation. The private sector is also participating in the discussions, and new partnerships and initiatives have arisen to address the issue. Standardized scientific methods are essential for detecting, identifying, quantifying, and characterizing MPs. Any treaty worth its salt will be grounded in scientific evidence [27].

18.2.2 National Policy Landscape

A rising awareness of the need to address MPs' sources and impacts characterizes the national policy landscape for these tiny pollutants. Legislation to speed up research, fill in data and knowledge gaps, and find and evaluate mitigation measures has been introduced or passed in a number of nations.

18.2.2.1 Legislative Frameworks

The legislation regarding MPs is a multifaceted matter that necessitates continuous refinement. The US Congress enacted the Microbead-Free Waters Act of 2015 to outlaw the sale of rinse-off cosmetic products that contain plastic microbeads. Nevertheless, there is a requirement for more extensive federal legislation to tackle the distinct adverse effects of MPs [28]. The ECHA's proposal to place restrictions on purposefully incorporated MP particles and to ensure that all plastic packaging is recyclable or reusable by 2030 serves as evidence that Europe is at the forefront of developing standardized regulatory guidelines. Global collaboration and standardization are essential for the worldwide detection, identification, measurement, and characterization of MPs [11].

18.2.2.1.1 Regulatory Mechanisms

MP pollution can be reduced through various economic strategies, such as extended producer responsibility (EPR), plastic taxes, deposit-return schemes, green public procurement, and the circular economy. EPR ensures producers are responsible for their products' entire lifecycle, encouraging sustainable design and recycling practices [29]. Plastic taxes discourage the use of plastic products, while deposit-return schemes charge for recycling. Green public procurement focuses on acquiring goods and services with a reduced environmental footprint, reducing plastic usage [30]. The circular economy aims to minimize waste and optimize resource utilization by prolonging material lifespans. These strategies require a coordinated effort from governments, producers, and consumers. In summary, these strategies can effectively mitigate MP pollution by advocating for sustainable alternatives and minimizing plastic waste [31].

18.2.2.2 Policy Variances among Nations

Each country's approach to MPs is unique, with its own set of priorities and measures. Many nations have outright forbidden the use of microbeads in cosmetics and personal care items. This includes the United States, Italy, Canada, France, and Germany. The use of rinse-off cosmetics containing MPs has been outlawed in Germany but not in France. A move to curb the release of MPs has led Norway to limit the use of recycled tires on athletic fields. Some nations have passed laws requiring or strongly recommending standardized procedures for MP sampling, detection, and characterization, while others have taken a more hands-off approach [32,33].

18.2.2.2.1 Challenges in Uniform Implementation

Different developed and developing nations have different policies regarding MPs. While wealthy nations put an emphasis on research, mitigation, and regulation, poor nations struggle with waste management, insufficient funding, and a lack of technical capability. Another obstacle to effective policies is the absence of regulation and standardized scientific methodologies. To help these nations establish effective policies and regulations regarding MPs, there needs to be a coordinated international effort, and this can only be achieved with global cooperation [34].

18.2.3 Current Policies and Regulations

Even though most regulations currently in place are focused on macroplastics, current policies and regulations regarding MPs acknowledge the necessity of handling their causes and effects. A growing awareness of the need for effective solutions is reflected in the significance of standardized scientific methods and regulations to mitigate their release and environmental impact [8,11].

18.2.3.1 European Union (EU) Single Use Plastics Directive

Targets for single-use plastics are recommended to fuel publicity and transparency and to institutionalize the city's goal of reducing their use [29]. Navarra, Spain, for instance, requires hotels, retail, and catering businesses to serve 80% of beer, 70% of soft drinks, and 40% of water in reusable containers by 2028 [35]. At a national level, countries including Spain, Sweden, and Greece have set targets to reduce single-use cups and food containers—Spain is aiming for a 50% reduction in these items by 2026 and 70% by 2030 (in terms of weight, compared with 2022 data) [35].

18.2.3.2 Global Initiatives and Collaborations

Global initiatives aim to address MP pollution through international cooperation, standardized scientific methods, and harmonized regulatory guidelines, acknowledging the cross-border nature of this issue and the need for coordinated action among nations [36].

18.2.3.3 Impact on MPs in Freshwater and Soil

Soil and freshwater ecosystems are particularly vulnerable to the detrimental effects of MPs. There is a correlation between aquatic animals and freshwater environments with high MP concentrations, suggesting that these contaminants may make their way up the food chain. When released into waterways, MPs pose a threat to aquatic life by entangling their tentacles and carrying toxic chemicals. MPs have the potential to influence soil pH, microbial processes, and faunal behavior in soil. Important for soil fertility and health, MP buildup can reduce microbial activity and change soil pH. Also, MPs have the potential to interact with soil fauna, which can impact their well-being and how they function. It is still unclear what the broader ecosystem will face as a result of these interactions and their long-term effects. According to the current research, additional research and actions are needed to reduce the effects of MPs on soil and freshwater ecosystems [11,26,33].

TABLE 18.1 Some Policies That Are Initiated for Addressing MP Pollution

No.	Policy	Target Area	Year(s)
1	Ban on microbeads	Several countries	2015–2020
2	Global Partnership on Marine Litter (GPML) at Rio +20	Worldwide	2012
3	MSFD	Europe	2008
4	UNEA Resolutions 1/6 and 2/11	Worldwide	2016
5	OSPAR/HELCOM/Barcelona Convention/ Associated Action Plans	Europe	2014–2020
5	SDG14	Worldwide	2017
6	ICRI call on microbeads	Worldwide	2016
7	Updated Basel Convention	Worldwide	2019
8	European Green Deal	Europe	2019
9	UN Plan to reduce MPs	Worldwide	2018
10	Oceans and the Law of the Sea A/74/19	Worldwide	2019
11	Antarctic Treaty	Antarctic	2019
12	ECHA restriction on MPs	Europe	2019

18.3 POLICY IMPLEMENTATION STRATEGIES

18.3.1 Monitoring and Assessment Protocols

Standardized protocols and methodologies are necessary for the surveillance and assessment of MPs in freshwater and soil environments to precisely ascertain their sources, fate, and consequences [37]. Extensive monitoring and analysis of environmental matrices, including soil, water, and sediment, are imperative in the scientific literature for comprehending the distribution, transport, and impacts of MPs. The implementation of standardized sampling and measurement methodologies is critical in order to generate monitoring results that are consistent and comparable across sites worldwide [38]. A review has been conducted on different pretreatment methods utilized in the monitoring of MPs in soil and freshwater sediment samples. This review emphasizes the criticality of sample processing, pretreatment, and characterization techniques in guaranteeing precise analysis [39]. To address this global environmental challenge, it is critical to establish comprehensive monitoring and assessment protocols, as MPs have been detected in Arctic regions and deep-sea sediments.

18.3.1.1 Surveillance Systems and Technologies

Monitoring MPs in freshwater and soil is predominantly accomplished through the implementation of pretreatment techniques that prepare samples for analysis. For accurate identification and quantification of MPs in environmental matrices, these techniques are vital. MPs analysis in freshwater and soil sediments frequently employs sample processing, pretreatment, and characterization techniques as pretreatment methods [40]. The implementation of standardized sampling and measurement methodologies is critical in order to generate monitoring results that are consistent and comparable across sites worldwide. Furthermore, it is imperative to undertake comprehensive monitoring and analysis of environmental matrices, including soil, water, and detritus, in order to ascertain the origins and ultimate destination of MPs [37]. The complex characteristics of MP transport and fate in freshwater and soil ecosystems emphasize the necessity for comprehensive protocols for monitoring and evaluation in order to effectively tackle this worldwide environmental dilemma [41].

18.3.1.1.1 Remote Sensing Applications

Emerging as a promising technology for monitoring MPs in freshwater and soil environments are remote sensing applications. Satellite sensors and synthetic aperture radar (SAR) are utilized by these applications to detect and monitor the presence of MPs in a variety of ecosystems [42]. Multispectral remote sensing employs wavelengths from the Near Infrared (NIR), visible spectrum, Infrared, and Microwave wavelengths to detect MPs at spatial resolutions. SAR is an environmental monitoring active satellite remote sensing system that measures fluid mechanics properties of surface water, including surface tension variations and sea surface roughness. Despite its promising potential, the use of satellite sensors for the detection of plastic debris in water is still in the research and development phase. The application of hyperspectral, thermal imaging, and multispectral sensors to monitor macroplastic debris has been effective [42].

Although remote sensing does indeed overcome certain limitations of conventional approaches, including the financial investment required and the challenge of monitoring expansive regions, it is critical to acknowledge that the development and implementation of this technology for MPs monitoring are still in their nascent phases. Hence, additional investigation and advancement are required to enhance the precision and dependability of these techniques to monitor MPs in soil and freshwater ecosystems [42].

18.3.1.1.2 Advanced Analytical Techniques

Advanced analytical methodologies are necessary for the surveillance of MPs in soil and freshwater ecosystems. Utilizing specialized instruments and methodologies, these processes detect, analyze, and quantify MP particles in samples that have been gathered. Frequently, electron and optical microscopy are applied to samples to visually identify and quantify MP particles. Particularly advantageous for larger MPs is this technique [40]. By employing methods such as Fourier-transform infrared (FTIR) spectroscopy and Raman spectroscopy, the chemical composition of MPs can be determined, thereby facilitating the classification of plastic varieties [43]. The thermal properties of MPs can be determined using methods such as differential scanning calorimetry (DSC) and thermogravimetric analysis (TGA), which can be utilized to aid in their identification [40].

18.3.1.2 Temporal and Spatial Monitoring

Particularly with respect to surveillance technologies and systems, temporal and spatial monitoring are vital elements of policy implementation strategies. To determine the efficacy of the implemented policies and technologies, these approaches employ rigorous monitoring and evaluation mechanisms that take temporal and spatial dimensions into account. Furthermore, flexibility and adaptability are emphasized as critical components of implementation strategies. This includes the capacity to observe and react to temporal and spatial fluctuations. Sustained stakeholder engagement and communication, which constitute an additional critical component, require temporal and spatial considerations to guarantee the continued effectiveness and relevance of policies and technologies. Consequently, temporal and spatial monitoring are essential for the effective execution of policies pertaining to surveillance technologies and systems [44].

18.3.1.2.1 Establishing Baselines

The establishment of baselines is an essential component of environmental surveillance, encompassing the monitoring of MPs in soil and freshwater. Baselines serve as a standard against which variations in dimension and temporal progression can be assessed in various spatial contexts [45]. In temporal monitoring, variations are tracked over time. This can provide insights into the efficacy of policies designed to mitigate MP pollution and aid in the identification of trends, such as fluctuations in MP concentrations [46]. In contrast, spatial monitoring entails the observation and analysis of alterations occurring in various geographical areas. This can aid in the identification of MP pollution hotspots and shed light on their origins and transport routes [47]. Regional environmental monitoring with enhanced spatial and temporal precision can be accomplished through the development and deployment of a highly precise spatiotemporal monitoring platform, thereby elevating the standard of regional environmental quality management and decision-making [44].

18.3.1.2.2 Detecting Hotspots and Trends

Various methodologies can be employed to identify concentrations and trends for temporal and spatial monitoring in the management and policy implementation of MPs in freshwater and soil. These include:

- **Bibliometrics-Based Review:** Analyze soil MP research to detect soil contamination hotspots and trends. This can reveal soil MPs' spatial and temporal distribution, helping build focused management methods [48].
- **Systematic Review and Meta-Analysis:** Use systematic review and meta-analysis to understand MP runoff and discharge into freshwater ecosystems. This method can identify freshwater contamination

hotspots and trends, influencing policy implementation and management [49].

- Utilize long-term soil monitoring programs and archival samples to track MP contamination fluctuations over time. This can identify agricultural topsoil hotspots and trends, aiding management policy creation [50].
- **Application of Management Intervention Efficacy:** Conduct a systematic review to compare MP control treatments in freshwater bodies. This can uncover intervention application and efficacy patterns, directing MP pollution management policy and regulation [51].

18.3.2 Regulatory Measures

Compliance with regulations is essential for the management of MPs in soil and freshwater. Policymakers in several OECD (Organisation for Economic Co-operation and Development) nations are becoming progressively more concerned with formulating policy responses to control accidental MP emissions. The aforementioned measures encompass prohibitions on commonly littered single-use plastic items as well as limitations on the production and distribution of specific personal care and cosmetic products that contain MPs [52]. Although they constitute a significant portion of total emissions into the environment, MPs emitted from vehicle tires and textile products continue to be largely excluded from current policy frameworks [52].

18.3.2.1 MPs Content Labeling

The incorporation of MP content labels in soil and freshwater management systems could be a significant regulatory measure. It provides insight into the environmental consciousness of businesses and has the potential to impact consumer decisions. Nevertheless, the efficacy of these identifiers might be contingent upon the regulatory clarity and the enforceability of measurement standards [53]. For example, regulations must be precise and enforceable, and they should be linked to risks in order to encourage both environmental safety and innovation [54].

18.3.2.1.1 Consumer Products

Consumer product labeling regarding the concentration of MPs can function as a regulatory mechanism to oversee the management of MPs in soil and freshwater. A great emphasis has been placed on product stewardship and safety, prompting numerous nations to implement regulations that specifically target MPs found in consumer goods [53]. For instance, the Microbead-Free Waters Act, which is enacted in the United States, forbids the incorporation of plastic microbeads into specific products with the intention of preventing their discharge into the environment. However, regulation regarding MPs contamination of food and potable water remains inadequate; this is a critical issue that requires coordinated international and national efforts to resolve [55].

18.3.2.1.2 Packaging Materials

Packaging material MP content labeling may serve as an efficient regulatory measure for the management of MPs in soil and freshwater. The packaging industry is a major contributor to the contamination of MPs. Secondary MPs are byproducts of the decomposition of larger plastic products and packaging materials, whereas primary MPs are deliberately incorporated into packaging materials for diverse functionalities [56]. Labels devoid of MPs can effectively communicate an organization's environmental stewardship and uphold the principle of communicating ecological standards. However, consumers might not consistently opt for MP-free products that bear the label if doing so comes with an additional expense [53].

18.3.2.2 Emission Standards

Emission standards are increasingly recognized as an essential regulatory measure in the management of MPs' presence in soil and freshwater. The primary objective of these standards is to restrict the release of MPs from a multitude of origins, encompassing consumer goods, industrial operations, and the reduction of larger plastic objects to more manageable pieces. The precise quantification and regulation of MP emissions present considerable obstacles owing to the wide array of origins and routes by which these particles infiltrate the ecosystem [38].

18.3.2.2.1 Industrial Discharges

Runoff and discharge pathways are components of industrial activities that can substantially contribute to the discharge of MPs into freshwater ecosystems. Untreated waste or discharge from industrial sources, for instance, has the potential to introduce MPs directly into aquatic ecosystems [49]. Moreover, industrial operations have the potential to discharge MPs into agricultural ecosystems, thereby posing a risk of contamination to adjacent freshwater sources and sediments [57]. The establishment and enforcement of emission standards are critical in order to effectively tackle this matter. By regulating the discharge of MPs from industrial sources, these standards would aid in mitigating their detrimental effects on freshwater and soil ecosystems. An expansion of the European Union's regulatory framework to encompass intentional MPs and biodegradable plastics may aid in the reduction of MP emissions from industry. In addition, the utilization of advanced measurement methodologies, including fluorescence-based approaches, can facilitate the identification and surveillance of MP discharges originating from industrial sources [38].

18.3.2.2.2 Agricultural Practices

The utilization of plastic mulching, water pipelines, and greenhouse covers, among other agricultural practices, contributes to the environmental accumulation of MPs. Furthermore, agricultural and horticultural practices frequently employ "intentionally added MPs," which are deliberately integrated into consumer and industrial goods to serve particular objectives. These MPs constitute a significant contributor to MP pollution. Agricultural fertilizers can significantly contribute to MP concentrations in soils, which may have long-term effects on the productivity and operation of agroecosystems, according to research. It is challenging to undo the damage caused by the use of plastic in agricultural soils, and the interactions that occur between MPs and soil fauna, vegetation, microbiota, and crop production are frequently detrimental [58].

18.3.3 Stakeholder Engagement

Stakeholder engagement is essential to MP control policy implementation in freshwater and soil. It involves the participation of scientists, policymakers, industry experts, NGOs, and the public. The Royal Society's MPs evidence synthesis study emphasizes consulting scientists, policymakers, industry, and NGO professionals to design policies that are inclusive [38]. The implementation of adaptive management approaches is recommended to continuously evaluate and refine policies and regulations, incorporating new scientific findings and adapting policies to address emerging MP challenges. Collaborating with industries, retailers, and manufacturers is crucial for promoting MP awareness and education. This participation can promote sustainable production, minimize MP consumption, and enlighten customers about MP-free options. Developing measures to raise awareness and educate the public about MP pollution is vital.

Public awareness campaigns, educational programs, and efforts that promote responsible consumption and behavior change can achieve this [59]. Stakeholder engagement is crucial owing to MPs in agriculture. To reduce MP use, alternative methods must be found and concerns about purposely introduced MPs in agriculture and horticulture addressed.

18.3.3.1 Industry Collaboration

Governments may educate consumers, support sustainable production, and limit MP use by collaborating with industries, retailers, and product producers. Collaboration can also build partnerships to achieve positive change at the source, share best practices, and regulate MPs norms and laws [60]. Furthermore, it is essential that the agricultural sector participate in efforts to resolve the challenges posed by MPs. This involves the exploration of viable alternatives to MPs in agricultural practices, as well as promoting optimal strategies to mitigate plastic pollution in agricultural regions [38].

18.3.3.1.1 Corporate Responsibility Programs

The promotion of sustainable production methods and the enforcement of the principle of communicating environmental consciousness can be accomplished through corporate responsibility programs. Despite this, it is critical to ensure that the adoption of MPs-free alternatives is not hindered by increased costs or consumer reluctance by addressing the obstacles associated with consumer choices [53]. Corporate responsibility programs and industry collaboration are important elements of stakeholder engagement in policy implementation strategies pertaining to the management of MPs in soil and freshwater. To effectively tackle the issue of MP pollution, governments can facilitate positive change, limit the utilization of MPs, and establish standardized practices through collaboration with industries and the promotion of corporate responsibility [53].

18.3.3.1.2 Industry Specific Guidelines

Industry-specific MP stakeholder engagement recommendations include many essential strategies:

- **Inclusive Consultation:** Include scientists, policymakers, industry, and NGO professionals to ensure policy creation considers diverse perspectives.

- Implement adaptive management approaches to continuously evaluate and modify policies and regulations, including new scientific findings and adapting policies to address MP issues [61].
- **Industry Engagement:** Collaborate with retailers, manufacturers, and industries to raise awareness and educate about MPs. Encourage them to implement sustainable production methods, decrease MPs in products, and advise consumers about MP-free alternatives [61].
- **Public Awareness and Education:** Create policies to educate the public about MP pollution. Promote behavioral change and responsible consumption through public awareness, education, and initiatives [61].
- **Collaborate and Share Knowledge:** Promote sector-level collaboration to share knowledge and scale up effective approaches. This can involve cross-sector alliances that use networks and knowledge [61].
- **Standardized Definitions and Clear Requirements:** Scientific evidence should guide requirements for microfiber capture and filtration systems. Standardized definitions and defined requirements enable technological solution creation [62].
- **Use-Oriented Policy Approaches:** Use methods to prevent synthetic polymer products from abrasion and MPs from entering the environment. This involves advocating best practices throughout use, like laundry and eco-driving [62].
- **Private Governance:** Use MBIs and certification processes to combat MPs globally. This requires non-fragmented governance and stakeholder participation.
- Conduct stakeholder analysis to identify and evaluate active participants in the MP regulatory debate. This can ensure that all important parties are involved in policymaking and their views are considered [62].

18.3.3.2 Public Participation

Public education and awareness campaigns are of the utmost importance to grab the attention of the public and promote responsible consumption and behavior modification. Support for these campaigns may be obtained through partnerships with academic institutions, non-governmental organizations (NGOs), and other community-based initiatives. Additionally crucial are adaptive management strategies that permit ongoing evaluation and modification of policies and regulations. By accommodating evolving policies and integrating recent scientific discoveries, governments have the ability to foster an atmosphere that encourages stakeholders, including the general public, to adopt proactive strategies in order to mitigate MP pollution [38].

18.3.3.2.1 Community Awareness Programs

Programs that raise community consciousness can mobilize and educate members of the public regarding MP pollution. This objective can be accomplished by means of public awareness campaigns, educational initiatives, and programs that promote responsible consumption practices and behavioral modification. Active involvement of industries, retailers, and product manufacturers is critical to raise awareness and facilitate education regarding MPs. These collaborations may facilitate constructive transformation at their origin, exchange optimal methodologies, and harmonize regulations and standards pertaining to MPs [62].

18.3.3.2.2 Citizen Science Initiatives

Citizen science offers a unique chance to raise awareness about MPs, educate the public about mitigation options, and engage volunteers in data-gathering programs. To promote varied and inclusive engagement in combating MP contamination, citizen science programs can be customized to the talents and concerns of certain groups. Addressing MPs requires global-scale data, analysis, and solutions. Citizen science projects can help collect MP data for understanding and addressing this issue [63]. Citizen science engagement enhances environmental awareness, particularly MP pollution. U.S. EPA and NOAA work on citizen science programs to reduce MPs. This shows that government-led citizen science programs can help solve environmental problems [64].

18.4 CHALLENGES IN POLICY IMPLEMENTATION

Many nations and areas have implemented management initiatives to address the problems of marine litter, and there is a basic regulatory and management framework in place. Despite this, marine litter is still becoming more prevalent everywhere, including on seafloors, in oceanic gyres, estuaries, and mangroves. This indicates that marine litter is still a concern, especially regarding MPS. This is due to a variety of intricate factors, some of which can be found in the existing framework and which make it impossible to effectively control MPs contamination [65].

18.4.1 Standardization Issues

Standardized and harmonized processes are necessary for reliable, widely recognized, and legally defensible data collection in environmental monitoring for regulatory reasons. Stakeholders at multiple conferences on plastics in the environment concurred that there is a significant, if not total, lack of standards for chemical analysis, biological effect analysis, sample preparation, sampling, and sampling in the field of plastic pollution. The fact that the problem of plastic materials in various environmental compartments is different from traditional environmental monitoring and evaluation issues presents a fundamental obstacle [59].

18.4.1.1 Lack of Unified Measurement Metrics

Results utilizing either mass concentration (mass of plastics/volume) or descriptors of abundance (items/volume) have been used to characterize MP concentrations in the environment, but the units have not been consistent [66]. Unfortunately, conducting unit discussions for experimental effects data is challenging, if not impossible, due to insufficient reporting of property features that are frequently supplied inside certificates of analysis from suppliers [67].

18.4.1.1.1 Diverse Size Classifications

The definition of a lower limit for particle size is now up for debate. Although it is widely acknowledged that all plastic objects smaller than 5 mm are considered MPs [68], other authors divide MPs into smaller groups according to size. A lower restriction on particle size would, on the one hand, encourage the development of regulatory standards by standardizing sample techniques [69]. A smaller size limit, however, would keep smaller particles out of the regulatory purview. Current MP particles are future NP particles since they are predicted to continually break down into smaller fragments on a sub-micrometer to nanometer scale. As a result, current MP regulatory measures will also affect future concentrations of NPs [70].

18.4.1.1.2 Sampling Method Variability

The physical property diversity of plastics and environmental factors influence changes in sampling and testing techniques. Various extraction techniques have been devised to isolate high-density MPs from sediments [71]. Additionally, biota-rich water samples can necessitate pre-filtration processing procedures [72]. Analyses of relative efficiency or possible consequences on the MPs are not always included when introducing new methods. Other researchers may also introduce adaptation through subsequent comparative analysis [73].

18.4.1.2 Inconsistencies in Regulatory Approaches

The MPs are a class of particles that have a wide range of physicochemical characteristics that influence their environmental fate and potential hazard to living things. It appears to be a field of study for scientists to explore unanswered issues about the breakdown process to the uptake by organisms and interactions with chemicals to the unique surface characteristics of old plastics. Although it is obvious that regulatory methods are needed immediately to reduce environmental MPs, there are still a lot of unanswered problems regarding how these strategies will be implemented in terms of monitoring and enforcement.

18.4.1.2.1 Definition Variations

Five widely used standards are used to define MP: (1) high polymer content synthetic materials; (2) solid particles; (3) less than 5 mm; (4) water-insoluble; and (5) non-biodegradable [74]. Nonetheless, there is ongoing disagreement on several issues. For example, experts cannot agree on whether tire abrasion counts as "MPs," given that the monitoring guidelines for marine litter classify rubber from tires apart from plastics [75].

18.4.1.2.2 Enforcement Discrepancies

The current lack of sufficient scientifically gathered data on the prevalence of plastic debris in the environment is due to a lack of implementation and enforcement of existing regulations and management actions [76], inadequate international cooperation, and insufficient state participation in regional initiatives [77]. However, there are further societal shifts that provide fundamental national and regional difficulties, such as urbanization, fast economic growth, and shifting patterns of production and consumption, which could compromise the effectiveness of initiatives to reduce the availability of these materials [78]. For example, the United Nations Environment Programme noted that only the nations in the Northwest Pacific and Wider Caribbean have national laws specifically addressing marine litter [79]. In nations like Germany, the amended MARPOL Annex

V has not yet been incorporated into national legislation. As a result, there is no legal foundation for its national implementation [77].

18.4.2 Resource Constraints

18.4.2.1 Funding Limitations

A lack of funding to implement the rules contributes to non-compliance with legal efforts made at the national and international levels to monitor marine pollution. Until all MP escape routes within waste management are closed, improving plastic waste management activities cannot provide the intended outcomes. To do this, it is necessary to study, examine, and evaluate the escape routes as well as the relative amounts of MP flows via them. This will enable prompt response by rerouting funds and focusing on obstructing the most harmful routes [80].

18.4.2.1.1 Research and Monitoring Budgets

Research and development expenditures must be significantly increased in order to create innovative technologies and plastic substitutes that can effectively manage plastic waste. Less than 1.0% of the gross domestic product is allocated to research and development in the majority of African nations [81]. Funding for scientific study that can lower the likelihood of MPS entering water bodies and releasing dangerous compounds that then impact aquatic life and human health should be set aside by governments and other stakeholders. For example, the European Union started initiatives to standardize analytical techniques for identifying MPS in aquatic environments [82].

18.4.2.1.2 Implementation Costs

Minimizing the amount of MPS released into the environment requires closing the loop between the manufacture of plastics and waste management. The centralized plastic waste management is difficult in low- and middle-income nations because of limited infrastructure and a lack of funding for creative recovery and recycling methods. In this situation, creating a circular economy in low- and middle-income nations may be successful if the initiatives are decentralized and localized plastic waste disposal, collecting, sorting, recovery, and reuse take place [83].

18.4.2.2 Technological Barriers

The ability to monitor and test the effects of MP pollution within wastewater, freshwater (rivers, streams, and lakes), and marine environments is currently lacking in international standards, and regional test methods are too costly and time-consuming. This is a result of the relative novelty of this field of study, the lack of pertinent reference materials, and the dearth of more extensive studies on the biodegradation of plastic materials in these settings [84].

18.4.2.2.1 Access to Advanced Detection Technologies

Advanced detection methods are needed for correct identification of MPS, particularly the smaller fraction of MPS, 1 μm–1 mm, is impractical for citizen science projects [85]. The fact that plastic waste is not evenly dispersed throughout the environment and among the samples submitted to the lab presents another important difficulty [86]. The percentage of a sample or dust filter that is analyzed can therefore have a big influence on the number of MPs and NPs that are found. Furthermore, it is challenging to assess MPs and NPs accurately because of their low ambient concentrations [82]. Asymmetric distribution of MP abundance and size in collected samples, as well as differences in sample pretreatment procedures and laboratory instrumentation, are other issues related to data quality [87].

18.4.2.2.2 Capacity Building Challenges

The general challenge remains in processing large sample amounts in a short amount of time and in an economical manner, especially for monitoring purposes in various environmental matrices, whether they be freshwaters to meet the objectives of the Water Framework Directive or marine environments to comply with the MSFD, aside from these specificities regarding the definition of polymer materials and improving analytical approaches. The development of methods for detecting MP contamination in "particle-rich environments," including soils, sediments, and compost will require a particular focus on finding time- and cost-effective techniques to reduce sample volume and remove MPS [88].

18.4.3 Transboundary Issues

The fact that marine debris crosses international borders emphasizes how widespread and impactful the issue is. To control marine trash, national actions alone are insufficient; instead, regional and international collaboration is needed. A worldwide or regional integration of ocean-based litter monitoring is necessary, as demonstrated by empirical long-term monitoring research conducted in the Southern Ocean.

18.4.3.1 Cross Border Movement of MPs

Since the majority of freshwater bodies in Africa are transboundary, effective regional environmental regulations and collaboration are necessary for their management [89]. However, competing political positions among the riparian nations seriously jeopardize the management of the majority of transboundary lakes and rivers ecosystems in Africa [90]. Tanzania, Kenya, and Uganda share Lake Victoria, which serves as an excellent illustration of this. Its management has proven difficult because there hasn't been adequate coordination or coordinated policies, mostly since the East African Community collapsed in 1977. But when laws, agreements, and collaborations do take place, more attention is frequently paid to the sharing of natural resources than to pollution prevention [91].

18.4.3.1.1 Atmospheric Transport

MPs have the ability to spread throughout various environments, such as the atmosphere, persist in temporal reservoirs, and possibly even penetrate trophic food webs. When particle dispersion modeling was done to take into account 10 and 1 μm MP particles, it was found that the majority of the MPs that were transported through the atmosphere and deposited on the glaciers under study came from Central Asia, Northern Africa in the fall, Central Europe, the Atlantic Ocean in the winter and spring, the northern Indian Ocean, and up toward Russia in the summer. The findings of the simulation suggest that since there is no human activity at higher elevations on the plateau, MPs that arrive at the Tibetan Plateau may have traveled both short- and long-distances [92].

18.4.3.1.2 Watershed Dynamics

The percentage of urban and suburban development within the watersheds and population density are positively connected with the presence of MPS. Understanding the movement of MPS and the biofilms they are associated with is essential for determining the spatial scales at which plastics are transported throughout a watershed. Research on this topic should be grounded on known principles of community and ecosystem ecology [65]. Economic activities that are vital to ensure the ecological functioning of river basins, estuaries, and nearby coastal waters have been disregarded when it comes to risk assessments for plastic contamination. In order to forecast the future of micro- and macroplastics in aquatic habitats, and the socioeconomic factors should be taken into account while developing modeling methodologies in order to inform management strategies [93].

18.4.3.2 Challenges in Policy Alignment

Inconsistencies between government agency mandates and actions discourage cooperation and communication, which are necessary for carrying out comprehensive management plans. In order to facilitate integrated ocean governance, provide mutually beneficial solutions, and take advantage of the economical and environmental advantages of ocean use, policies and laws at all governmental levels and within international organizations must be harmonized [94]. Therefore, in order to improve policy harmonization, environmental stakeholders and scientists must work together to address the difficulties of environmental contamination. Thus, to achieve the adoption of long-lasting and reliable methods to minimize MPS in the environment, ocean governance has to adopt a cross-sectoral approach, enhanced collaboration, a specified contribution framework, harmonization, and policy consistency [95].

18.5 FUTURE POLICY DIRECTIONS

18.5.1 Integration with Circular Economy

The circular economy is an economic paradigm that encourages the economy's use of resources in ways that reduce, reuse, recycle, and recover [96,97]. It's a regenerative and restorative economic model that emphasizes the design-based application of the circular economy's three fundamental principles in order to address global concerns including pollution, waste, biodiversity loss, and climate change [97]. These principles include minimizing resources as well as product waste and regenerating the natural systems. Policies for MP management can be included into the circular economy by decreasing the use of plastic, encouraging its recycling, and ensuring its sustainable use. The circular economy strategy for plastic seeks to save USD 200 billion annually, creating 700,000 net jobs, cut greenhouse gas emissions by 25%, and reduce the yearly volume of plastic waste entering seas by 80%. A recent research discusses strategies for effective management of MPs to achieve life cycle sustainability [96].

18.5.1.1 Extended Producer Responsibility (EPR)

The Europe and certain Asian nations are seeing a sharp increase in the adoption of ERP a policy option that requires producers to bear either financial or physical accountability for their products after their useful life. Producers must either manage spent products through remanufacturing, recycling, or reuse in accordance with EPR regulations, or assign this duty to an outside organization known as a producer responsibility organization (PRO), which is compensated by the producer for managing spent products. Encouraging manufacturers to take accountability for waste management can strongly incentivize them to redesign goods with reduced material consumption and enhanced recyclability, which is the fundamental principle of EPR [98]. One option to improve plastic packaging design and facilitate recycling is to refine the current EPR obligations. This will pave the road for sustainable use of plastic.

18.5.1.1.1 Incorporating MPs in EPR Programs

To facilitate and finance the avoidance of MP contamination at many stages of the value chain, integrating MPs into EPR projects is a necessary approach. The purpose of EPR programs is to give industry stakeholders a framework for collaboration in order to identify strategies to reduce the costs associated with pollution reduction [52]. The Organization for Economic Cooperation and Development (OECD) highlights the need for a national framework for the circular economy and has established steps to prevent MPs contamination in water, with an emphasis on textiles and tires [99]. The European Chemical Agency (ECHA) views the standardized ban on purposefully added MPs as both implementable and enforceable [100]. The OECD also addresses the potential for EPR to cover a broader variety of plastic items and the environmental effects that go along with them. This includes plastic products that are not just plastic packaging [99].

18.5.1.1.2 Industry Circular Practices

The circular economy offers industrial organizations a strong chance to improve the environment by implementing the 4 Rs i.e., recycle, reduce, reuse, and recover. This method entails a complete plan that covers all aspects of business, from supply chain management to product design and engineering, with an emphasis on reuse, which is the most effective way to cut plastic emissions. Through small modifications or maintenance, industrial organizations can set themselves apart as leaders in the new, environmentally friendly manufacturing paradigm. Furthermore, research indicates that investment in industry can facilitate and encourage circular economy activities, which are often regarded as essential to a sustainable economy [101]. Circular practices are extremely important in order to reduce the pollution caused by MPs in industries. However, in order to address this challenge, additional research and sustainable options are required, as there are concerns that the recycling process itself could result in the more production of MPs [102]. Therefore, even if circular economy methods present viable ways to manage plastic waste, it's crucial to carefully assess how they affect the production of MPs and make sure that the strategy as a whole helps to lessen the pollution caused by MPs in the environment.

18.5.1.2 Sustainable Product Design

The management of MP pollution through sustainable product design can be achieved through the implementation of EPR. This involves encouraging the design of products with lower environmental impact, such as eco-design, and guaranteeing efficient end-of-life collection to enhance collection rates [103]. The first step in establishing a circular economy is sustainable product design, where each component of a product should be a valuable resource [104]. An environmental sustainability framework known as Eco-design for Sustainable Products Regulation (ESPR) establishes energy performance and other eco-design criteria for particular product categories [105]. Pollution caused by MPs can be considerably reduced with the help of sustainable product design. Designers and companies can contribute to the prevention of MP production during the product lifetime by reducing plastic production, eliminating single-use plastic, and encouraging the use of alternative materials such as metal, paper, bamboo, and glass. Reducing the amount of MPs released into the environment can also be achieved by using recycled materials and making durable products [52]. Therefore, by emphasizing the use of materials that are less hazardous to the environment and considering the entire lifecycle of products, sustainable product design plays a key role in tackling the issue of MP pollution.

18.5.1.2.1 Reducing MPs Generation

The reduction in the generation of MP following the EPR policy can be done via the following different methods:

1. **Redesigning of Products:** Products can be redesigned in order to facilitate their handling after the product's retirement. Moreover, plastic can altogether be avoided for product manufacturing as it contributes to the MP generation, and in this way, MP generation can be minimized [101].
2. **Investing in Research and Technology:** Funds must be allocated to science and technology in order to create novel and effective

approaches to tackle the problem of MP contamination, such as upgrading stormwater and wastewater treatment systems [101].

3. **EPR Systems for Textile Products:** Establishment of EPR systems for textile products can solve the issue of MP shedding by using eco-modulated prices to promote activities that prioritize reuse of fabric.

4. **Integrated Waste Management:** One key strategy for lowering the production of MPs is to set up an integrated waste management system that will enhance the plastics' life cycle and reduce its negative effects on the environment [106].

18.5.1.2.2 Design for Disposal and Decomposition

One strategy that considers the management of a product after its use is known as the design for disposal and decomposition. In addition to assuring efficient end-of-life collection, reuse, and recycling, it entails designing goods for simple disassembly and material reuse. The "Design for Decomposition" effort of the Biomimicry Institute is focused on developing technologies that will change the fashion industry and minimize waste by mimicking natural decomposition processes to transform discarded textile products into new biocompatible materials [107]. This strategy encourages recycling, reuse, and the use of biocompatible materials in an effort to alleviate the problem associated with waste generation and resource consumption. MPs can be converted and removed using a variety of techniques and technologies such as recycling, burning, and biodegradation [108]. Furthermore, research has been done on the creation of catalytic conversion technologies that can convert MPs into nontoxic materials [109]. Additionally, the incorrect disposal of plastic products can result in the production of MPs, highlighting the necessity of efficient management techniques, such as EPR and design for disposal and decomposition [110].

18.5.2 Technological Advancements

A number of technical innovations are being developed to solve the problem of plastic pollution, including MPs.

18.5.2.1 Remote Sensing Applications

The process of gathering data remotely via the use of sensors and imaging technology is known as remote sensing. Remote sensing has shown promise in recent years as a method for tracking and identifying plastic pollution in freshwater and soil habitats, including MPs. Among the uses of remote sensing in this field are:

1. **Detection and Identification of Plastic Waste:** Plastic waste has been detected, identified, and measured in both terrestrial and marine habitats using data from remote sensing. For instance, the amount of agricultural plastic waste [111], marine plastic waste [112], and terrestrial aggregated plastic waste has been measured using the data obtained via remote sensing.
2. **Monitoring of Plastic Waste:** Monitoring the physical properties of the Earth's environmental components, including plastic contaminants, can be done at a reasonable cost via remote sensing [113].
3. **Mapping of Plastic Pollution:** Remote sensing is a promising technology for measuring plastic contamination in soil and freshwater settings. It provides regular observations and wide coverage, which makes it a useful tool for tracking the effects and spread of plastic pollution [114].

Overall, the use of remote sensing technologies can provide high-quality data to support current monitoring strategies in an economical manner. To detect and track MPs in soil ecosystems and freshwater, a deep understanding is necessary to develop and improve remote sensing techniques.

18.5.2.1.1 Satellite Imaging

Satellite imaging is an emerging technology that has the potential to monitor as well as control the levels of MP pollution [115]. MP pollution in oceans, rivers, and other water bodies can be easily identified and mapped using satellite images. The exceptionally low spatial resolution and signal-to-noise ratio of the sensors make the process of MP detection via satellite imaging challenging. In order to address this issue, deep learning and machine learning models are being developed that can automatically classify mixtures of natural organic matter using satellite and aerial images. To detect and categorize MPs in satellite images, these models make use of algorithms that can learn from enormous datasets of annotated images [116]. This approach has demonstrated positive results for the identification of MPs in waterbodies, including the Ganges River in India [117].

18.5.2.1.2 UAV Based Surveillance

Drone surveillance, also known as UAV-based surveillance, uses unmanned aerial vehicles (UAVs) to monitor and collect data from a high altitude in the form of still photos and videos. These drones are useful for monitoring, security, and surveillance applications since they are equipped with a variety of cameras, sensors, and modules that allow the transmittance of real-time imagery as well as its recording. A growing number of industries, including agriculture, security, environmental, and disease management are now exploring the use of UAVs for several types of surveillance. Using real-time data gathering of images, UAV-based surveillance can monitor MPs which is crucial for their monitoring and mapping. UAVs are able to detect and monitor MP debris in a variety of environments, including beaches, rivers, and aquatic ecosystems [118]. The monitoring of macro and MPs in aquatic ecosystems has been investigated with satellite sensors, and researchers have found that MPs may be monitored with SAR which is a type of remote sensing technology [42].

18.5.2.2 Emerging Analytical Techniques

Environmental protection is becoming increasingly concerned with increasing MP pollution, and a number of analytical methods are now being developed for the management of MPs. These approaches include ultra-high-resolution microscopy, Raman micro-spectroscopy, spectroscopic techniques, and modern analytical methods for MP research, such as processing and identification procedures in various environmental compartments [119]. The most frequently used method for the identification and quantification of MPs is the Fourier transform infrared (FTIR) spectroscopy [120]. This technique makes it possible to precisely characterize MPs and related pollutants. These analytical methods are used to measure the quantity of MPs in food or environmental systems, as well as to identify them and pollutants associated with them. These new analytical tools are essential for improving our knowledge of the prevalence and effects of MPs as well as creating efficient management plans to lessen the negative they have on the environment.

18.5.2.2.1 Nanoparticle Detection

Nanoparticles, including NPs can be easily detected using a variety of sophisticated methods. These are characterized by using a variety of techniques such as field flow fractionation, scanning electron microscopy

(SEM) and Raman spectroscopy [121]. Moreover, impurities adhered to the plastic particles have been identified and micro(nano)plastics (MNPs) have been characterized using ultra-high-resolution microscopic and spectroscopic techniques [122]. These techniques male it possible the identification of NPs both chemically and physically, which is essential for determining how they might affect the environment and public health. Since micro- and nano-plastics are ubiquitous, their detection is essential for understanding their impact on the environment.

18.5.2.2.2 High Resolution Imaging

The detection and analysis of MPs have been made possible by the development of high-resolution imaging techniques such as flow imaging microscopy, hyperspectral imaging, and polarization holographic imaging. Hyperspectral imaging has been refined to detect MPs quickly and to determine their composition as well as structural characteristics [123,124]. Automated real-time counting and morphological assessments of many materials, including MPs, can be achieved with a portable and affordable method called polarization holographic imaging [125]. High-resolution near-infrared imaging spectroscopy has been used for MP detection in various environmental settings [126]. Flow imaging microscopy, or FlowCam, has the benefit of completely automated MP investigation over conventional spectroscopic techniques [127].

18.5.3 Strengthening International Cooperation

MP pollution is a global problem, hence it requires a global solution. International cooperation is crucial for the management of MP pollution across the globe. Several developed countries have made efforts to address this issue. International measures to prevent plastic pollution include the Basel Convention, Paris Agreement, Stockholm Convention, UN SDG Goal 12, and many marine-focused programs [128]. Governments of different countries have employed various instruments for the management of plastic pollution. Cooperation between nations can improve the exchange and management of information and allow the monitoring of progress made in order to manage plastic waste.

18.5.3.1 Global Data Sharing Initiatives

To address the issue of MP pollution, several international data exchange efforts are now under development. These programs are designed in such a way that they facilitate the establishment of all-encompassing plans for the control of MPs by improving the transmission of data and information and management of knowledge. Some of the key initiatives for this purpose are listed below.

18.5.3.1.1 Collaborative Databases

In order to address the issue of MP pollution, various collaborative databases have been developed. The aim of these databases is to enhance knowledge management, data sharing, and dissemination of information for supporting the development of comprehensive strategies for the management of MPs. Among the most important cooperative databases and projects are bibliographic analysis of data sharing practices, public databases for MPs in the environment, and the National Oceanic and Atmospheric Administration (NOAA) National Centers for Environmental Information (NCEI) marine MPs database. Datasets on MPs are created quickly and kept in a variety of forms, spanning from kilobytes to terabytes [129]. A public database for MPs in the environment called the Marine Plastic Database (MPDB) is an openly available structured relational database for particle-based MP data that is extensible and adaptable. The purpose of the database is to reduce plastic waste and promote positive change. It was created to satisfy the need for an extensive database containing data on MPs [130]. The goal of the NOAA NCEI marine MPs database is to make a significant amount of data publicly accessible so that a thorough analysis of the environmental problem may be conducted [131].

18.5.3.1.2 Open Access Information Platforms

There are open access information platforms accessible for the exchange of knowledge about MPs. One such platform is the open access, international, peer-reviewed journal "MPs" in MDPI. Furthermore, the Toxicity of MPs (ToMEx) is a dynamic, open-access database that helps environmental managers and scientists around the world to keep up with the findings regarding the toxicity of MPs. Another open access platform is Plastiverse (https://www.plastiverse.org/tools) which is a digital platform that can gather structured data about plastic waste via crowdsourcing. It offers an accessible database containing quantitative metrics for company data on the production and consumption of plastic worldwide, such as the weight of plastic packaging, the percentage of reusable plastic packaging, and post-consumer recycled plastic.

18.5.3.2 Harmonization of Regulatory Approaches

Studies and interactions with stakeholders demonstrate that harmonizing regulatory methods for the control of MPs is a persistent global concern. Studies conducted in the Himalayas and India have brought attention to the serious and expanding issue of MP contamination in these areas [132,133]. Using sophisticated microscopic and spectroscopic techniques, attempts are being undertaken to identify MPs and related pollutants [122]. The significance of tackling MP pollution through targeted activities and policies, together with the necessity of following up on related studies and EU directives, has also been emphasized by stakeholders, such as the development of the second OSPAR Regional Action Plan on Marine Litter.

18.5.3.2.1 Standardizing Measurement Protocols

To provide accurate and consistent measurements of MP pollution, standardizing measuring procedures for MP management is essential. A number of research studies and publications have highlighted that standardized procedures are required for the collection, examination, and handling of MPs in a variety of environmental settings including, freshwater ecosystems, sediments, and soil ecosystems [134]. Furthermore, there is a demand for more focused and enforceable rules related to MP contamination in order to protect environmental safety and innovation [54].

18.5.3.2.2 Joint Policy Formulation

A comprehensive strategy that considers the different causes and effects of MP contamination is needed to develop joint policies for MP pollution management. Preventing and minimizing plastic waste, maintaining water quality, and management of plastics in a sustainable way are important components of these programs. Furthermore, legislative initiatives have also concentrated on tasks including enhanced waste management procedures, terminal strategies, and the development of uniform guidelines for tracking MPs in various environmental settings [52]. One example of a policy measure intended to avoid and reduce MP contamination is the recent prohibition on the intentional use of MPs for the manufacturing of different products by the EU [135]. In order to successfully address the global issue of MP pollution, joint policy formation should consider a variety of measures, such as waste management practices, regulatory actions, and the establishment of standardized monitoring techniques.

18.6 CONCLUSION

To address the widespread occurrence of MPs in freshwater and soil ecosystems, a comprehensive approach is needed that combines regulatory, technological, and social aspects. Well-designed policy frameworks are essential for reducing the detrimental effects of MP pollution on biodiversity, ecosystem function, and human health. Policymakers can create a cleaner and more sustainable future by enforcing strict monitoring procedures, encouraging sustainable alternatives to traditional plastics, and boosting international collaboration. Meaningful progress in MP management requires active involvement from all stakeholders, such as governments, companies, academics, and the public. Collaborative actions are crucial for reducing MPs contamination and protecting freshwater and soil ecosystems. Collaborating to achieve shared objectives allows us to successfully tackle the issues presented by MPs and move towards a more durable and environmentally friendly Earth.

REFERENCES

1. Microplastics in fresh water and Soil Royal Society 2019.
2. Ruiz A. 2023. 25 plastic waste statistics that will shock you. Available from: https://theroundup.org/plastic-waste-statistics/
3. UNEP. Microplastics: The long legacy left behind by plastic pollution.
4. Hale RC, Seeley ME, La Guardia MJ, Mai L, Zeng EY. A global perspective on microplastics. *Journal of Geophysical Research: Oceans*. 2020; 125(1): e2018JC014719.
5. Jambeck JR, Geyer R, Wilcox C, Siegler TR, Perryman M, Andrady A, et al. Plastic waste inputs from land into the ocean. *Science*. 2015; 347(6223): 768–71.
6. Microplastics: sources, effects and solutions. 2018. Available from: https://www.europarl.europa.eu/news/en/headlines/society/20181116STO19217/microplastics-sources-effects-and-solutions
7. Boucher J, Friot D. *Primary Microplastics in the Oceans: A Global Evaluation of Sources*: IUCN, Gland, Switzerland; 2017.
8. Seay J, Ternes ME. A review of current challenges and legal advances in the global management of plastic waste. *Clean Technologies and Environmental Policy*. 2022; 24(3): 731–8.
9. Abril Ortiz A, Sucozhañay D, Vanegas P, Martínez-Moscoso A. A regional response to a global problem: single use plastics regulation in the countries of the Pacific Alliance. *Sustainability*. 2020; 12(19): 8093.
10. Agamuthu P, Mehran S, Norkhairah A, Norkhairiyah A. Marine debris: a review of impacts and global initiatives. *Waste Management & Research*. 2019; 37(10): 987–1002.

11. Kumar R, Verma A, Shome A, Sinha R, Sinha S, Jha PK, et al. Impacts of plastic pollution on ecosystem services, sustainable development goals, and need to focus on circular economy and policy interventions. *Sustainability.* 2021; 13(17): 9963.
12. Chen T-L, Kim H, Pan S-Y, Tseng P-C, Lin Y-P, Chiang P-C. Implementation of green chemistry principles in circular economy system towards sustainable development goals: challenges and perspectives. *Science of the Total Environment.* 2020; 716: 136998.
13. Wahab SN, Lim FYJ. China's National Sword Policy Reaction to Southeast Asia's waste trade and the transition to a circular economy. *Journal of Technology and Operations Management.* 2022; 17(1): 1–10.
14. Kerdlap P, Low JSC, Ramakrishna S. Zero waste manufacturing: a framework and review of technology, research, and implementation barriers for enabling a circular economy transition in Singapore. *Resources, Conservation and Recycling.* 2019; 151: 104438.
15. Watkins E, Schweitzer J-P. *Moving Towards a Circular Economy for Plastics in the EU by 2030*: Institute for European Environmental Policy (IEEP); 2018.
16. Iacovidou E, Ebner N, Orsi B, Brown A. Plastic packaging-how do we get to where we want to be? 2020. Available from http://bura.brunel.ac.uk/handle/2438/26557
17. Villarin TS, Cuaresma P. Environmental governance and the ban on single use plastics. 2020. Available from https://www.researchgate.net/publication/345598566_Environmental_Governance_and_the_Ban_on_Single_Use_Plastics
18. Proshad R, Kormoker T, Islam MS, Haque MA, Rahman MM, Mithu MMR. Toxic effects of plastic on human health and environment: a consequences of health risk assessment in Bangladesh. *International Journal of Health.* 2018; 6(1): 1–5.
19. Chasse C. *Evaluation of legal strategies for the reduction of plastic bag consumption*: Doctoral dissertation, Harvard University; 2018.
20. Yang T, Percival RV. The emergence of global environmental law. *Ecology LQ.* 2009; 36: 615.
21. Zhao Y, Kim B. Environmental regulation and chronic conditions: evidence from China's air pollution prevention and control action plan. *International Journal of Environmental Research and Public Health.* 2022; 19(19): 12584.
22. Hecht AD, Fiksel J. Solving the problems we face: the United States Environmental Protection Agency, sustainability, and the challenges of the twenty-first century. *Sustainability: Science, Practice and Policy.* 2015; 11(1): 75–89.
23. Khan F, Ahmed W, Najmi A. Understanding consumers' behavior intentions towards dealing with the plastic waste: perspective of a developing country. *Resources, Conservation and Recycling.* 2019; 142: 49–58.
24. Da Costa JP, Mouneyrac C, Costa M, Duarte AC, Rocha-Santos T. The role of legislation, regulatory initiatives and guidelines on the control of plastic pollution. *Frontiers in Environmental Science.* 2020; 8: 104.

25. Conchubhair DÓ, Fitzhenry D, Lusher A, King AL, van Emmerik T, Lebreton L, et al. Joint effort among research infrastructures to quantify the impact of plastic debris in the ocean. *Environmental Research Letters*. 2019; 14(6): 065001.
26. Kentin E. Restricting microplastics in the European Union: process and criteria under REACH. *The European Physical Journal Plus*. 2018; 133: 1–12.
27. Tessnow-von Wysocki I, Le Billon P. Plastics at sea: treaty design for a global solution to marine plastic pollution. *Environmental Science & Policy*. 2019; 100: 94–104.
28. Strifling DA. The microbead-free waters act of 2015. *Journal of Land Use & Environmental Law*. 2016; 32(1): 151–66.
29. Fletcher S, March ALA, Roberts K, Shirian Y, i Canals LM, Cairns A, et al. Turning off the tap: how the world can end plastic pollution and create a circular economy. 2023.Available from https://www.unep.org/resources/turning-off-tap-end-plastic-pollution-create-circular-economy
30. Oke A, Osobajo O, Obi L, Omotayo T. Rethinking and optimising post-consumer packaging waste: a sentiment analysis of consumers' perceptions towards the introduction of a deposit refund scheme in Scotland. *Waste Management*. 2020; 118: 463–70.
31. Blomsma F, Tennant M. Circular economy: preserving materials or products? Introducing the resource states framework. *Resources, Conservation and Recycling*. 2020; 156: 104698.
32. An L, Liu Q, Deng Y, Wu W, Gao Y, Ling W. Sources of microplastic in the environment. In He D, Luo Y (Eds.), *Microplastics in Terrestrial Environments: Emerging Contaminants and Major Challenges*. 2020 (pp. 143–59). Springer.
33. Verschoor AJ, Van Gelderen A, Hofstra U. Fate of recycled tyre granulate used on artificial turf. *Environmental Sciences Europe*. 2021; 33(1): 1–15.
34. Marshall RE, Farahbakhsh K. Systems approaches to integrated solid waste management in developing countries. *Waste Management*. 2013; 33(4): 988–1003.
35. Lade C, Strickland P, Frew E, Willard P, Nagpal S, Osorio SC, et al. *International Tourism Futures: The Drivers and Impacts of* Change: Goodfellow Publishers Ltd; 2020.
36. Raubenheimer K, Urho N. *Possible Elements of a New Global Agreement to Prevent Plastic Pollution*: Nordic Council of Ministers; 2020.
37. Chen Y-N, Rani A, Chiang C-Y, Kim H, Pan S-Y. Monitoring, control and assessment of microplastics in bioenvironmental systems. *Environmental Technology & Innovation*. 2023; 32: 103250.
38. Muniz-Gonzalez A-B, Silva CJ, Silva ALP, Campos D, Pestana JL, Martínez-Guitarte J-L. Suborganismal responses of the aquatic midge Chironomus riparius to polyethylene microplastics. *Science of the Total Environment*. 2021; 783: 146981.
39. Sajjad M, Huang Q, Khan S, Khan MA, Liu Y, Wang J, et al. Microplastics in the soil environment: a critical review. *Environmental Technology & Innovation*. 2022; 27: 102408.

40. Lee H, Kim S, Sin A, Kim G, Khan S, Nadagouda MN, et al. Pretreatment methods for monitoring microplastics in soil and freshwater sediment samples: a comprehensive review. *Science of the Total Environment*. 2023; 871: 161718.
41. Cera A, Cesarini G, Scalici M. Microplastics in freshwater: what is the news from the world? *Diversity*. 2020; 12(7): 276.
42. Mukonza SS, Chiang J-L. Satellite sensors as an emerging technique for monitoring macro-and microplastics in aquatic ecosystems. *Water Emerg Contam Nanoplastics*. 2022; 1(4): 17.
43. Rossatto A, Arlindo MZF, de Morais MS, de Souza TD, Ogrodowski CS. Microplastics in aquatic systems: a review of occurrence, monitoring and potential environmental risks. *Environmental Advances*. 2023; 13: 100396.
44. Yang R, Hao X, Zhao L, Yin L, Liu L, Li X, et al. Design and implementation of a highly accurate spatiotemporal monitoring and early warning platform for air pollutants based on IPv6. *Scientific Reports*. 2022; 12(1): 4615.
45. Haldar S. (Ed.). Environmental system management of mineral resources and sustainable development. In *Mineral Exploration*. 2013 (pp. 267–85). Elsevier: Amsterdam, The Netherlands.
46. Mathes RW, Lall R, Levin-Rector A, Sell J, Paladini M, Konty KJ, et al. Evaluating and implementing temporal, spatial, and spatio-temporal methods for outbreak detection in a local syndromic surveillance system. *PLoS One*. 2017; 12(9): e0184419.
47. Zeunert S, Meon G. Influence of the spatial and temporal monitoring design on the identification of an instantaneous pollutant release in a river. *Advances in Water Resources*. 2020; 146: 103788.
48. Miao J, Huang W, Pan R, Zhou K. Research progress and hotspot analysis of soil microplastics: a bibliometrics-based review. *Frontiers in Environmental Science*. 2023; 11:1297646.
49. Wang J, Bucci K, Helm P, Hoellein T, Hoffman M, Rooney R, et al. Runoff and discharge pathways of microplastics into freshwater ecosystems: a systematic review and meta-analysis. *FACETS*. 2022; 7: 1473–1492.
50. Weber CJ, Bastijans J-E, Lügger K, Heller C. Microplastic in long-term soil monitoring: first spatial and temporal data on plastics in agricultural topsoils. *Journal of Soils and Sediments*. 2023; 24: 1–15.
51. Munien S, Adhikari PL, Reycraft K, Mays TJ, Naidoo T, Pruitt M, et al. Application and efficacy of management interventions for the control of microplastics in freshwater bodies: a systematic review. *Water*. 2024; 16(1): 176.
52. Buzzi E. *Policies to Reduce Microplastics Pollution in Water: Focus on Textiles and Tyres*: OECD Publishing; 2021.
53. Munhoz DR, Harkes P, Beriot N, Larreta J, Basurko OC. Microplastics: a review of policies and responses. *Microplastics*. 2022; 2(1): 1–26.
54. Mitrano DM, Wohlleben W. Microplastic regulation should be more precise to incentivize both innovation and environmental safety. *Nature Communications*. 2020; 11(1): 5324.

55. Usman S, Abdull Razis AF, Shaari K, Azmai MNA, Saad MZ, Mat Isa N, et al. The burden of microplastics pollution and contending policies and regulations. *International Journal of Environmental Research and Public Health*. 2022; 19(11): 6773.
56. Calero M, Godoy V, Quesada L, Martín-Lara MÁ. Green strategies for microplastics reduction. *Current Opinion in Green and Sustainable Chemistry*. 2021; 28: 100442.
57. Kallenbach EM, Rødland ES, Buenaventura NT, Hurley R. Microplastics in terrestrial and freshwater environments. In Bank MS (Ed.), *Microplastic in the Environment: Pattern and Process* (p. 87). Springer; 2022.
58. Cusworth SJ, Davies WJ, McAinsh MR, Gregory AS, Storkey J, Stevens CJ. Agricultural fertilisers contribute substantially to microplastic concentrations in UK soils. *Communications Earth & Environment*. 2024; 5(1): 7.
59. Brennholt N, Heß M, Reifferscheid G. Freshwater microplastics: challenges for regulation and management. In Wagner M, Scott Lambert S (Eds.), *Freshwater Microplastics: Emerging Environmental Contaminants?* 2018 (pp. 239–72). Springer.
60. Birnstiel S, Soares-Gomes A, da Gama BA. Depuration reduces microplastic content in wild and farmed mussels. *Marine Pollution Bulletin*. 2019; 140: 241–7.
61. Li J, Lusher AL, Rotchell JM, Deudero S, Turra A, Bråte ILN, et al. Using mussel as a global bioindicator of coastal microplastic pollution. *Environmental Pollution*. 2019; 244: 522–33.
62. Onyena AP, Aniche DC, Ogbolu BO, Rakib MRJ, Uddin J, Walker TR. Governance strategies for mitigating microplastic pollution in the marine environment: a review. *Microplastics*. 2021; 1(1): 15–46.
63. Litter Agaom. From pollution to solution. 2021.
64. Popa CL, Dontu SI, Savastru D, Carstea EM. Role of citizen scientists in environmental plastic litter research—a systematic review. *Sustainability*. 2022; 14(20): 13265.
65. McCormick AR, Hoellein TJ. Anthropogenic litter is abundant, diverse, and mobile in urban rivers: insights from cross-ecosystem analyses using ecosystem and community ecology tools. *Limnology and Oceanography*. 2016; 61(5): 1718–34.
66. Hidalgo-Ruz V, Gutow L, Thompson RC, Thiel M. Microplastics in the marine environment: a review of the methods used for identification and quantification. *Environmental Science & Technology*. 2012; 46(6): 3060–75.
67. Connors KA, Dyer SD, Belanger SE. Advancing the quality of environmental microplastic research. *Environmental Toxicology and Chemistry*. 2017; 36(7): 1697–703.
68. Rochman CM, Browne MA, Halpern BS, Hentschel BT, Hoh E, Karapanagioti HK, et al. Classify plastic waste as hazardous. *Nature*. 2013; 494(7436): 169–71.

69. Arthur C, Baker JE, Bamford HA. *Proceedings of the International Research Workshop on the Occurrence, Effects, and Fate of Microplastic Marine Debris*, September 9–11, 2008, University of Washington Tacoma, Tacoma, WA, USA. 2009.
70. Peters R, Brandhoff P, Weigel S, Marvin H, Bouwmeester H, Aschberger K, et al. Inventory of nanotechnology applications in the agricultural, feed and food sector. *EFSA Supporting Publications.* 2014; 11(7): 621E.
71. Stock F, Kochleus C, Bänsch-Baltruschat B, Brennholt N, Reifferscheid G. Sampling techniques and preparation methods for microplastic analyses in the aquatic environment–a review. *TrAC Trends in Analytical Chemistry.* 2019; 113: 84–92.
72. Cole M, Webb H, Lindeque PK, Fileman ES, Halsband C, Galloway TS. Isolation of microplastics in biota-rich seawater samples and marine organisms. *Scientific Reports.* 2014; 4(1): 4528.
73. Rist S, Hartmann NB, Welden NA. How fast, how far: diversification and adoption of novel methods in aquatic microplastic monitoring. *Environmental Pollution.* 2021; 291: 118174.
74. Verschoor A. Towards a definition of microplastics: considerations for the specification of physico-chemical properties. RIVM Letter Report 2015-0116. 2015.
75. Hanke G, Galgani F, Werner S, Oosterbaan L, Nilsson P, Fleet D, et al. Guidance on Monitoring of Marine Litter in European Seas: a guidance document within the common implementation strategy for the marine strategy framework directive. EUR 26113. Luxembourg (Luxembourg): Publications Office of the European Union; JRC83985. 2013.
76. Gold M, Mika K, Horowitz C, Herzog M. Stemming the tide of plastic litter: a global action agenda. *Tulane Environmental Law Journal.* 2013; 27: 165.
77. Chen C-L. Regulation and management of marine litter. In Bergmann M, Gutow L, Klages M (Eds.), *Marine Anthropogenic Litter.* 2015 (pp. 395–428). Springer.
78. da Costa JP, Duarte AC, Rocha-Santos TA. Microplastics–occurrence, fate and behaviour in the environment. *Comprehensive Analytical Chemistry.* 2017; 75: 1–24. Elsevier.
79. Warning UNEPDo E. *UNEP Year Book 2011: Emerging Issues in Our Global* Environment: UN; 2011.
80. Hettiarachchi H, Meegoda JN. Microplastic pollution prevention: the need for robust policy interventions to close the loopholes in current waste management practices. *International Journal of Environmental Research and Public Health.* 2023; 20(14): 6434.
81. Sanganyado E. Developing countries must fund local research. *Science.* 2021; 372(6549): 1403.
82. Li C, Busquets R, Campos LC. Assessment of microplastics in freshwater systems: a review. *Science of the Total Environment.* 2020; 707: 135578.
83. Joshi C, Seay J, Banadda N. A perspective on a locally managed decentralized circular economy for waste plastic in developing countries. *Environmental Progress & Sustainable Energy.* 2019; 38(1): 3–11.

84. Harrison JP, Boardman C, O'Callaghan K, Delort A-M, Song J. Biodegradability standards for carrier bags and plastic films in aquatic environments: a critical review. *Royal Society Open Science.* 2018; 5(5): 171792.
85. Löder MG, Gerdts G. Methodology used for the detection and identification of microplastics—a critical appraisal. In Bergmann M, Gutow L, Klages M (Eds.), *Marine Anthropogenic Litter.* 2015 (pp. 201–27). Springer.
86. Hartmann NB, Huffer T, Thompson RC, Hassellov M, Verschoor A, Daugaard AE, et al. Are we speaking the same language? Recommendations for a definition and categorization framework for plastic debris. *Environmental Science and Technology.* 2019; 53(3): 1039–1047. ACS Publications.
87. da Costa JP, Santos PS, Duarte AC, Rocha-Santos T. (Nano) plastics in the environment–sources, fates and effects. *Science of the Total Environment.* 2016; 566: 15–26.
88. Zarfl C. Promising techniques and open challenges for microplastic identification and quantification in environmental matrices. *Analytical and Bioanalytical Chemistry.* 2019; 411: 3743–56.
89. Botts L, Muldoon P, Botts P, von Moltke K. The great lakes water quality agreement: its past successes and uncertain future. In Hisschemöller M, Hoppe R, Dunn WN, Ravetz JR (Eds.), *Knowledge, Power, and Participation in Environmental Policy Analysis* (pp. 121–44). Routledge; 2018.
90. Khan FR, Mayoma BS, Biginagwa FJ, Syberg K. Microplastics in inland African waters: presence, sources, and fate. In Wagner M, Lambert S (Eds.), *Freshwater Microplastics: Emerging Environmental Contaminants?* 2018 (pp. 101–24). Springer.
91. Abila R, Onyango P, Odongkara K. *Socioeconomic Viability and Sustainability of BMUs: Case Study of the Cross-Border BMUs on Lake Victoria*: GLOW IV, Bagamoyo, Tanzania; 2006.
92. Zhang Y, Gao T, Kang S, Allen S, Luo X, Allen D. Microplastics in glaciers of the Tibetan Plateau: evidence for the long-range transport of microplastics. *Science of the Total Environment.* 2021; 758: 143634.
93. Pinheiro L, Agostini V, Lima A, Ward R, Pinho G. The fate of plastic litter within estuarine compartments: an overview of current knowledge for the transboundary issue to guide future assessments. *Environmental Pollution.* 2021; 279: 116908.
94. Stoett PJ. *Global Ecopolitics: Crisis, Governance,* and Justice: University of Toronto Press; 2019.
95. DasGupta R, Shaw R. Changing perspectives of mangrove management in India–an analytical overview. *Ocean & Coastal Management.* 2013; 80: 107–18.
96. Elsamahy T, Al-Tohamy R, Abdelkarim EA, Zhu D, El-Sheekh M, Sun J, et al. Strategies for efficient management of microplastics to achieve life cycle assessment and circular economy. *Environmental Monitoring and Assessment.* 2023; 195(11): 1361.
97. Wijkman A, Skånberg K. The circular economy and benefits for society. Club of Rome. 2015.

98. Hanisch C. Is extended producer responsibility effective? *Environmental Science & Technology*. 2000; 34(7): 170A–5A.
99. Brown A, Laubinger F, Börkey P. New aspects of EPR: extending producer responsibility to additional product groups and challenges throughout the product lifecycle. OECD Environment Working Papers, No. 225, OECD Publishing, Paris. 2023.
100. Nielsen MB, Clausen LPW, Cronin R, Hansen SF, Oturai NG, Syberg K. Unfolding the science behind policy initiatives targeting plastic pollution. *Microplastics and Nanoplastics*. 2023; 3(1): 3.
101. Dey PK, Malesios C, De D, Budhwar P, Chowdhury S, Cheffi W. Circular economy to enhance sustainability of small and medium sized enterprises. In Dey PK, Chowdhury S, Malesios C (Eds.), *Supply Chain Sustainability in Small and Medium Sized Enterprises*. 2022 (pp. 10–45). Routledge.
102. Syberg K, Nielsen MB, Oturai NB, Clausen LPW, Ramos TM, Hansen SF. Circular economy and reduction of micro (nano) plastics contamination. *Journal of Hazardous Materials Advances*. 2022; 5: 100044.
103. Komyakova V, Vince J, Haward M. Microplastics and the Australian marine environment: issues and options. Report to the National Environmental Science Program, Marine Biodiversity Hub. IMAS, University of Tasmania. 2020.
104. Manjunatheshwara K, Vinodh S. Sustainable electronics product design and manufacturing: state of art review. *International Journal of Sustainable Engineering*. 2021; 14(4): 541–51.
105. Malcolm R. Ecodesign laws and the environmental impact of our consumption of products. *Journal of Environmental Law*. 2011; 23(3): 487–503.
106. Prata JC, Silva ALP, Da Costa JP, Mouneyrac C, Walker TR, Duarte AC, et al. Solutions and integrated strategies for the control and mitigation of plastic and microplastic pollution. *International Journal of Environmental Research and Public Health*. 2019; 16(13): 2411.
107. Eadie L, Ghosh TK. Biomimicry in textiles: past, present and potential. An overview. *Journal of the Royal Society Interface*. 2011; 8(59): 761–75.
108. Anand U, Dey S, Bontempi E, Ducoli S, Vethaak AD, Dey A, et al. Biotechnological methods to remove microplastics: a review. *Environmental Chemistry Letters*. 2023; 21: 1–24.
109. Chen J, Wu J, Sherrell PC, Chen J, Wang H, Zhang W, et al. How to build a microplastics-free environment: strategies for microplastics degradation and plastics recycling. *Advanced Science*. 2022; 9(6): 2103764.
110. Arpia AA, Chen W-H, Ubando AT, Naqvi SR, Culaba AB. Microplastic degradation as a sustainable concurrent approach for producing biofuel and obliterating hazardous environmental effects: a state-of-the-art review. *Journal of Hazardous Materials*. 2021; 418: 126381.
111. Lanorte A, De Santis F, Nolè G, Blanco I, Loisi RV, Schettini E, et al. Agricultural plastic waste spatial estimation by Landsat 8 satellite images. *Computers and Electronics in Agriculture*. 2017; 141: 35–45.
112. Taggio N, Aiello A, Ceriola G, Kremezi M, Kristollari V, Kolokoussis P, et al. A combination of machine learning algorithms for marine plastic litter detection exploiting hyperspectral PRISMA data. *Remote Sensing*. 2022; 14(15): 3606.

113. Salgado-Hernanz PM, Bauzà J, Alomar C, Compa M, Romero L, Deudero S. Assessment of marine litter through remote sensing: recent approaches and future goals. *Marine Pollution Bulletin*. 2021; 168: 112347.
114. Lavender S. Detection of waste plastics in the environment: application of Copernicus earth observation data. *Remote Sensing*. 2022; 14(19): 4772.
115. Asokan A, Anitha J, Ciobanu M, Gabor A, Naaji A, Hemanth DJ. Image processing techniques for analysis of satellite images for historical maps classification—an overview. *Applied Sciences*. 2020; 10(12): 4207.
116. Iordache M-D, De Keukelaere L, Moelans R, Landuyt L, Moshtaghi M, Corradi P, et al. Targeting plastics: machine learning applied to litter detection in aerial multispectral images. *Remote Sensing*. 2022; 14(22): 5820.
117. Amrutha K, Warrier AK, Rangel-Buitrago N. Does the COVID-19 pandemic play a role in the spatial and temporal variations of microplastics? Evidence from a tropical river in southern India. *Marine Pollution Bulletin*. 2023; 192: 115088.
118. Maharjan N, Miyazaki H, Pati BM, Dailey MN, Shrestha S, Nakamura T. Detection of river plastic using UAV sensor data and deep learning. *Remote Sensing*. 2022; 14(13): 3049.
119. Yu H-W, Kim YS, Lee S, Yoo J, Choi J. A review on analytical methods and occurrences for microplastics in freshwater. *Journal of Environmental Analysis, Health and Toxicology*. 2020; 23(4): 180–93.
120. Veerasingam S, Ranjani M, Venkatachalapathy R, Bagaev A, Mukhanov V, Litvinyuk D, et al. Microplastics in different environmental compartments in India: analytical methods, distribution, associated contaminants and research needs. *TrAC Trends in Analytical Chemistry*. 2020; 133: 116071.
121. Schwaferts CW. *Characterization and Identification of Micro-and Nanoplastic by Raman Microspectroscopy, Scanning Electron Microscopy, Field-Flow Fractionation and Chemometrics*: Technische Universität München; 2022.
122. Melo-Agustín P, Kozak ER, de Jesús Perea-Flores M, Mendoza-Pérez JA. Identification of microplastics and associated contaminants using ultra high resolution microscopic and spectroscopic techniques. *Science of the Total Environment*. 2022; 828: 154434.
123. Faltynkova A, Johnsen G, Wagner M. Hyperspectral imaging as an emerging tool to analyze microplastics: a systematic review and recommendations for future development. *Microplastics and Nanoplastics*. 2021; 1(1): 1–19.
124. Zhu C, Kanaya Y, Tsuchiya M, Nakajima R, Nomaki H, Kitahashi T, et al. Optimization of a hyperspectral imaging system for rapid detection of microplastics down to 100 μm. *MethodsX*. 2021; 8: 101175.
125. Li Y, Zhu Y, Huang J, Ho Y-W, Fang JK-H, Lam EY. High-throughput microplastic assessment using polarization holographic imaging. *Scientific Reports*. 2024; 14(1): 2355.
126. Munz M, Kreiß J, Krüger L, Schmidt LK, Bochow M, Bednarz M, et al. Application of high-resolution near-infrared imaging spectroscopy to detect microplastic particles in different environmental compartments. *Water, Air, & Soil Pollution*. 2023; 234(5): 286.

127. Kim S, Hyeon Y, Park C. Microplastics' shape and morphology analysis in the presence of natural organic matter using flow imaging microscopy. *Molecules*. 2023; 28(19): 6913.
128. Löhr A, Savelli H, Beunen R, Kalz M, Ragas A, Van Belleghem F. Solutions for global marine litter pollution. *Current Opinion in Environmental Sustainability*. 2017; 28: 90–9.
129. Jenkins T, Persaud BD, Cowger W, Szigeti K, Roche DG, Clary E, et al. Current state of microplastic pollution research data: trends in availability and sources of open data. *Frontiers in Environmental Science*. 2022; 10: 912107.
130. Čerkasova N, Enders K, Lenz R, Oberbeckmann S, Brandt J, Fischer D, et al. A public database for microplastics in the environment. *Microplastics*. 2023; 2(1): 132–46.
131. Nyadjro ES, Webster JA, Boyer TP, Cebrian J, Collazo L, Kaltenberger G, et al. The NOAA NCEI marine microplastics database. *Scientific Data*. 2023; 10(1): 726.
132. Neelavannan K, Sen IS. Microplastics in freshwater ecosystems of India: current trends and future perspectives. *ACS Omega*. 2023; 8(38): 34235–48.
133. Talukdar A, Bhattacharya S, Bandyopadhyay A, Dey A. Microplastic pollution in the Himalayas: occurrence, distribution, accumulation and environmental impacts. *Science of the Total Environment*. 2023; 874: 162495.
134. Almuhtaram H, Andrews RC. Sampling microplastics in water matrices: a need for standardization. *ACS ES&T Water*. 2022; 2(8): 1276–8.
135. Hrustic A. EU restrictions on intentionally added microplastics-small plastic, low ambition. *Annals Fac LU Zenica*. 2022; 31: 7.

CHAPTER 19

Microplastics' Ecotoxicological Assessment and Quantification in the Soil Ecosystem

Sruthi S. Krishna, Indu M. Sasidharan Pillai, Priya K.L., and Suchith Chellappan

19.1 INTRODUCTION

Plastic pollution is a global concern due to its presence in both marine and terrestrial ecosystems (Zhang et al., 2020). Microplastics (MPs), the plastic in the size range less than 5 mm in diameter are considered an emerging pollutant with a longevity of hundreds or thousands of years (Meera et al., 2023). Environmental contamination of MPs is known to be an anthropogenically mediated threat to the biosphere and is becoming a significant concern worldwide (Sajjad et al., 2022). Landfills accommodate more than 79% of the plastic waste generated globally (Guo et al., 2020). These plastic wastes are subjected to physical, chemical, and biological degradation mainly by soil erosion, mechanical abrasion,

DOI: 10.1201/9781003487579-19

photolysis, and biodegradation and are converted to MPs, making soil a large microplastic sink (Meera et al., 2023; Ren et al., 2021; Wang et al., 2020). Almost 60% of MPs present in the soil are transported to aquatic ecosystems, thus soil acts as a source of MP also (Guo et al., 2020). Hence, MPs are found throughout the environment and are particularly prevalent in marine environments (Wu et al., 2020). Globally, MPs have been found in a range of soil types, including agricultural/farmland, greenhouse, home garden, coastal, industrial, and floodplain soils (Xu et al., 2020; Zhang et al., 2022). Their rapid build-up and delayed breakdown make them extremely persistent in the environment (Subair et al., 2023). Microplastics, which comprise plastic fibres, granules, and fragments, are heterogeneously mixed polymers (Guo et al., 2020). The main sources of MPs include municipal wastewater, worn tyres from cars, dust from the home and laundry, industrial activities (such as plastic deflashing and blasting), and deterioration of surfaces coated or composed of plastic, like polymeric paint and artificial turf (Guo et al., 2020; Subair et al., 2024a). The majority of these emissions take place in cities and in homes (Crossman et al., 2020). Soils are also constantly exposed to MPs from plastic mulching, compost, and sewage irrigation, but their impacts on the soil environment are mostly unknown (Yi et al., 2021).

MPs' strong hydrophobic nature and large specific surface area allow them to significantly adsorb pollutants such as organochlorine insecticides, polychlorinated biphenyls, plasticizers, antibiotics, polycyclic aromatic hydrocarbons (PAHs), and potentially toxic elements (PTEs). This will have an effect on how these compounds travel, change, and become bioavailable in the natural world (Subair et al., 2024b). MPs have the potential to contaminate subsurface water and be carried deeply into the soil (Sajjad et al., 2022). The movement of MPs through biopores has been suggested as a potential route for groundwater contamination (He et al., 2018). Colloids or nanoparticles can flow more quickly than pore-water because of the size exclusion phenomenon, which keeps larger particles out of narrow pores (Xu et al., 2020). Furthermore, as the plastics deteriorate, the additives and other harmful materials within them leak out. These compounds can quickly collect in soil and water and are relatively resistant to environmental degradation (Jiao Wang et al., 2019).

Microplastics may interact with biota in terrestrial systems, having an ecologically significant effect (De Souza MacHado et al., 2018). Microplastics serve as vectors for heavy metals, which may affect the microorganisms, and they could potentially harm soil biota such as earthworms

(Rillig et al., 2017). Reports indicate the presence of MPs in earthworm casts, chicken faeces, and snails (Guo et al., 2020; Rillig et al., 2017). This further increases the probability of bioaccumulation and trophic transfer. These aspects are studied in detail by researchers globally. This chapter intends to summarize these and provide an overview on the quantification and transport of MPs in soil and their impact on the soil ecosystem.

19.2 MICROPLASTIC CONTAMINATION ON SOIL ECOSYSTEM

The annual production of plastic is estimated to be 320 million tonnes (Yoon et al., 2024). Roughly 32% of all plastic generated, according to empirical calculations, is environmentally available in continental systems (Rillig et al., 2019a). These could be primary MPs or secondary MPs. Primary MPs that are intentionally created and utilized in a variety of industries and cosmetic items eventually end up accumulating in the environment because they are typically difficult to remove using current sewage disposal technology. Secondary MPs are created when bigger plastics gradually break down into smaller pieces due to a variety of intricate environmental factors, including temperature, UV light, wind, waves, and temperature (Guo et al., 2020). The main sources of secondary MPs in soil could be landfills, open dumps, sewage farming, and PMFs, or plastic mulch films. Plastic mulching is a widely used agricultural method for crop yield and quality. Despite their many advantages, they eventually break down into widely distributed MPs, which increase the likelihood of environmental issues (Wang et al., 2022; Yuan et al., 2022). Because of farming, a large amount of MPs are buried under topsoil and ploughed soil in dry or semi-arid settings (Li et al., 2020). The extensive use of wastewater sludge and greenhouses also makes agricultural land a major sink of MPs (Mahon et al., 2017; Jie Wang et al., 2019). Microplastic build-up in soil can have several negative consequences on soil ecosystems. A representation of the same is depicted in Figure 19.1.

Numerous alterations in the physicochemical properties of the soil, enzymatic and microbiological activity, plant growth, and yield are caused by MPs. Microplastics can affect soil quality and nutrient cycling by changing the physical characteristics of the soil, reducing soil fertility, and upsetting resident microbial communities (Xu et al., 2020). Drought and MPs in the soil may interact to impact a number of ecosystem processes, including soil aggregation, litter decomposition, and nutrient cycling (Lozano et al., 2021). The integration of MPs to soil depends on the shape

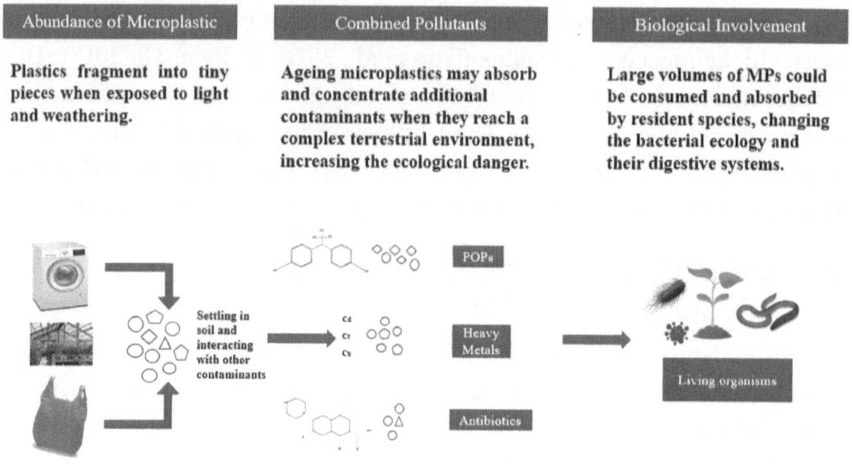

FIGURE 19.1 Sources and impact of MPs in soil.

characteristics of the MP and the soil characteristics (Guo et al., 2020). The distribution of MPs in various soil types across four countries is depicted in Guo et al. (2019). An updated report of the distribution in various soil types is provided in Table 19.1.

19.3 QUANTIFICATION TECHNIQUES FOR MICROPLASTIC IN SOIL

Detection and quantification of MPs in soil is challenging owing to the complexity of the soil matrix and the richness of organic matter (Wang et al., 2020). Till date there is no standardized method for the quantification of MPs. Three guidelines for MP analysis is provided by European commission (guideline issued in 2013) and National oceanic and atmospheric association (guideline issued in 2015). These guidelines focus more on MP detection and enumeration and less on identification and are provided different outcomes. Among these, only one guideline i.e. from UK water industry research is for the analysis of solid sample (Christian and Köper, 2023). There are three basic steps for the analysis of MPs. They include (1) sample collection (2) Pre-treatment/Extraction/Sample processing (3) sample analysis for identification and quantification.

During the sample processing, the separation of MPs from the soil matrix and other impurities is intended. Various processes include density separation, electrostatic separation, oil extraction, froth flotation,

TABLE 19.1 Microplastic Distribution in Various Soil Types

Soil Type	Extraction Method	Identification Technique	Concentration	References
Arable soils	Density separation using saturated salt solution and Fenton reaction	Visual identification using a dissection microscope	7,100–42,960 particles/kg	Zhang and Liu, (2018)
Beach soils	Floating separation and density separation	Visual identification, ATR-FTIR, and SEM	1.3–14,712.5 N/kg	Zhou et al. (2018)
Agricultural soils	Density separation and digestion	Fluorescence microscopy	1,320–8,190 particles/kg	Cusworth et al. (2024)
Agricultural soils and grasslands	Physical separation using sieving and Fenton reaction	Visual identification and ATR-FTIR	<6.7–80 MP/kg	Palazot et al. (2024)
Riparian soils	Density separation using saturated salt solution and Fenton reaction	Visual identification	428–799 pieces/kg	Krishna et al. (2023)
Cultivated soils	Density separation using saturated NaI and Fenton reaction	Visual identification using microscope	0.9×10^3 to 40.8×10^3 particles/(kg Ds)	Huang et al. (2021)
Garden soil	Physical separation using sieving	Optical microscopy, Micro-FTIR-ATR	35 pieces/2 kg sample	Brandolis et al. (2024)
Beach soils	Density separation using saturated NaCl and the Fenton reaction	visual identification using stereomicroscope and micro FTIR	188.3 ± 34.48 items/kg	Akkajit et al. (2021)

magnetic extraction, etc. (Kasa et al., 2022). Density separation using saturated salt helps in separating the MPs and inorganic soil particles. Low-density polymers like polyethylene and polypropylene can be easily separated using saturated NaCl solution. However, for high-density polymers like polyester and polyvinyl chloride, the recovery is low with NaCl. Hence, high-density salt solutions like NaI and ZnCl are required. However, these salts are costly, and some are toxic, limiting their application (Wang et al., 2020). After density separation using saturated salts also the separated MPs will be mixed with organic matter. The oxidation of the organic matter is achieved using oxidizing agents like H_2O_2, Fenton reagent, etc. or using enzymes (Wang et al., 2020). Electrostatic precipitation relies on separating particles based on their electrical conductivity and is size dependent. Also, it is not suitable for moist and cohesive soil samples (Kasa et al., 2022).

The most popular method of quantifying MPs is visual counting and categorization, usually according to the type, colour, size, and shape of the polymer. While visually sorting and identifying MPs under a dissecting microscope has been widely reported as a quantification method, its accuracy has numerous limits. The use of spectroscopy is crucial for verifying the existence of plastic and can improve the precision of visual counting (Hanvey et al., 2017). Numerous methods exist for spectroscopic identification, including near-infrared (NIR), Raman and Fourier transform infrared spectroscopy (FTIR), of which FTIR and Raman are the most commonly employed. The accuracy of FTIR and Raman technology to identify MPs is positively correlated with the degree of organic matter removal during the pre-digestion and extraction procedure (Yang et al., 2022). Furthermore, ATR-FTIR technology is an enhanced FTIR technology with an ATR mode that can independently identify each microplastic particle's polymer, resolving the microscope's inadequacies. Thermal analysis methods include pyrolysis-gas chromatography-mass spectrometry (Pyr-GC-MS), thermal extraction desorption-gas chromatography-mass spectrometry (TED-GC-MS), and thermogravimetric analysis-mass spectrometry (TGA-MS). Pyr-GC-MS is used to identify soil nanoplastics (Yang et al., 2022). Other techniques like flow cytometry, dynamic light scattering, scanning electron microscope and atomic force microscopy are also applied for this purpose (Christian and Köper, 2023). Although there are many techniques for MP analysis, till date there is no single technique for complete analysis. A detailed summary of commonly used techniques for the identification and quantification of MPs is provided in Wang et al. (2020). Based on a scaled comparison using four parameters for pretreatment and five parameters for sample analysis, the best suitable options among the various techniques available for solid samples was identified. It was indicated that floatation using saturated salt and Fenton reactions are equally good pretreatment process. Pyr-GC-MS was identified as the best analysis technique for the identification, quantification, and characterization (Wang et al., 2020). However, classification based on size and shape is not possible in this method (Christian and Köper, 2023). Another evaluation of the various techniques for separation of MP from the soil matrix was done using AGREE analysis in a 12-point scale. This analysis indicated electrostatic precipitation to be the best suited, followed by density separation with identification using FTIR (Kasa et al., 2022). Jia et al. (2022) also substantiates ATR-FTIR and LIDR as a reliable method for microplastic identification owing to the advantage of FTIR over Raman spectroscopy in the presence of organic matter (Jia et al., 2022).

19.4 MOVEMENT AND RETENTION OF MICROPLASTICS IN SOIL

The extent of vertical and horizontal distribution of MPs depends on various factors like soil biota, micropores, soil aggregation and cracking, agronomic practices, etc. Among these, the most favoured paths are root growing and uprooting. Soil characteristics have a greater influence on the MP migration capacity than the MP characteristics (Gui et al., 2022). In the case of the loam, MPs that have gotten into this area may migrate to a deeper soil layer; in the case of the clay loam, however, the stickiness of the clay particles will keep the MPs on the top layer. Soil organisms like collembola species, mite, earthworms, etc. are reported to transport MPs in soil (Li et al., 2020). Particles will eventually travel throughout soil systems and towards streams and rivers due to erosion caused by wind and water, which would reduce the MPs' abundance in soil surfaces (Cao et al., 2021; Hurley and Nizzetto, 2018). This migration rate varies with the characteristics of MP due to different interactions with soil aggregation (Guo et al., 2020). MP fibres are reported to be preferentially transported by wind, followed by fragments (Bullard et al., 2021). Low-density MPs of smaller size are susceptible to long-distance transport through erosion caused by water (Han et al., 2022). Animal disturbance, surface runoff, and water infiltration are some of the long-distance ways that soil MPs can go from the land into the groundwater environment. Microplastics can be moved and dispersed by soil animals and livestock, either through attachment to the exterior of the organism or transfer through consuming and excrement (R. Qi et al., 2020).

Leaching plays a significant part in promoting the vertical migration of MPs (Zhou et al., 2020). Studies indicate vertical penetration of 70 cm by MPs of average size 3.7 μm by leaching (Cey et al., 2009; Li et al., 2020). Earthworms are reported to transport MPs vertically at the rate of 10 cm in 21 days (He et al., 2018; Rillig et al., 2017). Other factors favouring vertical transport include bioturbation, farming activities like ploughing, harvesting of rhizome, etc. Climatic conditions can also affect the vertical transport of MPs with an alternate dry – wet cycle increasing the rate of vertical transport. During the dry phase, soil cracks are formed, and during the wet phase, MPs are leached to the deep soil through these cracks

(Li et al., 2020). Predicting the possibility of movement, uptake, and transformation within the soil profile would require an understanding of the spatial distribution and behaviour of MPs inside soil aggregates in both top and subsoils (R. Qi et al., 2020). Studies report the average penetration rate in China as 5.24 m (Guo et al., 2020). High-density MPs have a higher tendency to penetrate into the deeper soil layers and to stay deposited in the soil (Li et al., 2020; Xiang et al., 2022). Hydrophobicity of the MPs also affects their transport in soil (Li et al., 2020). Interaction with dissolved organic matter and microorganisms will turn MPs more hydrophilic, aiding the subsurface transport (Yu and Flury, 2021). Researchers conclude that smaller microplastic particles have greater mobility and that soil pores have retention effects on larger microplastic particles. MPs in the size range of 0.1–6 μm are reported to penetrate vertically in soil columns through the macro and meso soil pores (Li et al., 2020). In the meantime, the effective pore size will decrease due to the retention of MPs, increasing the future particle retention (Zhao et al., 2022).

Some researchers believe that natural places, such as green spaces, act as natural filters because of the retention capacity of permeable soils and plants. These regions may contain a large amount of MPs and could contribute to decreasing the movement of MPs via run-off to water bodies or wind erosion to other sites (Leitão et al., 2023). The retention mechanisms of MPs in soil include straining, pore exclusion, physicochemical deposition, and film straining (Li et al., 2020). Synthetic fibres made from sewage sludge were found to be kept in treated agricultural soil for up to 15 years following the last sludge treatment, according to a study on the retention of MPs in soils (Zubris and Richards, 2005). Accumulation hotspots can develop at depths as well, with fibres being identified at depths greater than 25 cm in regions with significant soil drainage flow downward (Horton and Dixon, 2018).

19.5 ECOTOXICOLOGICAL IMPACTS OF MICROPLASTICS ON SOIL ECOSYSTEM

The detrimental impacts of plastic on terrestrial ecosystems have been emphasized by numerous studies, which indicate that plastic pollution on land is far worse than it is in the aquatic environment. Different types of MPs have different effects on the characteristics of soil. Different plastic waste kinds, sizes, and contents may have different effects on the physicochemical and hydrological qualities of soil (Y. Qi et al., 2020). Microplastics have been shown to have detrimental impacts on a variety of soil species, including nematodes, collembolans, earthworms, and snails, according to

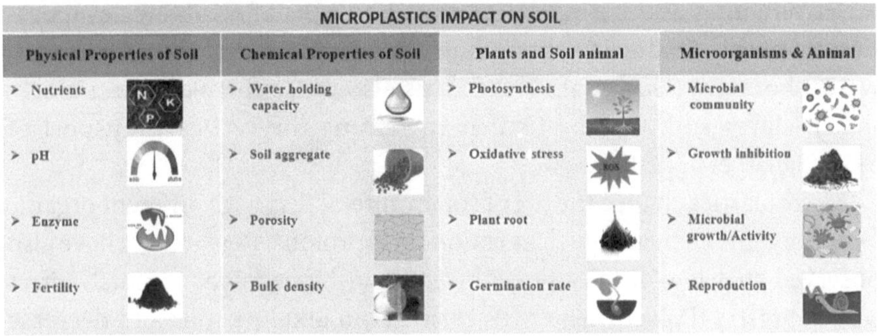

FIGURE 19.2 Effects of microplastics on soil physicochemical properties, microbes, plants, and animals.

recent studies. This is due to the possibility that MPs will be consumed and may have harmful effects on some types of soil fauna, disrupt the gut's symbiotic microbiota, or obstruct the migration of soil microarthropods by plugging soil pores (Lin et al., 2020). The effect of MPs on the physicochemical characteristics of soil, plants, animals, and microbes is depicted in Figure 19.2 and a detailed description of the same is provided in the following sections.

19.5.1 Effect on Soil Physiochemical Characteristics

The application of MPs has the potential to modify the physicochemical and hydrological properties of soil, hence impacting the microbial community composition and enzyme activity (H. Yu et al., 2021). Since MPs decrease the permeability (porousness) and aeration (entry of oxygen-containing air) of water in the soil, they may gradually alter the physical characteristics and structure of the soil by increasing the proportion of stable water aggregates (Sajjad et al., 2022). The effect of various types of MPs on soil varies with the soil characteristics. The chemical composition of the MPs also determines the effect (Guo et al., 2020). Depending on the kind of microplastic—like polyester, which lowers the bulk density of soil—the impacts differ. The water holding capacity is influenced differently by different MPs; polyester fibres, for instance, have a greater effect than polyethylene and polyacrylic in water holding capacity. Polyethylene was reported to decrease soil bulk density and increase water holding capacity (De Souza MacHado et al., 2018). On the contrary, another study indicated a decrease in water holding capacity and no detectable effect on the bulk density. The alteration of water retention and permeability of soil by MPs affects water evaporation from soil, leading to changes in the soil water cycle worsening soil water shortages (Guo et al., 2020). This also cause shrinkage and cracking in clayey soil, thereby further increasing the vertical transport of MPs (Sajjad et al., 2022).

Microplastics change the pH of soil by interacting with different organic and inorganic elements, such as cations and protons. Researchers have also observed that low-density polyethylene and biodegradable mulch affect soil chemistry. Polyethylene MPs raise pH in alkaline soils and decrease pH in acidic soils (Li et al., 2023). The organic matter content of soil can also be impacted by MPs. Certain MPs contain carbon (mostly inert carbon with long-term stability), which can raise soil organic carbon (SOC) and have a variety of effects on the carbon cycle, causing fertilized soil to release more CO_2, influencing soil microbial processes, plant growth, and the decomposition of litter (Yang et al., 2022). Microplastics also improve the soil quality by the accumulation of humic substances on the MP. The effect of MP on the soil fertility strongly depends on exposure time and MP concentration (Guo et al., 2020).

19.5.2 Effect on Terrestrial Plants

Apart from soil, the primary sources of MPs in plant communities are surface runoff, agricultural irrigation water, land application of municipal sludge, residual decomposition of plastic waste and agricultural mulching film, and atmospheric microplastic deposition. Among these, the airborne MPs have the potential to directly impact plant components above ground as well as penetrate the soil and affect roots (Z. Yu et al., 2021). Plant diversity and community composition may be impacted by MPs, and changes in these parameters may be caused by a number of ways (Rillig et al., 2019b). Plant type, leaf area, waxy leaves, trichomes, and the kind, size, and density of MPs particles all affected how much MPs accumulated and were deposited on leaves from the atmosphere (Allouzi et al., 2021). MPs may have a negative impact on terrestrial plants' physiological processes, namely on the rhizome and seeds (Ge et al., 2021).

Microplastics can have an immediate impact on plant growth or a secondary effect by altering the characteristics of the soil or microorganisms. The total root length of spring onions (Allium fistulosum) was found to be significantly increased by the six MPs of polyamide (PA), high-density polyethylene, polyester (PES), polyethylene terephthalate (PET), polypropylene, and polystyrene. Moreover, some of these MPs have been found to be responsible for changes in the average root diameter, root biomass, and root tissue density of spring onions (Yang et al., 2022). Microplastics have the potential to influence a plant community's horizontal organization by encouraging the gathering of plants within the same species (Z. Yu et al., 2021). Furthermore, MPs (0.2–1.5 mm) have been found in tissue sections of plants grown in commercial compost, and since these particles are comparable in size and density to the minerals that certain plants absorb, it is thought that plants could mechanically transport MPs (Yang et al., 2022). The way that microplastic particles migrate within plant communities is strongly influenced by their sizes, shapes, and chemical makeup. The size of plant cell wall pores, or 5–50 nm, is the range of microplastic particles that are more readily adsorbed to the root cell wall or seed epidermis (Z. Yu et al., 2021).The primary processes impacted by microplastic input may take place at the soil-plastic interface, also known as the microplastisphere, which is comparable to plant-soil interactions in the rhizosphere. Given that the majority of agricultural soils are contaminated with microplastic, the ecological significance of the unique niches occupied by the microorganisms within the microplastisphere (Zhou et al., 2021).

19.5.3 Effect on Soil Animals

MPs are consumed by soil animals and make their way into their digestive systems. Most plastics are excreted by soil organisms after ingestion. However, evidence suggests their retention in the intestine for long periods (Ya et al., 2021). Research revealed that PS microplastics produced intestinal barrier failure, bile acid metabolism disorders, and alterations in intestinal microbiota in mice. These findings suggest that the consumption of MPs and their prolonged presence in the gut may alter the feeding behaviours of animals (Yang et al., 2022). To investigate the effects of MPs on soil animal growth, survival, metabolism, gut microbiota, eating behaviour, and inflammatory response, a limited subset of invertebrates, including isopods, nematodes, collembolans, and oligochaetas, were studied (Kumar et al., 2020). Various studies used earthworm as a model organisms to depict the effect of MP on soil organisms for several types of MPs at varying concentrations and proved that the effect depends on the MP type and concentration (Sajjad et al., 2022; Ya et al., 2021). Animal reproduction can potentially be impacted by MPs. The reproduction and survival rates of the soil fauna steadily declined as concentrations increased, suggesting that MPs had detrimental impacts on soil animals (Wan et al., 2023). A decrease in snail TAOC was reported for snails which ingested PET, causing gastrointestinal damage. MPs are also reported to cause histological changes in digestive organs (Ya et al., 2021).

MPs attached to the body surface of soil organisms can impede their movement (Chang et al., 2022). The chemical additives, such as heavy metals, antibiotics, pesticides, and plasticizers, borne on the surface of MPs, may cause harm to soil animals and increase the chance of bioaccumulation through the food chain (Ya et al., 2021). Furthermore, the impact of the combination of hazardous chemicals and MPs on soil animals differs from that of the individual substances on soil animals (Yang et al., 2022). The concentration of MPs was thought to be a key factor influencing how the organisms behaved (Kumar et al., 2020). Trophic transfer of MPs also occurs through the food chain (Chang et al., 2022). In human beings, the exposure to MPs can lead to health issues like cancer, reproductive problems, organ damage, and development delay. As the body's levels of antioxidants and free radicals go out of balance, MPs will also produce oxidative stress. Additionally, it was shown that MPs can cause aberrant immunological responses and chemical reactions that affect the body's processes of catabolism and anabolism, both of which are detrimental to the health of living things (Allouzi et al., 2021).

19.5.4 Effect on Soil Microbes

Microplastic particles are widely identified as human-caused contaminants in soil, potentially detrimental to soil microorganisms (Lozano et al., 2021). Microplastics may have a direct or indirect impact on the makeup of microbial communities and microorganisms (Yang et al., 2022). Exposure of soil microorganisms, including fungi and bacteria, to excessive amounts of MPs can have negative effects (Kumar et al., 2020). According to reports, MPs are present in microorganisms' digestive systems because of their small sizes and shapes, which make it easy for them to be consumed and build up in the digestive systems of feeders that don't differentiate between them and food (Allouzi et al., 2021). Microplastics can alter how bacteria metabolize (Yang et al., 2022). The research on numerous variables, such as the polymer shape, type, concentration, and size of the MPs, made it challenging to draw broad conclusions about the harmful effects of MPs on microbiological activity in relation to specific variables. Based on the investigations, there was insufficient data to draw a general conclusion on the toxicity of MPs in relation to the composition of the microbial community and soil structure (Kumar et al., 2020). MPs are used as a carbon source by gram-positive bacteria. MPs also act as ecological habitats by soil microbes (Sajjad et al., 2022). In contrast to the microbial communities on other surfaces (rock and leaf), those colonized by MPs were distinct and clearly different (Yang et al., 2022). MPs particles have an impact on the structure of the soil and the cycling of nutrients. These alterations also have an impact on the soil's fertility, which lowers it and inhibits the development of microbial communities in the rhizosphere. Since these bacteria are crucial to the growth of plants, a decrease in their activity could have detrimental effects on crop growth rates and plant health (Allouzi et al., 2021).

19.6 RECOMMENDATION FOR FUTURE STUDIES

In freshwater systems, research regarding microplastic concentrations is expanding quickly. Most of the time, though, this knowledge has not yet been connected to ecological repercussions. It is challenging to determine the precise nature of the danger caused by MPs in these systems and how the presence of microplastic would affect these ecosystems due to a lack of quantitative data. In fact, compared to more thoroughly researched chemical pollutants, the long-term effects of MPs remain unknown in the field of MPs study (Horton and Dixon, 2018). Comparing the findings of research reporting on MPs presence in soils is currently difficult due to different sampling designs, missing fundamental data, and inconsistent methodologies (Hooge et al., 2023). Various research findings have clarified the intricate effects of soil contamination by distinct MPs while cautioning that additional investigation is necessary to fully understand the impact of MPs on soil water transport and retention (Zhao et al., 2022). The consequences of exposure to MPs on a wider variety of soil fauna under actual environmental situations need further investigation. Furthermore, it is still unclear whether MPs can promote the bioaccumulation and trophic transmission of pollutants generated from plastic, such as plastic additives and sorbed chemicals (Wang et al., 2020).

19.7 CONCLUSIONS

Microplastics are newly discovered contaminants that infiltrate terrestrial ecosystems through a variety of channels, build up in soil, and harm microbes, plants, animals, and soil. MP concentrations have the potential to alter the ecological and structural makeup of soil. There is a growing global pollution of MPs, and there are significant concerns linked with this that need to be addressed in terrestrial soils. Microplastic pollution affects the physiochemical properties, soil microorganisms, plant growth, and environmental hazards in a range of soils, plants, and people. There have been reports of ecological effects of MPs on human health, soil biota, and aquatic organisms. The major ways that MPs might get up in soils are by irrigation, fertilization, film mulching, atmospheric deposition, and land application of sewage sludge. Both aquatic and terrestrial organisms exhibited growth inhibition and decreased absorption efficiency. MPs may cause respiratory and digestive harm to humans who consume or inhale them. Although visual counting has proven a valuable method for gaining an understanding of MPs in environmental samples, analytical procedures need to advance because it is labour-intensive and prone to human error.

REFERENCES

Akkajit, P., Tipmanee, D., Cherdsukjai, P., Suteerasak, T., Thongnonghin, S., 2021. Occurrence and distribution of microplastics in beach sediments along Phuket coastline. *Mar. Pollut. Bull.* 169, 112496. https://doi.org/10.1016/J.MARPOLBUL.2021.112496

Allouzi, M.M.A., Tang, D.Y.Y., Chew, K.W., Rinklebe, J., Bolan, N., Allouzi, S.M.A., Show, P.L., 2021. Micro (nano) plastic pollution: The ecological influence on soil-plant system and human health. *Sci. Total Environ.* 788, 147815. https://doi.org/10.1016/j.scitotenv.2021.147815

Brandolis, I., Falconi, A., Mackay, M., Zafar, G., Holuszko, M.E., 2024. A perspective on plastics and microplastics contamination in garden soil in British Columbia, Canada. *Pollutants* 4, 153–173. https://doi.org/10.3390/POLLUTANTS4010010

Bullard, J.E., Ockelford, A., O'Brien, P., McKenna Neuman, C., 2021. Preferential transport of microplastics by wind. *Atmos. Environ.* 245, 118038. https://doi.org/10.1016/J.ATMOSENV.2020.118038

Cao, L., Wu, D., Liu, P., Hu, W., Xu, L., Sun, Y., Wu, Q., Tian, K., Huang, B., Yoon, S.J., Kwon, B.O., Khim, J.S., 2021. Occurrence, distribution and affecting factors of microplastics in agricultural soils along the lower reaches of Yangtze River, China. *Sci. Total Environ.* 794. https://doi.org/10.1016/j.scitotenv.2021.148694

Cey, E.E., Rudolph, D.L., Passmore, J., 2009. Influence of macroporosity on preferential solute and colloid transport in unsaturated field soils. *J. Contam. Hydrol.* 107, 45–57. https://doi.org/10.1016/J.JCONHYD.2009.03.004

Chang, X., Fang, Y., Wang, Y., Wang, F., Shang, L., Zhong, R., 2022. Microplastic pollution in soils, plants, and animals: A review of distributions, effects and potential mechanisms. *Sci. Total Environ.* 850, 157857. https://doi.org/10.1016/J.SCITOTENV.2022.157857

Christian, A.E., Köper, I., 2023. Microplastics in biosolids: A review of ecological implications and methods for identification, enumeration, and characterization. *Sci. Total Environ.* 864, 161083. https://doi.org/10.1016/J.SCITOTENV.2022.161083

Crossman, J., Hurley, R.R., Futter, M., Nizzetto, L., 2020. Transfer and transport of microplastics from biosolids to agricultural soils and the wider environment. *Sci. Total Environ.* 724, 138334. https://doi.org/10.1016/j.scitotenv.2020.138334

Cusworth, S.J., Davies, W.J., McAinsh, M.R., Stevens, C.J., 2024. A nationwide assessment of microplastic abundance in agricultural soils: The influence of plastic crop covers within the United Kingdom. *Plants People Planet* 6, 304–314. https://doi.org/10.1002/PPP3.10430

De Souza MacHado, A.A., Lau, C.W., Till, J., Kloas, W., Lehmann, A., Becker, R., Rillig, M.C., 2018. Impacts of microplastics on the soil biophysical environment. *Environ. Sci. Technol.* 52, 9656–9665. https://doi.org/10.1021/acs.est.8b02212

Ge, J., Li, H., Liu, P., Zhang, Z., Ouyang, Z., Guo, X., 2021. Review of the toxic effect of microplastics on terrestrial and aquatic plants. *Sci. Total Environ.* 791, 148333. https://doi.org/10.1016/j.scitotenv.2021.148333

Gui, X., Ren, Z., Xu, X., Chen, X., Chen, M., Wei, Y., Zhao, L., Qiu, H., Gao, B., Cao, X., 2022. Dispersion and transport of microplastics in three water-saturated coastal soils. *J. Hazard. Mater.* 424, 127614. https://doi.org/10.1016/J.JHAZMAT.2021.127614

Guo, J.J., Huang, X.P., Xiang, L., Wang, Y.Z., Li, Y.W., Li, H., Cai, Q.Y., Mo, C.H., Wong, M.H., 2020. Source, migration and toxicology of microplastics in soil. *Environ. Int.* 137, 105263. https://doi.org/10.1016/j.envint.2019.105263

Han, N., Zhao, Q., Ao, H., Hu, H., Wu, C., 2022. Horizontal transport of macro- and microplastics on soil surface by rainfall induced surface runoff as affected by vegetations. *Sci. Total Environ.* 831, 154989. https://doi.org/10.1016/J.SCITOTENV.2022.154989

Hanvey, J.S., Lewis, P.J., Lavers, J.L., Crosbie, N.D., Pozo, K., Clarke, B.O., 2017. A review of analytical techniques for quantifying microplastics in sediments. *Anal. Methods* 9, 1369–1383. https://doi.org/10.1039/c6ay02707e

He, D., Luo, Y., Lu, S., Liu, M., Song, Y., Lei, L., 2018. Microplastics in soils: Analytical methods, pollution characteristics and ecological risks. *TrAC Trends Anal. Chem.* 109, 163–172. https://doi.org/10.1016/j.trac.2018.10.006

Hooge, A., Hauggaard-Nielsen, H., Heinze, W.M., Lyngsie, G., Ramos, T.M., Sandgaard, M.H., Vollertsen, J., Syberg, K., 2023. Fate of microplastics in sewage sludge and in agricultural soils. *TrAC Trends Anal. Chem.* 166, 117184. https://doi.org/10.1016/j.trac.2023.117184

Horton, A.A., Dixon, S.J., 2018. Microplastics: An introduction to environmental transport processes. *Wiley Interdiscip. Rev. Water* 5, 1–10. https://doi.org/10.1002/WAT2.1268

Huang, B., Sun, L., Liu, M., Huang, H., He, H., Han, F., Wang, X., Xu, Z., Li, B., Pan, X., 2021. Abundance and distribution characteristics of microplastic in plateau cultivated land of Yunnan Province, China. *Environ. Sci. Pollut. Res.* 28, 1675–1688. https://doi.org/10.1007/s11356-020-10527-3

Hurley, R.R., Nizzetto, L., 2018. Fate and occurrence of micro(nano)plastics in soils: Knowledge gaps and possible risks. *Curr. Opin. Environ. Sci. Heal.* 1, 6–11. https://doi.org/10.1016/j.coesh.2017.10.006

Jia, W., Karapetrova, A., Zhang, M., Xu, L., Li, K., Huang, M., Wang, J., Huang, Y., 2022. Automated identification and quantification of invisible microplastics in agricultural soils. *Sci. Total Environ.* 844, 156853. https://doi.org/10.1016/J.SCITOTENV.2022.156853

Kasa, V.P., Thomas, A.P., Bordoloi, S., De Bhowmick, G., Dubey, B.K., Sarmah, A.K., 2022. Microplastics in soil: Current status and evaluation of the greenness of various analytical methods of identification. *Green Anal. Chem.* 3, 100038. https://doi.org/10.1016/J.GREEAC.2022.100038

Krishna, A.B.S., Madhu, M., Jayadev, A., 2023. Investigation of microplastics and microplastic communities in selected river and lake basin soils of Thiruvananthapuram District, Kerala, India. *Environ. Monit. Assess.* 196, 1–32. https://doi.org/10.1007/S10661-023-12219-0

Kumar, M., Xiong, X., He, M., Tsang, D.C.W., Gupta, J., Khan, E., Harrad, S., Hou, D., Ok, Y.S., Bolan, N.S., 2020. Microplastics as pollutants in agricultural soils. *Environ. Pollut.* 265, 114980. https://doi.org/10.1016/j.envpol.2020.114980

Leitão, I.A., van Schaik, L., Ferreira, A.J.D., Alexandre, N., Geissen, V., 2023. The spatial distribution of microplastics in topsoils of an urban environment - Coimbra city case-study. *Environ. Res.* 218. https://doi.org/10.1016/j.envres.2022.114961

Li, W., Wufuer, R., Duo, J., Wang, S., Luo, Y., Zhang, D., Pan, X., 2020. Microplastics in agricultural soils: Extraction and characterization after different periods of polythene film mulching in an arid region. *Sci. Total Environ.* 749, 141420. https://doi.org/10.1016/j.scitotenv.2020.141420

Li, Z., Yang, Y., Chen, X., He, Y., Bolan, N., Rinklebe, J., Lam, S.S., Peng, W., Sonne, C., 2023. A discussion of microplastics in soil and risks for ecosystems and food chains. *Chemosphere* 313. https://doi.org/10.1016/j.chemosphere.2022.137637

Lin, D., Yang, G., Dou, P., Qian, S., Zhao, L., Yang, Y., Fanin, N., 2020. Microplastics negatively affect soil fauna but stimulate microbial activity: Insights from a field-based microplastic addition experiment. *Proc. R. Soc. B Biol. Sci.* 287. https://doi.org/10.1098/rspb.2020.1268

Lozano, Y.M., Aguilar-Trigueros, C.A., Onandia, G., Maaß, S., Zhao, T., Rillig, M.C., 2021. Effects of microplastics and drought on soil ecosystem functions and multifunctionality. *J. Appl. Ecol.* 58, 988–996. https://doi.org/10.1111/1365-2664.13839

Mahon, A.M., O'Connell, B., Healy, M.G., O'Connor, I., Officer, R., Nash, R., Morrison, L., 2017. Microplastics in sewage sludge: Effects of treatment. *Environ. Sci. Technol.* 51, 810–818. https://doi.org/10.1021/acs.est.6b04048

Meera, G., Sasidharan Pillai, I.M., Reji, P.G., Sajithkumar, K.J., Priya, K.L., Chellappan, S., 2023. Coagulation studies on photodegraded and photocatalytically degraded polystyrene microplastics using polyaluminium chloride. *Waste Manag.* 170, 329–340. https://doi.org/10.1016/j.wasman.2023.09.018

Palazot, M., Soccalingame, L., Froger, C., Jolivet, C., Bispo, A., Kedzierski, M., Bruzaud, S., 2024. First national reference of microplastic contamination of French soils. *Sci. Total Environ.* 918, 170564. https://doi.org/10.1016/J.SCITOTENV.2024.170564

Qi, R., Jones, D.L., Li, Z., Liu, Q., Yan, C., 2020. Behavior of microplastics and plastic film residues in the soil environment: A critical review. *Sci. Total Environ.* 703, 134722. https://doi.org/10.1016/j.scitotenv.2019.134722

Qi, Y., Beriot, N., Gort, G., Huerta Lwanga, E., Gooren, H., Yang, X., Geissen, V., 2020. Impact of plastic mulch film debris on soil physicochemical and hydrological properties. *Environ. Pollut.* 266, 115097. https://doi.org/10.1016/j.envpol.2020.115097

Ren, Z., Gui, X., Xu, X., Zhao, L., Qiu, H., Cao, X., 2021. Microplastics in the soil-groundwater environment: Aging, migration, and co-transport of contaminants – A critical review. *J. Hazard. Mater.* 419, 126455. https://doi.org/10.1016/j.jhazmat.2021.126455

Rillig, M.C., De Souza Machado, A.A., Lehmann, A., Klümper, U., 2019a. Evolutionary implications of microplastics for soil biota. *Environ. Chem.* 16, 3–7. https://doi.org/10.1071/EN18118

Rillig, M.C., Lehmann, A., de Souza Machado, A.A., Yang, G., 2019b. Microplastic effects on plants. *New Phytol.* 223, 1066–1070. https://doi.org/10.1111/nph.15794

Rillig, M.C., Ziersch, L., Hempel, S., 2017. Microplastic transport in soil by earthworms. *Sci. Rep.* 7, 1–6. https://doi.org/10.1038/s41598-017-01594-7

Sajjad, M., Huang, Q., Khan, S., Khan, M.A., Liu, Y., Wang, J., Lian, F., Wang, Q., Guo, G., 2022. Microplastics in the soil environment: A critical review. *Environ. Technol. Innov.* 27, 102408. https://doi.org/10.1016/j.eti.2022.102408

Subair, A., Krishnamoorthy, P., Suchith, L., Chinglenthoiba, C., 2024b. Removal of polystyrene microplastics using biochar - based continuous flow fixed - bed column. *Environ. Sci. Pollut. Res.* 31, 13753. https://doi.org/10.1007/s11356-024-32088-5

Subair, A., Meera, G., Chellappan, S., Sajithkumar, K.J., Chinglenthoiba, C., Priya, K.L., Indu, M.S., 2023. Techniques for removal and degradation of microplastics, in: Wang, C., Babel, S., Lichtfouse, E. (Eds.), Microplastic Occurrence, Fate, Impact, and Remediation. Springer. pp. 127–153. https://doi.org/10.1007/978-3-031-36351-1_6

Subair, A., Priya, K.L., Chellappan, S., Rahuman, A.T., Hridya, J., Devi, P.S., Salkka, S.M., Indu, M.S., Pugazhendhi, A., Chinglenthoiba, C., 2024a. Evaluating the performance of electrocoagulation system in the removal of polystyrene microplastics from water. *Environ. Res.* 243, 117887. https://doi.org/10.1016/j.envres.2023.117887

Wan, L., Cheng, H., Liu, Y., Shen, Y., Liu, G., Su, X., 2023. Global meta-analysis reveals differential effects of microplastics on soil ecosystem. *Sci. Total Environ.* 867, 161403. https://doi.org/10.1016/j.scitotenv.2023.161403

Wang, Jiao, Liu, X., Li, Y., Powell, T., Wang, X., Wang, G., Zhang, P., 2019. Microplastics as contaminants in the soil environment: A mini-review. *Sci. Total Environ.* 691, 848–857. https://doi.org/10.1016/j.scitotenv.2019.07.209

Wang, Jie, Coffin, S., Sun, C., Schlenk, D., Gan, J., 2019. Negligible effects of microplastics on animal fitness and HOC bioaccumulation in earthworm Eisenia fetida in soil. *Environ. Pollut.* 249, 776–784. https://doi.org/10.1016/j.envpol.2019.03.102

Wang, Q., Adams, C.A., Wang, F., Sun, Y., Zhang, S., 2022. Interactions between microplastics and soil fauna: A critical review. *Crit. Rev. Environ. Sci. Technol.* 52, 3211–3243. https://doi.org/10.1080/10643389.2021.1915035

Wang, W., Ge, J., Yu, X., Li, H., 2020. Environmental fate and impacts of microplastics in soil ecosystems: Progress and perspective. *Sci. Total Environ.* 708, 134841. https://doi.org/10.1016/j.scitotenv.2019.134841

Wu, M., Yang, C., Du, C., Liu, H., 2020. Microplastics in waters and soils: Occurrence, analytical methods and ecotoxicological effects. *Ecotoxicol. Environ. Saf.* 202, 110910. https://doi.org/10.1016/j.ecoenv.2020.110910

Xiang, Y., Jiang, L., Zhou, Y., Luo, Z., Zhi, D., Yang, J., Lam, S.S., 2022. Microplastics and environmental pollutants: Key interaction and toxicology in aquatic and soil environments. *J. Hazard. Mater.* 422, 126843. https://doi.org/10.1016/j.jhazmat.2021.126843

Xu, B., Liu, F., Cryder, Z., Huang, D., Lu, Z., He, Y., Wang, H., Lu, Z., Brookes, P.C., Tang, C., Gan, J., Xu, J., 2020. Microplastics in the soil environment: Occurrence, risks, interactions and fate–A review. *Crit. Rev. Environ. Sci. Technol.* 50, 2175–2222. https://doi.org/10.1080/10643389.2019.1694822

Ya, H., Jiang, B., Xing, Y., Zhang, T., Lv, M., Wang, X., 2021. Recent advances on ecological effects of microplastics on soil environment. *Sci. Total Environ.* 798, 149338. https://doi.org/10.1016/j.scitotenv.2021.149338

Yang, H., Yumeng, Y., Yu, Y., Yinglin, H., Fu, B., Wang, J., 2022. Distribution, sources, migration, influence and analytical methods of microplastics in soil ecosystems. *Ecotoxicol. Environ. Saf.* 243, 114009. https://doi.org/10.1016/j.ecoenv.2022.114009

Yi, M., Zhou, S., Zhang, L., Ding, S., 2021. The effects of three different microplastics on enzyme activities and microbial communities in soil. *Water Environ. Res.* 93, 24–32. https://doi.org/10.1002/wer.1327

Yoon, J.-H., Kim, B.-H., Kim, K.-H., 2024. Distribution of microplastics in soil by types of land use in metropolitan area of Seoul. *Appl. Biol. Chem.* 67, 1–13. https://doi.org/10.1186/s13765-024-00869-8

Yu, H., Qi, W., Cao, X., Hu, J., Li, Y., Peng, J., Hu, C., Qu, J., 2021. Microplastic residues in wetland ecosystems: Do they truly threaten the plant-microbe-soil system? *Environ. Int.* 156, 106708. https://doi.org/10.1016/j.envint.2021.106708

Yu, Y., Flury, M., 2021. Current understanding of subsurface transport of micro- and nanoplastics in soil. *Vadose Zo. J.* 20, e20108. https://doi.org/10.1002/VZJ2.20108

Yu, Z.-F., Song, S., Xu, X.-L., Ma, Q., Lu, Y., 2021. Sources, migration, accumulation and influence of microplastics in terrestrial plant communities. *Environ. Exp. Bot.* 192, 104635. https://doi.org/10.1016/j.envexpbot.2021.104635

Yuan, Y., Qin, Y., Wang, M., Xu, W., Chen, Y., Zheng, L., Chen, W., Luo, T., 2022. Microplastics from agricultural plastic mulch films: A mini-review of their impacts on the animal reproductive system. *Ecotoxicol. Environ. Saf.* 244, 114030. https://doi.org/10.1016/j.ecoenv.2022.114030

Zhang, D., Ng, E.L., Hu, W., Wang, H., Galaviz, P., Yang, H., Sun, W., Li, C., Ma, X., Fu, B., Zhao, P., Zhang, F., Jin, S., Zhou, M., Du, L., Peng, C., Zhang, X., Xu, Z., Xi, B., Liu, X., Sun, S., Cheng, Z., Jiang, L., Wang, Y., Gong, L., Kou, C., Li, Y., Ma, Y., Huang, D., Zhu, J., Yao, J., Lin, C., Qin, S., Zhou, L., He, B., Chen, D., Li, H., Zhai, L., Lei, Q., Wu, S., Zhang, Y., Pan, J., Gu, B., Liu, H., 2020. Plastic pollution in croplands threatens long-term food security. *Glob. Chang. Biol.* 26, 3356–3367. https://doi.org/10.1111/gcb.15043

Zhang, G.S., Liu, Y.F., 2018. The distribution of microplastics in soil aggregate fractions in southwestern China. *Sci. Total Environ.* 642, 12–20. https://doi.org/10.1016/J.SCITOTENV.2018.06.004

Zhang, Z., Peng, W., Duan, C., Zhu, X., Wu, H., Zhang, X., Fang, L., 2022. Microplastics pollution from different plastic mulching years accentuate soil microbial nutrient limitations. *Gondwana Res.* 108, 91–101. https://doi.org/10.1016/j.gr.2021.07.028

Zhao, S., Zhang, Z., Chen, L., Cui, Q., Cui, Y., Song, D., Fang, L., 2022. Review on migration, transformation and ecological impacts of microplastics in soil. *Appl. Soil Ecol.* 176, 104486. https://doi.org/10.1016/j.apsoil.2022.104486

Zhou, J., Gui, H., Banfield, C.C., Wen, Y., Zang, H., Dippold, M.A., Charlton, A., Jones, D.L., 2021. The microplastisphere: Biodegradable microplastics addition alters soil microbial community structure and function. *Soil Biol. Biochem.* 156, 108211. https://doi.org/10.1016/j.soilbio.2021.108211

Zhou, Q., Zhang, H., Fu, C., Zhou, Y., Dai, Z., Li, Y., Tu, C., Luo, Y., 2018. The distribution and morphology of microplastics in coastal soils adjacent to the Bohai Sea and the Yellow Sea. *Geoderma* 322, 201–208. https://doi.org/10.1016/J.GEODERMA.2018.02.015

Zhou, Y., Wang, J., Zou, M., Jia, Z., Zhou, S., Li, Y., 2020. Microplastics in soils: A review of methods, occurrence, fate, transport, ecological and environmental risks. *Sci. Total Environ.* 748, 141368. https://doi.org/10.1016/j.scitotenv.2020.141368

Zubris, K.A.V., Richards, B.K., 2005. Synthetic fibers as an indicator of land application of sludge. *Environ. Pollut.* 138, 201–211. https://doi.org/10.1016/j.envpol.2005.04.013

CHAPTER 20

Microplastics in the Air

Zakia Gueboudji and Kenza Kadi

20.1 INTRODUCTION

Plastic waste has emerged as a significant global problem, prompting considerable scientific interest in its environmental impact (Kumar et al., 2023). In recent years, the pervasive presence of microplastics (MPs) in the environment has sparked widespread concern and garnered increasing attention from scientists, policymakers, and the general public alike (Thacharodi et al., 2024). While much focus has been directed towards understanding the impacts of MPs in aquatic ecosystems, another dimension of this complex issue has emerged: the ubiquitous presence of MPs in the air we breathe (Prata et al., 2022). The revelation of MPs being suspended in the atmosphere has unveiled a new frontier in the study of plastic pollution and its potential consequences for environmental and human health (Blackburn & Green, 2022).

The proliferation of MPs, defined as plastic particles smaller than 5 mm in size, has reached unprecedented levels, permeating virtually every corner of the planet (Costa & Duarte, 2017). Originating from a myriad of sources, including the fragmentation of larger plastic debris, abrasion of synthetic textiles, and breakdown of plastic packaging, these microscopic particles have become integral components of the atmospheric aerosol (Ahmad et al., 2023; Zhao et al., 2023). Their small size and buoyant nature enable them to be transported over vast distances by wind currents, leading to their widespread distribution across terrestrial and marine environments (Li et al., 2020; Akdogan & Guven, 2019).

DOI: 10.1201/9781003487579-20

The presence of MPs in the air presents a multitude of challenges and implications for both environmental and human health (Amato-Lourenço et al., 2020). As airborne particles, MPs can serve as carriers for a range of toxic pollutants, including heavy metals, polycyclic aromatic hydrocarbons (PAHs), and persistent organic pollutants (POPs) (Munyaneza et al., 2022; Yang et al., 2021). Adsorbed onto the surfaces of MPs, these contaminants can undergo long-range transport and deposition, potentially accumulating in remote regions far from their original sources (Behera et al., 2023).

Furthermore, the inhalation of MPs poses potential risks to respiratory health, as these particles can penetrate deep into the lungs and may trigger inflammatory responses or exacerbate existing respiratory conditions (Yao et al., 2022; Sridharan et al., 2021). Additionally, MPs in the air may act as vectors for the transmission of pathogens, further compounding concerns regarding public health (Facciolà et al., 2021).

The dynamics of MPs in the atmosphere are governed by a complex interplay of physical, chemical, and biological processes (Kek et al., 2023). Atmospheric deposition plays a crucial role in the deposition of MPs onto terrestrial and aquatic ecosystems, where they can exert cascading effects on biodiversity and ecosystem functioning (Zhang et al., 2020). Moreover, atmospheric transport facilitates the global dispersal of MPs, underscoring the interconnectedness of environmental compartments and the transboundary nature of plastic pollution (Gago et al., 2022).

In light of these multifaceted challenges, understanding the sources, fate, and impacts of MPs in the air requires a multidisciplinary approach that integrates expertise from atmospheric science, environmental chemistry, toxicology, and public health (Tiernan et al., 2022). Collaborative efforts aimed at characterizing the sources and pathways of MPs, quantifying their distribution and abundance in the atmosphere, and assessing their potential risks to human health and ecosystems are essential for informing evidence-based policies and mitigation strategies.

In this chapter, the intricate world of MPs suspended in the air is delved into, with their origins, fate, and impacts on environmental and human health being examined. Through a synthesis of current research findings and emerging insights, an attempt is made to shed light on this complex issue and highlight the urgent need for concerted action to address the growing threat of microplastic pollution in the atmosphere.

20.2 LITERATURE REVIEW

Microplastics, minute plastic particles measuring less than 5 mm, have emerged as a pressing environmental concern, infiltrating ecosystems worldwide (Enfrin et al., 2021). While considerable attention has been devoted to plastic pollution in oceans and water bodies, the presence of MPs in the atmosphere is an understudied facet of this global issue (Gago et al., 2022). The proliferation of plastics in modern society has led to their ubiquitous presence in the environment (Avio et al., 2017). Plastics, known for their durability, find applications in various industries, contributing to their widespread use (Kumar et al., 2023). However, the downside of this convenience is the persistent and pervasive nature of plastic pollution (Heidbreder et al., 2019). While marine plastic pollution has garnered significant attention, the airborne dispersion of MPs presents a novel and relatively uncharted territory (Akdogan & Guven, 2019). Investigations into airborne MPs constitute a dynamic and expanding realm in environmental science, reflecting a growing awareness of the potential ramifications of these particles on ecosystems and human health (Jin et al., 2022; VishnuRadhan et al., 2021). A thorough analysis of existing literature highlights the multifaceted nature of this phenomenon (Luo et al., 2024; Huppertsberg & Knepper, 2018). Studies examining the sources of airborne MPs reveal a an intricate interplay of factors, including the atmospheric deposition of particulates from terrestrial sources, the fragmentation of larger plastic items, and the release of MPs from various human activities, such as the abrasion of synthetic textiles (Ahmad et al., 2023; Sridharan et al., 2021; Prata, 2018).

Characterizing airborne MPs involves a complex evaluation of their prevalence, morphology, and composition in diverse environments, spanning from urban air to more secluded and untouched locations (dos Santos Galvão et al., 2022). These investigations not only disclose the widespread presence of MPs but also illuminate the factors influencing their spatial distribution and temporal variability (Priya et al., 2022). Research into transport mechanisms reveals the potential for long-range dispersal, with wind patterns and atmospheric processes playing crucial roles in the movement of MPs across geographical boundaries (Behera et al., 2023; Leonard et al., 2023).

Growing concerns revolve around the ecological and human health implications of inhaling airborne MPs (Yao et al., 2022). Studies have begun exploring the respiratory impacts on organisms exposed to these particles, considering the particles' role as carriers for contaminants (Yao et al., 2022; Pironti et al., 2021). However, despite advancements in understanding, significant knowledge gaps persist. The standardization of methodologies for sampling and analysis remains a challenge, impeding comparability across studies. Moreover, uncertainties persist regarding the long-term fate of airborne MPs in the atmosphere and their ecological consequences (Zhao et al., 2023).

This literature review offers a detailed examination of the current state of knowledge concerning MPs in the air. While substantial progress has been made in unraveling the complexities of this environmental issue, the review underscores existing gaps and emphasizes the necessity for ongoing research to comprehensively address these uncertainties. The evolving understanding of airborne MPs highlights the need for a holistic approach to assess their impact on ecosystems, human health, and the broader environment.

20.3 SOURCES OF AIRBORNE MICROPLASTICS

Various factors, encompassing transport, dispersion, and deposition mechanisms, exerted influence on the atmospheric movement of pollutants, as illustrated in Figure 20.1.

Airborne MPs can originate from various sources, and their presence in the atmosphere has raised concerns due to potential environmental and

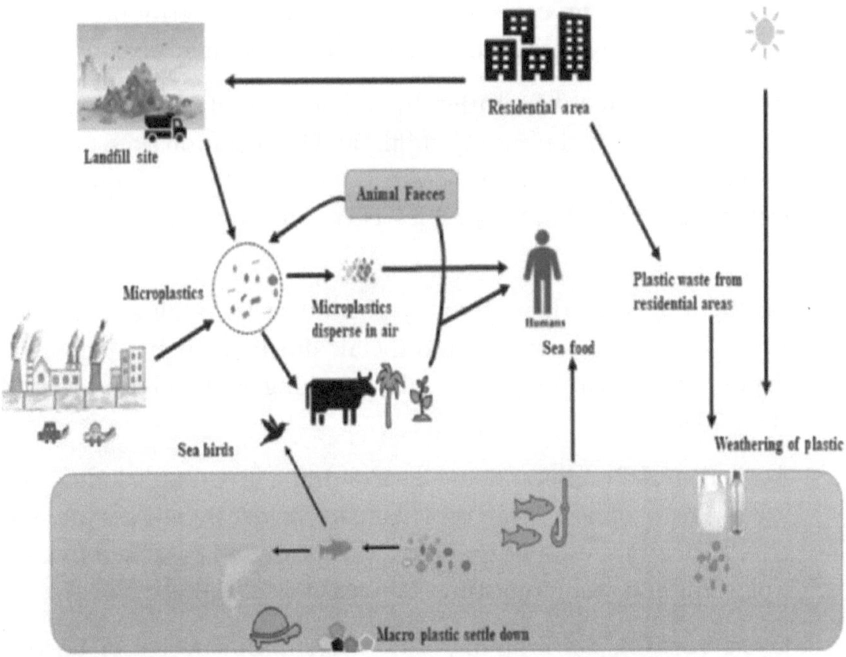

FIGURE 20.1 Origins, transport mechanisms, and fate of microplastic particles in the environment (Ahmad et al., 2023).

health impacts (Batool et al., 2022). Some common sources of airborne MPs include:

1. **Plastic Breakdown:** Larger plastic items, such as bottles, bags, and packaging materials, can break down into smaller particles through processes like weathering, UV radiation, and mechanical abrasion. These smaller particles, known as MPs, can become airborne (Gueboudji et al., 2024; Zhang & Wu, 2023).

2. **Tire Wear:** The wear and tear of tires on roads generates tiny particles, including MPs. As vehicles move, friction between the tires and the road surface releases small rubber particles containing plastic additives into the air (Gehrke et al., 2023).

3. **Road Paint:** Road markings and paints can contain plastic components. Over time, these materials can degrade, and the resulting particles can become airborne (Burghardt & Pashkevich, 2023).

4. **Industrial Processes:** Certain industrial activities, such as plastic manufacturing, processing, and recycling, can release MPs into the air. Additionally, processes like sanding, cutting, or grinding of plastic materials can generate airborne particles (Wieland et al., 2022).

5. **Waste Management Facilities:** Improper disposal and management of plastic waste, including at landfills and incineration sites, can lead to the release of MPs into the air. Wind and other environmental factors can carry these particles over long distances (Rajmohan et al., 2019).

6. **Synthetic Fibers:** Fibers from synthetic textiles, such as polyester and nylon, can be released into the air during activities like laundry or through natural processes like clothing wear and tear (Mishra et al., 2020).

7. **Agricultural Practices:** Some agricultural activities, including the use of plastic mulches, can contribute to the release of MPs into the air. Plastic mulches are thin plastic sheets used to cover soil to control weeds and retain moisture (Lwanga et al., 2023).

8. **Open Burning:** Burning of plastic waste in open areas or uncontrolled burning in landfill sites can release MPs and other pollutants into the air (Sridharan et al., 2021).

9. **Airborne Transport:** Microplastics present in water bodies, soil, or other environments can be transported into the air through wind and atmospheric processes (Enyoh et al., 2019).

10. **Urban Runoff:** Urban areas with impervious surfaces can contribute to the generation of MPs through runoff from roads and other surfaces during rainfall events (Müller et al., 2020).

Research is ongoing to better understand the sources, distribution, and potential impacts of airborne MPs. Monitoring and regulating these sources are essential to mitigate the environmental and health risks associated with airborne MPs. This section dissects the various pathways through which MPs enter the atmosphere, shedding light on their diverse sources.

20.3.1 Atmospheric Transport

This subsection delves into the mechanisms of atmospheric transport that facilitate the dispersal of MPs. It explores how MPs are lifted into the air, examining natural processes such as wind-driven suspension from surfaces, as well as anthropogenic activities that contribute to their release (Enyoh et al., 2019).

The dispersal of MPs in the atmosphere is facilitated by various atmospheric transport mechanisms. Once released into the air, MPs can be transported over long distances, affecting remote areas (Evangeliou et al., 2020). Some key atmospheric transport mechanisms include:

1. **Wind Transport:** Wind is a primary driver of atmospheric transport. Microplastics released into the air can be carried by wind currents over considerable distances. The smaller and lighter particles are more susceptible to wind transport (Evangeliou et al., 2020).

2. **Turbulence and Convection:** Turbulent air movements, which can occur near the Earth's surface and within the atmospheric boundary layer, play a role in the dispersal of MPs. Convection currents, driven by temperature differences, can also contribute to vertical transport (Xiao et al., 2023).

3. **Aerosolization:** Microplastics can become airborne through processes of aerosolization, where particles are lifted into the air as a result of mechanical or natural forces. This can include activities like

traffic on roads, agricultural practices, or breaking waves in aquatic environments (Zhang et al., 2020).

4. **Volatilization and Resuspension:** Some MPs may be associated with volatile organic compounds or other substances that can affect their buoyancy. Volatilization and resuspension processes can lift these particles into the air (Gueboudji, 2024a; Zhang et al., 2020).
5. **Rainfall and Deposition:** Microplastics in the atmosphere can be scavenged by rainfall and subsequently deposited on the Earth's surface. This process, known as wet deposition, can lead to the introduction of MPs into water bodies and soil (Zhang et al., 2020).
6. **Long-Range Transport:** Microplastics can be transported over long distances through atmospheric circulation patterns. This can result in the deposition of MPs in regions far from their original sources (Evangeliou et al., 2020).
7. **Particulate Matter Dynamics:** Microplastics can associate with particulate matter in the atmosphere. This association can affect their behavior, including settling and deposition, and may influence their transport patterns (Sridharan et al., 2021).
8. **Temperature and Humidity:** Atmospheric temperature and humidity conditions can influence the behavior of MPs. For example, high humidity may lead to the aggregation of MPs with other particles, affecting their transport dynamics (Yang et al., 2021).

20.3.2 Plastic Degradation

Plastic degradation refers to the process by which plastic materials break down into smaller fragments or particles over time. Microplastics in the air can be the result of various degradation processes, including:

1. **Mechanical Breakdown:** Exposure to physical forces such as sunlight, wind, and abrasion can lead to the mechanical breakdown of larger plastic items into smaller particles. This process is commonly observed in the natural environment, where plastic debris undergoes weathering and fragmentation (Zhang et al., 2021).
2. **Photodegradation:** Ultraviolet (UV) radiation from sunlight can cause photodegradation of plastics. The exposure to UV light breaks down the molecular structure of plastic polymers, leading to the

formation of smaller particles. This process is more pronounced in certain types of plastics, such as polyethylene and polypropylene (Fotopoulou & Karapanagioti, 2019).

3. **Thermal Degradation:** Elevated temperatures can accelerate the degradation of plastics. In the presence of heat, plastics may undergo thermal degradation, resulting in the release of volatile compounds and the formation of microplastic particles (Fotopoulou & Karapanagioti, 2019).
4. **Chemical Degradation:** Exposure to chemicals in the environment, including acids, bases, and other reactive substances, can contribute to the chemical degradation of plastics. This can lead to the breakdown of polymer chains and the formation of smaller plastic particles (Gunawardhana et al., 2023).
5. **Biodegradation:** While traditional plastics are not easily biodegradable, some additives or modifications can enhance their biodegradability. Biodegradation involves the breakdown of plastics by microorganisms into smaller components. However, complete biodegradation into non-toxic substances is not always achieved, and MPs may still persist (Corcoran, 2022; Liao & Chen, 2021).

Once MPs are generated through degradation processes, they can be transported into the air by various mechanisms, as discussed in the previous response. Wind, turbulence, and other atmospheric processes can lift these particles into the air, leading to the dispersal of airborne MPs.

It is important to note that the presence of MPs in the air raises concerns due to potential environmental and health impacts. Inhaling or ingesting airborne MPs may have adverse effects on human health and ecosystems (Sridharan et al., 2021). Therefore, understanding the sources, transport pathways, and fate of airborne MPs is critical for developing strategies to mitigate their impact on the environment.

20.3.3 Emissions from Human Activities

Human activities, ranging from industrial processes to daily practices, emit a myriad of pollutants, including MPs, into the environment (Prata et al., 2022).

Various sources contribute to the presence of MPs in the environment, spanning from their production to their disposal and eventual distribution.

Firstly, during the manufacturing of certain goods like personal care items and industrial abrasives, MPs are directly generated (Kumar et al., 2023). Secondly, improper disposal practices, including littering and inadequate waste management, lead to the release of macroplastics into the environment, which degrade into MPs over time (Rajmohan et al., 2019). In the textile industry, synthetic fibers like polyester and nylon shed microplastic particles during production, use, and washing, eventually finding their way into aquatic environments via wastewater (Mishra et al., 2020). Microplastics from roads, such as tire wear particles and plastic fragments, are carried by stormwater runoff into rivers and oceans (Gueboudji, 2024c). Plastic mulches used in agriculture for weed control and water conservation can also contribute to microplastic release into the soil and surrounding ecosystems as they degrade (Kumar et al., 2023). Additionally, MPs present in domestic and industrial wastewater can bypass treatment plants and enter water bodies through effluent discharge (Gueboudji & Kadi, 2023; Bui et al., 2020). Airborne MPs transported from various sources, such as industrial processes and road traffic, can be deposited over long distances onto land and water surfaces (Prata et al., 2022). Construction and demolition activities release MPs, especially when materials like paints and coatings containing MPs are used. Recreational activities, such as playing on synthetic turf fields made from recycled rubber tires, also contribute to microplastic release (Gueboudji, 2024c). Lastly, until recently, many personal care products contained microbeads made of plastic, which could enter wastewater during use (Thacharodi et al., 2024).

Efforts to mitigate the release of MPs involve better waste management, reducing plastic use, and developing alternative materials. Additionally, research is ongoing to understand the long-term impacts of MPs on ecosystems and human health.

20.4 DISTRIBUTION MECHANISMS

Understanding how airborne MPs distribute involves investigating their movement within the atmosphere, transport across distances, and eventual deposition in diverse environments (Batool et al., 2022). Numerous factors influence their dispersion. Wind plays a crucial role, lifting MPs into the air and carrying them over long distances, with particle size and weight affecting susceptibility to wind transport (Evangeliou et al., 2020). While turbulent processes and convection currents enable vertical transport within the atmosphere, most MPs remain in the lower atmosphere (Xiao et al., 2023). Microplastics can enter the air through aerosolization, propelled by activities like road traffic or breaking waves, and resuspension from surfaces (Luo et al., 2024). Deposition mechanisms include wet deposition, where rainfall removes MPs from the air, and dry deposition, where gravity causes particles to settle on surfaces (Zhao et al., 2023). Long-range transport occurs through atmospheric circulation patterns, potentially depositing MPs far from their sources. Particle size and density influence microplastic behavior, with smaller particles more easily transported by air currents (Behera et al., 2023). Areas with significant microplastic sources, like urban centers, may experience higher concentrations. Seasonal and weather variations, including temperature and wind patterns, impact dispersion and deposition. Microplastics deposited in various environments become potential sinks, requiring understanding of their distribution for assessing environmental and human exposure (Gueboudji, 2024b; Leonard et al., 2023).

Ongoing research is aimed at improving our understanding of these distribution mechanisms, assessing the potential risks associated with airborne MPs, and developing strategies to mitigate their impacts on ecosystems and human health.

20.5 DETECTION AND MEASUREMENT CHALLENGES

Detecting and measuring airborne MPs presents significant challenges, given their unique characteristics and the intricate environmental matrices in which they are embedded (Ahmad et al., 2023). Several key challenges are associated with assessing airborne MPs.

The small size range of airborne MPs, including submicron particles, poses a challenge for conventional detection methods, which may not be capable of capturing particles of such minuscule sizes (Munyaneza et al., 2022). Moreover, airborne MPs exhibit diverse types and compositions, ranging from fibers to fragments, making it difficult to develop standardized detection methods due to this variability (Zhang et al., 2020).

Low concentrations of airborne MPs, particularly when compared to larger plastic debris in the environment, require highly sensitive detection methods. Additionally, the presence of background contaminants in airborne environments can interfere with microplastic detection, complicating the distinction between plastic and non-plastic particles (Kek et al., 2023).

Temporal dynamics further complicate assessments, as the concentration of airborne MPs can vary over time due to weather conditions, human activities, and seasonal changes, necessitating frequent and systematic monitoring (Zhao et al., 2023). Furthermore, the lack of standardized protocols for sampling, extraction, and analysis hinders data comparison across studies and regions.

Developing effective and representative sampling strategies for airborne MPs is challenging due to the dynamic nature of airborne particles, requiring consideration of factors such as wind speed, particle size distribution, and sampling duration (Xiao et al., 2023). Additionally, airborne MPs may interact with various environmental matrices, such as air, water, soil, and sediments, altering their properties and complicating detection (Enfrin et al., 2021).

Some analytical techniques may lack the sensitivity required to detect MPs at low concentrations, necessitating the use of high-performance instruments to overcome these limitations (Huppertsberg & Knepper, 2018). The interdisciplinary nature of studying airborne MPs underscores the need for holistic approaches integrating expertise from atmospheric science, environmental science, chemistry, and materials science to develop comprehensive assessment methods.

Addressing these challenges involves ongoing research and the development of innovative detection and measurement techniques. Improved standardization, collaboration among researchers, and advancements in analytical technologies are essential for accurately assessing the presence and impact of airborne MPs.

20.6 TYPES OF AIRBORNE MICROPLASTICS

Airborne MPs manifest in diverse forms, originating from a variety of sources, and their classification is based on size, shape, and origin (Munyaneza et al., 2022). The common types of airborne MPs can be delineated as follows:

- **Microplastic Particles**
  Primary MPs, intentionally manufactured and typically measuring less than 5 mm, encompass microbeads in personal care products and industrial abrasives. Secondary MPs, on the other hand, originate from the fragmentation and breakdown of larger plastic items like bottles, packaging materials, and synthetic textiles, with weathering, UV radiation, and mechanical abrasion contributing to their formation (Chen et al., 2023).

- **Microplastic Fibers**
  Synthetic fibers shed from textiles, like polyester and nylon clothing, can become airborne during activities such as laundry and through natural wear and tear. These fibers represent a common type of airborne microplastic (Li et al., 2020).
- **Microplastic Fragments**
  Fragments constitute small pieces resulting from the breakdown of larger plastic items due to environmental processes. Examples include fragments from plastic bottles, containers, and other plastic products (Zhang et al., 2020).
- **Microplastic Films**
  Thin plastic films, commonly used in packaging and agricultural practices (e.g., plastic mulches), can become airborne as they degrade or break down. These films contribute to airborne microplastic pollution (Sridharan et al., 2021).
- **Microplastic Aggregates**
  Aggregates, or agglomerates, form when smaller microplastic particles bind together, either through natural processes or interactions with other substances in the environment (Sridharan et al., 2021).
- **Nurdles**
  Nurdles, also known as plastic pellets, are small pre-production plastic pellets used as raw material in plastic product manufacturing. Accidental spills during transportation or handling can release nurdles into the environment, including the air (Gunawardhana et al., 2023).
- **Biodegradable Microplastics**
  Some MPs are designed to be biodegradable, although the complete breakdown into non-toxic substances is not guaranteed. Residues from biodegradable plastics can still contribute to airborne microplastic pollution (Sridharan et al., 2021).
- **Microplastics Coated with Organic and Inorganic Matter**
  Microplastics in the environment can attract and absorb other substances, such as organic compounds and metals (Gueboudji, 2022). The coating of MPs with these substances may influence their transport and environmental impact (Ricardo et al., 2021).

20.7 CONSEQUENCES AND IMPACTS

The implications of airborne MPs in the environment extend across ecosystems, human health, and the global well-being. These repercussions encompass various dimensions.

20.7.1 Environmental Impact

Airborne MPs settling on both land and water surfaces contribute to ecosystem contamination, posing risks to a diverse range of organisms, including plants, animals, and microorganisms. The deposition of these particles into water bodies further exacerbates water pollution, impacting aquatic life and potentially leading to bioaccumulation within the food chain (Prata et al., 2022).

20.7.2 Human Health Impact

The inhalation of airborne MPs presents potential health risks, as these particles can be transported through the air and deposited in the respiratory system. The long-term health effects of inhaling MPs are still not fully understood and remain a subject of ongoing research. Additionally, MPs deposited in water bodies enter the food chain, indirectly exposing humans through the consumption of contaminated seafood and other food items (Chen et al., 2023).

20.7.3 Economic Impact

Agricultural fields experiencing the settling of airborne MPs may suffer consequences, affecting soil quality and potentially affecting crop growth. This could lead to economic challenges in agriculture, including reduced crop yields and increased costs for farmers (Enyoh et al., 2019).

20.7.4 Biodiversity Impact

The ingestion of MPs by wildlife, both terrestrial and aquatic, can result in physical harm, digestive issues, and potential long-term health effects. The pervasive presence of MPs in ecosystems contributes to declines in biodiversity, disrupting the delicate balance of various species (Loppi et al., 2021).

20.7.5 Ecological Disruption

The presence of MPs in the environment disrupts ecological processes and functions. This includes alterations in nutrient cycling, shifts in microbial communities, and changes in other dynamics within ecosystems (Zhao et al., 2023).

20.7.6 Pollutant Transport

Microplastics act as carriers for various pollutants, including POPs and heavy metals. This transport mechanism can contribute to the spread of contaminants through ecosystems, further amplifying environmental challenges (Munyaneza et al., 2022).

20.7.7 Social and Cultural Impact

The visible presence of airborne MPs has aesthetic impacts on landscapes and natural settings. Additionally, the contamination of recreational areas may have repercussions on outdoor activities and tourism (Kumar et al., 2021).

20.7.8 Regulatory and Management Challenges

The complexity of assessing and regulating airborne MPs presents challenges for policymakers and regulatory agencies. There is a pressing need for effective policies to comprehensively address microplastic pollution. Regulatory gaps and the technical challenges in monitoring and detecting airborne MPs make it challenging to establish baseline data and track changes over time. Addressing these multifaceted consequences and impacts requires a multidisciplinary approach that integrates scientific research, policy development, public awareness, and concerted efforts to reduce overall plastic use while improving waste management practices (Tiernan et al., 2022).

20.8 TOXIC IMPACTS

Following inhalation, MPs move into the lungs via the trachea, then enter the bloodstream and disperse throughout the body's circulatory system, leading to different levels of harm on cells, tissues, organs, and bodily systems.

20.8.1 Inhalation-Based Toxicity

Microplastics and nanoparticles (NPs) present in the air primarily enter the bodies of animals and humans through inhalation, initially interacting with the respiratory system. However, due to the body's clearance mechanisms within the respiratory barriers (such as the nasal cavity, trachea, bronchus, and alveolar macrophages), a significant portion of these plastic particles are unable to enter the body and instead form deposits (Ramsperger et al., 2023; Yang et al., 2021). The efficiency of deposition mainly depends on factors such as particle size and density. Smaller NPs (less than 2.5 μm) can evade these clearance mechanisms and penetrate the lung and respiratory barrier. Macrophages can either deposit plastic particles within the respiratory system, remove them, or assist in their migration within the respiratory system, allowing them to enter the circulatory system, a process known as dust metastasis (Amato-Lourenço et al., 2020).

Moreover, the presence of MPs and NPs in the respiratory system, with their large surface area, can trigger the release of chemokines affecting macrophage migration, potentially resulting in particle overload. For instance, studies on rats have shown that the influx and inflammation of neutrophils in the lungs, as well as the expression of pro-inflammatory genes in epithelial cells, are induced by polystyrene granules (64 nm), causing granule-induced oxidation (Rajput et al., 2022; Prata, 2018).

The large specific surface area of MPs enables them to adsorb numerous oxidizing substances, such as metals, which may lead to an overproduction of reactive oxygen species and subsequent oxidative stress in vivo. However, there is limited research on the toxic effects of atmospheric MPs on animals and humans. Inhalation of MPs poses the most significant toxic effects on the respiratory system (Yuan & Xu, 2023).

Studies have demonstrated that polyvinyl chloride (PVC) prepared by emulsion polymerization (2 mm) exhibits significant cytotoxic and hemolytic effects on rat and human lung cells in vitro. Additionally, research has shown that polystyrene nanoparticles (PS-NPs) can rapidly internalize into human alveolar epithelial cells, significantly reducing cell viability, affecting the cell cycle and apoptosis, and promoting an inflammatory response. The smaller the diameter of the particles, the faster their corresponding velocity (Yang et al., 2021).

Other findings suggest that PS-MPs induce cytotoxicity and inflammation in human lung epithelium by promoting the formation of reactive oxygen species. Furthermore, it has been observed that PS-MPs disrupt the lung barrier at low concentrations and decrease levels of α 1-antitrypsin at high concentrations, increasing the risk of chronic obstructive pulmonary disease. Studies have also shown that micro- and nanoplastics may exhibit strong genotoxicity when absorbed and internalized by the respiratory system (Mattioda et al., 2023).

Furthermore, respiratory lesions have been identified in workers exposed to synthetic textiles, fusing, vinyl chloride, or PVC, indicating distinct risks compared to the general population. Inhaled MPs and NPs are not efficiently cleared by human lungs and may remain in the lungs for a prolonged period, potentially triggering an inflammatory response. For instance, artificial glass fiber inhalation has been associated with DNA damage, potentially leading to cancer. In summary, inhaling MPs can exert toxic effects on both animals and humans (Facciolà et al., 2021).

20.8.2 Other Toxic Effects

Inhaled MPs can also exert adverse effects on various other systems within the body, primarily through diffusion or translocation. In vivo experiments have demonstrated that after inhalation, microplastic particles can enter the circulatory system via the migration of macrophages (Luo et al., 2024; Yao et al., 2022).

Once in the circulation, MPs may trigger inflammation, vascular occlusion, and other forms of blood toxicity. In vitro and in vivo studies have shown that polystyrene NPs induce red blood cell aggregation, whereas polypropylene particles increase hemolysis (Hwang et al., 2020).

Moreover, inhaled MPs may find their way into the gastrointestinal tract through the clearance of upper respiratory tract cilia. Within the digestive system, MPs can modify the permeability of intestinal epithelial cells and alter microbial composition (Huang et al., 2021).

Recent findings indicate that following maternal lung exposure to NPs, plastics can translocate to the placenta and the fetus, depositing in organs such as the fetal liver, kidney, nervous system, and circulatory system (Enyoh et al., 2023; Mortensen et al., 2019). However, there is a scarcity of research on the toxic effects of atmospheric MPs on systems beyond the respiratory system. Further investigation is crucial to elucidate their impacts on both animals and humans.

20.8.3 Joint Toxic Effects

Microplastics commonly contain various additives, including catalysts (such as organotin), flame retardants, polybrominated diphenyl ethers, antioxidants (like nonylphenol), antibacterial agents (such as triclosan), and plasticizers (such as phthalate PAEs). All of these additives pose harm to both animals and humans (Yang et al., 2021). Additionally, MPs adsorb numerous inorganic pollutants from the environment, such as gold (Au) and copper (Cu), as well as organic pollutants like polycyclic aromatic hydrocarbons and polychlorinated biphenyls. These pollutants are mutagenic and carcinogenic substances prevalent in the environment. Moreover, many plastic monomers themselves, such as polystyrene (PS) and PVC, exhibit mutagenic and carcinogenic toxic effects on animals and humans. Exposure to the additives or adsorbed pollutants in MPs results in combined toxicity (Rodrigues et al., 2019).

MPs in the atmosphere are prone to binding with other pollutants present in atmospheric environments, such as POPs and copper, acting as carriers for the long-distance transport of these pollutants in the atmosphere.

Some studies have investigated the adverse effects of MPs combined with these pollutants on animals and the human body, including oxidative stress, inflammatory reactions, and metabolic disorders, with associated risks of mutation and carcinogenesis. However, most of these studies have focused on oral ingestion or in vitro experiments (Luo et al., 2024; Yang et al., 2021). There is a lack of research on the toxicity mechanism of these bound substances after inhalation by animals and humans, necessitating more experimental data.

20.9 ENHANCED ANALYTICAL TECHNIQUES AND UNCOVERED DEFICIENCIES IN ASSESSING MICROPLASTICS IN THE ATMOSPHERE

Two primary sampling methods are commonly utilized for atmospheric MPs. One method involves passive collection of atmospheric fallout, where depositions are gathered using a non-plastic funnel, typically made of materials like stainless steel or glass. The collected material is then filtered to separate MPs from the sediment. Another method is active pump sampling and filtration, employing a system that collects air through a pump unit and filters it to retain plastic particles (Munyaneza et al., 2022; Chen et al., 2020; Zhang et al., 2020).

Following sample collection, various quantitative analysis methods can be applied to determine the types and sizes of particles (Chen et al., 2020).

Despite these established methods, several gaps in measuring atmospheric MPs persist. There is a pressing need for standardization and optimization of sampling techniques to ensure consistency and accuracy across studies. Additionally, the absence of standardized protocols for sample processing and analysis leads to variability in results, hindering comparability between studies. Furthermore, our understanding of the behavior, fate, and environmental impacts of atmospheric MPs remains limited, particularly regarding their sources and transport mechanisms. Addressing these gaps is essential for advancing our knowledge of atmospheric microplastic pollution and implementing effective mitigation strategies (Ahmad et al., 2023; Behera et al., 2023).

Notwithstanding the establishment of primary collection and analysis methods, several limitations and deficiencies persist. The current methods can only detect certain common types of plastic particles at or near ground level. Furthermore, accurately identifying the size of nanoplastics or the types of particles adsorbed by organic matter on their surfaces presents significant challenges.

Table 20.1 provides an overview of various methods utilized for the analysis of MPs and nanoplastics within the atmospheric environment. These methods encompass thermal analysis, spectral analysis, and other analytical techniques. Thermal analysis techniques such as Thermogravimetric Analysis coupled with Differential Scanning Calorimetry (TGA-DSC) involve the use of tools like TGA-DSC instruments to analyze plastic components like polyethylene (PE) and polypropylene (PP). While TGA-DSC offers simplicity and high accuracy, it faces limitations in distinguishing polymers with similar transition temperatures and identifying copolymers, besides destroying samples in the process. Pyrolysis-Gas Chromatography-Mass Spectrometry (Py-GC-MS) is another thermal analysis method offering high accuracy and simultaneous identification of microplastic types, weight, and additives without pretreatment. However, it also destructs samples and cannot discern morphology, size, or quantity of plastic particles. Spectral analysis methods such as Fourier Transform Infrared Spectroscopy (FT-IR) and Raman Spectroscopy (RM) offer advantages like simple pretreatment and non-destructive analysis. Still, they encounter challenges in identifying certain types of plastic particles and may face interference issues like fluorescence or background emission when utilizing lasers as the light source. Other analytical methods like Scanning Electron Microscopy coupled with Energy Dispersive X-ray Spectroscopy (SEM-EDS), Mass Spectrometry (MS), X-ray Photoelectron Spectroscopy (XPS), and Resonance Micro-Raman Spectroscopy (RMR) provide capabilities for imaging, elemental identification, and structural analysis. However, they have their respective drawbacks, such as high cost, limited applicability, or specific calibration requirements. Each method offers unique advantages and limitations, highlighting the need for careful consideration of the analytical approach based on the research objectives and constraints (Yang et al., 2021).

20.10 INTERDISCIPLINARY RESEARCH NEEDS

Addressing the complex challenges associated with airborne MPs requires interdisciplinary research efforts that involve collaboration among experts from various fields.

In the field of atmospheric science, researchers should investigate the sources, transport, and dispersion of airborne MPs in the atmosphere. Understanding the role of meteorological conditions in shaping the distribution of MPs is crucial, and developing models to simulate the behavior

TABLE 20.1 Prominent Methods for Analyzing Microplastics and Nanoplastics in the Atmospheric Environment (Yang et al., 2021)

Methods	Tools	Medium	Plastic Components	Optimal Size	Advantages	Disadvantages
Thermal analysis	TGA-DSC	NA	PE, PP, etc.	NA	• The operation is simple • Less sample is required (1–20 mg) • High accuracy	• Difficult to distinguish the polymers with similar transition temperatures • Difficult to identify copolymers • The samples are destroyed • It cannot identify the morphology, size, and quantity of the plastic particles
	Py-GC-MS	NA	PA, PC, PE, PS, PP, rubber, PET, PVC, PMMA	NA	• Less sample is required (5–200 μg) • The microplastic type and weight and additives can be identified simultaneously without pretreatment • The accuracy is high • It recognizes copolymers	• The samples are destroyed • It cannot identify the morphology, size, and quantity of the plastic particles
	TED-GC-MS	NA	PA, PE, PP, PS, PET	NA	• It involves simple pretreatment and operation	
Spectral analysis	FT-IR	Water, oil, air	RY, PE, PET, PAA	>20 μm	• Simple pretreatment • Does not destroy the sample • Plastic particles type can be determined	• It is difficult to identify the types of plastic particles that are aged or have contaminated surfaces
	RM	Water, air	PA, PC, PE, PP, PS, PET, PVC, PMP, PCL, PMMA	0.5–20 μm	• Does not destroy the sample • Supports nanosample imaging, and low sample amount identification • Environmentally friendly	• Long measurement time • Easy produce fluorescence interference • Low signal-to-noise ratio • Use of laser as the light source leads to background emission and sample degradation

(*Continued*)

TABLE 20.1 (*Continued*) Prominent Methods for Analyzing Microplastics and Nanoplastics in the Atmospheric Environment (Yang et al., 2021)

Methods	Tools	Medium	Plastic Components	Optimal Size	Advantages	Disadvantages
Other analytical methods	SEM-EDS	Majority	Majority	≥1 nm	• Imaging is at the nanoscale • Elements can be identified	• It is expensive • Work efficiency is low
	MS	Majority	Majority	≥1 nm	• It can identify the structure, molecular weight, degree of polymerization, functional group, and end group structure of the plastic particles	• Different samples require different ionizing reagents (poor applicability)
	XPS	Majority	Majority	>10 nm	• It can identify elemental composition and content, chemical state, molecular structure, and chemical bonds	• It cannot identify the nanoplastic types definitely
	RMR	Water, oil	Majority	>50 μm	• The cost is low • It is convenient for real-time field detection	• Only used to detect the concentration • It requires specific calibration samples

FT-IR, Fourier-transform infrared spectroscopy; RM, Raman spectroscopy; TGA-DSC, thermogravimetric analysis-differential scanning calorimeter; Py-GC-MS, pyrolysis gas chromatography-mass spectrum; TED-GC-MS, thermal desorption gas chromatography-mass spectrum; SEM-EDS, scanning electron microscope-energy dispersive spectrometer; MS, mass spectrometry; XPS, X-ray photoelectron spectroscopy; RMR, resonance microwave reflectometry; RY, rayon; PE, polyethylene; PET, polyethylene terephthalate; PAA, polyacrylic acid; PA, polyamide; PC, polycarbonate; PP, polypropylene; PS, polystyrene; PVC, polyvinyl chloride; PMP, polymethylpentene; PCL, polycaprolactone; PMMA, polymethylmethacrylate; NA, not available.

of MPs in the air is essential for accurate predictions (Behera et al., 2023; Tiernan et al., 2022).

Environmental scientists play a key role in studying the fate and impact of airborne MPs on terrestrial and aquatic ecosystems. Their work involves examining the interactions between MPs and other environmental pollutants and assessing the potential for bioaccumulation and biomagnification of MPs in food webs (Sridharan et al., 2021; Zhang et al., 2020; Enyoh et al., 2019).

Chemists and materials scientists contribute by analyzing the composition and characteristics of airborne MPs to distinguish them from natural particles. Investigating the degradation processes and transformation of MPs in different environmental matrices is crucial for understanding their lifecycle. Moreover, these researchers focus on developing innovative techniques for the detection and quantification of MPs (dos Santos Galvão et al., 2022; Yang et al., 2021; Chen et al., 2020; Rodrigues et al., 2019).

Toxicologists and health scientists assess the potential health effects of inhaling or ingesting airborne MPs. Their research aims to understand the bioavailability and toxicity of different types of MPs and to project the long-term health impacts on both humans and wildlife (Luo et al., 2024; Pironti et al., 2021).

Ecologists and biodiversity experts investigate the ecological consequences of microplastic exposure on plant and animal species. They assess the effects of MPs on biodiversity, ecosystem functioning, and community dynamics to understand the broader impact on natural environments (Thacharodi et al., 2024; Yao et al., 2022).

Engineers and technologists contribute by developing and optimizing air sampling technologies for collecting airborne MPs. They explore innovative waste management technologies to prevent the release of MPs into the environment and design filtration systems to reduce microplastic emissions from various sources (Bui et al., 2020).

Researchers in policy and social sciences examine the socio-economic impacts of airborne MPs on communities. They evaluate existing policies related to plastic use and waste management, proposing and assessing the effectiveness of regulatory measures to address microplastic pollution (Kumar et al., 2021; Rajmohan et al., 2019).

Communication and education experts play a crucial role in translating research findings to the public, policymakers, and industries. They develop educational programs to inform the public about the sources and impacts of MPs and collaborate with stakeholders to promote sustainable practices and behaviors (Kumar et al., 2021; Heidbreder et al., 2019).

In the realm of data science and modeling, researchers integrate data from various sources to create comprehensive models of microplastic distribution. They use machine learning and data analytics to analyze large datasets related to microplastic pollution and develop predictive models for assessing future trends in microplastic emissions (Yu & Hu, 2022).

Lastly, collaborative networks are essential for fostering communication and knowledge exchange among researchers, institutions, and industries. Establishing interdisciplinary research networks and consortia dedicated to studying airborne MPs is vital for addressing this form of pollution comprehensively.

20.11 CONCLUSIONS

The study on MPs in the air sheds light on a critical yet underexplored dimension of plastic pollution, emphasizing its global significance. The research has meticulously examined the sources, distribution, and consequences of airborne MPs, revealing their diverse origins and the intricate mechanisms governing their dispersion. The challenges associated with detection and measurement have been addressed, considering various collection methods and analytical approaches.

The identification of both primary and secondary forms of MPs, originating from direct release and larger plastic breakdown, provides a comprehensive understanding of the issue. The summary underscores the potential ecological and human health impacts, emphasizing the urgent need for interdisciplinary research to assess inhalation exposure risks and the far-reaching effects on terrestrial and aquatic ecosystems.

This study represents a crucial first step in comprehensively understanding plastic pollution, particularly in the context of airborne MPs. It serves as a clarion call for further investigation, urging a holistic approach to environmental management. As we navigate the challenges posed by this pervasive issue, collaboration among scientists, policymakers, and the public is essential to develop effective strategies for mitigating the impact of airborne MPs on our planet. Only through collective efforts can we hope to safeguard the health of ecosystems and protect human well-being from the consequences of this global environmental challenge.

REFERENCES

Ahmad, M., Chen, J., Khan, M. T., Yu, Q., Phairuang, W., Furuuchi, M., ... & Panyametheekul, S. (2023). Sources, analysis, and health implications of atmospheric microplastics. *Emerging Contaminants*, 100233. https://doi.org/10.1016/j.emcon.2023.100233

Akdogan, Z., & Guven, B. (2019). Microplastics in the environment: a critical review of current understanding and identification of future research needs. *Environmental Pollution*, 254, 113011. https://doi.org/10.1016/j.envpol.2019.113011

Amato-Lourenço, L. F., dos Santos Galvão, L., de Weger, L. A., Hiemstra, P. S., Vijver, M. G., & Mauad, T. (2020). An emerging class of air pollutants: potential effects of microplastics to respiratory human health? *Science of the Total Environment*, 749, 141676. https://doi.org/10.1016/j.scitotenv.2020.141676

Avio, C. G., Gorbi, S., & Regoli, F. (2017). Plastics and microplastics in the oceans: from emerging pollutants to emerged threat. *Marine Environmental Research*, 128, 2–11. https://doi.org/10.1016/j.marenvres.2016.05.012

Batool, I., Qadir, A., Levermore, J. M., & Kelly, F. J. (2022). Dynamics of airborne microplastics, appraisal and distributional behaviour in atmosphere: a review. *Science of the Total Environment*, 806, 150745. https://doi.org/10.1016/j.scitotenv.2021.150745

Behera, S. N., Yadav, M., Kumar, V., & Rout, P. R. (2023). Various perspectives on occurrence, sources, measurement techniques, transport, and insights into future scope for research of atmospheric microplastics. In Surampalli, R. Y. (Ed.), *Microconstituents in the Environment: Occurrence, Fate, Removal and Management* (pp. 203–225). Hoboken: John Wiley & Sons Inc. https://doi.org/10.1002/9781119825289.ch9

Blackburn, K., & Green, D. (2022). The potential effects of microplastics on human health: what is known and what is unknown. *Ambio*, 51(3), 518–530. https://doi.org/10.1007/s13280-021-01589-9

Bui, X. T., Nguyen, P. T., Nguyen, V. T., Dao, T. S., & Nguyen, P. D. (2020). Microplastics pollution in wastewater: characteristics, occurrence and removal technologies. *Environmental Technology & Innovation*, 19, 101013. https://doi.org/10.1016/j.eti.2020.101013

Burghardt, T. E., & Pashkevich, A. (2023). Road markings and microplastics–a critical literature review. *Transportation Research Part D: Transport and Environment*, 119, 103740. https://doi.org/10.1016/j.trd.2023.103740

Chen, G., Fu, Z., Yang, H., & Wang, J. (2020). An overview of analytical methods for detecting microplastics in the atmosphere. *TrAC Trends in Analytical Chemistry*, 130, 115981. https://doi.org/10.1016/j.trac.2020.115981

Chen, G., Li, Y., & Wang, J. (2023). Human health effects of airborne microplastics. *Comprehensive Analytical Chemistry*, 100, 185–223. https://doi.org/10.1016/bs.coac.2022.07.008

Corcoran, P. L. (2022). Degradation of microplastics in the environment. In Rocha-Santos, T., Costa, M. F., Mouneyrac, C. (Eds.), *Handbook of Microplastics in the Environment* (pp. 531–542). Cham: Springer International Publishing. https://doi.org/10.1007/978-3-030-39041-9_10

Costa, M. F., & Duarte, A. C. (2017). Microplastics sampling and sample handling. In *Comprehensive Analytical Chemistry* (Vol. 75, pp. 25–47). Elsevier. https://doi.org/10.1016/bs.coac.2016.11.002

dos Santos Galvão, L., Fernandes, E. M. S., Ferreira, R. R., dos Santos Rosa, D., & Wiebeck, H. (2022). Critical steps for microplastics characterization from the atmosphere. *Journal of Hazardous Materials*, 424, 127668. https://doi.org/10.1016/j.jhazmat.2021.127668

Enfrin, M., Hachemi, C., Hodgson, P. D., Jegatheesan, V., Vrouwenvelder, J., Callahan, D. L., ... & Dumée, L. F. (2021). Nano/micro plastics–challenges on quantification and remediation: a review. *Journal of Water Process Engineering*, 42, 102128. https://doi.org/10.1016/j.jwpe.2021.102128

Enyoh, C. E., Verla, A. W., Verla, E. N., Ibe, F. C., & Amaobi, C. E. (2019). Airborne microplastics: a review study on method for analysis, occurrence, movement and risks. *Environmental Monitoring and Assessment*, 191, 1–17. https://doi.org/10.1007/s10661-019-7842-0

Evangeliou, N., Grythe, H., Klimont, Z., Heyes, C., Eckhardt, S., Lopez-Aparicio, S., & Stohl, A. (2020). Atmospheric transport is a major pathway of microplastics to remote regions. *Nature Communications*, 11(1), 3381. https://doi.org/10.1038/s41467-020-17201-9

Facciolà, A., Visalli, G., Pruiti Ciarello, M., & Di Pietro, A. (2021). Newly emerging airborne pollutants: current knowledge of health impact of micro and nanoplastics. *International Journal of Environmental Research and Public Health*, 18(6), 2997. https://doi.org/10.3390/ijerph18062997

Fotopoulou, K. N., & Karapanagioti, H. K. (2019). Degradation of various plastics in the environment. In Takada, H., Karapanagioti, H. K. (Eds.), *Hazardous Chemicals Associated with Plastics in the Marine Environment* (pp. 71–92). Cham: Springer. https://doi.org/10.1007/698_2017_11

Gago, J., Booth, A. M., Tiller, R., Maes, T., & Larreta, J. (2022). Microplastics pollution and regulation. In *Handbook of Microplastics in the Environment* (pp. 1071–1096). Cham: Springer International Publishing. https://doi.org/10.1007/978-3-030-39041-9_52

Gehrke, I., Schläfle, S., Bertling, R., Öz, M., & Gregory, K. (2023). Mitigation measures to reduce tire and road wear particles. *Science of the Total Environment*, 166537. https://doi.org/10.1016/j.scitotenv.2023.166537

Gueboudji, Z. (2022). Effect of storage time on the quality of olive oil mill wastewaters obtained from the cold extraction of olive oil in the region of Khenchela and their biological properties (Doctoral dissertation, Abbes Laghrour University-Khenchela, Algeria).

Gueboudji, Z. (2024a). Waste valorization techniques. In: A. Anouzla, S. Souabi (Eds.), *Generation of Energy from Municipal Solid Waste*. Springer, Cham. https://doi.org/10.1007/978-3-031-74334-4_2

Gueboudji, Z. (2024b). Biogas generation and ecological implications: an overview. In: A. Anouzla, S. Souabi (Eds.), *Generation of Energy from Municipal Solid Waste*. Springer, Cham.

Gueboudji, Z. (2024c). From trash to treasure: Transforming landfill tyres, plastics, and glass into sustainable construction resources. In: S. Souabi, A. Anouzla (Eds.), *Technical Landfills and Waste Management. Springer Water*. Springer, Cham. https://doi.org/10.1007/978-3-031-55665-4_12

Gueboudji, Z., & Kadi, K. (2023). Physicochemical characterization and estimation of the pollution degree of olive oil mill wastewaters from the cold extraction system and the traditional system. In: S. Souabi, A. Anouzla (Eds.), Wastewater from Olive Oil Production. Springer Water. Springer, Cham. https://doi.org/10.1007/978-3-031-23449-1_6

Gueboudji, Z., Mahmoudi, M., Kadi, K., & Nagaz, K. (2024). Characteristics and Impacts of Municipal Solid Waste (MSW): A Review. In: *Anouzla, A., Souabi, S. (eds) Technical Landfills and Waste Management*. Springer Water. Springer, Cham. https://doi.org/10.1007/978-3-031-52633-6_4

Gunawardhana, G. M. S. S., Perera, U. L. H. P., & Ratnayake, A. S. (2023). Sources and fate of plastics into microplastics: degradation and remediation methods. In Vithanage, M., Alwis, A. D., Botheju, D. (Eds.), *Maritime Accidents and Environmental Pollution-The X-Press Pearl Disaster* (pp. 155–172). Boca Raton, FL: CRC Press.

Heidbreder, L. M., Bablok, I., Drews, S., & Menzel, C. (2019). Tackling the plastic problem: a review on perceptions, behaviors, and interventions. *Science of the Total Environment*, 668, 1077–1093. https://doi.org/10.1016/j.scitotenv.2019.02.437

Huang, Z., Weng, Y., Shen, Q., Zhao, Y., & Jin, Y. (2021). Microplastic: a potential threat to human and animal health by interfering with the intestinal barrier function and changing the intestinal microenvironment. *Science of the Total Environment*, 785, 147365. https://doi.org/10.1016/j.scitotenv.2021.147365

Huppertsberg, S., & Knepper, T. P. (2018). Instrumental analysis of microplastics—benefits and challenges. *Analytical and Bioanalytical Chemistry*, 410, 6343–6352. https://doi.org/10.1007/s00216-018-1210-8

Hwang, J., Choi, D., Han, S., Jung, S. Y., Choi, J., & Hong, J. (2020). Potential toxicity of polystyrene microplastic particles. *Scientific Reports*, 10(1), 7391. https://doi.org/10.1038/s41598-020-64464-9

Jin, L., Xie, J., He, T., Wu, D., & Li, X. (2022). Airborne transmission as an integral environmental dimension of antimicrobial resistance through the "One Health" lens. *Critical Reviews in Environmental Science and Technology*, 52(23), 4172–4193.

Kek, H. Y., Tan, H., Othman, M. H. D., Nyakuma, B. B., Ho, W. S., Sheng, D. D. C. V., … & Wong, K. Y. (2023). Critical review on airborne microplastics: an indoor air contaminant of emerging concern. *Environmental Research*, 118055. https://doi.org/10.1016/j.envres.2023.118055

Kumar, R., Verma, A., Shome, A., Sinha, R., Sinha, S., Jha, P. K., … & Vara Prasad, P. V. (2021). Impacts of plastic pollution on ecosystem services, sustainable development goals, and need to focus on circular economy and policy interventions. *Sustainability*, 13(17), 9963. https://doi.org/10.3390/su13179963

Kumar, V., Singh, E., Singh, S., Pandey, A., & Bhargava, P. C. (2023). Micro-and nano-plastics (MNPs) as emerging pollutant in ground water: environmental impact, potential risks, limitations and way forward towards sustainable management. *Chemical Engineering Journal*, 459, 141568. https://doi.org/10.1016/j.cej.2023.141568

Leonard, J., El Rassi, L. A., Samad, M. A., Prehn, S., & Mohanty, S. K. (2023). The relative importance of local climate and land use on the deposition rate of airborne microplastics on terrestrial land. *Atmospheric Environment*, 120212. https://doi.org/10.1016/j.atmosenv.2023.120212

Li, Y., Shao, L., Wang, W., Zhang, M., Feng, X., Li, W., & Zhang, D. (2020). Airborne fiber particles: types, size and concentration observed in Beijing. *Science of the Total Environment*, 705, 135967. https://doi.org/10.1016/j.scitotenv.2019.135967

Liao, J., & Chen, Q. (2021). Biodegradable plastics in the air and soil environment: low degradation rate and high microplastics formation. *Journal of Hazardous Materials*, 418, 126329. https://doi.org/10.1016/j.jhazmat.2021.126329

Loppi, S., Roblin, B., Paoli, L., & Aherne, J. (2021). Accumulation of airborne microplastics in lichens from a landfill dumping site (Italy). *Scientific Reports*, 11(1), 4564. https://doi.org/10.1038/s41598-021-84251-4

Luo, D., Chu, X., Wu, Y., Wang, Z., Liao, Z., Ji, X., … & Shang, X. (2024). Micro-and nano-plastics in the atmosphere: a review of occurrence, properties and human health risks. *Journal of Hazardous Materials*, 133412. https://doi.org/10.1016/j.jhazmat.2023.133412

Lwanga, E. H., Van Roshum, I., Munhoz, D. R., Meng, K., Rezaei, M., Goossens, D., … & Ritsema, C. (2023). Microplastic appraisal of soil, water, ditch sediment and airborne dust: the case of agricultural systems. *Environmental Pollution*, 316, 120513. https://doi.org/10.1016/j.envpol.2022.120513

Mattioda, V., Benedetti, V., Tessarolo, C., Oberto, F., Favole, A., Gallo, M., … & Giorda, F. (2023). Pro-inflammatory and cytotoxic effects of polystyrene microplastics on human and murine intestinal cell lines. *Biomolecules*, 13(1), 140. https://doi.org/10.3390/biom13010140

Mishra, S., Singh, R. P., Rath, C. C., & Das, A. P. (2020). Synthetic microfibers: source, transport and their remediation. *Journal of Water Process Engineering*, 38, 101612. https://doi.org/10.1016/j.jwpe.2020.101612

Mortensen, N. P., Johnson, L. M., Grieger, K. D., Ambroso, J. L., & Fennell, T. R. (2019). Biological interactions between nanomaterials and placental development and function following oral exposure. *Reproductive Toxicology*, 90, 150–165. https://doi.org/10.1016/j.reprotox.2019.08.016

Müller, A., Österlund, H., Marsalek, J., & Viklander, M. (2020). The pollution conveyed by urban runoff: a review of sources. *Science of the Total Environment*, 709, 136125. https://doi.org/10.1016/j.scitotenv.2019.136125

Munyaneza, J., Jia, Q., Qaraah, F. A., Hossain, M. F., Wu, C., Zhen, H., & Xiu, G. (2022). A review of atmospheric microplastics pollution: in-depth sighting of sources, analytical methods, physiognomies, transport and risks. *Science of the Total Environment*, 822, 153339. https://doi.org/10.1016/j.scitotenv.2022.153339

Pironti, C., Ricciardi, M., Motta, O., Miele, Y., Proto, A., & Montano, L. (2021). Microplastics in the environment: intake through the food web, human exposure and toxicological effects. *Toxics*, 9(9), 224. https://doi.org/10.3390/toxics9090224

Prata, J. C. (2018). Airborne microplastics: consequences to human health? *Environmental Pollution*, 234, 115–126. https://doi.org/10.1016/j.envpol.2017.11.043

Prata, J. C., Castro, J. L., da Costa, J. P., Cerqueira, M., Duarte, A. C., & Rocha-Santos, T. (2022). Airborne microplastics: concerns over public health and environmental impacts. In Rocha-Santos, T., Costa, M. F., Mouneyrac, C. (Eds.), *Handbook of Microplastics in the Environment* (pp. 177–201). Cham: Springer International Publishing. https://doi.org/10.1007/978-3-030-39041-9_37

Priya, A. K., Jalil, A. A., Dutta, K., Rajendran, S., Vasseghian, Y., Qin, J., & Soto-Moscoso, M. (2022). Microplastics in the environment: recent developments in characteristic, occurrence, identification and ecological risk. *Chemosphere*, 298, 134161. https://doi.org/10.1016/j.chemosphere.2022.134161

Rajmohan, K. V. S., Ramya, C., Viswanathan, M. R., & Varjani, S. (2019). Plastic pollutants: effective waste management for pollution control and abatement. *Current Opinion in Environmental Science & Health*, 12, 72–84. https://doi.org/10.1016/j.coesh.2019.08.006

Rajput, A., Kumar, R., Gupta, A., & Gupta, S. (2022). Microplastics in the air and their associated health impacts. In Ahamad, A., Singh, P., Tiwari, D. (Eds.) *Plastic and Microplastic in the Environment: Management and Health Risks* (pp. 84–102). Hoboken, NJ: John Wiley & Sons, Inc. https://doi.org/10.1002/9781119800897.ch6

Ramsperger, A. F., Bergamaschi, E., Panizzolo, M., Fenoglio, I., Barbero, F., Peters, R., … & Laforsch, C. (2023). Nano-and microplastics: a comprehensive review on their exposure routes, translocation, and fate in humans. *NanoImpact*, 29, 100441. https://doi.org/10.1016/j.impact.2022.100441

Ricardo, I. A., Alberto, E. A., Júnior, A. H. S., Macuvele, D. L. P., Padoin, N., Soares, C., … & Trovo, A. G. (2021). A critical review on microplastics, interaction with organic and inorganic pollutants, impacts and effectiveness of advanced oxidation processes applied for their removal from aqueous matrices. *Chemical Engineering Journal*, 424, 130282. https://doi.org/10.1016/j.cej.2021.130282

Rodrigues, M. O., Abrantes, N., Gonçalves, F. J. M., Nogueira, H., Marques, J. C., & Gonçalves, A. M. M. (2019). Impacts of plastic products used in daily life on the environment and human health: what is known? *Environmental Toxicology and Pharmacology*, 72, 103239. https://doi.org/10.1016/j.etap.2019.103239

Sridharan, S., Kumar, M., Singh, L., Bolan, N. S., & Saha, M. (2021). Microplastics as an emerging source of particulate air pollution: a critical review. *Journal of Hazardous Materials*, 418, 126245. https://doi.org/10.1016/j.jhazmat.2021.126245

Thacharodi, A., Hassan, S., Meenatchi, R., Bhat, M. A., Hussain, N., Arockiaraj, J., … & Pugazhendhi, A. (2024). Mitigating microplastic pollution: a critical review on the effects, remediation, and utilization strategies of microplastics. *Journal of Environmental Management*, 351, 119988. https://doi.org/10.1016/j.jenvman.2023.119988

Tiernan, H., Friedman, S., Clube, R. K., Burgman, M. A., Castillo, A. C., Stettler, M. E., … & De Nazelle, A. (2022). Implementation of a structured decision-making framework to evaluate and advance understanding of airborne microplastics. *Environmental Science & Policy*, 135, 169–181. https://doi.org/10.1016/j.envsci.2022.05.005

VishnuRadhan, R., Thresyamma, D. D., Eldho, T. I., & Bhagat, J. (2021). Atmospheric plastics-a potential airborne fomite with an emerging climate signature. *The Journal of Climate Change and Health*, 3, 100037. https://doi.org/10.1016/j.joclim.2021.100037

Wieland, S., Balmes, A., Bender, J., Kitzinger, J., Meyer, F., Ramsperger, A. F., … & Kress, H. (2022). From properties to toxicity: comparing microplastics to other airborne microparticles. *Journal of Hazardous Materials*, 428, 128151. https://doi.org/10.1016/j.jhazmat.2021.128151

Xiao, S., Cui, Y., Brahney, J., Mahowald, N. M., & Li, Q. (2023). Long-distance atmospheric transport of microplastic fibres influenced by their shapes. *Nature Geoscience*, 16(10), 863–870. https://doi.org/10.1038/s41561-023-01264-6

Yang, H., He, Y., Yan, Y., Junaid, M., & Wang, J. (2021). Characteristics, toxic effects, and analytical methods of microplastics in the atmosphere. *Nanomaterials*, 11(10), 2747. https://doi.org/10.3390/nano11102747

Yao, X., Luo, X. S., Fan, J., Zhang, T., Li, H., & Wei, Y. (2022). Ecological and human health risks of atmospheric microplastics (MPs): a review. *Environmental Science: Atmospheres*, 2, 921–942. https://doi.org/10.1039/D2EA00041E

Yu, F., & Hu, X. (2022). Machine learning may accelerate the recognition and control of microplastic pollution: future prospects. *Journal of Hazardous Materials*, 432, 128730. https://doi.org/10.1016/j.jhazmat.2022.128730

Yuan, Z., & Xu, X. R. (2023). Surface characteristics and biotoxicity of airborne microplastics. In Verma, S. K., Das, A. K. (Eds.), *Comprehensive Analytical Chemistry* (Vol. 100, pp. 117–164). Elsevier. https://doi.org/10.1016/bs.coac.2022.07.006

Zhang, K., & Wu, C. (2023). Formation of airborne microplastics. In Verma, S. K., Das, A. K. (Eds.), *Comprehensive Analytical Chemistry* (Vol. 100, pp. 1–16). Elsevier. https://doi.org/10.1016/bs.coac.2022.07.001

Zhang, K., Hamidian, A. H., Tubić, A., Zhang, Y., Fang, J. K., Wu, C., & Lam, P. K. (2021). Understanding plastic degradation and microplastic formation in the environment: a review. *Environmental Pollution*, 274, 116554. https://doi.org/10.1016/j.envpol.2021.116554

Zhang, Y., Kang, S., Allen, S., Allen, D., Gao, T., & Sillanpää, M. (2020). Atmospheric microplastics: a review on current status and perspectives. *Earth-Science Reviews*, 203, 103118. https://doi.org/10.1016/j.earscirev.2020.103118

Zhao, X., Zhou, Y., Liang, C., Song, J., Yu, S., Liao, G., … & Wu, C. (2023). Airborne microplastics: occurrence, sources, fate, risks and mitigation. *Science of the Total Environment*, 858, 159943. https://doi.org/10.1016/j.scitotenv.2022.159943

CHAPTER 21

Impact of Microplastics in Agricultural Soil on Nutrient Recycling and Fertility

Ajay Kumar Mishra, Pallavi Chattopadhyay, Somrita Nag, Anthony Fulford, and Sheetal Sharma

21.1 INTRODUCTION

The annual amount of plastic waste humans produce exceeds 350 million metric tons. The amount of plastic garbage generated worldwide is estimated to increase to one billion metric tons by 2060 if current practices remain unchanged (OECD, 2022). Macroplastics, like fishing nets or bottles, can also break down slowly and turn into microplastics (MPs) (Julienne et al., 2019), particles, fragments, or fibers with a diameter of less than 5 mm, and possibly even nanoplastics. This increases the chance of exposure and poses risks to the environment (Allen et al., 2022) and public health (Yee et al., 2021).

The emergence of microplastic pollution has become a significant environmental concern, especially in the context of agricultural soils. Microplastics, defined as plastic particles smaller than 5 mm, originate

DOI: 10.1201/9781003487579-21

from various sources, including agricultural inputs, waste disposal, and atmospheric deposition (Horton et al., 2017). Their presence in farm settings is particularly alarming due to the potential impacts on soil health and food security. In 2004 professor Richard Thompson coined "Microplastics" (Thompson et al., 2004). Plastic is ubiquitous, known as "white pollution," a matter of grave concern in the environmental ecosystem (Liu et al., 2014).

The introduction of MPs into soil disrupts the delicate balance of soil ecosystems. These particles can alter soil structure, water retention, and nutrient cycling processes (Weithmann et al., 2018). The interference with these fundamental soil functions has implications for agricultural productivity. For instance, changes in soil structure can affect root growth and water availability to plants, directly impacting crop yields.

Moreover, MPs in soil environments may act as carriers for other pollutants, including pesticides and heavy metals, posing additional risks to soil health and plant growth (Zhang et al., 2018). The interaction of MPs with soil biota further complicates this scenario. Soil microorganisms, crucial for nutrient recycling and organic matter decomposition, can be adversely affected by microplastic pollution (de Souza Machado et al., 2018).

Soils moderate many services, such as promoting carbon sequestration, biogeochemical cycling, and biodiversity. Soil microbes play a vital role in powering most of these services. They catalyze many biogeochemical transformations that yield human societal goods, such as increasing food security.

This study investigates microplastics' occurrence, properties, and effects on soil fertility and characteristics, interactions with soil biota, and wider environmental and health consequences. Understanding these aspects is vital for developing strategies to mitigate MPs' impacts on agriculture and ensure sustainable food production systems.

21.2 ORIGIN AND FATE OF MICROPLASTICS IN AGRICULTURE

The issue of MPs in agriculture is intricately linked to the broader narrative of global plastic production and its environmental ramifications. Since the advent of mass plastic production in the mid-20th century, the cumulative global production of plastics has soared to an astounding 8.3 billion tons, as reported by PlasticsEurope in 2019. This unprecedented production has led to the widespread environmental dispersion of plastics, with significant implications for agricultural ecosystems.

One of the critical challenges in managing plastic waste is its low recycling rate. As Letcher (2020) highlighted, only about 20% of produced plastics are recycled. The rest, amounting to a staggering 80%, finds its way into various environmental compartments, including soil, rivers, and oceans. The persistence of plastic waste in these environments leads to their gradual breakdown into smaller fragments through physical, chemical, and biological processes, eventually forming MPs, defined as plastic particles smaller than 5 mm in size (Thompson et al., 2004).

In agricultural contexts, the breakdown of plastic film residues is a significant MPs source. These residues are commonly subjected to weathering, ultraviolet radiation, and mechanical stresses from farming activities, resulting in their fragmentation into MPs. This process is particularly prevalent in mulched soils, where plastic films are widespread for moisture conservation and weed control. Studies by Zhou et al. (2020) have shown that mulched soils tend to have a higher concentration of plastic films than non-mulched soils.

Human activities have further exacerbated the spread of MPs in agricultural soils. Practices such as plastic mulching (Huang et al., 2020), sewage irrigation (Li et al., 2018), soil amendment application (Weithmann et al., 2018; Vithanage et al., 2021), fertilizer coatings (Bian et al., 2022), and littering (Yang et al., 2021) have been identified as key contributors. Environmental media like runoff (Nizzetto et al., 2016) and air (Dris et al., 2016) act as vectors for MP transport, further augmenting their presence in soil.

Phthalates (PAEs) are phthalic acid esters; they act as plastic additives and add plasticity to industrial polymers. An absence of a plasticizer renders a plastic brittle or inflexible. The reaction between phthalic anhydride and alcohol causes the formation of PAEs. Their chemical structure comprises a benzene ring and two carboxylic acids (Mankidy et al., 2013). The bonds are covalent, which allows PAEs to be released in a process known as lying in soil, freshwater, ocean, and sediments. PAEs have been widely used in occupational settings, mainly as plasticizers in various applications (Dutta et al., 2020).

Given the pervasiveness of MPs in agricultural soils, it is imperative to understand their environmental fate and potential impacts. MPs in soil can migrate, accumulate, and even act as vectors for contaminants such as heavy metals and organic pollutants. Their interaction with soil components and biota can have far-reaching implications for soil health, plant growth, food safety, and human health.

Thus, the origin and fate of MPs in agricultural settings present a complex and pressing environmental challenge. Their presence is not merely a result

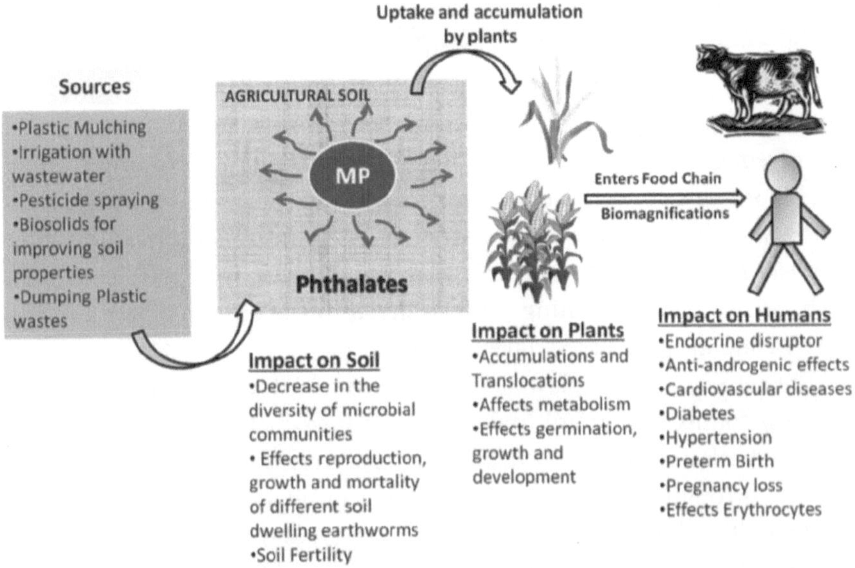

FIGURE 21.1 Fate and impact of phthalates from microplastics to biological systems. (Modified from Kumari and Kaur, 2020.)

of direct agricultural activities but is also significantly influenced by broader plastic use and disposal patterns in society. Addressing this issue requires a multifaceted approach that includes reducing plastic usage, improving waste management practices, and governmental regulations and necessary policies in the agricultural context and other areas of use (Figure 21.1).

21.3 MICROPLASTICS IN AGRICULTURE: OCCURRENCE AND CHARACTERISTICS

After global warming, microplastic contamination comes in second as an exponentially increasing ecological and environmental issue and a severe global concern (Azeem et al., 2021). The widespread occurrence of MPs in agricultural environments is a growing concern, reflecting a broader issue of environmental contamination. Microplastics, typically defined as plastic particles smaller than 5 mm, have been found in various ecosystems, with agriculture fields being significant reservoirs.

21.3.1 Occurrence of Microplastics in Agricultural Soils

Extensive studies have demonstrated the ubiquitous presence of MPs in agricultural soils across the globe. These particles predominantly originate from two types of plastics: polypropylene (PP) and polyethylene (PE), which are commonly used in agricultural activities and consumer products. Their prevalence of farming settings was first brought to attention by Pelamatti et al. (2021) and Koyuncuoglu and Erden (2021), who noted their frequent detection in soil samples from various agricultural regions.

Historically, concern about plastic pollution was raised as early as the 1970s, with seminal works like Carpenter and Smith (1972) warning about the risks associated with increasing plastic production and its uncontrolled disposal. This early warning is now echoed in the observed prevalence of MPs in aquatic and terrestrial ecosystems.

The extreme level of plastic pollution from wastewater treatment plants (WTPs) found on agricultural land is approximately 7.76 million tons (Peccia & Westerhoff, 2015). In 2012, agricultural systems added plastics to the soil through mulching, equivalent to around 4.4 million tons (Sintim & Flury, 2017; Luo et al., 2018).

Mulching is one of the most commonly used agriculture techniques to conserve soil moisture, increase water infiltration, and improve crop yields (Kader et al., 2017). Polyethylene (PE) mulching films are among the most widely used, cost-effective, and high-performance mulching materials (Qin et al., 2015). However, they are still one of the significant sources of plastic pollution in agricultural soil (Bläsing & Amelung, 2018).

Irrigation is a common practice in agriculture in water-deficient conditions, as it allows crops to grow even when precipitation is insufficient to cover plants' water demand. An area of 270 Mha is irrigated worldwide, accounting for 18% of total agricultural land (Bruinsma, 2003; Bläsing & Amelung, 2018). For irrigation, the direct use of partially treated or untreated wastewater from farm fields is rising and may become the only water source for many farmers. Thus, experts estimate that already 20 million ha (7% of the total irrigated land) are irrigated with untreated or partially treated wastewater (UNW-DPC, 2013) and that the food for 10% of the world population is produced by the use of contaminated wastewater (Corcoran, 2010). Untreated wastewater contains large amounts of plastic derived from, e.g., effluent of washing machines or care products like shampoos or peelings. Various plastic materials are widely used in agriculture, including soil fumigation film, irrigation drip tape/tubing, nursery pots, and silage bags. Li et al. (2018) concluded that 90% of MPs accumulate in wastewater sludge and that their concentration in sludge ranges from 1,500 to 56,400 particles/kg (Mintenig et al., 2017). Sludge is applied as fertilizer on agricultural land (Nizzetto et al., 2016).

Atmospheric deposition is another source of MPs and NPs entering surface soil. Atmospheric precipitation of plastic in the urban areas of Paris was around 2–35 particles/(m^2 days) (Dris et al., 2016). Allen et al. (2019) findings reported through air mass trajectory analysis that through atmospheric transportation, microplastic transport through the atmosphere at a distance of up to 95 km could reach and affect remote and sparsely populated areas. Microplastic detection in soil from remote/unsettled areas or restless high mountain areas has been reported (Scheurer & Bigalke, 2018).

UV-light irradiation degraded the soil environment's PS components (PS foam, single-use plates, and coffee cup caps) (Lambert & Wagner, 2016).

21.3.2 Characteristics of Microplastics in Soils

The characteristics of MPs in agricultural soils can vary widely based on their source, size, shape, and composition. Research has indicated that the concentration of MPs can be significantly higher in soil than in aquatic environments. Horton et al. (2017) estimated that terrestrial ecosystems might contain between 4 and 23 times more MPs than oceans. This difference is attributed to direct input from agricultural practices, atmospheric deposition, and runoff from urban and industrial areas.

In agricultural settings, MPs come from various sources, such as the breakdown of plastic mulch, synthetic fibers from clothing, and fragmented plastic equipment. These particles can range in size from visible fragments to microscopic fibers. Their physical properties, such as color, density, and shape, are influenced by their origin and the environmental processes they undergo, like UV degradation and mechanical breakdown.

Cey et al. (2009) reported that the average diameter of MPs that can leach down into deep soil of up to 70 cm is 3.7 mm. In addition, the penetration of water flow, such as rain or irrigation water from top to bottom in the ground, transmits MPs and NPs to the bottom with the soil vacuum and eventually leads to the groundwater (Panno et al., 2019; Wong et al., 2020).

Chemical properties of soil, such as pH, sorption and adsorption of heavy metals, and their redox potential play crucial roles in the bioavailability of plastic in soil (Ng & Obbard, 2006; Yu et al., 2020; Yu et al., 2021). Furthermore, iron–manganese oxide-bound fractions of nickel (Ni), copper (Cu), chromium (Cr), and exchangeable carbonated-bound fractions in the soil can be reduced by the presence of MPs (Yu et al., 2020). The presence of high-density polyethylene (HDPE), polystyrene (PS), polyethylene (PE), and polylactic acid (PLA) increases the diethylenetriamine pentaacetate (DTPA) extractable Cd concentration of soil (Wang et al., 2020a, 2020b; Adeel et al., 2018).

21.4 IMPACT ON SOIL FERTILITY AND NUTRIENT RECYCLING

Soil physicochemical, faunal, and microbiological properties represent the three leading indicators of soil health (Kibblewhite et al., 2016). The presence of MPs in soil triggers physical (stability of aggregates, soil bulk density, and water dynamics), chemical (nutrients availability, organic matter, and pH), and biological (microbial activity and soil fauna) changes.

21.4.1 Impact on Soil Fertility

21.4.1.1 Impact of Microplastics on Soil Properties

Plastic has been reported both in surface and subsurface soil. Numerous agricultural practices, such as tillage, irrigation, and soil organisms, promote plastic transport in different soil layers (Rillig et al., 2017c; Panno et al., 2019). Deep tillage, moldboard, and deep plowing methods disturb different soil layers and promote the deep penetration of MP and NP in subsoils (Rillig et al., 2017c; Baumhardt et al., 2008; Azeem et al., 2019). In addition, the pruning of rhizomes such as potatoes and carrots can also help MPs and NPs migrate downward (Zhou et al., 2020). A current study found that wet-dry circles can promote MPs to move downward. A dry climate induces soil cracks, which could facilitate the plastic to move in deep soil (O'Connor et al., 2019).

Microplastics in soils can profoundly impact soil properties. Studies have shown that MPs can affect soil texture, water retention capacity, and nutrient dynamics (Guo et al., 2022; Machado et al., 2019). The presence of MPs can alter the soil's physical structure, affecting its porosity and permeability. These changes can influence the soil's ability to hold and transmit water, potentially impacting plant growth and soil microbial activity.

Furthermore, the chemical composition of MPs can interact with soil chemistry. For example, plastic additives such as PAEs and bisphenol A (BPA) can leach into the soil, potentially affecting soil fertility and microbial communities. The interaction of MPs with other soil pollutants can also lead to complex environmental effects.

The occurrence and characteristics of MPs in agricultural soils represent a multifaceted environmental issue. Their widespread presence and the potential for significant impacts on soil health and agricultural

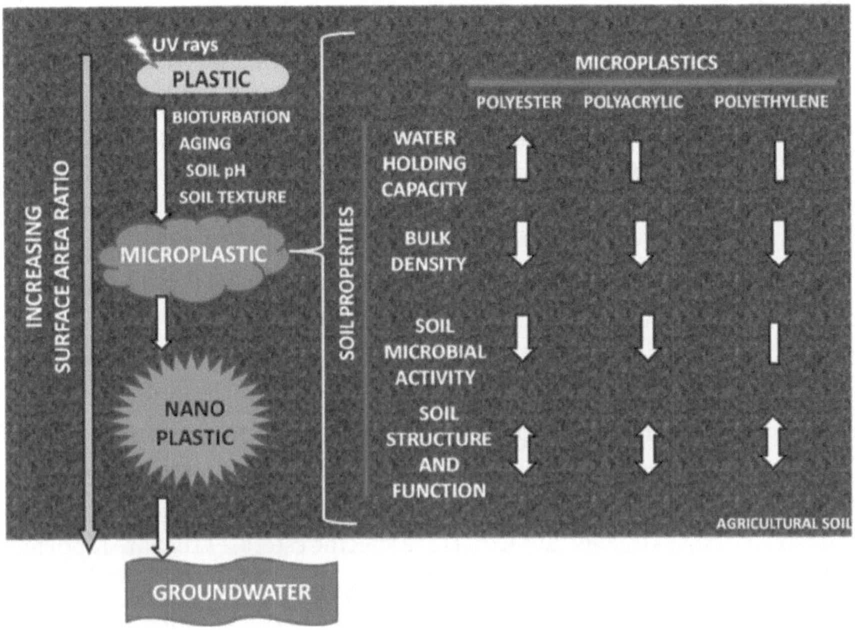

FIGURE 21.2 Impact of microplastic on soil properties (de Souza Machado et al., 2018).

productivity underscores the need for further research and effective management strategies to mitigate their impact on agricultural ecosystems.

Microplastics can alter carbon cycling in soil, creating variations in the decomposition process, which interferes with soil organic carbon (SOC) determination for soil quality checking (Kim et al., 2021; Rillig et al., 2021)—the impact of microplastic on soil properties (de Souza Machado et al., 2018) (Figure 21.2).

21.4.1.2 Microplastics as Carriers of Pollutants in Soil

This section discusses the role of MPs as vectors for harmful substances in soils, including pesticides, heavy metals, and other pollutants. This section also addresses how these pollutants can be transferred to plants and affect soil contamination.

The uptake of these pollutants takes place by the roots. The organic pollutant is first enriched at the root surface during the uptake process and then enters the plant via roots and water (Zhang et al., 2017). The contaminants, including PAEs, are reported to enter through the cuticle-free unsuberized cell wall (Müller & Kördel, 1993; Kvesitadze et al., 2006). Furthermore, the cell wall between the root cortex cells is porous; thus, contaminants can move freely before they reach the endodermis. Organic pollutants are reportedly translocated to different plant parts after their uptake (Lin et al., 2007). Various factors regulate uptake, like solubility, molecular mass of contaminant, pH, temperature phase of plant growth, etc. (Korte et al., 2000; Kvesitadze et al., 2006). In case the solubility of contaminants is low, they can also be absorbed by the roots via passive or active uptake (Inui et al., 2008). After plant uptake, PAEs may undergo enzymatic transformation to enhance their hydrophilicity and lower their toxicity. This process of contaminant transformation in plants is referred to as Sandermann's Green Liver Concept (Sandermann, 1994). Oxidation, reduction, hydrolysis, etc., are the main enzymatic reactions that mediate the conversion of hydrophobic contaminants into hydrophilic ones. This step increases the affinity of the formed intermediate towards the enzymes, and further transformations occur (Kvesitadze et al., 2006). Carboxylesterases (CXEs, EC 3.1.1.1) are specific esterases that are reported to be involved in the metabolism of PAEs in plants that display hydrolyzing activity against carboxylic esters (i.e., de-esterification), and are also documented to be involved in many functional roles in plants, including xenobiotics detoxification (Haslam et al. 2001; Gershater & Edwards, 2007; Sun et al., 2014).

PAEs induce stress in plants. Stress is a collective term used for both abiotic and biotic that contribute to limiting plant growth and development by affecting photosynthesis and reducing the carbon assimilation ability of plants. Decreasing crop productivity is a primary concern among people nowadays in order to meet the increasing food demands (FAO, 2009). Abiotic stresses are one of the major factors responsible for low productivity; worldwide, it is difficult to overcome such challenges. Mulching was introduced in agriculture, which helped overcome the challenges but came at the cost of high amounts of environmental additives (Huang et al., 2020; Qiang et al., 2023). There are several additives, but PAEs are the preferred plasticizers among them. Phthalates cause phytotoxicity, which has been reported (Mondal et al., 2022; Gao et al., 2019a). Different plants respond differently; some can withstand stress, while others are sensitive to pollutants. Plants are a crucial part of our ecosystem and provide numerous services to human beings (Beare et al., 1995; Kuzyakov & Blagodatskaya, 2015). Plants under phthalate stress showed adverse effects on germination, growth, and development along with biochemical and physiological processes by interfering with osmolytes, pigment content, and stress metabolites. Plastic agricultural films are the primary source of PAEs in agrarian soils (He et al., 2015). From soil, PAEs get into the plant system, where they accumulate and disrupt normal plant functions. During stressful conditions (phthalate stress), the plants show morphological and physiological responses and employ numerous strategies to maintain homeostasis. Strategies such as alterations in morphological and developmental patterns (i.e., growth plasticity) and biochemical and physiological processes (Tuteja, 2007; Saud et al., 2014).

In plants, pigments are crucial for regulating the photosynthesis rate. Abiotic stress affects photosynthesis. Photosynthetic capacity is directly proportional to their yield. To make carbohydrates, an energy source for heterotrophs, plants have to perform a series of complex reactions for carbon assimilation. The photosynthetic apparatus consists of pigments (Chl a, Chl b, pheophytins, and carotenoids), photosystems (PS), light reactions (for the generation of NADPH and ATP), and dark reactions (for CO_2 assimilation) (Singh & Thakur, 2018). The exposure of DEP (Diethyl phthalate) is also observed to enhance the expression of HSP (Heat Shock Proteins) in *Spirodela polyrhiza* and *Raphanus sativus*. Phthalates are observed to improve the level of ROS. The generation of reactive oxygen species (ROS) is an inevitable response of plants under stressed environments. They are produced in plants via partial oxygen reduction and are referred to as a collective term for oxygen and nonradical oxygen species (Ahmad, 2018). The presence of unpaired electrons is responsible for their high reactivity, which can even mediate the oxidation of cell structures and biomolecules and disturb cell integrity (Kanojia & Dijkwel, 2018). ROS formation also occurs under normal conditions because of various metabolic processes, but the level of ROS is observed to be accelerated during abiotic or biotic stresses. Cellular organelles like chloroplasts, mitochondria, and peroxisomes, which have high metabolic activity, act as main sites for ROS formation. Thus, to combat the enhanced levels of ROS, plants have a defense grid that relies on endogenous enzymatic and non-enzymatic antioxidants (Yu et al., 2019). The enzymatic antioxidative defense system is intricate and efficient and includes enzymes like superoxide dismutase (SOD), guaiacol peroxidase (POD), catalase (CAT), ascorbate peroxidase (APX), glutathione reductase (GR), etc. The non-enzymatic antioxidants include glutathione, ascorbate, carotenoids, tocopherols, polyphenolic compounds, etc.

Heavy metals such as dichlorodiphenyltrichloroethane (DDT), PAHs, and persistent organic pollutants (POPs) are adsorbed within MP and NP particles. Chemicals adsorbed on the surface of MPs and NPs are more toxic than those directly coming from plastics. For example, it can be seen that the sorption of PS MP on metals is in the order of $Pb^{2+} > Cu^{2+} > Cd^{2+} > Ni^{2+}$ (Yan et al., 2020). Under the same conditions, the rate of Cr released from MPs was faster than the rate of lead (Pb) release (Godoy et al., 2020). This may be due to different hydration ionic radii and differences in divalent cation complexing ability (Zou et al., 2020). The composition of MPs influencing the adsorption, diffusion, and release of heavy metals depends on their morphology, specific surface area, surface charge, and porosity (Godoy et al., 2019; Li et al., 2019; Rizwan et al., 2018). Wang et al. (2020a) reported that PE MPs adsorbed five pesticides (diflubenzuron, difenoconazole, carbendazim, malathion, and Diptera) in agricultural soil through exothermic and spontaneous processes; these results show that PE MPs can transfer different types of pesticides in agricultural soils.

Cr (VI) adsorption was inhibited at a $pH > 6$. At the same time, it increased at a $pH < 6$ due to adsorption sites that are available on PE MPs in increasing competition with sodium dodecyl benzene sulfonate. The four types of MPs [low-density polyethylene (LDPE), HDPE, polyvinyl chloride (PVC), and PE] were investigated using three types of heavy metals (Cd^{2+}, Pb^{2+}, and Cu^{2+}). Plastic adsorbed on metal was found in the order of PE > PVC > HDPE > LDPE, while metal adsorbed on MPs was $Pb^{2+} > Cu^{2+} > Cd^{2+}$ (Zou et al., 2020). The HDPE adsorption of Cd was increased with the increase in pH; in contrast, the efficiency was reduced with the increase in salinity. Desorption showed a high tendency towards the adsorbed Cd (Deshiikan et al., 1998) and posed a more significant threat to the biotic environment (Wang et al., 2019) (Table 21.1).

21.4.2 Impacts of Microplastics on Soil Nutrient Cycling

21.4.2.1 Effect of MPs on Nutrient Absorption and Recycling

In Wan et al.'s (2023) study, adding MPs significantly reduced soil organic carbon (SOC), pH, total nitrogen (TN), NH^{4+}-N, and bacterial diversity. Conversely, fungal diversity, dissolved organic carbon (DOC), and enzymatic activities—particularly those related to the biogeochemical cycle—were elevated.

Microplastics could significantly reduce soil phosphate content from 122.61 to 63.43 mg P/L. In contrast, MPs could significantly elevate soil ammonium content from 0.94 to 1.53 mg N/L.

The MP's can influence the *Arbuscular michorizha* fungi by altering the nutrient cycling and transportation to the roots as they are readily adsorbed (de Souza Machado et al., 2019). MP's such as polystyrene and polyethylene positively affect soil aggregation, thus increasing soil

TABLE 21.1 The Concentration, Duration, and Impact of Exposure of Microplastics on Various Crops

Plant	Microplastics	Concentration and Duration of Exposure	Impact	References
Brassica chinensis L. (Chinese cabbage pakchoi)	DBH and DEHP	(0–500 mg/kg) 72 hours	Inhibited seed germination	Ma et al. (2013)
Hordeum vulgare L. (Barley)	BBP	(0–1,600 g/L) 1 week		Kumari and Kaur (2018)
Vigna radiata (Mung bean)	DBP and DEHP	(0–500 mg/kg) 72 hours		Ma et al. (2014)
Hordeum vulgare L. (Barley)	DBP	(0–1,600 g/L) 7 days		Kumari and Kaur (2017)
Cucumis sativus L. (Cucumber)	DMP	(0–200 mg/L) 7 days		Zhang et al. (2016)
Chlorella emersonii and *Selenastrum capricornutum (microalgae)*	DBP	(10^{-3} to 10^{-7} M) 7 days	Inhibited the growth of algae	Melin and Egneus (1983)
Spirodela polyrhiza (Greater Duckweed/Duckmeat)	DAP	(0–400 mg/L) 7 days	Inhibit growth The concentration of dap is inversely proportional to algae growth	Sharma and Kaur (2019)
Phaseolus vulgaris (common bean), *Brassica campestris var. chinensis*, *Picea abies* (Norway spruce), *Trifolium repens* (white clover), *Plantago major* (broadleaf plantain) and *Holcus lanatus* (common velvet grass)	DBP	(0, 0.8, 1.5, 3.5, 10 μg/m^3) Exposure duration depends upon the lifecycle of the respective plants	Reduced biomass of shoots and roots	Dueck et al. (2003)

(*Continued*)

TABLE 21.1 (*Continued*) The Concentration, Duration, and Impact of Exposure of Microplastics on Various Crops

Plant	Microplastics	Concentration and Duration of Exposure	Impact	References
Chlorella vulgaris (Microalgae)	DBP	(0–100 mg/L) 2–10 days	Affects growth	Duan et al. (2018)
Raphanus sativus (Radish)	DEP and DEHP	(0–10^{-4} M) 24 hours	Decreased fresh weight	Saarma et al. (2003)
Scenedesmus obliquus (Microalgae), *Chlorella pyrenoidosa* (Microalgae)	DBP	(0–20 mg/L) 0–96 hours	Reduced growth	Gu et al. (2017)
Cultivars (V1, V2, V3, V4, V5) of *Ipomoea aquatic* (Water spinach)	DBP	(4.5, 10.3, 22.5 mg/kg) 39 days	Biomass declined	Cai et al. (2008)
Browallia speciosa (Sapphire flower), *Raphanus satvius* L. (Radish)	DBP	(0–4 µg/L) 14 days	Interfered in carotenoid synthesis	Virgin et al. (1981)
Raphanus sativus (Radish)	DBP	(0–120 ng/dm^3) 13 days	Inhibition in photosynthesis	Millar and Hannay (1986)
Spirodela polyrhiza (Common duckmeat), *Lemna minor* (Lesser duckweed)	DBP	(0–7.5 mg/L) 7 days	The decline in pigment content, which contributes to chlorosis and results in senescence	Huang et al. (2006)
Wheat	DBP and DEHP	(0–40 mg/kg) 14–40 days	Induces oxidative stress, which results in Chlorophyll content	Gao et al. (2019c)

(*Continued*)

TABLE 21.1 (*Continued*) The Concentration, Duration, and Impact of Exposure of Microplastics on Various Crops

Plant	Microplastics	Concentration and Duration of Exposure	Impact	References
Spirodela polyrhiza (Common Duckmeat)	DAP	(0–400 mg/L) 7 days	Proline content increased	Sharma and Kaur (2019)
Brassica chinensis L.	DBP and DEHP	(0–500 mg/kg) 72 hours and 14 days	Elevated MDA content	Ma et al. (2013)
Karenia brevis (Dinoflagellate)	DBP	(0–5.0 mg/L) 0–72 hours	Elevated MDA content	Li et al. (2015)
Wheat	DBP and DEHP	(0–40 mg/kg) 14–40 days	Elevated H_2O_2 content	Gao et al. (2019)
Cucumis sativus (Cucumber)	DBP	(0–200 mg/L) 1–7 days	Elevated MDA and H_2O_2 content with increasing duration and dose	Zhang et al. (2015)
Spirodela polyrhiza (Common Duckmeat), *Lemna minor* (Lesser Duckweed)	DBP	(0–7.5 mg/L) 7 days	Increased SOD and POD activities	Huang et al. (2006)
Wheat	DBP and DEHP	(0–40 mg/kg) 14–40 days	Elevated SOD, APX, CAT, and GPX activities	Gao et al. (2019)

nutrient retention (Lozano et al., 2021). Pot experiments conducted with loamy soils to analyze the onion roots showed significant changes in the nutrient uptake of various nutrients, such as a decrease in Sulfr concentrations and an increase in Mn when inoculated with the AMF (Jiménez et al., 2022). Also, the effects of MPs on the availability of soil nutrient distribution varied for various regions depending on the fertilizer regimes (Li & Liu, 2022).

21.4.2.2 Interaction of Microplastics with Soil Biota

This part highlights the interaction between MPs and soil organisms. It examines the effects of MPs on microbial diversity and activity and how these interactions can influence ecosystem services provided by soil biota.

The irregular or unorganized disposal of wastes on soils leads to potential risks to the biotic components, particularly microorganisms, nematodes, and plants (Bender & Heijden, 2015). Moreover, using untreated or partially reclaimed water for irrigation in some arid or semi-arid areas contributed to soil pollution. Sometimes, biosolids are also employed to improve the properties of the soil. The wastewater and biosolids contain a variety of harmful inorganic and organic contaminants. The soil acts as a primary sink for different types of waste, such as chemical, domestic, industrial, and agricultural wastes (Teng et al., 2014; Qing et al., 2015). These can also cause alterations in the physical and chemical characteristics of the soils.

Microplastics negatively impact different soil-dwelling earthworms' reproduction, growth, and mortality (Huerta Lwanga et al., 2017; Cao et al., 2017; Rodriguez-Seijo et al., 2017). Furthermore, besides plastic, its additives, like plasticizing agents, including PAEs, may harm soil-dwelling organisms when released from plastic. In many plastics, such additives are only loosely incorporated into the polymer structure and might thus be washed out. Some of these PAEs, like bis(2-ethylhexyl) phthalate, were found to inhibit soil microbiological activity (Wang et al., 2016).

There is growing evidence that soil organisms such as earthworms (*Lumbricus terrestris*), mites (*Hypoaspis aculeifer, Damaeus exspinosus*), and collembolans (*Folsomia candida*) can help migrate MPs and NPs from topsoil to deep soil (Rillig et al., 2017a; Lwanga et al., 2016; Zhu et al., 2018; Yu et al., 2019; Heinze, 2019; Adeel et al., 2020, 2019). Soil organisms can carry MPs and NPs through their casting, burrowing or ingestion, egestion and pushing, and adhesion to their exterior (Adeel et al., 2021). Rillig et al. (2017b) reported that earthworms (*Lumbricus terrestris*) added 35%–73%

of MPs (<50 and 63–150 μm in diameter) from surface debris to their burrows and also transported much smaller particles (<50 μm) into deep soil (Rillig et al., 2017a). In topsoil polyethylene, MPs were penetrated up to 10 cm, while NPs (710–850 μm) were primarily observed in the deepest layer. This result shows that earthworms transport plastic in terms of size, and NPs penetrate more profoundly in the soil than MPs. Earthworms can also pull large and microscopic plastic particles down into their burrows without eating the pieces of plastic (Zhang & Liu, 2018). The vertical movement of soil organisms creates micropores in the soil, which promotes the transportation of MPs and NPs by leaching. Yu et al. (2019) reported that MPs are moved up to 50 cm below due to the bioturbation of earthworms in sandy soil, and further leaching encouraged the MPs to reactivate even the most significant part (250–1000 μm), which was found in a 60 cm-high leachate column.

PS (polystyrene) microbeads inhibit Bacteroidetes, Proteobacteria, and Firmicutes due to their possible interaction with reduced soil nutrients, as shown in arsenic-polluted paddy soils (Dong et al., 2020). MPs and NPs change soil properties and bacterial communities. Rhizobia can change with changes in the soil matrix (Maity & Pramanick, 2020).

In the fertilized soil, MPs significantly enhanced the bacterial and fungal community. MPs indicate the selective impact on microbes and cause a severe hazard to biogeochemical cycles and microbes ecology (Ren et al., 2020). MPs change the AMF diversity and structure depending on their concentration level and type. Enriched with Ambispora (10% of PLA and PE), Archaeosporaceae (PLA 10%), and PLA hurt plant physiology, i.e., fresh/dry biomass, and chlorophyll content (Wang et al., 2020d). The ingestion of MPs by soil organisms could influence bacterial diversity e.g., Zhu et al. reported that in the collembolan (arthropod) gut, MPs potentially enhanced bacterial diversity, possibly due to a move in feeding after MP exposure (Zhu et al., 2018). Different sizes of plastic affect microbes due to their change in surface area (Brodhagen et al., 2017). NPs (<0.1 μm) especially could enter the cell membrane and cause cytotoxic effects (Lei et al., 2018) due to the bioaccumulation in the cell of filamentous and yeast fungi (de Souza Machado et al., 2018b; Miyazaki et al., 2015; Nomura et al., 2016). The surface of MP is a hotspot for some microbial development; MPs increased the soil microbial biomass (+43.6%). Shown positive effects on acid bacteria, bacteriocytes, and hydrolase and urease enzymes, while adverse effects are shown on Sphingomonadaceae, Fluorescein diacetate enzyme, decreased xylosidase and β-glucosidase activity by 16%–43%.

Soil organisms, including earthworms, mites, and collembolans, induce transportation through casting, ingestion, egestion, and adhesion to their exterior skin (Azeem et al., 2021). Brown et al. (2022), in a field experiment with the application of low-density polyethylene (LDPE), showed no significant changes in the soil microbial activity, including earthworm abundance, and the yield and growth of the wheat crop remained unchanged. Lu et al. (2023), a microcosmic experiment with the earthworms had varying results on the bacterial community and diversity with the application of polyethylene (PE) and polylactic acid (PLA).

21.5 DETECTION AND ANALYSIS OF MICROPLASTICS IN AGRICULTURAL SOILS

This section covers the methods and technologies used to detect and quantify soil MPs, discussing the current state and challenges in these analytical techniques.

Microplastics are present in diverse environments, and their potential nature can be determined with quantification, separation, and classification methods. Determination of MPs assists in checking the vulnerability of microplastic pollution and potential damage risks to living biota.

Singh et al. (2018) standardized a protocol for estimating different phthalate esters from soil (0–15 cm depths) and in MS broth. The recovery of different PAEs from soil and MS broth medium was also estimated using HPLC (Table 21.2).

TABLE 21.2 Techniques for Detecting Microplastics in Soil Samples (Yadav et al., 2022)

Source	Techniques for Detection	References
Manure application (manure and soil test)	FTIR spectroscopy	Wu et al. (2021)
PP and LDPE in organic soil (microbial activity determination)	DOC-DON ratios	Blöcker et al. (2020)
Sewage sludge-based fertilizers (soil test)	μ-FTIR	Zhang et al. (2020)
Plastic mulching in cropland with vegetables	Stereomicroscope	Beriot et al. (2021)
E-waste disposal site (soil test)	FTIR spectroscopy	Chai et al. (2020)
Domestic garbage (soil test)	Stereomicroscope and anatomical needle	Chen et al. (2020)
Agricultural plastics (soil test)	ATR-FTIR spectroscopy	Piehl et al. (2018)

(Continued)

TABLE 21.2 (*Continued*) Techniques for Detecting Microplastics in Soil Samples (Yadav et al., 2022)

Source	Techniques for Detection	References
Compost application (agricultural and horticulture soil sampling)	Density separation and light microscopy	Braun et al. (2021)
Lagooning sludge (co-compost, dry sludge, fresh sludge)	GC-MS; fluorescence staining microscopy; Raman spectroscopy	El Hayany et al. (2020)
Plastic waste (soil test, crops, earthworm cast, chicken feces)	Stereomicroscope and statistics	Huerta et al. (2017)
Long-term inorganic fertilizer application (soil test)	TOC analyzer, QIIME, FAPROTAX database, HT-qPCR	Li et al. (2021b)
Composite soil samples	HPLC	Singh et al. (2018)
Farm soil	SEM and LCSM (Liquid chromatography and mass spectrometry)	Mason et al. (2018)
Lettuce (*Lactuca sativa* L.)	Electron microscopy	Li et al. (2020)
Vegetables	Microscope	Chen et al. (2020)
Lettuce	SEM	Gao et al. (2021)
Sewage sludge	FAIR	Raju et al. (2020)
Tomato field	Raman Spectroscopy	Li et al. (2020), Sharma et al. (2023)
Cucumber	SEM	Li et al. (2021a)
Carrot (*D. carota*)	SEM and EDAX	Conti et al. (2020)
Apple (*M. domestica*)	SEM	
Lettuce		
Pear (*P. communis*)	EDAX and SEM	
Broccoli (*B. oleracea italica*)	EDAX and SEM	
Maize grown in hydroponics	IRMS	Urbina et al. (2020)

21.6 HUMAN HEALTH IMPLICATIONS

According to the World Health Organization (WHO), eating a fruit- and vegetable-rich diet is the best strategy for maintaining good health (Renzella et al., 2018). However, fruits grown in contaminated soil are unclean and pose health dangers to people.

This paper examines the potential pathways through which MPs can enter the food chain from agricultural soils and discusses the risks to human health.

Plastic additives include PAEs, dibutyl phthalate (DPB), and diethyl phthalate (DEP). Due to plastics and plastic additives, these tiny fragments of different shapes and colors are present in all environmental compartments. As regards their characteristics, PAEs can be introduced mainly through ingestion, inhalation, and dermal absorption. They can accumulate in the human body, where they have already been identified in blood, amniotic fluid, and urine (Arrigo et al., 2023).

Recent research results have led to defining several PAEs as "environmental hormones" because they interrupt biological hormones (Kambia et al., 2001) to impact living creatures intensely. The long-term accumulation of PAEs in humans may interfere with the secretion of hormones and the normal function of the nerve and immune systems, causing significant damage.

The reports on PAEs' impact on human health, particularly erythrocytes and the reproductive system, are significant. Although PAEs are considered very useful for their properties and are used in various fields, their disproportionate use must be reduced. Phthalates are found in all environments and, as mentioned above, can penetrate the body and cause damage to various organs and tissues. It would be essential to continue research to understand the individual effects of each class of PAEs on specific blood cells, such as granulocytes and isolated red blood cells (anucleated and nucleated), and thus on other organisms in addition to mammals. More scientific research is needed on some specific classes of PAEs, such as DBP, DMP, and DEHP, because they are more widely used and present in the environment (Arrigo et al., 2023).

Human exposure to PAEs has been detected in their metabolites in the body fluids and tissues, which have been reported to be associated with high blood pressure, pregnancy loss, preterm birth, diabetes, enhanced insulin resistance, anti-androgenic effects, cardiovascular disease, hypertension, etc. (Hoppin et al., 2013; Shiue, 2014; Sun et al., 2014; Trasande et al., 2014; Whyatt et al., 2014; Bai et al., 2017).

21.7 FUTURE PERSPECTIVES

This section offers suggestions for future research directions and discusses the implications for policy and management strategies.

Improvements in plastic waste collection methods and landfills are required. The government makes and implements policies on incorporating bio-based and biodegradable plastics and rewards tax rebates on recycled plastic. Improvement in e-waste management and recycling is also required (Prata et al., 2019; Jayan et al., 2019).

RICE-wheat is a dominant cropping system in Punjab. In addition to grains, they produce 22 and 24 m tons of rice and wheat straw per annum. Of the total rice residue, 82% of rice straw is burnt, 2% is sold commercially for commercial purposes, 11% is recycled in the field, and 0.5% is used for composting. Burning of a straw causes loss of nutrients, amounting to about 7.6 m tons of carbon, 0.08 m tons of nitrogen (N), and 0.008 m tons of sulfur. There is a need to recycle/use straw to increase crop production and conserve natural resources. The potential of straw to increase crop production has been evaluated extensively. Its use as a mulching material under supra-optimal conditions has been beneficial as it reduces maximum soil temperature and conserves water. Jalota et al. (2007) reviewed that straw mulching improved the yield of field, vegetable, and horticultural crops in irrigated and rain-fed environments of Punjab. This practice also saved irrigation water and fertilizer nitrogen. The benefits of mulching are more significant in summer/kharif season and soils with low water retentivity and cost-effective practices. Straw mulches such as paddy straws, wheat straws, and some long grasses can properly eliminate plastic use in the field and act as a better alternative to plastic mulches. Straw mulching of crops has been helpful in both irrigated and rain-fed environments. Biological mulching (straw mulching) as a potential and better alternative to plastic mulching in agricultural practices shown below (Figure 21.3):

Straw mulching is profitable because it escalates crop yields and minimizes the need for irrigation water and fertilizer N.

Chitosan, a biopolymer, is used in the formulation of green nanoparticles because of its simple production, inexpensive, and biodegradable,

FIGURE 21.3 Biological mulching (straw mulching) in agricultural practices.

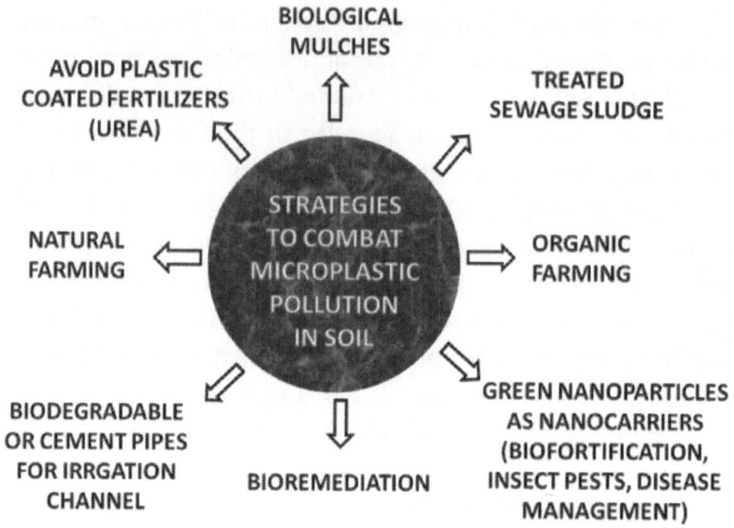

FIGURE 21.4 Strategies to combat microplastic pollution in agricultural soil (Sharma et al., 2023).

which contributes to the reducing micro and nanoplastics in crop production systems (Fakharzadeh et al., 2020). In horticulture, the use of green nanoparticles is an emerging sector, such as fertilization of nutrients and management of plant diseases and insect pests (Fakharzadeh et al., 2020; Deshpande et al., 2017; Campos et al., 2018) (Figure 21.4).

21.8 CONCLUSION

The study demonstrates that MPs significantly impair soil health, hinder nutrient recycling, and threaten agricultural productivity. The alteration in soil properties, such as decreased porosity and disrupted water dynamics, directly impacts plant growth and reduces crop yields. Furthermore, MPs facilitate the transfer of harmful pollutants into the soil ecosystem, complicating the interactions between soil biota and their natural functions. The findings underscore the necessity for immediate action to address microplastic pollution in agricultural landscapes. This involves enhancing waste management practices, developing biodegradable alternatives, and implementing stringent plastic-use regulations. Future research should focus on developing effective remediation techniques and exploring the long-term ecological impacts of MPs. These efforts are crucial for ensuring sustainable agricultural practices and the long-term health of soil ecosystems.

REFERENCES

Adeel, M., Ma, C., Ullah, S., Rizwan, M., Hao, Y., Chen, C., Jilani, G., Shakoor, N., Li, M., Wang, L., Tsang, D.C.W., Rinklebe, J., Rui, Y., & Xing, B. (2019). Exposure to nickel oxide nanoparticles insinuates physiological, ultrastructural and oxidative damage: A life cycle study on *Eisenia fetida*. *Environmental Pollution*, *254*, 113032.

Adeel, M., Shakoor, N., Ahmad, M.A., White, J.C., Jilani, G., & Rui, Y. (2021). Bioavailability and toxicity of nanoscale/bulk rare earth oxides in soil: Physiological and ultrastructural alterations in *Eisenia fetida*. *Environmental Science: Nano*, *8*, 1654–1666.

Adeel, M., Tingting, J., Hussain, T., He, X., Ahmad, M.A., Irshad, M.K., & Rui, Y. (2020). Bioaccumulation of ytterbium oxide nanoparticles insinuate oxidative stress, inflammatory, and pathological lesions in ICR mice. *Environmental Science and Pollution Research*, *27*, 32944–32953.

Adeel, M., Zain, M., Shafi, J., Ahmad, M., Rizwan, M., Shafiq, M., & Rui, Y. (2018). Conservation agriculture, a way to conserve soil carbon for sustainable agriculture productivity and mitigating climate change: A review. *Fresenius Environmental Bulletin*, *27*, 6297–6308.

Ahmad, R. (2018). Basics of free radicals and antioxidants. In: Ahmad, F. (ed) *Free Radicals, Antioxidants and Diseases*. IntechOpen, London, pp. 1–4.

Allen, D., Allen, S., Abbasi, S., Baker, A., Bergmann, M., Brahney, J., Butler, T., Duce, R.A., Eckhardt, S., Evangeliou, N., Jickells, T., Kanakidou, M., Kershaw, P., Laj, P., Levermore, J., Li, D., Liss, P., Liu, K., Mahowald, N., Masque, P., Materić, D., Mayes, A.G., McGinnity, P., Osvath, I., Prather, K.A., Prospero, J.M., Revell, L.E., Sander, S.G., Shim, W.J., Slade, J., Stein, A., Tarasova, O., & Wright, S. (2022). Microplastics and nanoplastics in the marine-atmosphere environment. *Nature Reviews Earth & Environment*, *3*, 393–405. https://doi.org/10.1038/s43017-022-00292-x

Allen, S., Allen, D., Phoenix, V., Le Roux, G., Jiménez, P.D., Simonneau, A., Binet, S., & Galop, D. (2019). Atmospheric transport and deposition of microplastics in a remote mountain catchment. *Nature Geoscience*, *12*, 339–344.

Arrigo, F., Impellitteri, F., Piccione, G., & Faggio, C. (2023). Phthalates and their effects on human health: Focus on erythrocytes and the reproductive system. *Comparative Biochemistry and Physiology Part C: Toxicology & Pharmacology*, *270*, 109645. https://doi.org/10.1016/j.cbpc.2023.109645

Azeem, I., Adeel, M., Ahmad, M.A., Shakoor, N., Jiangcuo, G.D., Azeem, K., Ishfaq, M., Shakoor, A., Ayaz, M., Xu, M., & Rui, Y. (2021). Uptake and accumulation of nano/microplastics in plants: A critical review. *Nanomaterials*, *11*(11), 2935. https://doi.org/10.3390/nano11112935

Azeem, I., Khan, A.A., Bibi, H., Azeem, K., Naz, F., Khan, K., Shah, S.A.A., & Shah, K.J.P. (2019). Nitrogen sources and Bioaab inclined yield and nutrients uptake specifically in *Triticum aestivum*. *Pure and Applied Biology*, *8*, 2311–2325.

Bai, P.Y., Wittert, G., Taylor, A.W., Martin, S.A., Milne, R.W., Jenkins, A.J., Januszewski, A.S., & Shi, Z. (2017). The association between total phthalate concentration and non-communicable diseases and chronic inflammation in South Australian urban dwelling men. *Environmental Research*, *158*, 366–372.

Bian, W., An, L., Zhang, S., Feng, J., Sun, D., Yao, Y., Shen, T., Yang, Y., & Zhang, M. (2022) The long-term effects of microplastics on soil organomineral complexes and bacterial communities from controlled-release fertilizer residual coating. *Journal of Environmental Management, 304*, 114193.

Baumhardt, R., Jones, O.R., & Schwartz, R.C. (2008). Long-term effects of profile-modifying deep plowing on soil properties and crop yield. *Soil Science Society of America Journal, 72*, 677–682.

Beare, M.H., Coleman, D.C., Crossley, D.A., Hendrix, P.F., & Odum, E.P. (1995). A hierarchical approach to evaluating the significance of soil biodiversity to biogeochemical cycling. *Plant and Soil, 170*, 5–22.

Bender, S.F., & van der Heijden, M.G.A. (2015). Soil biota enhance agricultural sustainability by improving crop yield and nutrient uptake and reducing nitrogen leaching losses. *Journal of Applied Ecology, 52*, 228–239.

Beriot, N., Peek, J., Zornoza, R., Geissen, V., & Lwanga, E.H. (2021). Low-density microplastics detected in sheep feces and soil: A case study from intensive vegetable farming in Southeast Spain. *Science of the Total Environment, 755*, 142653.

Bläsing, M., & Amelung, W. (2018). Plastics in soil: Analytical methods and possible sources. *Science of the Total Environment, 612*, 422–435. https://doi.org/10.1016/j.scitotenv.2017.08.086

Blöcker, L., Watson, C., & Wichern, F. (2020). Living in the plastic age-different short-term microbial response to microplastics addition to arable soils with contrasting soil organic matter content and farm management legacy. *Environmental Pollution, 267*, 115468.

Braun, M., Mail, M., Heyse, R., & Amelung, W. (2021). Plastic in compost: Prevalence and potential input into agricultural and horticultural soils. *Science of the Total Environment, 760*, 143335.

Brodhagen, M., Goldberger, J.R., Hayes, D., Inglis, D.A., Marsh, T.L., & Miles, C. (2017). Policy considerations for limiting unintended residual plastic in agricultural soils. *Environmental Science & Policy, 69*, 81–84.

Brown, R.W., Chadwick, D.R., Thornton, H., Marshall, M.R., Bei, S., Distaso, M.A., Bargiela, R., Marsden, K.A., Clode, P.L., Murphy, D.V., & Pagella, S. (2022). Field application of pure polyethylene microplastic has no significant short-term effect on soil biological quality and function. *Soil Biology and Biochemistry, 165*, 108496.

Bruinsma, J. (2003). *World Agriculture: Towards 2015/2030: An FAO Perspective*. FAO.

Cai, Q.Y., Mo, C.H., Zeng, Q.Y., Wu, Q.T., Férard, J.F., & Antizar-Ladislao, B. (2008). Potential of Ipomoea aquatica cultivars in phytoremediation of soils contaminated with di-n-butyl phthalate. *Environmental and Experimental Botany, 62*, 205–211.

Campos, E.V., Proença, P.L., Oliveira, J.L., Melville, C.C., Della Vecchia, JF, De Andrade, D.J., & Fraceto, L.F. (2018). Chitosan nanoparticles functionalized with β-cyclodextrin: A promising carrier for botanical pesticides. *Scientific Reports, 8*, 2067.

Cao, D., Wang, X., Luo, X., Liu, G., & Zheng, H. (2017). Effects of polystyrene microplastics on the fitness of earthworms in an agricultural soil. *IOP Conference Series: Earth and Environmental Science, 61*, 12148. https://dx.doi.org/10.1088/1755-1315/61/1/012148

Carpenter, E.J., & Smith Jr, K.L. (1972). Plastics on the Sargasso Sea surface. *Science, 175*(4027), 1240–1241.

Cey, E.E., Rudolph, D.L., & Passmore, J. (2009). Influence of macroporosity on preferential solute and colloid transport in unsaturated field soils. *Journal of Contaminant Hydrology, 107*, 45–57.

Chai, B., Wei, Q., She, Y., Lu, G., Dang, Z., & Yin, H. (2020). Soil microplastic pollution in an e-waste dismantling zone of China. *Waste Management, 118*, 291–301.

Chen, Y., Leng, Y., Liu, X., & Wang, J. (2020). Microplastic pollution in vegetable farmlands of suburb Wuhan, central China. *Environmental Pollution, 257*, 113449.

Conti, G.O., Ferrante, M., Banni, M., Favara, C., Nicolosi, I., Cristaldi, A., & Zuccarello, P. (2020). Micro-and nano-plastics in edible fruit and vegetables. The first diet risk assessment for the general population. *Environmental Research, 187*, 109677.

Corcoran, E. (2010). *Sick Water? The Central Role of Wastewater Management in Sustainable Development: A Rapid Response Assessment.* United Nations Environment Programme; Arendal, Norway.

De Souza Machado, A.A., Kloas, W., Zarfl, C., Hempel, S., & Rillig, M.C. (2018). Microplastics as an emerging threat to terrestrial ecosystems. *Global Change Biology, 24*(4), 1405–1416.

de Souza Machado, A.A., Lau, C.W., Kloas, W., Bergmann, J., Bachelier, J.B., Faltin, E., Becker, R., Görlich, A.S., & Rillig, M.C. (2019). Microplastics can change soil properties and affect plant performance. *Environmental Science & Technology, 53*(10), 6044–6052. https://doi.org/10.1021/acs.est.9b01339

De Souza Machado, A.A., Lau, C.W., Till, J., Kloas, W., Lehmann, A., Becker, R., & Rillig, M.C. (2018). Impacts of microplastics on the soil biophysical environment. *Environmental Science and Technology, 52*, 9656–9665.

de Souza Machado, A.A., Lau, C.W., Kloas, W., Bergmann, J., Bachelier, J.B., Faltin, E., Becker, R., Görlich, A.S., & Rillig, M.C. (2019). Microplastics can change soil properties and affect plant performance. *Environmental Science & Technology, 53*, 6044–6052.

Deshiikan, S.R., Eschenazi, E., & Papadopoulos, K.D. (1998). *Colloids and Surfaces A: Physicochemical and Engineering Aspects*, Elsevier: Amsterdam, The Netherlands, Volume 145, p. 93.

Deshpande, P., Dapkekar, A., Oak, MD, Paknikar, K.M., & Rajwade, J.M. (2017). Zinc complexed chitosan/TPP nanoparticles: A promising micronutrient nanocarrier suited for foliar application. *Carbohydrate Polymers, 165*, 394–401.

Dong, Y., Gao, M., Liu, X., Qiu, W., & Song, Z. (2020). The mechanism of polystyrene microplastics to affect arsenic volatilization in arsenic-contaminated paddy soils. *Journal of Hazardous Materials, 398*, 122896.

Dris, R., Gasperi, J., Saad, M., Miranda, C., & Tassin, B. (2016). Synthetic fibers in atmospheric fallout: A source of microplastics in the environment? *Marine Pollution Bulletin, 104*, 290–293.

Duan, K., Cui, M., Wu, Y., Huang, X., Xue, A., Deng, X., & Luo, L. (2018). Effect of dibutyl phthalate on the tolerance and lipid accumulation in the green microalgae Chlorella vulgaris. *Bulletin of Environmental Contamination and Toxicology, 101*, 338–343.

Dueck, T.A., Van Dijk, C.J., David, F., Scholz, N., & Vanwalleghem, F. (2003). Chronic effects of vapor phase di-n-butyl phthalate (DBP) on six plant species. *Chemosphere, 53*, 911–920.

Dutta, S., Haggerty, D.K., Rappolee, D.A., & Ruden, D.M. (2020). Phthalate exposure and long-term epigenomic consequences: A review. *Frontiers in Genetics, 11*, 499019. https://doi.org/10.3389/fgene.2020.00405

El Hayany, B., El Fels, L., Quénéa, K., Dignac, M.F., Rumpel, C., Gupta, V.K., Hafidi, M. (2020). Fluor staining-image analysis, Raman spectroscopy, and pyrolysis-GC/MS revealed microplastics from lagoons sludge to composts. *Journal of Environmental Management, 275*, 111249.

Fakharzadeh, S., Hafizi, M., Baghaei, M.A., Etesami, M., Khayamzadeh, M., Kalanaky, S., & Nazaran, M.H. (2020). Using nanochelating technology for biofortification and yield increase in rice. *Scientific Reports, 10*, 4351.

Gao, M., Liu, Y., & Song, Z. (2019a). Effects of polyethylene microplastic on the phytotoxicity of di-n-butyl phthalate in lettuce (Lactuca sativa L. var. ramosa Hort). *Chemosphere, 237*, 124482. https://doi.org/10.1016/j.chemosphere.2019.124482

Gao, M., Liu, Y., Dong, Y., & Song, Z. (2019b). Physiological responses of wheat planted in favor-aquic soils to di (2-ethylhexyl) and di-butyl phthalates. *Environmental Pollution, 244*, 774–782.

Gao, M., Xu, Y., Dong, Y., Song, Z., & Liu, Y. (2019c). Accumulation and metabolism of di (n-butyl) phthalate (DBP) and di (2-ethylhexyl) phthalate (DEHP) in mature wheat tissues and their effects on detoxification and the antioxidant system in grain. *Science of the Total Environment, 697*, 133981.

Gao, M., Xu, Y., Liu, Y., Wang, S., Wang, C., Dong, Y., & Song, Z. (2021). Effect of polystyrene on di-butyl phthalate (DBP) bioavailability and DBP-induced phytotoxicity in lettuce. *Environmental Pollution, 268*, 115870.

Gershater, M.C., & Edwards, R. (2007). Regulating biological activity in plants with carboxylesterases. *Plant Science, 173*(6), 579–588.

Godoy, V., Blázquez, G., Calero, M., Quesada, L., & Martín-Lara, M. (2019). The potential of microplastics as carriers of metals. *Environmental Pollutions, 255*, 113363.

Godoy, V., Martínez-Férez, A., Martín-Lara, M., Vellido-Pérez, J., Calero, M., & Blázquez, G. (2020). Microplastics as vectors of chromium and lead during dynamic simulation of the human gastrointestinal tract. *Sustainability, 12*, 4792.

Gu, S., Zheng, H., Xu, Q., Sun, C., Shi, M., Wang, Z., & Li, F. (2017). Comparative toxicity of the plasticizer dibutyl phthalate to two freshwater algae. *Aquatic Toxicology, 191*, 122–130.

Guo, Z., Li, P., Yang, X., Wang, Z., Lu, B., Chen, W., Wu, Y., Li, G., Zhao, Z., Liu, G., Ritsema, C., Geissen, V., & Xue, S. (2022). Soil texture is important in determining how microplastics affect soil hydraulic characteristics. *Environment International, 165*, 107293. https://doi.org/10.1016/j.envint.2022.107293

Haslam, R., Raveton, M., Cole, D.J., Pallett, K.E., & Coleman, J.O. (2001). The identification and properties of apoplastic carboxylesterases from wheat that catalyse deesterification of herbicides. *Pesticide Biochemistry and Physiology, 71*(3), 178–189.

He, L., Gielen, G., Bolan, N.S., Zhang, X., Qin, H., Huang, H., & Wang, H. (2015). Contamination and remediation of phthalic acid esters in agricultural soils in China: A review. *Agronomy for Sustainable Development*, *35*, 519–534.

Heinze, W.M. (2019). Nanoplastic transport in soils by advection and bioturbation. Mater's Thesis, Swedish University of Agricultural Sciences, Uppsala, Sweden.

Hoppin, J.A., Jaramillo, R., London, S.J., Bertelsen, R.J., Salo, P.M., Sandler, D.P., & Zeldin, D.C. (2013). Phthalate exposure and allergy in the US population: Results from NHANES 2005–2006. *Environmental Health Perspectives, 121*, 1129–1134.

Horton, A.A., Walton, A., Spurgeon, D.J., Lahive, E., & Svendsen, C. (2017). Microplastics in freshwater and terrestrial environments: Evaluating the current understanding to identify the knowledge gaps and future research priorities. *Science of the Total Environment*, *586*, 127–141.

Huang, Q., Wang, Q., Tan, W., Song, G., Lu, G., & Li, F. (2006). Biochemical responses of two typical duckweeds exposed to dibutyl phthalate. *Journal of Environmental Science and Health, Part A*, *41*, 1615–1626.

Huang, Y., Liu, Q., Jia, W., Yan, C., & Wang, J. (2020). Agricultural plastic mulching is a source of microplastics in the terrestrial environment. *Environmental Pollution*, *260*, 114096. https://doi.org/10.1016/j.envpol.2020.114096

Huerta Lwanga, E., Mendoza Vega, J., Ku Quej, V., de Los Angels Chi, A., del Cid, L.S., Chi, C., Segura, G.E., Gertsen, H., Salánki, T., van der Ploeg, M., Koelmans, A.A., & Geissen, V. Field evidence for transfer of plastic debris along a terrestrial food chain. *Scientific Reports*, *7*, 1–7.

Inui, H., Wakai, T., Gion, K., Kim, Y.S., & Eun, H. (2008). Differential uptake for dioxin-like compounds by zucchini subspecies. *Chemosphere*, *73*(10), 1602–1607.

Jalota, S.K., Khera, R., Arora, V.K., & Beri, V. (2007). Benefits of straw mulching in crop production: A review. *Journal of Research: Punjab Agricultural University*, *44*(2), 104–107.

Jayan, H., Leena, M.M., Sundari, S.S., Moses, J.A., & Anandharamakrishnan, C. (2019). Improvement of bioavailability for resveratrol through encapsulation in zein using electrospraying technique. *Journal of Functional Foods*, *57*, 417–424.

Jiménez, E.M., Leifheit, E.F., Plaza, C., Feng, L., Bergmann, J., Wulf, A., Lehmann, A., & Rillig, M.C. (2022). Effects of microplastics on crop nutrition in fertile soils and interaction with arbuscular mycorrhizal fungi. *Journal of Sustainable Agriculture and Environment*, *1*(1), 66–72. https://doi.org/10.1002/sae2.12006

Julienne, F., Delorme, N., & Lagarde, F. (2019). From macroplastics to microplastics: Role of water in the fragmentation of polyethylene. *Chemosphere*, *236*, 124409.

Kader, M.A., Senge, M., Mojid, M.A., & Ito, K. (2017). Recent advances in mulching materials and methods for modifying soil environment. *Soil and Tillage Research*, *168*, 155–166. https://doi.org/10.1016/j.still.2017.01.001

Kambia, K., Dine, T., Gressier, B., Germe, A.F., Luyckx, M., Brunet, C., Michaud, L., & Gottrand, F. (2001). High-performance liquid chromatographic method for the determination of di (2-ethylhexyl) phthalate in total parenteral nutrition and plasma. *Journal of Chromatography A*, *755*, 297–303.

Kanojia, A., & Dijkwel, P.P. (2018). Abiotic stress responses are governed by reactive oxygen species and age. *Annual Plant Reviews Online*, *1*, 1–32.

Kibblewhite, M.G., Chambers, B.J., & Goulding, K.W.T. (2016). How good is the evidence to support investment in soil protection? *Soil Use and Management*, *32*, 172–182.

Kim, S.W., Jeong, S.W., & An, Y.J. (2021). Microplastics disrupt accurate soil organic carbon measurement based on the chemical oxidation method. *Chemosphere*, *276*, 130178.

Korte, F., Kvesitadze, G., Ugrekhelidze, D., Gordeziani, M., Khatisashvili, G., Buadze, O., Zaalishvili, G., & Coulston, F. (2000). Organic toxicants and plants. *Ecotoxicology and Environmental Safety*, *47*, 1–26.

Koyuncuoğlu, P., & Erden, G. (2021). Sampling, pre-treatment, and identification methods of microplastics in sewage sludge and their effects in agricultural soils: a review. *Environmental Monitoring and Assessment*, *193*, 1–28.

Kumari, A. and Kaur, R. (2017). Germination and early growth toxicity to barley seedlings (Hordeum vulgare L.) under di-n-butyl phthalate (DBP) stress. *Journal of Pharmaceutical Sciences and Research*, *9*(12), 2361–2366.

Kumari, A., & Kaur, R. (2018). Evaluation of benzyl-butyl phthalate induced germination and early growth vulnerability to barley seedlings (Hordeum vulgare L.). *Indian Journal of Ecology*, *45*, 174–177.

Kumari, A., & Kaur, R. (2019). Modulation of biochemical and physiological parameters in Hordeum vulgare L. seedlings under the influence of benzyl-butyl phthalate. *PeerJ*, *7*, e6742.

Kumari, A., & Kaur, R. (2020). A review of morpho-physiological traits of plants under phthalates stress and insights into their uptake and translocation. *Plant Growth Regulation*, *91*(3), 327–347. https://doi.org/10.1007/s10725-020-00625-0

Kuzyakov, Y., & Blagodatskaya, E. (2015). Microbial hotspots and hot moments in soil: Concept and review. *Soil Biology and Biochemistry*, *83*, 184–199.

Kvesitadze, G., Khatisashvili, G., Sadunishvili, T., & Ramsden, J.J. (2006). *Biochemical Mechanisms of Detoxification in Higher Plants: The Basis of Phytoremediation*. Springer, New York.

Lambert, S., & Wagner, M. (2016). Characterisation of nanoplastics during the degradation of polystyrene. *Chemosphere*, *145*, 265–268.

Lei, L., Liu, M., Song, Y., Lu, S., Hu, J., Cao, C., & He, D. (2018). Polystyrene (nano) microplastics cause size-dependent neurotoxicity, oxidative damage and other adverse effects in *Caenorhabditis elegans*. *Environmental Science: Nano*, *5*, 2009–2020.

Letcher, T.M. (2020). Introduction to plastic waste and recycling. In: Letcher, T.M. (ed) *Plastic Waste and Recycling* (pp. 3–12). Academic Press.

Li, F.M., Wu, M., Yao, Y., Zheng, X., Zhao, J., Wang, Z.Y., & Xing, B.S. (2015). Inhibitory effects and oxidative target site of dibutyl phthalate on Karenia brevis. *Chemosphere*, *132*, 32–39.

Li, H., Liu, L., Xu, Y. and Zhang, J., 2022. Microplastic effects on soil system parameters: a meta-analysis study. *Environmental Science and Pollution Research*, 22, 11027–11038.

Li, Z., Li, Q., Li, R., Zhou, J., & Wang, G. (2021a). The distribution and impact of polystyrene nanoplastics on cucumber plants. *Environmental Science and Pollution Research, 28*, 16042–16053.

Li, H.Z., Zhu, D., Lindhardt, J.H., Lin, S.M., Ke, X., & Cui, L. (2021b). Long-term fertilization history alters the effects of microplastics on soil properties, microbial communities, and functions in diverse farmland ecosystems. *Environmental Science and Technology, 55*, 4658–4668.

Li, X., Chen, L., Mei, Q., Dong, B., Dai, X., Ding, G., & Zeng, E.Y. (2018). Microplastics in sewage sludge from the wastewater treatment plants in China. *Water Research, 142*, 75–85.

Li, X., Mei, Q., Chen, L., Zhang, H., Dong, B., Dai, X., He, C., & Zhou, J. (2019). Enhancement in adsorption potential of microplastics in sewage sludge for metal pollutants after the wastewater treatment process. *Water Research, 157*, 228–237.

Li, Z., Li, Q., Li, R., Zhao, Y., Geng, J., & Wang, G. (2020). Physiological responses of lettuce (*Lactuca sativa* L.) to microplastic pollution. *Environmental Science and Pollution Research, 27*, 30306–30314.

Li, Z., Li, Q., Li, R., Zhou, J., & Wang, G. (2021b). The distribution and impact of polystyrene nanoplastics on cucumber plants. *Environmental Science and Pollution Research, 28*, 16042–16053.

Lin, H., Tao, S., Zuo, Q., & Coveney, R.M. (2007). Uptake of polycyclic aromatic hydrocarbons by maize plants. *Environmental Pollution, 148*, 614–619.

Liu, E., He, W., & Yan, C.R. (2014). 'White revolution' to 'white pollution'—Agricultural plastic film mulch in China. *Environmental Research Letters, 9*, 091001.

Lozano, Y.M., Aguilar-Trigueros, C.A., Onandia, G., Maaß, S., Zhao, T., & Rillig, M.C. (2021). Effects of microplastics and drought on soil ecosystem functions and multifunctionality. *Journal of Applied Ecology, 58*, 1–9.

Lu, S., Hao, J., Yang, H., Chen, M., Lian, J., Chen, Y., Brown, R.W., Jones, D.L., Wan, Z., Wang, W., & Chang, W. (2023). Earthworms mediate the influence of polyethylene (PE) and polylactic acid (PLA) microplastics on soil bacterial communities. *Science of the Total Environment, 905*, 166959.

Luo, Y., Zhou, Q., Zhang, H.-B., Pan, X., Tu, C., Li, L., & Yang, J. (2018). Pay attention to research on soil microplastic pollution to prevent ecological and food chain risks. *Bulletin of the Chinese Academy of Sciences, 33*, 1021–1030.

Lwanga, E.H., Gertsen, H., Gooren, H., Peters, P., Salánki, T., Van Der Ploeg, M., Besseling, E., Koelmans, A.A., & Geissen, V. (2016). Microplastics in the terrestrial ecosystem: Implications for Lumbricus terrestris (Oligochaeta, Lumbricidae). *Environmental Science and Technology, 50*, 2685–2691.

Ma, T., Christie, P., Teng, Y., & Luo, Y. (2013). Rape (Brassica chinensis L.) seed germination, seedling growth, and physiology in soil polluted with di-n-butyl phthalate and bis(2-ethylhexyl) phthalate. *Environmental Science and Pollution Research, 20*, 5289–5298.

Ma, T.T., Christie, P., Luo, Y.M., & Teng, Y. (2014). Physiological and antioxidant responses of germinating mung bean seedlings to phthalate esters in soil. *Pedosphere, 24*, 107–115.

Maity, S., & Pramanick, K. (2020). Perspectives and challenges of micro/nanoplastics-induced toxicity with particular reference to phytotoxicity. *Global Change Biology, 26*, 3241–3250.

Mankidy, R., Wiseman, S., Ma, H., & Giesy, J.P. (2013). Biological impact of phthalates. *Toxicology Letters, 217*(1), 50–58. https://doi.org/10.1016/j.toxlet.2012.11.025

Mason, S.A., Welch, V.G., & Neratko, J. (2018). Synthetic polymer contamination in bottled water. *Frontiers in Chemistry, 6*, 407.

Melin, C., & Egneus, H. (1983). Effects of di-n-butyl phthalate on growth and photosynthesis in algae and isolated organelles from higher plants. *Physiologia Plantarum, 59*, 461–466.

Millar, D.J., & Hannay, J.W. (1986). Phytotoxicity of phthalate plasticizers: 1. Diagnosis and commercial implications. *Journal of Experimental Botany, 37*, 883–897.

Mintenig, S., Int-Veen, I., Löder, M., Primpke, S., & Gerdts, G. (2017). Identification of microplastic in wastewater treatment plants effluents using focal plane array-based micro-Fourier-transform infrared imaging. *Water Research, 108*, 365–372.

Miyazaki, J., Kuriyama, Y., Tokumoto, H., Konishi, Y., & Nomura, T. (2015). Cytotoxicity and behavior of polystyrene latex nanoparticles to budding yeast. *Colloids and Surfaces A: Physicochemical and Engineering Aspects, 469*, 287–293.

Mondal, T., Mondal, S., Ghosh, S.K., Pal, P., Soren, T., Pandey, S., & Maiti, T.K. (2022). Phthalates - A family of plasticizers, their health risks, phytotoxic effects, and microbial bioaugmentation approach. *Environmental Research, 214*(3), 114059. https://doi.org/10.1016/j.envres.2022.114059

Müller, J., & Kördel, W. (1993). Occurrence and fate of phthalates in soil and plants. *Science of the Total Environment, 134*, 431–437.

Ng, K., & Obbard, J. (2006). Prevalence of microplastics in Singapore's coastal marine environment. *Marine Pollution Bulletin, 52*, 761–767.

Nizzetto, L., Futter, M., & Langaas, S. (2016). Are agricultural soils dumps for microplastics of urban origin? *Environmental Science and Technology, 50*, 10777–10779.

Nomura, T., Tani, S., Yamamoto, M., Nakagawa, T., Toyoda, S., Fujisawa, E., Yasui, A., & Konishi, Y. (2016). Cytotoxicity and colloidal behavior of polystyrene latex nanoparticles toward filamentous fungi in isotonic solutions. *Chemosphere, 149*, 84–90.

O'Connor, D., Pan, S., Shen, Z., Song, Y., Jin, Y., Wu, W.M., & Hou, D. (2019). Microplastics undergo accelerated vertical migration in sand soil due to small size and wet-dry cycles. *Environmental Pollution, 249*, 527–534.

OECD. (2022). *Global Plastics Outlook: Policy Scenarios to 2060*. OECD Publishing. https://doi.org/10.1787/aa1edf33-en

Panno, S.V., Kelly, W.R., Scott, J., Zheng, W., McNeish, R.E., Holm, N., Hoellein, T.J., & Baranski, E.L. (2019). Microplastic contamination in karst groundwater systems. *Ground Water, 57*, 189–196.

Peccia, J., & Westerhoff, P. (2015). We should expect more out of our sewage sludge. *Environmental Science and Technology, 49*, 8271–8276.

Pelamatti, T., Rios-Mendoza, L.M., Hoyos-Padilla, E.M., Galván-Magaña, F., De Camillis, R., Marmolejo-Rodríguez, A.J., & González-Armas, R. (2021). Contamination knows no borders: Toxic organic compounds pollute plastics in the biodiversity hotspot of Revillagigedo Archipelago National Park, Mexico. *Marine Pollution Bulletin, 170*, 112623.

Piehl, S., Leibner, A., Löder, M.G., Dris, R., Bogner, C., & Laforsch, C. (2018). Identification and quantification of macro- and microplastics on agricultural farmland. *Scientific Reports, 8*, 1–9.

Pinstrup-Andersen, P., 2009. Food security: definition and measurement. *Food security, 1*, pp.5-7.

Prata, J.C., Silva, A.L.P., Da Costa, J.P., Mouneyrac, C., Walker, T.R., Duarte, A.C., & Rocha-Santos, T. (2019). Solutions and integrated strategies for the control and mitigation of plastic and microplastic pollution. *International Journal of Environmental Research and Public Health, 16*, 2411.

Qiang, L., Hu, H., Li, G., Xu, J., Cheng, J., Wang, J., & Zhang, R. (2023). Plastic mulching, and occurrence, incorporation, degradation, and impacts of polyethylene microplastics in agroecosystems. *Ecotoxicology and Environmental Safety, 263*, 115274. https://doi.org/10.1016/j.ecoenv.2023.115274

Qin, W., Hu, C., & Oenema, O. (2015). Soil mulching significantly enhances yields and water and nitrogen use efficiencies of maize and wheat: A meta-analysis. *Scientific Reports, 5*, 16210–16213. https://doi.org/10.1038/srep16210

Qing, X., Yutong, Z., & Shenggao, L. (2015). Assessment of heavy metal pollution and human health risk in urban soils of steel industrial city (Anshan), Liaoning, Northeast China. *Ecotoxicology and Environmental Safety, 120*, 377–385.

Raju, S., Carbery, M., Kuttykattil, A., Senthirajah, K., Lundmark, A., Rogers, Z., & Palanisami, T. (2020). Improved methodology to determine the fate and transport of microplastics in a secondary wastewater treatment plant. *Water Research, 173*, 115549.

Ren, X., Tang, J., Liu, X., & Liu, Q. (2020). Effects of microplastics on greenhouse gas emissions and the microbial community in fertilized soil. *Environmental Pollution, 256*, 113347.

Renzella, J., Townsend, N., Jewell, J., Breda, J., Roberts, N., Rayner, M., & Wickramasinghe, K. (2018). *What National and Subnational Interventions and Policies Based on Mediterranean and Nordic Diets Are Recommended or Implemented in the WHO European Region, and Is There Evidence of Effectiveness in Reducing Noncommunicable Diseases?* World Health Organization, Regional Office for Europe: Geneva, Switzerland.

Rillig, M.C., Ingraffia, R., & de Souza Machado, A.A. (2017a). Microplastic incorporation into soil in agroecosystems. *Frontiers in Plant Science, 8*, 1805.

Rillig, M.C., Leifheit, E., & Lehmann, J. (2021). Microplastic effects on soil carbon cycling processes. *PLoS Biology, 19*, 1–9.

Rillig, M.C., Muller, L.A., & Lehmann, A. (2017c). Soil aggregates as massively concurrent evolutionary incubators. *The ISME Journal, 11*(9), 1943–1948.

Rillig, M.C., Ziersch, L., & Hempel, S. (2017b). Microplastic transport in soil by earthworms. *Scientific Reports, 7*, 1–6.

Rizwan, M., Mostofa, M.G., Ahmad, M.Z., Imtiaz, M., Mehmood, S., Adeel, M., Dai, Z., Li, Z., Aziz, O., Zhang, Y., & Tu, S. (2018). Nitric oxide induces rice tolerance to excessive nickel by regulating nickel uptake, reactive oxygen species detoxification, and defense-related gene expression. *Chemosphere, 191*, 23–35.

Rodriguez-Seijo, A., Lourenço, J., Rocha-Santos, T.A.P., da Costa, J., Duarte, A.C., Vala, H., & Pereira, R. (2017). Histopathological and molecular effects of microplastics in Eisenia Andrei Bouché. *Environmental Pollution, 220*(Pt A), 495–503. https://dx.doi.org/10.1016/j.envpol.2016.09.092

Saarma, K., Tarkka, M.T., Itävaara, M., & Fagerstedt, K.V. (2003). Heat shock protein synthesis is induced by diethyl phthalate but not by di(2-ethylhexyl) phthalate in radish (Raphanus sativus). *Journal of Plant Physiology, 160*, 1001–1010.

Sandermann Jr, H. (1994). Higher plant metabolism of xenobiotics: the 'green liver' concept. *Pharmacogenetics and Genomics, 4*(5), 225–241.

Saud, S., Li, X., Chen, Y., Zhang, L., Fahad, S., Hussain, S., Sadiq, A., & Chen, Y. (2014). Silicon application increases drought tolerance of Kentucky bluegrass by improving plant water relations and morphophysiological functions. *The Scientific World Journal*, 2014, 1–10.

Scheurer, M., & Bigalke, M. (2018). Microplastics in Swiss floodplain soils. *Environmental Science and Technology, 52*, 3591–3598.

Sharma, R., & Kaur, R. (2019). Diallyl phthalate-triggered oxidative stress in Spirodela polyrhiza L. Schleiden: Physiological effects and role of the antioxidant defense system. *International Journal of Environmental Science and Technology, 17*(1). https://doi.org/10.1007/s13762-019-02491-4

Sharma, U., Sharma, S., Rana, V.S., Rana, N., Kumar, V., Sharma, S., Qadri, H., Kumar, V., & Bhat, S.A. (2023). Assessment of microplastics pollution on soil health and eco-toxicological risk in horticulture. *Soil Systems, 7*(1), 7. https://doi.org/10.3390/soilsystems7010007

Shiue, I. (2014). Higher urinary heavy metal, phthalate, and arsenic but not parabens concentrations in people with high blood pressure, US NHANES, 2011–2012. *International Journal of Environmental Research and Public Health, 11*, 5989–5999.

Singh, J., & Thakur, J.K. (2018). Photosynthesis and abiotic stress in plants. In: Vats, S. (ed) *Biotic and Abiotic Plant Stress Tolerance*. Springer, Singapore, pp. 27–46.

Singh, S.K.J., Prakash, G., & Bal, M. (2018). Assessment of diverse phthalate esters (PAEs) from irrigated agriculture soil under protected cultivation in IARI, New Delhi. *International Journal of Chemical Studies, 6*(3), 3432–3435.

Sintim, H., & Flury, M. (2017). Is biodegradable plastic mulch the solution to agriculture's plastic problem? *Environmental Science and Technology, 51*, 1068–1069.

Sun, Q., Cornelis, M.C., Townsend, M.K., Tobias, D.K., Eliassen, A.H., Franke, A.A., Hauser, R., & Hu, F.B. (2014). Association of urinary concentrations of bisphenol A and phthalate metabolites with risk of type 2 diabetes: A prospective investigation in the Nurses' Health Study (NHS) and NHSII Cohorts. *Environmental Health Perspectives, 122*, 616–623.

Teng, Y., Wu, J., Lu, S., Wang, Y., Jiao, X., & Song, L. (2014). Soil and soil environmental quality monitoring in China: A review. *Environment International, 69*, 177–199.

Thompson, R.C., Olsen, Y., Mitchell, R.P., Davis, A., Rowland, S.J., John, A.W., & Russell, A.E. (2004). Lost at sea: Where is all the plastic? *Science, 304*, 838.

Trasande, L., Sathyanarayana, S., & Trachtman, H. (2014). Dietary phthalates and low-grade albuminuria in US children and adolescents. *Clinical Journal of the American Society of Nephrology, 9*, 100–109.

Tuteja, N. (2007). Mechanisms of high salinity tolerance in plants. In: Häussinger, D., Sies, H. (eds) *Methods in Enzymology*. Academic Press, Cambridge, pp. 419–438.

UNW-DPC. (2013). *Proceedings of the UN-Water Project on the Safe Use of Wastewater in Agriculture*. United Nations University, Bonn.

Urbina, M.A., Correa, F., Aburto, F., & Ferrio, J.P. (2020). Adsorption of polyethylene microbeads and physiological effects on hydroponic maize. *Science of the Total Environment, 741*, 140216.

Vithanage, M., Ramanayaka, S., Hasinthara, S., & Navaratne, A. (2021). Compost as a carrier for microplastics and plastic-bound toxic metals into agroecosystems. *Current Opinion in Environmental Science & Health, 24*, 100297.

Virgin, H.I., Holst, A.M., & Mörner, J. (1981). Effect of di-n-butyl phthalate on the carotenoid synthesis in green plants. *Physiologia Plantarum, 53*, 158–163.

Wan, L., Cheng, H., Liu, Y., Shen, Y., Liu, G., & Su, X. (2023). Global meta-analysis reveals differential effects of microplastics on the soil ecosystem. *Science of the Total Environment, 867*, 161403.

Wang, J., Fang, R., Yuan, L., Yuan, G., Zhao, M., Zhu, S., Hou, J., Chen, G., & Wang, C. (2020d). Response of photosynthetic capacity and antioxidative system of chloroplast in two wucai (Brassica campestris L.) genotypes against chilling stress. *Physiology and Molecular Biology of Plants, 26*, 219–232.

Wang, J., Liu, X., Li, Y., Powell, T., Wang, X., Wang, G., & Zhang, P. (2019). Microplastics as contaminants in the soil environment: A mini-review. *Science of the Total Environment, 691*, 848–857.

Wang, J., Lv, S., Zhang, M., Chen, G., Zhu, T., Zhang, S., Teng, Y., Christie, P., & Luo, Y. (2016). Effects of plastic film residues on occurrence of phthalates and microbial activity in soils. *Chemosphere, 151*, 171–177.

Wang, F., Yang, W., Cheng, P., Zhang, S., Zhang, S., Jiao, W., & Sun, Y. (2019). Adsorption characteristics of cadmium onto microplastics from aqueous solutions. *Chemosphere, 235*, 1073–1080.

Wang, F., Zhang, X., Zhang, S., Zhang, S., & Sun, Y. (2020c). Interactions of microplastics and cadmium on plant growth and arbuscular mycorrhizal fungal communities in agricultural soil. *Chemosphere, 254*, 126791.

Wang, F., Zhang, X., Zhang, S., Zhang, S., Adams, C.A., & Sun, Y. (2020b). Effects of co-contamination of microplastics and Cd on plant growth and Cd accumulation. *Toxics, 8*, 36.

Wang, T., Yu, C., Chu, Q., Wang, F., Lan, T., & Wang, J. (2020a). Adsorption behavior and mechanism of five pesticides on microplastics from agricultural polyethylene films. *Chemosphere, 244*, 125491.

Weithmann, N., Möller, J.N., Löder, M.G.J., Piehl, S., Laforsch, C., & Freitag, R. (2018). Organic fertilizer is a vehicle for the entry of microplastic into the environment. *Science Advances*, *4*(4), eaap8060.

Whyatt, R.M., Perzanowski, M.S., Just, A.C., Rundle, A.G., Donohue, K.M., Calafat, A.M., Hoepner, L.A., Perera, F.P., & Miller, R.L. (2014). Asthma in inner-city children at 5–11 years of age and prenatal exposure to phthalates: The Columbia Center for Children's Environmental Health Cohort. *Environmental Health Perspectives*, *122*, 1141–1146.

Wong, J.K.H., Lee, K.K., Tang, K.H.D., & Yap, P.-S. (2020). Microplastics in the freshwater and terrestrial environments: Prevalence, fates, impacts, and sustainable solutions. *Science of the Total Environment*, *719*, 137512.

Wu, R.T., Cai, Y.F., Chen, Y.X., Yang, Y.W., Xing, S.C., & Di, L.X. (2021). Occurrence of microplastic in livestock and poultry manure in South China. *Environmental Pollution*, *277*, 116790.

Yadav, S., Gupta, E., Patel, A., Srivastava, S., Mishra, V.K., Singh, P.C., Srivastava, P.K., & Barik, S.K. (2022). Unravelling the emerging threats of microplastics to agroecosystems. *Reviews in Environmental Science and Bio/Technology*, *21*(3), 771–798.

Yan, W., Hamid, N., Deng, S., Jia, P.-P., & Pei, D.-S. (2020). Individual and combined toxicogenetic effects of microplastics and heavy metals (Cd, Pb, and Zn) perturb gut microbiota homeostasis and gonadal development in marine medaka (Oryzias melastigma). *Journal of Hazardous Materials*, *397*, 122795.

Yang, L., Zhang, Y., Kang, S., Wang, Z., & Wu, C. (2021). Microplastics in soil: A review on methods, occurrence, sources, and potential risk. *Science of the Total Environment*, *780*, 146546.

Yee, M.S.-L., Hii, L.-W., Looi, C.K., Lim, W.M., Wong, S.-F., Kok, Y.-Y., Tan, B.-K., Wong, C.-Y., & Leong, C.-O. (2021). Impact of microplastics and nanoplastics on human health. *Nanomaterials*, *11*(2), 496. https://doi.org/10.3390/nano11020496

Yu, H., Hou, J., Dang, Q., Cui, D., Xi, B., & Tan, W. (2020). Decrease in soil heavy metals' bioavailability caused by microplastics varies across aggregate levels. *Journal of Hazardious Materials*, *395*, 122690.

Yu, H., Zhang, Z., Zhang, Y., Fan, P., Xi, B., & Tan, W. (2021). Metal type and aggregate microenvironment govern the response sequence of speciation transformation of different heavy metals to microplastics in soil. *Science of the Total Environment*, *752*, 141956.

Yu, Y., Qin, W., Li, Y., Zhang, C., Wang, Y., Yang, Z., Ge, X., & Li, F. (2019). Red light promotes cotton embryogenic callus formation by influencing endogenous hormones, polyamines, and antioxidative enzyme activities. *Plant Growth Regulation*, *87*, 187–199.

Zhang, C., Feng, Y., Liu, Y.W., Chang, H.Q., Li, Z.J., & Xue, J.M. (2017). Uptake and translocation of organic pollutants in plants: A review. *Journal of Integrative Agriculture*, *16*, 1659–1668.

Zhang, G.S., & Liu, Y.F. (2018). The distribution of microplastics in soil aggregate fractions in southwestern China. *Science of the Total Environment*, *642*, 12–20.

Zhang, L., Xie, Y., Liu, J., Zhong, S., Qian, Y., & Gao, P. (2020). An overlooked entry pathway of microplastics into agricultural soils from applying sludge-based fertilizers. *Environmental Science and Technology*, *54*, 4248–4255.

Zhang, W., Zhang, S., Wang, J., Wang, Y., Mu, J., Wang, P., Lin, X., & Ma, D. (2018). Microplastic pollution in the surface waters of the Bohai Sea, China. *Environmental Pollution*, *236*, 619–625.

Zhang, Y., Du, N., Wang, L., Zhang, H., Zhao, J., Sun, G., & Wang, P. (2015). Physical and chemical indices of cucumber seedling leaves under dibutyl phthalate stress. *Environmental Science and Pollution Research*, *22*, 3477–3488.

Zhang, Y., Zhang, H., Sun, X., Wang, L., Du, N., Tao, Y., Sun, G., Erinle, K.O., Wang, P., Zhou, C., & Duan, S. (2016a). Effect of dimethyl phthalate (DMP) on germination, antioxidant system, and chloroplast ultrastructure in Cucumis sativus L. *Environmental Science and Pollution Research*, *23*, 1183–1192.

Zhou, Y., Wang, J., Zou, M., Jia, Z., Zhou, S., & Li, Y. (2020). Microplastics in soils: A review of methods, occurrence, fate, transport, ecological and environmental risks. *Science of the Total Environment*, *748*, 141368.

Zhu, D., Chen, Q.-L., An, X.-L., Yang, X.-R., Christie, P., Ke, X., Wu, L.-H., & Zhu, Y.-G. (2018). Exposure of soil collembolans to microplastics perturbs their gut microbiota and alters their isotopic composition. *Soil Biology and Biochemistry*, *116*, 302–310.

Zou, J., Liu, X., Zhang, D., & Yuan, X. (2020). Adsorption of three bivalent metals by four chemical distinct microplastics. *Chemosphere*, *248*, 126064.

CHAPTER 22

Microplastics as an Emerging Source of Particulate Air Pollution

Nimish Shah, Mamta Saiyad, Ankur Dwivedi, Shibu Pillai, and Milind Joshipura

22.1 INTRODUCTION

Microplastics(MPs) are small plastic particles that measure less than 5 mm in diameter. These tiny plastic particles can be found in various environments, including oceans, rivers, lakes, soils, and even the air. They originate from the fragmentation and breakdown of larger plastic items, as well as from the direct release of tiny plastic particles used in consumer products like cosmetics and textiles. MPs can also be generated through the degradation of synthetic fibers from clothing during washing. MPs pose significant environmental and health risks, as they can be ingested by marine life and subsequently enter the food chain, potentially harming both wildlife and humans. Efforts to mitigate MP pollution include improving waste management practices, reducing plastic consumption, and developing innovative technologies for capturing and removing MPs from the environment.

DOI: 10.1201/9781003487579-22

22.2 SOURCES OF MP

MPs originate from various sources, like the fragmentation of larger plastic items, synthetic textiles, microbeads in personal care products, industrial processes, tire wear and road runoff, paints and coatings, etc. Large plastic items such as bottles, bags, and packaging degrade over time due to exposure to sunlight, heat, and physical abrasion, breaking down into smaller fragments that become MPs. Synthetic fibers shed from clothing, textiles, and other synthetic materials during washing and wearing. Fabrics like polyester, nylon, and acrylic are major contributors to MP pollution in water bodies. Microbeads, tiny plastic particles used as exfoliants in cosmetics, toothpaste, and other personal care products, can enter waterways directly through wastewater effluent. Plastic pellets, or nurdles, used as raw materials in plastic manufacturing, are sometimes spilled during transportation and processing, contributing to MP pollution in rivers, oceans, and other environments. Tires degrade over time as they are used, releasing small rubber particles containing synthetic polymers onto road surfaces. These particles can be washed into water bodies by rainwater or runoff, contributing to MP pollution in aquatic environments. Small plastic particles are also released into the environment during the degradation of paints, coatings, and other products containing plastic polymers. These sources collectively contribute to the widespread distribution of MPs in the environment, posing environmental and health risks to ecosystems and organisms.

The degradation and fragmentation of plastics in the atmosphere vary depending on the polymer type. Polystyrene, for instance, rapidly generates secondary MPs when exposed to solar ultraviolet radiation due to easily breakable molecular chains [1]. Plastics with longer service lives persist longer in the environment but still contribute to MP generation [1]. Local emission rates of plastic particulates differ by location, influenced by polymer type, sources, and geography, including land distribution, water bodies, climate, and land use patterns such as agriculture and industries [2]. Airborne MP concentration and characteristics are impacted by wind patterns and distance from sources [3], with meteorological factors like wind speed, temperature, and boundary layer mixing influencing atmospheric MP transport [4]. Elevated synthetic MP concentrations in terrestrial and aquatic ecosystems during winter suggest a potential influence of precipitation, with monsoon-driven surface deposition of atmospheric particulates re-emitted post-monsoon, particularly in pollution hotspots like the Indo-Gangetic Plain [4,5].

22.3 SPREAD OF MP

MPs present in the atmosphere can undergo dry or wet deposition onto various surfaces such as soil, water bodies, vegetation, and buildings. Precipitation, including rain and snow, can effectively wash MPs out of the atmosphere and deposit them onto surfaces. Human activities such as industrial processes, agriculture, and construction can generate airborne particles containing MPs, dispersing them over large distances. This combination of natural and anthropogenic processes spreads MPs in the atmosphere, posing risks to human health and ecosystems. Understanding the sources, transport, and fate of airborne MPs is crucial for developing effective strategies to mitigate their environmental impacts.

Monsoon winds significantly contribute to the transboundary transport of airborne MPs. For example, the East Asian summer monsoon transports atmospheric MPs from Vietnam to southeastern China via the Philippines and Malaysia [6]. Rezaei et al. [7] found that wind-eroded sediments contained abundant light-density MPs, indicating wind's role in atmospheric MPs transport [7,8]. Mbachu et al. demonstrated the sensitivity of MPs to wind entrainment through wet and dry deposition, observing MPs in rainwater samples from 11 U.S. national parks. This study highlighted the long-range transport and precipitation of MPs into remote protected forests. Consequently, all atmospheric compartments can serve as potential sinks and sources of terrestrial and aquatic MP pollution due to airborne migration [9].

22.4 ROLE OF THE ATMOSPHERE IN THE GLOBAL TRANSPORT OF MPs

Airborne MPs possess lower densities than soil minerals, allowing them to remain suspended in the air longer than natural dust aggregates of similar size, facilitating transport to distant locations [10]. Consequently, the atmosphere serves as a potential vehicle for MPs' transport across terrestrial, aquatic, and aerial biomes. Cloud cover and meteorological factors modulate air-sea interactions, influencing the global transport of atmospheric particles [11]. Atmospheric currents, akin to oceanic gyres, play a crucial role in the global dispersal of MPs [12]. Wind stress and sea spray generate sea salt-MP aerosols (<0.2 to >2,000 μm), dispersing them to urban areas near coasts, although the extent of MP contamination in these areas remains unclear [13]. Field-based evidence from Liu et al. [14] confirms atmospheric transport of MPs from terrestrial to marine environments, with continuous cruise sampling in the western Pacific Ocean revealing high atmospheric MP abundance along coasts (0.13±0.24 items/m^3) compared to pelagic areas (0.01±0.01 items/m^3). However, Liu et al. [15] found inconsistencies in MP sizes and composition between aerosols and atmospheric deposition, suggesting limited atmospheric transport to the ocean compared to riverine inputs.

Rainfall, particularly freezing rain at 0°C, and snow in colder environments around −10°C, can scavenge organic vapors and particles from the atmosphere [16]. This scavenging effect may transport airborne MPs and their additives [17] or adsorbed co-pollutants [18] to remote locations. Bergmann et al. [19] identified MPs in polar snow, indicating substantial contamination in the Arctic atmosphere. They suggested a pathway where MPs traverse the atmosphere, bind with snow, and deposit in tundra, facilitated by scavenging effects (Figure 22.1). Atmospheric currents carry MPs, which are scavenged by clouds (wet/dry deposition) along with other particles and gases, precipitating them into terrestrial and aquatic environments. Remote Arctic lakes serve as potential sinks for MPs, possibly deposited through wind-driven atmospheric transport [20]. Klein and Fischer [21] observed increased accumulation of MP fragments in rural, fir forest, and urban areas of Hamburg, Germany, attributed to atmospheric transport. Zhou et al. [22] suggested that MPs present in riparian topsoil and sub-surface soil of remote mountains along the Yangtze River may have been transported by atmospheric deposition along with street runoff and wastewater discharge. While many studies worldwide document MPs

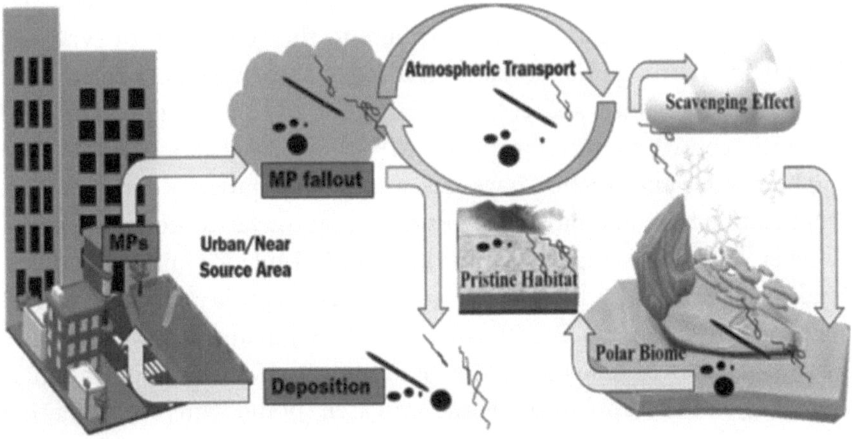

FIGURE 22.1 A simplified representation of atmospheric resuspension, transport, and scavenging of MPs.

in various environmental matrices and commercial food samples, estimating the contribution of aerial transport in MP deposition remains challenging. For example, Selvam et al. [23] identified natural and synthetic MPs in sea salt samples from the Tuticorin coast (Tamil Nadu), India, although their focus on sedimentation and resuspension from the ocean did not discount atmospheric transfer as a potential transmission route.

22.5 OCCURRENCE OF MPS IN AIR

Airborne MPs originate from various sources, including synthetic textile clothing, furniture, construction materials, rubber tire abrasion, and urban dust [24,25]. Additionally, MPs can arise from the resuspension of deposited dirt from automobiles [26]. Once suspended in the atmosphere, these MPs are carried away and can accumulate, further contaminating aquatic or terrestrial environments [27]. However, our understanding of the global concentrations of airborne MPs and microfibers (MFs) in both outdoor and indoor air remains limited.

22.5.1 MPs in Street Dust and Outdoor Air

A study examining the presence of MPs in street dust [28] revealed an underestimated source of airborne MPs. Street dust samples (<5 mm) from 15 sites in Asaluyeh, southern Iran, were predominantly composed of MPs (spheres, films, and fibers) and micro rubber (MR) particles, with averages of 900 MPs and 250 MRs per 15 g sample respectively [28]. To remove organic matter, the authors treated 15 g aliquots with 35 mL of 30% H_2O_2, inferring the identified particles to be synthetic. Increased vehicular activity and urban road traffic contribute to the generation of synthetic MR particles through tire abrasion, leading to suspended particulate matter (PM). Urban dust sources, including tire and paint aggregates, significantly contribute to atmospheric MP input [29]. Latex paints, which combine water and plastics, have seen a surge in popularity. Järlskog et al. [30] identified tire and bitumen MPs from road surfaces, stormwater, and sweeper sediment samples, identifying them as major emission sources. Studies suggest that vehicular tire abrasion is a significant source of MP aggregates in urban air [31,32]. Knight et al. [33] quantified synthetic tire wear aggregates in roadside drains of Plymouth, U.K., highlighting their contribution to MR pollution. Future research should focus on estimating urban outdoor MP pollution from tire abrasion. MPs emitted from tire abrasion are generally coarse particulate pollutants [34,35], with airflow and strong air currents from rotating vehicular tires contributing to PM emission [36]. Modeling approaches can help understand the abrasion and release of MPs from tires and other sources [37].

MPs can also originate from brake wear particles, primarily composed of metals and plastics (fibers, binders, fillers, etc., from brake linings) [38]. Paint emissions are another underestimated source of indoor and outdoor air pollution, with latex paints, combining water and plastics, experiencing a surge in popularity. Rist et al. [39] identified paint fragments on the Arctic coast, suggesting ship paint as an underestimated source of marine and airborne MPs. Disposable plastic bags can emit MPs into indoor or outdoor air [39], while plastic films used for agricultural mulching can also release high concentrations of MPs into the environment [40].

Incineration represents a potential but underexplored source of airborne MPs. Yang et al. [41] found significant levels of particles per metric ton of MPs (mainly PP and PS) in the bottom ash of municipal solid waste incinerators in Chinese cities, indicating incineration as a potential source of MP discharge. Atmospheric fallout from waste dumps and landfills also contributes to MP transfer in the air [42], with soil near electronic waste dumps found to be highly contaminated with MPs [43]. González-Pleiter et al. [44] sampled airborne MPs using aircraft, finding urban air predominantly polluted by fragments and rural/sub-rural air by fibers, with air mass trajectory analyses suggesting atmospheric transport of MPs over 1,000 km before deposition.

Synthetic textile fibers are recognized as a major source of MPs in atmospheric fallout [24,42], with polyethylene (PE), PP, and PS identified as the predominant types of urban wind-carried MPs in Dongguan City, China [45]. MP levels in London's outdoor air are estimated to be 20 times higher than in remote rural locations, indicating the influence of local sources such as human activities and vehicular movements [3].

Tracing the sources and types of airborne MPs is crucial despite limited studies, as their distribution varies globally due to regional anthropogenic activities, population density, and waste management practices [46]. Seasonal sampling gaps highlight the need for both wet and dry deposition sampling to understand atmospheric MP pollution and its impact on terrestrial and aquatic ecosystems [15].

Data on atmospheric deposition and MP transport is lacking outside a few megacities and specific regions like mountain catchments [26,27]. Improved standard methods are necessary to characterize and quantify MPs in air and environmental matrices [47,48], although minimizing background contamination remains a challenge [49,50].

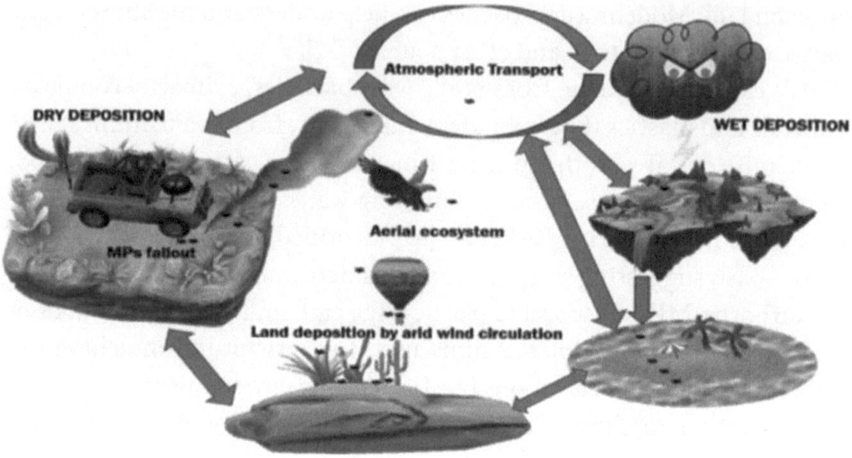

FIGURE 22.2 Overview of aerial transport of MPs through wet and dry deposition [5].

While many studies address atmospheric MP transport, much remains unknown, necessitating future research focusing on mechanisms and pathways. Understanding aerial deposition, transport, scavenging, and precipitation of MPs is vital for mapping sources and predicting ecosystem impacts (Figure 22.2). Identifying vulnerable regions can inform studies on MPs' contribution to particulate air pollution. Literature on MP loading rates from urban to remote areas is scarce, emphasizing the need for research on primary and secondary MP transfer mechanisms [8].

22.5.2 MPs in Indoor Air

Indoor airflow turbulence significantly influences the resuspension of airborne MPs and nanoparticles (NPs) [15]. Studies indicate that synthetic fibers and MPs in indoor air typically range from 50 to 250 μm [26,46], posing a higher risk of penetration into the human respiratory system [51]. Research in Paris highlighted the ubiquity of MPs in indoor and outdoor air, posing risks of human exposure through inhalation and settled dust ingestion [26].

Indoor concentrations of fibrous MPs were found to be higher than outdoor levels in France, suggesting air exchange as a transfer route from indoors to outdoors [26]. In a study at the East Normal China University, plastic particulates mainly sourced from clothing, bedding, and curtains were identified in indoor dust, with 60% being natural fibers. Documentation of MPs in indoor air from 12 countries showed increased concentrations of terephthalic acid and BPA in PET and polycarbonate (PC) MPs, respectively [15,52,53].

Laundry, particularly drying synthetic clothes in sunlight, is a significant source of indoor fibrous MPs [54–56]. E-waste dust may also emit MPs or additives indoors, contributing to indoor pollution [57]. Future research should focus on the understanding exposure and health impacts of MPs from ambient air in urban areas, as well as quantify the contribution of indoor MP fallout to outdoor particulate pollution. Understanding airborne MPs in indoor and outdoor environments is crucial for assessing human health risks.

22.5.3 Atmospheric MPs in the Vulnerable Ecosystems

The scavenging effect (Figure 22.3) can lead to the precipitation of MPs into delicate ecosystems when they deposit from the atmosphere to the lithosphere and hydrosphere. High-altitude ecosystems are particularly vulnerable to airborne exposure. For example, Napper et al. [58] identified MPs in remote areas of Mount Everest, suggesting sources such as trekkers' clothing and gear, as well as wind-driven atmospheric transport. While there are notable studies on polar ecosystems, Antarctic fauna's exposure to MPs is limited compared to the Arctic. MPs have been documented in pristine freshwater and marine ecosystems, with airborne

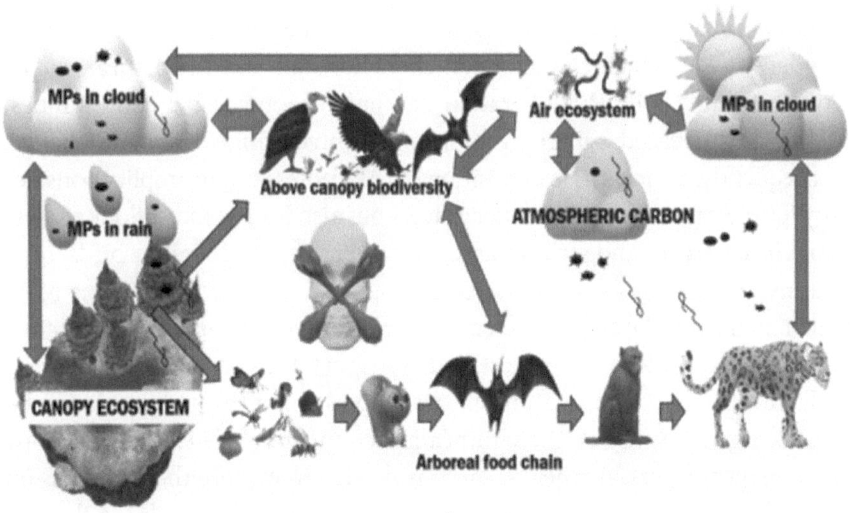

FIGURE 22.3 Abundance and trophic transfer of atmospheric MPs in canopy and aerial ecosystems [5].

deposition posing similar exposure risks to aquatic flora and fauna. Thus, emphasis on faunal and floral exposure to airborne MPs should focus on terrestrial ecosystems, notably forest ecosystems crucial for rainfall and climate regulation. Canopy ecosystems, especially in mountain and temperate forests, are susceptible to airborne MPs due to their biodiversity and height. However, understanding the impacts of MPs on canopy niches and species interactions requires further research.

Studies on airborne MPs primarily focus on their health implications, with comparatively fewer addressing their ecological impacts. Particularly lacking are discussions on their implications in remote canopy, above-tree canopy, and aerial habitats via wet/dry precipitation exposure. Knowledge on canopy, emergent layer, and airborne ecosystems and their community interactions is limited, despite their vital roles in atmospheric biogeochemical cycles, notably the carbon cycle. For example, tropical rainforest tree canopies influence carbon fluxes and forest dynamics, with canopy disturbances affecting deadwood decomposer communities and the natural carbon cycle. Canopy ecosystems are crucial for carbon and nitrogen biogeochemical cycles, highlighting the need for further research on their response to airborne MPs [13,59,60].

The emergent and canopy layers of forest ecosystems are directly impacted by the deposition of atmospheric MPs through precipitation or wind circulation. These suspended particles can adhere to plant surfaces upon deposition in the terrestrial environment. Studies on the transport and deposition of airborne MPs predominantly identify low-density polymers, such as low-density PE, which can alter soil characteristics and the natural microbiome of forest soil. This alteration can enhance anaerobic communities, leading to increased soil respiration and contributing to climate change. Consequently, atmospheric deposition of MPs poses serious implications not only for the canopy and understory layers but also for the forest floor ecosystem. The emergent and atmospheric ecosystems are inhabited by birds, bats, and insects. Migratory seabirds, for example, are known to consume MPs, which then enter the aerial food chain through predation by avian predators. Similarly, mosquitoes can ingest and transport MPs throughout their life cycle, ultimately contaminating the aerial food chain through predation by various organisms. Given the significant accumulation of MPs in the canopy and emergent layers, further research is needed to explore their impacts on atmospheric microbial communities and arboreal biodiversity [61–63].

Furthermore, carbon emissions stemming from atmospheric and ocean plastic particles can directly impact natural carbon sequestration processes, thus exacerbating climate change. To comprehensively understand the global plastic cycle, it is imperative to develop methodologies with high spatial resolution, alongside reliable modeling approaches incorporating factors like air-mass trajectories, dust dispersion, and wet/dry deposition rates, as well as population statistics. Mapping potential sources and vulnerable residential or forested areas susceptible to airborne MP fallout requires thorough documentation and exploration of the prevalence and distribution of MPs in outdoor and ambient air. While various studies have successfully identified MPs in both indoor and outdoor air, many have struggled to effectively trace the origins of this fallout [63].

22.5.4 Routes of Domestic and Industrial Exposure

Airborne MPs disseminate across terrestrial environments via wind erosion, posing a significant risk of exposure through direct inhalation of dust-borne MPs. Inhalation represents a primary pathway of aerial MP exposure, with dust serving as a notable medium for human respiratory intake. Inhalable fibers, including micro and nanofibers, suspended in the air, have the potential to release absorbed chemicals upon inhalation, potentially leading to various respiratory, cardiovascular, and even carcinogenic health issues. Studies have shown that household dust fallout may result in higher MP exposure compared to consuming contaminated seafood like mussels. In urban settings, children's daily exposure to PET through indoor dust can be considerable. However, research suggests that natural fibers may pose a greater inhalation risk due to their smaller size, despite synthetic fibers being more prevalent indoors. Discrepancies in pulmonary exposure parameters exist due to insufficient and uncertain data. Although ingestion exposure is generally considered higher than inhalation, airborne MPs can also be ingested orally during routine activities. Human exposure to plastic additives like phthalates through airborne transmission remains poorly understood. Additionally, dermal contact with airborne MPs, such as adherence to hair, represents an underestimated yet potential route of exposure [64].

In a study conducted locally in Iran, PE, PET, and PP fibers were consistently found in saliva, hair, and skin samples of residents, particularly in all hair samples, indicating a significant source of exposure. The study suggests that human head hair, with its high surface area and tortuosity, can accumulate electrostatic charge, attracting and retaining airborne MPs, especially fibers from clothing and soft furnishings. While the entry of these trapped MPs into the human system through dermal contact and absorption, particularly via the scalp, remains unproven, it underscores a potential exposure route. The composition of MFs in the air varies with seasonal clothing demands. To assess the health impacts of MPs on humans, understanding airborne pathways and realistic exposure levels is crucial. Regional, inter-regional, and local sources significantly influence human exposure to outdoor airborne MPs, highlighting the importance of tracking the spatial distribution and origins of tropospheric MPs to estimate inhalation exposure. Given that individuals residing near urban source areas are more susceptible to aerial MP exposure, investigating the potential contribution of airborne MPs to urban PM levels is essential.

Workers in synthetic textile industries are exposed to airborne MPs such as nylon, PE, PP, PS, PET, and others. Recent studies have highlighted concerning health impacts associated with exposure to these particles. For example, research has shown that inhalable MFs from polyester and nylon textiles can inhibit the development and repair of pulmonary airway organoids in both human and murine cells. Similarly, workers in industries dealing with electronics, packaging, automobiles, chemicals, pigments, and toys may encounter PS MPs, which have been associated with toxicological effects including morphological changes and inhibited cell proliferation. However, the findings from toxicity studies on airborne MPs have been inconsistent, largely due to limited knowledge of realistic industrial exposure scenarios [65,66].

Despite the potential risks, current occupational safety guidelines, such as those provided by the Occupational Safety and Health Administration (OSHA) in the United States, do not specifically address long-term exposure to plastics. Furthermore, there is a notable absence of legislation or occupational exposure limits for airborne MPs in any country worldwide. This lack of regulatory framework can be attributed to the relatively recent recognition of MPs as emerging contaminants and the limited availability of standardized characterization methods [67].

In countries like India, where guidelines exist for other occupational hazards, such as those related to the tobacco and construction industries, there is still a gap in addressing plastic exposure specifically. However, workers in the plastics and rubber industries have been observed to experience various health issues, including carcinomas, cytogenetic damage, and oxidative stress.

While some countries have updated regulations on outdoor and ambient PM, including MPs, there is a need for further research to inform comprehensive regulatory policies addressing the occupational hazards and long-term toxicity of airborne MPs. Additionally, more studies are needed to assess the occupational health risks associated with inhalable and non-inhalable MPs, as well as their potential dermal exposure routes, particularly among industrial workers.

Workers in industries such as flock, synthetic textile, plastic, and packaging are particularly susceptible to high levels of airborne MPs, which may lead to occupational diseases. Understanding the health impacts of airborne MPs requires consideration of the specific exposure scenarios encountered by workers in various industries [68].

22.6 EFFECTS OF MICROPLASTICS ON AIR QUALITY

The direct effect of MPs on air quality is relatively understudied compared to their impacts on aquatic environments. However, there are several potential ways in which MPs could affect air quality. MPs suspended in the air can contribute to PM pollution, specifically PM2.5 and PM10, which are particles with diameters less than 2.5 and 10 μm, respectively. Elevated levels of PM can worsen air quality and pose risks to human health, particularly respiratory and cardiovascular health. Inhalation of airborne MPs may pose health risks to humans and animals. While research on the health effects of inhaling MPs is still emerging, there is concern that inhalation could lead to respiratory problems and inflammation, as well as potential systemic effects if MPs enter the bloodstream. MPs deposited from the atmosphere onto terrestrial and aquatic ecosystems can have ecological consequences. They may accumulate in soil, water bodies, and vegetation, potentially affecting soil quality, water quality, and the health of organisms in these environments. MPs can adsorb and concentrate chemical pollutants from the air, such as polycyclic aromatic hydrocarbons (PAHs), polychlorinated biphenyls (PCBs), and heavy metals. When MPs are inhaled or deposited into ecosystems, these associated pollutants may be released, potentially causing additional harm to human health and the environment [13,26,69].

Overall, while the direct effects of MPs on air quality are still being researched, it is clear that MPs have the potential to impact both air quality and human health. Further studies are needed to better understand the sources, transport, and impacts of airborne MPs, as well as to develop strategies for mitigating their effects on air quality and ecosystems.

22.7 IMPLICATIONS OF MPs IN THE AQI

The Air Quality Index (AQI) serves as a standardized measure of air pollution levels, which can pose health risks upon exposure within relatively short periods, typically hours or days. It offers a normalized scale to gauge overall air quality within a region, incorporating various parameters such as PM, tropospheric ozone (O_3), sulfur oxides (SO_x), nitrogen oxides (NO_x), carbon monoxide (CO), ammonia (NH_3), and lead (Pb), among others, as outlined in the WHO Air Quality Guidelines of 2005.

Particulate matter, or PM, refers to fine particles suspended in the air, often comprising a mixture of liquid droplets and solid materials. These particles, including airborne MPs that are generally documented within the breathing zone of an average adult human (0–1.2 m above ground level), are typically classified under PM in air pollution guidelines rather than as a distinct category. The presence of PM in the atmosphere can reduce visibility, a parameter often assessed using a nephelometer to measure differences in light scattering, known as scattering potential [34,35,70,71].

Certain particulate plastics, such as MR from vehicles, MFs, and sea-spray MP aerosols, can contribute to light scattering, thereby potentially impacting visibility. Additionally, synthetic fibers, particularly polyamide (PA), with their hygroscopic surface properties, can absorb moisture, contributing to the formation of aerosols in windblown street dust, further affecting visibility. However, the specific influence of MPs on air visibility remains largely unexplored [72].

Despite the potential for airborne MPs to contribute to urban and industrial air contamination, their role in policy frameworks and assessment guidelines has received limited attention. Addressing this gap requires overcoming challenges in characterizing MPs, assessing their health impacts, and understanding realistic exposure levels before incorporating them into policy frameworks. While WHO has published comprehensive reports on the documentation, exposure, and potential health impacts of MPs in drinking water, comparable studies and policy standards for inhalable MPs and NPs in ambient air are currently lacking [73].

As per the AQI, plastic particulates primarily fall under the categories of PM, specifically PM10 (particles with a diameter of less than 10 μm, referred to as MPs) or PM2.5 (particles with a diameter of less than 2.5 μm, referred to as NPs). However, there is a notable absence of universally recognized ambient air quality standards specifically tailored for MPs/NPs, despite their emerging significance as contributors to PM10 and PM2.5 levels.

Many countries worldwide lack national standards for MPs/NPs in ambient air, primarily because the documentation and monitoring of airborne MPs have not been extensively undertaken, particularly in developing and industrialized nations. For example, in India, only a couple of studies have so far documented airborne particulate plastics in major metropolitan cities such as Chennai and Nagpur. Even during periods of significant PM elevation in cities like Delhi, particularly during winter months, there is a dearth of information regarding atmospheric MPs.

Global research on airborne MPs is fragmented, with no comprehensive studies correlating MP levels with their impact on AQI. There has been a lack of extensive research aimed at understanding the contribution of MPs/NPs to PM10 or PM2.5 emissions. Furthermore, the development of standard measurement methods for airborne MPs is crucial before establishing air quality standards specifically addressing this pollutant [6,26,74].

Elevated levels of PM with a diameter less than 10 μm (PM10) in the atmosphere contribute significantly to air pollution-related mortality, as outlined in the guidelines provided by the Environmental Protection Agency (EPA) and the WHO. Biomass burning activities, including bushfires, represent a significant source of PM pollution in the air. The smoke haze generated by uncontrolled biomass burning, along with forest and peat fires, detrimentally affects air quality, atmospheric visibility, the hydrologic cycle, ecosystems, climate, and human health.

Plastics utilized for mulching and land applications have the potential to release MPs and chemical vapors when subjected to combustion during bushfires. However, it remains uncertain whether the abundance of MP/NP particulates in the air heightens the risk of fatality. Recent findings by Dong et al. [75] suggest that PS MPs not only induce inflammation and cytotoxicity in BEAS-2B (human pulmonary epithelium) cells but also disrupt certain proteins known as *zonula occludens*, which are associated with chronic obstructive pulmonary disease (COPD). Their study concludes that PS particulates entering the lungs, even at low concentrations (1 $\mu g/cm^2$), have the potential to compromise pulmonary integrity and thereby contribute to serious lung disorders. Additionally, the toxicity of airborne MPs is influenced by weathering or aging processes [76].

Global guidelines have established the annual mean exposure standards for PM2.5 and PM10 at 10 and 20 $\mu g/m^3$, respectively, as outlined by the World Health Organization Air Quality Guidelines [35]. While these guidelines provide crucial benchmarks for assessing air quality, there is a notable absence of specific regulations for plastic particulates, including MPs, within the framework of air quality assessment.

Considering the increasing recognition of MPs as atmospheric particulates, there is a pressing need for further research to comprehensively understand their impacts and the extent of human exposure. Existing air quality frameworks have identified and established exposure limits for various chemical contaminants such as mercury (Hg), lead (Pb), phenols, and other organic compounds, as delineated by the Environmental Protection Agency Air Quality Index (AQI) Guidelines [34]. However, the absence of regulatory standards for plastic particulates is a significant gap, given their role in transporting such pollutants in the air [34,71].

It is imperative to investigate the presence of MPs and NPs within atmospheric aerosol PM and their potential to adsorb and desorb other airborne pollutants, including volatile organic compounds. Moreover, understanding the modes of transmission of toxic air pollutants facilitated by MPs is crucial for developing effective mitigation strategies. Comprehensive air quality assessment studies should aim to quantify the proportion of MPs within the overall PM composition to gain insights into their sources and implications for air pollution.

22.8 REMOVING MICROPLASTICS FROM THE AIR

Removing MPs from the air poses several challenges due to their small size and wide distribution. However, several methods are being explored for the removal of MPs from the air such as filtration, electrostatic precipitation, ionization, wet scrubbing, natural deposition, chemical degradation, etc. Air filtration systems equipped with high-efficiency particulate air (HEPA) filters can effectively capture airborne particles, including MPs. These filters use a dense mesh of fibers to physically trap particles as air passes through them. Electrostatic precipitators use an electric charge to attract and collect airborne particles, including MPs, onto collection plates or surfaces. This method can be effective for removing particles of various sizes from the air. Ionization systems release charged ions into the air, which attach to MP particles and cause them to become negatively charged. The charged particles can then be attracted to positively charged surfaces and removed from the air. Wet scrubbers use water or other liquids to capture and remove airborne particles, including MPs, by causing them to adhere to the liquid droplets. The contaminated liquid can then be treated and filtered to remove the MPs. MPs can settle out of the air over time through natural processes such as gravity and precipitation. Vegetation, soil, and bodies of water can act as sinks for airborne MPs, facilitating their removal from the atmosphere. Advanced oxidation processes (AOPs) and other chemical treatments can be used to degrade MPs in the air by breaking down their polymer chains into smaller, less harmful compounds. However, these methods are still in the early stages of development and may have limited applicability. Overall, while there is ongoing research into methods for removing MPs from the air, no single method is currently widely adopted or universally effective. A combination of approaches may be needed to address the challenge of airborne MP pollution effectively [77–79].

There have been relatively few case studies specifically focusing on MP air pollution. However, ongoing research is beginning to shed light on this emerging environmental issue. Here are a few notable case studies and research findings:

1. **Paris, France:** A study conducted in Paris found MPs in the air throughout the city, with higher concentrations observed in urban areas compared to rural areas. The researchers identified a variety of sources contributing to airborne MPs, including road traffic, industrial activities, and atmospheric deposition [80].
2. **Dongguan, China:** Researchers conducted a study in Dongguan, a major manufacturing hub in China, and found that MP particles were present in the air samples collected near industrial areas. The study suggested that industrial activities, such as plastic manufacturing and processing, were significant contributors to airborne MP pollution in the region [81].
3. **Western United States:** A study conducted in the western United States found MPs in atmospheric deposition samples collected in remote wilderness areas, far from urban centers. The presence of MPs in these remote areas suggests that airborne MPs can be transported over long distances and deposited in pristine environments [27].
4. **Mount Everest:** Recent research has detected MPs in snow samples collected from Mount Everest, the world's highest peak. The presence of MPs in this remote and pristine environment highlights the extent of plastic pollution and its potential to reach even the most isolated regions through atmospheric transport [82].
5. **Chennai, India:** Devipriya et al. investigated the levels of MP pollution in road dust samples collected from Chennai, the fifth-largest metropolitan city in India. Road dust samples were collected from 16 different locations across the city. The average abundance of MPs was found to be 227.94±91.37 particles/100 g of street dust sample. Nile Red dye was employed for the identification and quantification of MPs, revealing that 92.46% of the quantified MPs were fragments [83].

6. **Nagpur, India:** Narmadha et al. investigated multiple urban and rural sites in Nagpur, India, revealing an average total abundance of MP particles ranging from 50 to 120 particles/day. Fiber-shaped MPs were found to be the most prevalent, with transparent/white being the dominant color observed across all four sites. Chemical composition analysis using Fourier-transform infrared (FTIR) spectroscopy in Attenuated Reflectance mode identified several polymers present in the region, including low-density polyethylene, rayon, rubber fiber, polystyrene, polyaniline, polyolefin, and chlorinated polyvinyl chloride [84].

7. **London, UK:** Wright et al. present evidence of MP contamination in atmospheric deposition within a densely populated area, central London. Our findings reveal the presence of MPs in all samples, with deposition rates ranging from 575 to 1,008 MPs/m^2/day. Fibrous MPs were the most common shape observed, comprising 92% of the total MPs identified. Additionally, analysis of the samples identified 15 different petrochemical-based polymers [3].

8. **River Trent, UK:** Stanton et al. shed light on the prevalence of natural textile fibers in both upstream sections of the River Trent, UK, and atmospheric deposition within its catchment area, during a year-long MP sampling initiative. Out of 223 samples collected, natural textile fibers accounted for 93.8% of the total textile fiber population quantified. Surprisingly, extruded textile fibers, including synthetic ones, were notably absent from 82.8% of the samples, while natural textile fibers were absent from only 9.7% of the samples [85].

9. **Scotland, the UK:** Catarino et al. conducted an assessment of MP contamination in mussels gathered along the coastline of Scotland (UK) to characterize MP traits and evaluate the potential risk of human exposure to MPs through mussel consumption. The study involved the deployment of caged mussels (*Mytilus edulis*) in an urbanized estuary (Edinburgh, UK) to monitor seasonal variations in plastic pollution. Additionally, mussels (*Mytilus spp* and subtidal *Modiolus modiolus*) were collected from eight sampling stations across Scotland to analyze the types of MPs present at different locations. The investigation also included an assessment of potential human exposure to household dust fibers during meal consumption, which was compared with the levels of MPs found in edible mussels.

The findings revealed that the average number of MPs in *M. modiolus* was 0.086±0.031 (SE, $n=6$) per gram of wet weight (3.5±1.29 (SE) per mussel). For *Mytilus spp*, the mean number of MPs per gram of wet weight was 3.0±0.9 (SE, $n=36$) (3.2±0.52 (SE) per mussel), although this varied depending on the weight of the mussel. The study also predicted the ingestion of MPs by humans through the consumption of mussels, estimating an intake of 123 MP particles/year/capita in the UK, which could increase to 4,620 particles/year/capita in countries with higher shellfish consumption rates. However, the risk of plastic ingestion from mussel consumption was found to be minimal when compared to the exposure to fibers during a meal via household dust fallout, which ranged from 13,731 to 68,415 particles/year/capita [46].

10. Brahney et al. demonstrate that even the most remote regions of the United States, including national parks and wilderness areas, are not immune to MP pollution. Through wind and rain transport mechanisms, MP particles accumulate in these pristine environments. The study estimates that over 1,000 metric tons of MPs are deposited annually within protected areas in the south and central-western regions of the U.S. Predominantly, these plastic particles consist of synthetic MFs commonly used in clothing manufacturing. These findings underscore the critical importance of implementing measures to reduce pollution originating from such materials [8].

11. Gaston et al. conducted an investigation into airborne MPs within and outside buildings along the coastal regions of California. They employed glass fiber filters to filter known volumes of air, subsequently characterizing them using various microscopy techniques. The study revealed that MPs are prevalent in the air, with indoor environments (3.3±2.9 fibers and 12.6±8.0 fragments/m^3; mean±1 SD) containing approximately twice as much MPs as outdoor air (0.6±0.6 fibers and 5.6±3.2 fragments/m^3). While there was no significant difference in MP fiber length between indoor and outdoor environments, indoor MP fragments (58.6±55 μm) were observed to be approximately half the size of outdoor fragments (104.8±64.9 μm). Micro-Raman and FT-IR analyses presented slightly differing insights into airborne plastic compounds, with micro-Raman indicating a dominance of polyvinyl chloride indoors, followed by PE, whereas μFT-IR suggested polystyrene as the dominant compound

followed by PE and polyethylene terephthalate. The widespread presence of airborne MPs underscores significant potential sources of plastic inputs into terrestrial and marine ecosystems, raising substantial concerns regarding human inhalation exposure both indoors and outdoors [86].

12. Li et al. conducted a study in Beijing, collecting fiber particles from near-surface air, surface-deposited dust, and building materials. Analytical scanning electron microscopy was utilized for examination. The particles were initially categorized into two groups: organic and inorganic. Organic fibers included MPs and natural organic fiber particles, while inorganic fibers consisted mainly of man-made mineral fibers (MMMFs), asbestos (represented by chrysotile), calcium sulfate, and metal fiber particles. MPs and MMMFs fibers were the most abundant, accounting for 34.6% and 40.3% of the total, respectively, followed by asbestos (7.8%), calcium sulfate (7.2%), metal fibers (5.6%), and natural organic fiber particles (4.5%). The number concentration of these particles was approximately 16.7×10^{-3} fibers/mL at 1.5 m above the ground and about 14.1×10^{-3} fibers/mL at approximately 18 m, suggesting that the particles were primarily derived from the surface and were resuspended. Approximately 80% of the airborne fibers were smaller than 20 μm in length, which may be the critical size for fiber particles to resuspend into the air. Surface dust and construction sites were hypothesized to be the major contributors to the fiber particles [87].

13. Liu et al. conducted an investigation into the potential sources and spatial distribution of small airborne microplastics (SAMPs) in Shanghai in May 2018, utilizing an active suspended particulate sampler. The abundances of SAMPs from filtered air ranged from 0 to 4.18 n/m^3 (items per cubic meter of air). MFs comprised 67% of all SAMPs, followed by fragments and granules, which comprised 30% and 3% of SAMPs, respectively. Analysis using micro-FT-IR revealed that the SAMPs consisted of various polymers, including polyethylene terephthalate (PET), PE, polyester (PES), polyacrylonitrile (PAN), poly(*N*-methyl acrylamide) (PAA), rayon (RY), ethylene vinyl acetate (EVA), epoxy resin (EP), and alkyd resin (ALK). Synthetic compounds accounted for 54% of the observed particles, with PET, PE, PES, PAN, PAA, and RY comprising 91% of the MPs. The preliminary evaluation suggested that textile clothes are likely a major

source of airborne MPs. Modeling estimated that approximately 120.7 kg of SAMPs are annually transported through the air in Shanghai. Furthermore, an estimation of the ecological risk posed by SAMPs indicated a minor ecological consequence, highlighting the need for further evaluation of SAMPs pollution. Additionally, modeling estimated that approximately 21 particles of MPs are inhaled daily by individuals in Shanghai from outdoor environments [88].

14. Zhou et al. conducted a study on MPs in precipitation samples collected in Yantai, a coastal city in Shandong Province, East China. This study reports on the different shapes, deposition fluxes, and seasonal variations in MPs in the coastal urban atmospheric environment. The collected samples underwent pre-filtration using stainless steel sieves with mesh sizes of 5 and 1 mm. Residues such as insects and plant debris retained on a 5 mm sieve were discarded or archived, while residues retained on a 1 mm sieve were air-dried and examined under a stereomicroscope to select and count putative MPs with different morphologies and size fractions of 1–2, 2–3, and 3–5 mm. Samples that had passed through the 1 mm sieve were further filtered using a 5-μm pore size nitrocellulose filter for microscopic observation and enumeration of smaller MPs. Some samples containing organisms were digested with hydrogen peroxide (30%) to avoid interference during observation and selection of MPs, which were identified using Fourier transform infrared spectroscopy.

    Four shape types of MPs were found in the atmospheric samples: fibers, fragments, films, and foams. The majority were fibers of different colors (white, black, red, and transparent), accounting for about 95% of the MPs, while fragments accounted for about 4%, and films and foams accounted for less than 1%. The main polymers identified were PET for most fibers, polyvinyl chloride (PVC) for some fibers and films, PE for the fragments, and polystyrene for the foams. MPs with sizes below 0.5 mm accounted for more than 50% throughout the year, followed by the size ranges 0.5–1, 1–2, and then 2–3 mm. MPs with a size range of 3–5 mm appeared only in the spring samples at low abundance. Generally, the amount of MPs decreased dramatically with increasing particle size. The 100–300 μm size fraction formed the highest proportion of MPs below 0.5 mm in size, accounting for about 62%, while those below 50 μm showed the lowest abundance.

The annual site deposition flux of atmospheric MPs reached a maximum of 1.46×10^5 n/(m^2 a), with fibers up to 1.38×10^5 n/(m^2 a), and fragments, films, and foams up to 6.29×10^3, 7.65×10^2, and 2.45×10^2 n/(m^2 a), respectively. Deposition fluxes of different shape types ranged from 0.0 to 6.02×10^2 n/(m^{2}d), with fibers being the most common shape type. The deposition flux of MPs showed some seasonal variation, being higher in the spring, summer, and winter, and lowest in the autumn. Based on the observation site, the length of Yantai's urban coastline, and the annual deposition flux in this study, it was estimated that the total number of MPs deposited in the urban area was about 2.33×10^{13} particles, equivalent to about 0.9–1.4 tons [89].

15. Allen et al. reported observations of atmospheric MP deposition in a remote, pristine mountain catchment located in the French Pyrenees. Over a 5-month period, they analyzed samples representing atmospheric wet and dry deposition and identified fibers up to approximately 750 μm long and fragments ≤300 μm as MPs. They documented relative daily counts of 249 fragments, 73 films, and 44 fibers/m^2 deposited on the catchment. Air mass trajectory analysis revealed MP transport through the atmosphere over distances of up to 95 km. The study suggests that MPs can reach and affect remote, sparsely inhabited areas through atmospheric transport [27].

16. Dris et al. conducted sampling at three indoor sites, comprising two private apartments and one office, alongside outdoor air sampling at a single site. They estimated the deposition rate of fibers and their concentration in settled dust collected from vacuum cleaner bags. Indoor concentrations ranged from 1.0 to 60.0 fibers/m^3, while outdoor concentrations were significantly lower, ranging from 0.3 to 1.5 fibers/m^3. The deposition rate of fibers in indoor environments ranged between 1,586 and 11,130 fibers/day/m^2, leading to an accumulation of fibers in settled dust ranging from 190 to 670 fibers/mg. Analysis of fiber types revealed that 67% of the fibers in indoor environments were composed of natural materials, primarily cellulosic, while the remaining 33% contained petrochemicals, with polypropylene being predominant. Similar fibers have been observed in marine and continental studies dealing with MPs. Although the observed fibers were supposedly too large to be inhaled, exposure may occur through dust ingestion, particularly for young children [26].

17. Dris et al. investigated the atmospheric deposition of MPs in two distinct urban and suburban sites. MPs were continuously collected using a stainless steel funnel, and the collected samples were subsequently filtered and observed with a stereomicroscope. Fibers comprised almost all of the collected MPs. The study highlighted an atmospheric deposition rate ranging from 2 to 355 particles/m^2/day, with systematically higher fluxes observed at the urban site compared to the suburban site. Chemical characterization of the fibers revealed that approximately 29% of them were entirely synthetic (derived from petrochemicals) or a mixture of natural and synthetic materials. Extrapolation based on weight and volume estimates of the collected fibers suggested that between 3 and 10 tons of fibers are deposited annually through atmospheric fallout in the Parisian agglomeration, covering an area of 2,500 km^2 [24].

18. Thinh et al. documented a substantial presence of microplastics (MPs) in the atmospheric fallout at the Phuoc Hiep landfill, Cu Chi, Ho Chi Minh City. The study revealed that the concentration of MiPs in the atmospheric fallout at the Phuoc Hiep landfill was 1,801.2 items/m^2/day during the dry season and 913 items/m^2/day during the rainy season. The researchers also provided insights into the physical characteristics, including shape, color, and size, of the MP particles, along with detailed discussions on their implications [90].

19. Su et al. conducted a comprehensive study on roadside dust collected from rural and urban areas of Victoria, Australia, spanning two distinct seasons. It is noteworthy that any deposited fugitive dust and PM present in our samples could contribute to the observed MP abundance. Our analysis revealed an average MP abundance ranging from 20.6 to 529.3 items/kg (dry weight basis) across the two seasons, with fibers and items smaller than 1 mm being the predominant components. Polyester and polypropylene emerged as the dominant polymer types, constituting 26% of the total, while cotton and cellulose emerged as the most prevalent non-plastic items, accounting for 27% of the total. Remarkably, sampling sites exhibited consistent MP abundances over time and displayed similar patterns in terms of size, shape, and polymer composition. Through multiple correlation and principal component analyses, we identified urbanization and rainfall as critical factors influencing roadside MP accumulation. MP hotspots were predominantly situated in close proximity to

areas characterized by intensive urban land use and larger regional population sizes. Furthermore, we observed that MPS accumulated on roads and road verges during dry weather conditions and were subsequently flushed away during heavy storms, although the precise trigger values for this phenomenon remain unknown. Our findings underscore the potential of monitoring roadside dust as an initial and cost-effective approach for screening MP pollution in urban environments [91].

20. Abbasi et al. conducted a comprehensive study on street dust and suspended dust collected from the city and county of Asaluyeh, Iran. The samples underwent characterization using various microscopic techniques, including fluorescence, polarized light microscopy, and scanning electron microscopy (SEM), to quantify and classify MPs and MRs present in urban and industrial environments, which are potentially ingestible or inhalable by humans. In street dust samples retrieved from 15 sites, with particle sizes less than 5 mm, an average of 900 MPs and 250 MRs were observed per 15 g of sample. The MPs exhibited a wide range of colors and sizes, ranging from less than 100 to over 1,000 μm. The majority of street dust samples were dominated by spherical and film-like particles, while MRs primarily consisted of different sizes of black fragments and fibrous particulates.

    Airborne dust collected daily over an 8-day period at two locations revealed the widespread presence of fibrous MPs, with sizes ranging from about 2 to 100 μm, at an abundance of approximately 1 per m^3. These samples also contained small MR fragments, the precise characteristics of which were more challenging to define. Based on median concentrations observed in street dust, estimates of acute exposure through ingestion were approximately 5 and 15 MPs/day and 2 and 7 MRs/day for construction workers and young children, respectively. These findings highlight the potential health risks associated with exposure to MPs and MRs in urban and industrial environments [28].

21. Dehghani et al. conducted a comprehensive investigation into the presence, characteristics, and potential health risks associated with the ingestion of MP dust. The study focused on ten street dust samples, wherein the plastic load ranged from 88 to 605 MPs/30 g of dry dust. The dominant MPs observed were black and yellow granules, with sizes ranging from 250 to 500 μm, as determined using a binocular

microscope. Trace amounts of aluminum (Al), sodium (Na), calcium (Ca), magnesium (Mg), and silicon (Si) were detected, indicating the presence of additives from plastic polymers or adsorbed debris on MP surfaces.

Considering outdoor activities and workspaces with high concentrations of MPs as potential sources of acute exposure, the study estimated that children ingest an average of 3,223 MP particles/year, while adults ingest an average of 1,063 MP particles/year.

These findings underscore the significance of street dust as a potential source of MP contamination in urban environments, highlighting the necessity for control measures to mitigate the risks associated with MP ingestion [92].

22. Klein and Fischer conducted a study on MP concentrations in atmospheric deposition across the metropolitan region of Hamburg. The study spanned a period of 12 weeks, from December 2017 to March 2018, and involved six investigation sites equipped with three bulk precipitation samplers each. Sampling was conducted biweekly at these sites. The investigation sites comprised three locations in a rural area south of Hamburg, including one open field site and two throughfall sites situated under beech/oak and Douglas fir forest canopies, respectively. Additionally, three urban sites were selected along a transect from north to south within the city, representing varying degrees of urbanization in terms of population density, traffic, and industrial activity.

    The results revealed that MP particles were present ubiquitously across all sites. The median abundance ranged from 136.5 to 512.0 MP particles/m^2/day during the sampling period, with a mean MP abundance of 275 particles/m^2/day. Analysis using µRaman spectroscopy identified polyethylenes/ethylvinyl acetate copolymers as the dominant MP types, accounting for 48.8% and 22.0% of the total particles, respectively. Moreover, 16 particles (14.6%) were identified as contamination from PE samplers.

    In contrast to previous studies, fragments were found to significantly dominate over fibers. The spatial distribution of MP concentrations at urban sites followed a decreasing order from north to south: "north" (Henstedt-Ulzburg, low population density, suburb) – "center" (University; large population density) – "south" (Wilhelmsburg, middle population density, port and industrial facilities), with considerable variation observed within the time series. Surprisingly, the rural sites in the southern part of Hamburg exhibited the highest concentrations, with the Douglas fir site showing the highest concentration, followed by the open field and beech/oak sites [21].

While these case studies provide valuable insights into the presence and distribution of MPs in the air, further research is needed to fully understand the sources, transport mechanisms, and environmental impacts of airborne MP pollution. Ongoing monitoring efforts and interdisciplinary research collaborations will be essential for addressing this emerging environmental challenge.

Establishing standardized characterization methods for MPs is essential, not only for aquatic and terrestrial ecosystems but also for assessing MPs in the air. As evidence continues to accumulate, it becomes increasingly clear that airborne MPs may have significant implications for both the environment and human/animal health. Studies have measured substantial MP deposition in the atmosphere, such as observed in Dongguan with a range of 175–313 particles/m^2/day [45]. However, the contribution of airborne MPs to the overall increase in PM levels in the air remains largely unquantified.

Mapping the global distribution of airborne MPs is crucial for mitigating their levels and understanding the extent of pollution, especially in remote ecosystems affected by transboundary transport. Therefore, regardless of the environmental and health impacts, efforts to reduce aerial and terrestrial plastic waste generation are imperative. Moreover, research focusing on sustainable plastic waste management practices is essential to address this growing issue effectively [13,93,94]. Here are some critical recommendations for environmentalists, researchers, and policymakers in the field:

1. Future research efforts should prioritize elucidating the pathways of atmospheric transport of airborne MPs to identify susceptible hotspots associated with potential sources.

2. Investigate and quantify the contribution of airborne MPs and NPs to PM, including atmospheric aerosols.

3. Focus on understanding the atmospheric transport of MPs and mapping vulnerable ecosystems. For instance, analyze the impacts of MPs accumulating in polar snow on the polar ecosystem and trophic levels.

4. Develop standardized techniques and methods specific to sampling airborne MPs/NPs, complementing existing approaches for aquatic and terrestrial MPs.

5. Conduct ecotoxicological studies to understand the environmental and biological impacts of airborne MPs, including respiratory exposure levels.

6. Optimize in vivo and in vitro cell assays to explore cellular interactions, toxicity, genetic disruptions, and respiratory ailments caused by inhaled MPs.

7. Collect data on MP respiratory exposure among workers in industries dealing with plastics, including airborne MPs, and issue policy guidelines to address long-term exposure health effects.

8. Recognize MPs as an emerging air pollutant in air quality assessment guidelines, suggesting relevant standards and exposure limits to monitor air quality in vulnerable areas.

9. Conduct interdisciplinary research to gather evidence on the human health effects of airborne MPs, along with associated pollutants and additives.

10. Evaluate the impacts of atmospheric MPs on air and arboreal ecosystems, particularly in sensitive habitats like above tree canopies or aerial microbiota.

11. Conduct extensive studies to understand the potential links between atmospheric MPs/NPs and the atmospheric transport of virulent or antibiotic-resistant microbial strains, especially concerning the transmission of the ongoing COVID-19 pandemic.

REFERENCES

1. Biber NFA, Foggo A, Thompson RC. Characterising the deterioration of different plastics in air and seawater. *Mar Pollut Bull.* 2019;141:595–602.
2. Kawecki D, Nowack B. A proxy-based approach to predict spatially resolved emissions of macro-and microplastic to the environment. *Sci Total Environ.* 2020;748:141137.
3. Wright SL, Ulke J, Font A, Chan KLA, Kelly FJ. Atmospheric microplastic deposition in an urban environment and an evaluation of transport. *Environ Int.* 2020;136:105411.
4. Ojha N, Sharma A, Kumar M, Girach I, Ansari TU, Sharma SK, et al. On the widespread enhancement in fine particulate matter across the Indo-Gangetic Plain towards winter. *Sci Rep.* 2020;10(1):5862.
5. Sridharan S, Kumar M, Singh L, Bolan NS, Saha M. Microplastics as an emerging source of particulate air pollution: A critical review. *J Hazard Mater.* 2021;418:126245. Available from: https://www.sciencedirect.com/science/article/pii/S0304389421012097

6. Wang X, Liu K, Zhu L, Li C, Song Z, Li D. Efficient transport of atmospheric microplastics onto the continent via the East Asian summer monsoon. *J Hazard Mater.* 2021;414:125477.
7. Rezaei M, Riksen MJPM, Sirjani E, Sameni A, Geissen V. Wind erosion as a driver for transport of light density microplastics. *Sci Total Environ.* 2019;669:273–81. Available from: https://www.sciencedirect.com/science/article/pii/S0048969719308897
8. Brahney J, Hallerud M, Heim E, Hahnenberger M, Sukumaran S. Plastic rain in protected areas of the United States. *Science (80-).* 2020;368(6496):1257–60.
9. Mbachu O, Jenkins G, Pratt C, Kaparaju P. A new contaminant superhighway? A review of sources, measurement techniques and fate of atmospheric microplastics. *Water, Air, Soil Pollut.* 2020;231(2):85. https://doi.org/10.1007/s11270-020-4459-4
10. Rochman CM, Hoellein T. The global odyssey of plastic pollution. *Science (80-).* 2020 Jun 12;368(6496):1184–5. https://doi.org/10.1126/science.abc4428
11. Takahashi N, Hayasaka T. Air–sea interactions among oceanic low-level cloud, sea surface temperature, and atmospheric circulation on an intra-seasonal time scale in the summertime North Pacific based on satellite data analysis. *J Clim.* 2020;33(21):9195–212. Available from: https://journals.ametsoc.org/view/journals/clim/33/21/jcliD190670.xml
12. Gouin T, Roche N, Lohmann R, Hodges G. A thermodynamic approach for assessing the environmental exposure of chemicals absorbed to microplastic. *Environ Sci Technol.* 2011 Feb 15;45(4):1466–72. https://doi.org/10.1021/es1032025
13. Wright SL, Kelly FJ. Plastic and human health: A micro issue? *Environ Sci Technol.* 2017 Jun 20;51(12):6634–47. https://doi.org/10.1021/acs.est.7b00423
14. Liu K, Wu T, Wang X, Song Z, Zong C, Wei N, et al. Consistent transport of terrestrial microplastics to the ocean through atmosphere. *Environ Sci Technol.* 2019 Sep 17;53(18):10612–9. https://doi.org/10.1021/acs.est.9b03427
15. Liu K, Wang X, Song Z, Wei N, Ye H, Cong X, et al. Global inventory of atmospheric fibrous microplastics input into the ocean: An implication from the indoor origin. *J Hazard Mater.* 2020;400:123223. Available from: https://www.sciencedirect.com/science/article/pii/S0304389420312127
16. Lei YD, Wania F. Is rain or snow a more efficient scavenger of organic chemicals? *Atmos Environ.* 2004;38(22):3557–71. Available from: https://www.sciencedirect.com/science/article/pii/S1352231004003243
17. Hahladakis JN, Velis CA, Weber R, Iacovidou E, Purnell P. An overview of chemical additives present in plastics: Migration, release, fate and environmental impact during their use, disposal and recycling. *J Hazard Mater.* 2018;344:179–99. Available from: https://www.sciencedirect.com/science/article/pii/S030438941730763X
18. Akhbarizadeh R, Dobaradaran S, Amouei Torkmahalleh M, Saeedi R, Aibaghi R, Faraji Ghasemi F. Suspended fine particulate matter (PM2.5), microplastics (MPs), and polycyclic aromatic hydrocarbons (PAHs) in air: Their possible relationships and health implications. *Environ Res.* 2021;192:110339. Available from: https://www.sciencedirect.com/science/article/pii/S0013935120312366

19. Bergmann M, Mützel S, Primpke S, Tekman MB, Trachsel J, Gerdts G. White and wonderful? Microplastics prevail in snow from the Alps to the Arctic. *Sci Adv.* 2024 Mar 8;5(8):eaax1157. https://doi.org/10.1126/sciadv.aax1157
20. González-Pleiter M, Velázquez D, Edo C, Carretero O, Gago J, Barón-Sola Á, et al. Fibers spreading worldwide: Microplastics and other anthropogenic litter in an Arctic freshwater lake. *Sci Total Environ.* 2020;722:137904. Available from: https://www.sciencedirect.com/science/article/pii/S0048969720314170
21. Klein M, Fischer EK. Microplastic abundance in atmospheric deposition within the Metropolitan area of Hamburg, Germany. *Sci Total Environ.* 2019;685:96–103. Available from: https://www.sciencedirect.com/science/article/pii/S0048969719324581
22. Zhou Y, He G, Jiang X, Yao L, Ouyang L, Liu X, et al. Microplastic contamination is ubiquitous in riparian soils and strongly related to elevation, precipitation and population density. *J Hazard Mater.* 2021;411:125178. Available from: https://www.sciencedirect.com/science/article/pii/S0304389421001412
23. Selvam S, Manisha A, Venkatramanan S, Chung SY, Paramasivam CR, Singaraja C. Microplastic presence in commercial marine sea salts: A baseline study along Tuticorin Coastal salt pan stations, Gulf of Mannar, South India. *Mar Pollut Bull.* 2020;150:110675. Available from: https://www.sciencedirect.com/science/article/pii/S0025326X19308239
24. Dris R, Gasperi J, Saad M, Mirande C, Tassin B. Synthetic fibers in atmospheric fallout: A source of microplastics in the environment? *Mar Pollut Bull.* 2016;104(1):290–3. Available from: https://www.sciencedirect.com/science/article/pii/S0025326X16300066
25. Friot D, Boucher J. Primary microplastics in the oceans. IUCN Library System. 2017. 43 p. Available from: https://portals.iucn.org/library/node/46622
26. Dris R, Gasperi J, Mirande C, Mandin C, Guerrouache M, Langlois V, et al. A first overview of textile fibers, including microplastics, in indoor and outdoor environments. *Environ Pollut.* 2017;221:453–8. Available from: https://www.sciencedirect.com/science/article/pii/S0269749116312325
27. Allen S, Allen D, Phoenix VR, Le Roux G, Durántez Jiménez P, Simonneau A, et al. Atmospheric transport and deposition of microplastics in a remote mountain catchment. *Nat Geosci.* 2019;12(5):339–44. https://doi.org/10.1038/s41561-019-0335-5
28. Abbasi S, Keshavarzi B, Moore F, Turner A, Kelly FJ, Dominguez AO, et al. Distribution and potential health impacts of microplastics and microrubbers in air and street dusts from Asaluyeh County, Iran. *Environ Pollut.* 2019;244:153–64. Available from: https://www.sciencedirect.com/science/article/pii/S0269749118326290
29. Horton AA, Dixon SJ. Microplastics: An introduction to environmental transport processes. *WIREs Water.* 2018 Mar 1;5(2):e1268. https://doi.org/10.1002/wat2.1268

30. Järlskog I, Jaramillo-Vogel D, Rausch J, Gustafsson M, Strömvall A-M, Andersson-Sköld Y. Concentrations of tire wear microplastics and other traffic-derived non-exhaust particles in the road environment. *Environ Int.* 2022;170:107618. Available from: https://www.sciencedirect.com/science/article/pii/S0160412022005451
31. Lassen C, Hansen SF, Magnusson K, Noren F, Hartmann NIB, Jensen PR, et al. Microplastics-occurrence, effects and sources of releases to the environment in Denmark. The Danish Environmental Protection Agency 2015. https://backend.orbit.dtu.dk/ws/portalfiles/portal/118180844/Lassen_et_al._2015.pdf
32. Sundt P, Schulze P-E, Syversen F. Sources of microplastic-pollution to the marine environment. *Mepex Nor Environ Agency.* 2014;86:20.
33. Knight LJ, Parker-Jurd FNF, Al-Sid-Cheikh M, Thompson RC. Tyre wear particles: An abundant yet widely unreported microplastic? *Environ Sci Pollut Res.* 2020;27(15):18345–54. https://doi.org/10.1007/s11356-020-08187-4
34. US Environmental Protection Agency (EPA). Air quality index (AQI) - A guide to air quality and your health. 2021. Available from: https://www.airnow.gov/aqi/aqi-basics/
35. World Health Organization (WHO). WHO air quality guidelines for particulate matter, ozone, nitrogen dioxide and sulfur dioxide: Global update 2005. 2005. Available from: https://apps.who.int/iris/handle/10665/69477
36. Song J, Qian H, Zhao D, Jiang W, Zheng X. Particulate matter emission by an isolated rotating wheel. *Build Simul.* 2021;14(4):1163–73. https://doi.org/10.1007/s12273-020-0724-0
37. Unice KM, Weeber MP, Abramson MM, Reid RCD, van Gils JAG, Markus AA, et al. Characterizing export of land-based microplastics to the estuary - Part I: Application of integrated geospatial microplastic transport models to assess tire and road wear particles in the Seine watershed. *Sci Total Environ.* 2019;646:1639–49. Available from: https://www.sciencedirect.com/science/article/pii/S0048969718328638
38. Evangeliou N, Grythe H, Klimont Z, Heyes C, Eckhardt S, Lopez-Aparicio S, et al. Atmospheric transport is a major pathway of microplastics to remote regions. *Nat Commun.* 2020;11(1):3381. https://doi.org/10.1038/s41467-020-17201-9
39. Rist S, Vianello A, Winding MHS, Nielsen TG, Almeda R, Torres RR, et al. Quantification of plankton-sized microplastics in a productive coastal Arctic marine ecosystem. *Environ Pollut.* 2020;266:115248. Available from: https://www.sciencedirect.com/science/article/pii/S0269749120338902
40. Huang Y, Liu Q, Jia W, Yan C, Wang J. Agricultural plastic mulching as a source of microplastics in the terrestrial environment. *Environ Pollut.* 2020;260:114096. Available from: https://www.sciencedirect.com/science/article/pii/S0269749119362475
41. Yang Z, Lü F, Zhang H, Wang W, Shao L, Ye J, et al. Is incineration the terminator of plastics and microplastics? *J Hazard Mater.* 2021;401:123429. Available from: https://www.sciencedirect.com/science/article/pii/S0304389420314187

42. Chen G, Feng Q, Wang J. Mini-review of microplastics in the atmosphere and their risks to humans. *Sci Total Environ.* 2020;703:135504. Available from: https://www.sciencedirect.com/science/article/pii/S0048969719354981
43. Chai B, Wei Q, She Y, Lu G, Dang Z, Yin H. Soil microplastic pollution in an e-waste dismantling zone of China. *Waste Manag.* 2020;118:291–301. Available from: https://www.sciencedirect.com/science/article/pii/S0956053X20304967
44. González-Pleiter M, Edo C, Aguilera Á, Viúdez-Moreiras D, Pulido-Reyes G, González-Toril E, et al. Occurrence and transport of microplastics sampled within and above the planetary boundary layer. *Sci Total Environ.* 2021;761:143213. Available from: https://www.sciencedirect.com/science/article/pii/S0048969720367449
45. Cai L, Wang J, Peng J, Tan Z, Zhan Z, Tan X, et al. Correction to: Characteristic of microplastics in the atmospheric fallout from Dongguan City, China: Preliminary research and first evidence (*Environ Sci Pollut Res.* 2017;24(32):24928–35, https://doi.org/10.1007/S11356-017-0116-X). *Environ Sci Pollut Res.* 2019;26(35):36074–5.
46. Catarino AI, Macchia V, Sanderson WG, Thompson RC, Henry TB. Low levels of microplastics (MP) in wild mussels indicate that MP ingestion by humans is minimal compared to exposure via household fibres fallout during a meal. *Environ Pollut.* 2018;237:675–84. Available from: https://www.sciencedirect.com/science/article/pii/S0269749117344445
47. Enyoh CE, Verla AW, Verla EN, Ibe FC, Amaobi CE. Airborne microplastics: A review study on method for analysis, occurrence, movement and risks. *Environ Monit Assess.* 2019;191(11):668. https://doi.org/10.1007/s10661-019-7842-0
48. Wesch C, Elert AM, Wörner M, Braun U, Klein R, Paulus M. Assuring quality in microplastic monitoring: About the value of clean-air devices as essentials for verified data. *Sci Rep.* 2017;7(1):5424. https://doi.org/10.1038/s41598-017-05838-4
49. Miller E, Sedlak M, Lin D, Box C, Holleman C, Rochman CM, et al. Recommended best practices for collecting, analyzing, and reporting microplastics in environmental media: Lessons learned from comprehensive monitoring of San Francisco Bay. *J Hazard Mater.* 2021;409:124770. Available from: https://www.sciencedirect.com/science/article/pii/S0304389420327606
50. Prata JC, Reis V, da Costa JP, Mouneyrac C, Duarte AC, Rocha-Santos T. Contamination issues as a challenge in quality control and quality assurance in microplastics analytics. *J Hazard Mater.* 2021;403:123660. Available from: https://www.sciencedirect.com/science/article/pii/S0304389420316460
51. Hwang J, Choi D, Han S, Choi J, Hong J. An assessment of the toxicity of polypropylene microplastics in human derived cells. *Sci Total Environ.* 2019;684:657–69. Available from: https://www.sciencedirect.com/science/article/pii/S0048969719320832
52. Zhang J, Wang L, Kannan K. Microplastics in house dust from 12 countries and associated human exposure. *Environ Int.* 2020;134:105314. Available from: https://www.sciencedirect.com/science/article/pii/S016041201931952X

53. Zhang Y, Kang S, Allen S, Allen D, Gao T, Sillanpää M. Atmospheric microplastics: A review on current status and perspectives. *Earth Sci Rev.* 2020;203:103118. Available from: https://www.sciencedirect.com/science/article/pii/S001282521930621X
54. De Falco F, Cocca M, Avella M, Thompson RC. Microfiber release to water, via laundering, and to air, via everyday use: A comparison between polyester clothing with differing textile parameters. *Environ Sci Technol.* 2020 Mar 17;54(6):3288–96. https://doi.org/10.1021/acs.est.9b06892
55. O'Brien S, Okoffo ED, O'Brien JW, Ribeiro F, Wang X, Wright SL, et al. Airborne emissions of microplastic fibres from domestic laundry dryers. *Sci Total Environ.* 2020;747:141175. Available from: https://www.sciencedirect.com/science/article/pii/S0048969720347045
56. Song Z, Liu K, Wang X, Wei N, Zong C, Li C, et al. To what extent are we really free from airborne microplastics? *Sci Total Environ.* 2021;754:142118. Available from: https://www.sciencedirect.com/science/article/pii/S0048969720356473
57. Deng M, Han X, Ge J, Liang X, Du B, Li J, et al. Prevalence of phthalate alternatives and monoesters alongside traditional phthalates in indoor dust from a typical e-waste recycling area: Source elucidation and co-exposure risk. *J Hazard Mater.* 2021;413:125322. Available from: https://www.sciencedirect.com/science/article/pii/S0304389421002855
58. Napper IE, Davies BFR, Clifford H, Elvin S, Koldewey HJ, Mayewski PA, et al. Reaching new heights in plastic pollution—Preliminary findings of microplastics on Mount Everest. *One Earth.* 2020;3(5):621–30. Available from: https://www.sciencedirect.com/science/article/pii/S2590332220305509
59. Horton AA, Walton A, Spurgeon DJ, Lahive E, Svendsen C. Microplastics in freshwater and terrestrial environments: Evaluating the current understanding to identify the knowledge gaps and future research priorities. *Sci Total Environ.* 2017;586:127–41. Available from: https://www.sciencedirect.com/science/article/pii/S0048969717302073
60. Hartmann NB, Rist S, Bodin J, Jensen LHS, Schmidt SN, Mayer P, et al. Microplastics as vectors for environmental contaminants: Exploring sorption, desorption, and transfer to biota. *Integr Environ Assess Manag.* 2017 May 1;13(3):488–93. https://doi.org/10.1002/ieam.1904
61. Dehaut A, Cassone A-L, Frère L, Hermabessiere L, Himber C, Rinnert E, et al. Microplastics in seafood: Benchmark protocol for their extraction and characterization. *Environ Pollut.* 2016;215:223–33. Available from: https://www.sciencedirect.com/science/article/pii/S0269749116303979
62. Song YK, Hong SH, Jang M, Han GM, Rani M, Lee J, et al. A comparison of microscopic and spectroscopic identification methods for analysis of microplastics in environmental samples. *Mar Pollut Bull.* 2015;93(1):202–9. Available from: https://www.sciencedirect.com/science/article/pii/S0025326X15000314
63. Solomando A, Capó X, Alomar C, Álvarez E, Compa M, Valencia JM, et al. Long-term exposure to microplastics induces oxidative stress and a pro-inflammatory response in the gut of Sparus aurata Linnaeus, 1758. *Environ Pollut.* 2020;266:115295. Available from: https://www.sciencedirect.com/science/article/pii/S0269749120359832

64. Prata JC, da Costa JP, Lopes I, Duarte AC, Rocha-Santos T. Environmental exposure to microplastics: An overview on possible human health effects. *Sci Total Environ*. 2020;702:134455. Available from: https://www.sciencedirect.com/science/article/pii/S0048969719344468
65. Iñiguez ME, Conesa JA, Fullana A. Microplastics in Spanish table salt. *Sci Rep*. 2017;7(1):8620. https://doi.org/10.1038/s41598-017-09128-x
66. Hernandez E, Nowack B, Mitrano DM. Polyester textiles as a source of microplastics from households: A mechanistic study to understand microfiber release during washing. *Environ Sci Technol*. 2017 Jun 20;51(12):7036–46. https://doi.org/10.1021/acs.est.7b01750
67. Kosuth M, Mason SA, Wattenberg EV. Anthropogenic contamination of tap water, beer, and sea salt. *PLoS One*. 2018 Apr 11;13(4):e0194970. https://doi.org/10.1371/journal.pone.0194970
68. Venkataraman S, Suguna A, Surekha A, Sree Priyankha RS. Screening for respiratory morbidities and obstructive lung function among municipal waste handlers in Puducherry: A community-based cross-sectional study. *J Fam Med Prim Care*. 2022 Mar;11(3):1119–25.
69. Katsnelson A. Microplastics present pollution puzzle. *Proc Natl Acad Sci*. 2015 May 5;112(18):5547–9. https://doi.org/10.1073/pnas.1504135112
70. Horn SA, Dasgupta PK. The air quality index (AQI) in historical and analytical perspective a tutorial review. *Talanta*. 2024;267:125260. Available from: https://www.sciencedirect.com/science/article/pii/S0039914023010111
71. United States Environmental Protection Agency (EPA). Particulate matter (PM) pollution. 2021. Available from: https://www.epa.gov/pm-pollution
72. Darabi M, Majeed H, Diehl A, Norton J, Zhang Y. A review of microplastics in aquatic sediments: Occurrence, fate, transport, and ecological impact. *Curr Pollut Rep*. 2021;7(1):40–53. https://doi.org/10.1007/s40726-020-00171-3
73. Li J, Shan E, Zhao J, Teng J, Wang Q. The factors influencing the vertical transport of microplastics in marine environment: A review. *Sci Total Environ*. 2023;870:161893. Available from: https://www.sciencedirect.com/science/article/pii/S0048969723005089
74. Hurley RR, Woodward JC, Rothwell JJ. Ingestion of microplastics by freshwater tubifex worms. *Environ Sci Technol*. 2017 Nov 7;51(21):12844–51. https://doi.org/10.1021/acs.est.7b03567
75. Dong C-D, Chen C-W, Chen Y-C, Chen H-H, Lee J-S, Lin C-H. Polystyrene microplastic particles: In vitro pulmonary toxicity assessment. *J Hazard Mater*. 2020;385:121575. Available from: https://www.sciencedirect.com/science/article/pii/S0304389419315298
76. Xu X, Goros RA, Dong Z, Meng X, Li G, Chen W, et al. Microplastics and nanoplastics impair the biophysical function of pulmonary surfactant by forming heteroaggregates at the alveolar–Capillary interface. *Environ Sci Technol*. 2023 Dec 19;57(50):21050–60. https://doi.org/10.1021/acs.est.3c06668
77. Ahmed R, Hamid AK, Krebsbach SA, He J, Wang D. Critical review of microplastics removal from the environment. *Chemosphere*. 2022;293:133557. Available from: https://www.sciencedirect.com/science/article/pii/S0045653522000467

78. O'Brien S, Rauert C, Ribeiro F, Okoffo ED, Burrows SD, O'Brien JW, et al. There's something in the air: A review of sources, prevalence and behaviour of microplastics in the atmosphere. *Sci Total Environ.* 2023;874:162193. Available from: https://www.sciencedirect.com/science/article/pii/S0048969723008094
79. Shao L, Li Y, Jones T, Santosh M, Liu P, Zhang M, et al. Airborne microplastics: A review of current perspectives and environmental implications. *J Clean Prod.* 2022;347:131048. Available from: https://www.sciencedirect.com/science/article/pii/S0959652622006825
80. Liu M, Lu S, Song Y, Lei L, Hu J, Lv W, et al. Microplastic and mesoplastic pollution in farmland soils in suburbs of Shanghai, China. *Environ Pollut.* 2018;242:855–62. Available from: https://www.sciencedirect.com/science/article/pii/S0269749118317901
81. Yuan Z, Li H-X, Lin L, Pan Y-F, Liu S, Hou R, et al. Occurrence and human exposure risks of atmospheric microplastics: A review. *Gondwana Res.* 2022;108:200–12. Available from: https://www.sciencedirect.com/science/article/pii/S1342937X22000429
82. Imhof HK, Sigl R, Brauer E, Feyl S, Giesemann P, Klink S, et al. Spatial and temporal variation of macro-, meso- and microplastic abundance on a remote coral island of the Maldives, Indian Ocean. *Mar Pollut Bull.* 2017;116(1):340–7. Available from: https://www.sciencedirect.com/science/article/pii/S0025326X17300103
83. Patchaiyappan A, Dowarah K, Zaki Ahmed S, Prabakaran M, Jayakumar S, Thirunavukkarasu C, et al. Prevalence and characteristics of microplastics present in the street dust collected from Chennai metropolitan city, India. *Chemosphere.* 2021;269:128757. Available from: https://www.sciencedirect.com/science/article/pii/S0045653520329556
84. Narmadha VV, Jose J, Patil S, Farooqui MO, Srimuruganandam B, Saravanadevi S, et al. Assessment of microplastics in roadside suspended dust from urban and rural environment of Nagpur, India. *Int J Environ Res.* 2020;14(6):629–40. https://doi.org/10.1007/s41742-020-00283-0
85. Stanton T, Johnson M, Nathanail P, MacNaughtan W, Gomes RL. Freshwater microplastic concentrations vary through both space and time. *Environ Pollut.* 2020;263:114481. Available from: https://www.sciencedirect.com/science/article/pii/S0269749119367995
86. Gaston E, Woo M, Steele C, Sukumaran S, Anderson S. Microplastics differ between indoor and outdoor air masses: Insights from multiple microscopy methodologies. *Appl Spectrosc.* 2020 Apr 1;74(9):1079–98. https://doi.org/10.1177/0003702820920652
87. Li Y, Shao L, Wang W, Zhang M, Feng X, Li W, et al. Airborne fiber particles: Types, size and concentration observed in Beijing. *Sci Total Environ.* 2020;705:135967. Available from: https://www.sciencedirect.com/science/article/pii/S0048969719359625
88. Liu K, Wang X, Fang T, Xu P, Zhu L, Li D. Source and potential risk assessment of suspended atmospheric microplastics in Shanghai. *Sci Total Environ.* 2019;675:462–71. Available from: https://www.sciencedirect.com/science/article/pii/S0048969719316304

89. Qian Z, ChongGuo T, YongMing LUO. Various forms and deposition fluxes of microplastics identified in the coastal urban atmosphere. *Chinese Sci Bull.* 2017;62(33):3902–9.
90. Thinh TQ, Sang TTN, Viet TQ, Tam LTM, Dan NP, Strady E., et al. Preliminary assessment on the microplastic contamination in the atmospheric fallout in the Phuoc Hiep landfill, Cu Chi, Ho Chi Minh city. *Vietnam J Sci Technol Eng.* 2020 Sep 15;62(3 SE-):83–9. Available from: https://vietnamscience.vjst.vn/index.php/vjste/article/view/116
91. Su L, Nan B, Craig NJ, Pettigrove V. Temporal and spatial variations of microplastics in roadside dust from rural and urban Victoria, Australia: Implications for diffuse pollution. *Chemosphere.* 2020;252:126567. Available from: https://www.sciencedirect.com/science/article/pii/S0045653520307608
92. Dehghani S, Moore F, Akhbarizadeh R. Microplastic pollution in deposited urban dust, Tehran metropolis, Iran. *Environ Sci Pollut Res.* 2017;24(25):20360–71. https://doi.org/10.1007/s11356-017-9674-1
93. Orona-Návar C, García-Morales R, Loge FJ, Mahlknecht J, Aguilar-Hernández I, Ornelas-Soto N. Microplastics in Latin America and the Caribbean: A review on current status and perspectives. *J Environ Manage.* 2022;309:114698. Available from: https://www.sciencedirect.com/science/article/pii/S0301479722002717
94. Koelmans AA, Besseling E, Shim WJ. Nanoplastics in the aquatic environment. Critical review. In Bergmann M, Gutow L, Klages M (Eds.), *Anthropogenic Marine Litter.* 2015 (pp. 325–40). SpringerOpen.

CHAPTER 23

The Impact of Plastic Products on Air Pollution

A Comprehensive Review

Nimish Shah and Mamta Saiyad

23.1 INTRODUCTION

Plastic products have become an integral part of modern life, revolutionizing industries, and providing convenience and functionality in various applications. However, the widespread use and disposal of plastic products have led to a global environmental crisis, with far-reaching consequences for air quality. The impact of plastic products on air pollution is a pressing issue that demands attention from policymakers, scientists, and the public alike. Plastic pollution encompasses a wide range of environmental problems, from marine litter to landfills overflowing with discarded packaging. While much attention has been focused on the visible consequences of plastic pollution in oceans and terrestrial environments, the implications of plastic products on air quality are becoming increasingly evident. Plastic particles, ranging from microplastics to nanoplastics, are now ubiquitous in the atmosphere, with studies detecting their presence in urban air, remote environments, and even indoor settings.

DOI: 10.1201/9781003487579-23

The combustion of plastics, whether intentional or accidental, releases harmful emissions into the atmosphere, including particulate matter and toxic compounds. Additionally, plastic particles can serve as carriers for other pollutants, such as heavy metals and organic chemicals, further exacerbating air pollution. Inhalation of airborne plastic particles poses risks to respiratory and cardiovascular health, with potential long-term consequences for exposed populations. Despite growing awareness of the environmental and health impacts of plastic pollution, there remain critical gaps in our understanding of the sources, fate, and health effects of airborne plastic particles. Addressing these knowledge gaps requires interdisciplinary research efforts and holistic strategies that target both the sources and consequences of plastic pollution [1–4].

23.2 PLASTIC POLLUTION'S STEALTHY IMPACT ON AIR QUALITY

Plastic products have become ubiquitous in modern society, permeating nearly every aspect of daily life. While their adverse environmental effects, particularly in oceans and ecosystems, have garnered considerable attention, the impact of plastic pollution on air quality remains relatively understudied and underappreciated. This article delves into the often-overlooked consequences of plastic pollution on air quality, exploring how plastic products wield a profound influence that extends beyond their conspicuous ecological footprint. Plastic pollution emits a diverse array of pollutants into the atmosphere through various mechanisms. One significant contributor is the process of plastic degradation. Over time, exposure to sunlight and environmental factors causes plastic to degrade, releasing volatile organic compounds (VOCs) and other harmful chemicals into the air. Additionally, plastic waste incineration, a common disposal method in many regions, releases toxic gases such as dioxins and furans, further exacerbating air pollution. These pollutants contribute to the formation of smog, ozone depletion, and respiratory illnesses, posing significant risks to human health and environmental stability [5–7].

Microplastics, defined as plastic particles less than 5 mm in size, represent another insidious threat to air quality. These minute particles are ubiquitous in the environment, generated through the breakdown of larger plastic items and as additives in consumer products like cosmetics and textiles. Once airborne, microplastics can travel vast distances, infiltrating both urban and remote areas. Their small size enables them to penetrate deep into the respiratory system upon inhalation, potentially causing inflammation, oxidative stress, and respiratory ailments. Moreover, microplastics act as carriers for other pollutants, absorbing harmful chemicals from the environment and releasing them upon inhalation, amplifying their toxic effects [8].

The production and manufacturing of plastic products also contribute to air pollution. Petrochemical facilities, where raw materials are processed into plastic resins, emit significant quantities of greenhouse gases and air pollutants. These emissions not only contribute to climate change but also degrade air quality in surrounding communities, disproportionately impacting marginalized and vulnerable populations. Moreover, the transportation and distribution of plastic products further exacerbate air pollution, as fossil fuel-powered vehicles emit pollutants during transportation processes [9,10].

Addressing the complex interplay between plastic pollution and air quality requires multifaceted strategies. First and foremost, reducing plastic consumption and transitioning to sustainable alternatives is paramount. By adopting circular economy principles and investing in innovative recycling technologies, society can minimize the generation of plastic waste and mitigate its impact on air quality. Additionally, stringent regulations and enforcement mechanisms are necessary to curb emissions from plastic production facilities and waste management practices. Public awareness campaigns and education initiatives can also empower individuals to make informed choices and advocate for systemic change.

Plastic pollution poses a pervasive and multifaceted threat to air quality, with far-reaching consequences for human health and environmental integrity. As society grapples with the urgent need to address climate change and environmental degradation, acknowledging and mitigating the impact of plastic pollution on air quality must be prioritized. Through concerted efforts at the individual, community, and global levels, we can chart a path toward a cleaner, healthier future, free from the pervasive influence of plastic pollution on our air [11–13].

23.3 CONTRIBUTION OF PLASTICS TO AIR POLLUTION

Plastics, omnipresent in modern society, contribute to air pollution across their entire life cycle in multifaceted ways. From production to disposal, plastics release a plethora of pollutants into the atmosphere, exacerbating environmental degradation and human health risks. This article elucidates the intricate relationship between plastics and air pollution, highlighting the various stages in their life cycle where emissions occur such as production phase, transportation and distribution, use and consumption, and end-of-life management, etc. [14].

The manufacturing of plastics entails the conversion of raw materials, primarily fossil fuels, into polymers through processes like polymerization. Petrochemical plants, where this conversion occurs, emit copious amounts of greenhouse gases (GHGs), including carbon dioxide (CO_2) and methane (CH_4). Additionally, VOCs are released during polymer synthesis, contributing to ground-level ozone formation and smog. These emissions persist throughout the production chain, from extraction to processing, magnifying the environmental footprint of plastics. The transportation and distribution of plastic products further exacerbate air pollution. Fossil fuel-powered vehicles, employed in the transportation of raw materials, intermediates, and finished products, emit nitrogen oxides (NO_x), particulate matter (PM), and other air pollutants. Moreover, the globalized nature of the plastics industry necessitates extensive shipping and freight operations, leading to additional emissions. The combustion of marine fuels in shipping vessels releases sulfur oxides (SO_x) and nitrogen oxides into the atmosphere, contributing to regional and global air pollution [11,14,15].

During the use phase, plastics continue to contribute to air pollution through various mechanisms. Plastic products, such as polyvinyl chlorine (PVC) and polyurethane foams, release VOCs and phthalates into indoor air, posing health risks to occupants. Additionally, plastic combustion in incinerators and waste-to-energy facilities releases toxic gases, including dioxins, furans, and heavy metals, into the atmosphere. These emissions not only degrade air quality but also pose significant health hazards to nearby communities. At the end of their life cycle, plastics present significant challenges for waste management and disposal. Improper disposal practices, such as open burning and landfilling, release GHGs, VOCs, and other pollutants into the atmosphere. Landfills, where a significant portion of plastic waste ends up, emit methane—a potent greenhouse gas—during anaerobic decomposition. Furthermore, plastic litter, when exposed to environmental stressors like sunlight and moisture, undergoes degradation, releasing microplastics and chemical additives into the air [12].

Mitigating the air pollution associated with plastics demands comprehensive strategies that target each stage of their life cycle. Transitioning to renewable feedstocks and adopting energy-efficient manufacturing processes can reduce emissions during production. Promoting sustainable transportation practices and investing in alternative modes of freight transport can minimize emissions associated with transportation and distribution. Encouraging the use of plastic alternatives and promoting circular economy principles can mitigate emissions during the use phase and improve end-of-life management practices. Plastics contribute to air pollution throughout their life cycle, from production to disposal. Addressing this challenge requires concerted efforts from governments, industries, and individuals to adopt sustainable practices and technologies that minimize emissions and mitigate environmental impacts. By recognizing the multifaceted ways in which plastics contribute to air pollution, society can work toward a cleaner and healthier future [16,17].

23.4 EMISSION OF THE VOLATILE ORGANIC COMPOUNDS (VOCS) DURING MANUFACTURING

VOCs emitted during manufacturing processes represent a significant environmental and health concern, contributing to air pollution, smog formation, and adverse health effects. Understanding the sources, impacts, and mitigation strategies for VOC emissions is crucial for addressing this pervasive issue [16].

Sources of VOC Emissions: VOCs are a diverse group of organic chemicals that easily vaporize at room temperature, encompassing various compounds such as benzene, toluene, formaldehyde, and xylene. These compounds are emitted during the production of a wide range of materials and products, including plastics, paints, solvents, adhesives, and coatings. In the plastics industry, VOC emissions occur during polymerization, resin synthesis, and processing operations. Petrochemical plants, where raw materials are converted into plastic resins, are significant sources of VOC emissions. Additionally, VOCs are released during the formulation and application of coatings, paints, and adhesives used in plastic manufacturing processes.

Environmental and Health Impacts: VOC emissions contribute to the formation of ground-level ozone and smog, which have adverse effects on air quality and human health. Ozone, a major component of smog, can cause respiratory problems, exacerbate asthma, and increase susceptibility to respiratory infections. VOCs also react with nitrogen oxides (NO_x) in the presence of sunlight to form secondary organic aerosols, which contribute to haze and PM formation. Furthermore, certain VOCs, such as benzene and formaldehyde, are known carcinogens and pose long-term health risks, including cancer and neurological disorders. Chronic exposure to VOCs has been linked to respiratory issues, headaches, dizziness, and allergic reactions.

Mitigation Strategies: Addressing VOC emissions requires a multifaceted approach that incorporates technological advancements, regulatory measures, and industry best practices. Process optimization and engineering controls can help minimize VOC emissions during manufacturing processes. Switching to low-VOC or VOC-free alternatives, such as water-based coatings and solvents, can significantly reduce emissions without compromising product performance. Implementing effective ventilation systems and air pollution control technologies, such as catalytic oxidizers and adsorption systems, can capture and treat VOC emissions before they are released into the atmosphere. Additionally, regulatory agencies play a crucial role in establishing emission limits, monitoring compliance, and promoting pollution prevention measures through policies and standards.

Advancements in Green Chemistry: Advancements in green chemistry offer promising solutions for reducing VOC emissions in manufacturing processes. Green chemistry principles emphasize the design of products and processes that minimize environmental impacts, including the reduction or elimination of hazardous substances and emissions. Innovative approaches, such as solvent-free polymerization techniques, bio-based feedstocks, and recyclable materials, can help mitigate VOC emissions while promoting sustainability and resource efficiency. Research and development efforts focused on green chemistry continue to drive progress in creating safer and more environmentally friendly alternatives to conventional manufacturing practices. By reducing VOC emissions, we can improve air quality, protect public health, and mitigate the impacts of climate change [18–22].

23.5 THE DECOMPOSITION OF PLASTIC

The decomposition of plastic is a complex process influenced by environmental factors such as sunlight, temperature, humidity, and microbial activity. Plastic materials undergo physical, chemical, and biological transformations over time, leading to their eventual breakdown into smaller particles, including microplastics and nanoplastics.

Physical Degradation: Exposure to sunlight and mechanical forces like wind and waves can cause plastic materials to fragment into smaller pieces through a process known as photodegradation and mechanical degradation. Ultraviolet (UV) radiation from sunlight breaks down the chemical bonds in plastics, weakening their structure and causing them to become brittle. This makes plastics more susceptible to fragmentation into smaller particles. Additionally, physical abrasion from natural processes like erosion and weathering can further break down plastic debris into microplastics.

Chemical Degradation: Plastics can undergo chemical degradation through processes such as hydrolysis, oxidation, and polymer chain scission. Hydrolysis occurs when water molecules react with the polymer chains in plastics, leading to the cleavage of chemical bonds and the formation of smaller fragments. Oxidation, triggered by exposure to oxygen and UV radiation, causes the formation of free radicals in plastic polymers, leading to chain scission and the generation of smaller plastic particles. These chemical reactions contribute to the fragmentation and degradation of plastics over time.

Biological Degradation: Microorganisms play a crucial role in the decomposition of plastics through biodegradation. Certain bacteria, fungi, and algae possess the enzymatic machinery capable of breaking down plastic polymers into simpler compounds. This process, known as biodegradation, involves the enzymatic hydrolysis of polymer chains by microbial enzymes, leading to the assimilation of plastic fragments as carbon and energy sources by microorganisms. While biodegradation can reduce the size of plastic debris and facilitate their breakdown, it may also lead to the formation of microplastics and nanoplastics as intermediate products of microbial degradation.

The decomposition of plastic is a multifaceted process involving physical, chemical, and biological mechanisms that contribute to the generation of microplastics and nanoplastics. Understanding these processes is essential for assessing the environmental impacts of plastic pollution and developing strategies to mitigate its adverse effects on ecosystems and human health [23–25].

23.6 THE RELEASE OF MICROPLASTICS AND NANOPLASTICS DURING DECOMPOSITION

The release of microplastics and nanoplastics during decomposition is a growing concern with profound implications for environmental health and ecosystem integrity. As larger plastic items degrade over time due to environmental factors such as sunlight, temperature fluctuations, and mechanical forces, they break down into smaller fragments known as microplastics and nanoplastics. These tiny particles, measuring less than 5 mm and 100 nm in size, respectively, pose unique challenges due to their ubiquity, persistence, and potential toxicity. Understanding the mechanisms of microplastic and nanoplastics release, their environmental fate, and the consequences for ecosystems and human health is essential for developing effective mitigation strategies and safeguarding the environment [26–28].

Mechanisms of Microplastic and Nanoplastics Release: The decomposition of macroplastics into microplastics and nanoplastics can occur through several processes, including fragmentation, abrasion, and weathering. Mechanical forces such as wave action, wind erosion, and physical abrasion break down larger plastic items into smaller fragments over time. Additionally, exposure to UV radiation from sunlight causes polymer chains to degrade, weakening the structural integrity of plastics and facilitating fragmentation. Chemical weathering, including oxidation and hydrolysis, further accelerates the breakdown of plastics into microplastics and nanoplastics. As a result, plastic debris undergoes continuous degradation in the environment, releasing microplastics and nanoplastics into soil, water bodies, and the atmosphere [26,29].

23.7 ENVIRONMENTAL FATE OF MICROPLASTICS AND NANOPLASTICS

Once released into the environment, microplastics and nanoplastics exhibit complex transport and fate dynamics. In aquatic environments, buoyancy, water currents, and sedimentation play key roles in the dispersal and deposition of microplastics and nanoplastics. Microplastics may accumulate on the water surface, settle into sediments, or be transported over long distances by ocean currents. Nanoplastics, due to their smaller size and increased surface area, exhibit greater mobility and are capable of vertical transport through water columns. In terrestrial environments, microplastics and nanoplastics can migrate through soil profiles, affecting soil structure, nutrient cycling, and microbial communities. Atmospheric transport and deposition also contribute to the global distribution of microplastics and nanoplastics, with airborne particles settling on land and water surfaces [30,31].

23.8 ECOLOGICAL IMPACTS

Microplastics and nanoplastics pose a range of ecological risks to terrestrial and aquatic ecosystems. In marine environments, ingestion of microplastics by marine organisms, including fish, seabirds, and marine mammals, can lead to physical harm, gastrointestinal blockages, and toxicological effects. Nanoplastics, due to their small size and high surface area, may penetrate cell membranes and accumulate in tissues, potentially disrupting cellular function and metabolic processes. Microplastics and nanoplastics can also serve as vectors for pollutants, absorbing and desorbing hydrophobic contaminants such as polychlorinated biphenyls (PCBs), polycyclic aromatic hydrocarbons (PAHs), and heavy metals. These pollutants can bioaccumulate and biomagnify in the food web, posing risks to higher trophic levels and ecosystem health. In terrestrial ecosystems, microplastics and nanoplastics may alter soil properties, microbial communities, and plant physiology, with potential cascading effects on ecosystem functioning and productivity.

23.9 HUMAN HEALTH CONCERNS

The presence of microplastics and nanoplastics in the environment raises concerns about potential human health risks. Microplastics can enter the human body through ingestion of contaminated food and water, inhalation of airborne particles, and dermal exposure. Once ingested, microplastics may translocate across gastrointestinal barriers and accumulate in tissues, potentially causing inflammation, oxidative stress, and cellular damage. Nanoplastics, with their smaller size and increased surface area, may exhibit greater bioavailability and cellular uptake, raising concerns about systemic toxicity and adverse health effects. Furthermore, microplastics and nanoplastics may act as carriers for pathogens and harmful chemicals, increasing the risk of exposure to infectious agents and toxicants. Emerging research suggests potential links between microplastic exposure and adverse health outcomes, including gastrointestinal disorders, immune dysregulation, and carcinogenesis. However, further studies are needed to elucidate the mechanisms of microplastic toxicity and assess the long-term health effects of human exposure.

23.10 MITIGATION STRATEGIES

Addressing the release of microplastics and nanoplastics during decomposition requires a comprehensive approach that encompasses prevention, mitigation, and remediation measures. Strategies to reduce plastic pollution at the source include promoting sustainable consumption and production practices, implementing extended producer responsibility (EPR) schemes, and incentivizing the use of biodegradable and eco-friendly materials. Improved waste management and recycling infrastructure can help minimize plastic leakage into the environment and facilitate the recovery of plastic waste for recycling and reuse. Innovative technologies for plastic degradation and remediation, including enzymatic degradation, chemical recycling, and bioremediation, hold promise for reducing the accumulation of microplastics and nanoplastics in the environment. Furthermore, public awareness campaigns and education initiatives can raise awareness about the environmental impacts of plastic pollution and empower individuals to adopt more sustainable behaviors and lifestyle choices.

The release of microplastics and nanoplastics during decomposition represents a significant environmental challenge with far-reaching implications for ecosystems, human health, and planetary health. By understanding the mechanisms of microplastic and nanoplastic release, their environmental fate, and the associated ecological and health risks, society can develop effective strategies to mitigate plastic pollution and safeguard the environment for future generations [31,32].

23.11 COMBUSTION OF PLASTICS

The intentional or accidental combustion of plastics represents a significant source of air pollution, releasing harmful emissions that include PM and toxic compounds. Whether in industrial settings, waste incineration facilities, or open burning scenarios, the combustion of plastics contributes to air quality degradation and poses risks to human health and the environment. Understanding the mechanisms, impacts, and mitigation strategies associated with plastic combustion is essential for addressing this pressing environmental challenge.

When plastics are subjected to high temperatures, they undergo combustion, a chemical process involving the rapid oxidation of organic materials. During combustion, the heat generated breaks down the polymer chains in plastics, releasing energy in the form of heat and light. Combustion reactions also produce gaseous byproducts, such as carbon dioxide (CO_2), carbon monoxide (CO), and water vapor (H_2O), as well as solid residues in the form of ash and PM. The combustion of plastics can occur through various processes, including waste incineration, industrial combustion, and open burning of plastic waste.

The combustion of plastics releases a diverse array of emissions, including PM, VOCs), PAHs, dioxins, furans, and heavy metals. PM, consisting of fine particles suspended in the air, poses significant health risks when inhaled, contributing to respiratory problems, cardiovascular diseases, and premature death. VOCs, emitted during the combustion of plastic polymers, can react with nitrogen oxides (NO_x) in the atmosphere to form ground-level ozone and secondary organic aerosols, further exacerbating air pollution and smog formation. PAHs, dioxins, and furans, generated from incomplete combustion of plastics, are potent carcinogens and toxicants that pose long-term health risks to humans and wildlife. Heavy metals, present as additives or contaminants in plastics, can also be released into the environment during combustion, posing risks to ecosystems and human health.

The emissions from plastic combustion have adverse effects on both human health and the environment. Inhalation of PM and toxic compounds released during combustion can lead to respiratory problems, cardiovascular diseases, and neurological disorders. Children, the elderly, and individuals with pre-existing health conditions are particularly vulnerable to the health impacts of air pollution from plastic combustion. Additionally, exposure to toxic pollutants like PAHs, dioxins, and heavy metals can increase the risk of cancer, reproductive disorders, and developmental abnormalities in humans and wildlife. Environmental impacts include soil and water contamination, ecosystem disruption, and biodiversity loss, as well as the accumulation of plastic residues in terrestrial and aquatic environments.

Addressing the air pollution from plastic combustion requires comprehensive mitigation strategies that target both emissions reduction and waste management. Improving combustion technologies and adopting cleaner energy sources can reduce emissions from industrial processes and waste incineration facilities. Implementing emission control technologies, such as particulate filters, electrostatic precipitators, and scrubbers, can capture and treat pollutants before they are released into the atmosphere. Promoting alternative waste management practices, such as recycling, composting, and waste-to-energy conversion, can divert plastics from incineration and open burning, reducing air pollution and greenhouse gas emissions. Public awareness campaigns and education initiatives can also raise awareness about the health and environmental impacts of plastic combustion, encouraging individuals and communities to adopt sustainable practices and reduce plastic consumption.

The intentional or accidental combustion of plastics introduces harmful emissions, including PM and toxic compounds, into the atmosphere, contributing to air pollution and posing risks to human health and the environment. Addressing this environmental challenge requires concerted efforts from governments, industries, and communities to adopt cleaner technologies, improve waste management practices, and promote sustainable alternatives to plastic combustion. By mitigating the air pollution from plastic combustion, we can protect public health, safeguard the environment, and mitigate the impacts of climate change [33–36].

23.12 AIRBORNE PLASTIC PARTICLES

As airborne plastic particles increasingly infiltrate the atmosphere, concerns about their implications for respiratory and cardiovascular health are mounting. Microplastics and nanoplastics, generated from the degradation of larger plastic debris and a variety of human activities, are now ubiquitous in the air we breathe. Understanding the sources, transport mechanisms, health impacts, and mitigation strategies associated with airborne plastic particles is essential for addressing this emerging environmental and public health issue.

Airborne plastic particles originate from a variety of sources, including the fragmentation of larger plastic items, the breakdown of synthetic textiles, the abrasion of tires and road markings, and the release of microplastics from industrial processes and consumer products. Atmospheric transport mechanisms, such as wind and atmospheric deposition, can disperse these particles over long distances, leading to their widespread distribution in terrestrial and marine environments. Additionally, microplastics may be released directly into the air through activities such as waste handling, plastic recycling, and plastic manufacturing processes. Once airborne, plastic particles can be transported by atmospheric currents and weather patterns, dispersing them across local, regional, and even global scales. Atmospheric deposition, including dry and wet deposition, can deposit plastic particles onto terrestrial and aquatic surfaces, where they may accumulate in soil, water bodies, and sediments. However, some plastic particles remain suspended in the atmosphere for extended periods, undergoing atmospheric aging processes such as photodegradation and oxidation, which can alter their chemical and physical properties [37–39].

Exposure to airborne plastic particles poses potential risks to respiratory and cardiovascular health. Inhalation of microplastics and nanoplastics can deposit these particles deep within the lungs, where they may cause inflammation, oxidative stress, and tissue damage. Additionally, airborne plastic particles may translocate to other organs and tissues via the bloodstream, potentially exacerbating pre-existing health conditions and increasing the risk of respiratory and cardiovascular diseases. Furthermore, plastic particles may serve as carriers for other pollutants, such as heavy metals and organic contaminants, which can adsorb onto their surfaces and pose additional health risks upon inhalation. Addressing the issue of airborne plastic particles requires a multifaceted approach that encompasses pollution prevention, waste management, and regulatory measures. Strategies to reduce the release of plastic particles into the atmosphere include implementing measures to minimize plastic littering and promoting sustainable consumption and production practices. Improving waste management infrastructure, including recycling and waste-to-energy technologies, can help reduce the accumulation of plastic debris in the environment and mitigate emissions of airborne plastic particles from waste-handling activities. Additionally, regulatory measures, such as air quality standards and emission controls, can help limit emissions of plastic particles from industrial sources and combustion processes [27,39,40].

Furthermore, research efforts aimed at understanding the sources, fate, and health impacts of airborne plastic particles are essential for informing policy decisions and developing effective mitigation strategies. Long-term monitoring programs can track trends in airborne plastic pollution and assess the effectiveness of mitigation measures over time. Public awareness campaigns and education initiatives can also raise awareness about the health risks associated with airborne plastic particles and encourage individuals and communities to take action to reduce their exposure.

The increasing presence of airborne plastic particles in the atmosphere raises concerns about their implications for respiratory and cardiovascular health. Addressing this emerging environmental and public health issue requires collaborative efforts from governments, industries, researchers, and communities to reduce the release of plastic particles into the environment, mitigate their impacts on human health, and promote sustainable alternatives to plastic consumption and disposal. By taking proactive measures to address airborne plastic pollution, we can protect public health and safeguard the environment for future generations [27,37,38].

In recent years, the impact of plastic pollution on air quality has emerged as a growing concern as airborne plastic particles increasingly infiltrate the atmosphere. While much attention has been focused on the visible consequences of plastic pollution in oceans and terrestrial environments, the implications of plastic particles on respiratory and cardiovascular health are becoming increasingly apparent. Synthesizing existing knowledge underscores the urgency of further research and advocates for holistic strategies to mitigate the escalating impact of plastic products on air pollution. Existing research has revealed that plastic particles are ubiquitous in the atmosphere, with microplastics and nanoplastics being detected in urban air, remote environments, and even indoor settings. These particles originate from various sources, including the breakdown of larger plastic items, the abrasion of plastic surfaces, and the release of microfibers from textiles. Once airborne, plastic particles can travel long distances and remain suspended in the atmosphere for extended periods [39].

Studies have shown that airborne plastic particles can serve as carriers for other pollutants, including heavy metals, PAHs, and persistent organic pollutants (POPs). These contaminants can adhere to the surface of plastic particles, increasing their toxicity and potential health impacts. Additionally, plastic particles can act as nuclei for cloud formation, influencing weather patterns and atmospheric processes [37].

Exposure to airborne plastic particles poses risks to respiratory and cardiovascular health. Inhalation of these particles can lead to inflammation of the respiratory tract, exacerbation of asthma and other respiratory conditions, and impairment of lung function. Fine plastic particles, known as nanoplastics, have been shown to penetrate deep into the lungs and enter the bloodstream, where they can accumulate in various organs and tissues.

Moreover, the toxic chemicals associated with plastic particles, such as phthalates, bisphenol A (BPA), and flame retardants, have been linked to adverse health effects, including cardiovascular disease, neurodevelopmental disorders, and reproductive problems. These chemicals can leach from plastic particles and enter the body through inhalation, ingestion, and dermal absorption, posing risks to human health and wildlife. While significant strides have been made in understanding the presence and potential impacts of airborne plastic particles, there remain critical gaps in knowledge that warrant further research. Key areas for investigation include the sources, fate, and transport mechanisms of airborne plastic particles, as well as their interactions with other pollutants and their health effects on exposed populations [27,41].

Long-term studies are needed to assess the chronic health effects of exposure to airborne plastic particles and to identify vulnerable populations at increased risk. Furthermore, interdisciplinary research efforts involving atmospheric scientists, toxicologists, epidemiologists, and public health experts are essential to fully understand the complex interactions between plastic pollution and air quality. Addressing the escalating impact of plastic products on air pollution requires holistic strategies that target both the sources and consequences of plastic pollution. Efforts to reduce plastic production, consumption, and waste generation can help minimize the release of plastic particles into the environment. Transitioning to sustainable alternatives to plastic, promoting circular economy principles, and implementing extended producer responsibility schemes are critical steps in reducing the environmental burden of plastic pollution [42,43].

Furthermore, improving waste management practices, such as recycling, composting, and waste-to-energy conversion, can help prevent plastic waste from entering the environment and mitigate its impacts on air quality. In addition, regulations and policies aimed at reducing emissions of air pollutants from industrial processes, transportation, and waste incineration can help minimize the release of plastic particles and their associated contaminants into the atmosphere. Public awareness campaigns and education initiatives can also raise awareness about the health and environmental risks of airborne plastic particles, empowering individuals and communities to take action to reduce plastic pollution and protect air quality. By adopting holistic strategies that address the root causes of plastic pollution and mitigate its impacts on air quality, we can safeguard human health, preserve environmental quality, and ensure a sustainable future for generations to come.

23.12 CONCLUSION

In conclusion, the examination of plastic products' influence on air quality reveals a complex and far-reaching problem that extends beyond their visible environmental footprint. This comprehensive review has shed light on the various ways in which plastics contribute to air pollution across their life cycle, from the emissions of volatile organic compounds during manufacturing to the release of microplastics and nanoplastics during decomposition. Furthermore, the intentional or accidental combustion of plastics introduces harmful emissions, including PM and toxic compounds, further exacerbating air pollution.

As airborne plastic particles continue to infiltrate the atmosphere, concerns regarding their potential implications for respiratory and cardiovascular health intensify. Therefore, it is imperative to recognize the urgency of further research in this field and to advocate for holistic strategies aimed at mitigating the escalating impact of plastic products on air pollution.

Moving forward, interdisciplinary collaboration, technological innovation, and policy interventions will be essential in addressing this pressing issue. By fostering awareness, implementing sustainable practices, and promoting alternatives to plastic, we can work toward a cleaner and healthier environment for current and future generations.

REFERENCES

1. Smith O, Brisman A. Plastic waste and the environmental crisis industry. *Crit Criminol.* 2021;29:289–309.
2. Huang L, Yoshida Y, Li Y, Cheng N, Xue J, Long Y. Sustainable lifestyle: Quantification and determining factors analysis of household carbon footprints in Japan. *Energy Policy.* 2024;186:114016.
3. Zwicker M V, Brick C, Gruter G-JM, van Harreveld F. Consumer attitudes and willingness to pay for novel bio-based products using hypothetical bottle choice. *Sustain Prod Consum.* 2023;35:173–83.
4. Timiş C, Zoican Ş, Leah D, Şmuleac L, Paşcalău R. The importance of environmental measures in recycling in companies. *Res J Agric Sci.* 2023;55(3): 241–247.
5. Millican JM, Agarwal S. Plastic pollution: A material problem? *Macromolecules.* 2021;54(10):4455–69.
6. Iroegbu AOC, Ray SS, Mbarane V, Bordado JC, Sardinha JP. Plastic pollution: A perspective on matters arising: Challenges and opportunities. *ACS Omega.* 2021;6(30):19343–55.
7. Law KL, Narayan R. Reducing environmental plastic pollution by designing polymer materials for managed end-of-life. *Nat Rev Mater.* 2022;7(2):104–16.
8. North EJ, Halden RU. Plastics and environmental health: the road ahead. *Rev Environ Health.* 2013;28(1):1–8.
9. Forrest A, Giacovazzi L, Dunlop S, Reisser J, Tickler D, Jamieson A, et al. Eliminating plastic pollution: How a voluntary contribution from industry will drive the circular plastics economy. *Front Mar Sci.* 2019;6:627.
10. Rajmohan KVS, Ramya C, Viswanathan MR, Varjani S. Plastic pollutants: Effective waste management for pollution control and abatement. *Curr Opin Environ Sci Heal.* 2019;12:72–84.
11. Barrowclough D, Deere Birkbeck C, Christen J. Global trade in plastics: Insights from the first life-cycle trade database. UNCTAD Research Paper No. 53. 2020.

12. da Silva TR, de Azevedo ARG, Cecchin D, Marvila MT, Amran M, Fediuk R, et al. Application of plastic wastes in construction materials: A review using the concept of life-cycle assessment in the context of recent research for future perspectives. *Materials (Basel).* 2021;14(13):3549.
13. He Y, Kiehbadroudinezhad M, Hosseinzadeh-Bandbafha H, Gupta VK, Peng W, Lam SS, et al. Driving sustainable circular economy in electronics: A comprehensive review on environmental life cycle assessment of e-waste recycling. *Environ Pollut.* 2023;342:123081.
14. Hahladakis JN, Iacovidou E. Closing the loop on plastic packaging materials: What is quality and how does it affect their circularity? *Sci Total Environ.* 2018;630:1394–400.
15. Boucher J, Billard G. The challenges of measuring plastic pollution. *Field Actions Sci Reports J Field Actions.* 2019; SI19:68–75.
16. Ansar MA, Assawadithalerd M, Tipmanee D, Laokiat L, Khamdahsag P, Kittipongvises S. Occupational exposure to hazards and volatile organic compounds in small-scale plastic recycling plants in Thailand by integrating risk and life cycle assessment concepts. *J Clean Prod.* 2021;329:129582.
17. Klemeš JJ, Fan Y Van, Jiang P. Plastics: Friends or foes? The circularity and plastic waste footprint. *Energy Sources Part A Recover Util Environ Eff.* 2021;43(13):1549–65.
18. Ezeh E, Okeke O, Aniobi CC, Anya OU, Ndubuisi JO. Assessment of the levels of toxic volatile organic compounds and other substances in the air in a plastic recycling environment situated in scientific equipment development institute, Enugu. *IOSR J Appl Chem.* 2020;13(4):50–55.
19. Duan C, Liao H, Wang K, Ren Y. The research hotspots and trends of volatile organic compound emissions from anthropogenic and natural sources: A systematic quantitative review. *Environ Res.* 2023;216:114386.
20. Kwiatkowski S, Polat M, Yu W, Johnson MS. Industrial emissions control technologies: Introduction. In: Burns GL (Ed.), *Air Pollution Sources, Statistics and Health Effects.* Nova Science Publishers, Inc.; 2021; pp. 477–511.
21. Sapuan SM, Ilyas RA, Asyraf MRM, Sapuan SM, Ilyas RA, Asyraf MRM. Emission of hazardous air pollution in the composite production. In: Sapuan SM, Ilyas RA, Asyraf MRM (Eds.), *Safety and Health in Composite Industry.* Springer; 2022. pp. 35–66.
22. Stefaniak AB, Bowers LN, Cottrell G, Erdem E, Knepp AK, Martin Jr SB, et al. Towards sustainable additive manufacturing: The need for awareness of particle and vapor releases during polymer recycling, making filament, and fused filament fabrication 3-D printing. *Resour Conserv Recycl.* 2022;176:105911.
23. Casazza A, Spennati E, Converti A, Busca G. Study on the thermal decomposition of plastic residues. *Chem Eng Trans.* 2019;74:1141–6.
24. Litchfield SG, Schulz KG, Kelaher BP. The influence of plastic pollution and ocean change on detrital decomposition. *Mar Pollut Bull.* 2020;158:111354.
25. Yao D, Li H, Dai Y, Wang C-H. Impact of temperature on the activity of Fe-Ni catalysts for pyrolysis and decomposition processing of plastic waste. *Chem Eng J.* 2021;408:127268.

26. Pfohl P, Wagner M, Meyer L, Domercq P, Praetorius A, Hüffer T, et al. Environmental degradation of microplastics: How to measure fragmentation rates to secondary micro-and nanoplastic fragments and dissociation into dissolved organics. *Environ Sci Technol.* 2022;56(16):11323–34.
27. Tong H, Zhong X, Duan Z, Yi X, Cheng F, Xu W, et al. Micro-and nanoplastics released from biodegradable and conventional plastics during degradation: Formation, aging factors, and toxicity. *Sci Total Environ.* 2022;833:155275.
28. Lv S, Cui K, Zhao S, Li Y, Liu R, Hu R, et al. Continuous generation and release of microplastics and nanoplastics from polystyrene by plastic-degrading marine bacteria. *J Hazard Mater.* 2024;465:133339.
29. Du H, Xie Y, Wang J. Microplastic degradation methods and corresponding degradation mechanism: Research status and future perspectives. *J Hazard Mater [Internet].* 2021;418:126377. Available from: https://www.sciencedirect.com/science/article/pii/S0304389421013418
30. Wang L, Wu W-M, Bolan NS, Tsang DCW, Li Y, Qin M, et al. Environmental fate, toxicity and risk management strategies of nanoplastics in the environment: Current status and future perspectives. *J Hazard Mater.* 2021;401:123415.
31. Sharma VK, Ma X, Guo B, Zhang K. Environmental factors-mediated behavior of microplastics and nanoplastics in water: A review. *Chemosphere.* 2021;271:129597.
32. Pradel A, Catrouillet C, Gigault J. The environmental fate of nanoplastics: What we know and what we need to know about aggregation. *NanoImpact.* 2023;29:100453.
33. Velis CA, Cook E. Mismanagement of plastic waste through open burning with emphasis on the global south: A systematic review of risks to occupational and public health. *Environ Sci Technol.* 2021;55(11):7186–207.
34. Estrellan CR, Iino F. Toxic emissions from open burning. *Chemosphere.* 2010;80(3):193–207.
35. Cheng K, Hao W, Wang Y, Yi P, Zhang J, Ji W. Understanding the emission pattern and source contribution of hazardous air pollutants from open burning of municipal solid waste in China. *Environ Pollut.* 2020;263:114417.
36. Chavan D, Arya S, Kumar S. Open dumping of organic waste: Associated fire, environmental pollution and health hazards. In: Hussain C, Hait S (Eds.), *Advanced Organic Waste Management.* Elsevier; 2022. pp. 15–31.
37. Wright SL, Kelly FJ. Plastic and human health: A micro issue? *Environ Sci Technol [Internet].* 2017 Jun 20;51(12):6634–47. Available from: https://doi.org/10.1021/acs.est.7b00423
38. Wieland S, Balmes A, Bender J, Kitzinger J, Meyer F, Ramsperger AFRM, et al. From properties to toxicity: Comparing microplastics to other airborne microparticles. *J Hazard Mater.* 2022;428:128151.
39. Enyoh CE, Verla AW, Qingyue W, Ohiagu FO, Chowdhury AH, Enyoh EC, et al. An overview of emerging pollutants in air: Method of analysis and potential public health concern from human environmental exposure. *Trends Environ Anal Chem.* 2020;28:e00107.

40. Lu K, Lai KP, Stoeger T, Ji S, Lin Z, Lin X, et al. Detrimental effects of microplastic exposure on normal and asthmatic pulmonary physiology. *J Hazard Mater [Internet].* 2021;416:126069. Available from: https://www.sciencedirect.com/science/article/pii/S0304389421010335
41. Kacprzak S, Tijing LD. Microplastics in indoor environment: Sources, mitigation and fate. *J Environ Chem Eng [Internet].* 2022;10(2):107359. Available from: https://www.sciencedirect.com/science/article/pii/S2213343722002329
42. Cary CM, Seymore TN, Singh D, Vayas KN, Goedken MJ, Adams S, et al. Single inhalation exposure to polyamide micro and nanoplastic particles impairs vascular dilation without generating pulmonary inflammation in virgin female Sprague Dawley rats. *Part Fibre Toxicol [Internet].* 2023;20(1):16. Available from: https://doi.org/10.1186/s12989-023-00525-x
43. Yin X-H, Xu Y-M, Lau ATY. Nanoparticles: Excellent materials yet dangerous when they become airborne. *Toxics* 2022;10:50.

Index

Note: **Bold** page numbers refer to tables and *italic* page numbers refer to figures.